PENETRATING BARS THROUGH MASKS OF COSMIC DUST

ASTROPHYSICS AND SPACE SCIENCE LIBRARY

VOLUME 319

PENETRATING BARS THROUGH MASKS OF COSMIC DUST

The Hubble Tuning Fork strikes a New Note

Edited by

DAVID L. BLOCK
University of the Witwatersrand,
Johannesburg, South Africa

IVÂNIO PUERARI
Instituto Nacional de Astrofísica,
Óptica y Electrónica, Puebla, Mexico

KENNETH C. FREEMAN
The Australian National University,
Canberra, Australia

ROBERT GROESS
University of the Witwatersrand,
Johannesburg, South Africa

and

ELIZABETH K. BLOCK
Rand Afrikaans University (Soweto),
South Africa

A C.I.P. Catalogue record for this book is available from the Library of Congress.

ISBN 1-4020-2861-X (HB)
ISBN 1-4020-2862-8 (e-book)

Published by Springer,
P.O. Box 17, 3300 AA Dordrecht, The Netherlands.

Sold and distributed in North, Central and South America
by Springer,
101 Philip Drive, Norwell, MA 02061, U.S.A.

In all other countries, sold and distributed
by Springer,
P.O. Box 322, 3300 AH Dordrecht, The Netherlands.

Figure of Front Cover:
Elephants photographed in the late afternoon, as the sun sets in the African sub-continent. The effects of dust extinction and of dust scattering are both striking. Photograph supplied by Traci Hanson, Marketing Services Manager of Legacy Hotels and Resorts International. Their WWW site is www.legacyhotels.co.za.
Figure of Back Cover:
The logos of our principal conference sponsors (University of the Witwatersrand, Johannesburg and the Anglo American Chairman's Fund) are seen here, together with conference logo (centre) designed by R. Groess.

Printed on acid-free paper

springeronline.com

Printed in the Netherlands.

Table of Contents

Keynote review papers are identified by a † after the surname of the review speaker.

'Penetrating Bars Through Masks of Cosmic Dust: The Hubble Tuning Fork Strikes a New Note'

Members of the Scientific Organising Committee

David L. Block (South Africa) – SOC co-chair
Ken C. Freeman (Australia) – SOC co-chair

Bob Abraham (Canada)
Ron J. Allen (USA)
Ron J. Buta (USA)
Françoise Combes (France)
Bruce G. Elmegreen (USA)
Debbie M. Elmegreen (USA)
Antony Fairall (South Africa)
Paul Hodge (USA)
Garth Illingworth (USA)
Tom Jarrett (USA)
Johan H. Knapen (UK)
John Kormendy (USA)
Eija Laurikainen (Finland)
Duccio Macchetto (USA)
Daniel Pfenniger (Switzerland)
Ivânio Puerari (Mexico)
Vera Rubin (USA)
Heikki Salo (Finland)
Rogier A. Windhorst (USA)

SOC Secretary: Robert Groess

PREFACE

THE EDITORS: DAVID L. BLOCK AND KENNETH C. FREEMAN (SOC CO-CHAIRS), IVANIO PUERARI, ROBERT GROESS AND LIZ K. BLOCK

1. Harvard College Observatory, 1958

The past century has truly brought about an explosive period of growth and discovery for the physical sciences as a whole, and for astronomy in particular. Galaxy morphology has reached a renaissance..

The year: 1958.

The date: October 1.

The venue: Harvard College Observatory.

The lecturer: Walter Baade.

With amazing foresight, Baade penned these words:

"Young stars, supergiants and so on, make a terrific splash — *lots of light*. The total mass of these can be very small compared to the total mass of the system".

Dr Layzer then asked the key question:

"... the discussion raises the point of what this classification would look like if you were to ignore completely all the Population I, and just focus attention on the Population II ..."

We stand on the shoulders of giants. The great observer E.E. Barnard, in his pioneering efforts to photograph the Milky Way, devoted the major part of his life to identifying and numbering dusty "holes" and dust lanes in our Milky Way. No one could have dreamt that the pervasiveness of these cosmic dust masks (not only in our Galaxy but also in galaxies at high redshift) is so great, that their "penetration" is truly one of the pioneering challenges from both space-borne telescopes and from the ground.

It can be appropriately be remarked that "Minute dust grains have, in a very real sense, kept us in a scientific dark age about the true nature of galaxies."

The dust masks are highly complex: dust grains are invariably well embedded in the interstellar medium of the galaxies. Although negligible in mass, the "shrouding" or extinction effects of dust grains are extreme. A cosmic fog, in the truest sense, restricts our studies of galaxies to "tips of icebergs".

Dust also hides the bar-like structures which we now know lie near the centers of over 70% of galaxies and fundamentally affect their dynamics. What is the origin of these bars? Following the International Conference on bars in Tuscaloosa a decade ago, and another more recent meeting in Paris, the time was ripe for another major Conference – focusing almost exclusively on bars.

The Spitzer Space Telescope has been launched, and it is extraordinary powerful not only at penetrating dusty masks, but showing PAHs/dust grains in emission. We were extremely fortunate to have the PI of IRAC on-board Spitzer, G. Fazio, with us.

Penetration of dusty aerosols with infrared techniques has exposed the ubiquity of bars in a striking fashion, and we now realise that bars are much more common and influential than was previously believed. Furthermore we can now quantify the gravitational strengths (or torques) of spiral arms and of rotating bars from dust-penetrated images. We can seek answers to the question:

Do stronger bars drive stronger spirals?

Although much is missing (the speed of rotation of the bar, for example, is known for only a mere handful of galaxies), we can, for the first time, determine the observed distribution of gravitational torques of bars in Population II stellar disks. A crucial chapter has been missing from our story.

The actual birth process of galaxies such as our Milky Way still remains a mystery. What role does accretion of gas onto extragalactic disks play? Are the masses of disks constant with time, or has our Milky Way doubled its mass in one Hubble epoch? What is the influence of the dust in galaxies early in the life of the universe? At what epoch do gaseous disks become stabilized to the Fourier $m = 2$ component (spiral arms; bars)? And how far back in time do we need to look before we can no longer find massive (bulge-dominated) stellar disks?

We are now ready to ask: Has the Hubble tuning fork struck a New Note?

Are classifications in the Hubble fork transitory? Can one spiral galaxy support the formation/dissolution of possibly two/three bars in one Hubble time?

The Conference will focus on those scientific questions which must be answered if we are to truly exit our dark age in the understanding of bars and most importantly, of the principal driving mechanism(s) behind their formation/dissolution/rejuvenation.

Our scientific rationale included:

- 1. Morphological differences between galaxies in the early universe and today: HST ACS data, SIRTF, VLT, Gemini, ...

  a. Bar fraction as a function of look-back time

  b. Interacting galaxy morphology, then and now

  c. Galaxy luminosity functions, then and now.

d. Evolved stellar disks at high-z

e. Massive (bulge-dominated) stellar disks at high-z?

f. Penetrating high-z dust masks at restframe K: Challenges of L- and M- band imaging with NGST.

- 2. Properties of bars at low redshift

  Gravitational Torques from dust penetrated images: the bar/spiral torque separation

  a. Other measures of "bar strengths" (ellipticity, ...)

  b. The observed distribution of bar torques in spiral galaxies

  c. Bar torques in the SB0 (barred lenticular class)

  d. Pattern speeds of bars (Weinberg-Tremaine & other methodologies)

  e. Spiral/bar pattern speed coupling

  f. Rotating bars & their interaction with DM haloes

- 3. The morphology of HI and molecular gas distributions:

  a. Is HI born out of molecular hydrogen H_2?

  b. HI relative to spiral arms, bulge/disk distributions, general disk gradients

  c. photodissociation fronts in spiral arms, arm/interarm contrasts in gas, density wave triggered star formation: evidence

  d. phases of the ISM; variations with position.

  e. molecular gas and dust at high redshifts

- 4. Accretion of gas, as a major driving mechanism of galactic evolution

  a. Insight from theoretical models

  b. What can dust-penetrated images of stellar disks tell about gas accretion as a major driving mechanism in their evolution?

  c. Bars and galaxy environment (interactions, ...)

  d. Bars and nuclear inflows

  e. High Velocity Clouds / MW Disk Accretion

  f. From where does the gas come?

  g. HI "beards" in external galaxies

- 5. Infrared vs optical/uv morphology. The duality of spiral structure.

  a. Couplings vs decouplings between gaseous Pop I and stellar Pop II disks, as inferred from analyses of the shapes of rotation curves (form family & Hubble Type uncorrelated), as well as from near-IR imaging...

b. Results from 'The Large Galaxy Atlas': 2MASS

c. Results from recent deep surveys

d. Spiral arms, bulges, ovals, outer disks, rings ....; appearance of these at different wavelengths as tracers of DUSTY MASKS.

- 6. X-ray morphology of galaxies: diffuse x-rays from the interstellar medium, compact x-ray sources and supernova remnants in spiral arms and the general disk

- 7. Outer disk morphology:

a. Disk truncation radius

b. Star formation, spiral arms beyond the outer Lindblad resonance

c. Influence of dwarf companions

d. Tidal tails and bridges in interacting galaxies.

- 8. Dust-penetrated morphology: The Hubble Tuning Fork strikes a new note

a. Toward a new dust-penetrated classification scheme for the stellar disks of spiral galaxies

b. The duty cycle of bars: Bar dissolution/rejuvenation

c. the nature of those dust grains responsible for the EXTINCTION in optical morphologies: The 0.1 μ dust grains (i) Laboratory insights. (ii) Near-infrared albedo of 0.1 μ dust grains

d. cold dust: observations (sub-mm, near-IR, ...)

e. Unified images of dust: macromolecules and dust grains spanning the entire grain population diameter range

f. Insights from near- and mid-infrared minus optical images

g. Dusty ellipticals

- 9. On the origin of the Hubble sequence:

a. Hubble type evolution by internal processes (the role of viscosity, etc.)

b. Supermassive black holes and galaxy morphology: rotating Kerr-Newman black holes

c. gas infall & secondary bars

d. open vs. closed (non-accreting) systems: where have all the hot, axisymmetric disks gone?

e. dark halos and the Hubble sequence

- 10. The Hubble Sequence before $z \approx 1$

a. Insights from HST ACS

b. Insights from Spitzer, Gemini, VLT, Keck, ...

- 11. Masks in Astronomy – Lessons from our Milky Way

a. Fingerprinting the MW

b. The chemical mask of the MW

c. The mass mask of the MW

d. The dust mask [hot, warm, cold (20K), very cold (3K)] of the MW

The year 2004 marks the 40th anniversary of the famous density wave paper by C.C. Lin and Frank Shu published in 1964 (Figure 1). The paper was written while Frank was still an undergraduate at the Massachusetts Institute of Technology. It was a great joy to have Frank Shu personally with us (Figure 2).

DLB wishes to thank each invited review speaker, for their extremely generous gift of a mission bell, officially handed to him at the Gala Dinner (Figure 3) at Sun City, on the occassion of his 50th birthday.

2. Time for reflection: Quotable quotes from our Conference

Ivan King to Frank Shu:
"Is there something you can do with chaos, other than endure it?"

Frank Shu to David Block:
"The fact that galaxies are disordered in optical/blue light images may, ironically, be the best indicator that the background density (infrared) mode made in stars has been around for a long time."

Note : The order in which the papers appear in our table of contents follows the same order as that in our final conference program.

3. Morphology and Orthodoxy: Lessons from Saccheri

In moving forward in the arena of galaxy morphology, we may take heed from the genius Saccheri, who foreshadowed non-Euclidean geometry, but missed it.

Non-Euclidean geometry is now attributed to Gauss, Bolyai and Lobachevski. Why?

E.T. Bell, in his book *Development of Mathematics*, gives us the answer:

"[Saccheri's] brilliant failure is one of the most remarkable instances in the history of mathematical thought of the mential inertia induced by an education in obedience and orthodoxy, confirmed in mature life by an escessive reverence for the perishable works of the immortal dead [Euclid]. With two geometries, each as valid as Euclid's in his hand, Saccheri threw both away because he was willfully determined to continue in the obstinate worship of his idol, despite the insistent promptings of his own sane reason."

This remarkable quote by Bell also appears in Sandage, *Annual Review of Astronomy and Astrophysics*, volume 26 (1988).

The morphologist W.W. Morgan, writing in the same 'Annual Reviews' volume, summed it up thus:

" So much of the creative process takes place in the subconcious, in an area of incomplete communication... It is in such uncharted areas of mental space where some of the deepest science has its origins, and where the revoltionary philosophy of Wittgenstein's 'Philosophical Investigations' must have labored before birth."

As emphasized by Sandage (this volume), let us never understimate the fundamental role of imagination and intuition in the development of galaxy classification systems. Astronomers bravely penetrate masks of cosmic dust at higher and higher redshifts, and explore the diversity of the bar phenomenon. Vera Rubin (this volume) has it right: galaxy classification schemes are never final.

4. Post Venus Transit: A poem by Jacob Seloi

The Venus transit occurred during our Conference, and is forever etched in our minds. Jacob Seloi sent the Editors his poem, written as if Venus is penning her thoughts.

The Wonders of the Venus Transit

"Here I pass again after a century, two decades and two years. Your forefathers saw me and I am proud to pass across the disc of the Sun. A new generation of my twin sister are watching me. I have been passing here but what makes it different this time is that today different cultures are waiting for me to pass.

I am happy to know and to see that because of my beauty parade, people are able to come together and work together! Thank you for watching, Hope to see you again and I parade after eight years, off I go!"

– Jacob Seloi, S.G. Mafaesa Secondary School, Kagiso

5. Our Conference Spies: Cliff Brown and Robert Groess

Eack keynote review speaker has a caricature, reproduced in this volume. The caricature artist (who 'observed' many of the review speakers while 'in action') was Cliff Brown; we are enormously indebted to Mr Brown for his patience, hard work, and attention to detail.

Also 'observing' the delegates was one of us (RG), who secured over 900 digital images – many with the digital *nightshot* facility (to avoid flash interup-ptions). Almost all delegate photographs reproduced here come from RG. It was then up to Cliff Brown to do a detailed psycho-analysis of each of the reviewers!

6. Those who carried our vision...

We are deeply indebted to Mrs M. Keeton and the Board of Trustees of the Anglo American Chairman's Fund for their invaluable support and encourage-ment. Without their funding, this Conference would never have seen the light of day (see figure 8).

It was St Augustine who penned these words:

"Ex amante alio accenditur alius" – one loving spirit sets another on fire. Thank you Anglo American, for lighting our research candles.

Medallions commerating the 2004 transit of Venus, were struck in gold (generously to donated by AngloGold: we are hugely indebted to their Chief Excecutive Officer Bobby Godsell) and in silver, for all participants; a special vote of thanks is again owed to Anglo American.

It is finally a pleasure to thank Sean Summers, CEO of *Pick and Pay*, for his generous sponsoring of conference gifts.

To all our sponsors: astronomers thrive on vision. Without telescopic and theoretical vision, we perish. Your vision has made our vision possible.

Figure 1. The need to penetrate masks is exemplified by the Goroka Mudmen, photographed in the eastern highlands of Papua New Guinea. Masks may have negligible mass, but their effects can be awesome and dramatic: whether these be Population I dust and gas masks for galaxies, or human masks borne in the land of Papua New Guinea. Our Conference focussed much on how the Hubble tuning fork strikes a new note in the dust penetrated regime. Photograph by David Block.

ON THE SPIRAL STRUCTURE OF DISK GALAXIES

C. C. LIN AND FRANK H. SHU

Department of Mathematics, Massachusetts Institute of Technology

Received March 20, 1964

ABSTRACT

It is shown that gravitational instability is a plausible basis for the formation of the spiral pattern in disk galaxies. An explicit asymptotic formula is obtained for the form of the spiral. It gives reasonable numerical results for the galaxy, and qualitatively satisfactory trends for normal spirals of various types.

Figure 2. The year 2004 marks the 40th anniversary of the seminal density wave theory paper, by C.C. Lin and Frank Shu, published in the Atrophysical Journal in 1964. Courtesy the University of Chicago Press.

Figure 3. SOC co-chair David Block and Frank Shu enjoying the environs of the Pilanesberg Game Reserve and National Park.

Figure 4. At our Gala Dinner, Garth and Wendy Illingworth view a bell, cast from bronze in Nantes (France). The bell, estimated to be between 100-200 years old, comes from a mission station in California. It was paid for by 18 keynote review speakers elected by the SOC, as a gift to David Block for his 50th birthday. Bruce Elmegreen graciously facilitated the entire shipping process, of the mission bell to South Africa.

Figure 5. A giraffe camouflaged by (optically) thick acacia thorn trees reminded Garth Illingworth of the extinction effects of dust masks at high redshifts. Photograph courtesy Holland Ford.

Figure 6. The major Conference sponsors were the Anglo American Chairman's Fund and the University of the Witwatersrand, Johannesburg. The University crest is seen at left, the Conference Logo (designed by Robert Groess) is viewed in the centre, while the Anglo American logo is seen at right. Robert was inspired by the barred spiral NGC 1300 as a prototype for his design of the logo. The continent of Africa is focussed between the two spiral arms and is seen draped in the South African flag, which appropriately is Y-shaped to remind us of the Y-shape morphological schematic of Sir James Jeans (see article by Allan Sandage, this volume). The Y also represents the confluence of great minds in morphology from around the globe, deliberating, debating and merging their thoughts *'confluere'* at the tip of Africa.

A TRIBUTE TO COSMIC DUST PIONEER: J. MAYO GREENBERG

David L. Block
Cosmic Dust Laboratory, School of Computational and Applied Mathematics, University of the Witwatersrand, Johannesburg, South Africa

1. Personal Reflections

Allow me to make some personal reflections:

When I first decided to nominate Mayo for the Henry Norris Russell prize in 1997, letters in support of my nomination were received by Profesors L. Spitzer (Princeton University Observatory), H. vd Hulst, G. Miley and E. van Dishoeck (at Leiden), R.J. Allen (Baltimore), Bruce and Debbie Elmegreen (NY), D.A. Williams (University College London), P. Hodge (Editor of the *Astronomical Journal* in Seattle) and L. Allamandola, D. Cruikshank and Y. Pendleton (NASA-Ames). The letters bore stature of the greatness of the man. Sadly, both Lyman Spitzer and Henk vd Hulst passed on, shortly after they wrote their recommendation letters.

As fundamental as the Hertzsprung-Russell diagram is to our understanding of stellar evolution, so pivotal was the pioneering insight of Mayo Greenberg to our understanding of the chemical composition and evolution of dust grains in our Galaxy, and beyond.

The field in which Mayo played such an absolutely pioneering, leading role over the decades is now turning out to be one of the *most rapidly* developing areas in galactic and extragalactic astrophysics.

Not surprising. The distribution of dust tends to delineate the location of the material for future generations of stars and offers evidence of a history of past stellar processing of the ISM and metal enrichment.

As was recognised at our international conference on galaxy morphology and cold dust held at our University in January 1996 (attended by over 100 astronomers worldwide), the predictive powers of Mayo's work were absolutely remarkable. Remarkable both in depth, and in originality.

He did not develop theories to explain existing observations; rather, Mayo developed models and predicted scores of observations.

Conditions simulated by Professor Greenberg for the study of the photochemistry of low temperature ices related to the photoprocessing of interstel-

D. Block et al. (eds.), Penetrating Bars through Masks of Cosmic Dust, 1–12.

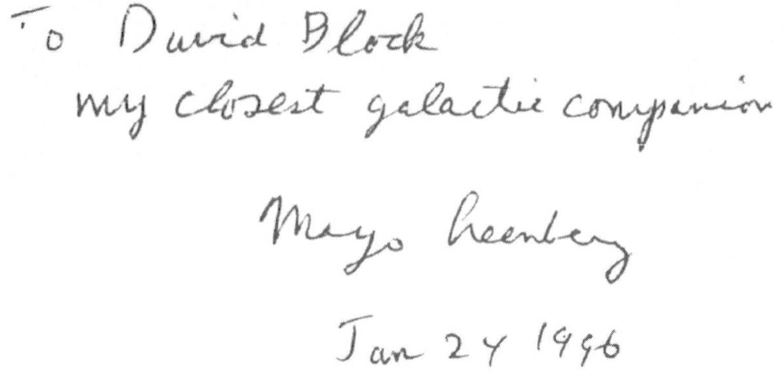

To David Block
my closest galactic companion
Mayo Greenberg
Jan 24 1996

Figure 1. The bond Mayo and I shared was exceptionally close. Mayo taught me to always look up. He always encouraged me. He once wrote this note in my copy of his book "Evoluzione della polvere interstellare e questioni attinenti" (International School of Physics: Enrico Fermi; Evolution of Interstellar Dust and Related Topics).

lar dust opened up fundamentally new paths: all of the chemical work that was done on molecules in giant molecular clouds prior to that assumed ion molecule chemistry, absolutely ignoring the fact that there was dust there. Researchers had ignored it from so many points of view, it would be impossible for me to enumerate them.

Greenberg's far-reaching insights into the pivotal role which the evolution of dust grains play in our understanding of star formation, are only now being fully appreciated. His now famous silicate core/organic refractory and ice mantle model for dust grains was developed *decades* before the advent of modern ground and space based instruments could test the models.

I think in particular, of the confirmation from spectroscopic studies of the silicate spectral features at $9.7\mu m$ and $18\mu m$ and the 3.4 μm organic feature in the diffuse interstellar medium.

But there is more. Dust is swept by stellar winds and explosions into clumps, clouds, shells, and filaments that can collapse to give rise to new stars. Such structure also mitigates the radiative transfer in a galaxy, shielding star forming clouds from photodissociation, or in its absence, leaving clouds mostly dissociated, as dense molecular cores inside large warm atomic shells. Greenberg was the first to *predict*, almost 30 years ago, the 'temperature fluctuations' or

Figure 2. Mayo Greenberg photographed in South Africa with Mr Michael O'Dowd, then Chairman of the Anglo American and de Beers Chairman's Fund. In the background is Professor James Lequeux, with the author at right.

Figure 3. Ready for a game drive in South Africa. Enjoying Life (capital L emphasized) to the utmost, is Mayo, seated in front of this 4×4. Seated behind Mayo is (L-R) his wife Naomi; Liz & Aaron Block, and Michelle Griffiths. In the back row are Ana-Maria Macchetto and Professor Richard Griffiths. Standing next to the 4×4 is Duccio Macchetto, Science Director at STSCI, while a game ranger keeps guard from the rear of the vehicle.

Figure 4. Mayo Greenberg in Pretoria, South Africa, standing on the steps of the Voortrekker Monument. A contrast in profiles: one in stone, the other in flesh. 'Voor'= front in Afrikaans; 'trekker' = 'traveller'. Photograph by the author.

Figure 5. Two master thinkers meet: Mayo Greenberg in deep discussion with former SA Cabinet Minister Ben Ngubane. Copyright reserved by the author & J. Waltham.

'temperature spiking' of very small dust grains. Again the master's vision has been soundly confirmed both by further theoretical and observational studies.

There is more. Cold (20K and colder) dust grains, acting as a cosmic mask or fog, may obscure a huge $\sim 95\%$ mass fraction of Pop II galactic backbones. Significant dust content can play havoc with attempts to accurately measure the light and color distributions in a galaxy, especially if it is embedded as opposed to a foreground screen. Moreover, dust can play havoc with inferences for the morphological classification of stellar Population II disks from gaseous Population I speciations, on which the Hubble tuning fork is based. A entirely new near-infrared classification scheme emerges when galaxies are *mask penetrated*; the Hubble tuning fork strikes a new note.

Greenberg was right on target. Over three decades ago, he *predicted* that 80-90 percent of the dust mass in a galaxy would consist of cold and very cold grains (see his pioneering paper 'Interstellar grains and spiral structure' in the 1970 volume edited by H. Habing) and therefore undetected by IRAS. It is precisely these 'large' (tenth micron) grains which are responsible for the visual extinction in galaxies; not the smaller, one-hundredth micron grains.

In January 1996, we could all salute the legendary Greenberg on the fact that optical minus near infrared imaging combined with radiative transfer codes, *now beautifully confirmed* his predictions: dust masses (and therefore dust-gas ratios) can increase by one order of magnitude. Interarm dust is everywhere. [See the volume which Mayo and I edited, entitled *'New Extragalactic Perspectives in the New South Africa'*, Kluwer, 1996].

Figure 6. Surrounded by his ever favourite 'yellow stuff' in this caricature (by Cliff Brown) is Mayo Greenberg.

Figure 7. Mayo Greenberg in Africa. (Top) with Matthias Steinmetz (left), Rogier Windhorst and B. Rocca-Volmerange. Photograph by Cliff Brown. (Bottom) On a tour visiting Soweto. Photograph courtesy Liz Block.

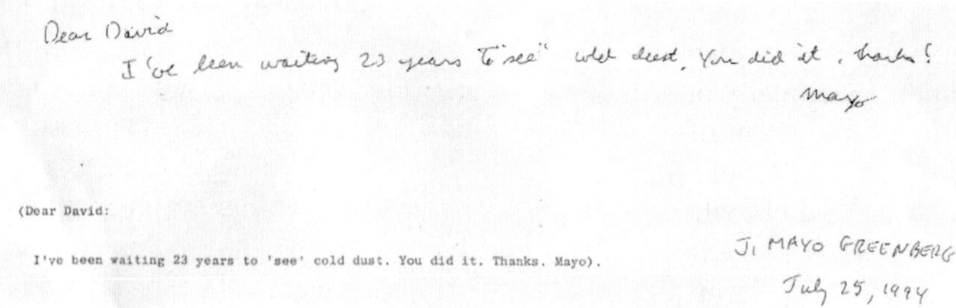

(Dear David:

I've been waiting 23 years to 'see' cold dust. You did it. Thanks. Mayo).

J. MAYO GREENBERG

July 25, 1994

Figure 8. Mayo had predicted the existence of 20K cold cosmic dust in our Galaxy, and beyond, decades ago. The observational challenge then was to find those grains which IRAS did not detect in galaxies outside of our own. Mayo handed this note to our team of researchers at a Conference in Cardiff in 1994: "Iv'e been waiting 23 years to 'see' cold dust. You did it. Thanks! Mayo." A greater mentor cannot be envisaged.

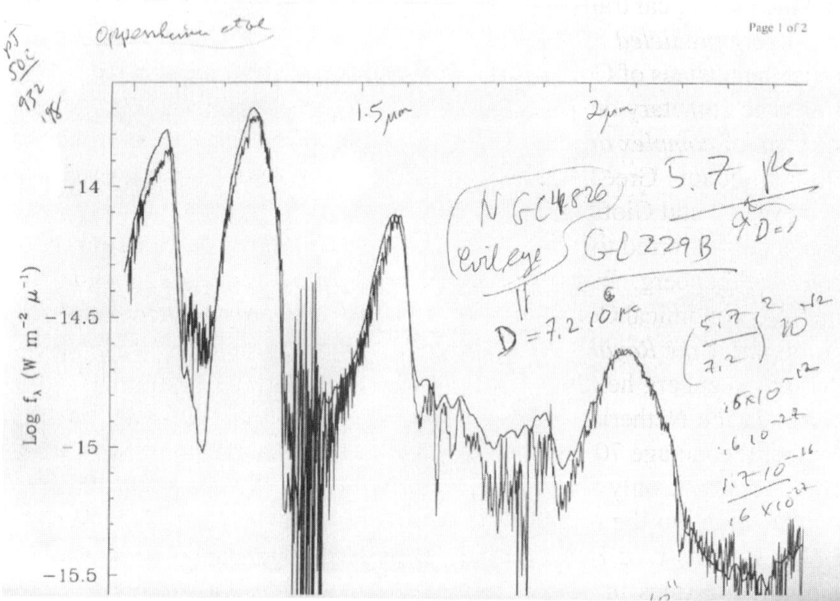

Figure 9. Mayo's 'office' knew no bounds. A few quick calculations (handwritten by Mayo in a hotel room in South Africa) appear on a Keck NIR spectrum of the brown dwarf Gliese 229B, published by B. R. Oppenheimer et al. in 1998. The words 'Oppenheimer et al.' appear, in Mayo's writing, at top left. Mayo – a man, heading for 80, with an unprecedented sweep and appreciation of current literature.

Greenberg's predictions were made some 30 years before 2D direct imaging could routinely become possible longward of I ($0.89 \mu m$). Sub-millimetre/mm observations had brought the issue of cold dust to the fore, but with controversial and diametrically opposed conclusions. Quantitative techniques to probe the full spatial extent of dust grains of all temperatures awaited the commissioning (circa 1990) of large format HgCdTe NICMOS arrays on telescopes at Mauna Kea and elsewhere. We generated optical minus near-infrared colour maps to establish the existence, at arcsecond resolution, of a cold and very cold population of (interarm) dust, and Greenberg's predictions were conclusively verified in 1994 and 1996.

As I reflect over the publication list of Greenberg, I must, in passing, salute the master in a paper co-authored by him in *Nature*, concerning the possibility of dust grains in in galactic haloes. 4096×4096 pixel CCD imaging of the galaxy NGC5907 with the Canada France Hawaii Telescope suggests – on observational grounds – that dust *may* be present in its halo. Colour excesses are at levels of ~ 0.1 magnitude for the halo NGC 5907, comparable to interarm colour excesses one finds in dusty galactic disks (eg. NGC4736). While it remains a great observational challenge to image galactic haloes, Greenberg blazed the theoretical trail.

Greenberg *predicted*

- that the nucleus of Comet Halley would be black (*Nature*, 321, 385, 1986)
- that the cometary dust would be $\sim$ one half rocky-silicate, the rest being in the form of *complex organic material.*

On both counts Greenberg was right again: predictions confirmed by the Vega 1, Vega 2 and Giotto missions.

I have only referred to a handful of the implications of the 300+ papers authored by Greenberg. In the citation survey conducted by P.C. van der Kruit entitled 'Astronomical Community in the Netherlands' [published in the *Quarterly Journal of the Royal Astronomical Society* (vol. 35. no. 4, pg 421, 1994)], Professor Greenberg held the highest number of cited publications of any astronomer in the Netherlands for the year 1991. No mean achievement for a researcher then at age 70! We indeed stand on the shoulders on giants.

Thus far, I have only focussed on some of the *publications* of Greenberg. I have not alluded to the *doctoral students* Professor Greenberg has mentored; scientists such as L. d'Hendecourt who are now, in their master's footsteps, playing crucial roles in our understandings of, for example, polycyclic aromatic hydrocarbons. Neither have I alluded to students who have worked with Greenberg in a postdoctoral capacity, and who later went on to establish research groups of the highest calibre. I think of Lou Allamandola, for example, who worked as a postdoc in Greenberg's Laboratory at Leiden.

What a fitting way it was to salute the man in 1997, when I invited astronomers from around the globe to send in their birthday wishes to Mayo on his 75th birthday. Scores, dozens, of emails, poured in. George Miley wrote the first congratulatory email:

"Your presence during the last two decades has been of immense benefit to the prestige of Leiden. Your immense enthusiasm and stimulating attitude to research is a joy to see. I cannot believe that you are three quarters of a century young. I regard it as a great privilege to have you as a colleague and friend. On behalf of Hanneke and myself, Happy Birthday!"

Next followed an email to my office from Sir Martin Rees, Astronomer Royal. This message was followed by emails from Ron Allen, the Elmegreens, Y. Terzian, G. Herbig, B. Draine, A. Li, F. Israel, J. Mather, and about 100 more researchers.

Walt Duley (University of Waterloo) pleaded:

"Happy 75th, Mayo! Please don't answer ALL the outstanding questions on dust before you retire. Leave a few for the rest of us to deal with! Your friend, Walt Duley."

Mayo always remembered his viewing of Comet Halley at the home of Bruce and Debbie Elmegreen. Bruce recalled:

"...you never failed to impress me with the breadth and originality of your work. I also have the fondest memories of viewing Halley's comet with you in our back yard. I wish you and your family many many more happy years. Bruce Elmegreen."

George Herbig, in his congratulatory email, included a small, but precise, calculation:

"Dear Mayo: May your life continue to be filled with dust, molecules, and other assorted interstellar debris! Best regards and congratulations on passing the 2.366×10^9 *s milestone, George Herbig."*

Kalevi Mattila recalls:

"Your visits to Finland and your presentations are well remembered still today – not least the magic tricks you presented to my (at that time) small children. The first international meeting I attended – the interstellar dust conference in Jena in 1969, was organized under your leadership. Ever since then it has always been a great pleasure and inspiration for me meeting you at different places and conferences..."

Martin Cohen (Berkeley) summed it up well:

"That yellow soup obviously agrees with you!"

From founding the microwave scattering laboratory at the Rensselaer Polytechnic Institute in New York, to the Laboratory of Astrophysics in Leiden, we continued to see a man not only characterized by a lifetime of trailblazing research, but there was more. Like his beloved hero Einstein, Mayo was, in his chosen field, a giant of a man whose profound *predictive* theoretical insights were truly astounding. As Professor John Kerridge (UCSD) emphasized:

"You have left an indelible and constructive mark in a host of fields, but of course most notably in the relationships between interstellar grains, comets, and primitive solar systems. Like so many others, I have learnt a great deal from your numerous studies in that area..."

Mayo was a family man, and Naomi was his closest earthly companion. The love they shared – how they still held hands whilst briskly walking the streets of Leiden – are especially etched in my mind. For me to be included in his innermost circle of galactic research companions was an immense privilege. When I first met Mayo, it seemed as if we had known one another for years. There was a mutual bond – a tie of *extraordinary friendship* (Figure 1).

The Greenberg home on the Rhine... the hospitality which Naomi and Mayo extended to my wife Liz and me at *Morsweg* over the years can, and will, never be forgotten. Taxis were 'forbidden'. On my last arrival at Schipol Airport, to speak at the Oort Centenary, Mayo and Naomi were both there to meet me. To quote George Miley: their hospitality was *legendary*.

Mayo had an exuberance for Life – with a captal L. He loved to LIVE. He loved to *encourage*. He visited me in South Africa on two occassions (figures 2-7), and we also walked the streets of Paris. We visited the Musee Rodin together. He marvelled at Camille Claudel's onyx and bronze *La Vague* and at Rodin's *The Walking Man, The Eternal Idol, La Tour du Travail* and many others. Each visit by him was filled with enough dreams to last for a lifetime.

When we went game viewing in our South African reserves, Mayo was always there – right in the front seat of the 4×4. His eye, always eager to spot a lion kill; a herd of elephant; a stalking cheetah.... At meal time, Mayo kept all enthralled. I recall staying at his home about a year or two before he passed away ... and to see a man, close to 80, carrying his back-pack, rushing off to catch the train (en route to the airport) to yet *another* Conference abroad, left an indelible impression on me.

There is a lot I will not say, because they are personal memories. But let me say this: Mayo enriched my life. He encouraged me, never to give up. He supported every facet of my research. Figure 8 shows but one example. We probably spoke on the telephone at least twice a month – often much more. If not by phone, we constantly emailed one another. We thoroughly enjoyed chatting about a huge range of topics, including the magnificent near-infrared spectra of the brown dwarf Gliese229B by Tom Geballe, B.R. Oppenheimer, S.R. Kulkarni and their collaborators (Figure 9).

I had already partly constituted a Scientific Organising Committee to organise a Conference here in South Africa for Mayo's 80th; much support was received from George Miley, Francoise Combes, Johan Knapen, Duccio Macchetto, Bruce Elmegreen and Ken Freeman, among others. But it was not to be. Fortunately George Miley, Ewine van Dishoeck and Willem Schutte were able to quickly organise a meeting in Mayo's honour, in Leiden; Mayo was physically still strong enough to attend.

I spoke to Mayo in his hospital bed in Belgium. We spoke just before he passed on, at his home in *Morsweg*. He was almost too weak to speak. But there came the voice, weak but *still speaking of future research plans and about dust in the high redshift universe*.

Mayo is no longer with us, but we are delighted to have his former student with us here in the Pilanesberg: Aigen Li. Mayo often spoke to me of Aigen, always in the fondest of terms; almost as in a 'father-son' relationship. Mayo was hugely proud of Aigen; he was particularly pleased to see Aigen move to Princeton, where Aigen conducted postdoctoral studies with Bruce Draine.

It is in his memory that we ask Prof. Jean-Loup Puget (Directeur d'Institut d'Astrophysique Spatiale, Orsay) to deliver the J. Mayo Greenberg Memorial Lecture (reproduced in this volume).

Mayo was one of my very closest research companions, and we miss him sorely.

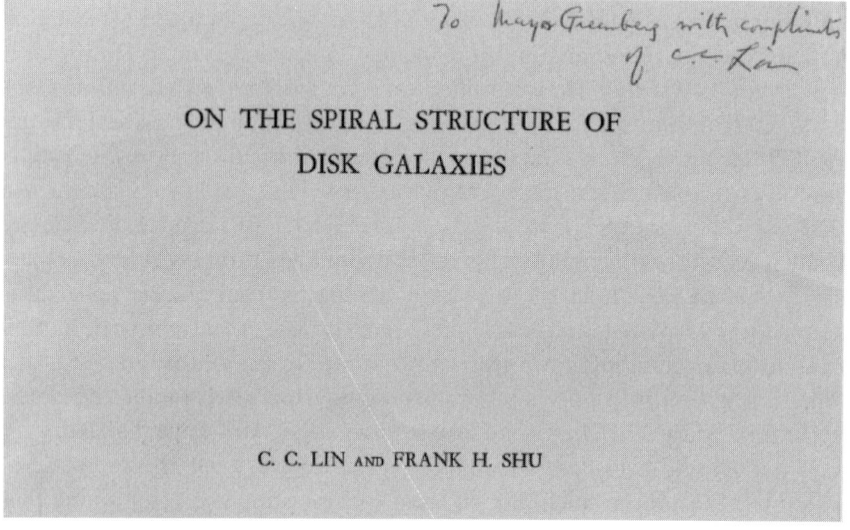

Figure 10. In the hand writing of C.C. Lin – "To Mayo Greenberg with compliments of C.C. Lin".

THE HUBBLE TUNING FORK STRIKES A NEW NOTE

D.L. Block[1], K.C. Freeman[2], I. Puerari[3], F. Combes[4], R. Buta[5], T. Jarrett[6], G. Worthey[7]

1. School of Computational and Applied Mathematics, University of Witwatersrand, South Africa; 2. RSAA, MSO, Australia; 3. INAOE, Mexico; 4. Obs. de Paris, France; 5. University of Alabama, USA; 6. IPAC, Caltech, USA; 7. University of Washington, USA

Abstract Astronomers have for decades referred to the *Hubble* Tuning Fork. We show that the actual originator of the Tuning Fork is that of Sir James Jeans. We next focus our attention on the duality of spiral structure and the temporal evolution of bars in our local Universe. Next, we present deep-IR observations of M33, which show gargantuan arcs of thermally pulsing asymptotic giant branch carbon stars; we attribute this intermediate age stellar population in the outer disk, to accretion of gas in the disk building process. We then consider power spectra of galaxies and Kolmogorov turbulence in gaseous Population I disks and the fractal nature of Population II disks. In particular, power spectra may serve as excellent diagnostics of the regime wherein vertical and horizontal gravitational instabilities in stellar disks are coupled. Finally, we argue that quantitative dust-penetrated templates for galaxies in our local Universe may also serve as excellent templates for galaxies at higher-z, because of the duality of spiral structure expected from gaseous to stellar disks.

Keywords: Galaxies: spiral, structure, kinematics and dynamics, high redshift — Methods: numerical

1. Renaming the Hubble Tuning Fork

Ever since Hubble published his famous book *The Realm of the Nebulae*, in which he presents a tuning fork diagram, astronomers have always referred to the *Hubble* Tuning Fork.

Who is the actual originator of the graphic tuning fork diagram?

Attention to this intriguing question was drawn to me by astronomy historian William Sheehan, and my interest in this matter deepened when I approached morphologists Allan Sandage and Sidney van den Bergh.

Curiously, no graphic description of a tuning fork is given in Hubble (1926). Indeed, the first time Hubble presents the tuning fork diagram is in his book, published in 1936.

D. Block et al. (eds.), Penetrating Bars through Masks of Cosmic Dust, 15–37.

The answer lies in a work published in 1929 by Sir James Jeans, entitled *Astronomy & Cosmogony*. In chapter XIII of that book, Jeans argues that a Y-shaped diagram would be appropriate to graphically depict the classifications suggested by Hubble in 1926.

"Hubble finds that it is not possible to place all observed nebulae in one continuous sequence; their proper representation demands a Y-shaped diagram such as shown in Figure 53" [of Jeans' book].

Jeans presented the first tuning fork diagram in 1929, seven years before *Realm of the Nebulae* appeared in print.

The reader is referred to an excellent historical perspective on these issues by Dr Allan Sandage (this volume).

May our Morphology Conference strike the correct historical note: while it is correct to speak of Hubble types and of the Hubble classification scheme (normal, barred), we must speak of the Jeans (or at least, Jeans-Hubble) Tuning Fork!

2. The Duality of Spiral Structure

Penetrating the dust mask: A ubiquity of low order modes

In Roget's Thesaurus, we find the following:

Masks:

[noun] screen, cloak, shroud. [verb] to camouflage, to make opaque, to disguise.

Optically thick dusty domains in galactic disks can completely camouflage or disguise underlying stellar structures. *Cosmic dust grains act as masks.* The dust masks obscure whether or not the dust lies in an actual screen or is well intermixed with the stars. The presence of dust and the morphology of a galaxy are inextricably intertwined: indeed, the morphology of a galaxy can dramatically change once the Population I disks of galaxies – the masks – are dust penetrated (e.g., Block and Wainscoat, 1991 Block et al., 1994).

The Hubble classification scheme of galaxies is based on their optical appearance. As one goes from early to late type spirals, both barred and unbarred, the optical appearance will be dominated more and more by the young Population I, i.e., blue stars and dust. Atlases reveal the rich variety of responses of the Population I component of gas to the underlying, older, stellar population. However, the gaseous Population I component, may only constitute 5 percent of the dynamical mass of the galaxy.

From a dynamical viewpoint, the disk of a spiral galaxy can be separated into two distinct components: the *gas–dominated* Population I disk, and the *star–dominated* Population II disk. The former component contains features of spiral structure (OB associations, HII regions, dust and cold interstellar HI gas), which are naturally fast evolving; dynamically, it is very active and responsive. The old stellar population betrays the underlying mass distribution - it is the 'backbone' of the galaxy. Infrared images of spirals are invariably not multi-moded, but show a ubiquity of global one and two armed structures in the underlying stellar disk (Jarrett et al. 2003; Block et al. 1994).

The degree to which dust penetrated near-infrared morphologies vary from gaseous Population I morphologies, varies from galaxy to galaxy. Many years before the advent of large format near-infrared camera arrays, it became increasingly obvious from rotation curve analyses that optical Hubble type is not correlated with the evolved Population II morphology. This was already evident in the pioneering work of Zwicky (1957) when he published his famous photographs showing the 'smooth red arms' in M51. In the *Hubble Atlas* and other atlases showing optical images of galaxies, we are looking at masks. At the gas. Not the stars, to which the properties of rotation curves are inextricably tied. Burstein and Rubin (1985) found three principal types of mass distribution, with Hubble type a and b classes being found *among all three types* more or less equally. Clearly, one needs to probe what lies behind the dusty masks.

We speak of a duality in spiral structure. Optically flocculent galaxies such as NGC 5055 (Elmegreen arm class 3) may present a radically different evolved disk morphology, almost identical to that of NGC 5861 (extreme grand design: Elmegreen class 12). Another optically flocculent with a *grand design* near-infrared disk, is NGC 4062. NGC 4062 even betrays spiral arm modulation in its stellar disk (Puerari et al. 2000), hitherto only detected in the optically grand design galaxies such as M51 and M81. Shu (this volume) attributes these dualities to very strongly coupled gaseous and stellar disks; so strongly coupled, he argues, that the gas actually follows a chaotic pattern. Studies of the 2MASS Large Galaxy Atlas (Jarrett et al. 2003) continues to reveal a rich duality in spiral structure.

To quote R.J. Allen (1996):

"We're now looking at a transition to a possible change in the way we look at galaxies. Sometimes ... we see disks that have a spiral structure that we couldn't have dreamt existed from looking at the optical picture ... we've got a possibility here of applying the morphology to a physical framework, perhaps in a way that none of us could have dreamt of before we had the capability of sweeping the dust away from the galaxy in a figurative sense.".

Dust Penetrated Disks betray ovals and bars

The ubiquity of bars in disk galaxies underlies the fundamental nature of the bar phenomenon. Evidence has been presented that the majority of spiral galaxies in our local Universe may be considered to be barred (Block et al. 2002), when these galaxies are examined in the near-IR. Grobol, Patsis and Pompei report that 26 / 30 galaxies classified optically as ordinary unbarred

(SA) present near-IR bars, down to their detection level at a relative amplitude of 3 percent. Moreover, there is a huge overlap in gravitational torque (defined below) between S and SB spirals (see e.g. Figure 4 in Block et al. 2001).

3. The temporal evolution of bars

Bars are a major perturbation of the gravitational potential and a major engine for the evolution of morphological and chemical structures. In purely stellar disks, bars are robust, long-lived structures (Combes & Sanders 1981). In reality, however, spiral galaxies contain reservoirs of gas (Sancisi 1983) which provoke the dissolution of bars (Bournaud & Combes 2002). The bar itself initiates an important radial gas inflow, which destroys the barred structure (Pfenniger & Norman 1990, Hasan et al. 1993).

The issue of whether *observations* support a galaxy dissolving its bar, and forming another, has awaited the completion of near-infrared surveys. In Block et al. (2002), we argued that bars are transient phenomena, undergoing a 'duty cycle' – dissolving and reforming themselves as many as four times during the age of the Universe.

The presence of gas in galactic disks is responsible for the destruction of bars in no more than 5 Gyrs. As elucidated by Bournaud & Combes (2002), gas is also responsible for bar renewal, when it is accreted from outside the disk. Gas accretion radically changes the temporal evolution of the bar strength: see for example Fig. 4 in Bournaud & Combes (2002). With accretion, once a bar is dissolved, the disk can become unstable again and a new bar may form. The disk rarely resides in axisymmetric states: even between bar episodes, spiral arms maintain a significant torque, for accretion also rejuvenates the spiral structure. Without accretion, the disks spend half of their lifetimes in nearly axisymmetric states with $Q_b < 0.05$: both bars and arms disappear. The number of galaxies in each class of the histogram in Fig.1 of Block et al. (2002) is interpreted as the time fraction galaxies spend in each class during their lives.

In summary, isolated galaxies (non-accreting systems) *cannot* reproduce the observed properties at all: they would become *unbarred*, nearly axisymmetric, after a few Gyrs (see also the contributions from Combes and Bournaud, this Volume). On the contrary, spiral galaxies appear to be *open* systems, that are still forming and continuously accreting mass today. We expect a doubling in disk mass every 10 billions years.

The origin of the accreted gas is most likely gaseous reservoirs observed outside the optical domains of nearly all spiral disks (Sancisi 1983, Katz et al. 2003). Pfenniger et al. (1994) have even postulated that the dark matter around spiral galaxies might be in the form of cold gas, mostly in molecular hydrogen form. Accretion from infalling dwarf satellites are not the main source, since these are only at a level of a few percent (Toth & Ostriker 1992).

The maximal torque over the entire disk is defined by

$$Q_{\rm T}(R) = \frac{F_{\rm T}^{\rm max}(R)}{F_0(R)} = \frac{\frac{1}{R}(\frac{\partial\Phi(R,\theta)}{\partial\theta})_{\rm max}}{\frac{d\Phi_0(R)}{dR}} \quad (1)$$

where $F_{\rm T}^{\rm max}(R)$ represents the maximum amplitude of the tangential force and $F_0(R)$ is the mean axisymmetric radial force, inferred from the axisymmetric component of the gravitational potential, Φ_0. The potential is inferred from the visible mass only, in both simulations and observations. We then measure $Q_{\rm b} = {\rm max}_R(Q_{\rm T}(R))$. $Q_{\rm b}$ is simply the bar strength in barred galaxies, or the maximal arms torque in unbarred or nearly unbarred galaxies.

The case of lenticulars will be treated elsewhere in this Conference. S0s are stellar systems without much gas. In the absence of gas, the dynamics of disks is different: pure stellar bars are very robust, and can endure for one Hubble time (Combes & Sanders 1981), contrary to bars in spiral galaxies. In SOs, bars are not destroyed, and no mechanism is needed to explain bar reformation.

4. The Building of Disks: Active evolution in the Triangulum Spiral M33

"That galaxies evolve is a trivial statement: galaxies are made of stars, stars evolve, and therefore galaxies evolve. But over and above such minimal evolution of the older stellar content of a galaxy, which is often called *passive* evolution, there exists the possibility of *active* galactic evolution..." Augustus Oemler, Jr.

Disks of galaxies appear to form from the inside out (Block et al. 2002) and appear to evolve in an *active* sense. When using the terminology *active galactive evolution*, I shall include evolution of carbon stars in a galaxy – specifically in the outer domains. Carbon stars are extremely important tracers of intermediate age stellar populations; they also make excellent test particles for investigating the kinematics of galaxies (Mould & Aaaronson, 1986). Carbon stars output such prodigious fluxes of light in the near-infrared that the K-band luminosity in galaxies containing a significant population of these intermediate age stars can be enhanced by up to a factor of 2 (Mouhcine and Lancon, 2003).

Relative to the inner regions of spiral galaxies, the mean ages of the outer regions are known to be somewhat younger and more metal-poor (Bell and de Jong 2000). We can therefore expect the contribution from the intermediate-age stars to be stronger in these outer regions, and the near-IR surface brightness of the outer disk will be preferentially enhanced by the presence of thermally pulsing- asymptotic giant branch (TP-AGB) stars. This brightening of

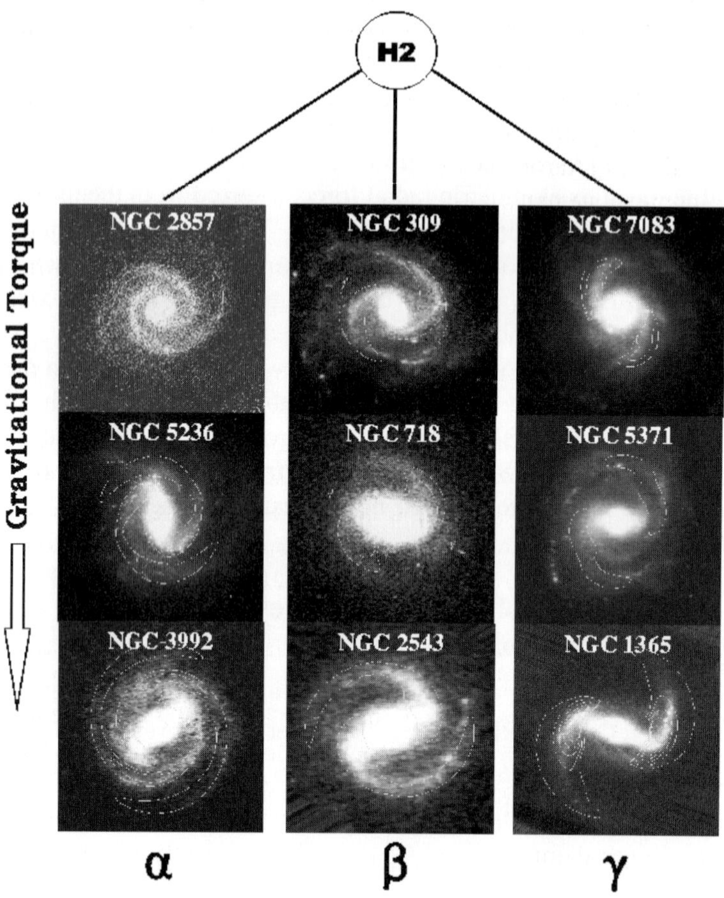

Figure 1. Spiral galaxies in the dust penetrated regime are binned according to three quantitative criteria: firstly, Hm, where m is the dominant Fourier harmonic (illustrated here are the two-armed H2 family). Next, follows the dust penetrated pitch angle families α, β or γ (for class α, the pitch angles range from $\sim$ 4-15°; for class β, the deprojected pitch angles range from $\sim$ 15-30°, while open-armed class γ spirals have pitch angles ranging between $\sim$ 35- 75°). Finally, we compute the gravitational torque, which is identical to the bar torque in galaxies presenting a bar. Bar torques are not derived from bar ellipticities but exploit the full gravitational potential of the disk within which it is embedded. Note that early type b spirals (NGC 3992, NGC 2543, NGC 7083, NGC 5371 and NGC 1365) are distributed within all three families (α, β and γ). Hubble type and dust penetrated class are uncorrelated.

the outer disk in the near-IR may well contribute to the apparent sharp radial truncation observed in the disks of many spiral systems (Kregel et al. 2002). Astronomers routinely use near-IR observations to minimize the effects of dust extinction, but it is precisely in this band that TP-AGB stars are so bright; the actual contribution of carbon stars to the near-IR light of spiral galaxies is in its infancy. Little has changed observationally since Aaronson noted the importance of the asymptotic giant branch for understanding the stellar content of nearer galaxies (Aaronson 1986).

In this context, we present deep new near-infrared images of the Triangulum Spiral M33 (Block et al. 2004). M33 was probably first discovered by Giovanni Hodierna in 1654, before being independently discovered by Messier almost a century later, in 1764. In his description of 'nebulae' and clusters photographed with the Crossley reflector at the Lick Observatory, Heber D. Curtis (1918) described the Triangulum Galaxy as 'a close rival to the Nebula of Andromeda as the most beautiful spiral known'.

The galaxy presents somewhat of an optically flocculent (fleece-like) appearance, a fact already attested to by Lord Rosse in his early drawings ('the whole nebula in [is] flocculi'). In a range of 1-12, Elmegreen and Elmegreen (1987) place the Triangulum Galaxy in their class 5 bin. Sandage and Humphresy (1980) note that ten spiral arms may be identified in deep optical images, although the galaxy is most famous for its two bright inner set of spiral arms. At longer, near-infrared wavelengths, the galaxy appears much smoother as the older underlying population dominates the light (Regan and Vogel 1994). M33 extends over one degree along the major axis and it has a plethora of star clusters, emission regions and supernova remnants (see Hodge, Skelton and Ashizawa 2002 for a complete atlas of these). Its distance modulus is 24.64^m, corresponding to a linear distance of 840 kpc (Freedman, Wilson & Madore 1991). Its linear diameter is $\sim$ 17 kpc, approximately one-half that of the Milky Way.

Although lessons learnt from carbon stars in the Magellanic Clouds is that they may be spatially extended in galactic disks (Blanco et al. 1978) – with about one-half of the bolometric luminosity in intermediate age Magellanic clusters coming from these TP-AGB stars (Persson et al. 1983) – literature discussing the morphological distribution of carbon stars in our closest spiral, the Triangulum Galaxy, is largely silent. One of the formidable difficulties (apart from the huge angular size of M33) is that the integrated surface brightness of carbon-bearing populations in the outer domain of galactic disks may, even in the near-infrared, be very low.

5. Mosaic

The near-infrared M33 images do *not* come from the original 2MASS survey (e.g., Jarrett et al. 2003), but from a special set of 2MASS observations in which the integration time was increased by a factor of six, extending approximately 1 mag deeper than the nominal survey.

The M33 region was surveyed in 2MASS with 0.14 $\times$ 1 degree tiles or "scans" (Skrutskie et al. 1997), forming 4 separate images per tile per band of size 512 $\times$ 1024 pixels (with resampled 1 arcsecond pixels). These images, 8.5 $\times$ 17 arcmin in angular size, are also known as "co-adds", since they comprise about six optimally dithered samples per pixel. The effective beam, or point-spread function FWHM, is 2-3$''$, depending on the atmospheric seeing. A total of 36 coadds per band comprise a full mosaic, covering approximately 1 deg^2.

Three mosaics are constructed for M33, corresponding to the J (1.2 μm), H (1.6μm), and Ks (2.2μm) bands. The typical 1σ background noise is 22.52, 21.24 and 21.05 mag arcsec^{-2} for J, H and Ks respectively.

We then constructed a "master" JHK mosaic, by combining the three individual mosaics, to enhance extremely faint surface brightness features. The combined mosaic image resolves features as small as 10 pc in M33, with 5σ sensitivities as faint as $\sim$ 20.2 mag arcsec^{-2} (K$_s$) for small, compact objects, and $\sim$22.5 mag arcsec^{-2} (K$_s$) for large- area features.

Removal of Foreground Stars

Foreground Milky Way stars were removed statistically, using the J-Ks histogram of two control star-fields to the East and West of M33 (but at the same galactic latitude) as a template. Of the $\sim$7000 point sources in the original image, $\sim$2300 were removed as foreground stars.

Results

The near-infrared images reveal remarkable arcs of red stars in the outer disk of M33, spanning over 120° in azimuth and delineating a swath up to 5$'$ (see Figure 2). The northern arc is dominant although a faint southern counterpart arc, forming a partial ring, can be seen in Figure 2. In order to better see these plume features visually, we can also subtract off an axisymmetric disk model. The $J - K_s$ colours of stars in these arcs are so red, that an M-giant population is excluded as the dominant light source. The arcs would be completely missed in conventional near-infrared surveys such as 2MASS.

Although M33 is optically famous for its bright inner pair of spiral arms, we show that, in the near-infrared, these arms surprisingly do *not* dominate the $m = 2$ Fourier mode.

To generate our Fourier spectra, we carefully determined the barycentre of the light (mass) distribution, and deprojected about that point. As in Deul & vd Hulst (1987) and Regan and Vogel (1994), we adopted the de Vaucouleurs values of log R = 0.23 (R is the ratio of the major-minor axis) and a position angle of 23 degrees.

In M33, we comment that the only dominant modes are $m = 1$ and $m = 2$; the galaxy does not present ten arms in the near-infrared, as it does optically. The theme of the duality of spiral structure is ever recurrent. We immediately note that the two inner 'grand-design' trailing spiral arms seen optically show up rather strikingly in the near-infrared regime. The computed pitch angle is $\sim$ 14 degrees. This inner set of near-infrared arms may also strikingly be seen in Regan and Vogel (1994) (see their Figures 1 and 3).

Far more important, however, is the *dominant* and very high peak for m=2; the peak does not occur at p_{max} of zero (as expected for bars, for example, with pitch angles $\sim$ 90 degrees, corresponding to spectral m=2 signatures with a peak at zero). Rather, the dominant m=2 peak corresponds to the set of giant arcs (very open two arm structures) on both the northern and southern sectors of the disk. The pitch angle of the dominant m=2 mode is $\sim$ 58 degrees. *It is these outer arcs (and not the inner grand design structure for which M33 is so famous) which dominates the m=2 Fourier spectra.*

M33 has a prominent central near-infrared bar/oval, whose gravitational torque class is two, in a range of class 0 - 7 (see also Groess, the Volume). The colour of the northern plume-like arc is estimated to range from J-Ks = 0.8 mag to 1.2 mag and redder, but it is exceedingly faint (the mean surface brightness of the plume lies between 20 and 21 mag arcsec^{-2} in Ks). In their study of M33 associations by Sandage & Humphreys (1980), associations 37, 39, 42, 43, 44 do lie in the vicinity of the arc. The northern arc of red stars is indeed seen – albeit very faintly – in the shallower near-infrared images of M33 (see Figure 3 of Regan and Vogel 1994).

The redness of the arcs is not produced by dust; numerous extinction estimates for M33 stars and associations are contained in the literature. Wilson (1991) obtained E(B-V) = 0.3 $\pm$ 0.1 mag, which included both foreground (Milky Way) and internal M33 extinction. The visual extinction in M33 is estimated to be A(V) $\sim$ 3 E(B-V) = 0.9 mag. The near-infrared extinction, A(K), is approximately one-tenth that in the optical (Rieke & Lebofsky 1985), so that our estimate of dust extinction at Ks is only $\sim$ 0.09 mag.

We therefore explore the possibility that the arcs may, in part, represent populations of very red *carbon stars*. The corner of parameter space occupied by individual carbon stars in our Milky Way may be studied in Figure 5 in Persson et. al. (1983). Ages of young clusters in the SMC have a turnoff mass which generates large populations of carbon stars that are very cool, with red J-Ks colours. The C stars themselves in Figure 3 have J-Ks colours in the range

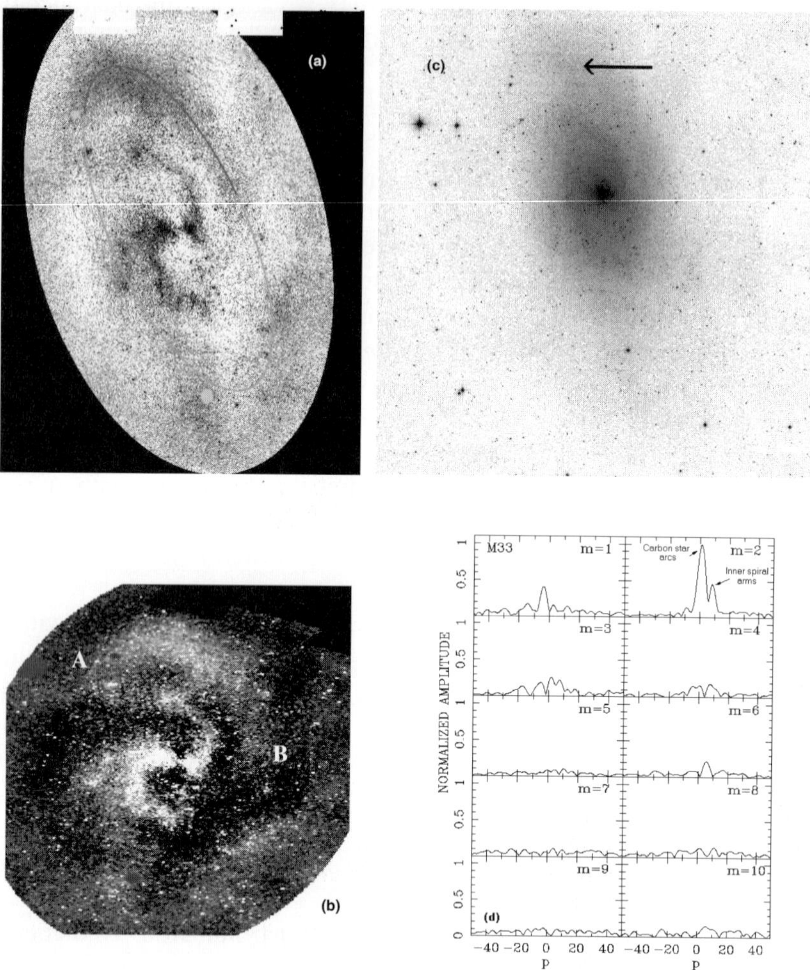

Figure 2. (a). A deep JHK_s mosaic of M33. A gargantuan plume-like ring of red stars stretches in a swath (up to $5'$ in width) for over 120 degrees, commencing at $\sim 14'$ north of the centre of M33. In order to best reveal the plumes and inner arms, we have subtracted an axisymmetric model of the disk. (b). A deprojected $JHK - s$ mosaic of M33, with an axisymmetric disk model subtracted. The northern plume is labelled A to B. (c). A deep 2MASS $H - band$ image of M33 (non-deprojected), reveals the northern arc (arrowed) and other red arcs in the outer disk of M33, (d). Fourier spectra generated from the deprojected near-infrared mosaic of M33. The dominant $m = 2$ mode in M33 arises from the giant outer ring or arcs of carbon stars.

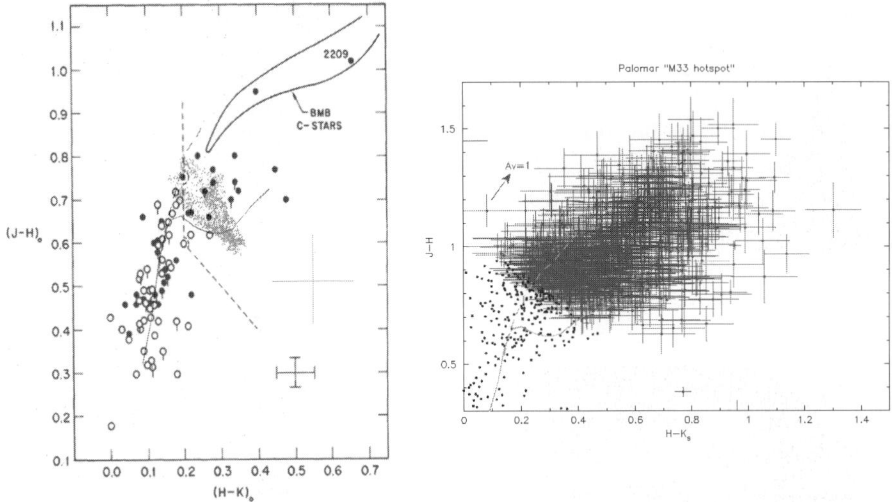

Figure 3. An overlay of the integrated cluster light in the Magellanic Clouds (from Figure 4 of Persson et al. 1983), with the integrated colours for plume stars (grey dots) in the SW sector of M33. The open symbols principally show the Searle, Wilkinson and Bagnuolo (SWB) Magellanic cluster types I, II and III; these do not contain carbon stars. The filled symbols are SWB types V, VI and VII. The red Magellanic carbon-star-bearing clusters lie above and to the right of the black dashed lines, in the Magellanic "IR-enhanced" cluster regime identified by Persson et al. (1983); these are predominantly SWB types V and VI. The thin error bars show our 1-sigma uncertainty in the color for each point in this figure; the thicker, blacker error bars are from the Persson et al. (1983) analysis. Also seen are the Bessell & Brett (1988) tracks for giants and dwarfs respectively. The overlay provides strong evidence for a carbon star population in the M33 plumes.

of $J - Ks \sim 1.2 - 1.9$. The integrated $J - Ks$ colours for clusters in the Large and Small Magellanic Clouds ranges from $\sim 0.4 - 1.2$.

Persson et al. (1983) comment:

"The effects of luminous carbon stars upon the infrared colors of the parent clusters are strong enough that metal-poor, intermediate age stellar populations may be detectable in the integrated light of more distant galaxies."

The colour of the northern arc extends to J-Ks > 1.1. Very old M giants of solar abundance can reach J-Ks ~ 1 (see e.g. Figure 2 in Bessell and Brett 1988), and even redder if they are super-metal-rich (Frogel and Whitford 1987). However, as reviewed by Pagel and Edmunds (1981), there is a strong radial abundance gradient (Searle et al. 1980) in M33 (-0.09 ± 0.02 dex/kpc in O/H); the outer regions are relatively metal-poor, and solar abundance is reached only within the inner $1''$ of M33. In regions of lower abundance, the giant branch is bluer. If stars with J-Ks > 1 are found in the outer low-metallicity regions, they cannot be M-giants. Figure 3 presents the integrated colours of the plumes in the SW sector of M33, overlayed with the integrated colors of the clusters in the Magellanic Clouds (1983). The overlay provides strong evidence for a carbon star population in the extended M33 plumes.

Follow-up observations with the Hale-5m at Palomar

One of us (TJ) recently imaged a section of the northern plume of M33 with the 5m-Hale reflector, using the 2048×2048 array near-infrared camera WIRC. The field of view is 8.5x8.5$'$ with 0.25 arcsec pixels. The seeing FWHM is $0.8''$ in J, and 0.7 $''$ in the K_s band. The telescope was centered at 01h34m28.1s, +30d54m00s (J2000). The total JHK_s integration time was ~ 9 minutes, reaching a limiting surface brightness at K_s of 23.7 mag per arcsec^{-2} (at 1σ, the RMS is 22.21 mag arcsec^{-2} per pixel; 4×4 pixels gives 1 square arsec), The point source photometry SNR=10 limits are 19.0, 18.0, 16.9 mag in JHKs, respectively. The figures (one with error bars, one without) shows the photometric $J - H$ and $H - K_s s$ colors for sources detected in the 76 arcmin^2 field. The sample has been cleaned of all low detection sources, with a $S/N < 10$. The extinction reddening vector is indicated with the red arrow. The evolved giant and main-sequence dwarf tracks are shown with green dashed and solid lines, respectively. In this color-color plot, no foreground MW stars were statistically removed, as in the other color-color plot. Here foreground stars appear with blue colors ($H - K_s < 0.2$ and $J - H < 0.8$; $J - K_s < 1.0$), while most of the M33 sources have colors that are redder than $H - K_s = 0.4$ and $J - H = 0.9$ ($J - K_s = 1.3$) mag. The color uncertainties for the red sources are less than 10A plethora of stars is seen toward the C-Star regime in the upper right corner of the figure, with colors exceeding $J - K_s \sim 1.8$ mag.

Discussion

The deep 2MASS images and some follow-up Palomar observations of our neighbouring spiral M33 identify for the first time the presence of large numbers of carbon stars in its outer regions. This directly emphasises the major contribution that carbon stars can make to the near-IR luminosity of a spiral galaxy with a sustained star formation history.

It is a sobering thought that only last year (Marigo 2003) have grids of asymptotic giant branch models become available which include carbon stars. All integrated-light models to date (e.g. Worthey 1994, Bruzual and Charlot 1993, Leitherer et al 1996, Vazdekis 2001) do *not* include carbon stars.

How young might carbon star population in the arcs be? TP-AGB stars have ages spanning at least 0.6 Gyr to 2 Gyr, but the interval could be wider. Inferring upper and low age limits for carbon stars is exceedingly difficult, simply because no calibrating SMC or LMC clusters exist in the two separate age regimes of (0.2 Gyr, 0.6 Gyr) and (2 Gyr, 4 Gyr) (see Figure 1 in Marigo et al 1996).

Where is accretion of gas in M33 most effectively expected to take place? M33 has a huge warped envelope of neutral hydrogen gas, with a notable asymmetric extension toward the North-West (eg. Corbelli and Schneider 1997). These authors suggest that this NW extension may be a betrayal of tidal interaction between M33 and M31, which are only $\sim$ 200 kpc distant.

In their modelling of the HI envelope, Corbelli and Schneider find that the phase changes at a radius of $\sim$ 20 arcmin, which is precisely the outer domain of the arcs reported in this study. It seems highly plausible therefore, that fresh, low-metallicity gas is being fed to the host galaxy M33, via external accretion. The outer disk from which mass is accreted is inclined to the inner disk of M33, and has a different angular momentum. We believe that it is the signature of gas flowing inward and accreting at $\sim$ 20 arcmin, from which the very red, and relatively metal-poor stars have been formed.

It is tempting to liken the ring in M33 to the one recently reported in the outer disk of our Milky Way (astro-ph/0301067). For the Milky Way ring, the rotation period is $\sim$ 600Myr. For M33, the rotation time-scale at the radius of the ring would be $\sim$ 240 Myr. Any carbon-bearing clusters of age 0.6 Gyr would have only undergone $\sim$ 2 orbits.

Our high z universe: when Carbon Stars were Rampant

Using the Large Magellanic Cloud as a guide, we have noted that carbon stars are produced in large numbers between ages of about 0.6 and 2 Gyr. Therefore galaxies that undergo a burst of star formation will have their IR light boosted 0.6 Gyr later, and this extra light will die away after about 2 Gyr. We know that star formation in the early universe (Bouwens at al. 2003) was already proceeding at redshifts $z \sim 6$. There may be an epoch starting about 0.6 Gyr after the onset of star formation in the universe (i.e. at redshifts $z \sim 4$) characterized by large numbers of carbon stars. The dominant output from the carbon stars will be redshifted into the mid-infrared where these stars could double the observed galactic flux. Instruments like the Mid-Infrared Instrument on board the JWST are needed to image these galaxies in their (rest-frame) 2.2 micron band.

6. Is the optical light in spiral galaxies the result of turbulence?

One of the major processes which structures the interstellar media in galaxies is mechanical feedback: for example, supersonic winds from massive stars and multiple supernovae from OB associations. Such mechanical feedback may generate wind-blown bubbles as seen, for example, in deep Schmidt plates of the Rosette Nebula (Block, 1990). Supernovae remnants as well super-bubbles (in both ionised as well as neutral gas) are two other well known examples.

The other major process that structures the interstellar medium on scales of 10 to $\sim$ 500 pc is *turbulence*. The standard reference for turbulent power spectra is the Kolmogorov model, applicable to homogeneous, isotropic, incompressible and adiabatic turbulence.

The correlated structure in a turbulent gas is with a power spectrum:

$$\hat{I}_c(k) = \sum_{n=1}^{N} \cos(k2\pi n/N)I(n)$$

$$\hat{I}_s(k) = \sum_{n=1}^{N} \sin(k2\pi n/N)I(n)$$

$$P(k) = \hat{I}_c(k)^2 + \hat{I}_s(k)^2$$

N is the number of pixels in the azimuthal scan and k is the wavenumber. Two dimensional power spectra of the neutral hydrogen (HI) emission of the Milky Way (e.g. Green, 1993; Dickey et al. 2001) are characterised by power laws, with slopes varying from -2.8 to -3, indicative of classical Kolmogorov turbulence. Elmegreen, Kim and Staveley-Smith (2001) probed spatial scales in the Large Magellanic Cloud, ranging over three orders of magnitude (30-4000pc), and found a 2D power law slope $\sim$ -2.7 (or equivalently -1.7 in 1D), again betraying Kolmogorov turbulence. Velocity and density spectra of the Small Magellanic Cloud are presented by Stanimirovic and Lazarian (2001) and are again found to be approximately Kolmogorov; the spectral indices (for their 3D power spectra) are -3.3 and -3.4, equivalent to a 1D power law slope of -1.4.

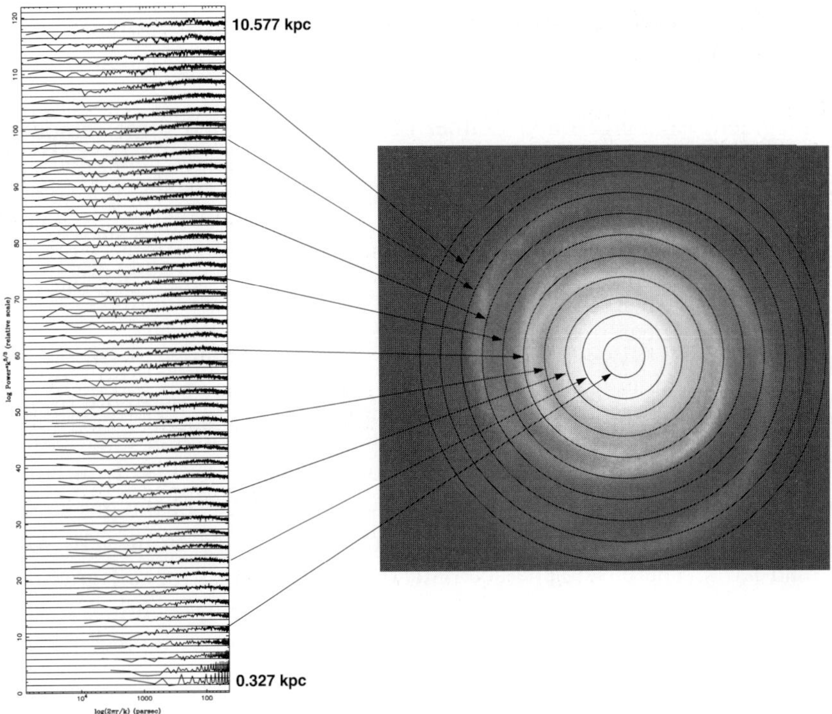

Figure 4. Forty-eight power spectra are generated from this HST image of the Sb galaxy NGC 4622. Of particular interest is the -5/3 power law slope within the domain of the probable thickness of its galactic disk (for the Milky Way, this is 325 pc). We believe that the 'hump' seen in the power spectra of NGC 4622 at high frequencies may well be a diagnostic of the coupling of vertical and horizontal gravitational instabilities in its stellar disk. HST image courtesy R. Buta.

For classical Kolmogorov turbulence, the implication is that the gas is fractal in nature, with the turbulent kinetic energy cascading to ever smaller scales, according to $E(k)\alpha k^{-5/3}$. This derives from the assumption of a constant energy transfer at at scales. Values steeper than -5/3 signify progressive energy losses, such as from compressible or shock dominated turbulence.

Why use power spectra instead of conventional 2D Fourier transforms? For galaxies, the 2D Fourier transform includes structure from the exponential disk, which is generally not wanted in analyses of interstellar clouds. Power spectra generated from azimuthal profiles are, however, ideal, since in any given azimuthal swath, there will be no systematic radial gradients from the disk.

Elmegreen, Elmegreen and Leitner (2003) have studied some famous examples of nearby spirals (such as NGC 3031=M81), and have generated power spectra of the optical light in passbands such as B, V and R. Their conclusion is that the optical light in spiral galaxies (even grand-design ones such as M81) is the result of turbulence; the implication therefore, is that young stars follow the gas as they form. They argue that large-scale turbulent motions may well be generated by sheared gravitational instabilities in the disk.

Elmegreen et al. (2003) suggest that *gravity* is the principal origin of turbulence, except in the diffuse interstellar medium, where the inertial term dominates the gravity term. Elsewhere, such as in self-gravitating giant molecular clouds, or on larger scales where the Toomre parameter Q$\sim$ 1, turbulence produces self-gravitating clouds at the crashing interfaces of the gas streams. Self-gravitating clouds of gas have a wide range of masses, from $\sim 10^7 M_\odot$ to less than $1 M_\odot$. There is no characteristic or dominant mass for self-gravitating clouds: most star-forming regions are similar, except for size. It is size which determines the velocity dispersion of the cloud, as well as its density, for a common background pressure.

Which comes first, the gravity or the turbulence, is a chicken and egg question: on the galactic scales we are studying, gravity helps drive turbulence and both produce cloud structure.

It might be surprising that it is *incompressible* Kolmogorov turbulence which appears to occur in the gaseous component of spiral galaxies, rather than compressible or shock-dominated turbulence. While there are no good theories for compressible turbulence, we do note that compressible turbulence gives a range of structures between incompressible turbulence and shocks. In a medium that is supersonically turbulent (shocks), the 2D power spectrum slope changes from -8/3 (-2.67) to -3; it is very difficult, therefore, to distinguish incompressible from shock dominated turbulence, as the differential signature is not large. Even the most extreme cloud formation scenarios, where all clouds represent shocks, would have a power spectrum similar to that of the classic Kolmogorov incompressible model.

Power spectra are extremely useful for much needed insight into the physical scale at which star formation in galaxies becomes coherent; in our Galaxy, we know this may range from several hundred parsecs, to a kpc (Gould's belt, of age $\sim$ 30 Myr, spans $\sim$ 1 kpc). If the scale height in the ambient medium is the typical scale for coherent star-forming structures, then self-gravity in the gas initiates the formation of star-forming clouds and star complexes.

7. The Fractal Nature of Dust Penetrated Disks

In this section, we generate power spectra on Population II disks, for the first time (see Figure 4).

As a test, forty-eight power spectra were generated from an HST image of the Sb galaxy NGC 4622. The galaxy was analysed in circular swaths, from a galactocentric radius of r=10 to r =490 pixels. The first circular swath covers (10, 19) pixels, the second one (20, 29) pixels, ..., (480, 489) pixels. Adopting a distance to NGC 4622 of 45.02 Mpc and using a HST/WFPC2 scale of 0.09993 arcsec pix^{-1}, gives a linear scale of 21.81 pc per pixel. The mean radius corresponding to the lowermost power spectrum is 15 pixels (or 0.327 kpc); that of the uppermost, 485 pixels or 10.577 kpc. Each power spectrum contains Fourier modes k=1 to k=1800. We have multiplied each power spectrum by $k^{5/3}$ so that power spectral slopes of -5/3 will appear horizontal here.

Of particular interest is the -5/3 power law slope within the domain of the probable thickness of its galactic disk (for the Milky Way, this is 325 pc). As soon as structures are formed with sizes comparable to, or smaller than, the disk thickness, the dynamics in the plane and vertically to it are no longer independent (see, e.g., Huber and Pfenniger, 2001). *We believe that the 'hump' seen in the power spectra of NGC 4622 at high frequencies may well be indicative of the coupling of vertical and horizontal gravitational instabilities in the stellar disk.*

8. Eyes to thc future: Restframe K-band images at high redshift

We have already noted that lessons learnt from studies of objects in our Local Universe is that stellar and gaseous disks can, and do, present the most striking dualities in morphology. Cold dust grains ($\sim$ 15-20 K) in our Local Universe increases the IRAS dust mass of a galaxy, by one order of magnitude (see e.g. Block et al. 1994) and these grains act as highly effective masks, obscuring (or partially obscuring) extragalactic stellar backbones. There is no reason why such dualities should not persist at higher redshift space, when the dust masks are penetrated. The evolved stellar disks of high-z galaxies in the Hubble Deep Field (HDF) have never been explored at restframe K-band. Images of galaxies with redshifts z $\sim 0.5 - 1$ or higher secured using the Hubble

Space Telescope and NICMOS never penetrate the dusty, gaseous Population I mask. At z values greater than 3, even the H-band 1.6μm observed flux stems from emission shortward of 4000 A. Examining HDF galaxies at restframe I (0.84μm) would not be sufficient; attenuation by dust even in the I band for some local field galaxies may still be at a level of 50 percent (e.g. Block 1996).

A poignant remark pertaining to those 'irregular' galaxies comprising the faint blue excess was made by Ellis (1997):

"It is tempting to connect the rapidly evolving blue galaxies in the redshift surveys with the irregular/peculiar/merger systems ... Could this category of objects not simply be an increasing proportion of sources rendered unfamiliar by redshift or other effects? ... it is not yet clear whether the available local data samples are properly represented in all classes... the precise distinction between late-type spiral and irregular/peculiar/merger may remain uncertain."

In their study, Abraham et al. (1996) commented that the "shape of the faint-end number count for peculiar objects is sensitive to the large systematic uncertainties inherent in the visual classification of these objects".

Submillimetre observations of galaxies at redshifts as large as $z \sim$ 4-5 show that dust masses do not decrease with redshift: dust masses at these redshifts may still be of order 10^8 $M_\odot$ (Norman and Braun 1996).

What would the dust penetrated images of spiral galaxies at high redshift z $\sim$ 1 look like, when the effects of dust are 'swept away' and when redshift as well as surface brightness dimming effects of $(1{+}z)^{-4}$ *are fully accounted for?*

We have begun an investigation of the capability of NGST (JWST) for extending local morphological studies to higher redshifts. Our simulations recreate a given image when moved out to higher redshifts, always in a pre-selected restframe. In a preliminary investigation (Block et al. 2001), our strategy is to focus on a single extreme case, NGC 922, which highlights the potential benefits of extending quantitative analyses of rest-wavelength near-IR morphology to higher redshifts. NGC 922 is particularly well-suited for our purposes, since in common with numerous high-redshift galaxies, "the morphology of NGC 922 is so peculiar as to be outside the classification system. It would be called a sport by nineteenth-century animal breeders" (Sandage and Bedke 1994; see the description to Panel 313). This galaxy bears a striking optical resemblance to chaotic objects seen at higher redshifts such as HDF 2-86 (van den Bergh 1998) at $z = 0.749$ in the HDF, which van den Bergh et al. (1996) suggest might be a spiral in the process of being assembled. Both NGC 922 and HDF 2-86 contain a prominent arc-like feature in the one-half of the galaxy in which star formation is apparently proceeding at a vigorous rate (see Cohen 1976 and Devereux 1989).

Our images of NGC 922 are simulated at redshifts $z = 0.7$ and $z = 1.2$ in the K$'$ (2.1 μm) restframe.

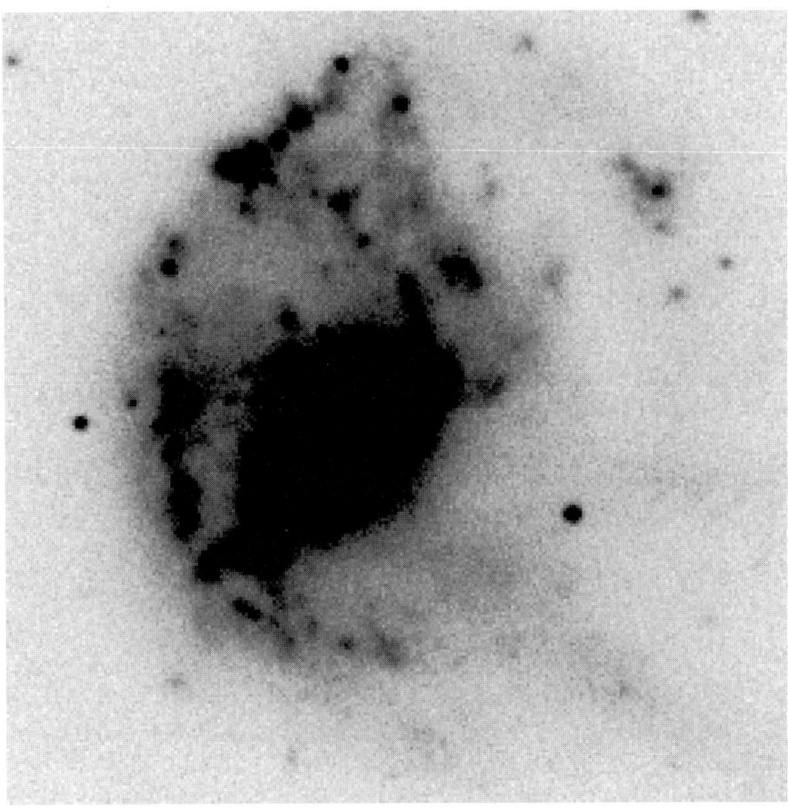

Figure 5. NGC 922 could well serve as a local Rosetta stone for morphologically peculiar systems in our higher redshift Universe (Block et al. 2001). Apparent chaos reigns supreme in this deep optical B-band image of NGC 922; dust-penetrated imaging, however, shows that a simple two-armed spiral (betraying the signature of arm modulation found in several grand design prototypes such as Messier 81) is largely responsible for the dynamics in the stellar backbone of NGC 922 whose Fourier spectra are presented in Figure 7.

Bi-dimensional Fourier analysis remains robust on these images (only $3''$ on a side) when the galaxy is moved out to redshifts $z = 0.7$ and $z = 1.2$, as seen in Figure 6. The analysis also confirms that only two low-order m components ($m = 1, 2$) are sufficient to describe the dust-penetrated morphology of NGC 922 (Fig. 7), even though this galaxy falls outside the Hubble classification scheme.

If NGC 922 is to serve as a local Rosetta stone for understanding optically chaotic systems at higher z, then the morphology of such optically peculiar objects may still be successfully described by our dust penetrated template for (spiral) galaxies at $z \sim 0$ (Figure 1).

References

Aaronson, M., *in Stellar Populations, Cambridge University Press*, 45, 1986

Abraham, R.G. et al., *Astrophysical Journal Supplement*, 107, 1, 1996

Allen, R.J., *in New Extragalactic Perspectives in the New South Africa, Eds. D.L. Block and J.M. Greenberg, Kluwer*, 50, 1996

Bell, E.F., de Jong, R.S., *Monthly Notices of the Royal Astronomical Society*, 312, 497, 2000

Bessell, M.S., Brett, J.M., *Publications of the Astronomical Society of the Pacific*, 100, 1134, 1988

Blanco, B.M., Blanco, V.M., McCarthy, M.F., *Nature*, 271, 638, 1978

Block, D.L., *Nature*, 347, 452, 1990

Block, D.L., *in New Extragalactic Perspectives in the New South Africa, Eds. D.L. Block and J.M. Greenberg, Kluwer*, 1, 1996

Block, D.L., Wainscoat, R.J., *Nature*, 353, 48, 1991

Block, D.L. et al., *Astronomy and Astrophysics*, 288, 365, 1994

Block, D.L. et al., *Astronomy and Astrophysics*, 371, 393, 2001

Block, D.L. et al., *Astronomy and Astrophysics*, 394, L35, 2002

Block, D.L. et al., *Astronomy and Astrophysics (Letters, in press)* (astro-ph/0406485), 2004

Bournaud, F., Combes, F., *Astronomy and Astrophysics*, 392, 83, 2002

Bouwens, R.J. et al., *Astrophysical Journal*, 595, 589, 2003

Bruzual, A.G., Charlot, S., *Astrophysical Journal*, 405, 538, 1993

Burstein, D., Rubin, V., *Astrophysical Journal*, 297, 423, 1985

Cohen, J.G., *Astronomical Journal*, 203, 587, 1976

Combes, F., Sanders, R.H., *Astronomy and Astrophysics*, 96, 164, 1981

Corbelli, E., Schneider, S.E., *Astrophysical Journal*, 479, 244, 1997

Curtis, H.D., *Publications of the Lick Observatory*, XIII, 11, 1918

Deul, E.R., van der Hulst, J.M., *Astronomy and Astrophysics Supplement*, 67, 509, 1987

Devereux, N.A., *Astrophysical Journal*, 346, 126, 1989

Dickey, J.M. et al., *Astrophysical Journal*, 561, 264, 2001

Ellis, R.S., *Annual Review of Astronomy and Astrophysics*, 35, 389, 1997

Elmegreen, D.M., Elmegreen, B.G., *Astrophysical Journal*, 314, 3, 1987

Elmegreen, B.G., Kim, S., Staveley-Smith, L., *Astrophysical Journal*, 548, 749, 2001

Elmegreen, B.G., Elmegreen, D.M., Leitner, S.N., *Astrophysical Journal*, 590, 271, 2003

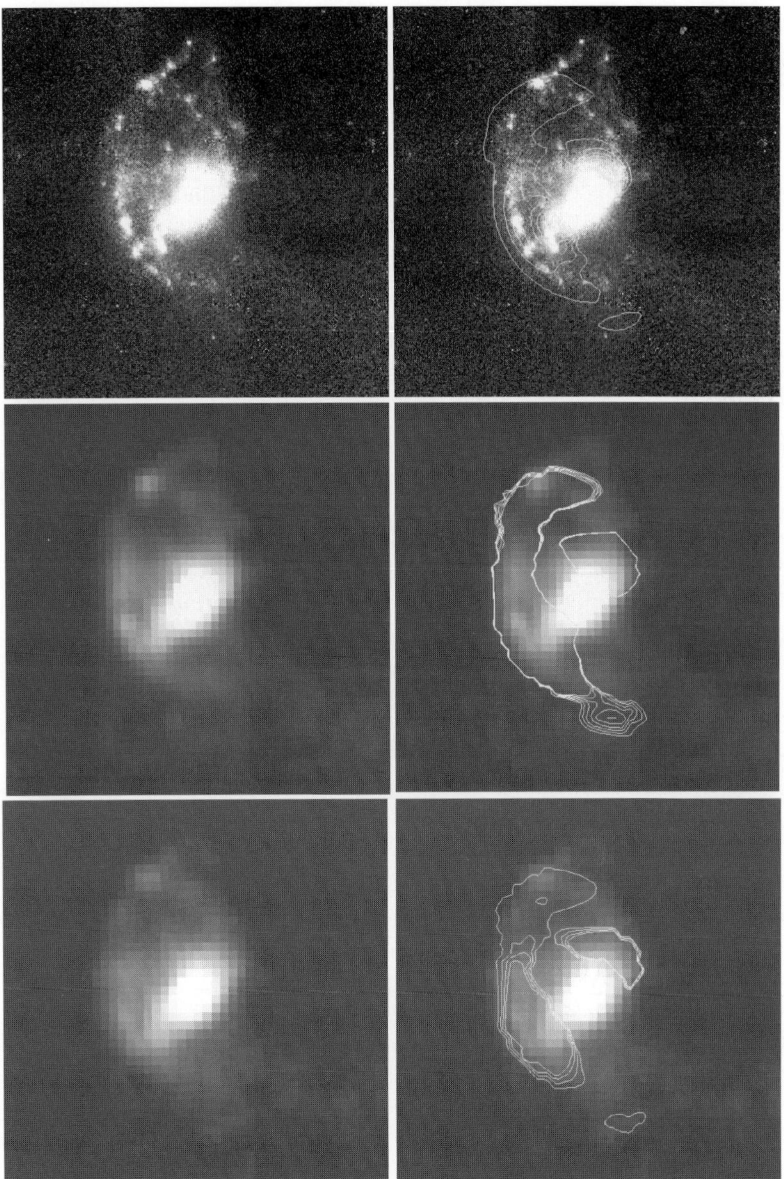

Figure 6. Dust penetrated images of NGC 922. Upper left: Groundbased K′ (2.1 μm) image. Middle left: simulated 1h exposure (assuming a 6m NGST), at redshift 0.7 (L band). Bottom left: NGC 922 simulated at redshift z=1.2 (M band) with an 6m NGST. The right hand panels show contour overlays of the low order m=1 and 2 inverse Fourier transforms. The cosmology we adopted assumes deceleration parameter $q_0 = 0.1$. Adapted from Block et al. (2001).

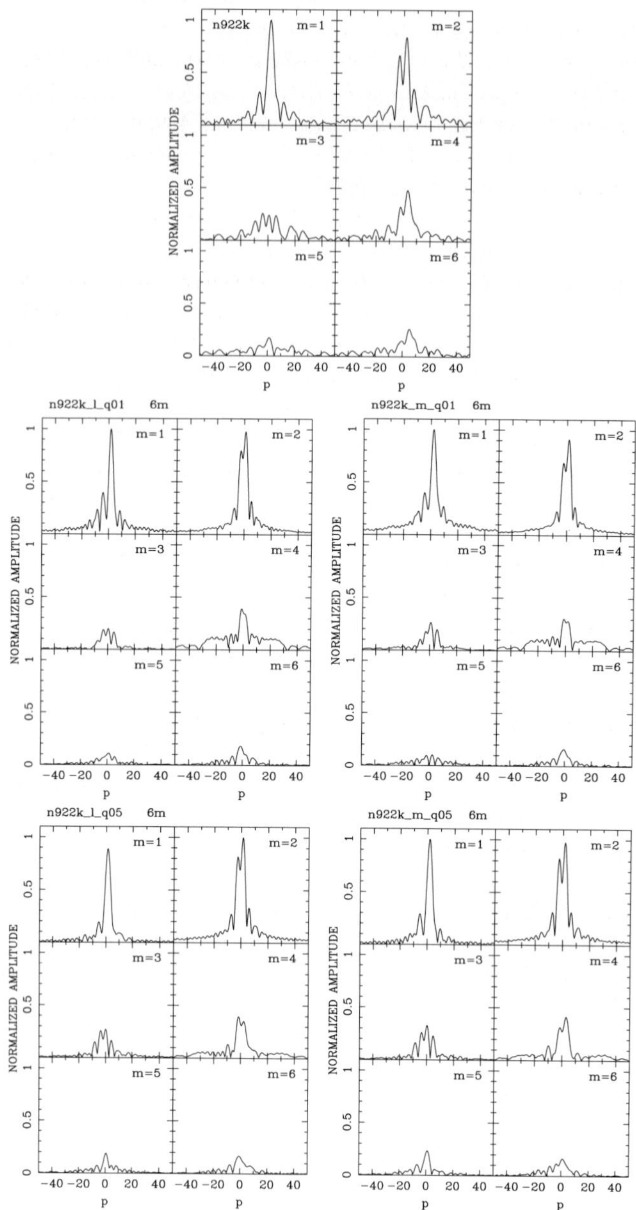

Figure 7. The Fourier spectra of the K′ (2.1 μm) image of NGC 922 at its original distance (top). The middle panels show the Fourier spectra viewed through a 6m NGST, when NGC 922 is moved to redshifts $z = 0.7$ (L-band) and $z = 1.2$ (M-band), where a deceleration parameter $q_0 = 0.1$ is used. The bottom panels are similar to the middle ones, except that a deceleration parameter of $q_0 = 0.5$ is adopted. A remarkable similarity in the restframe K′ images of NGC 922 is found, independent of redshift and of the deceleration parameter assumed.

Freedman, W.L., Wilson, C.D., Madore, B.F., *Astrophysical Journal*, 372, 455, 1991
Frogel, J.A., Whitford, A.E., *Astrophysical Journal*, 320, 199, 1987
Green, D.A., *Monthly Notices of the Royal Astronomical Society*, 262, 327, 1993
Grosbol, P., Patsis, P.A., Pompei, E., *Astronomy and Astrophysics (in press)*, 2004
Hassan, H., Pfenniger, D., Norman, C., *Astrophysical Journal*, 409, 91, 1993
Hodge, P.W., Skelton, B.P., Ashizawa, J., *An Atlas of Local Group Galaxies, Kluwer*, 2002
Hubble, E., *Astrophysical Journal*, 64, 321, 1926
Huber, D., Pfenniger, D., *Astronomy and Astrophysics*, 336, 840, 2001
Jarrett, T.H. et al., *Astronomical Journal*, 125, 525, 2003
Katz, N., Keres, D., Dave, R., Weinberg, D.H., *The IGM/Galaxy Connection: The Distribution of Baryons at z=0, Eds. J.L. Rosenberg and M.E. Putman, Kluwer*, 185, 2003
Leitherer, C. et al., *Publications of the Astronomical Society of the Pacific*, 108, 996, 1996
Mould, J., Aaronson, A., *Astrophysical Journal*, 303, 10, 1986
Marigo, P., Girardi, L., Chiosi, C., *Astronomy and Astrophysics*, 316, L1, 1996
Marigo, P., Girardi, L., Chiosi, C., *Astronomy and Astrophysics*, 403, 225, 2003
Mouhcine, M., Lancon, A., *Astronomy and Astrophysics*, 402, 425, 2003
Norman, C.A., Braun, R., *in Cold Gas at High Redshift, Eds. M.N. Bremer, P.P. van der Werf, H.J.A. Rottgering and C.L. Carilli, Kluwer*, 3, 1996
Pagel, B.E.J., Edmunds, M.G., *ARAstronomy and Astrophysics*, 19, 77, 1981
Persson, S.E. et al., *Astrophysical Journal*, 266, 105, 1983
Pfenniger, D., Norman, C., *Astrophysical Journal*, 363, 391, 1990
Pfenniger, D. Combes, F., Martinet, L., *Astronomy and Astrophysics*, 285, 79, 1994
Puerari, I. et al., *Astronomy and Astrophysics*, 359, 932, 2000
Regan, M.W., Vogel, S.N., *Astrophysical Journal*, 434, 536, 1994
Rieke, G.H., Lebofsky, M.J., *Astrophysical Journal*, 288, 618, 1985
Sancisi, R., *Internal Kinematics and Dynamics of Galaxies, IAU Symp. 100*, 55, 1983
Sandage, A., Humphreys, R.M., *Astrophysical Journal*, 236, L1, 1980
Sandage, A., Bedke, J., *The Carnegie Atlas of Galaxies, Carnegie Inst of Washington Pub. No. 638*, 1994
Searle, L. Wilkinson, A., Bagnuolo, W.G., *Astrophysical Journal*, 239, 803, 1980
Stanimirovic, S., Lazarian, A., *Astrophysical Journal*, 551, L53, 2001
Skrutskie, M.F. et al., *in The Impact of Large Scale Near-IR Sky Surveys, Eds. F. Garzon et al., Dordrecht: Kluwer*, 25, 1997
Toth, G., Ostriker, J.P., *Astrophysical Journal*, 389, 5, 1992
van den Bergh, S. et al., *Astronomical Journal*, 112, 359, 1996
van den Bergh, S., *Galaxy Morphology and Classification, Cambridge University Press*, 87, 1998
Vazdekis, A., *Astrophysics & Space Science*, 276, 921, 2001
Wilson, C.D., *Astronomical Journal*, 101, 1663, 1991
Worthey, G., *Astrophysical Journal Supplement*, 95, 107, 1994
Zwicky, F., *Morphological Astronomy*, Springer-Verlag, 1957

EPISODES IN THE DEVELOPMENT OF THE HUBBLE GALAXY CLASSIFICATION

Allan Sandage
The Observatories of the Carnegie Institution of Washington, 813 Santa Barbara Street, Pasadena, California, 91101, USA

Abstract In March, David Block telephoned from Johannesburg with an interesting invitation to comment on the provenance of Hubble's famous tuning fork galaxy classification diagram. Over the years he had often noticed that the diagram is not in Hubble's definitive 1926 classification paper (Astrophysical Journal, 64, 321), and that it had only appeared in his semi-popular book "Realm of the Nebulae", but with no antecedents mentioned. And then he made the central observation that a skeleton diagram, albeit turned 90 degrees, had first appeared in the 1929 edition of James Jeans' "Astronomy and Cosmogony". Could I comment? In Coda I, the question is addressed: Is there a duality of the classification depending on wavelength? No one can argue that there is no difference. The point is that the wavelength dependence of the morphology that is surely present at some level, is either dominantly serious for some problems but not for others. There also is a comment on the current thrust for new classifications by adding ever increasing detail. In Coda II, an insistance is made that there must be a strict separation between the pure morphologist and the theoretician. "Imagination" or "genius" or "intuition" provides the elusive opening with which to break the hermeneutical circle. Sherlock Holmes said it right in many places. "*It is a capital mistake to theorize before you have all the evidence. It biases the judgement.*" The discovery of a classification system, seeming so simple at the beginning is, in fact, extraordinarily complicated.

Keywords: Galaxies: spiral, structure, kinematics and dynamics

1. The Initial Development (1922-1936)

As the history had developed in the passage of time, I had known that the final tuning fork diagram had not been a part of Hubble's classification system as he had set it out in 1926. I also knew that the diagram had first appeared in "*The Realm of the Nebulae*", but had not traced its root to Jeans, as David Block had done. Accepting David's invitation to comment and to give an account of the subsequent development of the modern expression of that system,

D. Block et al. (eds.), Penetrating Bars through Masks of Cosmic Dust, 39–55.

I relate the following episodes that began in 1922 and progressed into the modern extension after Hubble had gone to his reward in 1953.

In his 1922 preliminary classification paper of nebulae (ApJ. 56, 162), Hubble makes no mention of Jeans. There is only a minimum outline of a classification what would be expanded four years later in his definitive 1926 paper. After this, the only other description of the developing classification was in "The Realm of the Nebulae".

What is clear is that between 1922 and 1926, Hubble had read Jeans' 1918 "*Problems of Cosmogony and Stellar Dynamics*". This essay was the winner of the Adams Prize for 1917. In it, Jeans puts forth the suggestion, based on the newly discovered rotation of the non-galactic nebulae (i.e. the galaxies), that the flattened structures were due to the evolving forms of gaseous, or even liquid, bodies as the rotation rate is increased, as in the Maclaurin and Jacoby ellipsoids.

Hubble mentions these suggestions by Jeans in the 1926 paper but he does not elaborate on why this theoretical discussion on the flattening of rotating gas or liquid should pertain to the galaxies which Hubble himself had shown in 1925 were composed of stars. In any case, Hubble writes in Astrophysical Journal 64, 321, 1926:

"*Although deliberate effort was made to find a descriptive classification which should be entirely independent of theoretical considerations, the results are almost identical with the path of development derived by Jeans* [and here a footnote to "Problems of Cosmology and Stellar Dynamics, 1919"] *from purely theoretical investigations. The agreement is very suggestive in view of the wide field covered by the data, and Jeans' theory might have been used both to interpret the observations and to guide research. It should be borne in mind however, that the basis of the classification is descriptive and entirely independent of any theory.*"

Despite this denial of any debilitating influence from theory, nevertheless, Hubble clearly was looking for a classification that he hoped could eventually be used for an ultimate theory of "origins". The mention of theories of flattenings of liquid/gaseous bodies under rotation shows this; he speaks of "development" which can only mean either "origins" or "evolution".

In his last published comments on the classification in the 1936 "Realm" he introduces the S0 class, but he had not yet made the redistribution of the SBa galaxies into an SB0 class as that redistribution was later described in the *Hubble Atlas* (Carnegie Publication 618, 1961) and in Chapter 1 of "Galaxies and the Universe" (Volume 9 of the Chicago Compendium series, University of Chicago Press).

Between the times of the 1926 paper and the writing of the "Realm", Jeans had published an update of his 1917 Adams Prize essay as a major book titled "Astronomy and Cosmogony" (Cambridge University Press, 1929). There it

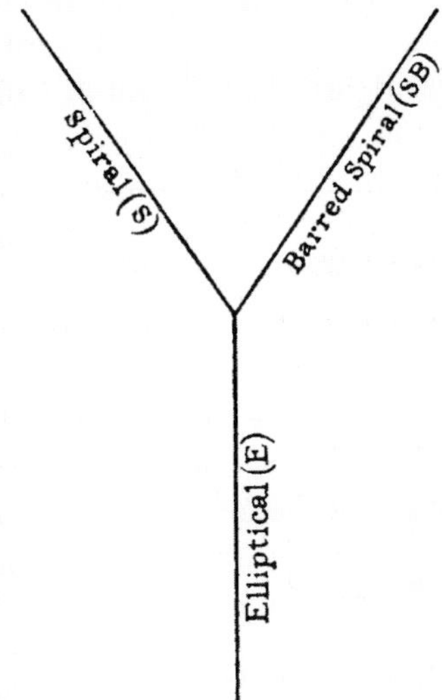

Figure 1. David Block (this Volume) made the central observation that a skeleton diagram, albeit turned 90 degrees, had first appeared in the 1929 edition of James Jeans' "Astronomy and Cosmogony". Seen here is the original Y diagram by Jeans in his 1929 book, page 332, his figure 53.

is evident that he had well understood Hubble's classification system of 1926. Although the descriptions of the classification bins had been well written there, Hubble had not given a diagram that illustrated the unifying connections between his galaxy types laid out in a linear sequence, the connections being the important advance in the classification scheme over that of Wolf (1908).

In the dry academic language of formal science, and with so few working extragalactic astronomers at the time (perhaps only 20 worldwide), without such a diagram there was the danger that the classification system, buried in the language of the Astrophysical Journal, might lay fallow. This was soon prevented by Jeans, as his connective diagram was adopted by Hubble in 1936.

To strengthen an understanding of the power of Hubble's linear connection of the classification bins, Jeans produced a Y shaped diagram that summarized the essence of the 1926 classification, shown here in Figure 1, reproduced from Figure 53 of "Astronomy and Cosmogony" (1929). Clearly, as emphasized by David Block, this is the provenance of the famous extended tuningfork diagram. Rotate the figure by 90 degrees, add pictures of the opening of the arms of the spiral sequences, add the flattening sequence of the E galaxies, and add the then (1936) "hypothetical" S0 class, and you have Hubble's diagram. The diagram appeared only in the "Realm" but that was enough to ensure it lasting fame.

Hubble mentions the influence of Jeans only once in the "Realm", and then only to justify his use of the terms "early" and "late" in the descriptions of the spiral types, not to describe the provenance of the diagram. He writes in "Realm":

"The terms "early" and "late" are used to denote relative position in the empirical sequence without regard to their temporal implications. These explanations emphasize the purely empirical nature of the sequence of classification. The consideration is important because the sequence closely resembles the line of development indicated by the current theory of nebular evolution as developed by Sir James Jeans." Then the footnote *"[The most recent statement of the theory is in Jeans, Astronomy and Cosmology (1928), XIII.]"*

There is no mention here or elsewhere of Jeans' inverted Y.

2. The Influence of the Diagram

Why did the diagram become so overwhelmingly important?. Despite the excellence of Hubble's 1926 word descriptions of the classification, the diagram is much easier to understand and to remember. It became the visual mnemonic. Indeed, we all learned to classify from it. Only later did we read the verbal descriptions in the 1926 fundamental paper. That was true in my generation. It is true now.

I learned the classification first from inspection of the diagram, and only later by direct instruction from Hubble. By then (1950), he had slightly rearranged the Sa, SBa, with the S0, and SB0 classes (which he had discovered after 1936, verifying his "hypothetical" S0 class). This redistribution had not been published as a research paper, nor had the characteristics of the S0 class been described by Hubble anywhere. These points, in fact, were not described in print until his notes were published in the front material of the Hubble Atlas in 1961, seven years after his death. But Hubble had described the S0 class to Baade privately, and that description had found its way into the literature (Spitzer and Baade 1951), but Hubble had not made a research paper about the S0 class. In 1936 the class was still "hypothetical".

3. The Extensions

Additions to the original Hubble sequence began when Holmberg came to Mount Wilson beginning in 1949 to observe with the 60-inch, and then a second time in 1957 in his program to measure the apparent magnitudes of 300 of the brightest galaxies in the northern hemisphere. Holmberg recognized that Hubble's Sa, Sb, and Sc classification bins could be usefully reduced in size by introducing intermediate classes which he called Sb+, Sb−, Sc+, Sc−. etc, following Shapley (1941) who had made the same suggestion, but it was unnoticed and not adopted at the time. de Vaucouleurs (1955), in an important survey of southern Shapley-Ames galaxies, changed Holmberg's plus and minus notation back to that of Shapley into the modern ab, and bc, notation for the intermediate types. de Vaucouleurs also added the later classes of Scd, Sd, as had Shapley, and added, as well, the transition of Sdm, Sm, and Im to the Magellanic Cloud types introduced earlier by Lundmark (1926, 1927). These additions completed the extension of the types along Hubble's original linear sequence for both regular and barred spirals.

Fine detail within the types then followed. By the time of the Hubble Atlas in 1961, we would recognize the (r) and (s) spiral arm subtypes as that added detail. The (r) type denoted arms that begin tangent to an internal ring. The (s) arm type are for arms that begin either from the central regions as in ordinary spirals, or from the ends of the bar in the barred spirals. Transition types were called either (rs) or (sr) depending on which dominates. de Vaucouleurs had visited Pasadena in August of 1955 to learn the extended Hubble system as it existed in 1954, and he also adopted the (r) and (s) arm detail in his 1955 classifications.

The concept of a "classification volume" had been developed in the preparations for the Hubble Atlas that had started in 1954. Here, the (r) and (s) subtypes were displayed on the surface of a rectangular box (Fig. 5, page 26 of the Hubble Atlas). The top and bottom of the box contained the ordinary spirals and the barred spirals respectively. Hodge (1966) recounts part of this history.

This rudimentary concept of 1954 was then brilliantly generalized by de Vaucouleurs into a continuous classification volume where all interior parts are filled by the (rs) and (sr) types and where the major Sa, Sab, Sb, Sbc, Sc, Scd, Sd, Sdm, Sm, and Im classes were changing along the long axis of an ellipsoid (the axis of the Jeans-Hubble tuning fork). The transition cases between the "ordinary" and the barred spirals were also accomodated in the "volume". Figures 2 and 3 show this development. de Vaucouleurs also introduced the R notation, such as RSBab(r) for galaxies that have an apparent external ring surrounding the main body that contains a bar and an Sab place in the sequence, and with a set of (r)-type arm connections to the bar.

The last significant extension of the Hubble classification was made in a major advance by van den Bergh (1960a,b) with his introduction of a "beauty" or "regularity" index which he called "luminosity class". The degree of symmetry and/or regularity of the image is the criterion for the assignment of the luminosity class. Hence, the notation became much more complex than the simple Hubble types of E, S0 Sa, Sb, or Sc, and their counterparts in the barred spirals.

Throughout, there had been no appeal to notions of any underlying physics that could "explain" the sequence with its variable bulge sizes, its bars or lack thereof, its (r) and (s) arm types in terms of the dynamics of resonances or bar

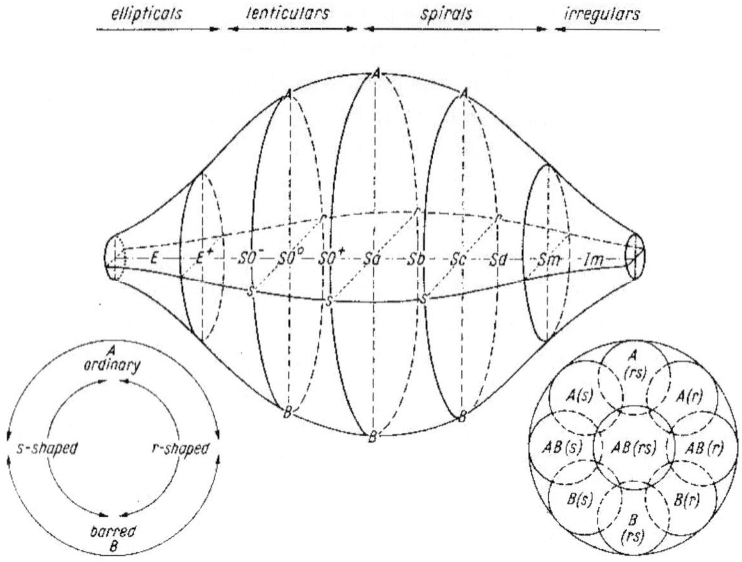

Figure 2. The classification volume by de Vaucouleurs from his review in Handbuch der Physik in 1959, generalizing a less complete three dimensional diagram from the Hubble Atlas.

instabilities, its variation of "beauty" whose physics must be related to rotation rates, or to any of the more subtle physical properties of color, metallicity, star formation rates, dust content, radio emission, etc. This vacuousness of "explanation" in the classification itself is, of course, its power.

Because it was devised with none of these physical properties as part of the classification criteria, any correlation of Hubble types with these physical processes (rotation rates, star formation rates, Lindblad resonances, etc.) shows that there is in fact an underlying physics to the classification.

The fact that there is no hermeneutical circularity with the underlying physics is the power of the classification.

4. The Next Step

The highest accolade that a morphologist can hope for has been given in the writings of one of the scheduled speakers here. In a review for *Annual Reviews of Astronomy and Astrophysics* he and his coauthor state:

"At the level of detail that we nowadays try to understand, the time has passed when we can make effective progress by defining morphological bins with no guidance from a theory. Years ago, people commonly reacted badly to a classification as complicated as (R)SB(r)b. The reason, we believe, was that the phenomenology alone didn't sell itself. People did not see why this level of detail was important. Now, we will show [in this essay] *that every letter in*

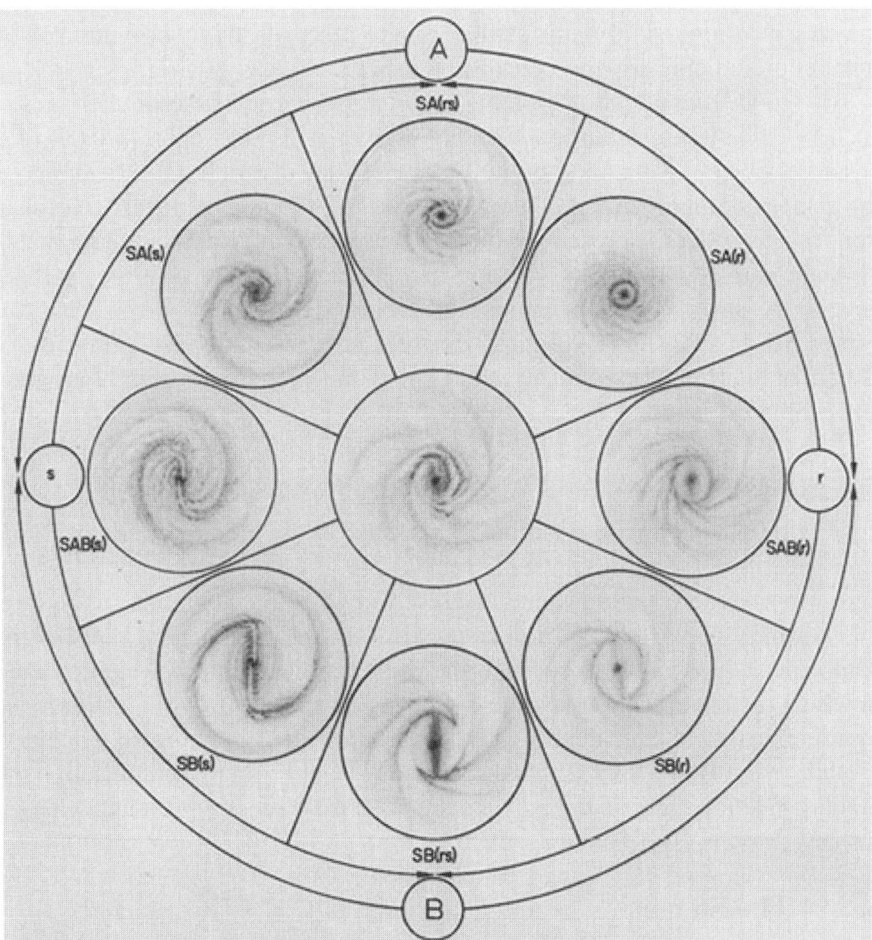

Figure 3. A cross section of the classification volume in the region of the Sb and SBb spirals. This diagram is from the frontispiece of the first Reference Catalogue of Bright Galaxies by G. and A. de Vaucouleurs (1964).

the above classification has a clearcut meaning in terms of formation physics. This is the goal of physical morphology."

This quote from Kormendy and Kennicutt (2004), that "every letter in the classification has a clearcut meaning in terms of the formation physics" certifies the promise of the Hubble system. Because the dross of the morphologist's suppositions of what the underlying physics "ought to be" has been removed, the classification that remains can be used to discover the underlying physics without prejudicing any theoretical conclusions.

All "good" classification systems have this property. The characteristics of such "ideal" systems and how they are derived is a philosophy in itself, discussed elsewhere (Chapter 4 of the Carnegie Atlas of Bright Galaxies). It was the genius of Hubble that produced the basic "morphological ideal" classification framework of the present system, as Kormendy and Kennicutt have noted.

And clearly, the "why" of what the "pure" morphologist sees is the next step beyond the sheer visual appearances of forms. That step is part of what this conference is about. The "origins" theorists here assembled have now, in the system of the Hubble classification, the silk with which to spin their "formation and evolution" webs.

5. Coda I

My original short essay, written in response to the request of David Block to comment on the provenance of Hubble's 1936 tuning fork diagram, ended here. However, on reading what I had sent, David pointed out that, "to be sure, this is what I requested, but it does not address the more central point of the conference which is ... how robust is the Hubble system to the wavelength band in which the classification is made? Could you comment on the claim made by some that there is a "duality" of the classification depending on whether it is made in the B or a near IR band such as H? Could you also comment on the S0 classification, both as to Hubble's unpublished discovery of it and where it fits now into the classification system?."

6. The S0 class

The simplest definition of an S0 galaxy remains "a disk galaxy more flattened than an E6 elliptical but with no trace of spiral arms or recent star formation." It is, of course, more detailed than this. The extended modern system has subclasses and degrees of S0-ness (e.g. the Carnegie Atlas Chapters 2 and 3 and panels 26 through 60). Nevertheless, the S0 types occur at the junction of the most flattened E galaxy (E6) and the disk spirals. The distribution of apparent flattenings shows that there is a predominance of true disks in the S0 type rather than ellipsoids of intermediate flattenings as in E galaxies (Sandage, Freeman, and Stokes 1970).

7. An Example of a Hermeneutical Circularity in an Incorrect Classification Scheme

I emphasized earlier that the power of the Hubble system is the absence of a hermeneutical circularity. By that I mean that there is no appeal to what the classification "ought" to be based on preconceived notions of the classifer of an underlying mechanism of how the galaxy got into the form it has. The S0 case is the prime example of how such a circularity has tainted the S0 classification system in some newly proposed replacements of the Hubble sequence.

Hubble had discussed his observational discovery of the S0 class with Walter Baade some time between 1936 and 1950, following his postulation of the existence of such a class between the E and spiral types, set out in the "Realm". Lyman Spitzer, in discussing with Baade the role of gas in galaxian disks and subsequent star formation, noted that stripping of gas from spiral disks, with the subsequent cessation of star formation, was a possibility in galaxy collisions. Hence, gas-free galaxies with disks like those along the Hubble spiral sequences (both ordinary and barred), but with no recent star formation, might be expected in dense galaxy environments such as occur in compact galaxy clusters where collisions might be expected. Indeed, Baade said he knew of observational cases. They were Hubble's S0 galaxies, and behold, they are more frequent in clusters than in the general field. Ipso facto, S0s are swept spirals. Notwithstanding, E galaxies are also more frequent in clusters than in the general field, and this surely is not due to sweeping. Rather, in the well known morphology/density relation, discovered by Hubble and Humason (1931), the S0s track the E galaxies, not spirals.

But there are many possibilities to explain the morphology/density relation, such as a hierarchy of density contrasts in the initial conditions where the original high density contrast protogalaxies make the E galaxies and the lower contrast blobs become disk galaxies. The classification bin in which a galaxy finally ends along the Hubble sequence depends on the strength of the initial fluctuation (Sandage 1986, 1990), or some other unknown mechanism that need not be gas sweeping.

The point is that a parallel S0 sequence of $S0_a$, $S0_b$, and $S0_c$ types with different bulge-to-disk ratios, as suggested by Spitzer and Baade in 1951, and by Baade in his 1963 Harvard lectures, does not exist. This conclusion is discussed at length in the S0 section of the Carnegie Atlas, and especially in the descriptions of NGC 1553 [$S0_{1/2}$(5)(P)] and NGC 3056 [$S0_{1/2}$(5)(P)] on Atlas panel 39 and NGC 5838 [$S0_2$(5)(P)] on Atlas panel 40. Part of the description for NGC 5838 reads "There are no examples in the RSA of S0 galaxies that could once have been like M33 or NGC 300 which have no large, bright central bulges that are characteristic of all S0 galaxies."

However, van den Bergh (1976) adopted Baade's hypothetical scheme in his suggestion for a major revision of the Hubble system by postulating a three-prong tuning fork with “anemic" spirals of all the Hubble types, plus the Baade $S0_a$, $S0_b$, and $S0_c$ types. The proposed third tuning-fork tyne for S0s is based on what “ought” to be, rather than what is. This is a hermeneutical circularity, with the consequence that the resulting classification is not morphologically pure because it was developed with a physical process in mind. More to the point, the third tyne does not exist in the sky as seen directly from the RSA data (opt cit).

Rather, morphologically, the S0 class forms a continuum between the E and the Sa galaxies, recognized by the “transition” classes of E/S0 on the one side and the S0/Sa class on the other of the S0 bin. It is this morphological continuity, found simply by inspection of the large scale images made with the Mount Wilson, Palomar, and Las Campanas reflectors, that puts the S0 class squarely between the E and the Sa types. The same continuity exists between the E, SB0 and SBa types of the barred sequence.

8. Is there a duality of the classification depending on wavelength?

Much is made in this conference of the difference in the appearance of galaxies along the Hubble sequence between images made in the blue and the near infrared spectral bands. No one can argue that there is no difference. Both Zwicky (1955) and Schweizer (1976) [see also Hackwell and Schweizer 1983, and Elmegreen 1981] showed, starting 50 years ago, that the contrast between the arm and the disk light changes dramatically with wavelength for the obvious reason that the Population I (young and blue) and the Population II (old and red) stars reside in different places in the image. Hence, arm vs. disk structures will show different dominance between the blue and red images. Dust differences will also play a role.

However, is this so serious that there will be different classifications assigned between the B and the H wavelengths? Yes and no, depending on the level of detail that any given classifier demands from whatever classification is being used.

By this is meant that we must examine the properties of the coarsest classification level of only say E, Sa, Sb, Sc, SBa, SBb, or SBc. At this level, do the B and H images give the same bin designation, as proved by Eskridge et al. (2002) to within one T class, or is there no correlation as claimed by Block and Puerari (1999)? (We come to the bar differences a paragraph below).

If the most detailed classification set out by de Vaucouleurs is used, for example of say $SAB(rs)_2(+)C_f$ for NGC 1232 (see Chapter 1 in Galaxies and the Universe), where the C denotes the arm character, the 2^+ denotes branching from two main arms of the (rs) types, and f denotes filamentary arms (flocculent in the current parlance) rather than "massive" in the Reynolds (1927a,b) sense, then there will be a classification change with wavelength at this level of detail. Add to this the 12-tone arm classification of Elmegreen (1981) and Elmegreen and Elmegreen (1982), and the six bar strength divisions by Buta and Block (2001) and the change between B and H will be even stronger. For example, it is said that the population I filamentary arms of say NGC 488 or NGC 5055 are changed into a underlying grand design arms of Population II stars of the old disk such as was showed by Zwicky (1955) in M51.

Is this the meaning of a "duality of classification" (a term often used by Block)? Buta and Block (2001) state that "A rich duality of spiral structure has been found from studies of optical and near-infared images; a spiral galaxy may present two completely different morphologies when examined optically and in the near infrared." However, a contrary statement is made by Eskridge, Frogel, Pogge et al. (2002) as: "This result [their near-IR images] does not support recent claims in the literature that the optical and near IR morphologies of spiral galaxies are uncorrelated."

Perhaps what is meant by Block and Puerari (1999), although not stated, is that there is no correlation of the detailed types between B and H, but that the overall wide Hubble bin types are the same in B and H as in Eskridge et al. (2002). In fact, Eskridge et al. show that the Hubble classes are identical in the B and H images for S0/a and Sm galaxies, and differ by only about one T class for intermediate Hubble types. Hence, there seems to be no fundamental difference in the Hubble types as a function of wavelength at the course level of the wide Hubble bins.

9. The bar strength

Only the strongest bars are recognized in the classifications in the Hubble Atlas, the Revised Shapley Ames Catalog (the RSA), and the Carnegie Atlas, whereas a finer division is made by de Vaucouleurs with his transition types of S$\underline{A}$B, and SA$\underline{B}$ for the barred Sa, Sb, and Sc galaxies, where the underline is the dominant mode. Hence, there is a higher percentage of barred spirals in the RC3 than in the RSA, which is the parent of the Carnegie Atlas. The finer division into six bar strengths by Buta and Block (2001) adds even more detail.

Yes! Indeed, when the "dust is unveiled", most galaxies show a bar strength of 2 or more on the Buta/Block scale of six, and weak bars are often missed in the blue because of dust veiling. What is the consequence?

Galaxy classifications are useful for a wide variety of problems in practical cosmology. The consequences vary with the type of problem they are used for. For a dynamisist who is attempting to understand the bar instabilities and their evolution with time in terms of classical dynamics coupled with gas hydrodynamics, the bar strength is all important and the change with wavelength of this detail of the classification is central.

However, for a statistical cosmologist interested in say the type-dependence of luminosity functions, binned into the Hubble classes, such detail is not important because the broad-bin Hubble types are sufficient. And again, for an evolutionist looking for galaxy change in the look–back time, minute details are probably not important, yet a strong wavelength dependence of spiral pattern with wavelength may dominate the answer and lead to incorrect conclusions.

All this is of course well known, and is a major subject of this conference. Suffice to repeat the argument, this requires that the observations in, say, the Hubble Deep Field (HDF) should be made at the same rest wavelength as for the nearby morphological samples. Block's insistence on this, using NGC 922 as a template example, is most important.

The observational problem at high redshifts is, of course, formidable because of the necessity to observe the HDF galaxies at very long IR wavelengths so as to compare with the B local band, or conversely we must determine the morphology of local galaxies in the far UV to compare with the HDF samples observed in the B and R bands. All this, of course, is obvious, but points to the necessity for a very wide wavelength coverage of the NGST as the next step.

My only point is that the wavelength dependence of the morphology that surely is present at some level, is either dominatingly serious for some problems but not for others.

10. The Current Thrust For New Classifications by Adding Ever Increasing Detail

The call now for making new classifications by introducing "physical morphology" into the classification, as urged by Kormendy and Kennicutt in the quote given earlier, seems to me to be correct as long as one can avoid the slippery slope of a hermeneutical circularity. The physics should be kept apart from the notation, i.e. the bin size should be kept large; this is the power of the Hubble system. Said differently, the current simplicity should be kept by not adding new symbols with increasing detail to the broad bin size in the current basic notation.

This is the identical problem encountered by Hubble (1926), in countering the criticisms by Reynolds (1927a, 1927b) who claimed that the Hubble classification is too simple because it does not cover the vast variety of detail shown in nature. But that is precisely the power of the Hubble system as emphasized by Hubble (1927) in his reply to Reynolds. The bin size in the current system is wide enough to accommodate a variety of different detail on the theme of say Sc. If one demands that the notation be so detailed to include the multitude of minute details (massive vs. filamentary arms, multiple arms rather than only two, bar strength, Fourier component multiplicities etc. etc.) then we are demanding such detail that we might as well revert to the Wolf descriptive system (as indeed did Danver, 1942, in preference to the Hubble broad types), or in more modern times to the vast descriptive notations of Vorontsov-Velyaminov et al. (1962-1968) in their catalog of 29,000 galaxies that have no backbone physics as, apparently, does the Hubble sequence. These details are important for the dynamistist, and can be used as the data necessary for any given problem, but they need not be included as part of the classification notation.

11. Evolution in a Coffee Cup

Evolution is a major theme of the conference. Do galaxies evolve along the Hubble sequence or do they arrive on the sequence from more pristine states? Much progress has been made in understanding the dynamics of bar and arm formation and their change with time. There is no doubt that galaxies change their appearance with time, but it is not likely that they wander much beyond their Hubble bin as they evolve as mature systems. Rather, it seems more reasonable to believe that they can only change within their types, albeit often from bar to ordinary as their bars are destroyed and are created anew in less than a Hubble time (Bournaud and Combes 2002).

Said differently, it is unlikely that an Scd galaxy with small central bulge, with small mass, and with a low rotation rate can (or will) evolve into a massive, large bulge, high rotating Sb or SBb galaxy. Nevertheless, an initial Sb galaxy is expected to be highly variable in appearance in its arm detail, and even in its van den Bergh luminosity class, many times in its life on the Hubble sequence, yet staying close to the Sb class.

An appreciation of how the appearance of a galaxy can change in a time short compared with one rotation period can be seen by making a galaxy in a coffee cup and watching it evolve.

The method is to start the coffee rotating by stirring with a spoon to set up a velocity field with shear. (The rotational velocity, of course, is zero at the walls of the cup and is a maximum at the center). Pour cream into the rotating differential velocity field. If it is poured near the center of the cup, an amazingly symmetrical spiral pattern forms with multiple arms. Depending on the strength of the shear field, one can produce a range of van den Bergh "beauty" variations. Indeed, one can call out both the Hubble types (eg. Sa,Sb, Sc, Sd) and the van den Bergh luminosity classes as the form decays with the decaying shear field as the rotational velocity of the coffee decays with time. (If also you know the local galaxies by name, you can even call out the NGC numbers as the coffee-cup galaxy evolves, impressing your companions).

By introducing the cream off center, one armed spirals result. By varying the initial rotational velocity of the coffee, one can make initial configurations that range from Sa through Sd and Sm. The "anemic" class of van den Bergh can also be produced by introducing a second influx of cream after the first has been stirred into the coffee. The contrast of the white cream with the blackness has been diluted.

But the most marvelous result of the demonstration is to see the "evolution" of the arms with time, with segments forming and dissolving in as short a time as a tenth of a revolution. Surely, this is the appearance of how the arms of a galaxy look progressively as they form and dissolve in times short compared with the rotation period.

As can be seen in the coffee cup, galaxies evolve, but not far along the Hubble sequence. They display their final Hubble types soon after formation, depending more on their mass, rotational velocities, and the initial density contrasts before their collapse to their present disk or ellipsodial forms, than by an evolution into their present forms along the Hubble sequence. Evolution is important, but it should not be confused with the formation processes of the initial galaxy soon after its collapse from a wider volume, or, alternately, from the coalescence of "fragments falling into equilibrium", which is ELS (1963) with noise (Sandage 1990).

12. Coda II

Upon reading to this point in the second edition of these notes, David Block reacted as will most readers in partial disagreement with the insistence here that there be a strict separation between the pure morphologist and the theoretician using the fruits of the morphologist's trade. But herein lies a seemingly insolvable epistimological problem. In the initial absence of knowledge of an underlying physical explanation for the origin of the objects, how does the classifier begin when there is no a prior knowledge of an "explanation".

Of course, once a classification system is so mature and "good" so as to suggest an underlying physics, it must be consistent in its final form with whatever the unknown physics turns out to be. But this is precisely the hermeneutical problem at the beginning of a quest for "goodness".

The problem is discussed at length elsewhere (Chapter 1 of the Carnegie Atlas of Galaxies) where the tension of the push and pull of the induction/deduction process gradually overcomes the initial ignorance. Yet even here the morphologist and the theoretician must be kept apart. I claim that the task of the morphologist is to speak only the truth, derived by what is seen, not supposed; "truth" defined here as "what is". And what "is" can only be found from observation, not a priori rational thought. (Of course this is an extreme position, not strictly true: no scientist can remain a pure morphologist if he/she is to go very far beyond the classification stage).

On the other hand, the theoretician, seeking "explanations" works on the basis of "faith", defined here as a "belief" in his/her "explanations". This tension between the morphologist and the theoretician at the onset of work can be diagrammed in the following way.

Construct a wall, on one side is the morphologist who is describing and connecting the objects by binning in some fashion. On the other side is the theoretician chomping at the bit to get the data from the morphologist to begin an explanation of why the binning is either good or is empty depending on how he/she can fit the binning into "explanations" consistent with an unknown physics.

There is a one-way door in the wall between the two domains. The door can transmit information only from the morphologist to the theoretician. The morphologist's side on the left is called "truth". The theoretician's side on the right is called "faith".

As in other contexts where "truth" and "faith" are also involved, it is essential to avoid confusion between the two. "Truth" must not contain "faith", but "faith", if it is not wholly myth, must contain "truth". This is the dilemma and tension between the morphologist and theoretician.

Sherlock Holmes said it right in many places. *"It is a capital mistake to theorize before you have all the evidence. It biases the judgement."* (A Study in Scarlet, Chapter 3), or [If one does so] *"before one has data, one begins to twist facts to suit theory instead of theory to fit facts"* (A Scandal in Bohemia), or [One must not] *"get into the habit of telling a story backward"* [from the theory to the facts] (The Problem of Thor Bridge).

How then does the morphologist begin the work of creating a "good" classification when faced with Aristotle's dilemma that "acquiring knowledge requires prior knowledge, which initially is unavailable"? This is called the problem of the hermeneutical circle, met before in these notes. Neither the pure initial induction by the Baconian method, or the pure deduction of Aristotle from initially sterile hypotheses, can work at first.

Edgar Allan Poe addressed this matter on how to start. In his long essay "Mellonta Tauta" and at the beginning of his remarkable "Eureka" (where he anticipates the expansion of the universe) he dismissed both the methods of Bacon and Aristotle as the initial paths to objective knowledge. He called his starting place "imagination"; we would call it intuition. This thing, "imagination", or "genius", or "intuition" provides the elusive opening with which to break the hermeneutical circle. Hubble used it; Wolf, Vorontsov-Velyaminov, and others using simple description, did not.

The discovery of a classification system, seeming so simple at the beginning is, in fact, extraordinarily complicated. But perhaps, after all, the Zen master has it right not to analyze the method too closely, but simply:

When eating do not sleep
When sleeping do not eat
When classifying do not theorize
rather
When eating, eat
When sleeping, sleep
When classifying, classify.

References

Baade, W. 1963, *Evolution of Stars and Galaxies*, ed. C. Payne-Gaposchkin (Cambridge: Harvard University Press)

Block, D.L. and Puerari, I. 1999, *Astronomy and Astrophysics*, 342, 627

Bournaud, F. and Combes, F 2002, *Astronomy and Astrophysics*, 392, 83

Buta, R., and Block, D.L. 2001, *Astrophysical Journal*, 550, 243

Danver, C.G. 1942, *Lund Obs. Ann.* Vol 10

de Vaucouleurs, G. 1959, *in Handbuch der Physik*, 53, 275

de Vaucouleurs, G. 1955, *Mem. Commonwealth Obs. (Mount Stromlo)*, Vol. 3, No. 13

de Vaucouleurs, G, and de Vaucolueurs, A. 1964, *Reference Catalogue of Bright Galaxies (Austin: University of Texas Press)*

Eggen, O.J., Lynden-Bell, D., and Sandage, A. 1963, *Astrophysical Journal*, 136, 748

Elmegreen, D.M., 1981, *Astrophysical Journal Supplement*, 47, 229

Elmegreen, D.M., and Elmegreen, B.G., 1982, *Monthly Notices of the Royal Astronomical Society*, 201, 1021

Eskridge, P.B., Frogel, J.A., Pogge, R.W. and 12 others, 2002, *Astrophysical Journal Supplement*, 143, 73

Hackwell, J.A. and Schweizer, F. 1983, *Astrophysical Journal*, 265, 643

Hodge, P.W. 1966 *The Physics and Astronomy of Galaxies and Cosmology (New York: McGraw-Hill Book Co.)*

Holmberg, E. 1950, *Medd. Lunds. Obs. Ser. 2.*, No. 128

Holmberg, E. 1958, *Medd. Lunds. Obs. Ser. 2.*, No. 136

Hubble, E. 1922, *Astrophysical Journal*, 56, 162

Hubble, E. 1926, *Astrophysical Journal*, 64, 321

Hubble, E. 1927, *Observatory*, 50, 276

Hubble, E. 1936, *Realm of the Nebulae (New Haven: Yale University Press)*

Hubble, H. and Humason, M.L. 1931, *Astrophysical Journal*, 228, 131

Jeans, J. 1919, *Problems of Cosmogony and Stellar Dynamics (Cambridge: Cambridge University Press)*

Jeans, J. 1929, *Astronomy and Cosmogony (Cambridge: Cambridge University Press)*

Kormendy, J, and Kennicutt, R. *Annual Review of Astronomy and Astrophysics*, 2004, vol. 45, *in press*

Lundmark, K. 1926, *Ark. Math. Astr. Phys., Ser. B.*, Vol. 19, No. 8

Lundmark, K. 1927, *Medd. Astr. Obs. Uppsala*, No. 30

Reynolds, J.H. 1927a, *Observatory*, 50, 185

Reynolds, J.H. 1927b, *Observatory*, 50, 308

Sandage, A. 1961, *The Hubble Atlas of Galaxies (Washington: Carnegie Institution of Washington)*, Pub. 618

Sandage, A. 1975, *in Galaxies and the Universe*, Eds. A. Sandage, M. Sandage, J. Kristian (Chicago: University of Chicago Press), Chapter 1.

Sandage, A. 1986, *Astronomy and Astrophysics*, 161, 89

Sandage, A. 1990, *Journal of the Royal Astronomical Society of Canada*, 84, 70

Sandage, A and Bedke, J. 1994, *The Carnegie Atlas of Bright Galaxies* (Washington: The Carnegie Institution of Washington), Chapter 4

Sandage, A., Freeman, K.C., and Stokes, N.R. 1970, *Astrophysical Journal*, 160, 831

Schweizer, F. 1976, *Astrophysical Journal Supplement*, 31, 313

Shapley, H. *in Galaxies* (1941 edition), (Philadelphia; The Blakiston Co.), p. 29

Spitzer, L. and Baade, W. 1951, *Astrophysical Journal*, 113, 413

van den Bergh, S. 1960a, *Astrophysical Journal*, 131, 215

van den Bergh, S. 1960b, *Astrophysical Journal*, 131, 558

van den Bergh, S. 1976, *Astrophysical Journal*, 206, 883

Vorontsov-Velyaminov, B.A., Krasnogorskaja, A., and Arkipova, V.P. 1962-1968 *Morphological Catalog of Galaxies (Moscow)*

Wolf, M. 1908, *Pub. Astroph. Inst. Hoenig. Heideberg*, Vol. 3, No. 5

Zwicky, F. 1955, *Publications of the Astronomical Society of the Pacific*, 67, 232

SECULAR EVOLUTION VERSUS HIERARCHICAL MERGING: GALAXY EVOLUTION ALONG THE HUBBLE SEQUENCE, IN THE FIELD AND RICH ENVIRONMENTS

Francoise Combes
LERMA, Observatoire de Paris, 61 Av. de l'Observatoire, F-75014, Paris, France

Abstract In the current galaxy formation scenarios, two physical phenomena are invoked to build disk galaxies: hierarchical mergers and more quiescent external gas accretion, coming from intergalactic filaments. Although both are thought to play a role, their relative importance is not known precisely. Here we consider the constraints on these scenarios brought by the observation-deduced star formation history on the one hand, and observed dynamics of galaxies on the other hand: the high frequency of bars and spirals, the high frequency of perturbations such as lopsidedness, warps, or polar rings. All these observations are not easily reproduced in simulations without important gas accretion. N-body simulations taking into account the mass exchange between stars and gas through star formation and feedback, can reproduce the data, only if galaxies double their mass in about 10 Gyr through gas accretion. Warped and polar ring systems are good tracers of this accretion, which occurs from cold gas which has not been virialised in the system's potential. The relative importance of these phenomena are compared between the field and rich clusters. The respective role of mergers and gas accretion vary considerably with environment.

Keywords: Galaxy – Evolution – Hubble sequence – accretion – star formation – Interactions – mergers

1. Introduction

Galaxies grow from small mass systems, the first structures to be unstable after recombination are of the order of a globular cluster mass ($\sim 10^6$ $M_\odot$). Then, according to the hierarchical scenario of galaxy formation, small systems merge to form larger and larger systems, and in the same time, small structures accrete gas mass and dark matter from larger gaseous structures in the shape of filaments in the cosmic web.

The relative importance of these two essential ways to build galaxies, through mergers or external gas accretion, is still unprecised in the numerical simula-

D. Block et al. (eds.), Penetrating Bars through Masks of Cosmic Dust, 57–74.

tions, since it depends on many unknown parameters of the baryonic physics, gas dissipation,, star formation efficiency, feedback, gas re-heating, etc.. However the dynamical history of galaxies strongly depends on these processes, and it might be possible to find constraints on them from the observations, locally and at various redshifts, of the dynamical states of galaxies.

Interaction with massive companions and mergers tend to heat the stellar component of galaxies, and to form the spheroids, either increase the mass of the central bulge, or even transform the system in a giant ellipticals in case of major mergers. On the contrary, gas accretion can replenish the young disk in spiral galaxies, and rejuvenate spiral or bar waves, and reduce the bulge-to-disk ratio. The thickness of galaxy disks, and their ability to maintain spiral structure and asymmetries is a tracer of their history, in terms of interactions or gas accretion. I will review here the recent progress in obtaining these constraints, both on the observational side, which provides statistics of the dynamical state of spiral galaxies, and on the theoretical side, which provides the interpretation through predictions of this dynamical state under several hypothesis about the past history of galaxies.

Section 2 reviews the star formation history of galaxies in the field, and Section 3 the constraints from the dynamics: bars, warps, polar rings and asymmetries. Section 4 considers the same phenomena in clusters, and conclusions about secular evolution as a function of environment are drawn in Section 5.

2. Star formation history

Models of the chemical evolution of the Milky Way suggest that the observed abundances of metals require a continuous infall of gas with metallicity about 0.1 times the solar value. This can solve the well-known G-dwarf problem, i.e. the observational fact that the metallicities of most long-lived stars near the Sun lie in a relatively narrow range ($-0.6 <$ [Fe/H] $< +0.2$, Rocha-Pinto & Maciel 1996). The infall of gas is also supported by the constant or increasing star formation rate (SFR) scenario inferred from the local distribution of stars (e.g. Haywood et al. 1997). Other abundance problems require also an infall rate integrated over the entire disk of the Milky Way of a few solar mass per year at least (Casuso & Beckman 2001). This infall dilutes the enrichment arising from the production of heavy elements in stars, and thereby prevents the metallicity of the interstellar medium from increasing steadily with time. Some of this gas could come from the High Velocity Clouds (HVC) infalling onto our galaxy disk (Wakker et al 1999).

The High-Velocity Clouds (HVCs) observed in the Galactic neighbourhood, at least those not included into the Magellanic Stream more akin to tidal debris, have been proposed to be remnants of the formation of the galaxies in the Local Group (Blitz et al 1999). With distances much larger than previously assumed (i.e. 1 Mpc instead of 100 kpc), their gas masses could represent a large fraction of the total baryonic mass of the Local Group. This hypothesis is supported by observational evidence that their kinematical centre is the Local Group barycentre (Blitz et al 1999). Within this hypothesis, HVCs can well explain the evolution of the light elements in the Galaxy, and the G-dwarf problem (Lopez-Corredoira et al 1999). The present infall rate of gas is estimated to be 7.5 Mo/yr (Blitz et al 1999). This is the right order of magnitude to double the baryonic mass of the Galaxy in 10 Gyr time-scale. The HVCs, whose properties are very similar to the higher redshift Lyα forest clouds, may thus form a significant constituent of baryonic, and of non-baryonic, dark matter.

It is now possible to investigate the star formation evolution in nearby galaxies as well: Worthey & Espana (2004) with HST colour-magnitude diagrams derived a stellar abundance distribution for M31. The results are quite similar to what is observed in the solar neighbourhood and they concluded that closed-box models in M31 suffer a G-dwarf problem even more severe than in the Milky Way.

The star formation rate in the Milky Way disk has remained of the same order of magnitude over the galaxy life-time (Rana & Wilkinson 1986), although some temporal fluctuations can be identified (Rocha-Pinto et al 2000). This appears to be a general results for spiral galaxies in the middle of the Hubble sequence (e.g. Kennicutt 1983, Kennicutt et al 1994).

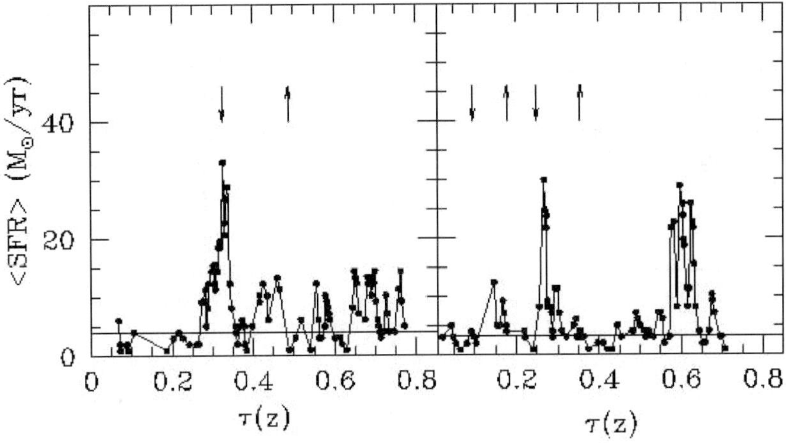

Figure 1. Star formation rate history of two galactic-like objects modelled by Tissera (2000). The arrival of a companion is indicated an arrow pointing up, and the merging of the baryonic cores by an arrow pointing down.

To maintain the same order of magnitude for star formation rate requires external gas accretion, since an isolated galaxy must have an exponentially decreasing star formation rate, even taking into account stellar mass loss. Nu-

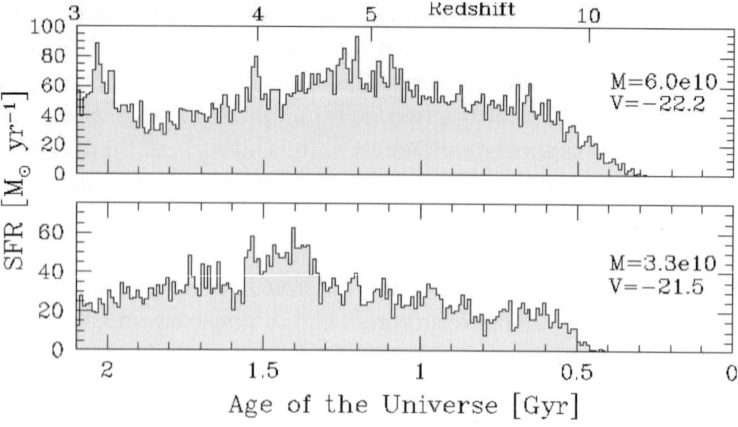

Figure 2. Star formation histories of Lyman-break galaxy models at $z = 3$, by Nagamine et al (2004). The mass of the galaxy in h^{-1} $M_{\odot}$ and its rest-frame V-band magnitude are indicated at right.

merical simulations of galaxy evolution, in a general cosmological frame, have shown that indeed a typical spiral galaxy in average environment, maintains a constant level of star formation rate, with superposed bursts (Tissera 2000, Nagamine et al 2004). While simulations of isolated galaxy mergers reveal how star formation is triggered by interaction and mergers (e.g. Mihos & Hernquist 1996), they cannot maintain a constant average level of star formation, as in cosmological simulations.

Figures 1 and 2 shows several histories of star formation for galaxy-like objects at $z = 0$, from Tissera (2000) and at $z = 3$, Nagamine et al. (2004). The history has been obtained by reconstructing the merging tree, in identifying the progenitors of the present galaxy with the particles it is made of in the simulation, a companion being defined by having at least 10% of the main galaxy mass (the accretion of baryonic clumps are assimilated to external gas accretion). Each time such a companion enters the main halo, an up arrow is drawn, and each merger is traced by a down arrow. Several phenomena can be emphasized: a large fraction of gas is accreted between mergers, and this contribute an essential part of the star formation. The tidal interaction triggers bursts of star formation, which in general are delayed with respect to the first passage, but the bursts are not only fuelled by the accreted companion gas, but mainly from the main galaxy gas driven in by the tidal forces. In that sense, the external gas accretion in between mergers also fuels the merger-triggered starbursts.

Observations of stellar populations in a large sample of local galaxies (SDSS, Heavens et al 2004, Jimenez et al 2004) have shown that massive galaxies have formed most of their stars at early times, while dwarf galaxies are forming their stars efficiently only now. Only intermediate mass galaxies could have in average maintained their star formation rate over a Hubble time. The high efficiency of star formation in massive galaxies underlines the role of environment, and the availability of high gas accretion. The rate of gas consumption is then high, and the galaxies run out of gas quickly. On the contrary, far from deep potential wells, dwarf galaxies need billion years to reach the threshold of star formation, and convert their gas only now.

In semi-analytic models, an essential free parameter to reconstruct merger trees and mass assembly in galaxies, in the fraction of accretion, and this can change considerably the final angular momentum of the system (Wechsler et al 2002). Figure 3 illustrates the time evolution of the baryonic content of halos that correspond to a "local group" ($V = 220$ km/s) sized halo at $z = 0$. This evolution corresponds to the best fit of the observations, i.e. the luminosity function of galaxies and the Tully-Fisher relation. The star formation rate has been calibrated, so that the star formation efficiency is constant with redshift (Somerville & Primack 1999). Note that the fraction of available gas in the form of cold phase remains quite high until $z = 0$.

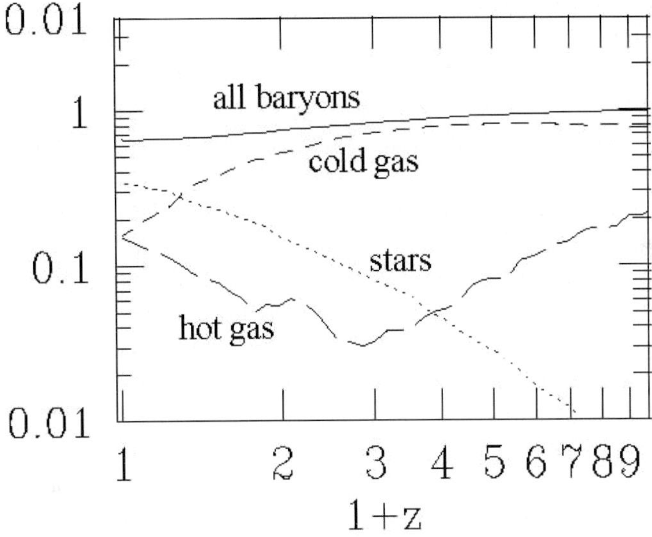

Figure 3. The fraction of cold and hot gas, and the history of star formation in halos of size $V_c = 220$ km/s at $z = 0$, in a standard CDM model, computed with semi-analytic models by Somerville & Primack (1999).

As far as the global gas content is concerned, constraints can be found in the frequency and column densities of the damped Lyman-α systems (DLA).

Already at high redshift, the models are limited by the observations, which then constrain the star formation rate (fig 4). A strong starburst mode is required in addition to quiescent star formation mode, as triggered by the frequent mergers around $z \sim 2-1$. Note that the observational points only correspond to HI gas of column density larger than 2 10^{20} cm^{-2}, ignoring molecular or ionised gas, that could bring the model closer to observations at low redshift.

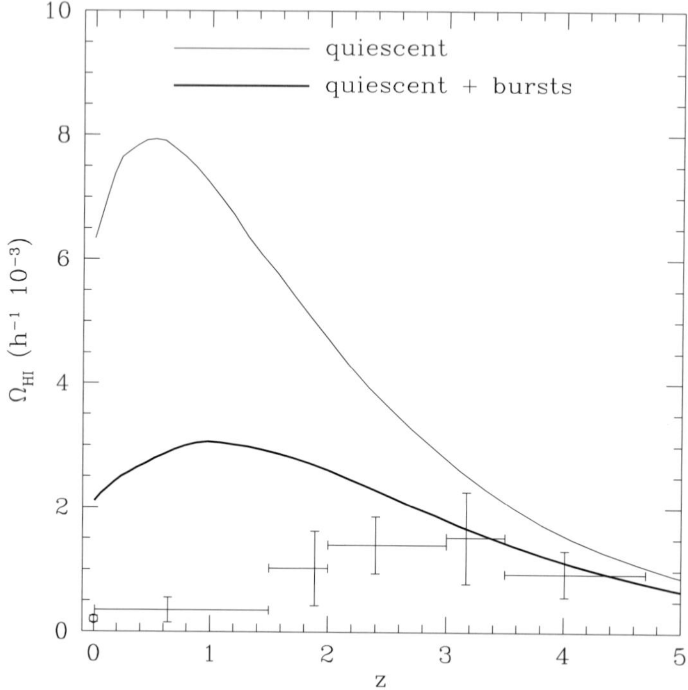

Figure 4. Cold gas density versus redshift, computed in semi-analytic models by Somerville & Primack (1998). The bold solid line includes starbursts, while the light line only quiescent star formation. Data points show the density in the form of HI estimated from observations of DLA (Storrie-Lombardi et al. 1996). The point at $z = 0$ is from local HI observations (Zwaan et al 1997).

3. Constraints from the dynamics

Several dynamical processes can be naturally explained with external gas accretion, and could be used as constraints on the amount of gas available, they are summarised in fig 5.

Bars and secular evolution

>From observations and numerical simulations in the last decades, secular evolution has been shown to be driven by bars and spirals (e.g. Kormendy & Kennicutt 2004). In particular, the bar gravity torques produce radial gas inflow, generating star formation. Dynamical instabilities then regulate themselves, since the gas inflow itself can destroy the bar. The bar is destroyed by two main mechanisms: first the central mass concentration built after the gas inflow, destroys the orbital structure sustaining the bar, scatter particles and push them on chaotic orbits (Hasan et al 1990, 1993, Hozumi & Hernquist 1999). Second, the gas inflow itself weakens the bar, since the gas loses its angular momentum to the stars forming the bar (Bournaud & Combes 2004). This increases the angular momentum of the bar wave, in decreasing the eccentricity of the orbits. This bar destruction is reversible, since the central mass concentration is then not strong enough to prevent bar formation.

Secular evolution then includes several bar episodes in a galaxy life-time. A spiral galaxy rich in gas (at least 5% of the disk mass) is unstable with respect to bar formation. Gravity torques are then efficient to drive the matter inwards. The galaxy morphological type evolves towards early-types, the mass is concentrated, the bulge is developped, through horizontal and vertical resonances. This weakens the bar, and when the galaxy becomes again axi-symmetric, gas can be accreted from the outer parts by viscosity (Bournaud & Combes 2002). The gas accretion, if significant with respect to the disk mass, can reduce the bulge-to-disk ratio (by replenishing the disk), and make the galaxy disk unstable again to a new bar.

Several bars can successively produce secular evolution of spiral galaxies, if there is enough external gas accretion. This can be used to quantify the amount of accretion in a galaxy life-time.

To estimate the number of bars that has occured in a typical galaxy, and the time spent in a bar phase, Block et al (2002) have estimated quantitatively the frequency of bars and $m = 2$ spiral components in a sample of 163 spiral galaxies, observed in the near-infrared (Eskridge et al 2002). In this band, it is more easy to identify the bars in the old stellar component, with limited dust extinction.

The bar strength Q_b has been estimated from the gravitational potential, derived from the NIR images, assuming constant mass-to-light ratios. The main surprising result in the histogram of bar strength is the hole at low values (lack of axi-symmetric galaxies), and a long tail at high values, corresponding to a large number of strongly barred galaxies (see also Whyte et al 2002, Buta et al 2004).

It can be concluded from the comparison of bar life-times given by numerical simulations, and the observation of bar frequency, that bars have to be reformed, and this can only be through significant accretion of external gas (see Bournaud & Combes, this volume).

Warps

Most spiral galaxies reveal a warped disk, particularly spectacular in the HI component (Briggs 1990), although it is still frequent in the optical disk (Reshetnikov & Combes 1998). Many physical mechanisms have been studied to explain the origin of these warps (see the review by Binney 1992), but the high frequency of warps is only accounted for by external gas and dark matter accretion. Matter accreted in the outer parts of a dark matter halo, with an angular momentum misaligned with that of the inner parts, car re-oriented the whole system in 7-10 Gyr, and the corresponding disk will show an integral-sign warp during the process (Jiang & Binney 1999). Intergalactic gas accretion is sufficient to produce a torque corresponding to the observed warps, with the infall amplitude already required by evolutionary chemical models (Lopez-Corredoira et al 2002). During some transient phase, a U-shape warp is predicted, combining m=0 and m=1 perturbations, and its frequency corresponds to the observations. The galaxy has just to move in an accretion flow of average density of 10^{-4} cm^{-3}, and velocity 100km/s.

Bullock et al (2001) have explored the angular momentum distribution in a large number of dark matter haloes, from a cosmological simulation. They found an almost universal form, with a moderate misalignment in the outer parts, that could also favor warps in the baryonic matter. However, both the halo growth through merger of smaller units, and quiescent and continuous accretion are able to produce the same universal profile, so this will not be discriminant.

Figure 5. Dynamical processes constraining the external gas accretion, illustrated by prototypical examples, from left to right: frequency of bars in galaxies (NGC 1365), frequency of warps (NGC 4013), polar rings (NGC 4650A), and lopsidedness (M101).

Polar rings

Polar ring galaxies are particular objects, composed of a primary galaxy in general of early-type (lenticular or elliptical), and a polar ring or disk, where the matter is orbiting in a nearly perpendicular plane. The polar material is quite gas-rich, but also composed of stars, of age larger than a few Gyrs, implying that the polar disk is stable. Indeed, in any of the various scenarios advanced to account for this structure, the bulk of the stars form later from the gas settled in the polar plane. The probability for a galaxy to have a polar ring has been estimated to $\sim 5\%$, given the detection biases, that these systems are better detected edge-on (Whitmore et al 1990).

The two main scenarios to form polar rings are mergers of two galaxies, or external gas accretion, this gas coming either from a passing-by companion, or from a cosmic gaseous filament. In the merging scenario (Bekki 1997, 1998), a special geometry is required for the two interacting system, it must be a head-on collision at low velocity, with the two galaxy planes nearly perpendicular to each other. In the accretion scenario, gas may be accreted from an unbound companion passing by (Schweizer et al. 1983, Reshetnikov & Sotnikova 1997), in a nearly polar orbit. The companion must be gas-rich, and may also be more massive than the primary. But gas could also come from a gaseous cosmic filament, during the long formation period of a giant galaxy, since the various filaments have not the same orientation. This is the extreme case of the warp formation mechanism, invoked in the last subsection, where the various accreted gas have not aligned angular momentum. Intermediate cases between warps and polar rings, where rings of gas orbit at around 45°, are observed (for instance NGC 660). Polar rings around galactic-like objects are frequently seen in cosmological simulations (Semelin & Combes 2004).

>From a large number of numerical simulations, exploring the various geometrical or mass parameters of the two scenarios, it has been shown that the accretion scenario is more likely to produce polar rings, corresponding to observations, than the merger scenario (Bournaud & Combes 2003). In addition to this statistical argument, some prototypical cases, like NGC 4650A, support the accretion hypothesis, since they do not possess a halo of very old stars, expected in the merger scenario.

Lopsidedness

The observation of extended disks of neutral hydrogen (through the HI line at 21cm) around most spiral disks, has revealed a large frequency of asymmetries and lopsidedness. In their compiled sample of 1700 galaxies, Richter & Sancisi (1994) found that at least 50% of spiral galaxies are lopsided, and have a characteristic signature in their global HI spectrum. This percentage is even higher in late-type galaxies, where Matthews et al (1998) found a frequency of

77% of HI distorted profiles. The asymmetries affect also the stellar disk, as shown by Rix & Zaritsky (1995) and Zaritsky & Rix (1997) on near-infrared images of galaxies. About 20-30% of galaxies have a significantly perturbed stellar disk, quantified by the $m = 1$ Fourier term in their potential or density distribution (Kornreich et al 1998, Combes et al 2004).

Since most of these galaxies are isolated without obvious sources of perturbations (e.g. Wilcots & Prescott 2004), the interpretation was searched in a possible long life-time of these features. Baldwin et al. (1980) proposed that the lopsidedness comes from $m = 1$ kinematic waves build from off-center elliptical orbits that may persist a long time against differential precession. Although this can prolonge the life-time significantly, since the winding out by differential precession is quite long in the outer parts of galaxies, they conclude that it is still not sufficient to explain the high frequency of lopsidedness in neutral gas disks.

Alternatively, the perturbations could come from recent minor mergers, explaining the non-correlation with the presence of companions (Zaritsky & Rix 1997). In this hypothesis, lopsidedness can be used to constrain the merger frequency, which appears very high. Indeed, Walker et al (1996) have estimated through N-body simulations that the life-time of perturbations are of the order of the Gyr. A high merging frequency may enter in conflict with other observations, for instance the presence of thin stellar disks in spiral galaxies (Toth & Ostriker 1992).

External gas accretion could solve the problem of the high frequency of lopsidedness, since it is likely that in many cases, the accretion is asymmetric.

4. Environmental effects

The relative role of galaxy merging and external gas accretion must depend strongly on environment. In particular, in rich clusters, mergers have considerably increased the fraction of spheroids and ellipticals, and tidal interactions and ram pressure have stripped and heated the cold gas around galaxies, so that external accretion will be reduced or suppressed. To quantify theses effects, we consider successively the influence of environment on star formation rate, morphological types, and dynamical features, such as bars or warps.

Star formation history

High resolution images with the HST, followed by spectroscopic surveys in a dozen galaxy clusters at redshift around 0.3-1. have allowed to follow galaxy evolution and in particular their star formation history as a function of time, and also position in the cluster (Oemler et al 1997, Dressler et al. 1999, Poggianti et al 1999).

Rich clusters of galaxies, and especially their dense cores, are well known to be dominated by early-type galaxies without star formation, or even passively evolving and anemic spirals, conspicuous for their low star-formation rate. To find star formation it is necessary to look back in time, at redshift at least larger than 0.2, where there exists a larger fraction of blue galaxies (B-O effect, Butcher & Oemler 1978, 1984). The recent HST images have shown that these blue galaxies are disk galaxies in majority perturbed by tidal interactions, strongly suggesting that galaxy interactions are still triggering star formation activity around $z = 0.5$ (Lavery & Henry 1988).

There is thus a strong evolution effect in clusters in a recent past. Moreover, detailed spectroscopic studies have shown that $z \sim 0.5$ clusters possess a large fraction of peculiar galaxies, likely post-starburst, called E+A (or k+a), devoid of emission lines (and therefore with no current star formation), but very strong Balmer absorption lines. (Dressler et al 1999, Poggianti et al 1999). This means that they have a large fraction of A stars, implying that the galaxy was experiencing a strong starburst that has just been abruptly interrupted. These galaxies are in majority disk-dominated. Their fraction is about 20%, much larger than in the field, and in the clusters at $z = 0$ (Dressler et al 1999).

That star formation is quenched in clusters is revealed by the low fraction ($\sim$ 10%) of Hα emitters; star formation and morphological evolution in cluster galaxies appear to be largely decoupled (Couch et al 2001). On the contrary, the star formation is continuing in groups. About 55% of galaxies have been cataloged in groups at $z < 0.1$ in the 2dF and SDSS surveys, and the Hα detection rate and equivalent width varies strongly and continuously with galaxy density (Balogh et al 2004).

The star formation rate, as traced by the Hα line, is strongly dependent on the density of galaxies (Lewis et al 2002, 2dF survey). There is a tight correlation between SFR and local projected density, as soon as the density is above Σ = 1 galaxy/Mpc^2, independent of the size of the structure. In clusters, the field star formation rate is only recovered at about 3 times the virial radius. Gomez et al (2003) find also a strong SFR-Σ relation with the early data release of the SDSS, the SF-quenching effect being even more noticeable for strongly star-forming galaxies. The same break of the SFR-Σ relation is observed at 1 galaxy/Mpc^2. This relation is somewhat linked to the morphological type-density (T-Σ) relation, but cannot be reduced to it,

The spatial distribution of star forming galaxies is also quite clear in clusters. There is a clear radial gradient, the star formation being more active in the outer parts (Balogh et al 1999). There is a smooth transition from a blue, disk-dominated galaxy population in the outskirts, to a red, bulge-dominated, galaxy population in the cluster cores. Although this is influenced by the various transformation processes occuring in clusters (tidal interaction, mergers, harrassment, ram-pressure..), this has been mainly interpreted in terms of the building up of the clusters themselves, i.e. gas-rich galaxies continue to fall into the cluster, and contribute to the star-formation activity there, but then their gas replenishment is stopped when these galaxies arrive into the core (Diaferio et al. 2001). In a model where the morphology of galaxies are assumed to depend only on their merger history, where mergers lead to spheroid and bulges, and subsequent gas cooling replenishes disks, it is possible to reproduce the morphological gradient, and star-formation gradient observed. Galaxies in clusters are much sooner devoid of their gas, and their gas fueling is stopped, as shown in Figure 6.

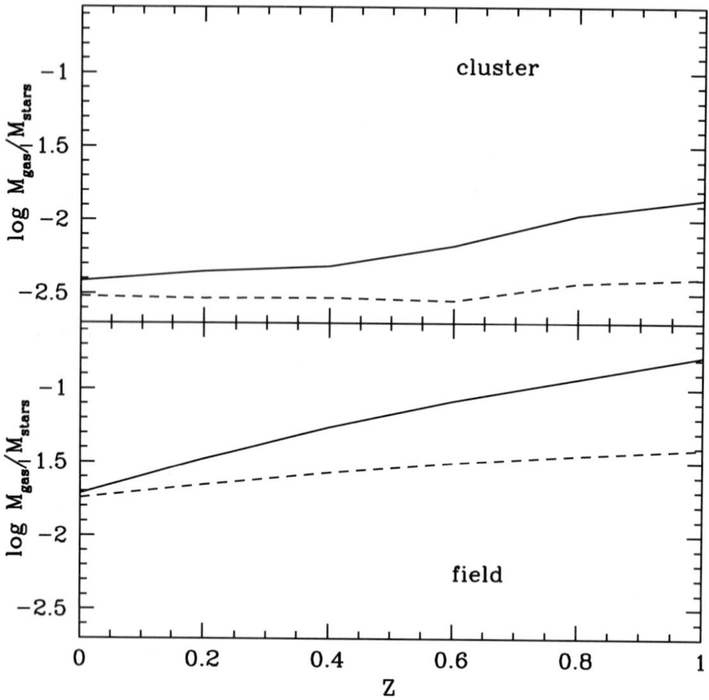

Figure 6. The ratio of gas mass to stellar mass as a function of redshift for galaxies larger than 3 10^{10} $M_{\odot}$ in clusters (top) and in the field (bottom), from semi-analytical models by Diaferio et al (2001). The dashed lines correspond to SFR proportional to the cold gas mass over the dynamical time. The solid lines include in addition a dependency of the SFR with redshift as $(1+z)^{-1.5}$.

Morphological types

There is a strong morphological segregation as a function of projected density Σ at $z = 0$ (Dressler 1980), essentially the fraction of spirals fall from two thirds to less than one third to the benefit of lenticulars and ellipticals. This segregation has also been observed at $z = 0.4$ (or 5Gyrs ago), where the fraction of ellipticals is the same as at $z = 0$ (Dressler at al 1997). The difference is in the fraction of lenticulars, which is less than at $z = 0$. This suggests that the formation of elliptical galaxies occurs before the formation of rich clusters, probably in the loose-group phase or earlier. Lenticulars are generated in large numbers only after cluster virialization, through the various galaxy transformation processes, tidal interactions, mergers, ram-pressure stripping and sudden halt in gas fueling (Poggianti et al 2001). These results have been confirmed and precised with the local Sloan survey by Goto et al (2003); at least two mechanisms are required to explain the morphological segregation: first at low density in the outskirts of the cluster, the gas supply is stopped, and star formation halted. Second at high density, in the cluster core, galaxy interactions and merging must explain the high frequency of early-types. The segregation is also found quite similar at $z = 0.5$ and locally, confirming that ellipticals in clusters have formed earlier. This corresponds to numerical simulations, which show that the halos located inside clusters form earlier than isolated halos of the same mass (Gottloeber et al 2001).

Many aspects of galaxy morphology in clusters can be explained by evolutionary processes (see the reviews by Moore 2003, Combes 2003), but the fact that the gas supply is halted plays in all these processes a major role (Shioya et al 2004 and fig 7): this gas shortening could happen in two ways, either gradual from gas stripping, and star formation rate dropping subsequently, or abruptly, through a starburst triggered by an interaction, and lack of gas supply afterwards. The second way is necessary to explain the high frequency of post-starburst systems (k+a) in distant clusters, and the low frequency of strong Hα line emitters.

Frequency of bars, warps

Spiral galaxies are globally less abundant in clusters, and are preferentially located in the outskirts. It has long been known that clusters are the site of a particular class of spirals, called "anemic", with low star formation, smooth arms, and low gas content (van den Bergh 1976). In average, anemic galaxies have an HI deficiency of a factor 4, but no CO emission deficiency, their spiral structure is fuzzy and short-lived, since new stars with low velocity dispersion are no longer formed out of the gas (Elmegreen et al 2002).

The frequency of bars in cluster spirals has been studied in order to test the influence of tidal interactions by Andersen (1996) for the Virgo cluster. Barred galaxies appear to be more centrally concentrated in the cluster core than unbarred galaxies. This is inferred from both velocity dispersions which are quite different. On the contrary, barred or unbarred lenticulars have the

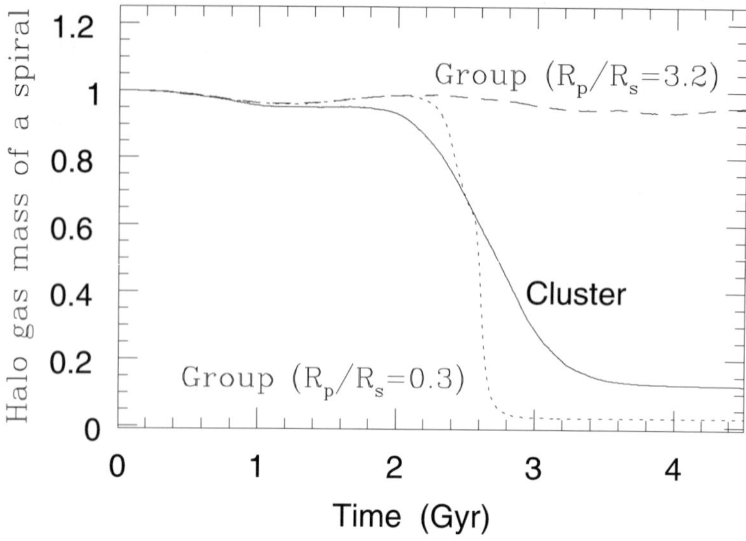

Figure 7. Evolution with time of the halo gas mass (normalized to the initial mass), for a spiral galaxy orbiting a cluster with a pericenter distance R_p equal to 3.2 the scale radius of the cluster R_s = 230kpc (solid line). By comparison, the dashed line corresponds to a comparable orbit in a group with R_s = 62kpc, and the dotted line, with R_p = 0.3 R_s in this group (from Bekki et al 2002).

same radial distribution in the cluster. Andersen (1996) suggests that tidal triggering by the cluster itself is the most likely source of the enhanced fraction of barred spirals in the cluster center. The same phenomenon is observed in Coma (Thompson 1981). Bars occur twice as often in the core region of Coma as they do in the outer parts of the cluster. This is certainly the consequence of tidal interactions triggering bar instability (Gerin et al. 1990, Miwa & Noguchi 1998, Berentzen et al 2004).

At a global scale however, there does not appear to be any significant difference between the frequency of bars in clusters and in the field (van den Bergh, 2002). Because bars are triggered by galaxy interactions in the cluster cores, this implies that the suppression of gas supply in clusters has a tendency to reduce bar frequency to compensate.

It is difficult to observe warps around spiral galaxies in clusters, because of their characteristic general HI deficiency. About two thirds of galaxy clusters are HI deficient (Solanes et al. 2001). Gas is particularly stripped in the outer parts of galaxies, where warps develop.

Polar rings and asymmetries

The environment of polar ring galaxies seems to be similar to that of normal galaxies (Brocca et al 1997). There is no evidence of more companions for example. Some polar ring galaxies are observed in clusters (Taniguchi et al. 1986). As for asymmetries, they are very frequent in clusters, but this has to be attributed to galaxy interactions; in a Hubble time, each galaxy suffers about 10 encounters with an impact parameter less than 10kpc, due to the importance of substructures (Gnedin 2003).

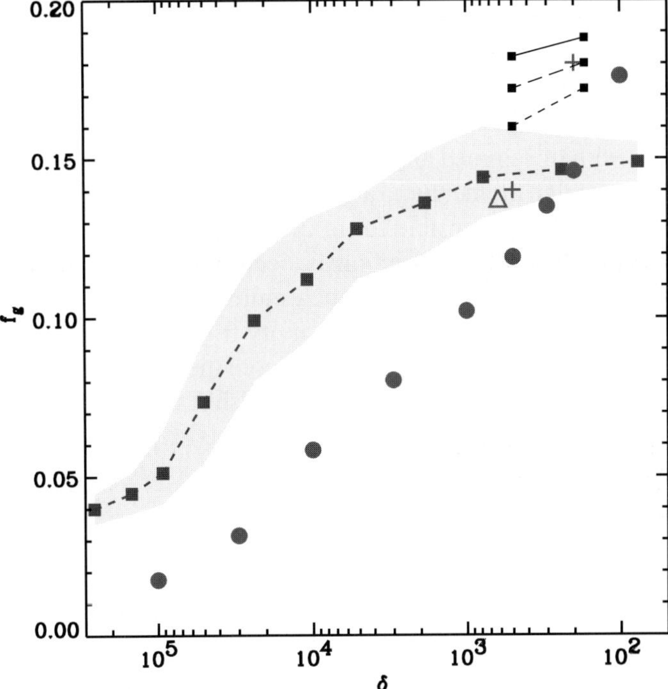

Figure 8. Distribution of the hot gas fraction f_g in clusters of galaxies as a function of δ, the average density inside the radius r (normalised to the critical density). The filled circles, empty triangle and crosses are the data, while the squares connected by a dashed line are the theoretical predictions (cosmological simulations from several groups), with an assumed baryon fraction of $f_b = \Omega_b/\Omega_m$ =0.16; other models (with or without stellar feedback) are shown with $f_b = \Omega_b/\Omega_m$ =0.20 (small squares), from Sadat & Blanchard (2001)

5. Summary

The mass assembly of galaxies occurs through two main processes: hierarchical merging of smaller entities, and more diffuse gas accretion. The relative importance of the two processes cannot be easily found by cosmological simulations, since many physical parameters such as gas dissipation, star formation and feedback, are still unknown. However, star forming histories in galaxies (age, kinematics and metallicity), and also the dynamical states of galaxies (bars, spirals, warps, polar rings, asymmetries..) can constrain the role of the two processes.

Chemical evolution of normal spiral galaxies requires gas infall of low-metallicity gas, in the proportion of a few solar masses per year. The developpment of bars and spirals in galaxies constrain the amount of external gas accretion at about 10 $M_{\odot}$/yr. Bars are transient features in galaxies, they exert gravity torques on the gas component, and drive mass to the center. A new bar can reform in disks that have been replenished in gas. The large frequency of gaseous warps in the outskirts of galaxies require also continuous gas accretion, with misaligned angular momentum. When the accreted gas has perpendicular angular momentum, a polar ring can be formed. Asymmetric gas accretion could be responsible for frequent lopsided galaxies.

In rich environments, the evolution is occuring at a much quicker pace. Galaxy interactions and mergers are much more frequent, and happen earlier in the Hubble time. The star formation activity triggered by the mergers must have occurred early in dense groups that coalesce then in galaxy clusters. Most of the star formation is then quenched after redshift $z \sim 1$. The only activity remaining comes from the infall of field spiral galaxies in the outer parts of the clusters. Secular evolution through external gas accretion is slowed down or halted, since galaxies are stripped from their gas, at large galactocentric radii. Most of the gas has been shock-heated at the virial temperature of the clusters, and is observed in X-rays. The mass fraction in hot gas reaches the baryonic fraction in the outer parts (e.g. Sadat & Blanchard 2001, fig 8, Valageas et al. 2002). Contrary to the field, it appears that hierarchical merging has dominated secular evolution in rich galaxy clusters.

Acknowledgments

I wish to thank David Block for the organisation of this exciting conference.

References

Andersen V.: 1996 AJ 111, 1805
Baldwin J., Lynden-Bell D., Sancisi R.: 1980, MNRAS 193, 313
Balogh, M., Eke, V., Miller, C. et al.: 2004, MNRAS 348, 1355
Balogh, M.L., Morris, S.L., Yee, H. K. C. et al: 1999, ApJ 527, 54

Bekki, K. 1997, ApJ, 490L, 37
Bekki, K. 1998, ApJ, 499, 635
Bekki, K., Couch W., Shioya Y.: 2002, ApJ 577, 651
Berentzen, I., Athanassoula, E., Heller, C. H., Fricke, K. J.: 2004 MNRAS 347, 220
Binney J.: 1992 ARA&A 30, 51
Blitz L., Spergel D., Teuben P. et al.: 1999, ApJ 514, 818
Block D., Bournaud F., Combes F., Puerari I., Buta R.: 2002, A&A 394, L35
Bournaud F., Combes F.: 2004, A&A Letter, in press
Bournaud F., Combes F.: 2003, A&A 401, 817
Bournaud F., Combes F.: 2002, A&A 392, 83
Briggs F.H.: 1990 ApJ 352, 15
Brocca, C., Bettoni, D., Galletta, G.: 1997 A&A 326, 907
Bullock, J. S., Dekel, A., Kolatt, T. S. et al.: 2001, ApJ 555, 240
Buta R., Laurikainen E., Salo H.: 2004, AJ 127, 279
Butcher, H., Oemler, A.: 1978, ApJ 219, 18
Butcher, H., Oemler, A.: 1984, ApJ 285, 426
Casuso E., Beckman J.E.: 2001, ApJ 557, 681
Combes F., Jog C., Bournaud F., Puerari I.: 2004, A&A in prep
Combes F.: 2003, IAU Symp. 217, Recycling intergalactic and interstellar matter, ASP Conf Series, ed. P-A. Duc et al. (astro-ph/0308293)
Couch W.J., Balogh, M.L., Bower, R.G. et al.: 2001, ApJ 549, 820
Diaferio A., Kauffmann G., Balogh M.L. et al.: 2001, MNRAS 323, 999
Dressler, A.: 1980 ApJ 236, 351
Dressler, A., Oemler A., Couch W.J. et al.: 1997, ApJ 490, 577
Dressler, A., Smail, I., Poggianti, B. et al.: 1999 ApJS 122, 51
Elmegreen D.M., Elmegreen B.G., Frogel J.A. et al: 2002 AJ 124, 777
Eskridge, P. B., Frogel, J. A., Pogge, R. W., et al.: 2000, AJ 119, 536
Gerin, M., Combes, F., Athanassoula, E.: 1990 A&A 230, 37
Gnedin, O.: 2003, ApJ 582, 141 and ApJ 589, 752
Gomez, P., Nichol, R., Miller, C. et al.: 2003, ApJ 584, 210
Goto, T., Yamauchi, C., Fujita, Y. et al.: 2003, MNRAS, 346, 601
Gottloeber, S., Klypin, A., Kravtsov, A. V.: 2001, ApJ 546, 223
Hasan, H., & Norman, C.A.: 1990, ApJ 361, 69
Hasan, H., Pfenniger, D., Norman, C.: 1993, ApJ 409, 91
Haywood M., Robin A.C., Creze M.: 1997, A&A 320, 428 & 440
Heavens, A., Panter, B., Jimenez, R., Dunlop, J.: 2004, Nature, 428, 625
Hozumi S., Hernquist L.: 1999, in "Galaxy Dynamics", ASP Conf Series, ed. D. R. Merritt, M. Valluri, and J. A. Sellwood, Vol 182, p.259
Jiang I-G., Binney J.: 1999 MNRAS 303, L7
Jimenez R., Panter B., Heavens A., Verde L.: 2004, MNRAS, preprint
Kennicutt, R. C., Tamblyn, P., Congdon, C. E.: 1994, ApJ 435, 22
Kennicutt, R. C.: 1983, ApJ 272, 54
Kormendy J., Kennicutt, R. C.: 2004, ARAA, in press
Kornreich, D. A., Haynes, M. P., Lovelace, R. V. E.: 1998, AJ 116, 2154
Lavery R.J., Henry J.P.: 1988, ApJ 330, 596
Lewis, I., Balogh, M., de Propris, R. et al. 2002: MNRAS 334, 673
Lopez-Corredoira, M., Beckman, J. E., Casuso, E.: 1999 A&A 351, 920
Lopez-Corredoira, M., Betancort-Rijo, J., Beckman, J. E.: 2002, A&A 386, 169
Matthews, L. D., van Driel, W., Gallagher, J. S.: 1998, AJ 116, 2196
Mihos J.C., Hernquist L.: 1996, ApJ 464, 641

Miwa, T., Noguchi, M.: 1998 ApJ 499, 149
Moore, B.: 2003, in Clusters of Galaxies: Probes of Cosmological Structure and Galaxy Evolution, Carnegie Observatories Symposium III. (astro-ph/0306596)
Nagamine K., Springel V., Hernquist L., Machacek M.: 2004, MNRAS 350, 385
Oemler, A., Dressler, A., Butcher, H.: 1997 ApJ 474, 561
Poggianti, B., Bridges T.J., Carter, D. et al. : 2001 ApJ 563, 118
Poggianti, B., Smail, I., Dressler, A. et al. : 1999 ApJ 518, 576
Rana, N. C., Wilkinson, D. A.: 1986 MNRAS, 218, 497
Reshetnikov V., Combes F.: 1998 A&A 337, 9
Reshetnikov, V., Sotnikova, N. 1997, A&A, 325, 933
Richter O., Sancisi R.: 1994, A&A 290, 9
Rix, H-W., Zaritsky D.: 1995, ApJ 447, 82
Rocha-Pinto, H. J., Scalo, J., Maciel, W. J., Flynn, C.: 2000, ApJ 531, L115
Rocha-Pinto H.J., Maciel W.J.: 1996, MNRAS 279, 447
Sadat R., Blanchard A.: 2001 A&A 371, 19
Schweizer, F., Whitmore, B. C., Rubin, V. C. 1983, AJ, 88, 909
Semelin B., Combes F.: 2004, A&A in prep.
Shioya Y., Bekki K., Couch W.: 2004, ApJ 601, 654
Solanes, J., Manrique, A., Garcia-Gomez, C. et al.: 2001, ApJ 548, 97
Somerville R.S., Primack J.R.: 1999, MNRAS 310, 1087
Somerville R.S., Primack J.R.: 1998, Proceedings of the Xth Rencontre de Blois
Storrie-Lombardi, L.J., McMahon, R.G., Irwin M.J.: 1996, MNRAS, 283, L79
Taniguchi, Y., Shibata, K., Wakamatsu, K.-I.: 1986 Ap&SS 118, 529
Tissera P.: 2000, ApJ 534, 636
Thompson L.A.: 1981 ApJ 244, L43
Toth G., Ostriker J.P.: 1992, ApJ 389, 5
Valageas, P., Schaeffer, R., Silk, J.: 2002 A&A 388, 741
van den Bergh, S.: 2002, AJ 124, 782
van den Bergh, S.: 1976 ApJ 206, 883
Wakker B.P., Howk J.C., Savage B.D. et al.: 1999 Nature, 402, 388
Walker, I. R., Mihos, J. C.; Hernquist, L.: 1996, ApJ 460, 121
Wechsler, R. H., Bullock, J. S., Primack, et al.: 2002, ApJ 568, 52
Wilcots, E. M., Prescott, M. K. M.: 2004, AJ 127, 1900
Whitmore, B. C., Lucas, R. A. 1990, AJ, 100, 1489
Whyte L.F., Abraham R.G., Merrifield M.R.et al.: 2002, MNRAS 336, 1281
Worthey G., Espana A.L.: 2004, in *Origin and Evolution of the Elements*, Carnegie Observatories Astrophysics Series, A. McWilliam and M. Rauch Ed.
Zaritsky D., Rix, H-W.: 1997, ApJ 477, 118
Zwaan M.A., Briggs F.H., Sprayberry D., Sorar E.: 1997, ApJ 490, 173

DENSE GAS AND STAR FORMATION IN BARS

Susanne Huttemeister
Astronomisches Institut, Ruhr-Universitat Bochum, 44780 Bochum, Germany

Abstract Bars serve as obvious funnels of gas towrard the cntral regions of galaxies, where the gas flow may then trigger activity. However, only a minority of bars is actively funnelling gas toward the nucleus and can thus be traced by dense, molecular gas. These objects offer an excellent laboratory for the study of changing gas properties, diagnosed by specific ratios between molecular lines. Ratios of CO isotopomer lines are of special importance and allow to identify a 'diffuse', unbound molecular phase as well as bound, almost disk-like clouds which may be the sites of star formation within bars. The diffuse medium which may dominate the region of the strongest bar shock is capable of offsetting the standard molecular mass determination from CO intensities. The examination of a small sample of gas-filled bars shows that not all specimens are equal – the shock strength and probably also the evolutionary stage of the (starburst) nuclei differ significantly. In a starburst nucleus, the dense gas should be more abundant. However, a few objects with very faint HCN emission have been identified, possibly indicating a dominating diffuse gas phase and/or a different star formation efficiency.

Keywords: Galaxies: Bars, Galaxies: Starbursts

1. Bars, gas flows and molecular line ratios

Bars are a frequent ingredient in galaxies: $\sim 1/3$ of all spirals in the local universe are strongly barred. Another third of all local spirals, including the Milky Way, show a weak bar, which is often confined to the inner few kpc of the galaxy. Bars represent an important transport mechanism of material toward the central regions of galaxies, fuelling nuclear starbursts and driving galaxy evolution through the concentration of mass close to the nucleus (e.g. Sakamoto et al. 1999, Combes, Dupraz, & Gerin 1990 and references therein). They efficiently transfer angular momentum outward and allow gas to flow into the central region along characteristic (x_1 and x_2) orbits.

To concentrate mass in the central region and thus affect galaxy evolution, obviously the gas flow inward through the bar has to be active. However, long bars that are filled with dense gas that can be traced in CO, the molecule most commonly used to map the distribution of the molecular phase of the Interstellar Medium, are considerably rarer than strong bars which show molecular gas

D. Block et al. (eds.), Penetrating Bars through Masks of Cosmic Dust, 75–79.

only at the bar ends or close to the center (if at all) and thus likely transient. The number of known 'active' bars in terms of gas transport depends on the amount of deep observations of molecular gas, but the overall fraction is likely to be no larger than $\sim 30\%$.

These bars differ significantly in their properties, i.e. velocity field, linewidth and derived shock structure. This is somewhat surprising, since the gas should in all cases be responding to a strong bar potential in a similar way. The dynamics of this response should then have consequences for the properties of the gas in the bar and the nucleus. For example, a diffuse component unbound from clouds should be produced (e.g. Das & Jog 1995). The evolutionary state of the bar and the degree of central mass concentration may be instrumental in regulating the conditions of the gas in the bar (see models by Athanassoula 1992). More specifically, the gas is funnelled from outer x_1-orbits to inner x_2-orbits as the bar evolves in time (see e.g. the simulations by Friedli & Benz 1993).

The investigation of the physical properties of the molecular gas requires more information than the lowest (1–0) transition of the most abundant (after H_2) molecule ^{12}CO alone can provide. If several CO lines have been observed (arising from higher J levels of ^{12}CO as well as from the rarer isotopomers ^{13}CO or even C^{18}O), it becomes possible to solve a simplified version of the radiative transfer problem, and obtain estimates of $n(H_2)$ and $T_{\rm kin}$.

Studying these models and relating them to know environments, it becomes clear that the the value of a diagnostic line ratio, e.g. ^{12}CO/^{13}CO(1–0) $= \mathcal{R}_{10}$, alone can already be an indication of the nature of the dominating gas component. Low values of $\mathcal{R}_{10} \approx 4 - 6$ are typical for spiral arms in galactic disks (Polk et al. 1988). In the starburst centers of galaxies, $\mathcal{R}_{10}$ is elevated to typical values of 10 – 15 (Aalto et al. 1995). Even higher values of > 15 are found in a few centers of luminous merging galaxies (Casoli, Dupraz & Combes 1992, Aalto et al. 1991). Physically, high values of $\mathcal{R}_{10}$ go along with low global $n(H_2)$ densities of $\leq 10^3$ cm-3. Alternatively, the ^{13}CO isotopomer, that, due its smaller abundance, is less able to shield itself against destructive UV radiation, may be selectively dissociated and thus physically removed from the gas, if the effective optical depth is low. This may be accomplished by a low (column) density and a high degree of turbulent motion, which may be expected in the environment of a starburst nucleus or a bar shock.

2. Examples of Gas-filled Bars

We have examined $\mathcal{R}_{10}$ as the most easily accessible diagnostic line ratio in a small sample of three galaxies with gas-filled bars. For two galaxies, high-resolution interferometric observations in ^{12}CO(1–0) are also available.

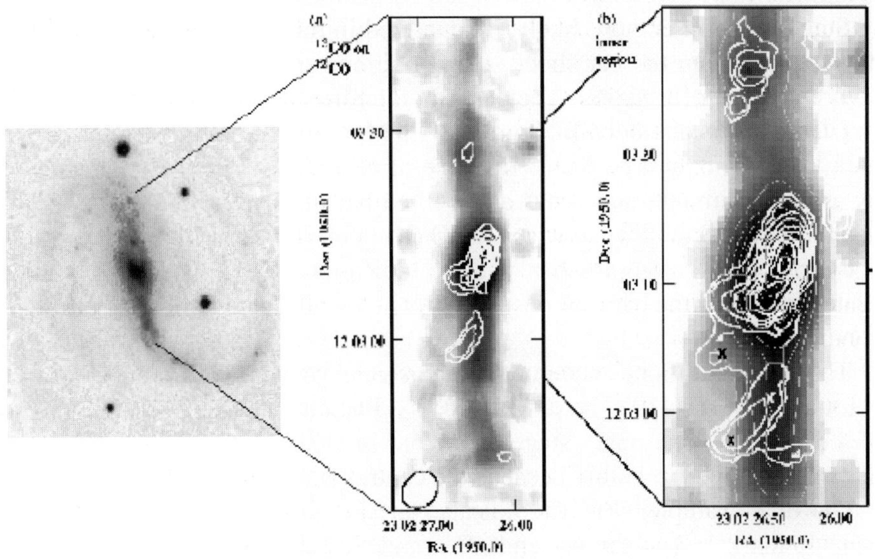

Figure 1. The gas-filled bar of NGC 7479. The optical image in the left panel was taken with HoLiCam at Hoher List Observatory in the V-band (courtesy M. Altmann). The middle panel shows the bar in the ^{12}CO(1–0) transition, the right panel depicts the inner part of the bar in ^{12}CO(1–0) (grayscale and thin gray contours) and ^{13}CO(1–0) (white contours, adapted from Huttemeister et al. 2000, based on OVRO data).

Despite being of the same Hubble type, SBc, the dominant physical properties of the gas in the bar of the three galaxies differ significantly.

NGC 7479 (Fig. 1) is an SBc starburst galaxy with a LINER (or possibly Seyfert 2) nucleus. Our study of $\mathcal{R}_{10}$ reveals strong variations within the bar and the starburst center of the galaxy: Within the central 1 kpc, $\mathcal{R}_{10}$ varies between 10 and 30, the global value for the central condensation is 22, i.e. somewhat higher than for typical starburst nuclei. Within the bar, we find variations between 4 and 40 within a few arcseconds (or a few 100 pc; $1'' \sim$ 155 pc at an assumed distance of 32 Mpc). It is striking that the low, disk-like values for $\mathcal{R}_{10}$ occur well away from the peak intensities in ^{12}CO, which in turn are well correlated with the dust lanes indicating the main bar shock.

Assuming that high values of $\mathcal{R}_{10}$ indicate low gas densities, we can infer that the shock region is dominated by low density molecular gas that is probably 'diffuse' in the sense of not being bound to clouds. This gas is expected to have an optical depth of (only) τ 1–2 in the ^{12}CO(1–0) transition, a property that has consequences for the way this transition traces the total gas mass: The gas mass can be overestimated by up to an order of magnitude if the standard CO–to-H_2 conversion factor is incorrectly applied. This gas component is unlikely to be involved in current star formation activity.

Star formation is more likely encountered in the region of low $\mathcal{R}_{10}$. Here, downstream from the bar shock, the conditions are more quiescent and thus favorable to star formation. The gas is in the form of bound clouds.

UGC 2855 is another SBc spiral, $\sim$ 20 Mpc distant and tidally interacting with UGC 2866, at a projected distance of 60 kpc. Its center is characterised by a weak starburst, and it has a gas-filled bar of $\sim$ 8 kpc length. The high resolution ^{12}CO map we took at OVRO (Huttemeister et al. 1999) shows solid body rotation all along the bar, with no indication of a bar shock. $\mathcal{R}_{10}$ in the center and along the bar is almost constant at 5 – 10, again indicating quiescent conditions.

Thus, the conditions encountered in this galaxy are very different from what is found in NGC 7479. The diagnostic $\mathcal{R}_{10}$ line ratio can serve as an indicator of a different evolutionary stage of the bar in UGC 2855: The bar shock has not yet developed, possibly because the central mass concentration is still low.

Our final example, *NGC 4123*, is also an SBc galaxy and member of a wide pair. Remarkably, at the bar end, $\mathcal{R}_{10}$ and $\mathcal{R}_{21}$ drop from central values of $\sim$ 25 to $\sim 4 - 5$. This is a strong argument against one possible explanation of changing $\mathcal{R}$, namely an inflow of gas that is underabundant in ^{13}CO. This scenario is very unlikely in a bar environment, which is expected to be well mixed, in any case, but the example of NGC 4123 clearly shows that the physical properties of the gas (possibly combined with selective photodissociation) dominate the changes in $\mathcal{R}$.

3. Faint HCN emission

Studies of HCN emission from (barred and unbarred) starburst galaxies show a generally good (and expected) correlation between $L_{\rm FIR}$ and $L_{\rm HCN}$ (Gao & Solomon 2004), indicating a higher dense gas fraction with rising star formation rate. However, in the regime of *moderate* starburst activity ($L_{\rm FIR}$ between $10^{10}\,{\rm L}_\odot$ and $10^{11}\,{\rm L}_\odot$), we have found a few galaxies that seem to defy the trend of an elevated dense gas fraction going along with a starburst. Fig. 2 shows three examples, and NGC 4123 may be a fourth specimen. UGC 2866 is the (barred) starburst companion of UGC 2855, and NGC 4194 (the Medusa) and Mrk 297 are moderate luminosity starburst mergers harbouring unusually extended bursts.

Since NGC 4123 is a barred galaxy and the CO distribution in UGC 2866 is, at least morphologically, also bar-like, one explanation for this unusual behaviour is a very dominant diffuse gas phase, possibly residing in the bar. However, the central values for $\mathcal{R}_{21}$ are normal for starburst nuclei in both galaxies, and the gas distribution in NGC 4194 does not seem to be dominated

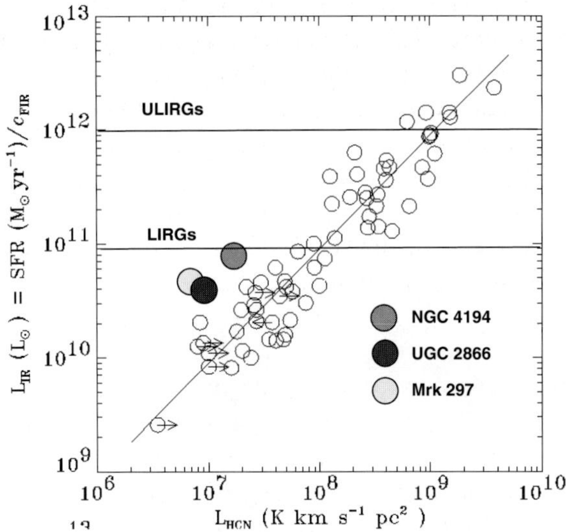

Figure 2. Three examples of galaxies which are underluminous in HCN, overlayed on the tight L(HCN) – L(IR) relation of Gao & Solomon 2004.

by a bar. Moreover, the galaxies in question do not appear to be overluminous in CO with respect to the FIR luminosity. Thus, it seems possible the dense gas fraction in these galaxies is low not just along the bar, but also in the nuclear region, despite an ongoing starburst. This may be related to the extent of the burst, which is possibly less concentrated than in, e.g., ULIRGs with high dense gas fractions. These galaxies may form stars very efficiently from a relatively small amount of dense gas. Clearly, further investigations are required to assess the importance of this class of objects.

References

Aalto. S., Black, J.H., Johansson, L.E.B., & Booth, R.S. 1991, A&A 249, 323
Aalto, S., Booth, R.S., Black, J.H., & Johansson, L.E.B. 1995, A&A 300, 369
Athanassoula, E. 1992, MNRAS 259, 345
Casoli, F., Dupraz, C., & Combes, F. 1992, A&A 264, 55
Combes, F., Dupraz, C., & Gerin, M. 1990, In: Wielen, R. (ed.) Dynamics and Interactions of Galaxies. Springer, Berlin, p. 305
Friedli, D., & Benz, W. 1993, A&A, 268, 65
Gao,Y., & Solomon, P.M. 2004, ApJS 152, 63
Huttemeister, S., Aalto, S., & Wall, W.F. 1999, A&A 346, 45
Huttemeister, S., Aalto, Das, M., S., & Wall, W.F. 2000, A&A 363, 93
Polk, K.S., Knapp, G.R., Stark, A.A.,& Wilson, R.W. 1988, ApJ 332, 432
Sakamoto, K., Okumura, S.K., Ishizuki, S., & Scoville, N.Z. 1999, ApJ 525, 691

ON THE ORIGIN OF S0 GALAXIES

Uta Fritze – v. Alvensleben
Universitätssternwarte Göttingen, Geismarlandstr. 11, 37083 Göttingen, Germany

Abstract I will review the basic properties of S0 galaxies in the local Universe in relation to both elliptical and spiral galaxies, their neighbours on the Hubble sequence, and also in relation to dwarf spheroidal (dSph) galaxies. This will include colours, luminosities, spectral features, information about the age and metallicity composition of their stellar populations and globular clusters, about their ISM content, as well as kinematic signatures and their implications for central black hole masses and past interaction events, and the number ratios of S0s to other galaxy types in relation to environmental galaxy density.
I will point out some caveats as to their morphological discrimination against other classes of galaxies, discuss the role of dust and the wavelength dependence of bulge/disk light ratios. These effects are of importance for investigations into the redshift evolution of S0 galaxies – both as individual objects and as a population. The various formation and transformation scenarios for S0 and dSph galaxies will be presented and confronted with the available observations.

Keywords: S0 galaxies, formation and evolution

1. Introduction

Until some years ago, S0 galaxies were thought to be simple and well-understood: old passively evolving purely stellar systems in which everything had happened in the distant past, which are boring now and bound to go on fading into a dark and unspectacular future. In recent years a series of surprising observations led to a vividly renewed interest in S0 galaxies, their formation, transformation and evolution.

2. Conventional Wisdom and Recent Puzzles

Located on the Hubble sequence between **E**llipticals (**E**s) and **Sp**irals (**Sp**s) because of their strong $r^{1/4}$–bulges and their small and smooth exponential disks with **B**ulge-to-**T**otal light ratios **B/T**$\sim$ 0.6 in the B-band, S0 galaxies hold keys to 3 fundamental issues: **1.** does Hubble's classification describe a continuous sequence from Es through Sd and Irr galaxies with S0s being the transition types or are Es and Sps fundamentally different and, if so, where

D. Block et al. (eds.), Penetrating Bars through Masks of Cosmic Dust, 81–99.

do the S0s belong? **2.** What determines the properties of S0 galaxies: initial conditions or environmental effects, nature or nurture? **3.** The rate of evolution of S0 (and E) galaxies can test cosmological structure formation scenarios and constrain their parameters.

Classically, S0 galaxies were thought to have formed in a monolithic initial collapse with most of their **S**tar **F**ormation (**SF**) on a short timescale in the early Universe. Their relatively unimportant stellar disks (as compared to spirals) formed from leftover or accreted gas shortly after the bulge as both components do contain red stellar populations, little gas and essentially no SF. S0 galaxies follow relations established for ellipticals, like the color–magnitude relation, the black-hole mass – velocity dispersion, Faber–Jackson, **F**undamental **P**lane (**FP**), and luminosity–metallicity relations. They also follow the Tully–Fisher relation for spirals, although with larger scatter. The observations that S0s are the dominant galaxy population in the central regions of nearby galaxy clusters while scarce in clusters at intermediate redshift and in the field provided a major trigger into renewed interest in the S0s and their formation or transformation histories. Recent detections of unexpected amounts of gas, dust and SF, kinematic peculiarities, fine-structure, inward central color gradients, etc. added to this new interest. Various scenarios have been suggested for the transformation accompanying the infall into the gradually forming galaxy clusters of the spiral rich field galaxy population into the S0-, dSph-, and dE-rich population of todays rich clusters. Timescales for the various aspects of the transformation have been discussed and possible progenitors or early stages of these transformation processes are being looked for in high and intermediate redshift clusters. Probably more than one evolutionary path has led to the observed manifold of S0 and dSph systems.

3. S0s on the Hubble Sequence

S0 follow many trends along the Hubble sequence originally defined by morphological appearance, i.e. the bulge component getting less prominent and the disk component and spiral structure getting more pronounced from early to late-type galaxies: the colors of S0s are almost as red on average as those of ellipticals and redder than those of spirals that get increasingly bluer towards later Hubble types, their spectra, as well, have properties that fit in between those of Es and Sa's. The content in neutral and molecular gas in most S0s is low to negligible similar to ellipticals while it increases systematically towards later spiral types. IRAS though has detected 50 % of all E and S0 galaxies (Jura *et al.* 1987), later investigations revealed that cold gas is particularly common among peculiar S0s that also show relatively high HI content and H_α. HI structures and kinematics often suggest an external origin. Dust masses from FIR measurements are typically a factor ~ 10 higher than those from optical extinction, indicating that the dust distribution is more extended than that of the star light. CO and HI are more strongly concentrated in the centers than for spirals (Bregman *et al.* 1998, Pogge & Eskridge 1993, Sadler *et al.* 2000, Welch & Sage 2003). However, even the most gas-rich S0s contain much less gas than expected if they had formed all their stars in an early burst and evolved as a closed-box since then (Faber & Gallagher 1976). As opposed to massive ellipticals, S0s do not show luminous X-ray halos. Present-day star formation rates SFR_o are low in most S0s, their ratio of time averaged past SFR $\langle SFR \rangle$ to present-day SFR_o falls between the respective values for Es and the earliest type spirals (e.g. Sandage 1986). Their SF efficiences, i.e. SFRs normalised to gas content, are higher than for spirals, for which they decrease further from Sa through Sd. In terms of average luminosity and M/L−ratio, S0s also range between Es and Sa's.

All these observations indicate that the Hubble sequence is a continuous sequence, not only in terms of B/T−light ratios, but also in terms of colors and spectral properties, gas content, SFR_o, $\langle SFR \rangle$, and, hence, stellar population ages. On the other hand, one might look at the S0s in the light of the apparent dichotomy among ellipticals: Low-luminosity Es are fast rotators, have power-law surface brightness profiles towards the resolution limit, disky isophotes, and approximately isotropic velocity dispersions, all compatible with the idea that dissipation has played a role in their formation and dynamical evolution. Luminous Es, however, show slow or no rotation, are triaxial, have cores, boxy isophotes, and anisotropic velocity dispersions (Kormendy & Illingworth 1982). Kormendy & Bender (1996) redefine the Hubble sequence as extending from boxy Es via disky Es to S0s, Sa, ..., Sd, Irr. van den Bergh (1994) argued in favor of the S0s sharing the dichotomy among the ellipticals with faint S0s being flattened and prolate and bright S0s being the ones that are truely intermediate between Es and Sa's. Faber *et al.* 's (1997) observations that, with one exception (NGC 524), S0s all have power-law surface brightness profiles support this view.

It has long been known that in terms of central parameters (central surface brightness, central velocity dispersion) and absolute luminosity S0s and bulges follow the relations among ellipticals while **d**warf **E**llipticals (**dE**s) and Globular Clusters follow different relations in the famous Kormendy (1985) diagrams. From these diagrams and the knowledge from stellardynamical modelling that mergers cannot increase the central phase space density by more than a factor of two, it is clear that S0s or normal/giant Es cannot be produced in mergers of dEs or dSphs. Ryden *et al.* (1999) found no distinct structural differences between dEs and dSphs/dS0s.

S0s harbour nuclear black holes like bulges and Es, as revealed by absorption or, in some cases, emission line analyses of high spatial resolution HST spectra. The two S0s in Pinkney *et al.* 's (2003) sample follow the black hole mass – velocity dispersion relation $M_{BH} - \sigma_e$ for Es and bulges found by Gebhardt *et al.* (2000) and Tremaine *et al.* (2002). If all S0s have central BHs still seems to be an open question, e.g. in view of NGC 524's core.

S0s follow the central $Mg_2 - \sigma_o$ relation of ellipticals and have metallicity gradients shallower by factors 2–3 than spirals, therewith fitting into the trend of increasing metallicity gradients from early to late-type spirals. In contrast to spirals and ellipticals, however, many S0s apparently get older from inside out, i.e. their bulge stellar populations are younger than those of their disks. S0s follow the anticorrelation between EW(H_β) and velocity dispersion σ_o of ellipticals – indicating that more massive galaxies have older stellar populations (Fisher *et al.* 1996). The age of a stellar population in this context means the time elapsed since the last epoch of significant SF. Concerning the relative stellar population ages in the disks and bulges of field S0s, observations have revealed discrepant results: BJHK photometry with surface brightness profile decomposition of 35 S0s showed that because of similar $J - K$ and $H - K$ colors both components probably have similar metallicities and the different $B - H$ colors indicate that disks are $\sim 3 - 5$ Gyr younger than bulges. Some S0s show evidence for AGB light contributions implying active SF until $2 - 3$ Gyr ago (Caldwell 1983, Bothun & Gregg 1990). NGC 7332 which has emission line gas kinematically decoupled from the stellar component and an A-type bulge spectrum (Bertola *et al.* 1992, Fisher *et al.* 1994, Hibbard & Rich 1990), indicative of recent accretion and starburst events, is found by Bender & Paquet (1999) to have a bulge star population spectroscopically older than its disk population. NGC 5102 and NGC 404, resolved into individual stars with HST, both show blue inward color gradients and significant populations of young stars $\gtrsim$ 15 and $\gtrsim$ 300 Myr, respectively, in their centers (Deharveng *et al.* 1997, Tikhonov *et al.* 2003). Investigations into stellar population ages of 33 S0s and 19 Es in the Coma **cluster** over a wide range in luminosities $-20.5 \leq M_B \leq -17.5$ showed that 1. ≥ 40 % of the S0s (and none of the Es!) had significant SF in their **central** regions during the last 5 Gyr and that 2. the fraction of S0s with recent SF increases with decreasing luminosity (Poggianti *et al.* 2001a). This luminosity dependence of the luminosity weighted ages is independently found from optical and NIR photometry of S0s in Abell 2218 (z= 0.17) by Smail *et al.* (2001). It explains the discrepant results of earlier studies and support the idea of a dichotomy between low and high-luminosity S0s.

S0s as well as ellipticals in Virgo, Coma and other nearby clusters – both rich and poor – follow a **C**olor–**M**agnitude **R**elation (**CMR**) in the sense that luminous galaxies are redder than fainter ones (Bower *et al.* 1992, Andreon 2003). This CMR is primarily due to higher metallicities in brighter galaxies, as evidenced by spectroscopic observations yielding similar relations between Mg_2 and luminosity or Mg_2 and σ. In addition to metallicity differences, age differences may also contribute. E.g. Kuntschner & Davis (1998) and Kuntschner (2000) find from line strength analyses of Fornax Es and S0s that the luminous Es & S0s have formed their stars earlier or stopped their SF earlier than the lower luminosity ones. The tightness of the CMR is conventionally assumed to imply uniform old stellar population ages and, within hierarchical galaxy formation scenarios, that more massive galaxies formed earlier from more massive building blocks.

Schweizer *et al.* (1990) found fine structure, i.e. ripples, shells, plumes, boxiness, X-structure, etc. in $\sim$ 50% of the **field S0s**. The amount of fine structure as quantified by his fine structure parameter Σ correlates with deviations from the CMR towards bluer colors, with increasing $H_\beta-$ EWs, and with decreasing Mg_2 (Schweizer & Seitzer 1992) in the sense that $\sim$ 50 % of the field S0s had major mergers with significant starbursts about $3-8$ Gyr ago (Schweizer 1993, 1999).

E/S0s form a homologous 3–parameter family with low scatter described by the apparantly universal **F**undamental **P**lane (**FP**) relation between effective radius r_e, velocity dispersion σ, and effective surface brightness $\langle I_e \rangle$ within r_e (Djorgovski & Davis 1987, Dressler *et al.* 1987, Jorgensen *et al.* 1993, 1996, Scodeggio *et al.* 1998a, b)

$$\log r_e = a \cdot \log\sigma + b \cdot \log\langle I_e \rangle + c$$

The scatter is generally $\lesssim$ the measurement errors, smaller in clusters than in the field and strongest for the low-luminosity S0s (van Dokkum *et al.* 2001). In combination with the Virial Theorem the FP relations imply a weak mass dependence of the mass-to-light ratio $M/L \sim M^{0.2}$ (Faber *et al.* 1987). A salient feature of the FP is that it indicates an intimite relation between spectral and dynamical evolution aspects of galaxies by coupling tightly structural and stellar population parameters. Observations of the FP in the NIR show that 1.) the scatter is independent of passband, implying that variations in age are balanced by variations in metallicity, and 2.) that the slope increases steadily from U through K, pointing to systematic changes in metallicity **and** age **and** DM content/homology breaking (Pahre *et al.* 1998). Conventionally, the tight FP is interpreted in terms of E/S0s being very homogeneous old stellar systems. However, it has been shown that stellardynamical mergers conserve the FP and its small scatter (Capelato *et al.* 1995, Evstigneeva *et al.* 2004) and that E+Sp merger remnants soon after merging come to lie on a *FP of changes* that is only tilted by 15° relative to the observed FP (Levine 1995). Observations by Lake & Dressler (1986) of a dozend well-known Sp – Sp merger remnants from Arp's (1966) and Arp & Madore's (1985) catalogues still featuring long tidal tails like NGC 7252 or NGC 3921 have shown that – very surprisingly – these objects ≤ 1 Gyr after their merger and strong starburst well fit into the $L-\sigma$- and FP–relations, casting serious doubt upon the idea that the very existence and the tightness of these relations necessarily imply uniform old ages for E and S0 galaxies. Apparently violent relaxation very fast transforms the **inner** regions within $1r_e$.

S0 galaxies not only follow relations for elliptical galaxies, they also obey the **T**ully–**F**isher **R**elation (**TFR**) for spirals: $L \sim V_{max}^{\alpha}$ with $\alpha \sim 4$ and V_{max} the maximum rotation velocity (Tully & Fisher 1977, Pierce & Tully 1992). Mathieu *et al.* (2002) find 6 S0s to follow the H-band TFR with small scatter ~ 0.3 mag while being offset by ~ 1.1 mag to fainter magnitudes in I, consistent with some fading after SF truncation (see below). Results are still somewhat controversial with Hinz *et al.* (2001, 2003) reporting a small offset ≤ 0.2 mag from the H-band TFR and a large scatter for massive ($V_c \geq 200 kms^{-1}$) S0s in Virgo and Coma and Neistein *et al.* (1999) finding a smaller offset in I and a large scatter for 18 nearby field S0s, in particular at the faint end. The luminosity range of the sample, like for the stellar population ages, or the field versus cluster environment might be a key to reconcile the apparently discrepant results. Note that rotation curves are much more difficult to measure for S0s than for spirals because they largely have to rely on absorption lines. Moreover, about 50 % of the Virgo S0s show weird rotation curves that do not allow to derive a V_{max} at all (Rubin *et al.* 1999). Different formation paths may well lead to S0s with diverse properties.

4. Classification Problems

S0s are distinguished against Sa galaxies either by their less prominent and featureless disks or, in quantitative analyses, by some limiting B/T–light ratio. Face-on S0s are hard to distinguish against ellipticals, the only reliable method being long-slit absorption line spectroscopy that allows to detect an S0's rapidly rotating disk. In a magnitude-limited Coma sample Jorgensen & Franx (1994) found only 12 % Es among the huge E/S0 population. For distant clusters often only the combined E/S0 population is considered. Using evolutionary synthesis models for bulge and disk components separately with two different SF histories – const. SFR as appropriate for disks and exponentially declining SFRs ($t_* \sim 1$ Gyr) for bulges – and combining bulges and disks to obtain after a Hubble time the observed average B-band B/T–light ratios of various Hubble types we could study both the wavelength and the redshift dependences of B/T–light ratios (Schulz *et al.* 2003). We investigate three different scenarios for the onset of bulge and disk SF: 1.) both start in the early universe at $z \sim 3-5$, 2.) bulge SF starts at $z \sim 3-5$ and disk SF later at $z \sim 1$, and 3.) disk SF starts at $z \sim 3-5$ and bulge SF later at $z \sim 1$.

We found a significant wavelength dependence of B/T–light ratios (Fig.1a): for S0s B/T increases by a factor $2-3$ from U through K (for Sd's by a factor $3-4$) in agreement with H-band B/T– determinations by Eskridge *et*

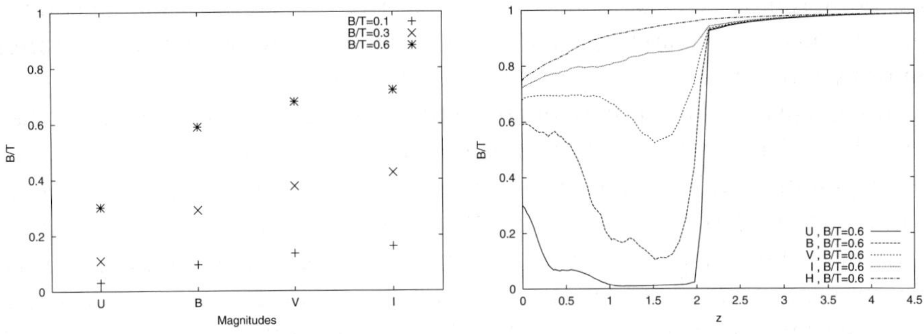

Figure 1. (a) Wavelength dependence, (b) redshift dependence of the B/T-light ratio for S0s. Bulge and disk SF both starting at z $\sim 3-5$.

al. (2002). The amount of increase slightly depends on the relative ages of bulge and disk stars (cf. Schulz *et al.* for details). In addition, we found a strong redshift evolution of B/T−light ratios that cannot be compensated for by switching from B to V, R, I in a comparison with galaxies at redshifts z=0.3, 0.5, 0.7. Models show that in any of the three scenarios B/T−light ratios of S0s at higher redshift determined from surface profile decomposition in bands that correspond to restframe B are significantly overestimated. S0s at higher redshift therefore have a high chance to be misclassified as Es (Fig.1b). The reason is the difference in time evolution of the disk and bulge components due to their different SF histories. We therefore decided to give cosmological and evolutionary corrections separately for bulge and disk components and the 3 scenarios that we explored: bulge and disk of equal age, bulge older or younger than the disk (cf. Schulz *et al.* 2003). It is true that changes get smaller towards longer wavelengths and that therefore galaxy classification in I, as e.g. done for the HDF galaxies by Marleau & Simard (1998), or in a NIR band is less affected. For S0s and Es, unfortunately, the NIR B/T−light ratios are very similar and do not allow to separate them from each other.

5. Formation and Transformation of S0s

In the central regions of local rich virialised galaxy clusters S0 galaxies are the dominant galaxy type, making up as much as $\sim 60\%$ of the bright galaxy population. Dwarf galaxies (dEs, dS0s, dSphs) are about twice as numerous as the luminous ones in Coma (de Lucia *et al.* 2004). Distant clusters, on the other hand, contain increasing fractions of blue galaxies, as first discovered by Butcher & Oemler (1978, 1984). Van Dokkum (2001) and Dahlen et al. (2004) find a factor $\sim$ 5 increase in the blue galaxy fraction from z$\sim$ 0.5 to z= 0, i.e. over the last 5 Gyr (using $H_o = 65$, $\Omega_m = 0.3$, $\Omega_\lambda = 0.7$).

Most of the blue galaxies in distant clusters are Sps and Irrs with ongoing SF (Smail *et al.* 1997), some are having starbursts, others show Balmer absorption lines indicative of recent starburst (Dressler & Gunn 1983). Some of the red galaxies also have strong Balmer absorption lines (E+A, k+a), i.e. are slightly older post-starbursts (spectroscopic BO−effect). Fasano *et al.* (2000, 2001) and Couch *et al.* (1998) show that while the fraction of ellipticals in the cluster population remains approximately constant from redshift z ~ 0.5 to z $= 0$, the fraction of spirals decreases by a factor ~ 5, and the fraction of S0s increases by the same factor. This suggests a significant transformation of spirals into S0s from z ~ 0.5 to z $= 0$, i.e. over the last 5 Gyr. At the same time, the faint-to-luminous galaxy ratio in clusters increased from ~ 1 at z $= 0.75$ to ~ 2 at z $= 0$ (de Propris *et al.* 2003).

The field galaxy population also being spiral-rich and poor in S0s and Es, the transformations of spirals into S0s and of luminous into dwarf galaxies seem to be linked with processes during the continuous infall of field spirals towards increasingly rich clusters. Formation and evolution of galaxy clusters and Large Scale Structure hence appear to go hand in hand with morphological and spectral transformation of important parts of their galaxy population.

A large variety of formation and transformation scenarios have been proposed for S0s over the years and it seems that more than one of them is needed to account for the observed diversity of the S0 population. Which one(s) is (are) realised or prevalent may depend on environment and epoch; massive and low-mass S0s may have different formation and evolution histories.

Like bulges in later-type spirals the spheroidal componets of S0s could have formed in the classical *fast initial collapse with rapid SF* scenario – at least for those 50% of the field S0s that do not show any fine structure or signs of central rejuvenation.

Internal instabilities in spiral disks can form bars that efficiently funnel gas and stars into the central parts and bars, in turn, can dissolve to form a bulge component. The huge bulges of S0s require several circles of bar formation and destruction as seen in numerical simulations (Combes *et al.* 1990, Raha *et al.* 1991). Depending on the relative amounts of gas and disk stars funneled into the central regions, stellar populations in the bulge might be older or younger than those in the disk.

Major spiral – spiral mergers have been shown to result in elliptical galaxies since the Toomres' pioneering work. Incomplete violent relaxations transforms stellar disks into de Vaucouleurs profiles while saving small gradients. The delayed and protracted backfall of HI from the tidal tails onto the main body, as e.g. observed in NGC 7252, can rebuild a gaseous disk on timescales of few Gyr that by and by transforms into a stellar disk, resulting in a disky elliptical or an S0 galaxy as seen in combined N-body and hydrodynamical simulations by Hibbard & Mihos (1995) and in Barnes' (2002) models. Few Gyr after the global or nuclear starbursts triggered respectively in prograde and retrograde encounters (Bekki 1995) the merger remnants have been shown to reach typical S0 – and somewhat later elliptical – galaxy colors and spectra in our evolutionary synthesis models (Fritze – v. Alvensleben & Gerhard 1994a, b). It is clear that this major merger scenario is only viable for massive and luminous S0s. Only in field or group environments galaxy encounter velocities are low enough for efficient merging. Hence the luminous S0s in clusters must have been preprocessed in the field or within infalling groups – in agreement with observations that **luminous** S0s have stellar population ages as old as ellipticals. In this scenario, the stellar population in the bulge will be older than that in the disk. Observations of the rich cluster MS 1054-03 at z= 0.83 reveal a substantial number of major mergers, 17% in total, not only in the center but also in the outer parts, consistent with the idea of enhanced merging in infalling groups (van Dokkum *et al.* 1999). Gavazzi *et al.* (2003) observationally caught a collective starburst among the members of a galaxy group falling into the cluster Abell 1387.

Minor mergers (e.g. mergers with galaxy mass ratios 3:1) or accretion events have been shown by Barnes (1996) and Bekki (1998) to be viable routes towards intermediate or low-luminosity S0s, again in the field or in groups rather than in dense clusters where galaxies have too high relative velocities. The nature and gas content of the objects involved determines the outcome. A Sp+dE or a Sp+(d)Irr merger will result in an S0 with a bulge older or younger than the disk depending on the strength of a possibly triggered starburst.

Major mergers among gas-rich galaxies form populous systems of new star clusters, many of which are compact and strongly enough bound to survive for Gyrs (Schweizer 2002, Fritze – v. Alvensleben 1998, 1999). Their enhanced metallicities and younger ages, when determined from spectroscopy or multi-wavelength photometry, are much more precise tracers of past violent SF events than the integrated galaxy light (Anders *et al.* 2004a, Fritze – v. Alvensleben 2004a, b). If the young star clusters forming in minor mergers are also long-lived still is an open question (Anders *et al.* 2004b, Fritze – v. Alvensleben 2004b). S0 galaxies with bimodal **G**lobular **C**luster (**GC**) color distributions in any case testify back to a major merger origin. Both the luminous S0 NGC 1380 (Kissler – Patig *et al.* 1997) and the low-luminosity S0 NGC 3115 show bimodal GC color distributions. The field stars in NGC 3115 also show the same bimodal color distribution (Elson 1997). Kundu & Whitmore (2001) estimate that $\gtrsim 10-20$ % of the S0s have bimodal GC color distributions and, hence, a clear major merger origin. Age differences between the blue metal-poor and the red metal-rich GCs are small ($\lesssim$ 3 Gyr). Hence, the mergers producing the red GC population in the starbursts they triggered must have happened early. On average, S0s have less populous GC systems than ellipticals of comparable luminosity and are closer to spirals in this respect, although with large scatter.

Harassment is an important process in dense cluster environments, where fast galaxy – galaxy encounters destabilise the disks of infalling spirals, drag out strong short-lived tidal tails and tear away stars from their outer disks, leaving the inner parts as dE, low-luminosity S0 or dSph galaxies on timescales of few Gyr while releasing up to 50% of the original stellar mass of the incoming spiral to the cluster potential (Moore *et al.* 1995, 1998). As the remnants of this process consist of the former central parts of much more massive galaxies, they are to be expected to deviate from the $Mg_2 - \sigma -$ relation towards too high metallicities for their mass. Poggianti *et al.* (2001b) report a significant population of low-luminosity galaxies in Coma with exactly these properties: enhanced metallicities for their luminosities. Apart from being harassment products they could also be Tidal Dwarf Galaxies, i.e. recycling galaxies forming out of gas and stars torn out into a tidal tail from a disk galaxy in interaction and becoming self-gravitating and dynamically independent there (cf. Duc & Mirabel 1999, Weilbacher *et al.* 2002, 2003).

Beyond these different types of galaxy – galaxy interactions the presence of a hot dense ICM as observed at X-wavelengths in the centers of nearby rich clusters will affect the gaseous components of infalling spirals. Processes like ram pressure stripping or sweeping of their HI disks have been predicted as the density and pressure of the hot ICM are observed to be comparable or higher than those of the ISM within galaxies, and they can directly be seen in terms of increasingly truncated and displaced HI disks in spirals towards the center of the Coma cluster (Cayatte *et al.* 1990, Bravo – Alfaro *et al.* 2000). It is clear that SF in these anemic spirals (van den Bergh 1976) will also be truncated. If the instability intrigued by the rapid removal of considerable amounts of HI will lead to a starburst consuming all the molecular gas and part of the central HI in one shot before final SF truncation and if it can also affect the stellar configuration is less clear. Observations that the galaxy population in clusters starts to deviate from the field galaxy population in terms of their SF activity at unexpectedly large distances from cluster centers (around $3 - 4\ R_{vir}$) (Lewis *et al.* 2002, Gomez *et al.* 2003, Balogh *et al.* 2004, Gerken *et al.* 2004) raised the idea that the ICM densities that far out might still be high enough to drive away the low density gas from infalling galaxy halos that otherwise could serve as a reservoir for disk accretion and, hence, for SF (Larson *et al.* 1980, Bekki *et al.* 2002). If realised by nature, this halo gas stripping or starvation would lead to SF strangulation on a rather long timescale of order Gyr.

We have investigated the effect of SF truncation without or after a preceeding starburst on the photometric and spectral evolution of various spiral types in Bicker *et al.* (2002). Using evolutionary synthesis models that successfully describe undisturbed galaxies of types Sa, Sb, Sc, and Sd, we added bursts of various strengths and/or SF truncation at various evolutionary ages. Figs. 2a, b show the time evolution of $(B-V)$ colors and of B-band luminosities M_B for starbursts and/or SF truncation occuring at z = 1 in a 6 Gyr old Sc galaxy. Unless the SFR does not go to zero, all models reach S0 galaxy colors fairly rapidly: within 2 Gyr after SF truncation and within $4-5$ Gyr after a starburst, depending on its strength, in case of $B-V$. The luminosity evolution, of course, strongly depends on whether or not a starburst occured before SF truncation: between pure SF truncation and strong burst models there is a difference in final luminosity of 2 mag in M_B.

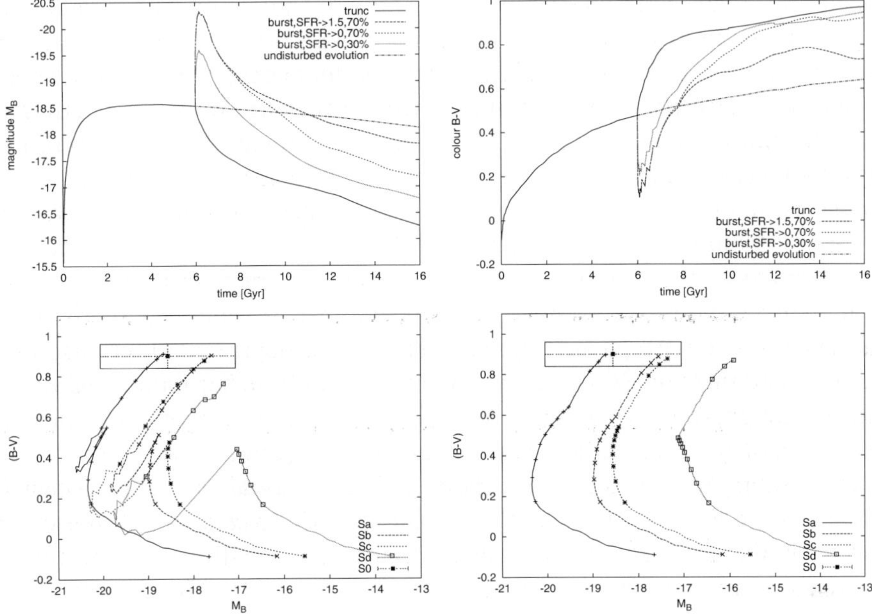

Figure 2. Luminosity (a) and color (b) evolution of an Sc spiral model with starbursts and/or SF truncation occuring at an age of 6 Gyr. CMD for various spiral models with starbursts occuring at 6 Gyr (c) and with SF truncation at 9 Gyr (d). Ticks mark 1 Gyr steps. The box gives average observed S0 color and luminosity with 1σ ranges.

The photometric evolution in $UBVRIJHK$ of these models is compared with observed S0 properties in terms of **C**olor – **M**agnitude **D**iagrams (**CMD**s). Note that our undisturbed models are calibrated in mass as to yield after a Hubble time the average observed M_B for the respective spiral type. As an example, Figs. 2c, d show in a CMD the evolution of various spiral types with strong bursts and/or SF truncation occuring at ages of 6 Gyr and 9 Gyr, corresponding to redshifts z = 1 and z = 0.5, respectively. The box in the upper left indicates the observed average color and luminosity of S0s with their 1σ ranges. About 4 Gyr after their strong starbursts Sa, Sb, Sc models all reach the color range of

S0s, Sd models remain too blue. Only the Sa model reaches the average luminosity of S0s. Sb and Sc models after strong starbursts are able to account for the fainter S0s. Only 1.5 Gyr after SF truncation without a preceeding starburst all models show S0 galaxy colors, but again only the Sa progenitors reach average S0 luminosities, the Sb snd Sc progenitors turn into low-luminosity S0s, and Sd galaxies with truncated SF end up with luminosities of dSphs. SF truncation in 3 Gyr young Sa and Sb progenitors, i.e. at $z = 2$, would produce galaxies too red for S0s. At such young ages, however, the dense hot ICM of today's rich cluster centers that is thought to be responsible for SF truncation by sweeping out the gas from infalling spirals might not yet have accumulated. We also considered mergers of equal spiral types that double the mass in stars and gas and the luminosity while not changing the colors. After strong bursts accompanying these mergers, Sa progenitors reach the bright end of the S0 galaxy luminosity distribution, and even Sd-galaxies, when merging with each other, reach average S0 luminosities. As far as the color evolution is concerned, our results are confirmed by an independent study by Shioya *et al.* (2002). Including also the luminosity evolution, however, our models allow for tighter constraints on the manifold of possible S0 progenitors than theirs. To investigate the recently renewed suggestion that SF strangulation by halo gas starvation might truncate SF on a longer timescale of ≥ 1 Gyr (Larson *et al.* 1980, Bekki *et al.* 2002), we recalculated models under these conditions and we also analyse the spectral evolution, e.g. the evolution of emission and absorption lines (Fritze & Bicker *in prep.*).

We find that SF truncation on a short timescale of order 10^8 yr is followed by a phase of moderate $H_\delta-$ strength, while SF strangulation on a timescale of 10^9 yrs does not develop enhanced H_δ absorption. SF truncation after a preceeding burst results in an $H_\delta-$ strong phase of $\sim$ 1.5 Gyr duration (see also Barger *et al.* 1996, Poggianti *et al.* 1999, and Shioya *et al.* 2004). Note that red $H_\delta-$ strong galaxies are also called E+A or k+a types.

We conclude that the progenitors of low-luminosity S0s can be Sa through Sc galaxies that experienced a starburst > 3 Gyr ago, as well as Sa/Sb galaxies with SF truncation at ages between 6 and 9 Gyr, i.e. at $1 \leq z \leq 0.5$, or Sc galaxies with SF truncation as late as $z = 0.5$, i.e. after 9 Gyr of undisturbed evolution. The strength of the burst does not make much difference. The progenitors of luminous S0s can only be early-type spiral – spiral or multiple mergers with starbursts occuring at ages ≤ 9 Gyr, i.e. at $z \geq 0.5$.

The progenitors of low- and high-luminosity S0s that had experienced a starburst have gone through an $H_\delta-$ strong phase, first blue and then red, of $\sim$ 1.5 Gyr duration. The progenitors of those low-luminosity S0s and dSphs that had their SF truncated without a starburst can, at maximum, have gone through a phase of intermediate-strength Balmer absorption provided the SF truncation happened on a short timescale of order 10^8 yr.

Barger *et al.* (1996) and Kodama & Smail (2001) quantitatively investigated the possibility of a spiral-to-S0 transformation in terms of galaxy numbers and morphologies. Couch & Sharples (1987) report a high proportion ($\sim 30\%$) of actively SFing and starbursting galaxies in 3 rich clusters at z= 0.31, many of which show signs of interactions while still having disk morphologies, and a high fraction of $H_\delta -$ strong galaxies, most of which are regular spheroids, indicating that the timescale for the photometric transformation is somewhat shorter than that for the morphological transformation, but that the strong Balmer absorption features after strong starbursts still are detectable after the morphological transformation is accomplished. Maybe in mergers which are probably the origin of strong starbursts the morphological transformation is faster than for other transformation processes?

The decrease of the blue galaxy fraction towards $z = 0$ probably is a consequence of the interplay between a decline in the galaxy infall rate with cosmic time, the overall decrease in gas content and SFR in the field galaxy population, and the increasing cluster richness together with the successive build-up of their dense hot ICM content. While at intermediate and high redshift, the starburst and E+A galaxies in clusters were bright high-mass objects, the actively SFing, starburst and postburst galaxies in local clusters are predominantly low-luminosity late-type dwarf galaxies. This trend is called *downsizing* effect (Bower *et al.* 1999, Poggianti 2004) and also reported for the field galaxy population (Cowie *et al.* 1996). Duc *et al.* (2002) conclude from ISOCAM mid-IR observations that up to 90% of the SF in a cluster $z = 0.18$ is hidden. If this would selectively apply for the most massive and metal-rich galaxies it might explain an apparent downsizing effect simply by a strong prevalence of dust at low redshift.

6. Redshift Evolution of S0 Galaxy Populations

Van Dokkum *et al.* (1998) investigate the **redshift evolution of the CMR** for 194 E/S0 galaxies in DL 1358+62 at z= 0.3. Thcy find the scatter, i.e. the age spread, for Es very small and independent of clustercentric radius R_{cl}. The scatter for the S0s is equally small in the center, however, it increases considerably towards larger R_{cl} with the S0s being increasingly offset towards bluer colors at large R_{cl}. This shows that while ellipticals have terminated their SF well before accretion, S0s stop SF in the outer parts of the cluster. Stanford *et al.* (1998) extend this study to 19 clusters at $z\sim 0.9$ using NIR photometry and find the local slope conserved, the scatter, i.e. the apparent age spread among E/S0s still small and the degree of evolution independent of cluster richness or L_X. The fact that the slope for these young galaxy populations is not different from the local one implies that it results from a correlation of galaxy mass with metallicity, rather than with age.

Studies of the **redshift evolution of the FP** aim at 3 aspects: the redshift evolution of its slope tells about size and mass evolution, i.e. about accretion and evolution in orbital structure, changes in the zero-point indicate evolution in M/L, due in part to the fading of the stellar population and possibly also affected by rejuvenation through accretion and SF, the redshift evolution of the scatter is a measure of the homogeneity of the E/S0 population at various lookback times. Semianalytical ΛCDM structure formation models predict that high density environments should collapse first and fastest. Hence, they lead us to expect cluster galaxies to be older than field galaxies and therefore to expect differences in the redshift evolution of the FPs of field and cluster galaxies (Kauffmann 1996, Diaferio *et al.* 2001). Remember again that age in this context means time since the last epoch of significant SF.

The redshift evolution of the FP for **cluster E/S0s** in the redshift interval $0.3 \leq z \leq 0.8$ as studied by Kelson *et al.* (1997), Bender *et al.* (1998), van Dokkum *et al.* (1998), Jorgensen *et al.* (1999), and Wuyts *et al.* (2004) indicates that the scatter remains close to local until $z \sim 0.5$ and only increases at $z = 0.8$, the zero-point shift indicates ~ 1 mag brightening to $z = 0.8$. The evolution in M/L is slow, implying a galaxy formation redshift $z_f > 2.8$ consistent with results from the redshift evolution of the CMR. The redshift evolution of the FP for **field galaxies** was studied by means of Keck-LRes spectroscopy by Gebhardt *et al.* (2003) on 21 Es and 15 S0s from the DEEP sample in the redshift range $0.3 \leq z \leq 1$ with $\langle z \rangle = 0.8$ and by van Dokkum & Ellis (2003) on 10 E/S0s from the HDF-N with $0.56 \leq z \leq 1.02$. Both groups find slope and scatter similar to local, and a zero-point offset by 2.4 mag in *I* (corresponding to rest-frame B at $z = 1$), marginally in agreement with passive evolution of simple old populations. Comparison with the evolution of cluster E/S0s indicates an age difference between cluster and field E/S0s of less than 2 Gyr, which poses a problem for hierarchical structure formation models that predicted a larger age difference.

The **redshift evolution of the TF relation** remains inconclusive at present for the S0s due to the difficulties in measuring stellar velocity dispersions from absorption lines in high redshift spectra.

7. Conclusions and Open Questions

Located between ellipticals and spirals on the Hubble sequence, S0 galaxies play a key role in 3 fundamental and still largely open questions: as to the nature of Hubble's sequence, to the respective roles of nature versus nurture in shaping galaxy properties, and as valuable test cases for cosmological galaxy formation scenarios. Are S0s transition types on one continuous line from E through Sd galaxy types or do they belong to either of two fundamentally different classes of objects? I reviewed arguments for and against boths options and conclude that current evidence seems to favor a dichotomy among S0s similar to the one for ellipticals. In this sense, the luminous S0s have formed their stars earlier than the low-luminosity S0s and may have had different formation scenarios.

S0 galaxies share many properties with other dynamically hot stellar systems, as e.g. the relation between central black hole mass and velocity dispersion, the color–magnitude and Fundamental Plane relations, while also following the Tully–Fisher relation for spirals. We have seen that constraints from the FP and CMR on the ages and formation histories of S0s are less tight than originally thought, and that 50% of the field S0s show evidence for important intermediate age stellar contributions in their central regions together with morphological and kinematic traces of interactions events. Detections of HI, CO, HII, dust and SF activity in quite a number of S0s adds to this view. Together with the strong redshift evolution of the S0 and spiral fractions in galaxy clusters this provides evidence that nurture rather than nature, environmental effects rather than initial conditions have shaped S0 galaxies, that they are **transformation** rather than formation products. It is not clear at present if any classical *early collapse + short timescale SF* S0s do at all exist in the field – for sure not in clusters. It also seems clear that more than one transformation scenario must be at work to produce the observed manifold of S0s. Major spiral–spiral/Irr mergers with starbursts may result in luminous S0s, as shown by dynamical simulations as well as evolutionary spectral synthesis, with the protracted backfall of HI from the tidal tail(s) leading to the build-up of a secondary stellar disk on timescales of 3 Gyr. Minor mergers (3:1) or accretion events on the one hand, harassment, tidal stripping of stars and ram-pressure sweeping of gas on the other hand may result in lower-luminosity S0s and dSphs. Any kind of merging or accretion requires low relative velocities and, hence, is more probable within groups before or during their infall into clusters or in very early stages of cluster evolution. It may leave kinematic peculiarities, fine structure, and positive color gradients that can survive for a few Gyrs. If the bursts are strong enough the resulting new star cluster population with its age and metallicity may be better suited to trace back the SF history of its parent galaxy than the integrated light. Pixel-by-pixel analyses provide spatially resolved information about the respective contributions of old and younger stars. Harassment through fast galaxy–galaxy encounters is probably the dominant process in dense clusters, its end products will be low-luminosity S0s, dSphs or dEs that can be expected to deviate from the luminosity–metallicity relation because they are the leftover central parts of originally much more massive galaxies. Interactions with the dense hot X-ray emitting ICM in nearby rich galaxy clusters is observed to efficiently remove HI disks from spirals plunging into it, cutting their gas supply and, hence, truncating their SF while destabilising the stellar disks. The timescale for the morphological transforma-

tion seems to be longer than that for the photometric transformation, the strong Balmer absorption features after strong bursts, however, seem to survive the morphological transformation, at least in some cases.

If the much less dense halo gas is removed from spirals by the same process at larger cluster-centric radii already – with the effect of cutting the SFing disk off from its accretion reservoir causing SF strangulation on a relatively long timescale seems less clear. Detailed statistics of the various progenitor and transition types (normal and strong emission line galaxies, blue and red H_δ–strong galaxies) in clusters of varying richness and degree of relaxation and at various cluster-centric radii should ultimately allow to identify the relative importance of all these different formation paths. The fact that the morphology–density relation seems to be continuous from the densest cluster centers out to the field probably indicates that, beyond the direct interaction with the clusters' central ICM and potential, more localised processes, e.g. in groups, must also play an important role.

It seems for sure that the S0 galaxies and the questions as to their origin will keep us excited for years to come and may still hold further surprises.

Acknowledgments

I gratefully acknowledge generous travel support from the organisers of this conference without which I could not have attended.

References

Anders, P., Bissantz, N., Fritze – v. Alvensleben, U., de Grijs, R., 2004a, MN 347, 196
Anders, P., de Grijs, R., Fritze – v. Alvensleben, U., Bissantz, N., 2004b, MN 347, 17
Andreon, S., 2003, A&A 409, 37
Arp, H., 1966, *Atlas of Peculiar Galaxies*
Arp, H., Madore, B. F., 1985, *A Catalogue of Southern Peculiar Galaxies*
Balogh, M., Eke, V., Miller, C. *et al.* , 2004, MN 348, 1355
Barger, A. J., Aragon – Salamanca, A., Ellis, R. S. *et al.* , 1996, MN 279, 1
Barnes, J. E., 1996, IAU Symp. 171, 191
Barnes, J. E., 2002, MN 333, 481
Bekki, K., 1995, MN 276, 9
Bekki, K., 1998, ApJ 499, 635
Bekki, K., Couch, W. J., Shioya, Y., 2002, ApJ 577, 651
Bender, R., Saglia, R. P., Ziegler, B. *et al.* , 1998, ApJ 493, 529
Bender, R., Paquet, A., 1999, Ap&SS 267, 283
Bertola, F., Buson, L. M., Zeilinger, W. W., 1992, ApJ 401, L79
Bicker, J., Fritze – v. Alvensleben, U., Fricke, K. J., 2002, A&A 387, 412
Bower, R. G., Lucey, J. R., Ellis, R. S., 1992, MN 254, 601
Bower, R. G., Kodama, T., Terlevich, A., 1998, MN 299, 1193
Bothun, G. D., Gregg, M. D., 1990, ApJ 350, 73
Bravo – Alfaro, H., Cayatte, V., van Gorkom, J. H., Balkowski, C., 2000, AJ 119, 580
Bregman, J. N., Snider, B. A., Grego, L., Cox, C. V., 1998, ApJ 499, 670

Butcher, H., Oemler, A., 1978, ApJ 219, 18
Butcher, H., Oemler, A., 1984, ApJ 285, 426
Caldwell, N., 1983, ApJ 268, 90
Capelato, H. V., de Carvalho, R. R., Carlberg, R. G., 1995, ApJ 451, 525
Cayatte, V., van Gorkom, J. H., Balkowski, C., Kotanyi, C., 1990, AJ 100, 604
Combes, F., Debbasch, F., Friedli, D., Pfenniger, D., 1990, A&A 233, 82
Couch, W. J., Sharples, R. M., 1987, MN 229, 423
Couch, W. J., Barger, A. J., Smail, I. *et al.* , 1998, ApJ 497, 188
Cowie, L. L., Songaila, A., Hu, E. M., Cohen, J. G., 1996, AJ 112, 839
Dahlen, T., Fransson, C., Østlin, G., Naslund, M., 2004, MN 350, 253
Deharveng, J.-M., Jedrzejewski, R., Crane, P. *et al.* , 1997, A&A 326, 528
de Lucia, G., Poggianti, B. M., Aragon – Salamanca, A. *et al.* , 2004, *astro-ph/0404084*
de Propris, R., Colless, M., Driver, S. *et al.* , 2003, MN 342, 725
Diaferio, A., Kauffmann, G., Balogh, M. L. *et al.* , 2001, MN 323, 999
Djorgovski, S., Davis, M., 1987, ApJ 313, 59
Dressler, A., Gunn, J. E., 1983, ApJ 270, 7
Dressler, A., Lynden – Bell, D., Burstein, D. *et al.* , 1987, ApJ 313, 42
Duc, P.-A., Mirabel, I. F., 1999, IAU Symp. 186, 61
Duc, P.-A., Poggianti, B. M., Fadda, D. *et al.* , 2002, A&A 382, 60
Elson, R., 1997, MN 286, 771
Eskridge, P. B., Frogel, J. A., Pogge, R. W. *et al.* 2002, ApJS 143, 73
Evstigneeva, E. A., de Carvalho, R. R., Ribeiro, A. L., Capelato, H. V., 2004, MN 349, 1052
Faber, S. M., Dressler, A., Davies, R. L. *et al.* , 1987, *Nearly Normal Galaxies*, Springer, New York, p. 175
Faber, S. M., Tremaine, S. *et al.* , 1997, AJ 114, 1771
Fasano, G., Poggianti, B. M., Couch, W. J. *et al.* , 2000, ApJ 542, 673
Fasano, G., Poggianti, B. M., Couch, W. J. *et al.* , 2001, Ap&SS 277, 417
Fisher, D., Franx, M., Illingworth, G., 1996, ApJ 459, 110
Fisher, D., Illingworth, G., Franx, M., 1994, AJ 107, 160
Fritze – v. Alvensleben, U., 1998, A&A 336, 83
Fritze – v. Alvensleben, U., 1999, A&A 342, L25
Fritze – v. Alvensleben, U., 2004a, A&A 414, 515
Fritze – v. Alvensleben, U., 2004b, in *The Young Local Universe*, eds. A. Chalabaev, Y. Fukui, T. Montmerle, *in press*
Fritze – v. Alvensleben, U., Gerhard, O. E., 1994a, A&A 285, 751
Fritze – v. Alvensleben, U., Gerhard, O. E., 1994b, A&A 285, 775
Gavazzi, G., Cortese, L., Boselli, A. *et al.* , 2003, ApJ 597, 210
Gebhardt, K., Bender, R., Bower, G. *et al.* , 2000, ApJ 539, L13
Gebhardt, K., Faber, S. M., Koo, D. C. *et al.* , 2003, ApJ 597, 239
Gerken, B., Ziegler, B., Balogh, M. *et al.* , 2004, *astro-ph/0403652*
Gomez, P. L., Nichol, R. C., Miller, C. J. *et al.* , 2003, ApJ 584, 210
Hibbard, J. E., Mihos, C. J., 1995, AJ 110, 140
Hibbard, J. E., Rich, R. M., 1990, *ESO/CTIO Workshop on Bulges of Galaxies*, ESO Garching, p. 295
Hinz, J. L., Rix., H.-W., Bernstein, G. M., 2001, AJ 121, 683
Hinz, J. L., Rieke, G. H., Caldwell, N., 2003, AJ 126, 2622
Jorgensen, I., Franx, M., 1994, ApJ 433, 553
Jorgensen, I., Franx, M., Kjaergaard, P., 1993, ApJ 411, 34
Jorgensen, I., Franx, M., Kjaergaard, P., 1996, MN 280, 167
Jorgensen, I., Franx, M., Hjorth, J., van Dokkum, P. G., 1999, MN 308, 833

Jura, M., Kim, D. W., Knapp, G. R. *et al.* 1987, ApJ 312, L11
Kauffmann, G., 1996, MN 281, 487
Kelson, D. D., van Dokkum, P. G., Franx, M. *et al.* , 1997, ApJ 478, L13
Kissler – Patig, M., Richtler, T., Storm, J., della Valle, M., 1997, A&A 327, 503
Kodama, T., Smail, I., 2001, MN 326, 637
Kormendy, J., 1985, ApJ 295, 73
Kormendy, J., Bender, R., 1996, ApJ 464, L119
Kormendy, J., Illingworth, G., 1982, ApJ 256, 460
Kundu, A., Whitmore, B. C., 2001, AJ 122, 1251
Kuntschner, H., 2000, MN 315, 184
Kuntschner, H., Davis, R. L., 1998, MN 295, L29
Lake, G., Dressler, A., 1986, ApJ 310, 605
Larson, R. B., Tinsley, B. M., Caldwell, C. N., 1980, ApJ 237, 692
Levine, S., 1995, *Interacting Galaxies*, ed. G. Longo, p. 129
Lewis, I., Balogh, M., de Propris, R. *et al.* , 2002, MN 334, 673
Marleau, F. R., Simard, L., 1998, ApJ 507, 585
Mathieu, A., Merrifield, M. R., Kuijken, K., 2002, MN 330, 251
Moore, B., Katz, N., Lake, G. *et al.* , 1995, Nat 379, 613
Moore, B., Lake, G., Katz, N., 1998, ApJ 495, 139
Neistein, E., Maoz, D., Rix, H.-W., Tonry, J. L., 1999, AJ 117, 2666
Pahre, M. A., de Carvalho, R. R., Djorgovski, S. G., 1998, AJ 116, 1606
Pierce, M. J., Tully, R. B., 1992, ApJ 387, 47
Pinkney, J., Gebhardt, K., Bender, R. *et al.* , 2003, ApJ 596, 903
Poggianti, B. M., Smail, I., Dressler, A. *et al.* , 1999, ApJ 518, 576
Poggianti, B. M., Bridges, T. J., Carter, D. *et al.* , 2001a, ApJ 563, 118
Poggianti, B. M., Bridges, T. J., Mobasher, T. J. *et al.* , 2001b, ApJ 562, 689
Poggianti, B. M., Bridges, T. J., Yagi, M. *et al.* , 2004, IAU Symp. 195, *in press*
Pogge, R. W., Eskridge, P. B., 1993, AJ 106, 1405
Raha, N., Sellwood, J. A., James, R. A., Kahn, F. D., 1991, Nat 352, 411
Rubin, V. C., Waterman, A. H., Kenney, J. D. P., 1999, AJ 118, 236
Ryden, B. S., Terndrup, D. M., Pogge, R. W. *et al.* , 1999, ApJ 517, 650
Sadler, E. M., Oosterloo, T. A., Morganti, R., Karakas, A., 2000, AJ 119, 1180
Sandage, A., 1986, A&A 161, 89
Shioya, Y., Bekki, K., Couch, W. J., de Propris, R., 2002, ApJ 565, 223
Shioya, Y., Bekki, K., Couch, W. J., 2004, ApJ 601, 654
Schulz, J., Fritze – v. Alvensleben, U., Fricke, K. J., 2003, A&A 398, 89
Schweizer, F., 1993, *Dynamics and Interactions of Galaxies*, Springer, p. 60
Schweizer, F., 1999, Ap&SS 267, 299
Schweizer, F., 2002, IAU Symp. 207, 630
Schweizer, F., Seitzer, P., Faber, S. M. *et al.* 1990, ApJ 364, L33
Schweizer, F., Seitzer, P., 1992, AJ 104, 1039
Scodeggio, M., Giovanelli, R., Haynes, M. P., 1998a, AJ 116, 2728
Scodeggio, M., Giovanelli, R., Haynes, M. P., 1998b, AJ 116, 2738
Smail, I., Dressler, A., Couch, W. J. *et al.* , 1997, ApJS 110, 213
Smail, I., Kuntschner, H., Kodama, T. *et al.* , 2001, MN 323, 839
Tikhonov, N. A., Galazutdinova, O. A., Aparicio, A., 2003, A&A 401, 863
Terlevich, A. I., Caldwell, N., Bower, R. G., 2001, MN 326, 1547
Tremaine, S., Gebhardt, K., Bender, R. *et al.* , 2002, ApJ 574, 740
Tully, R. B., Fisher, J. R., 1977, A&A 54, 661
van den Bergh, S., 1976, ApJ 206, 883

van den Bergh, S., 1994, AJ 107, 153
van den Bergh, S., 2004, ApJ 601, L37
van Dokkum, P. G., Ellis, R. S., 2003, ApJ 592, L53
van Dokkum, P. G., Franx, M., Kelson, D. D. *et al.* , 1998, ApJ 500, 714
van Dokkum, P. G., Franx, M., Fabricant, D. *et al.* , 1999, ApJ 520, L95
van Dokkum, P. G., Franx, M., Kelson, D. D., Illingworth, G. D., 2001, ApJ 553, L39
Weilbacher, P. M., Fritze – v. Alvensleben, U., Duc, P.-A., Fricke, K. J., 2002, ApJ 579, L79
Weilbacher, P. M., Duc, P.-A., Fritze – v. Alvensleben, U., 2003, A&A 397, 545
Welch, G. A., Sage, L. J., 2003, ApJ 584, 260
Wuyts, S., van Dokkum, P. G., Kelson, D. D. *et al.* , 2004, ApJ 605, 677

GRAVITATIONAL BAR TORQUES IN THE SPIRAL/S0 DIVIDE

R. Buta

Department of Physics & Astronomy, University of Alabama, Box 870324, Tuscaloosa, AL 35487-0324

Abstract The properties of bars in S0 galaxies and early-type spirals is only beginning to be explored. This paper discusses some preliminary results on maximum relative gravitational bar torques and relative Fourier intensity amplitudes in SB0 and SB0/a galaxies and how these compare with similar quantities estimated for spirals.

Keywords: galaxies: barred; galaxies: structure

1. Introduction

S0 galaxies were introduced to the Hubble sequence by Hubble (1936) as a means of bridging the apparently catastrophic gap between E7 and Sa galaxies. After real examples were discovered (see Sandage 1961), the hallmark of the class became a disk shape (definitely "later than" E7) and an absence of spiral arms. SB0 galaxies were originally classified as SBa by Hubble (1926) even though they also lacked arms. This inconsistency was corrected in Sandage (1961).

Barred S0 galaxies are extremely interesting because they help to take some of the mystery out of S0s in general: a bar is a disk feature that is closely related to spiral structure in many galaxies (see, for example, Kormendy and Norman 1979). In addition, ring features are directly related to bars in spirals and SB0s may show vestiges of similar rings. One could therefore ask whether bars, if studied properly, might provide any clues as to how S0s and spirals might be related in an evolutionary sense.

As a class, S0s are now recognized to be a rather mixed bag of possibly unrelated objects (van den Bergh 1990). Van den Bergh (1994) discussed the issue of whether S0s are truly intermediate between elliptical and Sa galaxies, and concluded that they are not on the basis of luminosity functions and axial ratio distributions. Also, the classification of the most featureless S0s, which

D. Block et al. (eds.), Penetrating Bars through Masks of Cosmic Dust, 101–110.

have virtually no structure other than a bulge and a disk, can be confused if based solely on visual inspection.

The origin of S0s is still a mystery. The most popular idea, that S0s represent spirals stripped of their interstellar gas by ram-pressure forces (or other interactions) in the cluster environment, can account perhaps for the high frequency of S0s in clusters, but can't fully account for field S0s and the bulge properties of S0s (see Larson, Tinsley, and Caldwell 1980, and references therein). Dressler et al. (1997) show that the morphology/density relation, where the fraction of Es and S0s correlate with local surface galaxy density, may have proceeded in a time-dependent, hierarchical manner. Shioya, Bekki, and Couch (2004) use red Hδ-strong galaxies to argue for truncated starbursts as an explanation for some S0s in clusters, while Bekki, Couch, and Shioya (2002) discuss halo gas stripping and conversion of a passive early-type spiral into an S0 as another mechanism. Welch and Sage (2003) use measures of the molecular gas content of S0s to argue that no current model of S0 formation can account for the low gas content as observed. However, bars have not really been factored into any of these previous studies, and indeed barred S0s may sometimes be ignored in the analysis of S0 properties (e.g., van den Bergh 1994).

2. Quantification of Bar Strengths in Disk Galaxies

One way to examine the properties of bars in S0 galaxies as compared to spirals is to use an indicator of bar strength that can be applied in a consistent manner for either type of galaxy. The gravitational torque method (GTM, Buta and Block 2001; see also Combes and Sanders 1981) provides one such indicator; it uses near-infrared images to infer approximate stellar gravitational potentials in galaxies under the assumptions of a constant mass-to-light ratio and a simple model of the vertical light distribution (see, e.g., Quillen, Frogel, and Gonzalez 1994). From the resulting planar force field, one derives the maximum ratio of the tangential force to the axisymmetric part of the radial force as a function of radius: $Q_T(r) = |F_T(r)/F_{0R}(r)|_{max}$. The maximum of this function, Q_g, provides a single number that characterizes the strength of nonaxisymmetric features in a galaxy. Q_g is equivalent to the maximum gravitational torque per unit mass per unit square of the circular speed in a galaxy. In strongly-barred galaxies, the *bar strength* $Q_b \approx Q_g$ while in pure spirals the *spiral strength* $Q_s \approx Q_g$ (Buta, Block, and Knapen 2003=BBK). In many spirals, Q_g measures a combination of bar and spiral strength.

The GTM has been applied to large samples of spiral galaxies in several recent studies (Block et al. 2001, 2002; Laurikainen and Salo 2002; Laurikainen, Salo, and Rautiainen 2002=LSR), and was greatly refined and improved by Laurikainen et al. (2004b). With the refined GTM (which allows for bulge shape, a type dependence in the ratio of radial to vertical scalelengths, and improved orientation parameters), Buta, Laurikainen, and Salo (2003=BLS) and Laurikainen, Salo, and Buta (2004a=LSB) derived Q_g values for the statistically well-defined Ohio State University Bright Galaxy Survey (OSUBGS, Eskridge et al. 2002).

3. Bars and the Evolution of Galaxies

The distribution of bar strengths in disk galaxies is a potentially powerful way of studying galaxy evolution, and in particular the origin and evolution of bars themselves (Sellwood 2000). For example, it has been shown by Bournaud and Combes (2002) that while the buildup of a central mass concentration in a barred spiral galaxy can destroy a bar, accretion of external gas can regenerate a bar, leading to recurrent bar formation. In this idea, the distribution of bar strengths simply reflects the relative amounts of time that a galaxy stays in a given bar state (strong, weak, or nonbarred). For spirals, the observed distribution of bar strengths shows an extended "tail" towards strong bars, and a deficiency of weak bars, that are only expected to be seen if accretion is a significant process (Block et al. 2002; see Figure 1, which is from BLS).

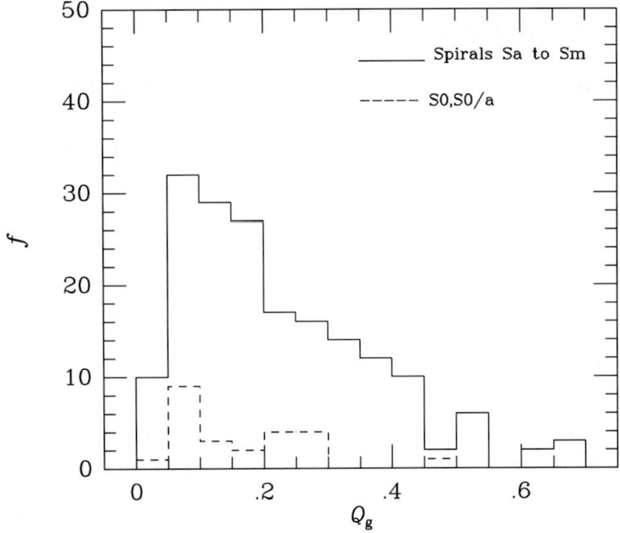

Figure 1. The distribution of maximum relative gravitational torques for 158 OSUBGS spiral galaxies, supplemented by 22 2MASS galaxies (solid histogram, from BLS) and for 24 S0,S0/a galaxies (dashed histogram), from new K_s-band observations.

In contrast, the bars of S0 galaxies are thought to be virtually primordial, fossil relics in systems that have not accreted any new gas for a significant fraction of a Hubble time and whose initial gas was either swept out or used up before the bar could be destroyed. If accretion is not important in these systems, then the distribution of bar strengths would reflect some primordial distribution that could be very different from that currently found for nearby spirals. Spirals which don't accrete any new gas but which drive much of their residual gas to the center can lose their bars, and the distribution of their bar strengths should be very different from that of accreting spirals: there should be no extended "tail" and a preponderance of nonbarred cases (Block et al. 2002). We should see something like this in S0s only if they were able to keep their gas long enough for the process of natural bar dissolution to run its course. *The history of S0 galaxies ought to be impressed on their distribution of bar strengths, and could be deciphered if we could measure this distribution as reliably as it has been measured for spirals.*

E. Laurikainen and I have carried out an extensive near-infrared imaging survey of a significant sample of S0 and S0/a galaxies in order to apply the GTM and determine the range of bar strengths in such galaxies as well as the distribution function of these strengths as compared to spirals. This survey (which we call the "Near-IR S0 Survey", or NIRS0S) is meant to mimic the OSUBGS for S0 galaxies in order to provide a statistically well-defined sample. Here I give some preliminary results from that survey. First I summarize what is known about torque properties of bars in spirals.

4. Gravitational Torques in Spiral Galaxies

Studies of maximum relative torques in spiral galaxies have centered around several issues: (1) quantitative dust-penetrated bar classification (Buta and Block 2001); (2) the distribution of these torques (Block et al. 2002; BLS); (3) a comparison of torques between active and nonactive galaxies (LSR; LSB); (4) the correlation between bar torque strength and inner ring shape (Buta 2002), and (5) the relation, if any, between maximum bar torques and spiral torques in the same galaxy (BBK; Block et al. 2004). Block and Puerari (1999) discuss a dust-penetrated classification of spirals based on pitch angle and dominant harmonic classes, a system supplemented with bar torque classes by Buta and Block (2001). Block et al. (2002) derived the distribution of Q_g values based on a preliminary analysis of OSUBGS H-band images of 163 galaxies. These authors did not allow for bulge shape and used a fixed relation between the radial scalelength and the vertical scaleheight: $h_z = h_r/12$, which partly led to a deficiency of low Q_g values. BLS used the more refined application of the GTM (Laurikainen et al. 2004b) to get the distribution in Figure 1, which shows similar characteristics to that obtained by Block et al. (2002), but has more low Q_g values. Block et al. (2002) argued that the observed distribution favors that spiral galaxies double their mass in 10^{10} yr, based on a comparison with numerical simulations that were analyzed with the same approximations as the observations. The BLS analysis still favors accretion but perhaps at a lesser rate.

LSR analyzed a sample of 2MASS images and showed that active galaxies seem to prefer weaker bars. This is in part due to a preference of early-type spirals as AGN hosts. LSB show that early-type galaxies have long and massive bars which at same time have weak maximum relative torques. This effect can explain some of the differences found between bar properties in active and nonactive galaxies. Buta (2002) analyzed deprojected red and Hα images of a sample of ringed galaxies and showed that there is little correlation between bar torque strength and intrinsic inner ring axis ratio.

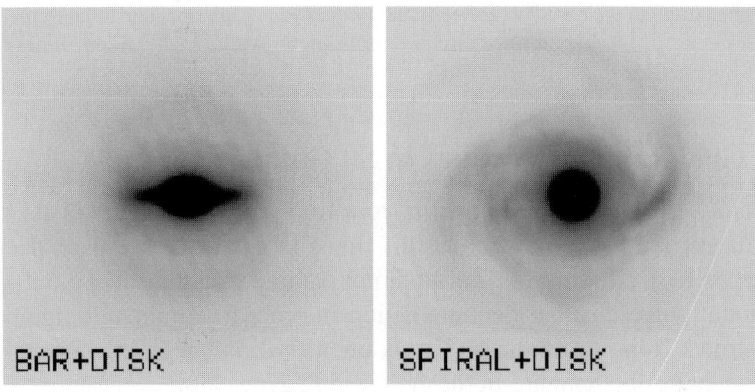

Figure 2. Separated bar and spiral images for the SAB(rs)bc spiral NGC 6951 (from BBK).

The issue of the relation between bars and the spirals that appear with them was addressed by Block et al. (2004), who used a Fourier-based technique (BBK) to separate the near-IR intensity distributions of bars and spirals (Figure 2). The method depends on the assumption that relative Fourier intensity amplitudes I_m/I_0, where m is an even integer, in a bar decline past a maximum in the same or a similar manner as they rise to that maximum (Figure 3, left panel). This is known as the symmetry assumption, and it allows the extrapolation of the bar into the spiral region. From such separated images, Block et al. (2004) derived bar strengths Q_b and spiral strengths Q_s for 17 intermediate- to late-type spiral galaxies. It was found that bar and spiral strength are largely uncorrelated except for $Q_b > 0.4$ (Figure 3, right panel). Block et al. concluded that the apparent correlation for $Q_b > 0.4$ indicates that a strong bar/spiral pattern results from a combined instability that has a single pattern speed and growth rate in a relatively massive disk.

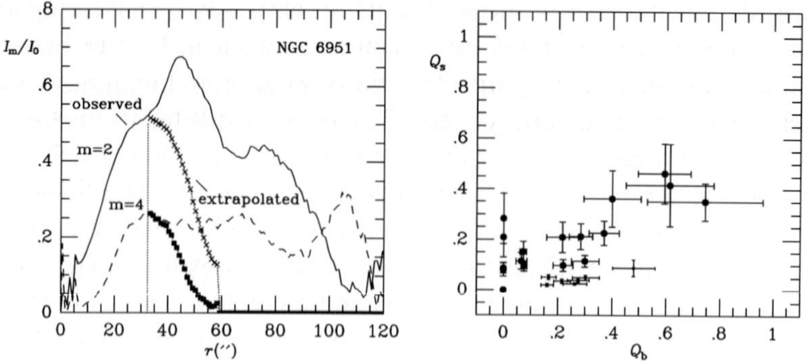

Figure 3. left: Bar extrapolations (symbols) for the m=2 and m=4 Fourier components of NGC 6951, based on the symmetry assumption (from BBK). The vertical line shows the radius r_2 around which the rising m=2 Fourier amplitudes are reflected. right: Plot of spiral strength Q_s versus bar strength Q_b based on gravitational torques for 17 intermediate to late-type spirals (from Block et al. 2004). Larger filled circles refer to spirals while smaller filled circles refer to SB0 and SB0/a galaxies.

5. Gravitational Torques in S0 Galaxies

Our sample of S0s has been drawn from RC3 (de Vaucouleurs et al. 1991), and 24 have been fully analyzed at this time. In each case we used deep optical images to derive improved orientation parameters, and carried out a two-dimensional bulge/disk/bar decomposition in order to eliminate bulge deprojection stretch from the analysis (Laurikainen et al. 2004b). The near-infrared images we have obtained are in the K_s-band (2.2μm).

One of the first results of our survey is that some galaxies classified as SB0^o or SB0$^+$ in RC3 are actually early-type spirals, based on our higher quality imaging. Such galaxies therefore cannot be considered true S0s in the Hubble (1936) sense but belong instead in the "spiral/S0 divide", at least in blue light. Another issue is that some galaxies classified as SB0 in RC3 are likely to be edge-on S0s with significant bulges, giving the false appearance of a bar in a face-on disk.

With such a small sample at hand, we cannot yet derive a meaningful distribution of bar strengths for S0s that can be directly compared to spirals. (Our preliminary distribution is shown as the dashed histogram in Figure 1, which emphasizes how much further we have to go to match the data for spirals.). However, one thing we can do is determine how strong bars in SB0 and SB0/a galaxies can get. Also, near-infrared images of SB0 and SB0/a galaxies provide us with an opportunity to study the properties of stellar bars uncomplicated by effects of both spiral arms and dust. In particular, we can use these galaxies to evaluate the symmetry assumption in bar/spiral separation.

I have addressed these questions in a preliminary way by analyzing several of our S0 sample galaxies in the same manner as the spirals were analyzed in Block et al. (2004). In this approach, relative Fourier amplitudes are derived from deprojected K_s-band images, and the bars are separated from other structure such as faint rings and weak spiral arms using the method of BBK. Figure 4 shows the relative m = 2 and 4 amplitudes for eight SB0 and SB0/a galaxies. For five of these galaxies (NGC 936, 1440, 2217, 4596, and 4608), the two profiles are remarkably symmetric (and the same is true for higher even Fourier terms to m = 20). For NGC 1452 and 2859, the m = 4 profile is relatively symmetric, but the m = 2 profile is not. In the case of NGC 2859, this asymmetry is due in part to a faint inner ring around the ends of the bar. For NGC 1452, decomposition uncertainties contribute to some of the asymmetry. Both profiles for the very sharp-ended bar in NGC 4643 are asymmetric, and this asymmetry carries into the higher-order terms as well. However, in no case is the asymmetry so extreme that we cannot do a reasonable representation of the bar with the symmetry assumption.

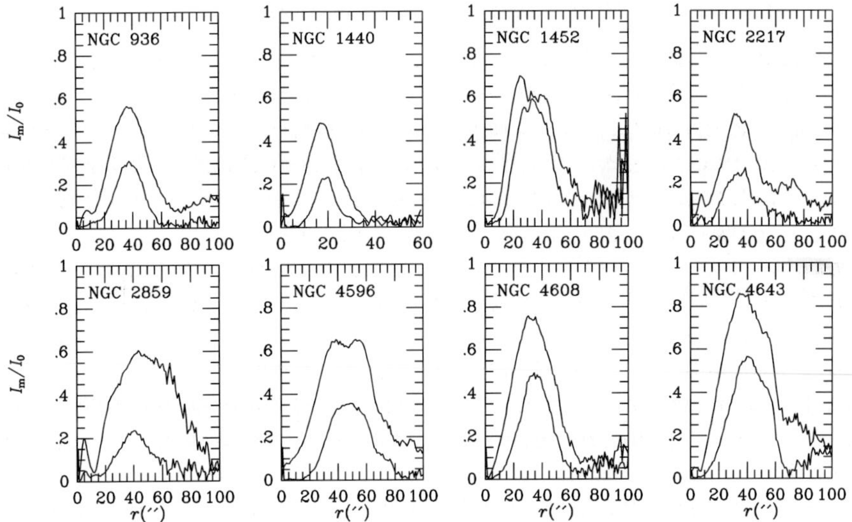

Figure 4. Relative amplitudes of the m=2 (upper curve) and m=4 Fourier components in the K_s-band light distribution of 8 barred S0, S0/a galaxies.

I believe these results verify the symmetry assumption in bar/spiral separation, if the bars in spirals are similar in behavior to those in these SB0 and SB0/a galaxies. BBK verified this using the SB(r)b spiral NGC 4394. The assumption does not necessarily apply to very late-type galaxy bars because these can have considerable asymmetry (de Vaucouleurs and Freeman 1972), and separating them from their spirals would require examining the behavior of odd Fourier terms in the bar (see Block et al. 2004 for one example).

Using the method of Quillen, Frogel, and Gonzalez (1994), I have transformed the bar and spiral images from the BBK analysis into force ratio maps and derived $Q_T(r)$ for these maps as well as for the total image. For this purpose, either 2D decompositions or elliptically-averaged profiles were used to estimate the disk radial scalelength, and then the vertical scaleheight was derived as $h_z = h_r/4$ (de Grijs 1998). Figure 5 compares the $Q_T(r)$ functions for the SB0/a galaxy NGC 4596 with the same functions for NGC 6951. In NGC 4596, the spiral is very weak, while the bar is moderately strong. The bar in NGC 6951 is of similar strength, but the spiral is also very strong.

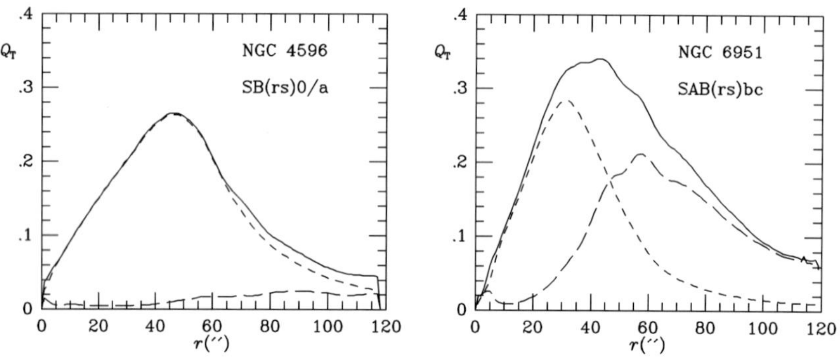

Figure 5. These plots show $Q_T(r)$ curves for the SB0/a galaxy NGC 4596 and for the SAB(rs)bc spiral NGC 6951 (from BBK). Short-dashed curves are for the bar, long-dashed curves are for the spiral, and solid curves are for the total images, Fourier-smoothed to m=20.

Table 1 summarizes preliminary values of Q_b and Q_s for seven of the galaxies in Figure 4. These are preliminary in the sense that they are based on approximate estimates of the radial scalelengths and also I have ignored the shape of the bulge in the force calculation (i.e., I have assumed it to be as flat as the disk). In each case, Q_s measures what is left after the symmetry assumption is applied to the bar. Most of these values are affected by noise outside the bar region and may not be significant. Figure 3 (right panel) shows these seven galaxies plotted with the spirals from Block et al. (2004). The results show that bars in these early types can still get strong in spite of the significant bulge components. NGC 1452 (see Figure 6) appears to be the strongest; its bar is long, thin, and has an $m = 20$ Fourier term that is still significant. It lies in the area in Figure 3 where no spirals were found in the Block et al. (2004)

study. The remaining cases have strong-looking bars, but these do not have exceptional torques, probably owing to the significant bulges.

Table 1. Maximum Relative Torques in the Spiral/S0 Divide[a]

Galaxy	Type[a]	h_z(pc)	Q_b	$r(Q_b)/r_o$	Q_s	$r(Q_s)/r_o$
NGC 936	SB(rs)0$^+$	740	0.22	0.28	0.04:	0.57:
NGC 1452	(R$'$)SB(r)0/a	570	0.48	0.52	0.09:	0.68:
NGC 2859	(R)SB(r)0$^+$	940	0.16	0.31	0.02:	0.46:
NGC 4245	SB(r)0/a	360	0.17	0.32	0.05:	0.46:
NGC 4596	SB(rs)0/a	890	0.26	0.41	0.03:	0.79:
NGC 4608	SB(r)0°	630	0.28	0.42	0.04:	0.55:
NGC 4643	SB(rs)0/a	850	0.30	0.46	0.05:	0.91:

[a]Types from RC3, except for NGC 4596 which is based on new optical images. Vertical scaleheight h_z is inferred from radial scalelength estimates h_r as h_z=h_r/4. $r(Q_b)$ and $r(Q_s)$ are the radii of the maximum relative bar and spiral torques, respectively. r_o is the extinction-corrected isophotal radius, also from RC3.

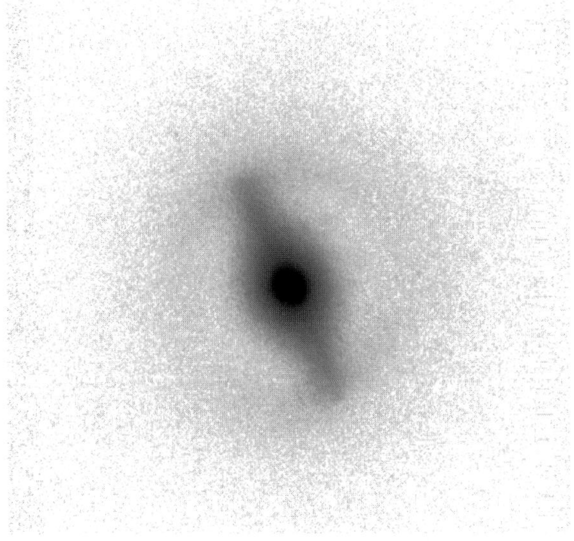

Figure 6. Deprojected K_s-band image of NGC 1452, the currently strongest known bar in our SB0 and SB0/a sample. The bar is surrounded by a nearly circular stellar inner ring.

6. Future Work on this Topic

E. Laurikainen and I are continuing our survey with the goal of accumulating at least a comparable number of relatively nearby S0 and S0/a galaxies as

OSUBGS spirals. The NIRS0S will provide a wealth of information on bulge and disk parameters and on bar strengths, shapes, morphologies, and Fourier amplitudes in early-type disk galaxies.

Acknowledgments

I thank E. Laurikainen for providing the deprojected K_s-band images and many of the Q_g values used for this presentation. These observations are based on several runs with the Nordic Optical Telescope in 2003 and 2004. I also thank Dave Block and the Anglo-American Chairman's Fund for partial travel support. This work has been supported by NSF Grant AST-0205143 to the University of Alabama.

References

Bekki, K., Couch, W.J., and Y. Shioya. (2002). ApJ, 577, 651
Block, D. L., and I. Puerari. (1999). A&A, 342, 627
Block, D. L., Puerari, I., Knapen, J. H. et al. (2001). A&A, 375, 761
Block, D. L., Bournaud, F., Combes, F. et al. (2002). A&A, 394, L35
Block, D. L., Buta, R., Knapen, J. H. et al. (2004). AJ, in press
Bournaud, F., and F. Combes. (2002). A&A, 392, 83
Buta, R. (2002). ASP Conf. Ser., 275, p. 185
Buta, R., and D. L. Block. (2001). ApJ, 550, 243
Buta, R., Block, D. L., and J. H. Knapen. (2003). AJ, 126, 1148 (BBK)
Buta, R., Laurikainen, E., and H. Salo. (2004). AJ, 127, 279 (BLS)
Combes, F. and R. H. Sanders. (1981). A&A, 96, 164
de Grijs, R. (1998). MNRAS, 299, 59
de Vaucouleurs, G. and K. C. Freeman. (1972). Vistas in Astronomy, 14, 163
de Vaucouleurs, G. et al. (1991). *Third Reference Catalogue of Bright Galaxies* New York: Springer (RC3)
Dressler, A., Oemler, A., Couch, W. J. et al. (1997). ApJ, 490, 577
Eskridge, P., Frogel, J. A., Pogge, R. W. et al. (2002). ApJS, 143, 73
Hubble, E. (1926). ApJ, 64, 321
Hubble, E. (1936). *The Realm of the Nebulae* New York: Dover Publications
Kormendy, J., and C. Norman. (1979). ApJ, 233, 539
Larson, R. B., Tinsley, B. M., and N. Caldwell. (1980). ApJ, 237, 692
Laurikainen, E., and H. Salo. (2002). MNRAS, 337, 1118
Laurikainen, E., Salo, H., and P. Rautiainen. (2002). MNRAS, 331, 880 (LSR)
Laurikainen, E., Salo, H., and R. Buta. (2004a). ApJ, in press (LSB)
Laurikainen, E., Salo, H., Buta, R., and S. Vasylyev. (2004b), in preparation
Quillen, A. C, Frogel, J. A. and R. Gonzalez. (1994). ApJ, 437, 162
Sandage, A. (1961). *The Hubble Atlas of Galaxies*. Carnegie Inst. of Wash. Pub. No. 618
Sellwood, J. (2000). ASP Conf. Ser., 197, p. 3
Shioya, Y., Bekki, K., and W. J. Couch. (2004). ApJ, 601, 654
van den Bergh, S. (1990). ApJ, 348, 57
van den Bergh, S. (1994). AJ, 107, 153
Welch, G. A. and L. J. Sage. (2003). ApJ, 584, 260

DIRECT MEASUREMENT OF PATTERN SPEEDS IN DOUBLE-BARRED SB0'S

E. M. Corsini[1], V. P. Debattista[2] & J. A. L. Aguerri[3]
[1]*Dipartimento di Astronomia, Universita di Padova, vicolo dell'Osservatorio 2, I-35122 Padova, Italy,* [2]*Institut fur Astronomie, ETH Honggerberg, HPF G4.2, CH-8093 Zurich, Switzerland,* [3]*Instituto de Astrofisica de Canarias, Via Lactea s/n, E-38200 La Laguna, Spain*

Abstract NGC 2950 is a nearby and undisturbed SB0 galaxy hosting two nested stellar bars. We used the Tremaine-Weinberg method to measure the pattern speed of the primary bar. This also permitted us to establish that the two nested bars are rotating with different pattern speeds (at better than 99% confidence level). In particular, the rotation frequency of the secondary bar is higher than that of the primary one.

Keywords: galaxies: individual (NGC 2950) — galaxies: kinematics and dynamics — galaxies: elliptical and lenticular, cD — galaxies: photometry — galaxies: structure

1. Introduction

Secondary bars are present in some $1/3$ of barred galaxies (Erwin 2004). Interest in secondary stellar bars is motivated by the hypothesis that they are a mechanism for driving gas to small radii, but it is still unclear whether they are efficient agents of radial mass transport. This uncertainty is largely due to the current lack of knowledge of the relation between the bar pattern speed of the primary, $\Omega_{\rm p}$, and secondary, $\Omega_{\rm s}$. Three cases are possible: (1) $\Omega_{\rm s} > \Omega_{\rm p}$, as favored by some hydrodynamical simulations (*e.g.*, Friedli & Martinet 1993) and theoretical models (*e.g.*, Maciejewski & Sparke 2000). (2) $\Omega_{\rm s} = \Omega_{\rm p}$ occurs in other hydrodynamical simulations (*e.g.*, Knapen et al. 1995), although the apparent arbitrary angle between the principle axes of the primary and secondary bars may contradict this. (3) $\Omega_{\rm p} > 0$ and $\Omega_{\rm s} < 0$, i.e. the two bars counter-rotate, as may happen if counter-rotating material is accreted (Sellwood & Merritt 1994). Each of these 3 cases predicts different radial mass transport rates (*e.g.*, Shlosman & Heller 2002). Measuring $\Omega_{\rm p}$ and $\Omega_{\rm s}$ for a sample of double-barred galaxies is a crucial step to address this issue.

D. Block et al. (eds.), Penetrating Bars through Masks of Cosmic Dust, 111–115.

2. The Tremaine-Weinberg Method

A model-independent method for measuring pattern speeds is the Tremaine-Weinberg (1984, hereafter TW) method; this gives the pattern speed Ω of a bar as $\mathcal{X}\,\Omega\,\sin i = \mathcal{V}$. where $\mathcal{X} = \int X\,\Sigma\,dX/\int\Sigma\,dX$, is the luminosity-weighted average of the position X and $\mathcal{V} = \int V_{\rm los}\,\Sigma\,dX/\int\Sigma\,dX$ is the luminosity-weighted average of the line-of-sight velocity $V_{\rm los}$ measured parallel to the major axis of the galaxy disk, respectively. Σ and i are the surface brightness and disk inclination, respectively. Slit observations parallel to the major axis of the disk measure all the quantities needed by the TW equation.

When a secondary bar is present, for a slit passing through both bars, the TW equation is modified to $(\mathcal{X}_{\rm p}\Omega_{\rm p} + \mathcal{X}_{\rm s}\Omega_{\rm s})\sin i = \mathcal{V}_{\rm p} + \mathcal{V}_{\rm s} \equiv \mathcal{V}$ and it is a consequence of the linearity of the continuity equation if the two bars are rigidly rotating through each other. The equation can be solved for $\Omega_{\rm s}$ by first measuring $\Omega_{\rm p}$ with slits which avoid the secondary bar, and then modelling the photometry of the primary bar to obtain $\mathcal{X}_{\rm s} = \mathcal{X} - \mathcal{X}_{\rm p}$. It has to be noticed that when $\Omega_{\rm p} \neq \Omega_{\rm s}$ the two bars are not likely to rotate rigidly through each other (Louis & Gerhard 1988; Maciejewski & Sparke 2000; Rautiainen et al. 2002). This require additional modelling to obtain $\Omega_{\rm s}$. Nonetheless, when $\Omega_{\rm p} = \Omega_{\rm s}$ the equation is satisfied exactly.

Therefore the application of the TW method to derive the pattern speed of two nested bars needs modelling in order to measure $\Omega_{\rm s}$, but it *does not* require any further assumption to test whether $\Omega_{\rm p} = \Omega_{\rm s}$.

3. The Double-Barred Galaxy NGC 2950

An ideal target for this purpose is the double-barred SB0 galaxy NGC 2950, which meets all the requirements for the TW analysis. It has an intermediate inclination, both bars have intermediate position angles between the major and minor axes of the disk and no evidence of spirals, dust lanes and/or patches, star-forming regions, small-scale structures (such as nuclear disks or rings) or significant companions (Fig. 1a,b). The stellar secondary bar of NGC 2950 (Fig. 1c,d) has been studied in optical and near-infrared bandpasses using ground-based and *Hubble Space Telescope* images (Wozniak et al. 1995; Friedli et al. 1996; Erwin & Sparke 2002).

We obtained *BVI* imaging of NGC 2950 at the Jacobus Kapteyn Telescope (FWHM $\simeq 1''$). We measured the length of the bars ($a_{\rm p} = 34.3'' \pm 2.1''$, $a_{\rm s} = 4.5'' \pm 1.0''$) by means of the photometric decomposition and Fourier analysis as done in Aguerri et al. (2003). The inclination ($45.6^\circ \pm 1.0^\circ$) and PA of the galaxy disk ($116.1^\circ \pm 1.0^\circ$) were determined by averaging the values measured between $65''$ and $100''$ in the $I-$band profile. Long-slit spectroscopy of NGC 2950 was carried out at the Telescopio Nazionale Galileo ($\lambda\lambda$ 4660 – 6820 A,

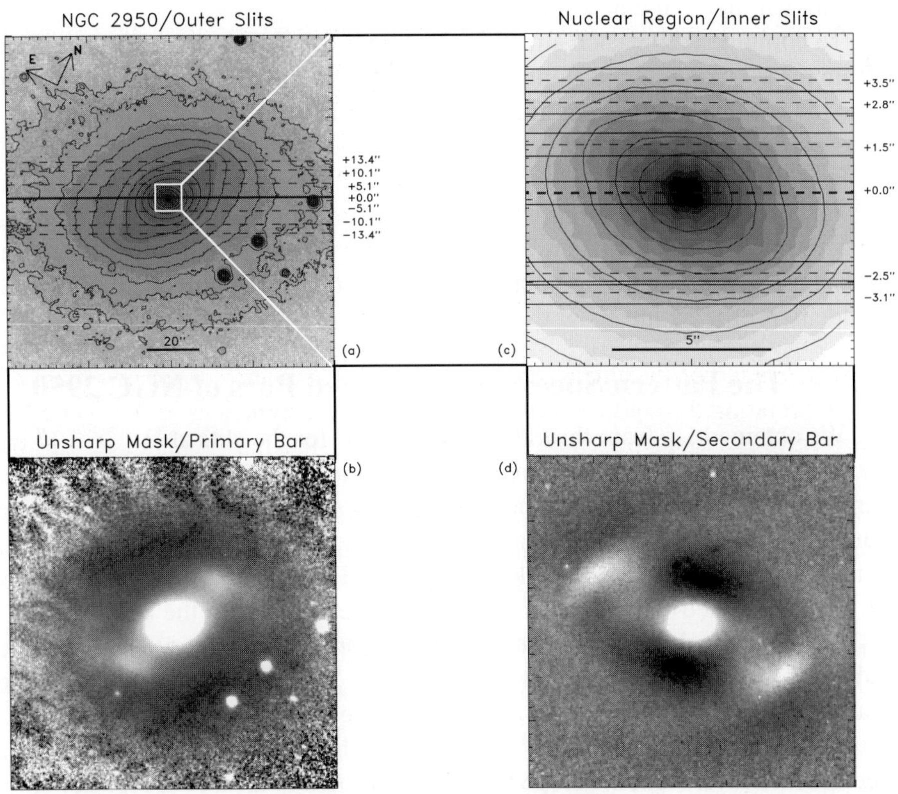

Figure 1. The double-barred SB0 galaxy NGC 2950. *(a)* $I-$band image showing the primary bar and disk and slit positions overlaid. Contours are spaced at 0.5 mag arcsec^{-2} and the outermost corresponds to $\mu_I = 23.0$ mag arcsec^{-2}. The solid and dashed lines correspond to the position of the spectra obtained along the disk major axis and at large offsets ($|Y| \geq 5.1''$), respectively. For each slit position the offset, Y, is given. Orientation is specified in the upper left corner and it is the same for all the panels. *(b)* The primary bar unveiled by the unsharp masking of the $I-$band image. We divided the image by itself after convolution by a circular Gaussian of width $\sigma = 12''$. This procedure enhanced any surface-brightness fluctuation and non-circular structure extending over a spatial region comparable to the σ of the smoothing Gaussian. *(c)* A zoom into the central region showing secondary bar and slit positions overlaid. Contours are spaced at 0.5 mag arcsec^{-2}, with the outermost corresponding to $\mu_I = 19.5$ mag arcsec^{-2}. For each spectrum obtained at $|Y| \leq 3.5''$ the solid and dashed lines mark the edges and center of the slit, respectively. *(d)* The secondary bar in the unsharp masking of the archival WFPC2/F814W image. The smoothing Gaussian has $\sigma = 0.7''$.

$\sigma_{st} \simeq 80$ km s^{-1}, FWHM $= 0.9'' - 1.5''$). We obtained 2 major axis spectra with the slit crossing the nucleus, and 11 offset spectra across the galaxy with the slit parallel to the disk major axis (Fig. 1a,c).

To compute $\mathcal{X}$ for each slit (Fig. 2b), we extracted the luminosity profiles from the V−band image along the position of the slit after convolving the image to the seeing of the spectrum. We used the V−band profiles to compute $\mathcal{X}$ because they are less noisy than those extracted from the spectra, particularly at large radii. To measure $\mathcal{V}$ for each slit (Fig. 2a), we first collapsed each two-dimensional spectrum along its spatial direction obtaining a one-dimensional spectrum. The value of $\mathcal{V}$ was then derived by fitting the resulting spectrum with the convolution of a template stellar spectrum and a Gaussian line-of-sight velocity profile by means of the Fourier Quotient Correlation method.

4. The Pattern Speeds of the Nested Bars of NGC 2950

We obtained $\Omega_{\rm p}$ from the values of $\mathcal{X}$ and $\mathcal{V}$ for the slits at $|Y| \geq 3.1''$ and at $Y = 0''$, which constrains only the zero point. Since the slits at $|Y| \geq 3.1''$ are not dominated by the secondary bar, we assume $\mathcal{X} = \mathcal{X}_{\rm p}$ and $\mathcal{V} = \mathcal{V}_{\rm p}$ for them and obtain $\Omega_{\rm p} \sin i$ with a straight line fit. This gives $\Omega_{\rm p} = 11.2 \pm 2.4$ km s^{-1} arcsec^{-1} (99.2 ± 21.2 km s^{-1} kpc^{-1}, Fig. 2c,d).

The photometric and kinematic integrals measured with the innermost slits ($|Y| \leq 2.8''$) include a strong contribution from the secondary bar. In particular, $|\mathcal{V}| \gg |\mathcal{V}_{\rm p}|$ for the slits at $Y = -2.5'', +2.8''$ if we extrapolate $\mathcal{V}_{\rm p}$ from large $|Y|$. A straight line fit to $\mathcal{X}$ and $\mathcal{V}$ for the slits at $|Y| \leq 2.8''$ has a slope ($= 63.7 \pm 7.1$ km s^{-1} arcsec^{-1}, Fig. 2c) which is different at better than 99% confidence level from the slope ($= 8.0 \pm 1.7$ km s^{-1} arcsec^{-1}, Fig. 2c) of the straight line fit for the primary bar. This may be because $\Omega_{\rm p} \neq \Omega_{\rm s}$. Estimating $\Omega_{\rm s}$ is model dependent. We illustrate this by considering two extreme cases. First, we assumed that the secondary bar dominates at $|Y| \leq 2.8''$ obtaining $\Omega_{\rm s,1} = 89.2 \pm 9.9$ km s^{-1} arcsec^{-1} (Fig. 2c). In the second case, we rewrote the TW equation for nested bars as $\mathcal{X}_{\rm s}(\Omega_{\rm s} - \Omega_{\rm p}) \sin i = \mathcal{V} - \Omega_{\rm p}\mathcal{X} \sin i$. The observed quantities are $\mathcal{X}$ and $\mathcal{V}$, while $\Omega_{\rm p}$ was measured above. To obtain $\mathcal{X}_{\rm s}$, first we derived the values of $\mathcal{X}_{\rm p}$ in the region of the nuclear bar by fitting a straight line to the $\mathcal{X}$ values at $|Y| \geq 10.1''$, and obtained $\mathcal{X}_{\rm s} = \mathcal{X} - \mathcal{X}_{\rm p}$. Then plotting $(\mathcal{V} - \Omega_{\rm p}\mathcal{X} \sin i)$ versus $\mathcal{X}_{\rm s}$, we obtain $(\Omega_{\rm s} - \Omega_{\rm p}) \sin i$ as the slope of the best fitting line; the result is $\Omega_{\rm s,2} = -21.2 \pm 6.2$ km s^{-1} arcsec^{-1}, *i.e.*, the secondary bar is counter-rotating relative to the primary one.

5. Conclusions

We conclude that in NGC 2950 the two bars must have different pattern speeds (at better than 99% confidence level), with the secondary bar having a larger pattern speed. However, since the two bars cannot be in exact solid body rotation, a more accurate measurement of $\Omega_{\rm s}$ will require careful modeling and

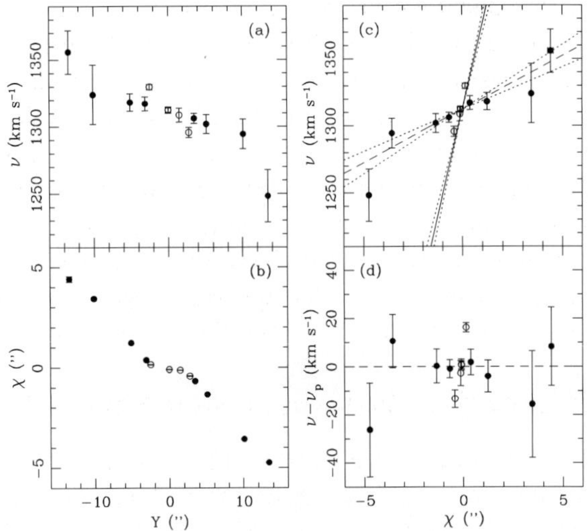

Figure 2. Pattern speeds of the nested bars of NGC 2950 (from Corsini et al. 2003). *(a)* The kinematic integrals $\mathcal{V}$ as a function of the slit offset Y with respect to the major axis ($Y = 0''$). Open and filled circles correspond to slits crossing the secondary bar ($|Y| \leq 2.8''$) and only the primary bar ($|Y| \geq 3.1''$), respectively. *(b)* The photometric integrals $\mathcal{X}$ as a function of the slit offset Y. *(c)* $\mathcal{V}$ as a function of $\mathcal{X}$ with different straight-line fits. These were obtained including the slits at $Y = 0''$ and only the slits at $|Y| \geq 3.1'$ (dashed line, slope $\Omega_p \sin i = 8.0 \pm 1.7$ km s^{-1} arcsec^{-1}) or the innermost slits at $|Y| \leq 2.8''$ (solid line, slope $\Omega_s \sin i = 63.7 \pm 7.1$ km s^{-1} arcsec^{-1}). The very different slope of the two straight lines suggests that $\Omega_p \neq \Omega_s$. *(d)* The residuals from the straight line fit to the slits at $|Y| \geq 3.1''$ and at $Y = 0''$.

comparison with simulations to account for such effects. Second, it may be that secondary bars oscillate about an orientation perpendicular to the primary bar, possibly accounting for $\Omega_s < 0$. This can be tested by repeating our measurements on a sample of double-barred galaxies.

References

Aguerri, J. A. L., Debattista, V. P., & Corsini, E. M. 2003, MNRAS, 338, 465

Corsini, E. M., Debattista, V. P., & Aguerri, J. A. L. 2003, ApJ, 599, L29

Erwin, P. 2004, A&A, 415, 941

Erwin, P., & Sparke, L. S. 2002, AJ, 124, 65

Friedli, D., & Martinet, L. 1993, A&A, 277, 27

Friedli, D., Wozniak, H., Rieke, M., Martinet, L., & Bratschi, P. 1996, A&AS, 118, 461

Knapen, J. H., Beckman, J. E., Heller, et al. 1995, ApJ, 454, 623

Louis, P. D., & Gerhard, O. E. 1988, MNRAS, 233, 337

Maciejewski, W., & Sparke, L. 2000, MNRAS, 313, 745

Rautiainen, P., Salo, H., & Laurikainen, E. 2002, MNRAS, 337, 1233

Sellwood, J. A., & Merritt, D. 1994, ApJ, 425, 530

Shlosman, I., & Heller, H. C. 2002, ApJ, 565, 921

Tremaine, S., & Weinberg, M. D. 1984, ApJ, 282, L5
Wozniak, H., Friedli, D., Martinet, L., Martin, P., & Bratschi, P. 1995, A&AS, 111, 115

GAS FLOWS, STAR FORMATION AND GALAXY EVOLUTION

John E. Beckman[1,2], Emilio Casuso[1], Almudena Zurita[3,4] and Mónica Relaño[1]
[1] *Instituto de Astrofísica de Canarias, 38200–La Laguna, Spain,*
[2] *Consejo Superior de Investigaciones Científicas, Spain*
[3] *Universidad de Granada, Dept. de Física Teórica y del Cosmos, 18071–Granada, Spain*
[4] *Isaac Newton Group of Telescopes, 38700–La Palma, Spain*

Abstract In the first part of this article we show how observations of the chemical evolution of the Galaxy: G- and K–dwarf numbers as functions of metallicity, and abundances of the light elements, D, Li, Be and B, in both stars and the interstellar medium (ISM), lead to the conclusion that metal poor H I gas has been accreting to the Galactic disc during the whole of its lifetime, and is accreting today at a measurable rate, $\sim 2M_{\odot}$ per year across the full disc. Estimates of the local star formation rate (SFR) using methods based on stellar activity, support this picture. The best fits to all these data are for models where the accretion rate is constant, or slowly rising with epoch. We explain here how this conclusion, for a galaxy in a small bound group, is not in conflict with graphs such as the Madau plot, which show that the universal SFR has declined steadily from $z = 1$ to the present day. We also show that a model in which disc galaxies in general evolve by accreting major clouds of low metallicity gas from their surroundings can explain many observations, notably that the SFR for whole galaxies tends to show obvious variability, and fractionally more for early than for late types, and yields lower dark to baryonic matter ratios for large disc galaxies than for dwarfs. In the second part of the article we use NGC 1530 as a template object, showing from Fabry–Pérot observations of its Hα emission how strong shear in this strongly barred galaxy acts to inhibit star formation, while compression acts to stimulate it.

Keywords: Galaxy: evolution, Galaxy: accretion, galaxies: ISM, galaxies: kinematics galaxies: star formation

1. Introduction

The role of mergers in galaxy evolution has become increasingly recognized recently, stimulated by the central role of CDM cosmology (Navarro, Frenk & White 1994, 1995; Power et al. 2003). The importance of mergers was realized from purely observational arguments by Toomre & Toomre (1972); they argued that as tidal encounters generate short lived features, to yield today's

D. Block et al. (eds.), Penetrating Bars through Masks of Cosmic Dust, 119–138.

"peculiar" galaxies a population of binary galaxies with highly eccentric orbits is required. If these have a flat binding energy distribution, their merger rate must have declined with time as $t^{-5/3}$ so that the ten obviously merging objects in the New General Catalogue must be the tail end of 750 remnants (see also Toomre, 1977). Zepf & Koo (1989), prior to the Hubble Deep Fields, and Abraham (1999) using their contents, inferred that galaxy pair density grows as $(1+z)^3(\pm 1)$, while Brinchmann et al. (1998) showed that irregular galaxies form 10% of the total at $z \sim 0.4$ and 30% at $z \sim 0.8$. Peculiar morphologies, characteristic of mergers and interactions dominate at high redshift. The route to ellipticals via major mergers of spirals was first explored by Toomre & Toomre (1972), by Toomre (1977), and by Verraraghavan & White (1985).

Although most studies focused on the more spectacular major mergers, studies of the Galaxy have brought out the importance of minor mergers and accretions. For example Gilmore & Feltzing (2000) in a major review, using arguments based on metallicity and kinematics, discuss its structural components: bulge, thin disc, thick disc (Freeman 1993; Gilmore & Wyse 1998), and stellar halo, showing that the thin disc and bulge are relatively young, while the thick disc and halo are relatively old. In particular the thick disc formed at a specific epoch, some 10 Gyr ago (Gilmore & Wyse 1998) as a result of a minor merger, and the Galaxy is accreting material now as it disrupts its smaller neighbours (Lynden–Bell 1976; Mirabel 1982; Savage et al. 2000). Here we will point up a process which is playing a key role in the evolution of the Galaxy, but which tends to be overlooked: the accretion of gas clouds of sub–galactic mass. This must be important in galaxy groups, small clusters, and the outskirts of rich clusters in general. We will show here, as the main thrust of the article, that the Galactic accretion rate has not declined during the disc lifetime and while this may seem surprising, it is a natural consequence of a plausible model for the Local Group. To maintain the star formation rates (SFR's) observed in late type galaxies now a steady inflow of gas from outside the galaxy is maintained. A key difference between a gas–rich and a gas–poor galaxy is that the capture cross–section for gas cloud accretion is higher in the former; this alone could account for the lower mean SFR's seen in late type galaxies but also for their greater relative scatter. The distribution of star formation across a galaxy disc is also affected strongly by gas flows, but here by in–plane flows. We bring this out in a semi–quantitative way in the final part of the paper, where we present observations of the relation between the velocity gradient of the gas flow in a strongly barred galaxy, NGC 1530, and the local formation rate of massive stars, inferring that compression enhances the local SFR, while shear reduces it.

2. Direct and indirect evidence for continuous gaseous inflow to the Galactic disc

Long before galaxy mergers were considered important, Larson (1972a,b) suggested that long term infall of low metallicity gas to the Galactic disc could explain the observed relative dearth of metal poor stars (the G–dwarf problem). An advantage of this scenario over its rivals is that it offers an explanation for the presence of measurable deuterium abundance in the Galaxy, notably near the centre (Audouze et al. 1976, and more recently Lubowich et al. 2000). With no known source of D within the Galaxy, and as astration destroys it, continuous replenishment by infall can explain its presence. High redshift D abundances from the Lyα forest (e.g. Kirkman et al. 2000; Pettini & Bowen 2001) are $\sim$2-4$\times 10^{-5}$, while recent FUSE values for the local ISM (e.g. Lehner et al. 2002; Oliveira et al. 2003) are $\sim$1-2$\times 10^{-5}$, and to complete the argument for infall the D abundance recently measured with FUSE in the high velocity cloud (HVC) "complex C" (Sembach et al. 2004) is 2.2$\times 10^{-5}$, a value intermediate between the first two. The idea is that some HVC's are supplying metal–poor non–astrated gas continuously to the Galaxy: they are the infall component mentioned. Some early models of Galactic inflow (Hunt & Sciama, 1972; Tinsley 1977), which derived infall mass rates similar to the Galaxy–wide SFR, entailed a sweeping up of Local Group gas, while Larson (1976) suggested collapsing remnants of the pregalactic nebula. Muller, Oort & Raimond (1963) detected HVC's in H I at 21 cm; Oort (1970) noted that some of them have radial velocities higher than the local Galactic escape velocity, and might be primordial. Recent observations that their metallicities are $\sim$ 0.1 solar (e.g. Lehner et al. 2002; Wakker et al. 2003; Savage et al. 2003; Sembach et al. 2004), rule out primordiality, but also rule out a Galactic origin, so they are candidates for low metallicity infall. Evidence that some HVC's may well belong to the Local Group but not the Galaxy had been accumulating (Mirabel 1982; Lepine & Duvert 1994; Blitz et al. 1999; López–Corredoira et al. 1999; Casuso & Beckman 2001; Gibson et al. 2001; Putman et al. 2002), and recently Braun et al. (2002) in a sensitive H I survey, estimated that a gas mass of a few times $10^9 M_\odot$ could have accreted to both the Milky Way and M31 during their lifetimes. Direct evidence for infall for other galaxies is small. Phookun et al. (1993) detected H I falling onto NGC 5254, while recently Fraternali et al. (2004) detected flows perpendicular to the planes of nearby spirals using Hα as the tracer. Observations of infall to external galaxies are few so far because the detection sensitivity has been limited at $\sim 10^8 M_\odot$ of gas, while it is probable that the bulk of these flows are in clouds of lower mass. As sensitivities have improved, signs of accretion are being reported, see e.g. van der Hulst & Sancisi (2003).

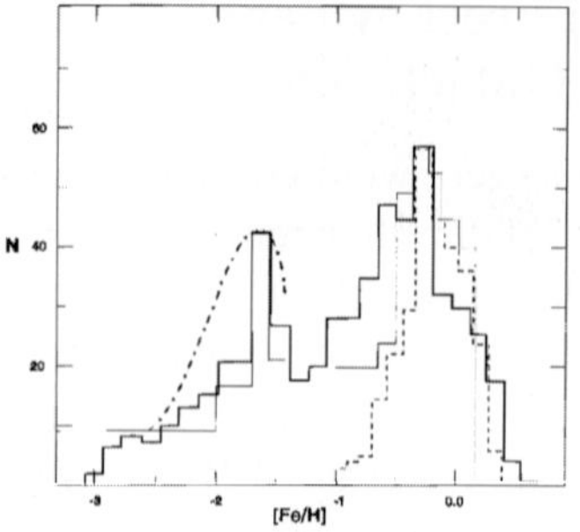

Figure 1. Early frequency histogram of local G dwarfs as a function of metallicity (using [Fe/H] as parameter) from CB97. Two peaks, for halo and thin disc stars, are clearly seen. The thick disc contribution, less obvious, lies between them and widens the higher metallicity peak (for the disc stars).

The G–dwarf problem is a strong pointer to infall for the Milky Way (van den Bergh 1962; Schmidt 1963; Pagel & Pachett 1975; Tinsley 1980; and Pagel 1987). Closed box models for chemical evolution of the Galactic disc predict far more low metallicity stars than those observed, and integrated population studies of external galaxies point to the same conclusion (Worthey et al. 1996; Espana & Worthey 2002; Bellazzini et al. 2003). Late F and G stars may be nearly as old as the disc, and preserve a record of its evolution though the oldest have evolved off the main sequence, so as improving technique brought K–dwarf statistics within range (Flynn & Morrell 1997) their local metallicity distribution has given an more accurate history to follow. Excluding halo and thick disc stars kinematically, a frequency v. metallicity plot for G, or better K dwarfs show low numbers for low disc metallicity ([Fe/H]< -1) rising sharply to a narrow peak between [Fe/H]=-0.6 and -0.2, then falling off to higher metallicities. The earlier G dwarf statistics (Carney et al. 1990; Rocha–Pinto & Maciel 1996) are well explained in a scenario with near constant infall of low metallicity gas to the Galactic plane (Casuso & Beckman 1997). The K dwarfs show the same metallicity pattern (Flynn & Morrell 1997; Favata et al. 1997; Kotoneva et al. 2002) and we will discuss this and how it is modelled in Sect. 4 below. Fig. 1 shows an early stellar metallicity–frequency plot, taking in thin disc, thick disc and halo G–dwarfs from Casuso & Beckman (1997) (hereinafter CB97).

Further chemical evidence explained well by low metallicity infall is the evolution of beryllium and boron, measured v. iron as standard in local disc stars. A recent compilation of observational data on this is shown in Fig. 2 (left and right). The "loop–back" seen for disc stars in both plots is well explained by infall, and best explained if the infall rises slowly with time. Recent models for Be and B evolution in the Galaxy (Fields et al 2000; Ramaty et al. 2000) have focused on their evolution in the halo, i.e. for [Fe/H]< -1,

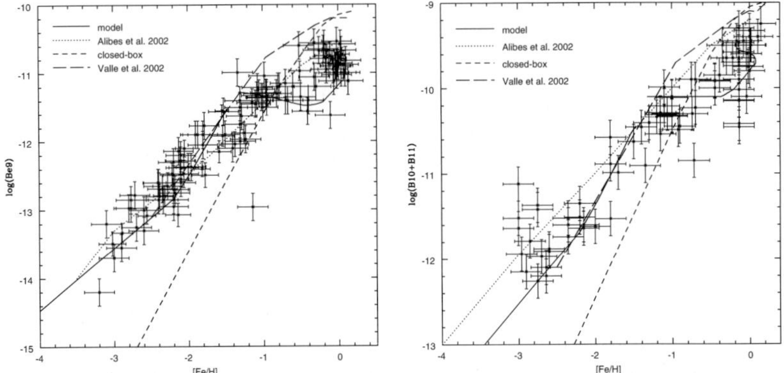

Figure 2. **(left)** Compilation of Be abundances measured in local disc and halo stars (taken from CB04). Model predictions shown are for a closed box model, for two models with infalling H I but at rates which decline with time (Alibés et al. 2002; Valle et al. 2002) and one model, for the disc, with slowly rising infall: CB97, solid line. For more details see text. **(right)** Compilation of B abundances measured in local disc and halo stars, taken from CB04. Model predictions are from the same sources as in Fig. 2 (left).

invoking specific mechanisms for their production by high cosmic ray fluxes in star forming zones to explain the observed linear relations of both Be and B with Fe. But their treatment of the disc evolution of Be and B lacks finesse, notably in the chemical evolution model used, mainly because the authors are not really aiming to explain the disc observations, even though over 90% of Be and B were made in the disc. Alibés et al. (2002) use a much better disc model, but even they do not predict very well the observations in Fig. 2. In CB97 we had managed to model earlier versions of these data sets. To our surprise, even concern, the model entailed H I infall with a rising rate over the disc lifetime. In 1997 the B observations were sparse but as shown in Fig. 2 the trends for B and Be v. Fe have been strongly confirmed as data has accumulated. It was these B and Be observations which first pushed us towards an increasing inflow rate scenario, but other types of evidence have accumulated to support this as we now describe.

3. The history of the local star formation rate (SFR)

Since Vaughan & Preston (1980) brought out the possible use of chromospheric activity in late type dwarfs as a chronometer, a number of surveys within 1 kpc of the Sun have used activity indices to investigate the time dependence of the local SFR. Barry (1988) found using this technique that the SFR, averaged in bins of 10^9yr showed evidence of a secular increase during the disc lifetime, but also large amplitude excursions in this general trend. From a much wider stellar sample Rocha–Pinto et al. (2000) confirmed both

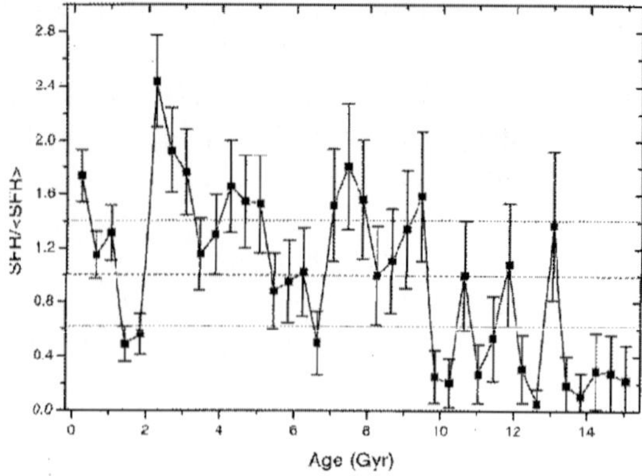

Figure 3. Star formation rate as a function of epoch, from Rocha–Pinto et al. (2000) based on stellar activity, and from observations of stars within 1 kpc of the Sun. Although absolute ages are progressively less accurate as they increase, the general run of SFR has clearly increased with increasing time, on which major shorter term variations are superposed.

findings, as we show here in Fig. 3. Although both samples were taken locally to the Sun, Wielen (1977, 1996) showed that the diffusion of stellar orbits implies that such a sample must include stars which have drifted radially from their birthplaces by over 2 kpc, both inwards and outwards. Thus any large excursion in the SFR must have occurred simultaneously over a major fraction of the Galactic disc. Interaction could cause this, but not only with another galaxy; accretion of a large gas cloud will also work. The data in Fig. 3 are well interpreted in a scenario of H I infall of clouds with a range of sizes. We can estimate a time scale for single cloud accretion, rough as it depends on distance estimates, using measured properties of extragalactic HVC's. From Blitz et al. (1999) or Putman et al. (2002) we set a lowish scale size for a large cloud at $\sim$5 kpc. Measured radial velocity of $\sim$200 km s^{-1} then gives an accretion time of a few times 10^8 yr, which is consistent with the widths of the surge peaks in Fig. 3. Also from Fig. 3 we can see a long term slowly rising trend, which is clearly worth further comment.

Is this apparently rising SFR compatible with cosmological trends? Since the Madau et al. (1996) plot it is known that the global SFR in the universe has been falling steadily from $z = 1$ to the present epoch. We must note though that the Local Group is gravitationally bound; interaction rates between galaxies have been steady since the group condensed. The HVC accretion rate to the Milky Way must depend on the density of the clouds within the Local Group, the gravitational range of the Galaxy, and its accretion cross–section. While accretion causes the first parameter to fall with time, it causes the second and third to rise as gas accumulates in the plane. In López–Corredoira et al. (1999) we computed that for the current accretion rate to be rising (or constant in the limiting case) the present mean cloud density must have a lower limit, which can be converted to an estimate of the fractional mass of the Local Group in the form of these intracluster clouds. The value is ~50%, which is fully plausible, comparing the masses of its galaxies (the Milky Way and M31 as the smaller objects contribute little) with dynamical estimates of the Local Group mass, as first done by Kahn & Woltjer (1959); for a recent estimate see Whiting (1999). (Note that these conclusions depend on only one assumption about dark matter: that the baryon to dark matter density does not vary much from object to object, which we cannot discuss in detail here, but we take as reasonable). We have also shown, in López–Corredoira et al. (2002) that if the net long term vector of this inflow is directed to the Local Group barycentre (Blitz et al. 1999) its dynamic parameters are those required to yield the known Milky Way H I warp amplitude and direction. A global check on the inflow rate can come from measurements of the SFR. For example SN II rates, of order a few per century in the disc, (e.g. Dragicevich et al. 1999) agree with an accretion rate of a few $M_\odot$ yr^{-1}, using a Salpeter IMF with mass limits 0.1$M_\odot$ and 100$M_\odot$, and that type II SNe come from stars with M>8$M_\odot$. Chemical evolution models require mean accretion rates during the disc lifetime of ~2$M_\odot$ yr^{-1}, consistent with estimates of HVC accretion rates (Blitz et al. 1999; Braun et al. 2002).

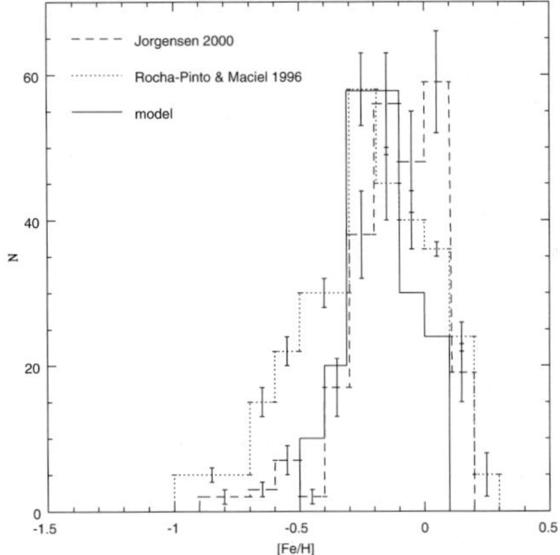

Figure 4. Frequency histograms of metallicity for G dwarfs in the local Galactic disc, from Rocha–Pinto & Maciel (1996), and from Jorgensen (2001). In the latter a scale height criterion has been used to minimize the thick disc contribution. Compared with these data is a model based on chemical evolution with increasing H I infall from CB04.

4. The K-dwarf problem

The dearth of low metallicity K dwarfs in the Galactic disc is a more severe test of chemical evolution models than previous similar data for G dwarfs. In a recent article (Casuso & Beckman 2004) we describe the observational and theoretical developments in this field and this section gives an overview of these. The G–dwarf data by Rocha–Pinto & Maciel (1996) already showed a much narrower peak in the metallicity frequency distribution than that in Fig. 1. Using a chemical evolution model (CB97) developed to explain the Be and B data, in Casuso & Beckman (2001) (CB01) we modelled this narrow peak well. The model included a rising infall rate, and predicted even better the G–dwarf data of Jorgensen (2001), published while CB01 was submitted, so this was a genuine prediction and not a post–diction as one so ofter finds. It is significant that we selected the best model for its Be, B predictions, although improvements were possible using only the Rocha–Pinto & Maciel (1996) data, so the improved agreement with Jorgensen was remarkable. One feature of Jorgensen's data was the careful exclusion of thick disc stars from the sample. In Fig. 4 we compare the two data sets with the model from CB01; these fits,especially to Jorgensen's data, strengthened our "increasing infall rate" hypothesis.The next step was to use K-dwarf metallicity distributions. In Fig.5 we show two data sets, by Favata et al. (1997) and Kotoneva et al. (2002), compared with our model prediction.We can see that agreement is good in the metallicity range close to solar, but less good outside this range. Of the two data sets,one lies above and the other below our prediction, at lower metallicities. The cause in both cases is the thick disc population. Favata et al. (1997) cut off their graph where the thick disc numbers become appreciable, while Kotoneva et al. (2002) did not take special steps to exclude thick disc stars. We conclude that the K–dwarf metallicity distribution in the thin disc gives some support to our model, but further measurement and analysis are needed here.

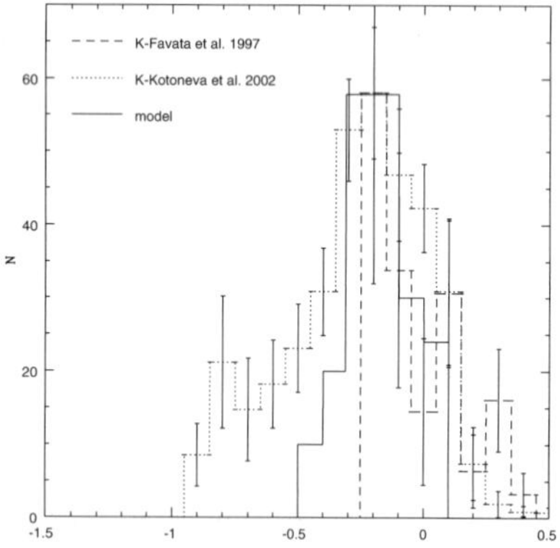

Figure 5. Frequency histograms of metallicity for K dwarfs in the local Galactic disc, from Favata et al. (1997) and Kotoneva et al. (2002). Compared with these data we show a model based on chemical evolution with increasing H I infall from CB04. Agreement is not as good as for the G–dwarfs in Fig. 4, but no attempt to limit the sample to thin disc stars was made in either observational paper, while the model is strictly for thin disc evolution.

5. The light element abundances in the Galactic Disc

The light elements: hydrogen, helium, lithium, beryllium and boron, are exceptional in that their main production sites are principally non-stellar. Hydrogen (H, D and T) most present day helium (^{4}He and ^{3}He), and some 10% of lithium are primordial, while the remaining lithium, beryllium and boron were formed in the interstellar medium (ISM) in processes involving Galactic Cosmic Rays (GCR). Reeves (1974) reviewed light element nucleosynthesis setting the scene for all subsequent developments. For the present article the importance of the light elements is that they present sensitive tests of abundance dilution by infall of element–poor gas into the ISM where they are being produced. We have presented the Be and B v. Fe data plots in Fig. 2, and in Fig. 6 we show a similar compilation for Li. For Be and B there has been much interest in why they show linear dependences on metallicity; they are spallation products of CNO so initial predictions were for a quadratic dependence, as secondary elements. Without entering in depth in this complex field, it is clear from Fig. 2 that this linear relation holds only in the halo metallicity range ([Fe/H]$<$-1.5) while for disc metallicities the dependence is far from linear. Clearly halo and disc behaviour are different, and this can be explained independently of the specific GCR production mechanisms for Be and B by differences in the surrounding gas flow conditions. For the halo stars, not the subject of this article, a model in which the star forming gas is flowing inwards to the Galactic centre as the initial spherical collapse occurs, gives a satisfactory explanation of the pseudo linear Be and B v. Fe observations. For the disc stars we show four model comparisons, a closed box model and three inflow models. The best fits are for secularly slowly increasing inflow. The same model, with no fine tuning, was used to fit the G- and K–dwarf distributions shown in Figs. 4 and 5, and is used as a framework for predicting the Li v. Fe plot in Fig. 6, though here the importance of both production and destruction mechanisms is such that the goodness of fit is not a simple test of the gas flow model (see Casuso & Beckman 2000 for more details). Finally we can return to the case of deuterium which we discussed briefly above. As deuterium is destroyed in stars, there are no equivalent plots to those in Fig. 2 for this nuclide. However the mere presence of D in the ISM of the Galaxy today, especially near the Galactic centre, implies very recent infall, and we discussed implications of this in Casuso & Beckman (1999). To summarize this section, the best fits to the observed evolutionary plots of light element abundances in the Galactic disc are provided by models in which the infall of H I has occurred at a constant or slowly rising rate.

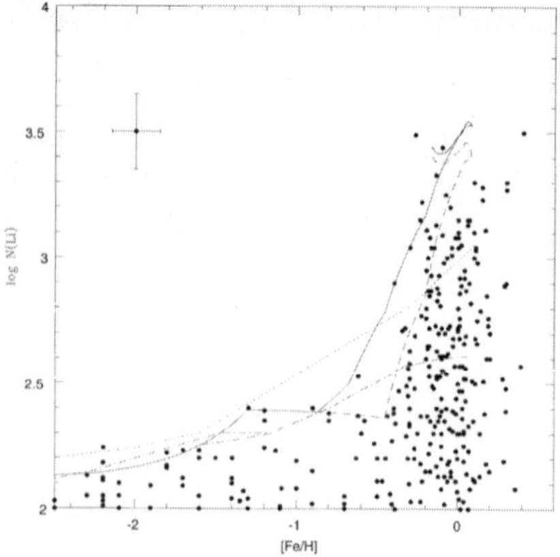

Figure 6. Lithium abundance as a function of metallicity ([Fe/H]) for local disc stars; compilation from Casuso & Beckman (2000). The upper envelope represents minimum Li depletion in the star observed, and is the datum for model fitting. Compared are a number of models, also from Casuso & Beckman (2000). The complexity of the production and destruction processes implies that model fits do not tell us too much about the time dependence of gaseous infall for Li, but all models showing fair fits to the upper envelope use constant or increasing infall assumptions.

6. Gas accretion in the general process of galaxy evolution

Far less attention has been paid to gas accretion as a driver of galaxy evolution than to mergers, because the latter are more spectacular and much easier to observe. The lower end of the galaxy mass function should contain clouds whose baryonic mass could be as high as $10^8 M_\odot$, and would range downwards from there. Their numbers should be high, conforming to general laws of mass fractionation, which Elmegreen (2002) has characterized as fractal in his discussion of Galactic gas clouds. Estimates at the low end of the galaxy mass function by Bell et al. (2003) confirm that the number of objects is rising as the gas to stars ratio is rising, for the faintest dwarf galaxies, and should keep rising for "failed galaxies" which are the gas clouds we "need". Rosenberg & Schneider (2002) find a slope of -1.5 for the H I mass function for galaxies. λCDM model cosmologies (see Kauffmann, White & Guiderdoni 1993, for an early model including λ) do predict a steep mass function at the low mass end, though they also overpredict satellite galaxy numbers, which is a problem. So we would expect galaxy groups and clusters to contain clouds in the mass range proposed here, but most such objects lie below present H I detectability limits ($\sim 10^7 M_\odot$) and can be picked up only in nearby galaxy groups.

One result of considerable relevance is that by Kennicutt, Tamblyn & Congdon (1994). In a survey they found that early type spirals have much lower current SFRs than late types (as expected) but a much higher fractional variability in the SFR. Tomita et al. (1996) inferred that their results imply SFR variability on timescales of 10^8 years, and Hirashita & Kamaya (2000) modelled this using a scenario with quasi–periodic limit cycles. One problem with this mechanism is to explain how SFR surges can occur over a whole galaxy. We have shown how the SFR excursions in Rocha–Pinto et al. (2000) imply Galaxy–wide changes, which are consistent with triggering by external H I clouds if these are large enough to affect a large part of the Galaxy: sizes of several kiloparsecs for the "external" HVC's were estimated by Blitz et al. (1999) and Putman et. al (2002). The variability difference between early and late type galaxies noted above is because a gas rich disc presents a larger cross–section for cloud accretion, and a higher probability for a star formation surge when a cloud is captured. Late types respond only to the most massive incoming clouds, and do so in a less dramatic way.

The current gas accretion rate of $2M_\odot$ yr^{-1} cited in Sect. 3, assuming constant infall, would imply $2\times10^{10}M_\odot$ of gas accreted in the disc lifetime; an exponentially rising rate starting from zero would give $1.4\times10^{10}M_\odot$. Using a dark to baryonic matter ratio of 5 the total mass accreted would be between 7 and 10 times $10^{10}M_\odot$. These total masses will be upper limits as the dark matter component of any HVC with incoming velocity higher than escape velocity will not be retained (this would give a natural explanation for possible low dark matter: baryonic matter ratios in massive late type spirals). These predictions are just compatible with observationally based estimates. Nikiforov et al. (2000) estimate a Galaxy disc mass of $6\times10^{10}M_\odot$ within a 20.5 kpc radius, in a total Galaxy mass of $3.3\times10^{11}M_\odot$. Even if these figures are rough, we find that a major fraction of the disc has been accreted by gas infall. This is qualitatively compatible with "inside–out" evolution of the disc, in which growing gas column density as a function of time and radius have produced increased SFRs at smaller radii, which in turn yields a radial metallicity gradient in the disc. Inside–out disc formation occurs in many λCDM based scenarios (e.g. Samland & Gerhard 2003). In any model where the disc is built up by gas accretion the column density threshold for quiescent star formation will give higher SFR's in the inner disc, although diffusion, especially if impelled by a bar, will tend to blur out the radial distribution of stellar age and metallicity. As it is extremely hard, as yet to estimate stellar population age distributions from outside a galaxy, the local measurements we have used to predict the time dependence of infall must be used as the best benchmark for some time to come.

7. And now for something completely different: how gas flows affect star formation in a barred galaxy

7.1 Non–circular velocities around the bar

So far we have been discussing accretion inflow perpendicular to the planes of galaxies. In this shorter final section we will see how gas flows within the

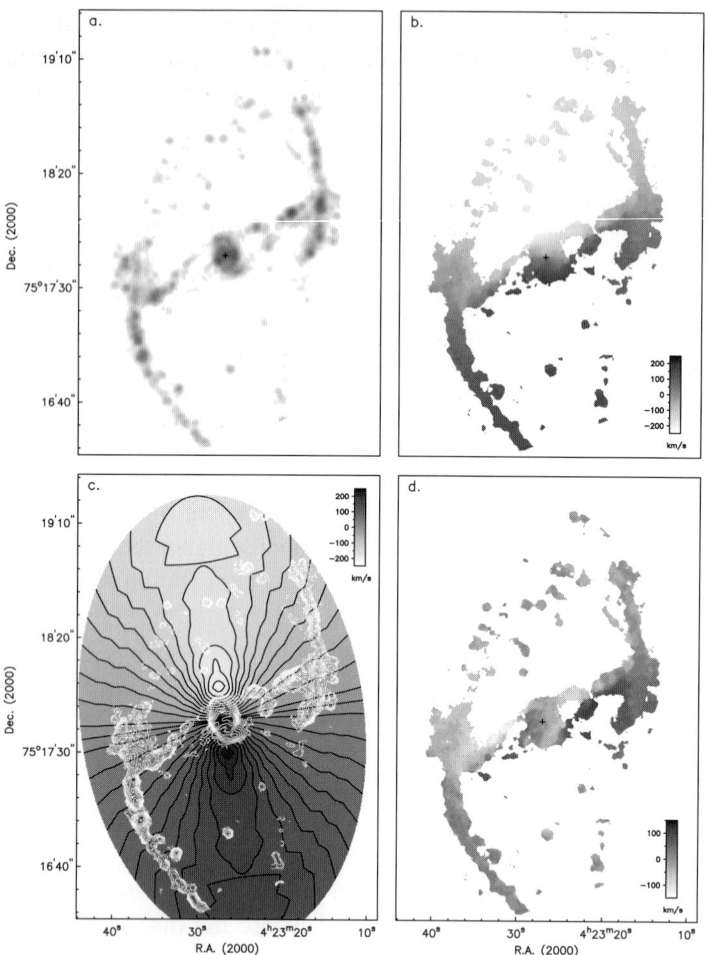

Figure 7. **(a)** Surface brightness map in Hα of NGC 1530, from a TAURUS data cube from the 4.2m WHT La Palma. **(b)** Radial velocity map of ionized gas from the peaks of the Hα emission lines across the face of the galaxy, using the same data cube. **(c)** Contours of Hα surface brightness superposed on a two–dimensional projection of the rotation curve derived from the velocity field shown in (b). **(d)** Map of the residual, non–circular velocity field, obtained by subtracting off the projected rotation curve in (c) from the complete velocity field in (b). The strong non–circular velocity field aligned with the bar is clearly seen here, as the galaxy inclination causes the flow along one side to be directed towards us, and away from us along the other side of the bar (see text for more details, also see Zurita et al. 2004).

plane affect the local SFR on short timescales. For this we use our recently measured velocity field of the strongly barred galaxy NGC 1530. In Zurita et al. (2004) we describe the TAURUS Fabry–Pérot interferometer on the 4.2m

WHT, La Palma (Spain) with which we made the two dimensional velocity map of the ionized hydrogen using Hα emission, and give details of the observations and reduction. The map has a velocity interval of $\sim$18.6 km s^{-1} and an angular resolution of $\sim$1$''$, which mark improvements on previous optical and radio maps by Regan et al. (1996), by Downes et al. (1996) and by Reynaud & Downes (1997, 1998, 1999). We derived first the classical rotation curve, and used its two dimensional projection to subtract off from the observed velocity field, yielding a residual field of non–circular projected velocity over the full face of the galaxy. In Fig. 7 we show how this was done and illustrate our results. NGC 1530 is an excellent subject for kinematic analysis. Its arms, bar and circumnuclear disc are very well separated; the major axis and bar position angle are well separated, and the inclination permits good velocity and intensity measurement. In Fig. 7c, streaming in the arms due to the disc density wave system shows up as ripples in the two dimensional velocity field, with amplitude 20-30 km s^{-1}, which we will not analyze further here. In Fig. 7b the non–circular motions round the bar show up clearly as distorted isovels. These motions are well seen in Fig. 7d: the residual non–circular velocity field, after subtracting off the rotational field seen in Fig. 7c. Cross–sections through this field are shown in Fig. 8, revealing counterflow with amplitude $\gtrsim \pm 100$ km s^{-1} top either side of the bar.

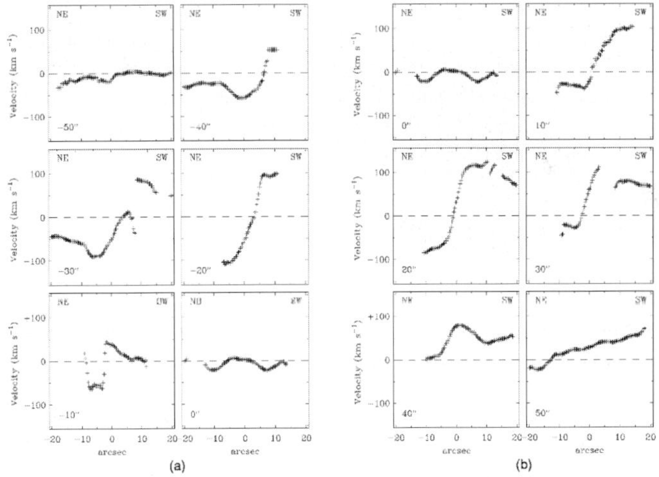

Figure 8. Cross–sections perpendicular to the bar of the non–circular velocity field illustrated in Fig. 7d, showing the amplitude of the flows, >100 km s^{-1} in opposite directions at either side of the bar.

Although Hα gives patchy velocity fields compared with H I, we have high spatial resolution to compensate. This allows us to produce maps of non–circular velocity gradient across the whole galaxy from a few hours' observa-

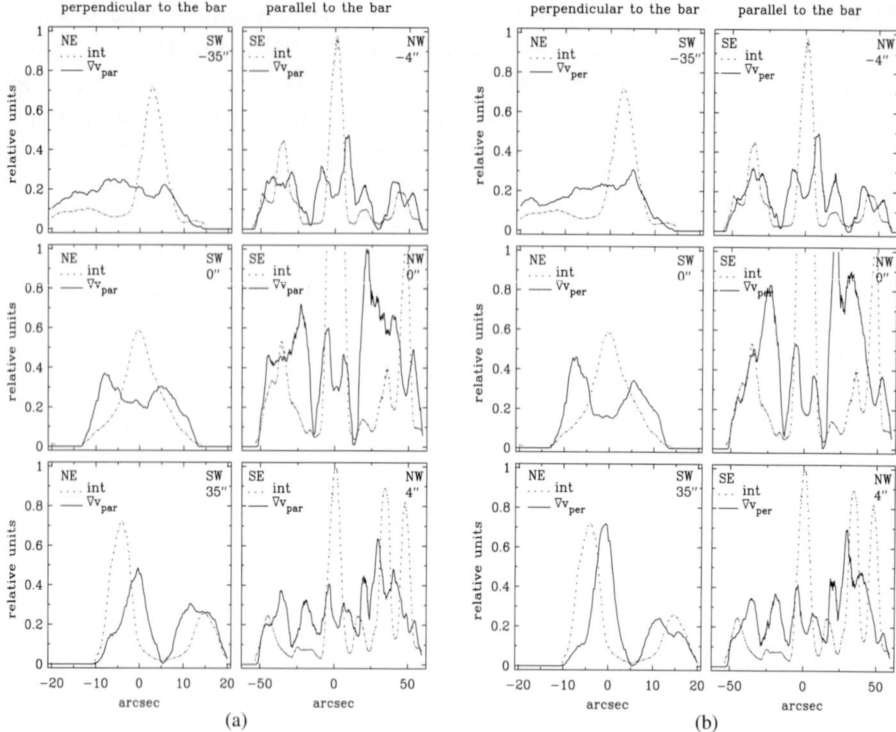

Figure 9. **(a)** Cross–sections, in the directions indicated, of the surface brightness in Hα and of the gradient in velocity parallel to the bar. **(b)** Cross–sections in the directions indicated, of the surface brightness in Hα and of the gradient in velocity perpendicular to the bar. In both graphs we clearly see local maxima in brightness (SFR) at local minima in velocity gradient. For further explanation see text, also Zurita et al. (2004).

tions. In a poster complementary to this article (Zurita et al. 2004a) we show a beautiful result of this work: the complete coincidence of the strong dust lanes along the bar with the loci of maximum velocity gradient perpendicular to the bar. This shows that lines of maximum shear, which are also lines of net zero non–circular velocity, are zones to which the dust migrates. We can draw more powerful inferences by superposing cross–sections of our maps in velocity gradient on cross–sections across the surface brightness map (Fig. 7a); the results are shown in Fig. 9. These show clearly an anticorrelation between local SFR (Hα surface brightness) and local velocity gradient,but in two distinct regimes. In zones of high shear (gradient perpendicular to the flow along the bar) star formation is suppressed, a conclusion inferred by Reynaud & Downes (1998) from CO mapping. This is as expected if shear disrupts large molecular clouds on timescales shorter than those for massive star formation (see Kennicutt 1998 for a discussion of this and related points). In zones of high compression star formation is enhanced, but occurs offset by hundreds of parsecs from the zone of maximum present compression. The offset relation; compression–high local SFR is seen on scales of hundreds of parsecs. On smaller scales we see the effects of outflow from luminous H II regions which impinges on the surrounding ISM, yielding "walls" of velocity gradient around these regions. These can be seen in Fig. 1 of Zurita et al. (2004a), and show up here in Fig. 9, where each peak in the Hα surface brightness due to a bright H II region coincides with a dip surrounded by symmetrically offset peaks, in the velocity gradient. These outflows have been reported by Relaño et al. (2003), and have amplitudes of 40–80 km s^{-1} from the centre of the H II region. We will present a quantitative treatment of these outflows and some initial models in Relaño et al. (2004).

7.2 Gas flow spiralling to the nucleus in the inner disc

In the inner 5$''$ of our original rotation curve (Zurita et al. 2004) we found a steep gradient which relaxed to a slightly less steep gradient outside this radius. Assuming this to be caused by a bulge, we used an HST-NICMOS near-IR image to check this, but as seen in Fig. 10a the central spheroidal component is diminutive, and could not cause the steep inner gradient. Assuming that we are detecting projected non–circular motion, we extrapolated the outer part of the rising rotation curve linearly to the origin, and subtracted this from the observed inner velocity field, yielding a map of projected non–circular velocity in the inner disc, as shown in Fig. 10c. This "yin/yang" pattern reveals spiral inflow to the nucleus. To show this we took an image of the zone in J–K from Pérez-Ramírez et al. (2000), see Fig. 10b, which shows interlocking spiral dust lanes. Assuming that these indicate inward flow lines we computed the projected velocity we would see, taking for simplicity a constant velocity amplitude along the flow. The result, in Fig. 10d agrees nicely with the observational field in Fig. 10c, except at the north and south ends of the inner disc, where we know that the Hα data do not allow us to plot the true circular rotation curve, so the residual map in Fig. 10c will have systematic errors here. This inward spiralling flow is not due to an inner bar. In Zurita et al. (2004, 2004a) we show using unsharp masking of the HST image that any circumnuclear bar must be less than 0.5$''$ in length, and with negligible dynamical effect over the inner disc. However, as modelled by Englmeier & Shlosman (2000) in a galaxy with an inner Lindblad resonance at a radius within the length of a major bar, as the dominant stellar x_1 orbits in the bar give way to perpendicular x_2 orbits within the ILR the gas is pulled round into opposing spiral trajectories, very much as seen in Fig. 10, which take it down to within a couple of hundred parsecs of the nucleus. Fabry–Pérot velocity fields from ionized gas emission give us an excellent tool for exploring these dynamical effects.

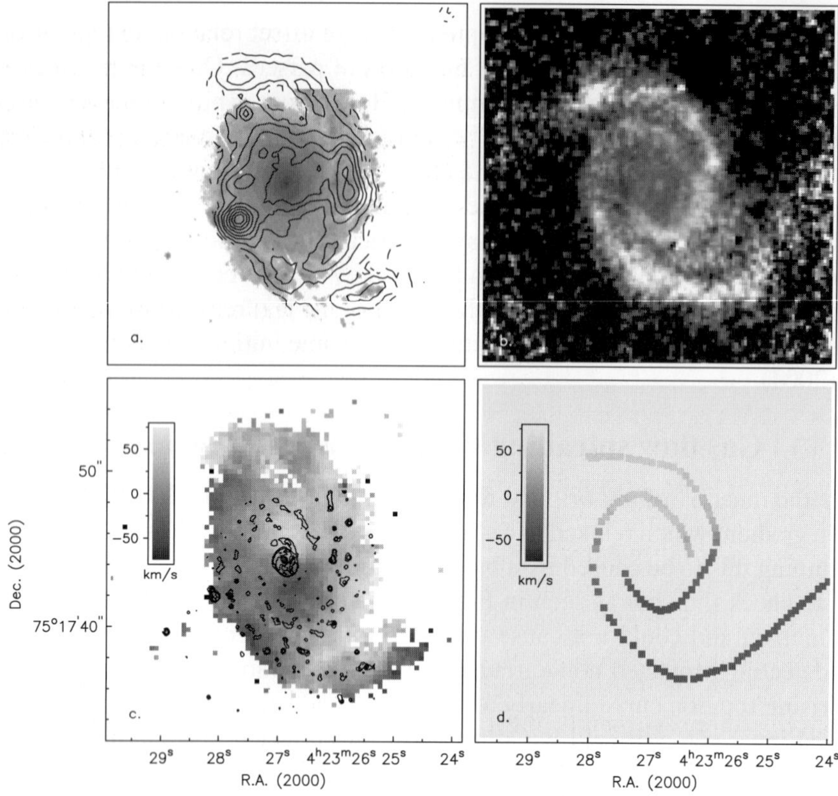

Figure 10. **(a)** HST–NICMOS H band image of the central disc of NGC 1530, with Hα contours superposed. **(b)** J–K image of the zone, from Pérez–Ramírez et al. (2000), clearly showing the interlocking spiral dust lanes around the nucleus. **(c)** Residual non–circular velocity map, projected along the line of sight, for the circumnuclear disc of NGC 1530, obtained from Fabry–Pérot observations. **(d)** Model predicted velocity field, based on constant amplitude flow vectors directed along the paths defined by the dust lanes in (b). The predictions match very well the observed pattern, notably within 5$''$ of nucleus, and show the power of the Fabry–Pérot method for velocity field diagnostics.

8. Some general conclusions

The message of this article is the same for the two different astrophysical systems considered: galaxy evolution cannot be well understood without a knowledge of gas flows and their effects on star formation. For the Milky Way our reasoning has been indirect. Chemical evidence is used to show that gas has been accreting to the disc during its whole lifetime, and at rates which do not appear to have declined during this period. This does not contradict the general picture in which the SFR in the universe has been declining steadily from $z = 1$ to the present epoch, as the overall density has fallen. We note that the ongoing global disc SFR is of order 2M$_\odot$ yr^{-1}, while during the formation of the initial spherical component (stellar halo) the rate must have been much higher. This means that a constant, even slowly increasing SFR in the discs of galaxies and small groups, regulated by inflow and feedback processes, need not contradict inferences about the universal SFR from the Madau et al. (1996) plot and its subsequent refinements. One feature of star formation in galactic discs is its apparently low efficiency averaged over time. In a model where the disc mass has grown slowly via accretion, this low efficiency is only apparent, an effect of time averaging. Any gas which arrives as infalling H I suffers relatively rapid conversion to stars, subject to feedback from the massive stars themselves. This scenario of gas persistently raining down onto galaxy discs, with the occasional sharper shower, a more massive H I cloud, is finding support as H I detection sensitivities increase, and more H I clouds falling in to other galaxies are detected. And infall can explain rather well what other mechanisms (major interactions apart) cannot: how the star formation rate can increase and decrease simultaneously over the whole, or the major part of a Milky Way sized disc galaxy. We have considered here the evidence that this does happen, both for the Milky Way and for galaxies in general.

In the second part of the paper we have shown how gas flow is a key parameter determining the star forming pattern in NGC 1530. While shocks, strong velocity gradients along the line of flow, yield massive stars, shear, strong velocity gradients perpendicular to the flow, inhibit massive star production. NGC 1530 is a specially favourable case for study, for the essentially geometrical and morphological reasons explained in Sect. 7, but we should treat it here as an example revealing processes of general importance. Here we have been able to present a semi–quantitative picture, but to disentangle fully the effects produced by gas flow on star formation we will need, as observers, to combine similar information from ionized, neutral (atomic) and molecular gas. Only then will we have the basis for full quantitative understanding, aimed at the major prize: a physical theory of star formation in different galactic environments.

The research discussed in this article was supported by grants AYA2001-0435 (Spanish Ministry of Science and Technology) and AYA2004-08251-C02-01 (Spanish Ministry of Education and Science). A. Zurita acknowledges support by the Consejería de Educación y Ciencia de la Junta de Andalucía, Spain. The authors thank the organizers of the conference for the invitation to present this paper in an excellent context, both scientific and social.

References

Abraham, R. G. 1999 in "Galaxy Interactions at Low and High Redshifts, (Eds. J.E. Barnes and D.B. Sanders), Dordrecht, Kluwer, p.11

Alibés, A., Labay, J., & Canal, R. 2002, ApJ, 571, 326

Audouze, J., Lequeux, J., Reeeves, H., & Vigroux, L. 1976, ApJ, 208, L51

Barry, D. C. 1988, ApJ, 324, 436

Bell, E. F., McIntosh, D. H., Katz, N., & Weinberg M. D. 2003, ApJL, 585, 117
Bellazzini, M., Cacciari, C., Federici, L., Fusi Pecci, F., & Rich, M. 2003, A&A , 405, 867
Blitz, L., Spergel, D. N., Teuben, P. J., Hartmann, D., & Burton, W.B. 1999, ApJ, 514, 818
Braun, R., de Heij, V., & Burton, W. B. 2002, BAAS, 200, 3304
Carney, B. W., Latham, D. W., & Laird, J. B. 1990, AJ, 99, 572
Casuso, E., & Beckman, J. E. 1997, ApJ, 475, 155 (CB97)
Casuso, E., & Beckman, J. E. 1999, AJ, 118, 1907
Casuso, E., & Beckman, J. E. 2000, PASP, 112, 942
Casuso, E., & Beckman, J. E. 2001, ApJ, 557, 681 (CB01)
Casuso, E., & Beckman, J. E. 2004, A&A, 419, 181 (CB04).
Downes, D., Reynaud, D., Solomon P. M., & Radford, S.J.E. 1996, ApJ, 461, 186
Dragicevich, P. M., Blair D. G., & Burman, R. R. 1999, MNRAS, 302, 693
Elmegreen, B. G. 2002, ApJ, 564, 773
Englmeier, P., & Shlosman I. 2000, ApJ, 528, 677
Espana, L., & Worthey, G. 2002, AAS, 201, 1408
Favata, F., Micela, G., & Sciortino, S. 1997, A&A, 323, 809
Fields, B. D., Olive, K. A., Vangioni–Flam, E., & Cassé, M. 2000, ApJ, 540, 930
Flynn, C ., & Morell, O. 1997, MNRAS, 286, 617
Fraternali, F., Osterloo, T., & Sancisi, R. 2004, A&A (in press)
Freeman, K. C. 1993 in "Galaxy Evolution: the Milky Way perspective"(Ed. S. R. Majewski) ASP Converence Series, Vol. 49, ASP, San Francisco, 1993
Gibson, B. K., Penton, S. V., Giroux, M. L., Stocke, J. T., Shull, J. M., & Tumlinson, J. 2001, AJ, 122, 3280
Gilmore, G., & Feltzing S. 2000 in "The Evolution of Galaxies on Cosmological Timescales" (Eds. J. E. Beckman & T. J. Mahoney), ASP Conference Series, Vol. 187, ASP, San Francisco, p. 20
Gilmore, G., & Wyse, R. F. G. 1998, AJ, 116, 748
Hirashita, H., & Kamaya, H. 2000, AJ, 120, 728
Hunt, R., & Sciama, D. W. 1972, MNRAS, 157, 335
Jorgensen, B. R. 2001, A&A, 363, 947
Kahn, F. D., & Woltjer, L. 1959, ApJ, 130, 70
Kauffman, G., White, S. D. M., & Guiderdoni, B. 1993, MNRAS, 264, 201
Kennicutt, R. C. 1998, ARA&A, 36, 189
Kennicutt, R. C., Tamblyn, P., & Congdon, C. W. 1994, ApJ, 435, 22
Kirkman, D., Tytler, D., Burles, S., Lubin, D., & O'Meara, J. M. 2000, ApJ, 529, 655
Kotoneva, E., Flynn, C., Chiappini, C., & Matteucci F., 2002, MNRAS, 336, 879
Larson, R.B. 1972a, Nature, 236, 21
Larson, R.B. 1972b, NPhS, 236, 7
Larson, R.B. 1976, MNRAS 176, 31
Lehner, N., Gry, C., Sembach, K., Hébrard, G., Chayer, P., Moos, H. W., Howk, J. C., Desert, J. M. 2002, ApJS, 140, 81
Lepine, J. R. D., & Duvert, G. 1994, A&A, 286, 60
López–Corredoira, M., Beckman, J. E., & Casuso, E. 1999, A&A, 351, 920
López–Corredoira, M., Betancort–Rijo, J., Beckman, J. E. 2002, A&A, 386, 169
Lubowich, D. A., Pasachoff, J. M., Balonek, T. J., Millar, T. J., Tremonti, C., Roberts, H., & Galloway, R. P. 2000, Nature, 405, 1025
Lynden–Bell, D. 1976, MNRAS, 174, 695
Madau, P., Ferguson, H. C., Dickinson, M. E., Giavalisco, M., Steidel, C. C., Fruchter, A. 1996, MNRAS, 283, 1388
Mirabel, F. 1982, ApJ, 256, 112

Muller, C. A., Oort ,J. H., & Raimond, E. 1963, Comptes Rendus Acad. Sci. Paris 257, 1661
Navarro, J. F., Frenk, C. S, & White, S. 1994, MNRAS, 267, 1
Navarro, J. F., Frenk, C. S, & White, S. 1995, MNRAS, 275, 56
Nikiforov, I. I., Petrovsky, I. V., & Ninkova, S. 2000, in "Small Galaxy Groups" ASP conference series 209, (Eds. M. Valtonen & C. Flynn), p. 399
Oliveira, C. M., Hébrard, G., Howk, J. C., Kruk, J. W., Chayer, P., & Moos, H. W. 2003, ApJ, 587, 235
Oort, J. H. 1970, A&A, 7, 181
Pagel, B. E. J. 1987 in "The Galaxy" (Eds. G. Gilmore & R. Carswell) D.Reidel, p. 341
Pagel, B. E. J., Patchett, B. E. 1975, MNRAS, 172, 13
Pérez–Ramírez, D., Knapen, J. H., Peletier, R. F. et al. 2000, MNRAS, 317, 234
Pettini, M., & Bowen, D. V. 2001, ApJ, 560, 41
Phookun, B., Vogel, S. N., & Mundy, L. G. 1993, ApJ, 418, 113
Power, C., Navarro, J. F., Jenkins, A., Frenk, C. S., White, S., Springel, V., Stadel, J., & Quinn, T. 2003, MNRAS, 338, 14
Putman, M. E. et al. 2002, AJ, 123, 873
Ramaty, R., Scully, S. T., Lingenfelter, R. E., & Kozlovsky, B., 2000, ApJ, 534, 747
Reeves H. 1974, ARA&A, 12, 437
Regan, M., Teuben, P., Vogel, S., & van der Hulst, T. 1996, AJ, 112, 2549
Relaño, M., Beckman, J. E., Zurita, A., & Rozas, M. 2003, Rev. Mex.Astr Astrofis., 15, 205
Relaño, M., Beckman, J. E., Zurita, A., & Rozas, M. 2004, A&A, (submitted)
Reynaud, D., & Downes D. 1997, A&A, 319, 737
Reynaud, D., & Downes D. 1998, A&A, 337, 671
Reynaud, D., & Downes D. 1999, A&A, 347, 37
Rocha–Pinto, H. J., &Maciel, W. J. 1996, MNRAS, 279, 447
Rocha–Pinto, H. J., Scalo, J., Maciel, W. J., & Flynn, C. 2000, A&A, 358, 869
Rosenberg, J. L., & Schneider, S. E. 2002, ApJ, 567, 247
Samland, M., & Gerhard, O.E. 2003, A&A, 399, 961
Savage, B. D. et al. 2000, ApJS, 129, 563
Savage, B. D. et al. 2003, ApJS, 146, 165
Schmidt, M. 1963, ApJ, 137, 758
Sembach, K. R. et al. 2004, ApJS, 150, 387
Tinsley, B. 1977, ApJ, 216, 548
Tinsley, B. 1980, Fund. Cosm. Phys., 5, 287
Tomita, A., Tomita, Y., & Saito, M. 1996, PASJ, 48, 285
Toomre, A. 1977 in "The Evolution of Galaxies and Stellar Populations" (Eds. B. Tinsley & R. B. Larson), New Haven, Yale U. Obs., p. 401
Toomre, A., & Toomre, J. 1972, ApJ, 178, 623
Valle, G., Ferrini, F., Galli, D., & Shore, S. N. 2002, ApJ, 566, 252
van den Bergh, S. 1962, AJ, 67, 486
van der Hulst, T. J. M., & Sancisi, R. 2003, in "Recycling Intergalactic and Interstellar Matter", IAU Symposium 217, p. 140
Vaughan, A., & Preston, G. W. 1980, PASP, 92, 385
Veeraraghavan, S. & White, S. D. M. 1985, ApJ, 296, 376
Wakker, B. P. et al. 2003, ApJS, 140, 91
Whiting, A. B. 1999, Proc. IAU Symposium 192 "The Stellar Content of the Local Group of Galaxies" (Eds. P. Whitelock & R. Cannon), ASP, p. 420
Wielen, R. F. 1977, A&A, 60, 263
Wielen, R. F., Fuchs, B., & Dettbarn, C. 1996, A&A, 314, 438
Worthey, G., Dorman, B., & Jones, L. A. 1996, AJ, 112, 948

Zepf, S. E., & Koo, D. C. 1989, ApJ, 337, 34
Zurita, A., Relaño, M., Beckman, J. E., & Knapen, J. H. 2004, A&A, 413, 73
Zurita, A., Relaño, M., Beckman, J. E., & Knapen, J. H. 2004a (these proceedings)

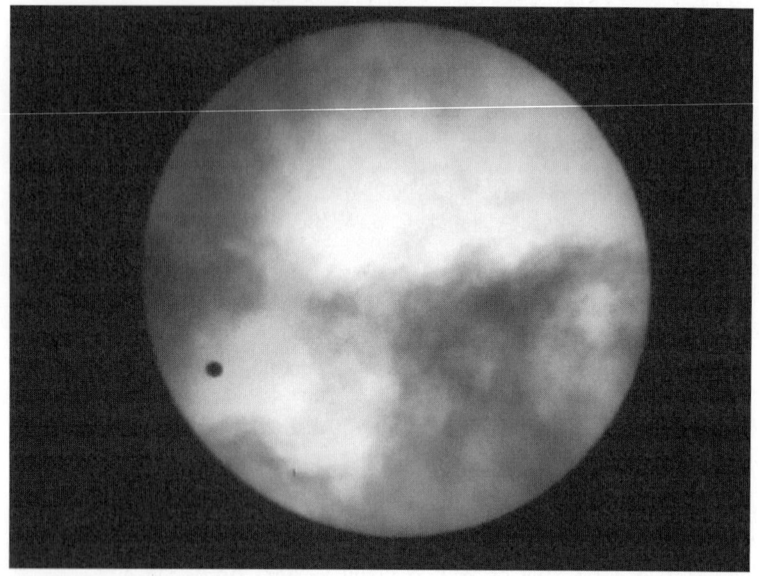

BAR-DRIVEN EVOLUTION AND 2D SPECTROSCOPY OF BULGES

M. Bureau[1], E. Athanassoula[2], A. Chung[3] and G. Aronica[4]

[1]*Hubble Fellow, Columbia Astrophysics Laboratory, 550 West 120th Street, 1027 Pupin Hall, MC 5247, New York, NY 10027, U.S.A.,*

[2]*Observatoire de Marseille, 2 place Le Verrier, F-13248 Marseille Cedex 4, France,*

[3]*Department of Astronomy, Columbia University, 550 West 120th Street, 1411 Pupin Hall, MC 5246, New York, NY 10027, U.S.A.,*

[4]*Astronomisches Institut, Ruhr-Universität Bochum, D-44780 Bochum, Germany*

Abstract A multi-faceted approach is described to constrain the importance of bar-driven evolution in disk galaxies, with a special emphasis on bulge formation. N-body simulations of bars are compared to the stellar kinematics and near-infrared morphology of 30 edge-on spirals, most with a boxy bulge. The N-body simulations allow to construct stellar kinematic bar diagnostics for edge-on systems and to quantify the expected vertical structure of bars. Long-slit spectra of the sample galaxies show characteristic double-hump rotation curves, dispersion profiles with secondary peaks and/or flat maxima, and correlated h_3 and V profiles, indicating that most of them indeed harbor edge-on bars. The stellar kinematics also suggests the presence of cold, quasi-axisymmetric central stellar disks. The ionized-gas distribution and kinematics further suggests that those disks formed through bar-driven gaseous inflow and subsequent star formation, which are absent in our simulations. Minimally affected by dust and dominated by Population II stars, K-band imaging of the same galaxies spectacularly highlights radial variations of the bars' scaleheights, as expected from vertical disk instabilities. The light profiles also vary radially in shape but never approach a classic deVaucouleurs law. Filtering of the images further isolates the specific orbit families at the origin of the boxy structure, which can be directly related to periodic orbit calculations in generic 3D barred potentials. Bars are thus shown to contribute substantially to the formation of both large-scale triaxial bulges and embedded central disks. Relevant results from the SAURON survey of the stellar/ionized-gas kinematics and stellar populations of spheroids are also briefly described. Specific examples supporting the above view are used to illustrate the potential of coupling stellar kinematics and linestrengths (age and metallicity), here specifically to unravel the dynamical evolution and related chemical enrichment history of bars and bulges.

Keywords: galaxies: bulges – galaxies: evolution – galaxies: kinematics and dynamics – galaxies: photometry – galaxies: spiral – galaxies: structure

D. Block et al. (eds.), Penetrating Bars through Masks of Cosmic Dust, 139–148.

1. Introduction

Conventional wisdom states that the bulges of spiral galaxies are analogous to low-luminosity elliptical galaxies residing at the centers of disks (Davies et al. 1983), and thus probably formed through rapid collapse or merging. There is however mounting evidence against significant merger growth in many bulges and numerous studies argue for the importance of slower (i.e. secular) processes (e.g. Kormendy 1993; Andredakis, Peletier, & Balcells 1995). Given the rapid decrease of the merger and star formation rates since $z \approx 1$, secular evolution mechanisms have probably been non-negligible for some time already, and their relative importance will only increase with time. Most works emphasize the potentially crucial role of bars and other asymmetries in disks (e.g. Friedli et al. 1996; Erwin & Sparke 2002). Theoretical models often involve the growth of a central mass through bar-driven inflow and recurring bar destruction and formation (e.g. Pfenniger & Norman 1990; Friedli & Benz 1993). The efficiency of bar dissolution mechanisms remains however uncertain and bar re-formation requires substantial external gas accretion over the lifetime of a galaxy (e.g. Bournaud & Combes 2002; Shen & Sellwood 2004). In this paper, we thus focus on slow processes which can occur in isolation.

Beside bar formation, which is well-studied and documented, N-body simulations of cold disks systematically show that, soon after their formation, bars should thicken and settle with an increased vertical velocity dispersion, appearing boxy or peanut-shaped (B/PS) when viewed edge-on (e.g. Combes & Sanders 1981; Combes et al. 1990). The large but constant fraction of B/PS bulges across the Hubble sequence ($\approx 45\%$) supports both the importance of this mechanism for real galaxies and its association with bars (e.g. Lütticke, Dettmar, & Pohlen 2000). Although bars are not readily identifiable in edge-on systems and other mechanisms can give rise to axisymmetric B/PS bulges (e.g. Binney & Petrou 1985; Rowley 1988), the particular gaseous kinematics of B/PS bulges further supports the view that they are simply thick bars viewed edge-on (e.g. Kuijken & Merrifield 1995; Merrifield & Kuijken 1999; Athanassoula & Bureau 1999; Bureau & Freeman 1999). Considering that, if proven true, this link would argue that at least 50% of all bulges formed through bar-driven processes, it is crucial to extend those tests to earlier type galaxies, where one might naively think that merging is more important.

In this paper, we thus present generic N-body simulations of bar-unstable disks developing a B/PS bulge (§ 2), and positively compare them to the stellar kinematics (§ 3) and K-band morphology (§ 4) of a sample of 30 spiral galaxies, most with a B/PS bulge. We also present SAURON integral-field observations of a few B/PS bulges which show not only the kinematic and structural signatures of bars, but also rather homogeneous stellar populations, as expected

(§ 5). We conclude by supporting the importance of bar-driven evolution (§ 6).

2. N-Body Simulations of Bars

We have run a large number of standard N-body simulations of bar-unstable disks and discuss below a few generic results. Figure 1 contrasts our results for a typical strongly barred case viewed edge-on (at late times) from various viewing angles. The initial conditions consist of a cold luminous exponential disk with constant Q and a live spherical dark halo (see Athanassoula 2003 for more details). No luminous spheroidal component was initially included in the simulations. The prominent thick central component, which would normally be identified with a bulge both morphologically and from the major-axis surface brightness profiles, is thus composed entirely of disk material. It has acquired a large vertical extent through disk vertical instabilities and, as expected, appears round when seen end-on, boxy-shaped at intermediate viewing angles, and peanut-shaped when seen side-on (e.g. Combes et al. 1990).

Based on a large number of similar simulations and Gauss-Hermite fits (van der Marel & Franx 1993), edge-on barred disks typically show the following kinematic signatures along their major-axis (Bureau & Athanassoula 2004): 1) a major-axis light profile with a quasi-exponential central peak and a plateau at moderate radii (Freeman Type II profile); 2) a "double-hump" rotation curve; 3) a rather flat central velocity dispersion peak with a plateau at moderate radii and, occasionally, a local central minimum and secondary peak; 4) an $h_3 - V$ correlation over the projected bar length. h_4 is rather featureless and is in any case hard to measure observationally, so we do not discuss it further. Those kinematic features are all spatially correlated and can be understood from the orbital structure of barred disks. They therefore provide a reliable and practical tool to identify bars in edge-on disks. Contrary to popular belief, so-called "figure-of-eight" position-velocity diagrams (Kuijken & Merrifield 1995; Merrifield & Kuijken 1999) do not occur in the stellar kinematics, as expected for realistic orbital configurations. However, while they are not uniquely related to triaxiality, line-of-sight velocity distributions with a high velocity tail (i.e. an $h_3 - V$ correlation) do appear to be particularly useful tracers of bars. All the characteristic kinematic features identified grow in strength as the bar evolves and do not change significantly for small departures from edge-on. Most can provide useful measurements of the bar length.

3. Stellar Kinematics of Boxy Bulges

As shown in Chung & Bureau (2004), we have obtained similar long-slit stellar kinematics along the major-axis of the 30 edge-on spirals from the sample of Bureau & Freeman (1999), most of which have a B/PS bulge. Comparing

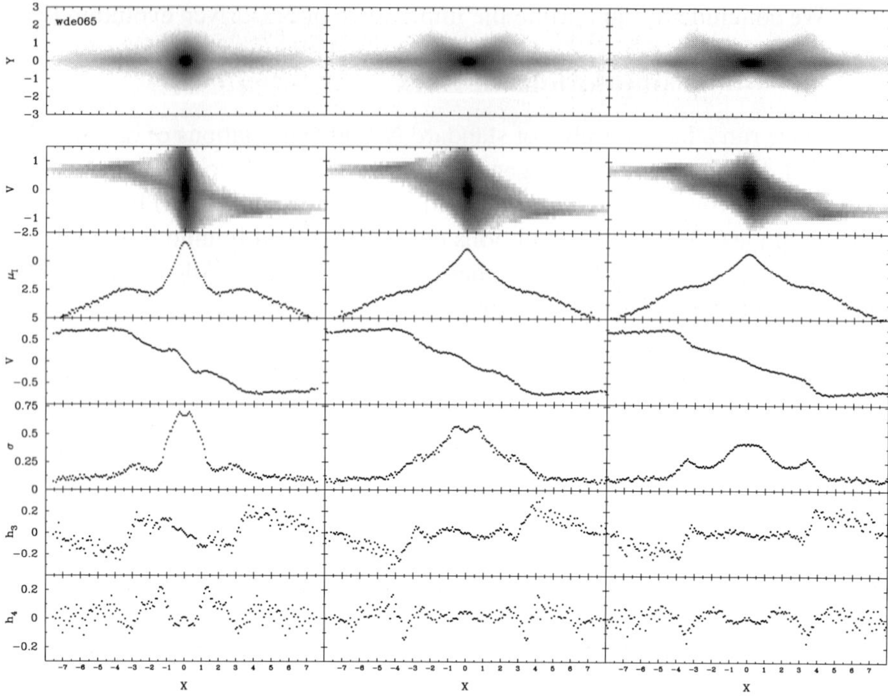

Figure 1. Bar diagnostics as a function of viewing angle for a strong bar. From left to right: simulation seen end-on, at intermediate viewing angle, and side-on. From top to bottom, each panel shows an edge-on view of the simulation, the position-velocity diagram along the major-axis, the major-axis surface brightness profile (μ_I), and the derived Gauss-Hermite coefficients V (mean velocity), σ (velocity dispersion), h_3 (skewness), and h_4 (kurtosis). All grayscales are plotted on a logarithmic scale. Adapted from Bureau & Athanassoula (2004) with permission.

those profiles with the N-body bar diagnostics (§ 2), we find bar signatures in 80% of the galaxies, including the S0s. The diagnostics thus appear robust and support the formation of most B/PS bulges through bar thickening.

As predicted, galaxies with a B/PS bulge frequently show a double-hump rotation curve with an intermediate dip or plateau, they often display a rather flat central velocity dispersion profile with a secondary peak or plateau, and a significant fraction of the objects have a local central σ minimum ($\geq$ 40%). The h_3 profiles display up to three slope reversals and, most importantly, h_3 is normally correlated with V over the presumed bar length, contrary to expectations from an axisymmetric disk. Figure 2 shows the derived stellar kinematics for 4 objects. The characteristic bar signatures strengthen the case for an intimate relationship between B/PS bulges and bars, even for early-type systems, and they leave little room for competing explanations of the bulges' shape (e.g. Binney & Petrou 1985). We also find that h_3 is anti-correlated with V in the

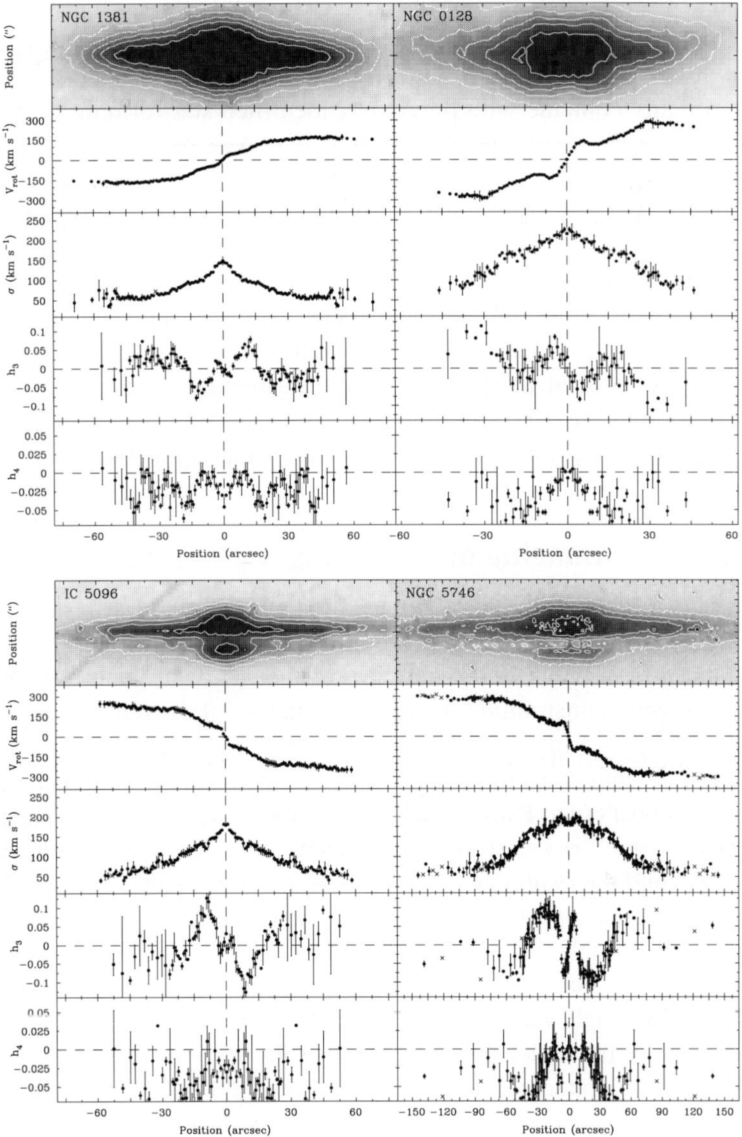

Figure 2. Stellar kinematics of spiral galaxies with a boxy bulge. The galaxies shown are the gas-poor S0 galaxies NGC1381 (top-left) and NGC128 (top-right) and the gas-rich intermediate-type spirals IC5096 (bottom-left) and NGC5746 (bottom-right). From top to bottom, each panel shows an optical image of the galaxy from the Digitized Sky Survey and the registered V, σ, h_3, and h_4 profiles along the major-axis. The measurements were folded about the center and the errors represent half the difference between the approaching and receding sides. Adapted from Chung & Bureau (2004) with permission.

very center of most galaxies ($\geq 60\%$), suggesting that those objects additionally harbor cold and dense (bright) (quasi-)axisymmetric central stellar disks.

Those disks may be related to the steep central light profiles observed (§ 4), and they roughly coincide with previously identified star-forming ionized-gas disks (Bureau & Freeman 1999). They thus may well have formed out of gas accumulated by the bar at its center through inflow. As suggested by N-body models, the skewness of the velocity profile (h_3) appears to be a reliable tracer of asymmetries, allowing to discriminate between axisymmetric and barred disks seen in projection. Based on their kinematics, B/PS bulges (and thus a large fraction of all bulges) appear to be made-up mostly of disk material which has acquired a large vertical extent through bar-driven instabilities, although we have not yet probed the potentials of the galaxies out of the disk plane systematically (but see § 5). Our observations are nevertheless consistent with standard bar-driven evolution models, and the formation of B/PS bulges does appear to be dominated by secular evolution processes rather than merging.

4. *K*-Band Imaging of Boxy Bulges

We have also obtained K-band images for all the galaxies in our sample. The K-band observations penetrate the prominent dust lanes present in many galaxies and offer a much improved view of their structure and morphology, apparently directly constraining the orbital structure of the objects. Indeed, as illustrated in Figure 3 for NGC128, unsharp-masking of those images reveals features entirely analogous to those expected from the orbital families thought to dominate 3D bars (e.g. Patsis, Skokos, & Athanassoula 2002). In particular, the x_1 family "tree" is clearly seen and many galaxies show secondary enhancements along the major-axis (see Skokos, Patsis, & Athanassoula 2002).

Moreover, the key aspect of bar thickening mechanisms to form B/PS bulges is that the disk material is rearranged vertically (as well as radially) by instabilities, rather than new material being added (as expected for accretion scenarios). This process is strongly supported by our observations, as most galaxies show a clear increase of the scaleheight where the B/PS bulge reaches its maximum extent. This is illustrated again in Figure 3 for NGC128, where we fitted the vertical profiles at each (projected) radial position with a generalized Gaussian (equivalent to a Sersic law with $n = \lambda^{-1}$), following Athanassoula & Misiriotis (2002). The shape of the profiles also changes with radius, the profiles being shallower where the peanut shape is maximum. We note that the profiles never approach a deVaucouleurs law ($n = 4$), even in the center, arguing again against violent relaxation and merging. Furthermore, in many galaxies, the steep part of the light profile is much shorter than the vertically-extended component, so that those two definitions of a bulge are clearly inconsistent (as is that of a kinematically hot sub-system, since B/PS bulges are most likely

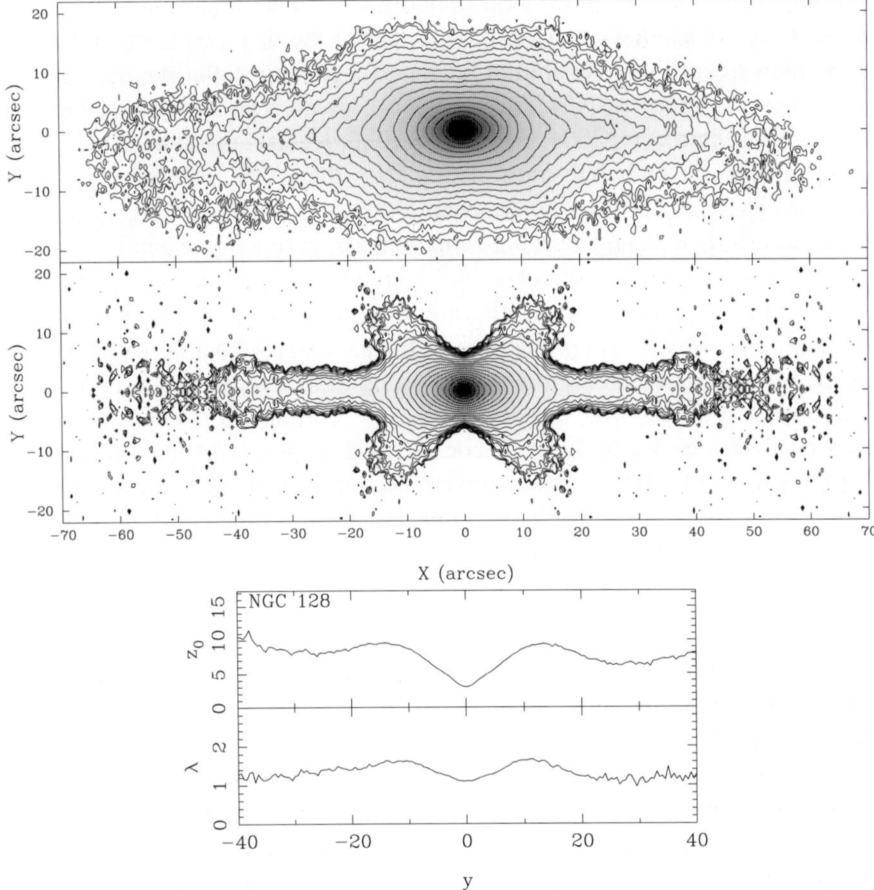

Figure 3. Morphology and structure of spiral galaxies with a boxy bulge. Top: K-band image of the S0 galaxy NGC128, showing the strong peanut-shape of the bulge. The isophotes are separated by 0.5 mag arcsec^{-2}. Middle: Symmetrized unsharp-masked (median-filtered) image of NGC128, revealing the underlying orbital structure. Bottom: Scaleheight (z_0) and shape (λ) of the fitted vertical profiles of NGC128 as a function of projected radius (registered). The radially varying thickening of the disk/bulge material is clearly visible. Adapted from Aronica et al. (2005) and Athanassoula et al. (2005) with permission.

rotationally supported). The steep part of the profiles is thus probably more closely related to the central disks discussed in the previous section.

5. SAURON Observations of Boxy Bulges

The SAURON team has conducted a survey of the 2D stellar kinematics, ionized-gas kinematics, and absorption linestrengths of a representative sample of nearby early-type galaxies (e.g. de Zeeuw et al. 2002). The SAURON data on edge-on early-type spirals thus offer a unique opportunity to extend the kinematic tests described above (§ 2) out of the disk plane and to test the predictions of bar-driven evolution scenarios regarding stellar populations.

The best examples are NGC7332 (Falcón-Barroso et al. 2004) and NGC4526 (Fig. 4; Emsellem et al. 2004; Sarzi et al. 2005; Kuntschner et al. 2005). Both S0 galaxies have a boxy bulge and clearly show the stellar kinematic signatures of a bar. They also possess homogeneous stellar populations (age and metallicity) across the disk and bar/bulge components, except in the central parts, as expected from simple bar-driven evolution models (e.g. Friedli & Benz 1995). NGC4526 also clearly shows a cold central stellar disk embedded in its triaxial bulge, as traced by a strong pinching of the stellar isovelocities in the center, a wide local central σ minimum, and a strong central $h_3 - V$ anti-correlation (while the rest of the bulge displays an $h_3 - V$ correlation as expected from a thick bar). This central stellar disk coexists with an ionized and molecular gas disk (Young et al. 2005), presumably formed through bar-driven inflow (and ensuing star formation). NGC4526 thus beautifully illustrates and confirms most aspect of bar-driven secular evolution models in spiral galaxies.

6. Conclusions

As secular processes for the evolution of galaxies gain in respect and popularity, developing practical tools to test the various models and gain novel insights into the structure and dynamics of real galaxies becomes increasingly pressing. The stellar kinematic bar diagnostics presented in § 2 allow to identify edge-on bars easily and thus to test the origin of B/PS bulges through bar-driven vertical instabilities. The spectroscopic observations of a large sample of spiral galaxies with a B/PS bulge shown in § 3 vindicate the usefulness of those diagnostics and confirm that B/PS bulges are generally consistent with the presence of a thick bar viewed edge-on, even for the earliest types. They also suggest the presence of rapidly rotating central stellar disks at the center of the bars, which may well have formed through bar-driven inflow and subsequent star formation. Analysis of K-band observations of the same sample reveals the orbital backbone of B/PS bulges (§ 4), in agreement with expectations from periodic orbit calculations of 3D bars. As expected, the scaleheight (and shape) of the material varies rapidly with (projected) radius, and it is largest where the extent of the B/PS bulge is maximum.

The emerging bar-driven evolutionary scenario is beautifully confirmed by synthesis CO observations and SAURON integral-field spectroscopy (§ 5), which further allow to study the distribution of the stellar populations (luminosity-weighted age and metallicity). As the quality of stellar population data is rapidly improving, a parallel improvement of the model predictions (which remain rudimentary) is urgently needed. Interestingly, although clear observa-

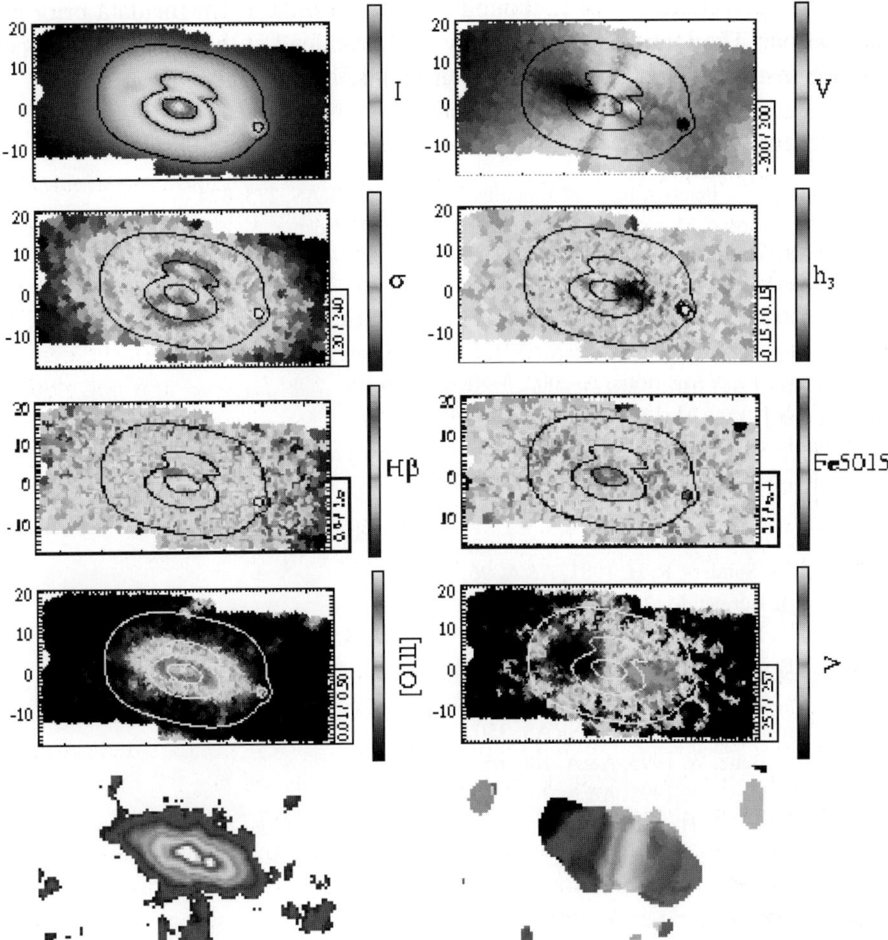

Figure 4. SAURON and BIMA observations of the S0 galaxy NGC4526. From left to right, top to bottom: Reconstructed image, mean stellar velocity V, stellar velocity dispersion σ, asymmetry of the stellar velocity profile h_3, Hβ linestrength (age), Fe5015 linestrength (metallicity), [OIII] intensity, ionized-gas velocity, total CO flux, and CO velocity field.

tional diagnostics of such scenarios are still largely inexistent, our observations do not seem to require nor indicate that bars may be destroyed (and possibly reformed). Specific tests would thus also be appreciated.

Acknowledgments

M.B. acknowledges support by NASA through Hubble Fellowship grant HST-HF-01136.01 awarded by the Space Telescope Science Institute. The authors wish to thank J.C. Lambert, K.C. Freeman, and E. Emsellem for support and useful discussions, and L. Young and the SAURON Team for data prior to publication. The Digitized Sky Survey was produced at the Space Telescope Science Institute under U.S. Government grant NAG W-2166.

References

Andredakis, Y. C., Peletier, R. F., & Balcells, M. 1995, MNRAS, 275, 874

Aronica, G., Bureau, M., Athanassoula, E., Dettmar, R.-J., Bosma, A., & Freeman, K. C. 2005, MNRAS, in preparation.

Athanassoula, E. 2003, MNRAS, 341, 1179

Athanassoula, E., Aronica, G., & Bureau, M. 2005, MNRAS, in preparation.

Athanassoula, E., & Bureau, M. 1999, ApJ, 522, 699

Athanassoula, E., & Misiriotis, A. 2002, MNRAS, 330, 35

Binney, J., & Petrou, M. 1985, MNRAS, 214, 449

Bournaud, F., & Combes, F. 2002, A&A, 392, 83

Bureau, M., & Athanassoula, E. 2004, ApJ, in press.

Bureau, M., & Freeman, K. C. 1999, AJ, 118, 126

Combes, F., Debbasch, F., Friedli, D., & Pfenniger, D. 1990, A&A, 233, 82

Combes, F., & Sanders, R. H. 1981, A&A, 96, 164

Chung, A., & Bureau, M. 2004, AJ, 127, 3192

Davies, R. L., Efstathiou, G., Fall, S. M., Illingworth, G., & Schechter, P.L. 1983, ApJ, 266, 41

Emsellem, E., et al. 2004, MNRAS, in press.

Erwin, P., & Sparke, L. S. 2002, AJ, 124, 65

Falcón-Barroso, J., et al. 2004, MNRAS, 350, 35

Friedli, D., & Benz, W. 1993, A&A, 268, 65

Friedli, D., & Benz, W. 1995, A&A, 301, 649

Friedli, D., Wozniak, H., Rieke, M., Martinet, L., & Bratschi, P. 1996, A&AS, 118, 461

Kormendy, J. 1993, in Galactic Bulges, eds. H. Dejonghe, & H. J. Habing (Dordrecht: Kluwer), 209

Kuijken, K., & Merrifield, M. R. 1995, ApJ, 443, L13

Kuntschner, H., et al. 2005, MNRAS, submitted.

Lütticke, R., Dettmar, R.-J., & Pohlen, M. 2000, A&AS, 145, 405

van der Marel, R. P., & Franx, M. 1993, ApJ, 407, 525

Merrifield, M. R., & Kuijken, K. 1999, A&A, 443, L47

Patsis, P. A., Skokos, Ch., & Athanassoula, E. 2002, MNRAS, 337, 578

Pfenniger, D., & Norman, C. 1990, ApJ, 363, 391

Rowley, G. 1988, ApJ, 331, 124

Sarzi, M., et al. 2005, MNRAS, in preparation.

Shen, J., & Sellwood, J. A. 2004, ApJ, 604, 614

Skokos, Ch., Patsis, P. A., & Athanassoula, E. 2002, MNRAS, 333, 847

Young, L. M., et al. 2005, AJ, in preparation.

de Zeeuw, P. T., et al. 2002, MNRAS, 329, 513

BAR-DRIVEN FUELING OF GALACTIC NUCLEI: A 2D VIEW

Eric Emsellem
Centre de Recherche Astronomique de Lyon, 9 av. Charles Andre, 69561 Saint Genis Laval, France

Abstract I briefly discuss evidences for bar-driven gas fueling in the central regions of galaxies, focusing on scales down to about 10 pc. I thus mention the building of inner disks, and the link with resonances, as well as the corresponding kinematic signatures such as σ-drops and counter-rotating nuclear disks as probed via integral-field spectroscopy.

Keywords: Galaxies, inner bars, inner disks, fueling, σ-drops

1. Introduction

Before embarking onto this short report on bar-driven fueling of the central regions of galaxies, let me define what I mean by "fueling galactic nuclei". Going from the kpc scale down to the presumed black hole of a galaxy, we progress through regions where the physical scales and regimes of the involved processes vary considerably. We should therefore not expect a direct link between the large-scale dynamical processes (e.g. the presence of a bar), and the central engine (accretion disk surrounding the black hole). I would also like to follow the nomenclature advocated by Jean-Luc Nieto, and thus only use the words "nucleus" and "nuclear" for structures at a scale of the order of $\sim$ 10 pc (an excellent illustration being the nearby nucleus of M 31). I will therefore focus here on the question of "how to accumulate mass at the scales of galactic nuclei, i.e. in the central tens of parsecs" (see also Emsellem 2004).

We know that bars can be efficient at redistributing gas within the stellar disk of a galaxy: outwards when the gas is in between the outer Lindblad resonance and Corotation, and inwards when it is inside Corotation (and outside the inner Lindblad resonance - ILR - if there is one). We therefore expect structures to build up at a scale related to the presumed ILR. The questions are therefore: do we observe such features, and are they small enough to be considered as "nuclear" structures?

D. Block et al. (eds.), Penetrating Bars through Masks of Cosmic Dust, 149–153.

2. Building inner disks

There are numerous examples of galaxy structures which have obviously been formed under the influence of a bar, the most generic ones being galactic rings (see e.g. work by Buta and collaborators). I would however like to focus on another type of dynamically cold systems, namely the building of inner disks. Although the inner disk of e.g., the Sombrero galaxy qualifies as a secularly evolved structure (Emsellem 1995), one of the best case to date remains the photometric features exhibited by the nearly edge-on S0 galaxy NGC 4570: there is strong evidence that the two ring-like structures and the 100 pc inner disk of that galaxy are the result of bar-driven secular evolution (van den Bosch & Emsellem 1998). Numerical simulations by Friedli, Benz & Kennicutt (1994) predicted that a radially decreasing initial abundance gradient should evolve due to the presence of a bar, with a strong flattening of the gradient outside cororation, a weakened one inside the bar region and a possible starburst at the very centre. This seems consistent with the observed colour gradients in NGC 4570 along the minor-axis, which retains the original vertical gradient, and along the major-axis which shows a clear correlation with the presumed location of the bars and its resonances (Fig. 1). The jump

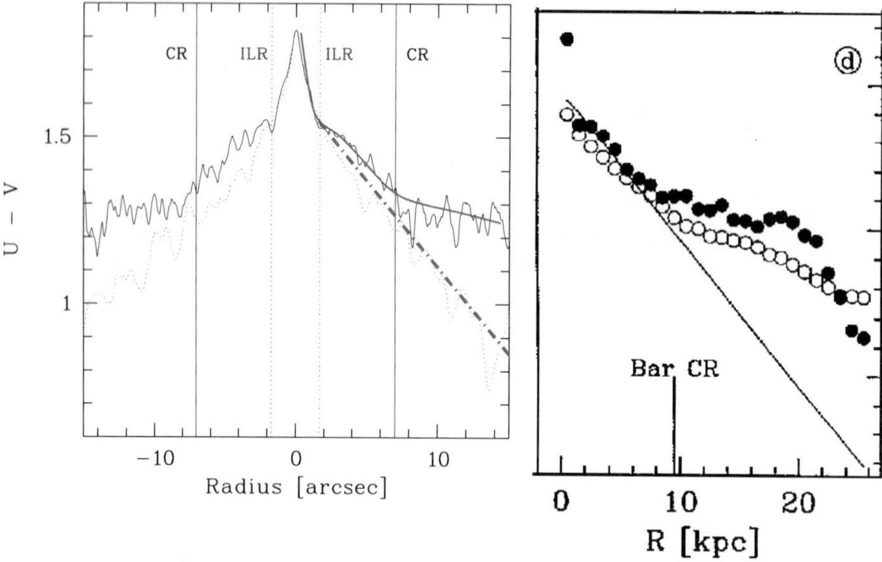

Figure 1. Left panel: mean abundance gradients obtained after redistribution due to the presence of a bar (from Friedli, Benz & Kennicutt 1994). The slopes are clearly different inside and outside the bar. Right panel: $V - I$ colour gradients along the minor and major axis of NGC 4570. The observed gradients are linked with the presumed locations of the resonances as expected.

from gas to stellar abundances, and the use of broad band colours to assess these gradients should forbid us to conclude too hastily. However, we recently obtained line-strength maps of NGC 4570 within the course of the SAURON survey (Bacon et al. 2001, de Zeeuw et al. 2002, Emsellem et al. 2004), in which we see a direct correspondence between the previously observed structures and the metal enrichment (as probed here by e.g. the Mgb index, Fig. 2).

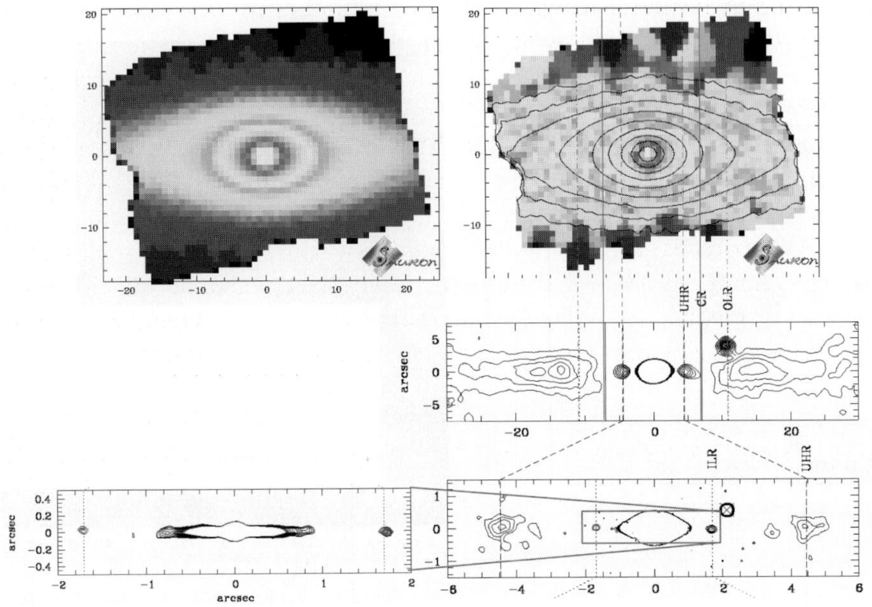

Figure 2. Top panels: SAURON reconstructed intensity (left) and Mgb (right) maps (to appear in a forthcoming paper of the SAURON team). The structures observed in the Mgb map are clearly correlated with the presumed resonances, emphasized in the isophotes of an unsharp masking image of NGC 4570 (contours plots extracted from van den Bosch & Emsellem 1998).

3. Towards the nucleus

Inner disks with sizes ranging from 100 to 500 pc, such as the one observed in NGC 4570, are quite common in early-type galaxies and could indeed be the result of bar-driven accretion followed by star formation (e.g. see the case of NGC 3115). This would in fact require inner bars with diameter from $\sim$ 200 to 1000 pc, similar to the ones now routinely observed in disk galaxies (Laine et al. 2002, Erwin & Sparke 2002). In some cases, such as the S0/a galaxy NGC 2974, the size of the fueled region (ILR) is less than 20 pc in diameter, a scale at which we can start making the link with the nucleus itself (Emsellem, Goudfrooij & Ferruit 2003).

Another clear case of gas fueling of the nuclear regions is provided by the high resolution ^{12}CO(2-1) map of NGC 6946 (Schinnerer et al., in preparation; Fig.3). The spiral-like distribution of the molecular gas is reminiscent of the dust lanes observed in barred galaxies, and indeed a small inner bar has been detected via K band imaging in this galaxy (Elmegreen et al. 1998). The amount of gas which is fueled within the central 20 pc is uncertain, but star formation is already ongoing there, which may thus lead to the formation of a flattened nuclear disk.

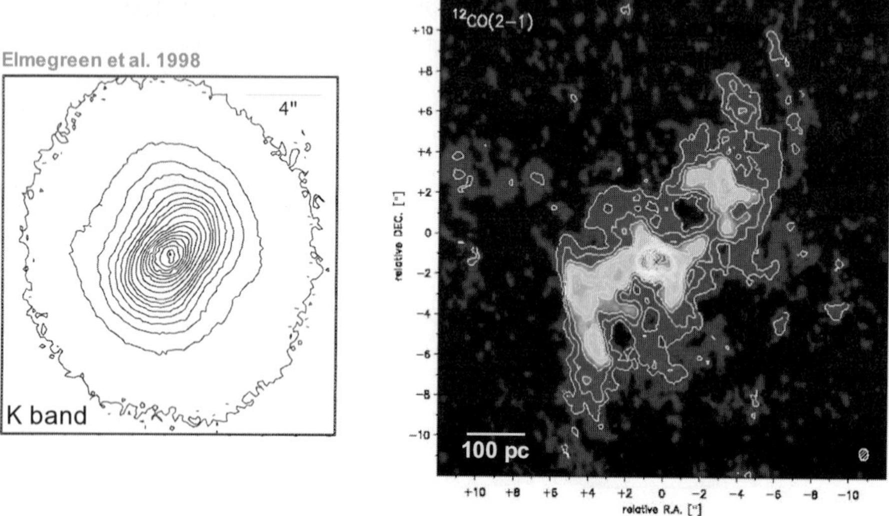

Figure 3. Right panel: ^{12}C0(2-1) map obtained with tha IRAM interferometer (from Schinnerer et al., in preparation) of the central kpc of NGC 6946, showing a two-arm spiral-like structure reminiscent of dust lanes in bars. Left panel: K band image (from Elmegreen et al. 1998) showing the presence of an inner bar in NGC 6946.

4. Signatures and the importance of bars

Looking for signatures of past accretion events, we should turn away from purely photometric features. The dynamical status of the central regions of galaxies may help us to probe such accretion events long after the nucleus itself has been formed. The so-called σ-drops (the DEBCA project: Emsellem et al. 2001), a central depression in the stellar velocity dispersion profile, are now routinely detected in disk galaxies (e.g. Marquez et al. 2004) and could indeed be the signatures we are looking for (Wozniak et al. 2003). It would be interesting to examine if there is any link between the blue nuclei observed in most spiral galaxies (e.g. Bøker et al. 2004) and the presence of such σ-drops.

It should finally be made clear that bars are not the only way gas can be transported inside the central 10 pc or so. The first ingredient for a successful fueling is obviously the availability of gas. In this context, interactions and/or external accretion have certainly a significant role in galaxy evolution (either with companions, or from large-scale structures; see Bournaud's, and Combes' contributions, these Proceedings). This is now clearly witnessed in NGC 7332, for which two-dimensional SAURON spectrography has been obtained (Falcon-Barroso et al. 2004). The ionized gas distribution and kinematics does not leave any doubt on the external origin of part of the dissipative component which is counter-rotating with respect to the stars. Evidences for the presence of a strong bar are also very strong: e.g. a boxy bulge, a cylindrical stellar velocity field. An hybrid scenario emerges in this case where the interaction with a companion provides the gas which is then fueled to the central inner 50 pc with the help of a bar (the formation of which could have been triggered by the interaction). It is finally important to note that the central stellar kinematics shows the presence of a counter-rotating stellar disk, less than 100 pc in diameter, a possible remnant of a past accretion episode.

I would like here to thank my collaborators, and more specifically Torsten Boker, Eva Schinnerer, Ute Lisenfeld, as well as the DEBCA and SAURON teams.

References

Bacon, R., et al. 2001, MNRAS, 326, 23

Boker, T., Sarzi, M., McLaughlin, D. E., van der Marel, R. P., Rix, H., Ho, L. C., & Shields, J. C. 2004, AJ, 127, 105

Elmegreen, D. M., Chromey, F. R., & Santos, M. 1998, AJ, 116, 1221

Emsellem, E., 1995, A&A, 303, 673

Emsellem, E., 2004, in The interplay between black holes, Stars, the ISM in galactic nuclei, IAU 222, Gramado, in press

Emsellem, E., Greusard, D., Combes, F., Friedli, D., Leon, S., Pecontal, E., & Wozniak, H. 2001, A&A, 368, 52

Emsellem, E., Goudfrooij, P., & Ferruit, P. 2003, MNRAS, 345, 1297

Erwin, P. & Sparke, L. S. 2002, AJ, 124, 65

Falcon-Barroso, J., Peletier, R. F., Emsellem, E., Kuntschner, H., Fathi, K., Bureau, M., Bacon, R., Cappellari, M., Copin, Y., Davies, R., L., de Zeeuw, T., 2004, MNRAS, 350, 35

Friedli, D., Benz, W.. Kennicutt, R., 1994, ApJL, 430, 105

Laine, S., Shlosman, I., Knapen, J. H., & Peletier, R. F. 2002, ApJ, 567, 97

Marquez, I., Masegosa, J., Durret, F., Gonzalez Delgado, R. M., Moles, M., Maza, J., Perez, E., & Roth, M. 2003, A&A, 409, 459

van den Bosch, F. C. & Emsellem, E. 1998, MNRAS, 298, 267

Wozniak, H., Combes, F., Emsellem, E., & Friedli, D. 2003, A&A, 409, 469

de Zeeuw, P. T., et al. 2002, MNRAS, 329, 513

Pilanesberg
National Park
ABSA

DUST PENETRATED ARM CLASSES: INSIGHT FROM RISING AND FALLING ROTATION CURVES

Marc S. Seigar[1], David Block[2] and Ivanio Puerari[3]
[1]*Joint Astronomy Centre, 660 N. A'ohoku Place, Hilo, HI 96720*
[2]*School of Computational and Applied Mathematics, University of the Witwatersrand, P.O. Box 60, Wits, Gauteng 2050, South Africa*
[3]*Instituto Nacional de Astrofisica, Optica y Electronica, Calle Luis Enrique Erro 1, 72840 Tonantzintla, Puebla, Mexico*

Abstract We present near-infrared K-band images of 15 galaxies. We have performed a Fourier analysis on the spiral structure of these galaxies in order to determine their pitch angles and dust-penetrated arm classes. We have also obtained rotation curve data for these galaxies and calculated their shear rates. We show that there is a correlation between pitch angle and shear rate and conclude that the main determinant of pitch angle is the mass distribution within the galaxy. This correlation provides a physical basis for the dust-penetrated classification scheme of Block & Puerari (1999).

Keywords: galaxies: fundamental parameters — galaxies: spiral — galaxies: structure — infrared: galaxies

1. Introduction

The classification of galaxies, i.e. Hubble type, has traditionally been inferred in the optical regime, where dust extinction still has a large affect, and the light is dominated by the young Population I stars. Infrared arrays offer opportunities for deconvolving the Population I and Population II morphologies, because in the K-band (2.2 μm), dust extinction is minimal, and the light is dominated by old Population II stars. The extinction at this wavelength is only 10% of that in the V-band (Martin & Whittet 1990).

Hubble type is not correlated with the Population II morphology, as confirmed by the near-infrared studies of de Jong (1996) and Seigar & James (1998a, b). Also, it has been shown that near-infrared morphologies of spiral galaxies are often vastly different from their optical morphologies (Block & Wainscoat 1991; Block et al. 1994a; Thornley 1996; Seigar & James 1998a, b; Seigar, Chorney & James 2003). Often galaxies with flocculent spiral structure

D. Block et al. (eds.), Penetrating Bars through Masks of Cosmic Dust, 155–163.

in the optical appear to have Grand-Design spiral structure in the near-infrared (Thornley 1996; Seigar et al. 2003). This suggests that the optical morphology bears little resemblance to underlying stellar mass distribution.

Burstein & Rubin (1985) showed that spiral galaxies can have one of three different types of rotation curve, rising, flat or falling. From these rotation curve types they derived three principle types of mass distribution and found that Hubble types Sa and Sb were amongst all three types in approximately equal amounts. This supports the idea that the optical morphology is not correlated with the underlying mass distribution in spiral galaxies.

The disk of a spiral galaxy can be separated into two distinct components: the *gas-dominated* Population I disk, and the *star-dominated* Population II disk. The former component contains features of spiral structure (OB associations, HII regions, and cold interstellar HI gas). In contrast, the Population II disk contains the old stellar population highlighting the underlying stellar mass distribution (Lin 1971). One might expect, even in the absence of appreciable optical depths, for the two morphologies to be very different, since the near-infrared light comes from mainly giant and supergiant stars (Rix & Rieke 1993; Frogel et al. 1996).

One therefore needs a near-infrared classification scheme, such as the dust-penetrated class (Block & Puerari 1999; Block et al. 1999) to describe the Population II disk, as well as Hubble type to describe the Population I disk. The dynamic interplay between the two components (via a feedback mechanism) is crucial, and has been studied extensively (Bertin & Lin 1996). A central aspect here is the likely coupling of the Population I disk with that of the Population II disk via a feedback mechanism (Pfenniger et al. 1996).

The theoretical framework to explain the co-existence of completely different morphologies within the same galaxy when it is studied optically and in the near-infrared is described by Bertin & Lin (1996). A global mode (Bertin et al. 1989a, b) is composed of spiral wavetrains propagating radially in opposite directions. Thus a feedback of wavetrains is required from the center. The return of wavetrains back to the corotation circle is guaranteed by refraction, either by the bulge or because the inner disk is dynamically warmer. In the stellar disk, such a feedback can be interrupted by the Inner Lindblad Resonance (ILR), which is a location where the stars meet the slower rotating density wave crests in resonance with their epicyclic frequency (Mark 1971; Lynden-Bell & Kalnajs 1972). In the gaseous disk, the related resonant absorption is only partial, so that some feedback is guaranteed. Once the above described wavecycle is set up, a self-excited global mode can be generated.

The tightness of the arms in the modal theory comes from the mass distribution and rate of shear. Galaxies with more mass concentration, i.e. higher overall densities (including dark matter) and higher shear, should have more tightly wound arms. If the disk is very light (low σ where σ is the disk density) the mode can be very tight, and one is in the domain of small epicycles. If one increases the mass of the disk one finds a trend towards more open structures, but soon one runs the risk of a disk that is too heavy and a bar mode results.

The goal of this paper is to highlight the expected correlation between the shear rate in spiral galaxies (as derived from their rotation curves) and the near-infared spiral arm pitch angle. The dust penetrated morphology depends on the near-infared spiral arm pitch angle, and so, such a correlation would provide a physical basis for the dust penetrated classification scheme of Block & Puerari (1999).

2. Decomposition and identification of modes

We have observed a sample of 15 galaxies in the near-infrared K-band (2.2 μm). These objects were taken from the study of Mathewson et al. (1992), who observed rotation curves for 1355 southern hemisphere spiral galaxies. Our sample includes galaxies with different rotation curve types (rising, falling and flat) and span as wide a range of optical Hubble Types as possible. The images were observed at the United Kingdom Infrared Telescope (UKIRT) using the UKIRT Fast Track Imager (UFTI) between 1–4 August 2001 and 11–12 March 2002.

The 2-D Fast Fourier decomposition of all the near-infrared images in this study, employed a program developed by I. Puerari (Schroeder et al. 1994). Logarithmic spirals are assumed in the decomposition.

The amplitude of each Fourier component is given by:

$$A(m,p) = \frac{\Sigma_{i=1}^{I}\Sigma_{j=1}^{J} I_{ij}(\ln r,\theta)\exp -(i(m\theta + p\ln r))}{\Sigma_{i=1}^{I}\Sigma_{j=1}^{J} I_{ij}(\ln r,\theta)} \tag{1}$$

where r and θ are polar coordinates, $I(\ln r, \theta)$ is the intensity at position $(\ln r, \theta)$, m represents the number of arms or modes, and p is the variable associated with the pitch angle P, defined by $\tan P = -\frac{m}{p_{max}}$.

Our Fourier spectra corroborate earlier observational indications (Block et al . 1994a, 1999; Block & Puerari 1999) that there is indeed a ubiquity of m=1 and m=2 modes. Block & Puerari (1999) proposed three principle archetypes for the evolved stellar disk of such galaxies. The first of these, designated dust-penetrated class α, covers the pitch angle range $4^\circ < P < 15^\circ$, the second, designated β, covers $18^\circ < P < 30^\circ$ and the third, designated γ covers $36^\circ < P < 76^\circ$.

Those lopsided galaxies (where m=1 is a dominant mode) are designated Lα, Lβ and Lγ according to the dust penetrated pitch angle. Evensided galaxies (where m=2 is the dominant Fourier mode) are classified into classes Eα, Eβ and Eγ. Higher order harmonics are classified as H3 (for m=3) and H4 (for m=4).

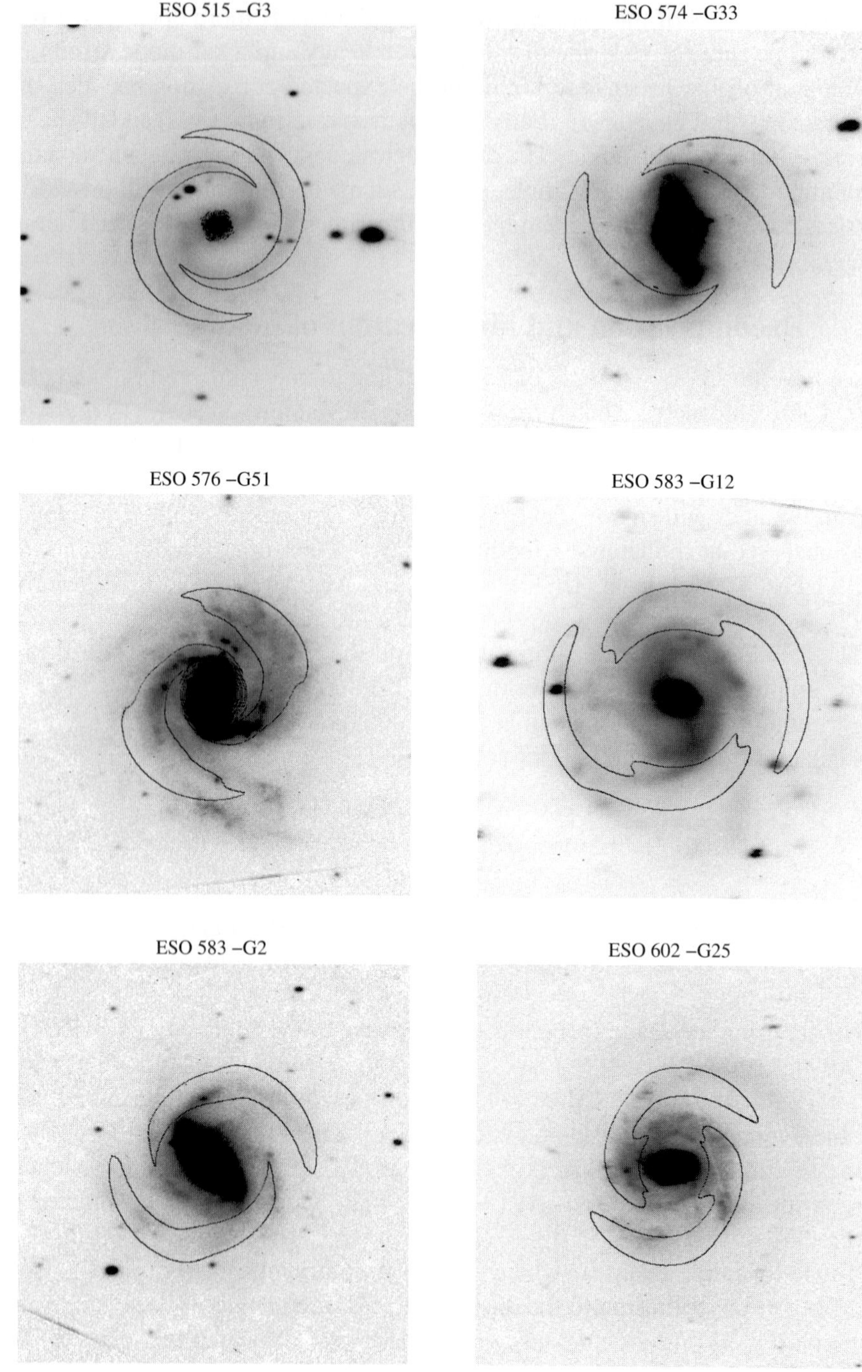
ESO 515 –G3
ESO 574 –G33
ESO 576 –G51
ESO 583 –G12
ESO 583 –G2
ESO 602 –G25

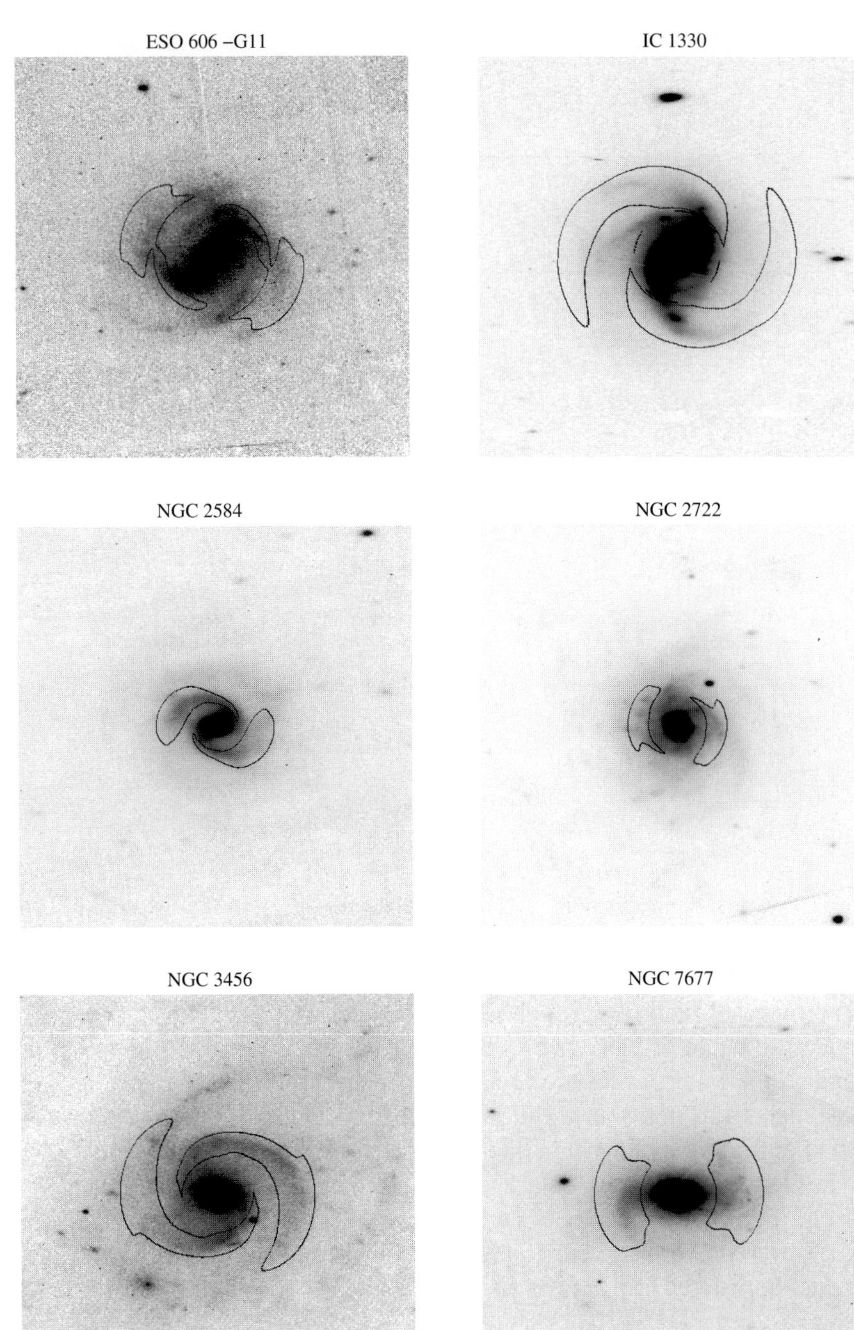
ESO 606 –G11
IC 1330
NGC 2584
NGC 2722
NGC 3456
NGC 7677

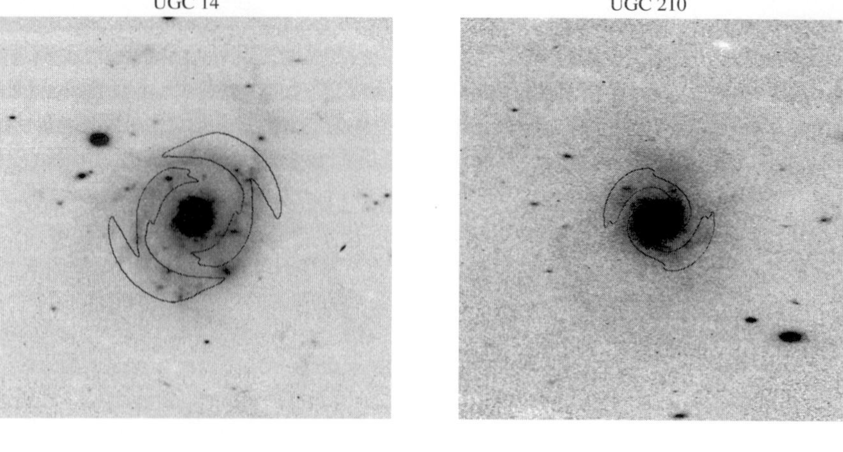

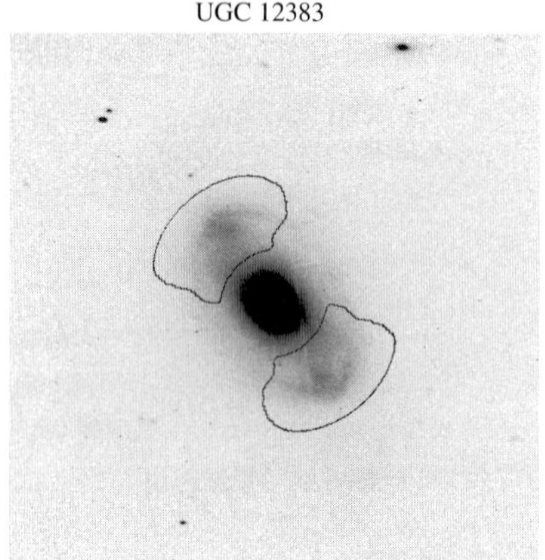

Figure 1. Greyscale images of the galaxies for which the FFT analysis was performed. The overlaid contours represent the FFT fit to the spiral structure.

The range of radii over which the Fourier fits were applied are selected to exclude the bulge or bar (where there is no information about the arms) and extend to the outer limits of the arms in our images. Pitch angles are then determined from peaks in the Fourier spectra, as this is the most powerful method to find periodicity in a distribution (Considere & Athanassoula 1998; Garcia-Gomez & Athanassoula 1993). The images were first deprojected to face-on. Figure 1 shows the images of the spiral galaxies observed for this project, overlaid with contours representing the results of the Fourier analysis. The dust-penetrated arm classes and pitch angles are listed in Table 1.

3. Discussion

Block et al. (1999) showed the first evidence that the pitch angle (and therefore dust penetrated arm class) of a spiral galaxy depends upon the shear rate as derived from rotation curves, consistent with the theoretical predictions of the modal theory (Bertin et al. 1989a, b; Bertin & Lin 1996; Fuchs 1991, 2000). The work presented by Block et al. (1999) consisted of just 4 galaxies. Here we present a further 15 galaxies, all with measured rotation curves. Their shear rates are derived from their rotation curves as follows

$$\frac{A}{\omega} = \frac{1}{2}\left(1 - \frac{R}{V}\frac{dV}{dR}\right) \tag{2}$$

where A is the first Oort Constant, ω is the rotational velocity, and V is the velocity measured at radius R. The value A/ω gives the shear rate.

Table 1. Results from the Fourier analysis and rotation curve analysis of 15 spiral galaxies. Column 1 shows the name of the galaxy; Column 2 shows the derived dust penetrated class; Column 3 shows the Hubble type; Column 4 shows the pitch angle of the K-band spiral arms and column 5 shows the derived shear rate.

Galaxy	Dust-penetrated class	Hubble type	P_K	A/ω
ESO 515 G3	Lγ	SBc	47.8±0.7	0.27±0.02
ESO 574 G33	Eγ	SBbc	39.9±1.0	0.37±0.02
ESO 576 G51	Eβ	SBbc	30.4±1.9	0.47±0.02
ESO 583 G2	Eβ	SBbc	28.4±1.0	0.47±0.02
ESO 583 G12	Lβ	SBc	17.7±2.0	0.54±0.02
ESO 602 G25	Lβ	Sb	21.8±2.0	0.45±0.02
ESO 606 G11	H4β	SBbc	25.2±2.4	0.50±0.02
IC 1330	Eγ	Sc	37.8±1.1	0.35±0.02
NGC 2584	Eβ	SBbc	29.7±3.0	0.50±0.02
NGC 2722	Eβ	Sbc	32.8±3.4	0.46±0.02
NGC 3456	Eγ	SBc	38.0±0.6	0.31±0.02
NGC 7677	Eα	SABbc	17.0±0.8	0.66±0.02
UGC 14	Lβ	Sc	20.9±1.3	0.53±0.02
UGC 210	Lβ	Sb	16.0±0.8	0.55±0.02
UGC 12383	Eα	SABb	15.0±1.6	0.66±0.02

Figure 2 shows a plot of the shear rate of spiral galaxies versus the spiral arm pitch angle. As well as showing a good correlation, it is also interesting to note how the galaxies seem to fall into 3 distinct areas on this plot, according to both their shear rates and pitch angles, and possibly their mass distributions. Galaxies, with high shear rates (rising rotation curves) and tightly wound spiral structure are found in the bottom right and are designated α in the dust penetrated class. Galaxies with shear rates around 0.5 (flat) and moderately wound spiral structure are in the middle and are designated β. The top left contains those galaxies with loosely wound structure and low shear rates (falling). These are designated γ. Figure 3 shows a plot between rotation curve

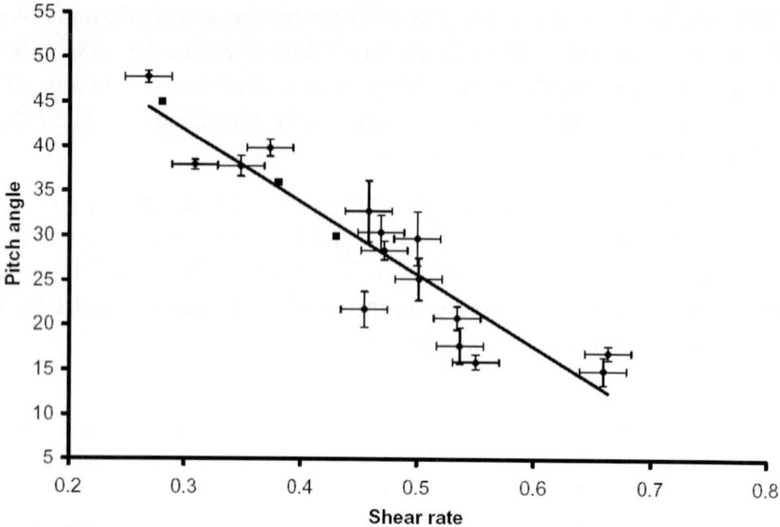

Figure 2. Pitch angle versus shear rate. Circles represent the 15 galaxies presented here. Squares are 3 galaxies presented in Block et al. (1999).

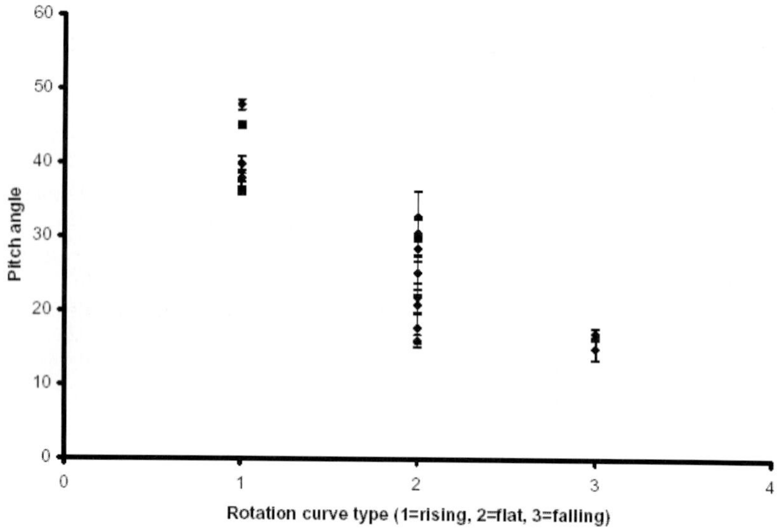

Figure 3. Pitch angle versus rotation curve type. The symbols are the same as Figure 2

type (Burstein & Rubin 1985), i.e. rising, flat or falling, versus spiral arm pitch angle, also showing a good correlation.

The shape of a rotation curve is determined largely by the distrbution of luminous and dark mass contained in a spiral galaxy. The correlation found between pitch angle and shear rate, therefore suggests that the main factor determing the tightness of spiral structure, is in fact the central mass concentration. Essentially, a declining rotation curve is indicative of a large central bulge. The correlation between mass concentration and spiral arm pitch angle has been suggested by many theoretical models (e.g. Fuchs 1991, 2000; Bertin et al. 1989a, b; Bertin & Lin 1996).

Acknowledgments

The United Kingdom Infrared Telescope (UKIRT) is operated by the Joint Astronomy Centre on behalf of the U.K. Particle Physics and Astronomy Research Council (PPARC). The authors would like to thank David Block, Ken Freeman and the scientific organising committee for the opportunity to present this work at the Bars Congress 2004.

References

Bertin G., Lin C.C., Lowe S.A., Thurstans R.P., 1989a, ApJ, 338, 78
Bertin G., Lin C.C., Lowe S.A., Thurstans R.P., 1989b, ApJ, 338, 104
Bertin G., Lin C.C., 1996, *Spiral Structure in Galaxies: A density wave theory*, MIT Press, Cambridge, MA
Block D.L., Wainscoat R., 1991, Nature, 353, 48
Block D.L., et al., 1994a, A&A, 288, 365
Block D.L., et al., 1994b, A&A, 288, 383
Block D.L., Puerari I., 1999, A&A, 342, 627
Block D.L., et al., 1999, Ap&SS, 269-270, 5
Burstein D., Rubin V.C., 1985, ApJ, 297, 423
Considere S., Athanassoula E., 1988, A&AS, 76, 365
de Jong R.S., 1996, A&A, 313, 45
Frogel J.A., Quillen A.C., Pogge R.W., in *New Extragalactic Perspectives in the New South Africa*, eds D.L. Block, J.M. Greenberg, Kluwer, Dordrecht, p251
Fuchs B., 1991, in *Dynamics of Disk Galaxies*, ed B. Sundelius, Chalmers University of Technology, p359
Fuchs B., 2000, in *Galaxy Dynamics: From the Early Universe to the Present*, eds F. Combes & G. Mamon, ASP Conf. Ser. Vol. 197, p53
Garcia–Gomez C., Athanassoula E., 1993, A&AS, 100, 431
Lin C.C., 1971, in *Highlights of Astronomy Vol. 2*, ed. C. de Jager, Reidel, Dordrecht, p88
Lynden-Bell D., Kalnajs A.J., 1972, MNRAS, 157, 1
Mark J.W-K., 1971, Proc. Natl. Acad. Sci., 68, 2095
Martin P.G., Whittet D.G.B., 1990, ApJ 357, 113
Mathewson D.S., Ford V.L., Buchhorn M., 1992, ApJS, 81, 413
Pfenniger D., Martinet L., Combes F., 1996, in *New Extragalactic Perspectives in the New South Africa* eds D.L. Block, J.M. Greenberg, Kluwer, Dordrecht, p291
Rix H.-W., Rieke M.J., 1993, ApJ, 481, 123
Schroeder M.F.S., Pastoriza M.G., Kepler S.O., Puerari I., 1994, A&AS, 108, 41
Seigar M.S., James P.A., 1998, MNRAS, 299, 672
Seigar M.S., James P.A., 1998, MNRAS, 299, 685
Seigar M.S., Chorney N.E., James P.A., 2003, MNRAS, 342, 1
Thornley M.D., 1996, ApJ, 469, 45

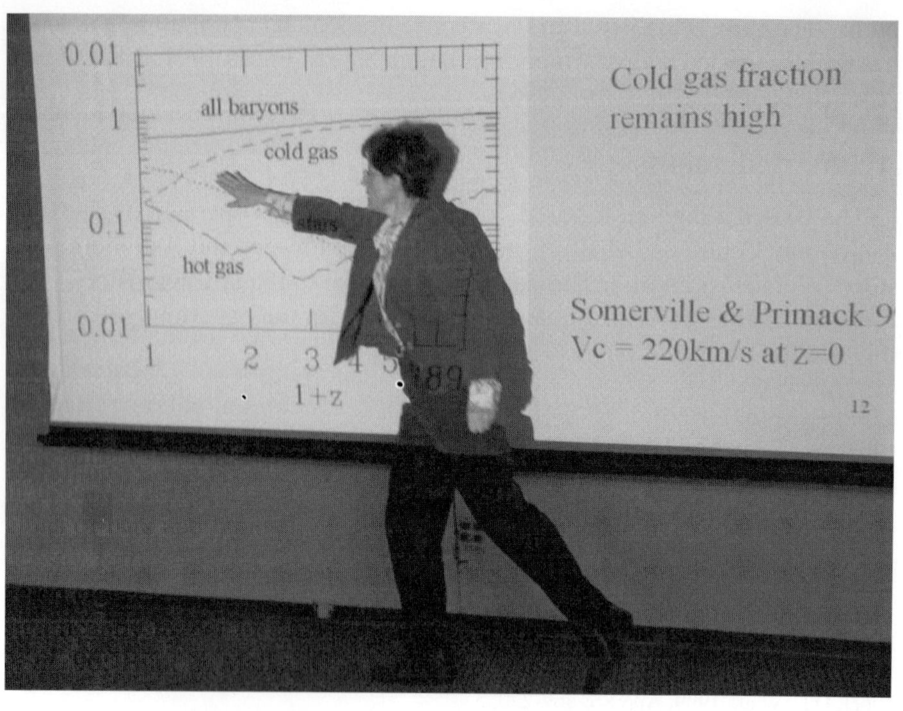
0.01
1
0.1
0.01
all baryons
cold gas
hot gas
1
2
3
4
5
1+z
Cold gas fraction
remains high
Somerville & Primack 9
Vc = 220km/s at z=0
12

BAR DISSOLUTION AND REFORMATION MECHANISMS

Frederic Bournaud and Francoise Combes
LERMA, Observatoire de Paris, 61 av. de l'Observatoire, F-75014 Paris, France

Abstract We detail the bar dissolution mechanisms. Bars in spiral galaxies with a normal gas content are transient features, because of the effects of growing central mass concentrations, gravity torques, and decoupled nuclear bars. Their life-time is 1 to 4 Gyrs. As bars were already present several billions years ago, the ubiquity of bars in the Local universe requires that bars are reformed. The most efficient process to reform bars is the accretion of large amounts of gas by spiral galaxies.

Keywords: galaxies: evolution – galaxies: spiral – galaxies: kinematics and dynamics

1. Introduction: are bars long-lived or transient features?

Bars are very common features in the Local Universe. Near infrared observations have shown that more than two thirds of galaxies are barred (e.g., Eskridge et al. 2002). Have they always existed? Basically, one can think that several billion years ago, the velocity dispersion of stars was smaller, galaxies contained a larger gas mass fraction, and bulges were much lighter. Thus, barred instabilities should have appeared more frequently and be stronger than in the Local Universe. Observationally, bars at high redshift can easily be missed, for the red-NIR light in which they are unveiled from dust is shifted to larger wavelengths. Accounting for this effect, Sheth et al. (2003) have reported that bars are common at redshift $z > 0.6$. They even find that these bars can be strong, long compared to the disk radius, with a high axis ratio, as expected from the simple arguments mentioned above.

That bars were already present in spiral galaxies several billion years ago may suggest that they are long-lived features (Miller 1996). An alternative scenario is that bars are dissolved and reformed (Sellwood 1996). It is then fundamental to know whether bars are robust or short-lived. We here present a detailed study of the bar dissolution mechanisms, and argue that bars cannot be long-lived, excepted in gas-poor spiral galaxies. With physical parameters of normal spiral galaxies, we find that the bar life-time is 1 to 4 Gyrs, and may

D. Block et al. (eds.), Penetrating Bars through Masks of Cosmic Dust, 165–174.

even have been smaller in the past. Then, the large fraction of barred galaxies observed today implies that bars have been reformed. We discuss the reformation mechanisms of bars, and show that the most efficient one is the accretion of large amounts of gas by spiral galaxies: this reforms bars, maintains spiral arms, and may even explain warps, thick disks, and lopsidedness.

In the following, we mainly use N-body simulations where the ISM is modelled by a sticky-particles code: dissipative collisions between gas clouds are computed on a fixed Lagrangian grid. Star formation is included. The code is described in Bournaud & Combes (2002). The bar strength P2 is defined as the maximal bar gravity torque in the $m = 2$ mode over the whole disk.

2. The response of gas to barred potentials: gravity torques and gas inflows.

Gas in barred galaxies is observed to be concentrated on the leading side of the bar (de Vaucouleurs & de Vaucouleurs 1963). This is reproduced in hydrodynamical simulations (e.g., Athanassoula 1992), as well as in our sticky-particles models (see Fig. 1). Because of this phase shift, the bar gravity torques make gas lose angular momentum, which initiates a gas inflow.

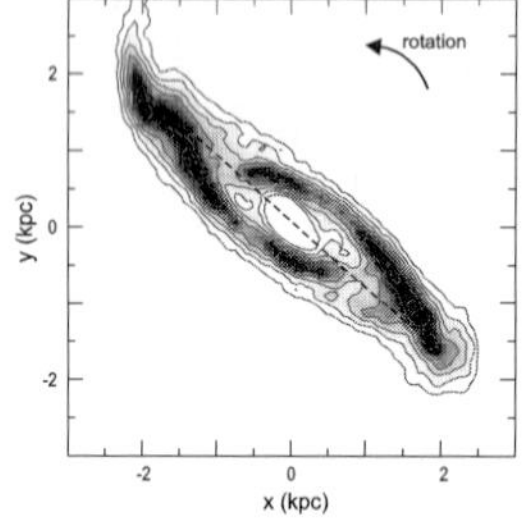

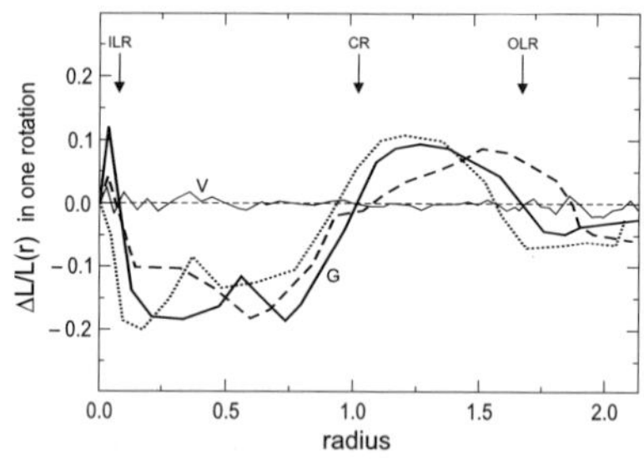

Figure 1. Left: Gas surface density (log scale map and contours) in the simulation of an Sb galaxy, when the bar reaches its maximum amplitude. Rotation is counter-clockwise. The dashed line represents the bar major axis. Between the ILR and the CR, gas forms two arms on the leading side of the bar, which initiates strong gravity torques – Right: Gravity torques exerted by the stellar bar on gas, in units of relative loss of angular momentum in one rotation. The three curves labeled 'G' correspond to three different instants, at the beginning of the bar dissolution. Inside the CR, the gravity torques make mass loss 10 to 20 % of its initial angular momentum in one rotation. The viscous torques 'V' are much smaller and do not directly contribute the gas inflow.

The amplitude of gravity torques will be fundamental when we study the bar dissolution (next Section). Viscous torques are negligible with respect to gravity torques (see Fig. 1), but the viscosity of gas is responsible for the gas/bar misalignment (van Albada & Roberts 1981). A too large viscosity may result in an over-estimated phase shift, which would induce unrealistically large gravity torques.

Then, we have checked that gravity torques between the bar and interstellar gas are correctly reproduced in numerical simulations. The gravity torques simulated using the sticky-particles code are shown in Fig. 1: they make gas lose 10 to 20 percent of its angular momentum in one rotation. To check that these torques are realistic, we made different tests:

- We used the hydrodynamical code of Junqueira & Combes (1996) that produced gravity torques of the same amplitude or even slightly stronger.
- We varied the parameters of the sticky-particles code (see next Section) and checked that they had no major influence on the results.
- We estimated the gravity torques in observed systems, using several tracers for the gas distribution (Hα, CO, B-V, dust emission). The results are shown in Fig. 2 for three systems. For these ones and many other ones, the amplitude of gravity torques is the same as in our simulations.

These arguments insure that the amplitude of gravity torques between the bar and gas is correctly reproduced in our sticky-particles simulations.

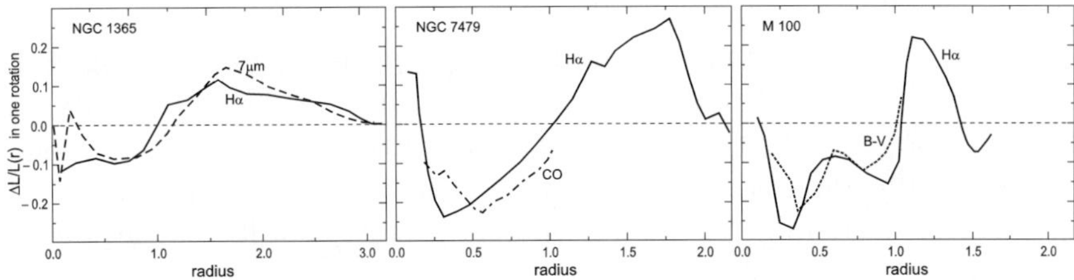

Figure 2. Gravity torques exerted by the bar on gaseous arms in observed barred galaxies, with the same units as Fig. 1. NIR and R images were used to derivate the bar potential, while the gas has been traced by the Hα emission, the dust emission at 7 μm, the CO millimetric emission, and the B-V index, according to the curves labels. In these systems, as in several others, gravity torques make gas lose 10 to 20 percent of its momentum in one rotation: this is correctly reproduced in our sticky-particles simulations.

3. Bar dissolution: central mass concentrations and gravity torques

The dissolution of bars in galaxy simulations: the role of central mass concentrations

Our numerical simulations of barred spiral galaxies (Bournaud & Combes 2002 and 2004, hereafter BC02 and BC04), show that bars are dissolved in a few dynamical times in spiral galaxies with bulges parameters and gas masses typical of Sa to Sd galaxies. We here detail the mechanisms for bar dissolution and show that the bar life-time in our simulations is realistic.

The dissolution of bars is usually attributed to the growth of central mass concentrations (CMCs). The bar themselves, through their gravity torques, fuel these concentrations over a few dynamical times. The growth of a CMC, through a process of escaping orbits (e.g., Pfenniger & Norman 1990), can strongly weaken the bar. Many works have reported the dissolution of bars with CMC masses of 0.5 to 2% of the disk mass (see Shen & Sellwood 2004, hereafter SS04, for a review). Yet, SS04 have recently claimed that bars are most robust versus the growth of CMC than found before. They find that realistic CMCs are not massive and/or concentrated enough to fully dissolve bars but only partly weaken them. They also point out that one must be very careful regarding the integration time step in numerical simulations, for a too long time step may trigger the effects of the CMCs in an unrealistic way.

We now analyze in detail one of our simulations, using 10^6 particles, a resolution of 75 pc, and physical parameters typical of Sbc galaxies. In this simulation, the bar is fully dissolved 1 Gyr after its formation (Fig. 3). It fuels a CMC of mass 1.4% of the disk mass and radius 90 pc. According to SS04, such a CMC should not fully dissolve the bar. Having checked that the bar dissolution is not caused by a too long time step (see Fig. 3), we reproduced the model of SS04: in a purely stellar simulation, we add an analytical CMC with the same mass and radius as in the complete simulation, growing at the same speed. This model confirms that the CMC is not sufficient to fully dissolve the bar (Fig. 3). Moreover, as seen on this figure, the bar is already much weakened before the CMC begins to grow. This leads to the conclusion that the CMC effects are limited, and that another process contributes much to the bar dissolution.

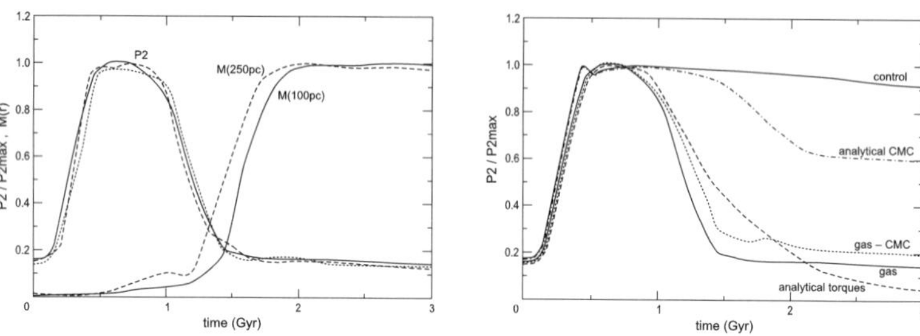

Figure 3. Left: Evolution of the bar strength P2 for the simulation analyzed in Sect. 3. The dashed and dotted curves correspond to a 3 times and 6 times shorter time step: the bar dissolution is not an artifact of a too long time step. The increase in the masses inside radii 250 pc and 100 pc occur only at the end of the bar dissolution, suggesting that another process than the CMC growth weakens the bar. – Right: Evolution of the bar strength in the purely stellar run ('control'), the complete simulation ('gas'), the simulation were the CMC is artificially removed ('gas-CMC'), the model with the CMC alone ('analytical CMC'), and the model with gravity torques alone ('analytical torques'). Gravity torques contribute to the bar dissolution, and are even more efficient than the CMC growth.

The role of gravity torques

The complete dissolution of the bar is not an effect of the CMC alone, but is caused by the presence of gas. Indeed, the bar is not dissolved in the purely stellar simulation, but is dissolved in a simulation where gas is present but the associated CMC has been artificially removed (Fig. 3). The CMC growth is not the only effect of gas: gas also exerts gravity torques. The phase shift between gas and stars initiates torques from the stellar bar to the gas arms, but there are also torques exerted by gas on the stellar bar. To check whether these torques can efficiently weaken the bar, we have measured them in the simulation, and applied them in the purely stellar run. As shown in Fig. 3, the gravity torques exerted by gas can fully dissolve the bar, in a much more efficient way than the CMC growth. Indeed, the torques undergone by stars in the bar are positive, which tends to make orbits become rounder. They also vary with radius, which can break the orbit alignment of the bar. An interpretation in terms of the bar angular momentum can be given. We have seen that gravity torques make gas lose about 15 percent of its angular momentum in one rotation. Since the gas mass is only 5 to 10 percent of the stellar mass, the torques exerted by gas on stars will make they earn about 1% of their momentum in one rotation. The barred wave has a negative angular momentum, typically -10 to -20 % of the stars momentum. The gravity torques exerted by gas will then compensate this negative momentum, and hence dissolve the bar, in 10 to 20 rotations. This corresponds to the life-time of bars in our simulations.

In the simulation that we have detailed above, the gravity torques have nearly dissolved the bar even before the CMC begins to grow. In other simulations, the effect of gravity torques can be less efficient, only leading to a partial (but important) weakening of the bar. Then, the growth of the CMC finishes to fully dissolve the bar (see details in BC04).

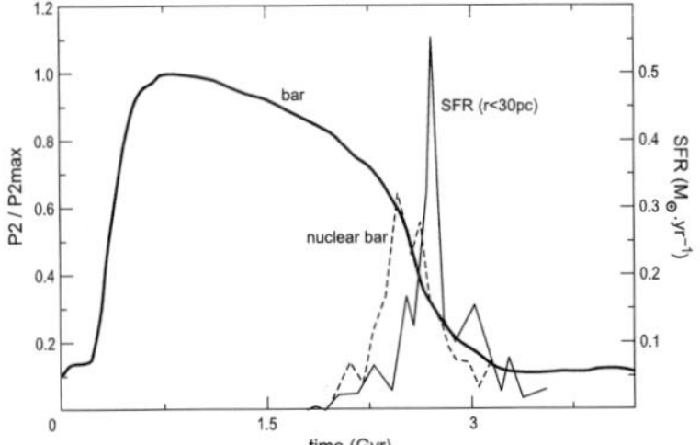

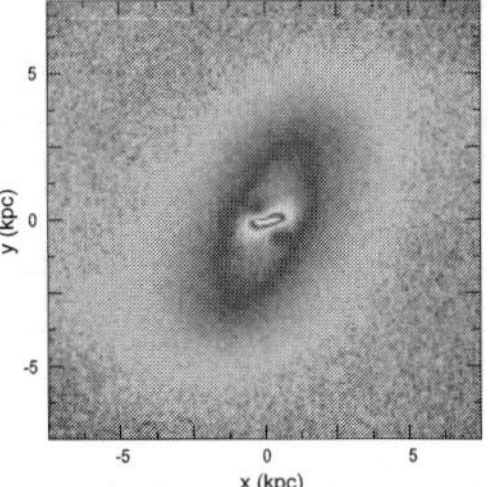

Figure 4. Nuclear bar in a barred early-type spiral galaxy. Left: Evolution of the bar strength P2 (thick line), the nuclear bar strength (dotted line), and central SFR (thin line). The nuclear bar fuels a nuclear starburst, why may initiate an AGN phase. The development of the nuclear bar around $t = 2.4$ Gyr triggers the dissolution of the main bar and provokes a rapid decrease in the bar strength. – Right: snapshot of the stars showing the two embedded bars.

The effects of nuclear bars

A third process of bar weakening occurs in early-type spirals, or late-type spirals after a significant slowing-down of the bar. In such galaxies, the bar has an ILR inside which kinematically cold disks are observed. The properties of these cold disks can be matched in simulations of bar-driven gas inflows (Wozniak et al. 2003). Nuclear arms and bars are observed in these nuclear disks (e.g., Emsellem et al. 2003), and form in numerical simulations, too. The pattern speed of the nuclear bar is different from the pattern speed of the main bar. This means that orbits inside the ILR leave the main bar alignment, which is likely to weaken the bar. Moreover, the nuclear bar initiate a nuclear gas infall, which makes the mass be more concentrated than without nuclear bar, and enhances the effects of the CMC. Indeed, in the simulation shown in Fig. 4, the bar weakening is observed to be faster when the nuclear bar has developed, providing an example to the weakening of bars by nuclear bars.

Bars in spiral galaxies are transient features

Because of the CMC growth, the gas gravity torques, and the effects of nuclear density waves, bars are found to be transient features. As discussed in Section 2, the gravity torques have the same amplitude than in many observed systems. Their effects on bars, as well as the fueling of a CMC, are thus believed to be realistic in our simulations. We then conclude that bars are transient features that are dissolved in a few dynamical times. In late-type systems, the bar dissolution is the result of gravity torques and CMC growth, and typically takes 1–3 Gyrs. These two processes need more time too dissolve bars in early-type galaxies, for they contain less gas and their bulge reduces the gravity torques. However, nuclear bars are common and they also affect the main bar. Finally, the bar life-time rarely excess 4 Gyrs, excepted in unusually gas poor spiral galaxies, with not more than 4% of gas in their disk, and massive bulges: in these galaxies, bars can last a Hubble Time. At the opposite, a few billion years ago, galaxies were more gas-rich and disk dominated than today. Gravity torques were then higher, and could fuel more massive central concentrations. Bars at high redshift are thus expected to be shorter-lived than today. The bars observed at redshifts larger than 0.6 cannot have lasted for several billion years. The ubiquity of bars today then implies that they have been reformed after their dissolution, or that an external process has prevented their dissolution.

The influence of viscosity in the sticky-particle code

The main concern in our results is the influence of viscosity. The viscous torques in our simulations is much smaller than gravity torques (Fig. 1), and do not directly fuel an unrealistic CMC, contrary to the assumption of Regan & Teuben (2004) on our results. Yet, as explained in Sect. 2, a too large numerical viscosity may cause an unrealistically large gravity torque to develop. We have already mentioned that gravity torques in our simulations are comparable to observed systems. In addition, we also checked that our results are not dependent on the box size used to compute particles collisions in the sticky-particles scheme: this parameter controls the distance along which particles interact, influences the numerical viscosity, thus the gas/bar phase shift and the gravity torques may depend on it. We used box sizes ranging from 10 pc to 350 pc: this does not lead to large changes in the gravity torques, and the bar life-time is not dependent on this parameter at all, as shown in Fig 5.

We also changed the mean free path of gas clouds by varying the frequency of collisions, and changed the elasticity coefficient of collisions to keep the overall energy dissipation constant. Reducing the mean free path of gas clouds reduces the viscosity, but this does not significantly change our results on the gravity torques amplitude and bar life-time. We thus believe that the life-time of bars in our sticky-particles simulations are realistic, for the gravity torques and associated CMC fueling are correctly reproduced.

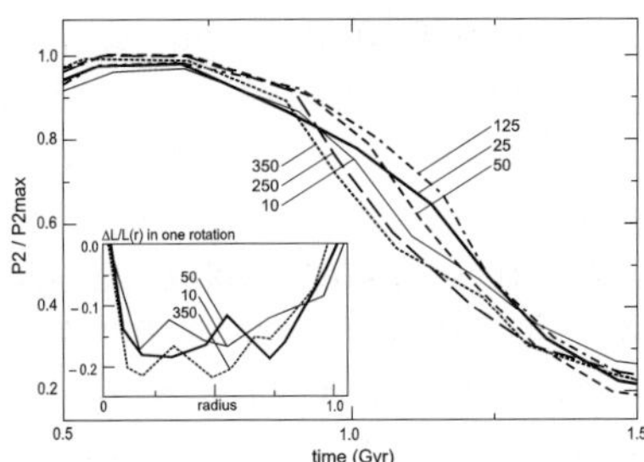

Figure 5. Evolution of the bar strength P2 for the same simulation as Fig. 3 with several values of the cell size of the collision grid in the sticky-particle scheme. The cell size has been varied from 10 to 350 pc, and the bar life-time remains unchanged. The thick line represents the main simulation detailed in Fig. 3 (cell size 50 pc). The inset, to be compared with Fig. 1, shows the gravity torques inside the bar CR, for cell sizes of 10, 50 and 350 pc: the torques amplitude remains nearly unchanged.

4. The reformation of bars: gas accretion on spiral galaxies

Since bars are dissolved in a few billion years, they must have been reformed to be frequent at low redshift. Two mechanisms may reform bars: galaxy interactions and accretion of large amounts of gas (Sellwood 1996). Berentzen et al. (2004) have shown that interactions can trigger bars in gas poor galaxies, but cannot reform bars that have been dissolved in galaxies that contain several percent of gas. Thus, interactions have only limited effects on bars at $z = 0$, and were even less efficient to reform bars a few billion years ago, when the galaxies contained more gas. We also led simulations of interactions that confirm this conclusion: galaxy interactions are not efficient enough to explain the ubiquity of bars in the Local Universe.

We have studied the accretion of gas by spiral galaxies in BC02. In our simulations, fresh gas can fuel the disk mainly when the first bar is dissolved: during the barred period, accreted gas is repelled outside the CR. After the bar dissolution, this gas increases the disk-to-bulge mass ratio, and reduces the stellar velocity dispersion through star formation. Then, a new bar forms. We have observed up to 4 cycles of bar formation/destruction over a Hubble time, as illustrated by Fig. 6. In some cases, gas fuels the disk in a continuous way, and maintains the bar for a Hubble time; such a long-term bar maintenance is however less frequent than the complete dissolution/reformation cycle.

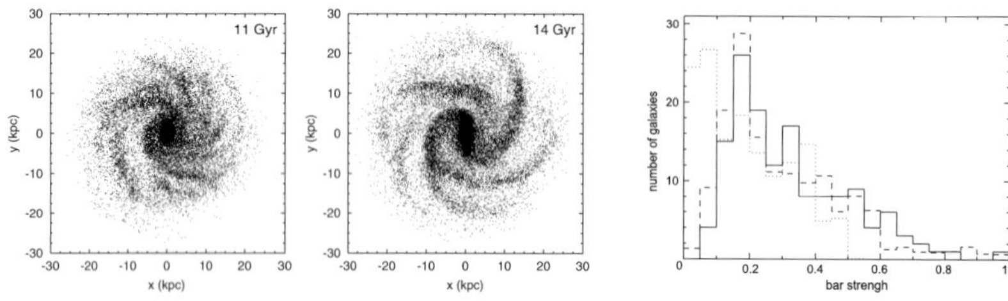

Figure 6. Left: Simulation of bar reformation with gas accretion. At 11 Gyr, this spiral galaxy has already been barred twice, and the second bar has been dissolved. At 14 Gyr, a third bar has formed. In the same time, a grand-design spiral is maintained, while such structures disappear in a few dynamical times in simulations of isolated galaxies. – Right : Distribution of bar strengths in observed galaxies and simulations (Block et al. 2002). Solid lines represent observations, dashed lines simulations with gas accretion, and dotted lines simulations without gas accretion.

This scenario of bar reformation through gas accretion is in good agreement with observations. Our simulations are able to reproduce the distribution of bar strength observed in the Local Universe (see Fig. 6 and Block et al. 2002), provided that the gas accretion rate on spiral galaxies is of the order of 10 $M_{\odot}.yr^{-1}$. Such an accretion rate between $z = 1$ and $z = 0$ is expected from cosmological models (e.g., Semelin & Combes 2002). Gas accretion on spiral galaxies not only reforms bars, but also maintains spiral arms. Bars also change their pattern speed when they are reformed.

A major issue is to know how gas can enter inside the disk in a few dynamical times. Several solutions can be proposed:

- infalling gas has a low angular momentum, thus directly reaches the inner disk without requiring any specific process

- accreted gas has a large velocity dispersion (for instance made-up of massive clouds that have various velocities) which triggers its viscosity. This makes gas enter the disk in the simulations of BC02.
- gas is accreted outside the disk plane: differential precession, collisions between different planes, and energy dissipation, will make gas fall into the disk plane and reduce its angular momentum, which can naturally fuel the inner regions.

The exact mechanism depends on the properties of infalling gas, which we are studying using cosmological simulations.

5. Conclusion: the cycle of bars in spiral galaxies

Contrary to what is usually believed, the dissolution of bars is not only a consequence of the growth of CMCs: gravity torques and nuclear bars also play an important role. The gravity torques, and the gas infall that they initiate, have been shown to be realistic in our sticky-particles simulations, so that the life-time of bars is 1 to 4 billions years in most spiral galaxies. In the past, this life-time could even have been shorter. In the same time, higher accretion rates may have reformed bars more rapidly.

The ubiquity of bars in the Local Universe can they be explained only if bars are reformed, which requires the accretion of large amounts of gas by galaxies, that double their mass in a few billion years. Only a small part of this accretion can come from infalling companions, but cosmological simulations predict that spiral galaxies accrete large amount of gas along cosmic filaments (Semelin & Combes 2002).

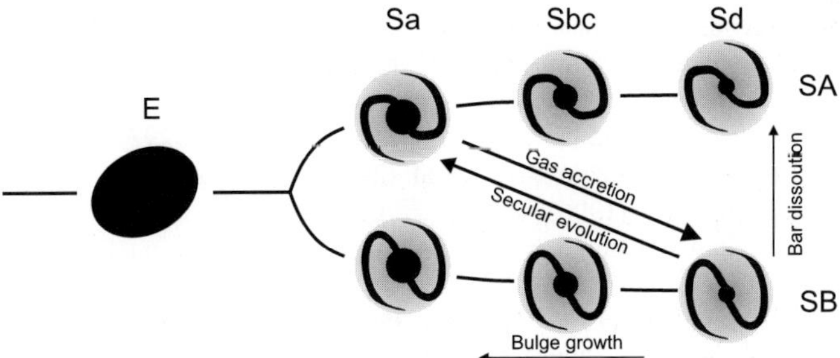

Figure 7. Evolution of spiral galaxies along the Hubble sequence, under the effects of secular evolution (bar dissolution, bulge growth – minor mergers may also contribute to the later one) and gas accretion. The secular disk thickening is not shown. The reverse effects of gas accretion can regulate the presence of bars, the disk thickness, and the bulge-to-disk mass ratio.

The cycle of bar dissolution and reformation, through the effects of secular evolution and gas accretion, is illustrated in Fig. 7. Gas accretion on spiral galaxies has many other effects. It explains the star formation history (Semelin

& Combes 2002). The combined effects of galaxy mergers and gas accretion can explain the formation of thick disks around younger thin disks. Accretion of gas form outside the disk plane can form and maintain disk warps. Asymmetrical accretion may be responsible for lopsidedness morphologies. The growth of bulges during bar episodes can be regulated by the disk fueling. The bar dissolution/reformation can provoke a succession of active phases in the nucleus, and each galaxy may have undergone several bar-induced AGN phases, at the end of each bar episode. Thus, the accretion of gas by spiral galaxies, predicted by the standard cosmological theories, not only influences bars, but more generally the whole evolution of spirals along the Hubble sequence.

Finally, spiral galaxies are open systems. Their Hubble type can change in a few dynamical times. Inner mechanisms tend to make them evolve toward earlier-types, but the cosmological accretion compensate these effects and make them return to late-type classes.

Acknowledgments

We thank D. L. Block, I. Puerari, and R. Buta, who have contributed to a part of this work. We are grateful to B. Elmegreen for constructive comments. Numerical simulations were computed at CNRS computing center, at IDRIS. FB acknowledges financial support from ENS, Paris, France.

References

van Albada, G. D., and Roberts, W. W. (1981) ApJ, 246, 740
Athanassoula, E. (1992) MNRAS, 259, 345
Berentzen, I., Athanassoula, E., Heller, C. H., and Fricke, K. J. (2004) MNRAS, 347, 220
Block, D. L., Bournaud, F., Combes, F., Puerari, I., and Buta, R. (2002) A&A, 394, L35
Bournaud, F., and Combes, F. (2002) A&A, 392, 83 (BC02)
Bournaud, F., and Combes, F. (2004) A&A, in prep. (BC04)
Emsellem, E., Goudfrooij, P., and Ferruit, P. (2003) MNRAS, 345, 1297
Eskridge, P. B., Frogel, J. A., Pogge, R. W., et al. (2002) ApJS, 143, 73
Junqueira, S., and Combes, F. (1996) A&A, 312, 703
Miller, R. H. (1996), in ASP Conf. Ser. 91, Barred Galaxies, ed. R. Buta, D. A. Crocker, and B. G. Elmegreen, 569
Pfenniger, D., and Norman, C. (1990) ApJ, 363, 391
Regan, M. W., and Teuben, P. J. (2004) ApJ, 600, 595
Sellwood, J. A. (1996), in ASP Conf. Ser. 91, Barred Galaxies, ed. R. Buta, D. A. Crocker, and B. G. Elmegreen, 259
Semelin, B., and Combes, F. (2002) A&A, 388, 826
Shen, J., and Sellwood, J. A. (2004) ApJ, 604, 614 (SS04)
Sheth, K., Regan, M. W., Scoville, N. Z., and Strubbe, L. E. (2003) ApJ, 592, L13
de Vaucouleurs, G., and de Vaucouleurs, A. (1963) AJ, 68, 278
Wozniak, H., Combes, F., Emsellem, E., and Friedli, D. (2003) A&A, 409, 469

DYNAMICS OF DOUBLY BARRED GALAXIES, ALSO WITH THE INNER BAR RETROGRADE

Witold Maciejewski
Obserwatorium Astronomiczne Uniwersytetu Jagiellonskiego, Orla 171, 30-244 Krakow, Poland

Abstract I use a method developed by Maciejewski & Sparke (2000) to find regular orbits that can support doubly barred galaxies. In such systems, double-frequency orbits play the same role as closed periodic orbits do in unchanging potentials. Orbits in double bars can be best studied through their maps. Here I use such maps in order to determine the fraction of the phase-space occupied by ordered motions. I also examine whether the recent kinematical observations of NGC 2950 can be explained in terms of two prograde bars, or whether they require retrograde inner bar.

Keywords: stellar dynamics — galaxies: kinematics and dynamics — galaxies: structure

1. Introduction

Bars within bars appear to be a common phenomenon in galaxies. Recent surveys indicate that up to 30% of early-type barred galaxies contain nested bars (Erwin & Sparke 2002). The relative orientation of the two bars is random, therefore it is likely that the bars rotate with different pattern speeds. Inner bars, like large bars, are made of relatively old stellar populations, since they remain distinct in near infrared (Friedli et al. 1996). Galaxies with two independently rotating bars do not conserve the Jacobi integral, and it is a complex dynamical task to explain how such systems are sustained. To account for their longevity, one has to find sets of particles that support the shape of the potential in which they move. Closed periodic orbits are separated in phase-space in such systems, and therefore they are unlikely to provide orbital support for nested bars. However, in systems with two pattern speeds, double-frequency orbits play a fundamental role. Thus in double bars a large fraction of particle trajectories gets trapped around a class of double-frequency orbits. Although such orbits do not close in any reference frame, they can be conveniently mapped onto the loops (Maciejewski & Sparke 1997), which are an efficient descriptor of orbital structure in a pulsating potential. Orbital support for nested bars can be provided by placing particles on the loops.

D. Block et al. (eds.), Penetrating Bars through Masks of Cosmic Dust, 175–178.

2. Double-frequency orbit and its corresponding loop

An example of a double-frequency orbit is given in Fig.1. Fourier spectrum of this orbit has two clear maxima at frequencies related to the pattern speeds of the two bars. The double-frequency orbit is not closed, and its appearance depends on the rotating frame in which it is drawn. However, if one creates a map of this orbit by plotting the particle position at every alignment of the bars, such a set of points is frame-independent. Moreover, for a double-frequency orbit these points populate closed curves, called loops (Maciejewski & Sparke 1997). Loops in the potential of a doubly barred galaxy pulsate with the relative period of the bars (Fig.2).

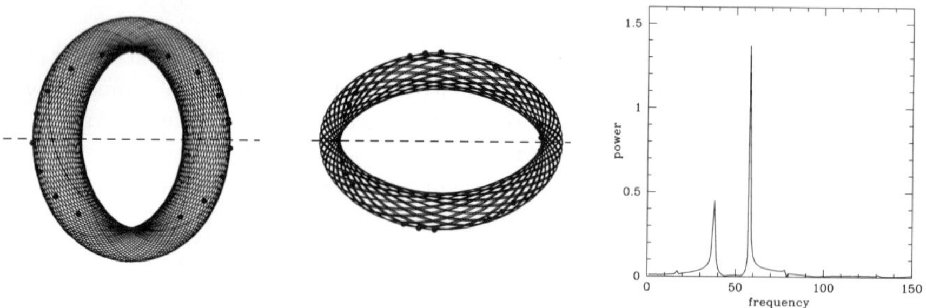

Figure 1. Double-frequency orbit in doubly barred potential, written in the frames corotating with the big and the small bar (horizontal), together with its Fourier spectrum. Dots mark positions of the particle at every alignment of the bars – they populate the loop.

Double-frequency orbits play crucial role in providing orbital support for the pulsating potential of double bars. No closed periodic orbits have been proposed as candidates for the backbone of such a potential. If in a given potential of two bars there are loops that follow the inner bar, and other loops that follow the outer bar, then one may expect that such a potential is dynamically possible. An example of such a potential has been constructed by Maciejewski & Sparke (2000).

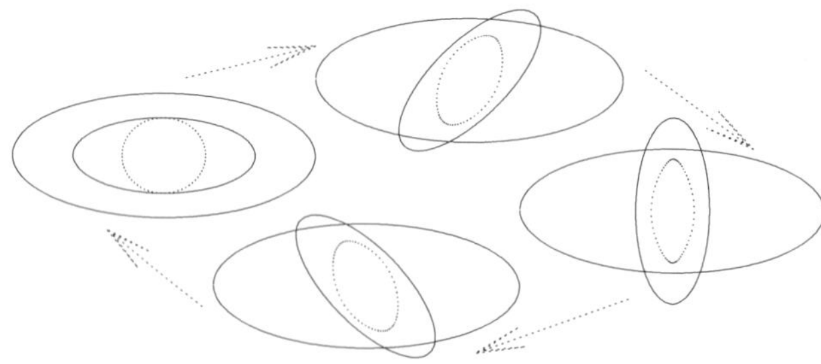

Figure 2. Evolution of the loop from Fig.1 during one relative period of the bars. The bars are outlined with solid lines.

3. Trapping particles around the loops

Double-frequency orbits in double bars are surrounded by regular orbits in the same way as are the closed periodic orbits in a single bar. In both cases, the trapped regular orbits oscillate around the parent orbit. How much of the phase space in double bars is occupied by orbits trapped around double-frequency orbits? If an orbit is trapped around a double-frequency one, the points in its map defined above should gather in a ring around the loop, which is the map of the double-frequency orbit. The width of this ring can serve as an indicator of how well an orbit is trapped around the loop.

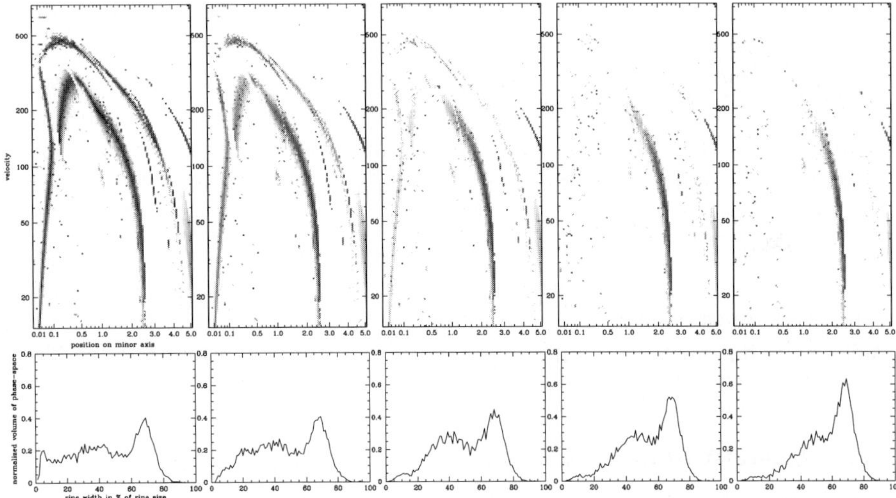

Figure 3. **Top panels:** The width of the ring in maps of orbits in double bars. Starting velocity is tangential for the diagram on the left, and departs from it by 3.6° , 7.2° , 10.8° , and 14.4° for the diagrams towards right. Darker color means smaller width. **Bottom panels:** Corresponding histograms of the ring width.

For the potential of Model 2 defined in Maciejewski & Sparke (2000), the widths of the rings are displayed in top panels of Fig.3 as a function of 3 parameters: the starting position (particle starts on the minor axis of the aligned bars), and starting velocity (its value and direction). With these 3 parameters one can describe all orbits that ever pass through the minor axis of the bars when the bars are aligned (Maciejewski & Athanassoula 2004). As expected, there are areas of small width, particularly for starting velocities perpendicular to the minor axis of the bars (symmetric loops). Two stripes of low width appear on the diagrams, which correspond to the x_1 and x_2 orbits in a single bar (the outer and the inner arch, respectively). There are possible regions of chaos in double bars (white areas), but overall loops in double bars and periodic orbits in single bars trap similar volumes of phase-space around them.

Lower panels of Fig.3 show histograms of ring width. Characteristic quasi-Gaussian profile on the right of each histogram indicates chaotic orbits. Low-width rings to the left represent regular orbits which are trapped around the loops. These diagrams confirm that regular orbits constitute considerable fraction of the phase-space, and this fraction is maximal for orbits starting with no radial velocity component.

4. Counter-rotating inner bar?

Recent direct kinematical observations have indicated that two stellar bars in the doubly barred galaxy NGC 2950 rotate with two different pattern speeds (Corsini et al. 2003). One can generalize the Tremaine-Weinberg method to multiple pattern speeds, but problems occur when one wants to evaluate the centroid and the luminosity-weighted velocity for each bar (Maciejewski 2004). However, in the formalism described above, these quantities are well defined for each loop, and one can look whether their combination can reproduce the observed data. So far I found it impossible to reproduce the data with loops in two prograde bars. This may indicate that the inner bar in NGC 2950 counter-rotates with respect to the primary bar. It is peculiar, given that the rotation curve does not indicate any significant fraction of stars on retrograde orbits. Dynamics of two counter-rotating bars made only out of stars on prograde orbits has not been studied yet, and it may significantly differ from the already investigated scenarios.

5. Conclusions

In a potential of two independently rotating bars, a large fraction of phase space can be occupied by trajectories trapped around parent regular orbits. These are double-frequency orbits, with the two frequencies corresponding to the forcing frequencies of the two bars, and they do not close in any reference frame. The structure of these orbits can be mapped using the loop approach, which allows us to single out dynamically possible double bars, and to extract their observable kinematical signatures.

References

Corsini, E. M., Debattista, V. P. & Aguerri, J. A. L. 2003. ApJL, 599, L29
Erwin, P. & Sparke, L. S. 2002, AJ, 124, 65
Friedli, D., Wozniak, H., Rieke, M., Martinet, L., & Bratschi, P. 1996, A&AS, 118, 461
Maciejewski, W., in preparation
Maciejewski, W. & Athanassoula, E., in preparation
Maciejewski, W. & Sparke, L. S. 1997, ApJL, 484, L117
Maciejewski, W. & Sparke, L. S. 2000, MNRAS, 313, 745

A COORDINATED EPISODE OF AGB STAR PRODUCTION AT LARGE GALACTOCENTRIC DISTANCES IN THE ANDROMEDA GALAXY?

Guy Worthey
Astronomy Program, Washington State University, Pullman, WA 99164-2814, USA

Abstract Stars in the halo of M31 are enriched with heavy elements a factor of ten more than the Milky Way halo. Many M31 globular clusters show high metallicities and enhanced nitrogen abundance, as does the M31 nucleus and near-nuclear regions. The bulk of the nitrogen comes from mass-loss in intermediate-mass asymptotic giant branch stars, a fairly quiescent process incapable of driving enriched gas out of a galaxy. We weigh the relative merits of supernova-wind-driven turbulent mixing versus in-situ large-galactocentric-radius enrichment to explain the observations.

Keywords: Chemical evolution, spiral galaxies, abundances, galaxy formation.

1. Introduction

Posters of the beautiful Andromeda galaxy and its companions M32 and NGC 205 adorn many of our office walls, and we have knowledge of this nearby spiral that rivals and in some ways surpasses our knowledge of the Milky Way. In particular, the Andromeda galaxy promises to test theories of galaxy development and history that were developed to explain observations in the Milky Way, that is, the modern versions of the Eggen, Lynden-Bell, and Sandage (1962) gaseous collapse and the Searle & Zinn (1978) falling fragments.

2. Contrasting M31 and the Milky Way

The Andromeda Galaxy (M31; NGC 224) has about twice the mass and twice the number of globular clusters as does the Milky Way. Gas phase material comprises 2% of the baryonic total in M31, but about twice this in the Milky Way (van den Bergh 2000). Further, the Milky Way is forming stars at a greater rate, with about a factor of 4 greater mass in ionised hydrogen. Most of the gas in M31 is concentrated in a 10-kpc “ring of fire”, although OB associations appear throughout the disk. Van der Kruit (1989) finds that

D. Block et al. (eds.), Penetrating Bars through Masks of Cosmic Dust, 179–187.

the bulge of M31 contributes 25% of the total light while Milky Way's bulge contributes more like 12%. (One should temper this result with the realization that the Milky Way emits more light per unit mass than does M31; in terms of mass the percentage contributions may be more nearly equal.) The nuclear regions appear gas- and dust-free, but the nucleus is double. One component is located at the bulge photocenter, but the brighter component is 0.5 arcsec away. Explanations for this observation usually involve the words "black" and "hole." The M31 nucleus contrasts greatly with Milky Way's, the latter having active star formation while the former is presently quiescent. A mean age for the stars in the M31 nuclear region can be guessed at via Lick system absorption lines. Sil'chenko et al. (1998) find that the nuclear region is younger than the bulge, and diagrams in Worthey (1998) show a mean age of around 3 Gyr. The interpretation of this can vary between a fairly strong star formation burst 3 Gyr ago to a very mild starburst much more recently with all intermediate possibilities allowed.

Globular cluster systems differ in the two galaxies despite having about the same number of clusters per unit mass. With a dividing line between metal poor and metal rich placed at [Fe/H] $= -1.0$, Milky Way metal rich globular clusters are somewhat concentrated toward the center of the galaxy, but they are not strongly disky despite the fact that they are often referred to ask "disk globulars". M31 has more metal-rich globular clusters than the Milky Way, and the most metal-rich of these *do* appear to have disk like kinematics (van den Bergh 1969). As in the Milky Way, the metal rich M31 globular clusters lie preferentially toward the center of the galaxy, but unlike the Milky Way, very metal-rich globulars (of nearly solar abundance) are found quite far from the center.

Halo field stars in the Milky Way have an abundance distribution similar to that of the globular clusters spatially coresident there: with a peak at [Fe/H] ~ -1.5 and a broad dispersion that matches the simplest of chemical evolution models, the "Simple model," that assumes a closed box, full mixing, instantaneous recycling, and a constant heavy element yield in every generation of stars. Halo field stars in M31 are quite different. Figure 1 shows a series of abundance distributions based on Hubble Space Telescope images of red giant stars at a variety of sampling points in M31 from the inner disk out to the halo. The distributions are fairly broad (but not as broad as the Simple model just mentioned) and also show a progression in median abundance that goes from metal-rich in the inner disk to metal-poor in the halo. But Figure 2 shows that the outer disk and halo abundance holds steady at [Fe/H] ≈ -0.5, a full factor of ten more heavy element enriched than the Milky Way halo. This is almost alarming: why should a galaxy only a factor of two larger have a halo abundance a factor of ten higher?

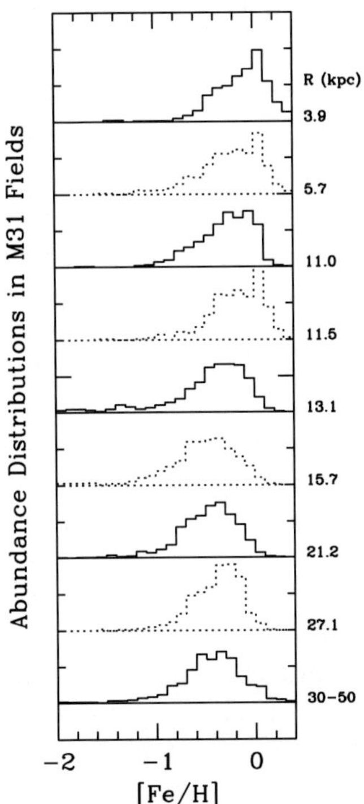

Figure 1. Stellar abundance distributions as a function of galactocentric radius in the Andromeda galaxy. The widths of the distributions are similar at all radii but the peak drifts modestly more metal poor toward the outskirts. Data from Worthey et al. (2004).

In the case of M31, the abundance of the metal-poor globular clusters (and also the several metal-poor dwarf spheroidal galaxies that orbit M31) does not match the abundance of the halo they occupy, making them look like intruders whose chemical enrichment occurred quite separately from the bulk of the halo. Having precise ages for neither the clusters nor the halo stars, it is impossible to establish the time order of events. Brown et al. (2003) have established the presence of at least a trace of few-Gyr stars in one halo field by detecting main sequence turnoff stars at 30th magnitude, but this came at the cost of many orbits of HST time and therefore is not trivially extensible to the rest of the halo. If one takes a close look at the chemistry of Milky Way globular clusters and dwarf spheroidals, one finds clear differences in abundance ratios of the dwarf spheroidals compared with halo field stars (Shetrone 2004) but the globular cluster abundances, on average, match the halo abundance pattern

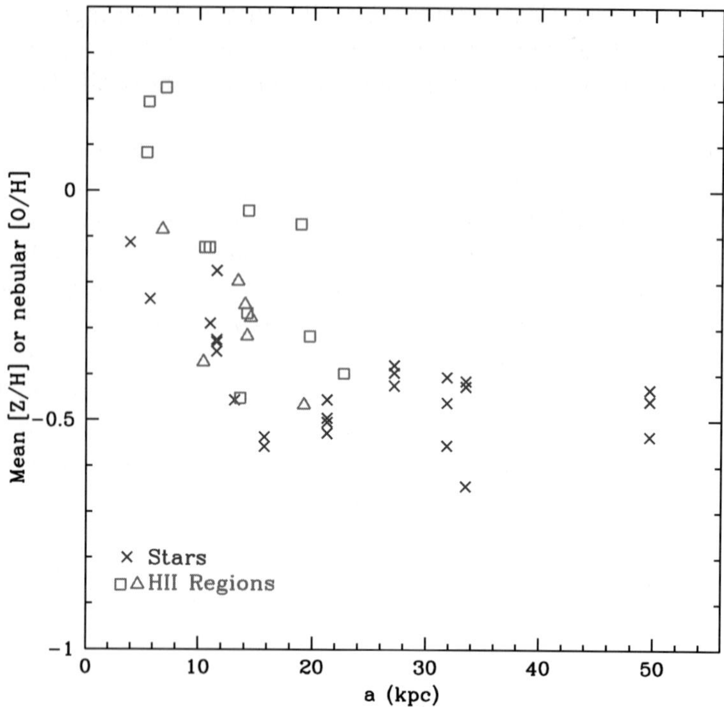

Figure 2. Abundance gradients in M31. Nebular abundances from HII regions (Blair et al. 1982; Dennefeld & Kunth 1981) lie parallel to but slightly higher than the (median) stellar abundances in the range of overlap, which is the inner 25 kpc of the galaxy. Outside this radius, the stellar abundances flatten, scattering about a mean of −0.5 dex.

(Sneden et al. 2004). So for the Milky Way, the dwarf spheroidals look like intruders, but the globular clusters look like they are part of the halo.

By counting red giants in the vicinity of M31, Ferguson et al. (2002) have established the presence of dynamical streams in the halo. The Milky Way also has qualitatively similar streams, the most blatant of which is the Sagittarius dwarf, which has been tidally stretched to almost 270 degrees of arc (Figure 3 of Majewski et al. 2003).

3. Nitrogen

The fact that M31 is composed of stars that have very strong absorption lines was observed by Morgan and Mayall (1957), who observed in particular the CN feature at λ4216. By the 1970's it was realized that this implied both that giant stars dominated the integrated light and that the giants in M31 must be as strong-lined as the strongest-lined of local giants, implying a high metallicity. Later work (Worthey et al. 1992; Worthey 1998) strongly indicated that elements could decouple from scaled solar, so that M31's strong cyanogen could be due to elevated nitrogen abundance, not necessarily all elements in lockstep.

Subsequent analysis has not changed this conclusion at all. Figure 3 shows good evidence that the M31 nucleus and bulge has at least a factor of two enhancement in the [N/Fe] ratio. Figure 3 contains a lot of information; let me summarize. Figure 3 is an example of an index-index diagram that is nearly age-metallicity degenerate. That is, any model of a given age and abundance will land very close to the solid lines drawn on the plot (except for very young ages, which will drift left; a 1.5-Gyr sequence is drawn as a short line segment to illustrate this). But the models are built with Milky Way stars and Milky Way abundance ratios, which are nearly solar for pretty much the whole span of abundance. So enhancements in [N/Fe] will move objects to the right in this diagram. If we knew the object's age exactly then we could measure an [N/Fe] value with very little ambiguity.

The various objects plotted are Milky Way and M31 globular clusters, and E and S0 galaxies of various sizes and density environments. There is model ambiguity at the very metal-poor end, but of the intermediate and metal-rich globular clusters, Milky Way objects are split between scaled-solar and [N/Fe] enhanced, but M31 globulars are all N-enhanced. The M31 nuclear region is also plotted as an asterisk at high [N/Fe]. Solar neighborhood environments would plot along the model sequence, as does the other asterisk, compact elliptical M32. There is a systematic among the galaxies in that the large ones tend to have higher [N/Fe]. Parenthetically, they also have higher [Mg/Fe] and presumably other supernova type II enrichment in a pattern that the M31 globular clusters do *not* share.

In terms of nucleosynthetic origins, it is well established that most nitrogen comes from the CNO cycle in upper-main sequence stars. The nitrogen thus produced requires the presence of carbon in the star first, so it is thus termed *secondary* nitrogen rather than *primary* nitrogen produced by supernovae. An important characteristic of secondary N production is that its effectiveness increases with increasing abundance (cf. Henry & Worthey 1999). This means that we would naturally expect a higher [N/Fe] ratio in metal rich environments.

The dispersal of nitrogen is fairly unique in that supernovae are minimally involved. Rather, most of the nitrogen is re-injected into the interstellar medium via relatively slow mass-loss during the asymptotic giant branch phase of evolution. This process is energetically incapable of generating galaxy-scale winds by itself; unless stirred by some other mechanism, nitrogen enrichment must be a purely local phenomenon.

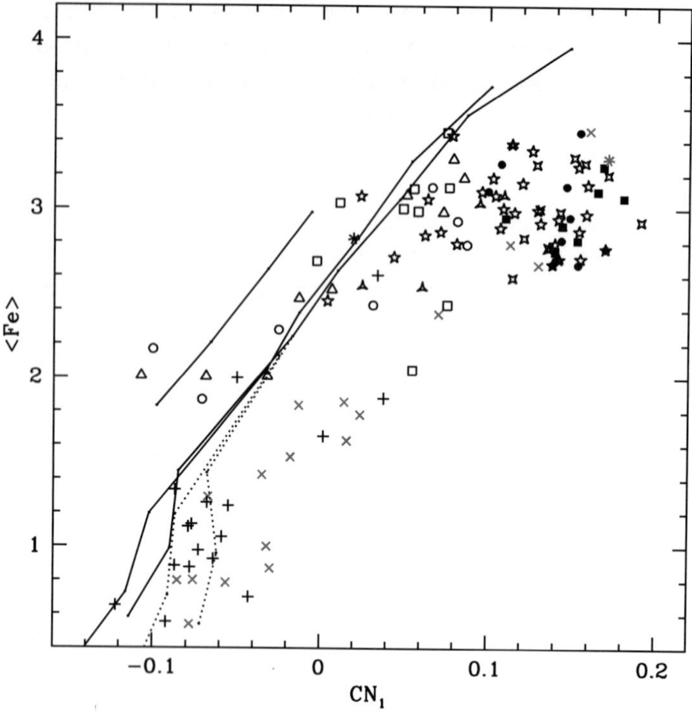

Figure 3. A variety of objects are plotted in an integrated light index-index diagram. Index <Fe> measures something between iron abundance and total heavy element abundance. The CN index measures mostly nitrogen abundance. The "x" symbols are M31 globular clusters, but "+" crosses are Milky Way globular clusters. Other symbols are E and S0 galaxy nuclei, with M32 and M31 marked with asterisks. The solid lines are model predictions for ages 9 and 18 Gyr (and 1.5 Gyr for the shorter line) and a variety of abundances with [Fe/H] = −2 in the bottom left, and supersolar in the upper right. The models have scaled-solar abundances, so objects that lie to the right are interpreted to have a high nitrogen abundance.

4. Formation

The main puzzle I seek to explain is why M31 has a factor of two more nitrogen than the Milky Way. The near-nuclear regions are enhanced, almost all metal-rich globular clusters are enhanced, and at least the inner parts of the gas disk are enhanced (as measured from N/O ratios in HII regions; Dennefeld & Kunth 1981). The outer (20-25 kpc) disk may have solar N/O ratios. And, parenthetically, the satellite elliptical M32 has quite normal abundance ratios, so M32 did not participate in M31's enrichment pattern.

An obvious, but unappealing answer would be that there was a bump in the initial mass function in M31 around 5 solar masses. More 5-solar-mass stars would leave the supernova enrichment unaffected, but would produce more nitrogen. The trouble with this answer is that it is unphysical and would require such major fine-tuning of star formation parameters that Occam would cringe. The explanation I proposed in Worthey (1998) is that M31 underwent a very major galactic wind episode after it was already established as a galaxy, with plenty of heavy elements from both supernova types and also nitrogen pooled in the gas phase. This material was blown to very large galactocentric radius by a second wave of star formation, where it remained available for incorporation into globular clusters and failed clusters that evaporated into halo stars. The "second wind" could also be caused in part by an especially violent merging event. Either before this phase or after, accreted satellite galaxies would bring their own metal-poor stars and clusters with them, adding metal-poor objects to the halo. The Milky Way lacked this "second wind" phase.

One problem with this picture is that, in contrast to M31, Milky Way bulge stars probably have $[N/Fe] < 0$ (McWilliam & Rich 1994), and the disk has $[N/Fe] \approx 0$, so the Milky Way simply seems not to have made the N at all. In this case, I think it might be best to wait until the decade-old McWilliam & Rich result can be confirmed.

Perhaps, a more appealing explanation is that the N was created in the outer disk of M31 in regions that have now faded to very low surface brightness. That this *can* happen is shown by the startling carbon-star-rich arcuate feature in local spiral M33 (cf. Block et al., this volume). At the present epoch, the M33 carbon stars appear distributed very widely and smoothly (not clumpily) and yet they must have been coeval to within about half a Gyr. The van den Bergh (1969) result that the metal-rich globular clusters have disk like kinematics lends support to this argument, although the "second wind" scenario could also form metal-rich clusters with disky kinematics if the post-wind gas was allowed to settle for a while. In the disk-origin model, the M31 bulge may be decoupled from the globular cluster enrichment, and may simply follow the trend seen in elliptical galaxies that larger velocity dispersions lead to light-element preferential enrichment.

One problem with this picture is that, in gas phase at the present epoch, there is a strong N/O gradient in the M31 disk, with more N toward the center of the galaxy (Dennefeld & Kunth 1981). If the N production was a disk phenomenon, then the M31 disk should also be polluted throughout by extra nitrogen. Another problem with this scenario is to explain how the Milky Way escaped the same disk-generated nitrogen enrichment. Speaking for myself, obvious mechanisms do not spring to mind. A final objection is the overall high abundance seen in the M31 halo. Since it is kinematically decoupled from the disk, it is very difficult to think of any mechanism by which the halo could be enriched by the outer disk.

An interesting speculation is that M31 has a "true" bulge while the Milky Way has a "pseudobulge," meaning a bulge-like dense region that was created by secular processes such as bar dynamics from the inner disk (Kormendy 1993). This could explain the M31/Milky Way disparity and also offers a prediction: that pseudobulge galaxies generally will not have metal-rich halo stars or globular clusters outside a few kiloparsecs galactocentric radius (since nearly all of the mass accretion for these galaxies will have occurred as quiet gaseous disk accretion) while true bulges and elliptical galaxies will exhibit at least some globular clusters with high metallicities at large galactocentric radius. The latter statement is true, but the former has yet to be confirmed.

In summary, we are left with an unsolved dichotomy between the Milky Way and the Andromeda galaxy, whose halo metallicities differ by a factor of ten, whose globular cluster systems seem perfectly disparate at the metal-rich end, and whose [N/Fe] ratios differ by a factor of two in all parts of the galaxies except the outer disk.

Acknowledgments

I would like to acknowledge support from the organisers of this meeting, Washington State University, and Space Telescope Science Institute. John Kormendy deserves credit for the idea of true bulge versus pseudobulge possibly giving rise to some of the differences between M31 and the Milky Way.

References

Blair, W. P., Kirshner, R. P., & Chevalier, R. A. 1982, ApJ, 254, 50

Brown, T. M., Ferguson, H. C., Smith, E., Kimble, R. A., Sweigart, A. V., Renzini, A., Rich, R. M., VandenBerg, D. A. 2003, ApJL, 592, L17

Dennefeld, M., & Kunth, D. 1981, AJ, 86, 989

Eggen, O. J., Lynden-Bell, D., & Sandage, A. R. 1962, ApJ, 136, 748

Ferguson, A. M. N., Irwin, M. J., Ibata, R. A., Lewis, G. F., & Tanvir, N. R. 2002, AJ, 124, 1452

Henry, R. B. C., & Worthey, G. 1999, PASP, 111, 919

Kormendy, J. 1993, in IAU Symp. 153, Galactic Bulges, ed. H. Habing & H. Dejonghe (Dordrecht: Kluwer), 209

Majewski, S. R., Skrutskie, M. F., Weinberg, M. D., & Ostheimer, J. C. 2003, ApJ, 599, 1082

McWilliam, A., & Rich, R. M. 1994, ApJS, 91, 749

Morgan, W. W., & Mayall, N. U. 1957, PASP, 69, 291

Searle, L., & Zinn, R. 1978, ApJ, 225, 357

Shetrone, M. D. 2004, in Carnegie Observatories Astrophysics Series, Vol. 4: Origins and Evolution of the Elements, ed. A. McWilliam and M. Rauch (Cambridge: Cambridge Univ. Press), in press

Sneden, C., Ivans, I. I., & Fulbright, J. P. 2004, in Carnegie Observatories Astrophysics Series, Vol. 4: Origins and Evolution of the Elements, ed. A. McWilliam and M. Rauch (Cambridge: Cambridge Univ. Press), in press

Trager, S. C., Worthey, G., Faber, S. M., Burstein, D., & Gonzalez, J. J. 1998, ApJS, 116, 1

van den Bergh, S. 1969, ApJS, 19, 145

van den Bergh, S. 2000, The Galaxies of the Local Group (Cambridge: Cambridge University Press)
van der Kruit, P. C. 1989, in The Milky Way as a Galaxy, ed. G. Gilmore, I. R. King, and P. C. van der Kruit (Sauverny: Geneva Observatory), 331
Worthey, G., Faber, S. M., & Gonzalez, J. J. 1992, ApJ, 398, 69
Worthey, G. 1998, PASP, 110, 888
Worthey, G., Espana, A. L., Courteau, S. C., & MacArthur, L. A. 2005, in preparation

Paul Eskridge: Given the uncertainties, do you really think the nebular abundances in Figure 2 are higher than the stellar values?

Guy Worthey: No.

Comments from J. Kormendy: It is tempting to interpret the different metallicities of the M31 and Galactic halos in the context of bulge formation mechanisms. M31 has a classical bulge, which like an elliptical, is thought to form via a major merger. Mergers are violent. Lots of stars in the halo get flung there from much smaller radii. This dilutes metallicity gradients and makes halo stars – on average – be more similar in their metallicities to stars in the inner bulge. In contrast, the "bulge" of our Galaxy is box-shaped and interpreted as a pseudobulge. Pseudobulges are believed to form secularly out of disks, with much less violence and almost no "splashing". This allows the Galactic halo to be very different – for example, lower in metallicity – than the bulge. So your observations seem very consistent with our developing picture of bulge and pseudobulge formation.

FUELLING STARBURSTS AND AGN

Johan H. Knapen
Centre for Astrophysics Research, University of Hertfordshire, Hatfield, Herts AL10 9AB, UK

Abstract There is considerable evidence that the circumnuclear regions of galaxies are intimately related to their host galaxies, most directly through their bars. There is also convincing evidence for relations between the properties of supermassive black holes in the nuclei of galaxies and those of their host galaxy. It is much less clear, however, how stellar (starburst) and non-stellar (AGN) activity in the nuclear regions can be initiated and fuelled. I review gas transport from the disk to the nuclear and circumnuclear regions of galaxies, as well as the statistical relationships between the occurrence of nuclear activity and mechanisms which can cause central gas concentration, in particular bars and interactions. There are strong indications from theory and modelling for bar-induced central gas concentration, accompanied by limited observational evidence. Bars are related to activity, but this is only a weak statistical effect in the case of Seyferts, whereas the relation is limited to specific cases in starbursts. There is no observational evidence for a statistical connection between interactions and activity in Seyferts, and some evidence for this in starbursts, but probably limited to the extremes, e.g., ULIRGs. Some interesting hints at relations between rings, including nuclear rings, and the presence of nuclear activity are emerging. It is likely that the connection between the inflow of gaseous fuel from the disk of a galaxy on the one hand and the activity in its nuclear region on the other is not as straightforward as sometimes suggested, because the spatial- or time-scales concerned may be significantly different.

Keywords: galaxies: kinematics and dynamics – galaxies: spiral – galaxies: structure – galaxies: active – galaxies: starburst

1. Introduction

Black holes are ubiquitous in the nuclei of both active and non-active galaxies (e.g., Kormendy & Richstone 1995), and are thought to be at the direct origin of non-stellar nuclear activity (e.g., Lynden-Bell 1969; Begelman, Blandford & Rees 1984). The fact that the mass of the central supermassive black hole (SMBH) in a galaxy is correlated with the velocity dispersion of the bulge, and hence with its mass (Ferrarese & Merritt 2000; Gebhardt et al.

D. Block et al. (eds.), Penetrating Bars through Masks of Cosmic Dust, 189–206.

2000) provides the most tangible link between the nuclear regions and their host galaxies. But because not all galaxies with SMBHs have AGN characteristics, the presence of an SMBH in itself cannot be enough to make a galaxy "active", at least not continuously, and additional mechanisms must be considered which can ignite the nuclear activity.

In the case of starburst galaxies, defined rather loosely as galaxies which show abnormally enhanced massive star formation activity in their central regions (or in some more extreme cases throughout the galaxy), a similar question can be posed, namely what ignites the starburst. In both the AGN and the starbursts, the availability of fuel at the right place and at the right time must play a critical role. Such gaseous fuel is plentiful in the disks of galaxies, but must lose significant quantities of angular momentum in order to move radially inward. In fact, the "fuelling problem" is not the amount of fuel that is available, but how to get it to the right place, as graphically illustrated by Phinney (1994, his fig. 1). Estimates for the mass accretion rate needed to fuel AGN vary from around 10^{-4} $M_{\odot}$/year for low-luminosity AGN such as LINERs, up to around 10 $M_{\odot}$/year for high-luminosity AGN such as QSOs, or, over a putative lifetime of 10^8 years for the AGN activity, only 10^4 to 10^9 $M_{\odot}$.

Large stellar bars, as well as tidal interactions and mergers, have some time ago been identified as prime candidates to drive gas efficiently from the disk into the inner kpc (see next Section). In this review we will concentrate on the observational evidence, mostly statistical in nature, for the effectiveness of these gravitational mechanisms, concentrating on the effects of bars in Section 2, and on those of interactions in Section 3. Galactic rings are considered in Sections 4 and 5, and summarising remarks are given in Section 6. Related reviews considering the fuelling of primarily AGN include those by Shlosman, Begelman & Frank (1990), Beckman (2001), Combes (2001), Shlosman (2003), Wada (2004), and Jogee (2004).

2. The effects of bars

Theoretically and numerically, bars are expected to concentrate gas in the central regions of spiral galaxies because the torqued and shocked gas within the bar loses angular momentum which allows the gas to move further in (e.g., Schwarz 1984; Combes & Gerin 1985; Noguchi 1988; Shlosman, Frank & Begelman 1989; Knapen et al. 1995a). The dynamics of bars and their influence on the circumnuclear regions has most recently been reviewed by Kormendy & Kennicutt (2004), and previously by, e.g., Sellwood & Wilkinson (1993) and Shlosman (2001). The general theoretical and numerical formalism of bars is now well understood, and different aspects of it are continuously being confirmed by observations. For instance, we recently investigated the well-known numerical result that stronger bars will lead to straight dust lanes along the leading edges of the bar, whereas the dust lanes will be more curved in weak bars (Athanassoula 1992). Using a small number of barred galaxies for which we had adequate data, we could indeed confirm observationally that there is an anti-correlation between the amount of curvature of the dust lanes and the gravitational bar torque, or bar strength (Knapen, Pérez-Ramírez & Laine 2002; see Fig. 1). In another study (Zurita et al. 2004), we used Hα Fabry-Pérot data of the strongly barred galaxy NGC 1530 to show in a graphic, two-dimensional way that indeed, as predicted by theory, large velocity gradients are found at the position of the dust lanes. Within those lanes, directly tracing enhanced concentrations of dust and thus gas, but indirectly tracing the location of shocks in the gas, the large velocity gradient prohibits massive star formation, which we observe to be located just outside the regions of largest shear or velocity gradient (Zurita et al. 2004; see Regan, Vogel & Teuben 1997 for an Hα Fabry-Pérot map at lower resolution which nevertheless indicates the shocks in the velocity field).

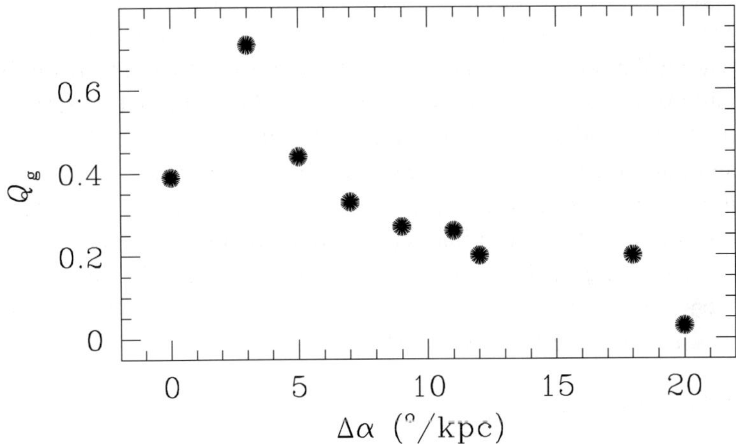

Figure 1. Gravitational bar torque $Q_{\rm g}$, an indicator of bar strength, as a function of the curvature of the dust lanes $\Delta\alpha$ in a sample of 9 barred galaxies. Small values of $\Delta\alpha$ indicate straight dust lanes, which are seen to occur in strong bars, thus confirming theoretical and numerical predictions. Data from Knapen, Pérez-Ramírez & Laine (2002).

Considering now specifically the theoretical and numerical view that bars can instigate radial inflow of gas, and thus lead to gas accumulation in the central regions of barred galaxies, several pieces of observational evidence to fit this picture have been forthcoming in recent years, both from observations of gas tracers in barred and non-barred galaxies (e.g., Sakamoto et al. 1999; Jogee, Scoville, & Kenney 2004; Sheth et al. 2004), and from other, less direct, measures of the gas concentration (e.g., Maiolino, Risaliti, & Salvati 1999; Alonso-Herrero & Knapen 2001). These results have been reviewed in somewhat more detail by Knapen (2004a), and we limit ourselves here to the conclusion that there is increasing observational support for the theoretical suggestion that bars lead to gas accumulation in the central regions of galaxies.

One must keep in mind that in all cases the observed correlation between the presence of a bar and increased central gas concentration is *statistical*, and not very strong, that there is a large overlap region in which the properties of barred and non-barred galaxies are very similar indeed (for instance, in the study by Sakamoto et al. 1999 just over half of the 19 sample galaxies are in the overlap range of gas concentration parameter $t_{\rm con}$, which is inhabited by both barred and non-barred galaxies), and that in the CO studies the X factor which gives the transformation of CO luminosity to mass is assumed to have the same value in the circumnuclear regions and in the disk. One can also question whether the statistical gas accumulation by bars is in fact related to the occurrence of nuclear activity of the non-stellar or stellar variety, and a careful consideration of both spatial- and timescales must be made to connect gravitationally driven inflow to fuelling of the starburst and/or the AGN. Finally, there is as yet no convincing direct observational evidence of inflow in a barred galaxy, mainly because the inflow rates are so low that they may be unobservable in practive (see above), and because most of the gas in bars moves around the bar, and will thus move inward during a part of its orbit, but then move outward again on a subsequent part (see discussion in Knapen 2001).

In the remainder of this Section, we will review the question of whether there is observational evidence that bars are related to the occurrence of nuclear activity.

2.1 Bars and starburst activity

There is a clearly observed trend for nuclear starbursts to occur preferentially in barred hosts (e.g., Hummel 1981; Hawarden et al. 1986; Devereux 1987; Dressel 1988; Puxley, Hawarden, & Mountain 1988; Arsenault 1989; Huang et al. 1996; Martinet & Friedli 1997; Hunt & Malkan 1999; Roussel et al. 2001). For example, Hummel (1981) reported that the central radio continuum component is typically twice as strong in barred as in non-barred galaxies; Hawarden et al. (1986) found that barred galaxies dominate the group of galaxies with a high 25μm/12μm flux ratio; and Arsenault (1989) found an enhanced bar+ring fraction among starburst hosts. Huang et al. (1996) revisited IRAS data to confirm that starburst hosts are preferentially barred, but did point out that this result only holds for strong bars (SB class in the RC3 catalogue, de Vaucouleurs et al. 1991) and in early-type galaxies, results confirmed more recently by Roussel et al. (2001). In contrast, Isobe & Feigelson (1992) did not find an enhanced far-IR to blue flux ratio among barred galaxies, and Ho, Filippenko & Sargent (1997) found only a very marginal increase in the detection rate of H II nuclei (indicative of starburst activity) among the barred as compared to non-barred galaxies in their sample, only among the late-type spirals (Sc-Sm), and most likely resulting from selection effects rather than bar-induced inflow (Ho et al. 1997). All results mentioned above rely on optical catalogues such as the RC3 to derive the morphological classifications, whereas it is now well-known that the presence of a bar can be deduced more reliably from near-IR imaging (e.g., Scoville et al. 1988; Knapen et al. 1995b). Although near-IR imaging leads to enhanced bar fractions as compared to optical imaging (e.g., Knapen, Shlosman & Peletier 2000; Eskridge et al. 2000), it is not clear how it would affect the results on bars and starbursts.

The statistical studies referred to above thus seem to show that bars and starbursts are connected, but that the results are subject to important caveats and exclusions. Further study is needed, determining bar parameters from near-IR imaging, using carefully defined samples, and exploring more direct starburst indicators than the IRAS fluxes which have often been used. Higher-resolution imaging of the starburst galaxies is also needed, to confirm the possible *circum*nuclear nature of the starburst, already suggested back in 1986 by Hawarden et al.

2.2 Bars and Seyfert activity

Seyferts are almost ideal for a study of AGN host galaxies: they are relatively local and occur predominantly in disk galaxies (see Fig. 2 for a nice example). One of the aspects of the host galaxy – Seyfert activity connection that has received a good deal of attention over the years is the question of whether Seyfert hosts are more often barred than non-Seyferts. Starting with the work of Adams (1977), many authors have dedicated efforts to resolve this question, without finding conclusive evidence (e.g., Adams 1977; Simkin, Su, & Schwarz 1980; Balick & Heckman 1982; MacKenty 1990; Moles, Márquez, & Pérez 1995; Ho et al. 1997; Crenshaw, Kraemer, & Gabel 2003). Unfortunately, many of these investigations are plagued by the absence of a properly matched control sample, by the use of the RC3 classification or, worse perhaps, ad-hoc and non-reproducible classification criteria to determine whether a galaxy is barred; and all of them are based on optical imaging.

Near-IR imaging is much better suited for finding bars (see Section 2.1), and a small number of studies have combined the use of high-quality, near-IR imaging with careful selection and matching of samples of Seyfert and quiescent galaxies. One such study, by Mulchaey & Regan (1997), reports identical bar fractions, but Knapen, Shlosman, & Peletier (2000), using imaging at higher spatial resolution and a rigorously applied set of bar criteria, find

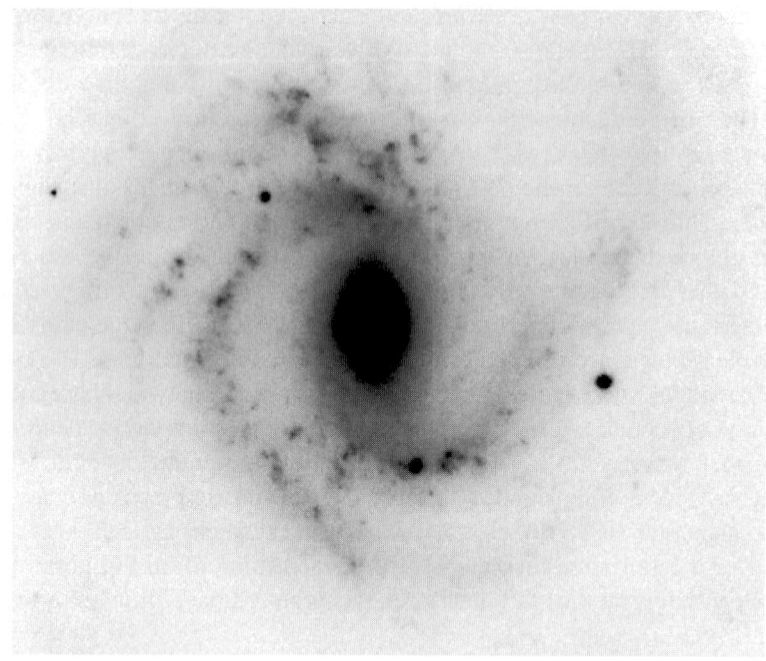

Figure 2. Near-IR K_s image of the Seyfert host galaxy NGC 4303, showing a prominent bar and a well-structured set of spiral arms. North is up, East to the left, and the size of the image shown is approximately four minutes of arc. The image was taken with the INGRID camera on the William Herschel Telescope, see Knapen et al. (2003) for more details.

a marginally significant difference, with a higher bar fraction in a sample of CfA Seyferts than in a control sample of non-Seyferts (approximately 80% vs. 60%). The results found by Mulchaey & Regan (1997) can be reconciled with those reported by Knapen et al. (2000) by considering the lower spatial resolution employed by the former authors.

Laine et al. (2002) later confirmed this difference at a 2.5 σ level by increasing the sample size and using high-resolution *HST* NICMOS near-IR images of the central regions of all their active and non-active sample galaxies. Laine et al. (2002) also found that almost one of every five sample galaxies, and almost one of every three barred galaxies, has more than one bar. The nuclear bar fraction, however, is not enhanced in Seyfert galaxies as compared to non-Seyferts (see also Erwin & Sparke 2002).

In a recent paper, Laurikainen, Salo & Buta (2004) study the bar properties of some 150 galaxies from the Ohio State University Bright Galaxy Survey (Eskridge et al. 2002) in terms of their nuclear properties, among other factors. Using several bar classification methods on optical and near-IR images, they find that only a Fourier method applied to near-IR images leads to a significant excess of bars among Seyferts/LINERs as compared to non-active galaxies: a bar fraction of 71%±4% for the active/starburst galaxies, versus 55%±5% for the non-active galaxies (at a significance level of 2.5σ)[1].

The Fourier method employed by Laurikainen et al. (2004) is objective but sensitive largely to classical bars with high surface brightnesses, and in this respect similar to the strict criteria applied by Knapen et al. (2000) and Laine et al. (2002), who basically rely upon significant a rise and fall in a radial ellipticity profile for bar identification. The results of the Fourier analysis by Laurikainen et al. (2004) are remarkably similar to those from Knapen et al. and Laine et al., apparently because all trace prominent bars in near-IR images. Laurikainen et al. find that the excess of bars among Seyferts/LINERs does not manifest itself in an analysis of optical images, which agrees with the general lack of excess found by the many authors who relied upon optical imaging for their bar classification (see references above). Laurikainen et al. (2004) also find that the bars in active galaxies are weaker than those in non-active galaxies, a result which confirms earlier indications to this effect by Shlosman, Peletier & Knapen (2000) and by Laurikainen, Salo, & Rautiainen (2002).

We can thus conclude that there is a slight, though significant, excess of bars among Seyfert galaxies as compared to non-active galaxies. This result is found only when using near-IR images, and only when applying rigorous and objective bar classification methods. Even so, there remain important numbers of active galaxies without any evidence for a bar, and, on the other hand, many non-active galaxies which do have apparently suitable bars. Given that any fuelling process must be accompanied by angular momentum loss, most easily achieved by gravitational non-axisymmetries, either the timescales of bars (or interactions, see below) are different from those of the activity, or the non-axisymmetries are not as easy to measure as we think, for instance because they occur at spatially unresolvable scales, and could be masqueraded to a significant extent by, e.g., dust or star formation (Laine et al. 2002), or because they occur in the form of weak ovals (e.g., Kormendy 1979) which will not necessarily be picked up by ellipse fitting or Fourier techniques. Additional work is clearly needed, but it is not clear whether this should be aimed primarily at the large-scale bars described in this Section, or perhaps better at the kinematics and dynamics of the very central regions of active and non-active galaxies. In any case, the use of carefully matched samples and control samples is of paramount importance.

[1]It is an interesting exercise to add the numbers found by Laurikainen et al. (2004) to those found by Laine et al. (2002), which would give largely the same overall result in terms of bar fractions, but with smaller error bars thanks to the increased sample sizes, and an overall significance level of more than 3σ. Formally this is not allowed though because the original samples have been selected using different methods, and should not simply be added.

3. The effects of interactions

Galaxy interactions can easily lead to non-axisymmetries in the gravitational potential of one or more of the galaxies involved, and as such can be implicated in angular momentum loss of inflowing material, and thus conceivably in starburst and AGN fuelling (e.g., Shlosman et al. 1989, 1990; Mihos & Hernquist 1995).

3.1 Interactions and starburst activity

It is well known that there is ample anecdotal evidence for the connection between galaxy interactions and starburst activity. This is perhaps clearest for the most extreme infrared sources, specifically the Ultra-Luminous InfraRed Galaxies (ULIRGs). They are powered mainly by starbursts (Genzel et al. 1998), and it has been known since briefly after their discovery that they occur in galaxies with disturbed morphologies, presumably as a result of recent interactions (e.g., Joseph & Wright 1985; Armus, Heckman, & Miley 1987; Sanders et al. 1988; Clements et al. 1996; Murphy et al. 1996; Sanders & Mirabel 1996). Given that the ULIRGs are both among the most extreme starbursts known, and are occurring in interacting galaxies, one can infer that such massive starbursts are in fact powered by gas which has lost angular momentum in galaxies which are undergoing a major upheaval, i.e., are merging or interacting.

More in general, and considering galaxies less extreme than those in the ULIRG class, there is considerable evidence for a connection between interactions and enhanced star formation in galaxies, often measured using galaxy colours which are bluer in the case of current star formation (see, e.g., the seminal paper by Larson & Tinsley 1978). But even so, a more detailed consideration can expose possible caveats. We mention the recent paper by Bergvall, Laurikainen, & Aalto (2003), who considered two matched samples of nearby interacting (pairs and clear cases of mergers) and non-interacting galaxies, and measured star formation indices based on UBV colours. From this analysis, Bergvall et al. do *not* find evidence for significantly enhanced star-forming activity among the interacting/merging galaxies, although they do report a moderate increase in star formation in the very centres of their interacting galaxies. Interesting in this respect are also recent results from a combination of Sloan Digital Sky Survey and 2dF Galaxy Redshift Survey data, presented by Balogh et al. (2004). These authors study the equivalent width of Hα emission, a measure of starburst activity, and find no correlation between its distribution among the star-forming population of galaxies and the environment.

So although mergers can undoubtedly lead to massive starbursts, they appear to do so only in exceptionally rare cases. Bergvall et al. (2003) estimate that only about 0.1% of a magnitude limited sample of galaxies will host massive starbursts generated by interactions and mergers. Most interactions between galaxies may not lead to any increase in the starburst activity. Those that do may be selected cases where a set of parameters, both internal to the galaxies and regarding the orbital geometry of the merger, is conducive to the occurrence of starburst activity (e.g., Mihos & Hernquist 1996). To further illustrate this point, we quote the results published by Laine et al. (2003), who find very little evidence for trends in starburst activity from detailed *HST* imaging of the Toomre sequence of merging galaxies.

3.2 Interactions and Seyfert activity

Interactions and mergers have long been suspected of triggering high-luminosity AGN such as QSOs (e.g., Disney et al. 1995; Bahcall et al. 1997), although many of such AGN seem to lie in entirely undisturbed elliptical systems. In fact, Dunlop et al. (2003) show that the host galaxy properties of radio-loud and radio-quiet AGN are indistinguishable from those of quiescent but otherwise comparable galaxies, and Floyd et al. (2004) find no correlation between the luminosity of a quasar and the presence of any morphological disturbance in the host.

Seyfert activity is known to occur in interacting and merging galaxies, and several rather spectacular examples are well known (for instance NGC 2992, or a number of the closest ULIRGs such as Mrk 273). To check statistically whether there is a connection between interactions and the occurrence of this type of nuclear activity, authors have considered the numbers of companions to Seyfert galaxies as compared to non-active control galaxies (e.g., Fuentes-Williams & Stocke 1988; de Robertis, Yee, & Hayhoe 1998; Schmitt 2001), or, alternatively, have searched for different fractions of Seyfert or AGN activity among more or less crowded environments (e.g., Kelm, Focardi, & Palumbo 1998; Miller et al. 2003). The conclusion from this substantial body of work must be that no unambiguous evidence exists for a direct connection between the occurrence of Seyfert activity and interactions. Some earlier work did report claims of a statistical connection, but this work was unfortunately plagued by poor control sample selection (see Laurikainen & Salo 1995 for a detailed review), and most early studies, but also some of the recent ones, are not based on complete sets of redshift information for the possible companion galaxies. In addition, Laine et al. (2002) have shown that the bar fraction among both the Seyfert and non-Seyfert galaxies in their sample is completely independent of the presence of companions (interacting galaxies were not considered by Laine et al.). We thus conclude that interactions and Seyfert activity may well be linked in individual cases, but that as yet the case that they are statistically connected has not been made convincingly.

4. Bars and nuclear rings

Apart from concentrating gas in the central regions of galaxies, as discussed in Section 2, bars also set up resonances which can act as focal points for the gas flow. As reviewed by, e.g., Shlosman (1999), gas concentrates there in limited radial ranges, where it can become gravitationally unstable and form stars. Rings in disk galaxies are mostly identified by their star formation, either by their blue colours or by Hα emission, and are intimately linked to the internal dynamics and the evolution of their hosts (see Buta & Combes 1996 for a comprehensive review on galactic rings).

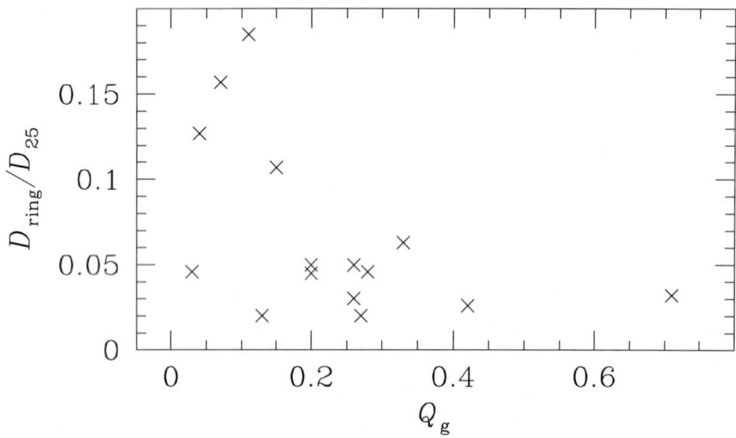

Figure 3. Relative size (ring diameter divided by host galaxy diameter) for a sample of 15 nuclear rings as a function of the gravitational torque Q_g, or strength, of the bar of its host galaxy. Data from Knapen, Pérez-Ramírez, & Laine (2002) and Knapen (2004b).

Nuclear rings are those on scales of less than one to roughly two kiloparsec in radius. They are rather common, and occur in about 20% of nearby spiral galaxies (Knapen 2004b). They can be directly linked to inner Lindblad resonances (Knapen et al. 1995a; Heller & Shlosman 1996; Shlosman 1999), and are in fact found almost exclusively in barred galaxies (e.g., Buta & Combes 1996; Knapen 2004b; but see below for a counterexample). Individual gas clouds in a nuclear ring can undergo a Jeans-type collapse either spontaneously (Elmegreen 1994), or after compression by density waves set up by the bar (Knapen et al. 1995a; Ryder, Knapen, & Takamiya 2001). Nuclear rings are thus not only excellent tracers of massive star formation in starburst regions, but also of the dynamics of their host galaxies. The latter point is illustrated by Fig. 3, which shows that nuclear rings with a large size relative to their host galaxy can only occur in bars with a low gravitational torque, or:

large rings cannot occur in strong bars. This confirms the results from earlier theory and modelling that the extent of the perpendicular $x2$ orbits needed to sustain the nuclear ring is limited as the bar gets stronger, i.e., as the $x1$ orbits become more elongated (see Knapen et al. 1995a; Heller & Shlosman 1996; Knapen, Pérez-Ramírez, & Laine 2002; Knapen 2004b), and graphically shows how intricately bars and nuclear rings are related.

A small number of rings, or pseudo-rings, apparently occur in non-barred galaxies. Some of these hosts, although classified as non-barred from optical imaging in the major catalogues, are obviously barred when imaged in the near-IR (e.g., NGC 1068, Scoville et al. 1988, and NGC 4725, Shaw et al. 1993; Möllenhoff, Matthias, & Gerhard 1995). In other cases, a non-barred host galaxy may either have an oval distortion, or be undergoing the effects of an interaction with a companion galaxy. In either of these cases, the gravitational potential of the galaxy could be disturbed, and the non-axisymmetric potential could lead to ring formation, in much the same way as in the presence of a bar potential (Shlosman et al. 1989).

A nice example is that of NGC 278, a small, nearby and isolated spiral galaxy ($v_{\rm sys} = 640\,{\rm km\,s^{-1}}$; $D = 11.8$ Mpc; $D_{25} = 7.2$ kpc; $M_B = -18.8$). Although classified as SAB(rs)b in the RC3, there is no evidence for the presence of a bar in this galaxy from either *HST* WFPC2 or ground-based NIR imaging (Knapen et al. 2004a). The optical disk of NGC 278 shows two distinct regions, an inner one with copious star formation and clear spiral arm structure, shown in Fig. 4, and an outer one ($r > 27$ arcsec or $r > 1.5$ kpc) which is almost completely featureless, of low surface brightness, and rather red. NGC 278 has a large H I disk, which is morphologically and kinematically disturbed, as seen from H I data (Knapen et al. 2004a). These disturbances suggest a recent minor merger with a small gas-rich galaxy, perhaps similar to a Magellanic cloud.

The scale and morphology of the region of star formation in NGC 278 indicate that this is in fact a nuclear ring, albeit one with a much larger *relative* size with respect to its host galaxy than practically all other known nuclear rings (the absolute radius of the nuclear ring is about a kiloparsec, normal for nuclear rings). Knapen et al. (2004a) postulate that it is in fact the past interaction which has set up a non-axisymmetry in the gravitational potential, which in turn, in a way very similar to the action of a classic bar, leads to the formation of the nuclear ring. The case of NGC 278 illustrates how in apparently non-barred galaxies rings can be caused by departures from axisymmetry induced by interactions, but also shows how difficult it can be to uncover this: in the case of NGC 278 only through detailed H I observations.

Figure 4. Real-colour image of the galaxy NGC 278 produced from archival $F450W$, $F606W$, and $F814W$ HST images. Area shown is about 100 arcsec on a side, or about 5.7 kpc. The blue region, apparently made up of spiral arm fragments, indicates the nuclear ring, with a radius of 1.1 kpc. Image data from Knapen et al. (2004a).

5. Rings and nuclear activity

Although nuclear rings and especially nuclear activity have as separate topics received considerable attention in the literature, their possible interrelation has not been much studied. Many nuclear rings, of course, are anecdotally known to occur in galaxies which also host a nuclear starburst or a prominent AGN (often of Seyfert or LINER type, given the typical parameters of the host galaxies involved), and some famous examples include NGC 1068 and NGC 4303.

In a recent paper, we explored the correlations between nuclear activity (both of the non-stellar and starburst variety) and the occurrence of nuclear rings in a sample of 57 nearby spiral galaxies (Knapen 2004b). Using information on the activity from the NASA/IPAC Extragalactic Database (NED) and ring parameters as derived from new Hα imaging (Knapen et al. 2004b), we found not only that nuclear rings significantly more often than not occur in galaxies which also host nuclear activity (only two of the 12 nuclear rings occur in a galaxy which is neither a starburst nor an AGN host; 30 of the 57 sample galaxies overall would fall into this category), but also that the circumnuclear Hα emission morphology of the AGN and starbursts is significantly more often in the form of a ring than in non-AGN, non-starburst galaxies (38% of AGN, 33% of starbursts, 11% of non-AGN, and 7% of non-AGN non-starburst galaxies have circumnuclear rings in our sample of galaxies).

Although the number of galaxies in this initial study is rather small for detailed statistical analyses, we did find this most interesting correlation between the occurrence of nuclear rings and that of nuclear activity. Our initial interpretation of this effect is that both nuclear rings, as traced by the massive star formation within them, and starbursts and AGN (of the Seyfert or LINER variety) trace very recent gas inflow. Although it is not clear *a priori* why the kpc-scale fuelling of nuclear rings and the pc-scale fuelling of activity might be so closely related, nuclear rings do seem to show a potential as tracers of AGN fuelling. These findings may also be related to the reported higher incidence of rings among Seyfert and LINER hosts than among non-AGN galaxies (Arsenault 1989 for nuclear rings; Hunt & Malkan 1999 for inner and outer rings). All these aspects of rings and nuclear activity need further scrutiny.

6. Concluding remarks

From theory and modelling, and increasingly also from observations, it is clear that bars can remove angular momentum from gaseous material, and thus drive it from the disk into the central kpc-scale regions of a galaxy (Section 2). In contrast, the evidence that this centrally condensed gas directly and immediately leads to AGN or starburst activity remains rather elusive. The relevant results, reviewed more in depth elsewhere in this paper as indicated below, have been summarised in Table 1, where bars and interactions have been labelled as primary indicators for links between the inflow-provoking mechanisms and the possibly resulting AGN or starburst activity. Also listed in Table 1 is a small number of so-called secondary indicators, which have received attention as outlining possible links between inflow and activity, but which may well be a result of one of the primary indicators. The information summarised in Table 1 can be related to the content of this paper as follows:

Table 1. Summary of observational evidence for relations between various host galaxy features and Seyfert/LINER and starburst activity

Feature	*Seyferts/LINERs*	*Starbursts*
	Primary indicators	
Bars	yes (but 2.5σ)	yes (but not in general?)
Interactions	no	yes (but extremes only?)
	Secondary indicators	
Nuclear bars	no	N/A
Rings	yes, some	yes (nuclear rings at least)

- Starbursts can be provoked by bars or interactions, at least in some cases (Section 2,1, 3.1).
- There is an increasing body of evidence that Seyfert activity preferentially occurs in barred host galaxies, but the effect is a statistical one, and not very pronounced (Section 2.2).
- There is no convincing evidence that AGN hosts are interacting more often than non-AGN (Section 3.2).
- Nuclear bars (Section 2.2) have great theoretical potential for bringing fuel very close to the centre of a galaxy (the "bars within bars" scenario, Shlosman et al. 1989) but have so far not lived up to that potential in terms of their detections in imaging surveys, where no higher nuclear bar fractions have been found in Seyferts as compared to non-Seyferts. As far as we aware, the possible statistical connections between nuclear bars and starbursts have not yet been studied.
- There is some interesting evidence that rings, both nuclear and non-nuclear, may be related to the presence of low-luminosity AGN activity (Section 5). This issue needs further exploration, but in any case the rings will most likely have formed under the influence of either a bar or another form of non-axisymmetry in the gravitational potential of the host, hence the inclusion of rings as secondary indicators in Table 1.
- Finally, there is a direct link between starburst activity and the presence of nuclear rings, since small nuclear rings with significant massive star formation can be classified as starburst, and since many starbursts might in fact be circumnuclear rather than nuclear, albeit with small ring radii of tens to hundreds of parsec (e.g., González-Delgado et al. 1998), or

possibly even smaller. Statistical links between inner/outer rings and starburst activity have not yet been explored.

Since we have known for quite some time that net radial gas inflow must be accompanied by the loss of significant quantities of angular momentum, and that the kind of deviations from axisymmetry in the gravitational potential of the host galaxy set up by bars and interactions is well suited to lead to such angular momentum loss (see reviews by Shlosman et al. 1989, 1990; Shlosman 2003), we must be missing some part of the puzzle. One possibility is that we are not looking at the right things at the right time: the spatial- as well as the time-scales under consideration may not be correct. So far, the spatial scale considered observationally has been from tens of kpc down to, roughly, a few hundred parsec. Whereas this range may be wholly adequate for the study of a major starburst, which can span up to a kpc, it may well be wholly *in*adequate for AGN fuelling, which is expected to be related to accretion to a SMBH, on scales of AUs. As far as timescales are concerned, what has been considered in the studies reviewed here is generally a rather long-lived phenomenon influencing kpc-scale regions (bar or galaxy-galaxy interaction). Starburst or AGN activity, on the other hand, occurs on essentially unknown timescales (somewhere around $10^6 - 10^8$ years could be expected for most AGN or starbursts). If the starburst or AGN activity is indeed short-lived, and possibly also periodic, the connection between the presence of activity at the currently observed epoch and any parameter of the host galaxy is not necessarily straightforward (as pointed out, e.g., by Beckman 2001).

The fact that we see any correlations at all, such as those of bar fractions with the presence of starburst or Seyfert activity, indicates that bars and interactions do have a role, presumably by establishing a gas reservoir in the central kpc region. In the coming years, we must start to disentangle the effects of gravitationally induced gas inflow, which brings gas to the inner kpc region at least, from those of possible other mechanisms which can transport the gas further in, and from the time scales and duty cycles of the activity. We seem to have reached the limits of purely morphological studies of the central regions of active and non-active galaxies (e.g., Laine et al. 2002), and must start to worry about the effects of the AGN or starburst on their immediate surroundings as we push the observations to smaller spatial scales, of tens of parsecs. One must, hence, move on to careful studies of the gas and stellar kinematics and dynamics. Integral field spectroscopy (e.g., Bacon et al. 2001), especially when used in conjunction with adaptive optics techniques, should allow a good deal of progress here, giving simultaneous high-resolution mapping of the distributions of stellar populations and dust, as well as of the gas and stellar kinematics. In combination with detailed numerical modelling, this could lead to the detection of the dynamical effects of, e.g, nuclear bars on gas flows which may be more directly related to the fuelling process of starbursts and/or AGN.

Acknowledgements I am indebted to my collaborators on the various aspects of the work described here, especially John Beckman, Shardha Jogee, Seppo Laine, Reynier Peletier, and Isaac Shlosman. Valuable comments by conference participants have helped improve this paper.

References

Adams, T. F. 1977, ApJS, 33, 19

Alonso-Herrero, A. & Knapen, J. H. 2001, AJ, 122, 1350

Armus, L., Heckman, T., & Miley, G. 1987, AJ, 94, 831

Arsenault, R. 1989, A&A, 217, 66

Athanassoula, E. 1992, MNRAS, 259, 345

Bacon, R., et al. 2001, MNRAS, 326, 23

Bahcall, J. N., Kirhakos, S., Saxe, D. H., & Schneider, D. P. 1997, ApJ, 479, 642

Balick, B. & Heckman, T. M. 1982, ARA&A, 20, 431

Balogh, M., et al. 2004, MNRAS, 348, 1355

Beckman, J. E. 2001, in ASP Conf. Ser. 249, The Central Kiloparsec of Starbursts and AGN: The La Palma Connection, eds. J. H. Knapen, J. E. Beckman, I. Shlosman, & T. J. Mahoney (San Francisco: Astronomical Society of the Pacific), 11

Begelman, M. C., Blandford, R. D., & Rees, M. J. 1984, Reviews of Modern Physics, 56, 255

Bergvall, N., Laurikainen, E., & Aalto, S. 2003, A&A, 405, 31

Buta, R. & Combes, F. 1996, Fundamentals of Cosmic Physics, 17, 95

Clements, D. L., Sutherland, W. J., McMahon, R. G., & Saunders, W. 1996, MNRAS, 279, 477

Combes, F. 2001, in Advanced Lectures on the Starburst-AGN Connection, eds. I. Aretxaga, D. Kunth, & R. Mújica (Singapore: World Scientific), 223

Combes, F. & Gerin, M. 1985, A&A, 150, 327

Crenshaw, D. M., Kraemer, S. B., & Gabel, J. R. 2003, AJ, 126, 1690

de Robertis, M. M., Yee, H. K. C., & Hayhoe, K. 1998, ApJ, 496, 93

de Vaucouleurs G., de Vaucouleurs A., Corwin J. R., Buta R. J., Paturel G., Fouque P., 1991, Third reference catalogue of Bright galaxies, 1991, New York : Springer-Verlag (RC3)

Devereux, N. 1987, ApJ, 323, 91

Disney, M. J., et al. 1995, Nature, 376, 150

Dressel, L. L. 1988, ApJ, 329, L69

Dunlop, J. S., McLure, R. J., Kukula, M. J., Baum, S. A., O'Dea, C. P., & Hughes, D. H. 2003, MNRAS, 340, 1095

Elmegreen, B. G. 1994, ApJ, 425, L73

Erwin, P. & Sparke, L. S. 2002, AJ, 124, 65

Eskridge, P. B., et al. 2000, AJ, 119, 536

Eskridge, P. B., et al. 2002, ApJS, 143, 73

Ferrarese, L. & Merritt, D. 2000, ApJ, 539, L9

Floyd, D. J. E., Kukula, M. J., Dunlop, J. S., McLure, R. J., Miller, L., Percival, W. J., Baum, S. A. & O'Dea, C. P. 2004, MNRAS, submitted (astro-ph/0308436)

Fuentes-Williams, T. & Stocke, J. T. 1988, AJ, 96, 1235

Gebhardt, K., et al. 2000, ApJ, 539, L13

Genzel, R., et al. 1998, ApJ, 498, 579

González-Delgado, R.M., Heckman, T., Leitherer, C., Meurer, G., Krolik, J., Wilson, A.S., Kinney, A., Loratkar, A. 1998, ApJ, 505, 174

Hawarden, T. G., Mountain, C. M., Leggett, S. K., & Puxley, P. J. 1986, MNRAS, 221, 41P

Heller, C. H. & Shlosman, I. 1996, ApJ, 471, 143

Ho, L. C., Filippenko, A. V., & Sargent, W. L. W. 1997, ApJ, 487, 591

Huang, J. H., Gu, Q. S., Su, H. J., Hawarden, T. G., Liao, X. H., & Wu, G. X. 1996, A&A, 313, 13

Hummel, E. 1981, A&A, 93, 93

Hunt, L. K. & Malkan, M. A. 1999, ApJ, 516, 660

Isobe, T. & Feigelson, E. D. 1992, ApJS, 79, 197

Jogee, S., Scoville, N. Z., & Kenney, J. 2004, ApJ, in press (astro-ph/0402341)

Jogee, S. 2004, in AGN Physics on All Scales, eds. D. Alloin, R. Johnson, & P. Lira, (Berlin: Springer), in press

Joseph, R. D. & Wright, G. S. 1985, MNRAS, 214, 87

Kelm, B., Focardi, P., & Palumbo, G. G. C. 1998, A&A, 335, 912

Knapen, J. H. 2001, in ASP Conf. Ser. 249, The central kiloparsec of starbursts and AGN: The La Palma connection. eds. J. H. Knapen, J. E. Beckman, I. Shlosman, & T. J. Mahoney (San Francisco: Astronomical Society of the Pacific) 249, 37 (astro-ph/0108349)

Knapen, J. H. 2004a, in Proc. of The neutral ISM in starburst galaxies, eds. S. Aalto, S. Hüttemeister, & A. Pedlar (San Francisco: Astronomical Society of the Pacific), in press (astro-ph/0312172)

Knapen, J. H. 2004b, A&A, submitted

Knapen, J. H., Beckman, J. E., Heller, C. H., Shlosman, I., & de Jong, R. S. 1995a, ApJ, 454, 623

Knapen, J. H., Beckman, J. E., Shlosman, I., Peletier, R. F., Heller, C. H., & de Jong, R. S. 1995b, ApJ, 443, L73

Knapen, J. H., Shlosman, I., & Peletier, R. F. 2000, ApJ, 529, 93

Knapen, J. H., Pérez-Ramírez, D., & Laine, S. 2002, MNRAS, 337, 808

Knapen, J. H., de Jong, R. S., Stedman, S., & Bramich, D. M. 2003, MNRAS, 344, 527 (Erratum MNRAS 346, 333)

Knapen, J. H., Stedman, S., Bramich, D. M., Folkes, S. F., & Bradley, T. R. 2004b, A&A, in press

Knapen, J. H., Whyte, L. F., de Blok, W. J. G., & van der Hulst, J. M. 2004a, A&A, in press (astro-ph/0405107)

Kormendy, J. 1979, ApJ, 227, 714

Kormendy, J., & Kennicutt, R. C. 2004, ARA&A, in press

Kormendy, J., & Richstone, D. 1995, ARA&A, 33, 581

Laine, S., Shlosman, I., Knapen, J. H., & Peletier, R. F. 2002, ApJ, 567, 97

Laine, S., van der Marel, R. P., Rossa, J., Hibbard, J. E., Mihos, J. C., Böker, T., & Zabludoff, A. I. 2003, AJ, 126, 2717

Larson, R. B. & Tinsley, B. M. 1978, ApJ, 219, 46

Laurikainen, E. & Salo, H. 1995, A&A, 293, 683

Laurikainen, E., Salo, H., & Rautiainen, P. 2002, MNRAS, 331, 880

Laurikainen, E., Salo, H., & Buta, R. 2004, ApJ, 607, 103

Lynden-Bell, D. 1969, Nature, 223, 690

MacKenty, J. W. 1990, ApJS, 72, 231

Maiolino, R., Risaliti, G., & Salvati, M. 1999, A&A, 341, L35

Martinet, L. & Friedli, D. 1997, A&A, 323, 363

Mihos, J. C. & Hernquist, L. 1996, ApJ, 464, 641

Miller, C. J., Nichol, R. C., Gómez, P. L., Hopkins, A. M., & Bernardi, M. 2003, ApJ, 597, 142

Moles, M., Marquez, I., & Perez, E. 1995, ApJ, 438, 604

Möllenhoff, C., Matthias, M., & Gerhard, O. E. 1995, A&A, 301, 359

Mulchaey, J. S. & Regan, M. W. 1997, ApJ, 482, L135

Murphy, T. W., Armus, L., Matthews, K., Soifer, B. T., Mazzarella, J. M., Shupe, D. L., Strauss, M. A., & Neugebauer, G. 1996, AJ, 111, 1025

Noguchi, M. 1988, A&A, 203, 259

Phinney, E. S. 1994, in Mass-Transfer Induced Activity in Galaxies, ed. I. Shlosman (Cambridge: Cambridge University Press), 1

Puxley, P. J., Hawarden, T. G., & Mountain, C. M. 1988, MNRAS, 231, 465

Regan, M. W., Vogel, S. N., & Teuben, P. J. 1997, ApJ, 482, L143

Roussel, H., et al. 2001, A&A, 372, 406

Ryder, S. D., Knapen, J. H., & Takamiya, M. 2001, MNRAS, 323, 663

Sakamoto, K., Okumura, S. K., Ishizuki, S., & Scoville, N. Z. 1999, ApJ, 525, 691

Sanders, D. B., Soifer, B. T., Elias, J. H., Madore, B. F., Matthews, K., Neugebauer, G., & Scoville, N. Z. 1988, ApJ, 325, 74

Sanders, D. B. & Mirabel, I. F. 1996, ARA&A, 34, 749

Schmitt, H. R. 2001, AJ, 122, 2243

Schwarz, M. P. 1984, MNRAS, 209, 93

Scoville, N. Z., Matthews, K., Carico, D. P., & Sanders, D. B. 1988, ApJ, 327, L61

Sellwood, J.A. & Wilkinson, A. 1993, Rep. Prog. Phys. 56, 173

Shaw, M. A., Combes, F., Axon, D. J., & Wright, G. S. 1993, A&A, 273, 31

Sheth, K., Vogel, S. N., Regan, M. W., Teuben, P. J., Harris, A. I., Thornley, M. D., & Helfer, T. T. 2004, ApJ, submitted

Shlosman, I. 1999, in ASP Conf. Ser. 187, The evolution of galaxies on cosmological timescales, eds. J. E. Beckman, & T. J. Mahoney (San Francisco: Astronomical Society of the Pacific), 100

Shlosman, I. 2001, in ASP Conf. Ser. 249, The central kiloparsec of starbursts and AGN: The La Palma connection. eds. J. H. Knapen, J. E. Beckman, I. Shlosman, & T. J. Mahoney (San Francisco: Astronomical Society of the Pacific) 249, 55

Shlosman, I. 2003, in ASP Conf. Ser. 290, Active Galactic Nuclei: From central engine to host galaxy, eds. S. Collin, F. Combes, & I. Shlosman (San Francisco: Astronomical Society of the Pacific), 427

Shlosman, I., Frank, J., & Begelman, M. C. 1989, Nature, 338, 45

Shlosman, I., Begelman, M. C., & Frank, J. 1990, Nature, 345, 679

Shlosman, I., Peletier, R. F., & Knapen, J. H. 2000, ApJ, 535, L83

Simkin, S. M., Su, H. J., & Schwarz, M. P. 1980, ApJ, 237, 404

Wada, K. 2004, in Coevolution of black holes and galaxies, ed. L. C. Ho (Cambridge: Cambridge University Press), 187

Zurita, A., Relaño, M., Beckman, J. E., & Knapen, J. H. 2004, A&A, 413, 73

PENETRATING DUST TORI IN AGN

Gabriela Canalizo[1], Claire Max[2,3], Robert Antonucci[4], David Whysong[4], Alan Stockton[5], Mark Lacy[6]

[1] *Department of Earth Sciences and IGPP, University of California, Riverside, CA 95521,* [2]*IGPP, Lawrence Livermore National Laboratory,* [3]*Center for Adaptive Optics, University of California, Santa Cruz,* [4]*Physics Department, University of California, Santa Barbara,* [5]*Institute for Astronomy, University of Hawaii,* [6]*Spitzer Science Center, MS 220-6, Caltech*

Abstract We present preliminary results from high resolution (~ 0.05") adaptive optics observations of Cygnus A. The images show a biconical structure strongly suggestive of an obscuring torus around a quasar nucleus. A bright ($K' = 18.5$) point source is found near the expected position of the nucleus. We interpret this source as the hot inner rim of the torus seen through the opening of the torus. Using high angular resolution K-band spectroscopy, we measure the ratio of molecular to recombination hydrogen lines as a function of distance to the center of the putative torus. These measurements place constraints on the properties of the torus and indicate a projected diameter of $\sim$ 600 pc.

Keywords: Active galaxies, infrared galaxies, adaptive optics

1. Introduction

The interplay between dust and radio emission has been the subject of vigorous research in recent years. We now know that radio loud sources have a high incidence of dust in their central regions (e.g., de Koff et al. 2000) and that there are several correlations between the properties of the dust and those of the radio source. However, this same dust has hampered the study of the of the nuclear regions of radio galaxies, particularly since most of the high resolution imaging studies of the hosts have been carried out at optical wavelengths.

We are conducting a Keck and Lick adaptive optics (AO) imaging and spectroscopic survey of nearby radio galaxies and other AGN in the near infrared that will allow us to pierce the dust in these objects and study the nuclear regions. One of the goals of this survey is to study the obscuring dust tori (if present) in these objects. Here we present preliminary results for one of the objects in our sample, the prototype radio galaxy Cyg A. For details on the observations, see Canalizo et al. (2003, hereafter Paper I; 2004, in preparation).

D. Block et al. (eds.), Penetrating Bars through Masks of Cosmic Dust, 207–211.

2. The Torus in Cygnus A

Because of its proximity ($z = 0.056$) and extreme characteristics, Cyg A has played a fundamental role in the study of virtually every aspect of powerful radio galaxies. Different lines of evidence indicate that Cyg A harbors a heavily extincted quasar (e.g., Ogle et al. 1997). According to unification schemes, this quasar would be hidden by an optically and geometrically thick dust torus. Figure 1 shows a Keck NIRC2 AO K' image of the central region of Cyg A with a resolution of ~ 0.05" (or 50 pc for $H_0 = 70$ km s^{-1} Mpc^{-1}). The two edge-brightened cones clearly seen in this image are strongly suggestive of a dust torus casting a shadow onto the surrounding gas. There are two unresolved point sources in the central region: the "primary" source, near (but slightly off-center) the vertex of the cones, and the "secondary", which appears to be the tidally stripped core of a lower mass merging galaxy (Paper I).

HST NICMOS observations of the primary point source show that it is highly polarized ($P_k \sim 25\%$) with a polarization angle roughly perpendicular to the jets (Tadhunter et al. 2000; 1999). Overlaying these data with HST FOC polarization data indicates that the primary is within 0.1" of the scattering center (M. Kishimoto, personal communication). While it is tempting to associate this point source with the quasar nucleus, its flux is considerably higher than that predicted by X-ray observations. Ward et al. (1991) predict a continuum flux density at 2.2 μm of 3.6×10^{-15} erg cm^{-2} s^{-1} A^{-1}. Assuming a normal dust-to-gas ratio, Ueno et al. (1994) estimate $A_V = 170$ from the X-ray absorption column. Combining both results, we expect a quasar continuum emission of 9.6×10^{-21} erg cm^{-2} s^{-1} A^{-1} at 2.2 μm. However, we measure a 2.2 μm flux of 1.53×10^{-18} erg cm^{-2} s^{-1} A^{-1}. From our NIRC2 K-band spectrum (see below; Fig. 2) we estimate that 36% of the total flux comes from emission lines, so the observed continuum flux density is 9.8×10^{-19} erg cm^{-2} s^{-1} A^{-1}, two orders of magnitude greater than expected. It is certainly possible that the extinction may be overestimated if the mean dust grain size is larger than assumed (e.g. Maiolino et al. 2001). However, the K-band spectrum of this point source (Fig. 2) does not show obvious broad lines, indicating that the broad line region is indeed at least partially hidden at these wavelengths.

A more likely interpretation for the K' point source is that it is the hot inner rim of the dust torus seen through the upper opening of the torus. The quasar itself is hidden behind the SE half of the torus, which is closer to us than its NW counterpart (recall that the NW jet points toward us and the SE jet away from us; see e.g., Carilli & Barthel 1996). Using observations of the 9.7 μm silicate dust absorption feature toward the nucleus, Imanishi & Ueno (2000) estimate an inner radius for the torus of less than 10 pc. Since the resolution of our images is 50 pc, a torus opening of this order would certainly appear unresolved.

Figure 1. False-color NIRC2 AO K' image of the core of Cyg A. North is up and East is to the left. White arrows indicate the direction of the jets and parallel lines the slit positions.

We obtained K-band Keck AO spectroscopy at the two slit positions indicated in Fig. 1. Slit A corresponds to the NIRC2 spectrum shown in Fig. 2, with total integration of 3600 s. The slit was placed roughly perpendicular to the jets, presumably along the torus. The plate scale in the cross-dispersion direction is 0.04" pixel^{-1} (or $\sim$ 40 pc for Cyg A). Slit B corresponds to a NIRSPEC spectrum with a total integration time of 2400 s. This slit was placed as close to the axis of the jets as possible, given the constraints imposed by the AO system (Paper I). The data were binned in the cross dispersion direction to match the spatial resolution of the NIRC2 data.

Figure 3 shows, for each slit position, a plot of the flux ratio of the H_2 $\nu = 1 - 0$ S(1) line to Paα, as a function of the distance from the primary point source. For Slit A (left panel), the ratio is clearly lower in the region

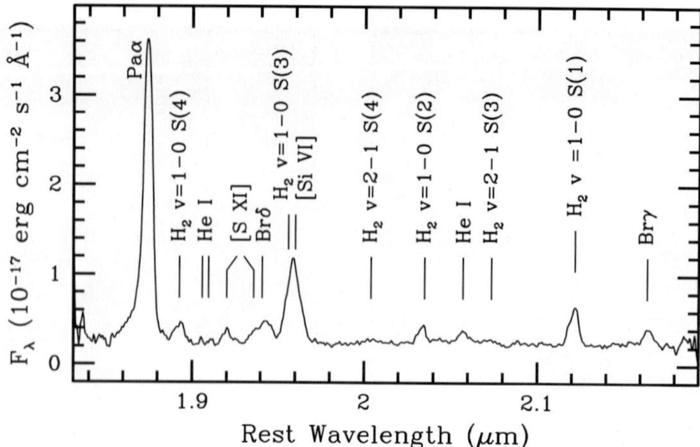

Figure 2. Keck NIRC2 AO spectrum of the 640 central parsecs of Cyg A. The slit was placed roughly perpendicular to the direction of the jets (Slit A in Fig. 1)

centered around the point source. The change in the ratio is unlikely to be due to extinction, since the ratio of Brγ to Paα remains constant over this region. A plausible explanation for this trend is that the molecular hydrogen is shielded from the continuum source on either side of the torus, leading to a higher $H_2/$Paα ratio. Since the torus is inclined some 30 to 60 degrees with respect to the line of sight, the emission we observe in this region actually comes from the region *above* the torus. Here the molecular hydrogen is no longer shielded and is thus photodissociated, so that the only emission we observe comes from the foreground; hence its ratio to Paα drops. From this plot, it may be inferred that the diameter of the torus is between 500 and 600 pc.

In the case of Slit B (Fig. 3, right panel), we are measuring $H_2/$Paα in the direction of the ionization cones, so that H_2 is mostly photodissociated within the cones. However, in the region SE of the opening of the torus (at distance < 0), the ratio increases since this region is shielded by the SE half of the torus. The ratio decreases again at roughly -300 pc, presumably beyond the shadow of the torus. The projected size of the torus would then be roughly $300 \times 2 = 600$ pc. To obtain an actual diameter of the torus from this value we must take into consideration the scale height as well as the inclination angle, and the angle between the slit and the axis of the jets.

Thus, we can place an upper limit of 600 pc to the diameter of the torus in Cyg A. This value is consistent with values inferred using other methods, e.g., < 800 pc from ~ 0.8" resolution optical images (Vestergaard & Barthel 1993) and "a few hundred parsecs" from infrared observations of the 9.7 μm silicate

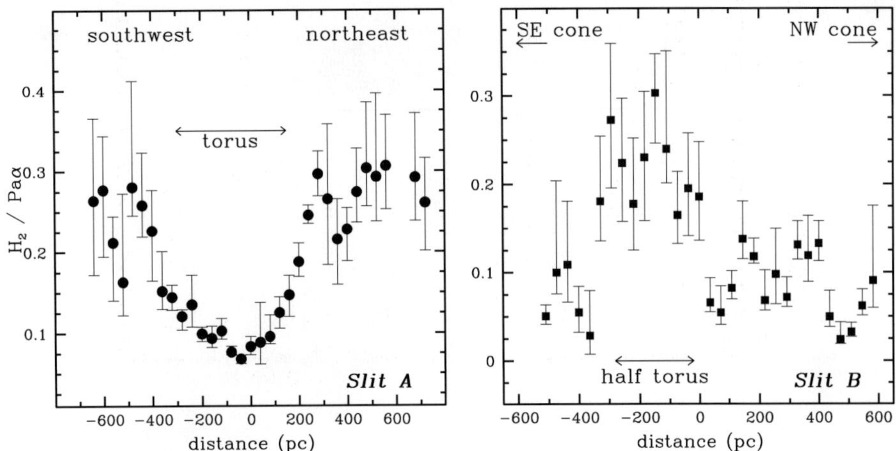

Figure 3. The ratio of the flux of H_2 $\nu = 1 - 0$ S(1) to Paα plotted as a function of distance from the primary point source for Slit A (left) and Slit B (right).

dust absorption feature (Imanishi & Ueno 2000). Detailed modeling should allow us to obtain better estimates for the size and inclination of the torus.

Acknowledgments

This work was supported in part under the auspices of the U.S. Department of Energy, National Nuclear Security Administration by the University of California, Lawrence Livermore National Laboratory under contract No. W-7405-Eng-48.

References

Canalizo, G., Max, C. E., Whysong, D., Antonucci, R., Dahm, S. E. 2003, ApJ, 597, 823

Carilli, C. L., & Barthel, P. D. 1996, A&A Rev., 7, 1

de Koff, S. et al. 2000, ApJ, 129, 33

Imanishi, M., Ueno, S. 2000, ApJ, 535, 626

Ogle, P., Cohen, M., Miller, J. S., Tran, H. D., Fosbury, R. A. E., Goodrich, R. W. 1997, ApJ, 482, L37

Tadhunter, C. N., Packham, C., Axon, D. J., Jackson, N. J., Hough, J. H., Robinson, A., Young, S., Sparks, W. 1999, ApJ, 512, 91

Tadhunter, C. N., Sparks, W., Axon, D. J., Bergeron, L., Jackson, N. J., Packham, C., Hough, J. H., Robinson, A., Young, S. 2000, MNRAS, 313, 52

Ueno, S., Koyama, K., Nishida, M., Yamauchi, S., Ward, M.J. 1994, ApJL, 431, 1

Vestergaard, M., Barthel,P.D. 1993, AJ, 105, 456

Ward, M. J., Blanco, P. R., Wilson, A. S., Nishida, M. 1991, ApJ, 382, 115

BARS FROM THE INSIDE OUT: AN HST STUDY OF THEIR DUSTY CIRCUMNUCLEAR REGIONS

Paul Martini

Harvard-Smithsonian Center for Astrophysics; 60 Garden Street, MS 20; Cambridge, MA 02138; USA

Abstract The results of bar-driven mass inflow are directly observable in high-resolution *HST* observations of their circumnuclear regions. These observations reveal a wealth of structures dominated by dust lanes, often with a spiral-like morphology, and recent star formation. Recent work has shown that some of these structures are correlated with the presence or absence of a bar. I extend this work with an investigation of circumnuclear morphology as a function of bar strength for a sample of 48 galaxies with both measured bar strengths and "structure maps" computed from *HST* images. The structure maps for these galaxies, which have projected spatial resolutions of 2 – 15 pc, show that the fraction of galaxies with grand-design (GD) circumnuclear dust spirals increases significantly with bar strength, while tightly wound dust spirals are only present in the most axisymmetric galaxies. In the subset of galaxies classified SB(s), SB(rs), or SB(r), GD structure is only found at the centers of SB(s) or SB(rs) galaxies, and not SB(r). Bar strength measurements of 45 SB(s), SB(rs), and SB(r) galaxies show that SB(s) galaxies have the strongest bars, while SB(r) galaxies have the weakest bars. As SB(s) galaxies are also observed to most commonly possess dust lanes along their leading edges, this is further support of a connection between GD structure and bar-driven inflow on larger scales. There is also a modest increase in the fraction of loosely wound dust spirals at later morphological types, and a corresponding decrease in the fraction of chaotic structures. This trend may reflect an increase in the fraction of galaxies with circumnuclear, gaseous disks. The trend appears to reverse at type Scd, where the fraction of galaxies with chaotic circumnuclear dust structure increases dramatically, although these data are of poorer quality.

Keywords: Barred galaxies, galaxy classification, circumnuclear structure

1. Introduction

Bars are the most effective means of driving gas toward the centers of isolated galaxies, inflow which is often invoked to explain circumnuclear star formation and secular evolution (e.g. Kormendy & Kennicutt 2004). Observations of many barred galaxies show evidence for dust lanes along the leading edges

D. Block et al. (eds.), Penetrating Bars through Masks of Cosmic Dust, 213–222.

of the bar and these dust lanes likely trace the shocks and inflow driven by gravitational torques. The structure of circumnuclear dust within the semiminor axis of the bar can provide important information about the effectiveness of bar-driven mass transport.

It is now possible to quantitatively study the connection between bars and their circumnuclear region due to a combination of three factors: near-infrared surface photometry of a large number of nearby galaxies, a relatively straightforward measure of bar strength Q_b from near-infrared images (Buta & Block 2001) through application of the gravitational torque method of Combes & Sanders (1981), and *HST* images of many of these galaxies. In this contribution I begin with a brief overview of the classification of circumnuclear dust structure. I then apply this system to a large sample of nearby galaxies and investigate correlations between bar strength and circumnuclear structure.

2. Circumnuclear Structure in Galaxies

Martini et al. (2003a) conducted an imaging survey of 123 nearby galaxies with the NICMOS and WFPC2 cameras on *HST* to study circumnuclear dust. These data were used to develop a purely empirical classification system based on common features in the dust distribution and without regard to either the larger scale or nuclear properties of the galaxy. This system is thus complementary to the subject of this conference as it is a dust classification scheme, rather than a *dust-penetrated* classification scheme. The classification system has six categories:

Grand design (GD): Two dominant and coherent dust spirals

Tightly wound (TW): Coherent and large pitch angle dust spirals

Loosely wound (LW): Coherent and small pitch angle dust spirals

Chaotic spiral (CS): Multiple, fragmented dust lanes implying the same sense of rotation.

Chaotic (C): Dust structure without a well-defined morphology

No Structure (N): No evidence for nuclear dust structure

An example of each of these classes is shown in Figure 1.

The initial sample was culled of all galaxies with $v > 5000$ km s^{-1} and inclinations $R_{25} > 0.30$ and then each unbarred galaxy was matched with a barred galaxy of approximately the same morphological (T) type, blue luminosity, heliocentric velocity, inclination, and angular size. This resulted in an extremely well-matched set of 19 barred and 19 unbarred galaxies. The distribution of these 38 galaxies into the six circumnuclear dust classes is shown

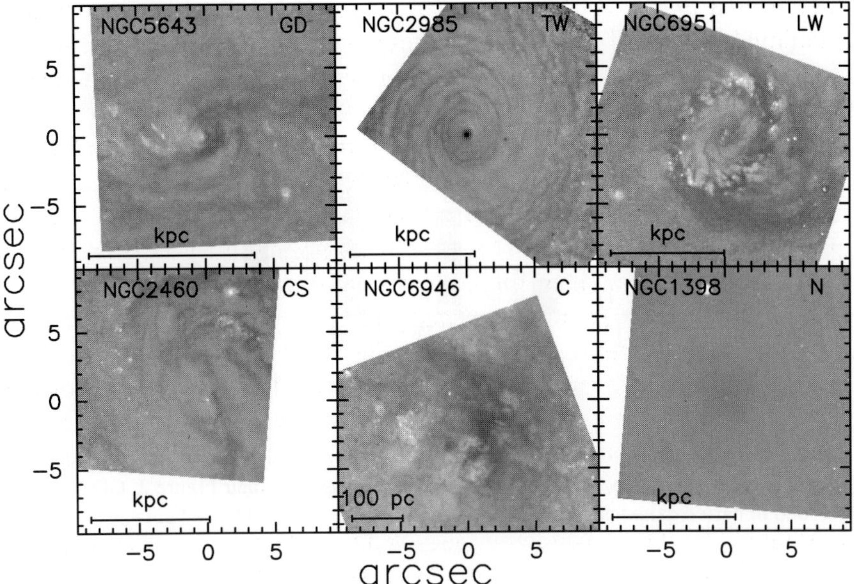

Figure 1. $(V - H)$ color maps of prototypes for the six circumnuclear morphology classes proposed by Martini et al. (2003a) and reproduced from their Figure 3.

in Figure 2. This figure clearly demonstrates two connections between bars and their circumnuclear region: GD structure is only found in barred galaxies, while TW structure avoids barred galaxies (Martini et al. 2003b). In addition, GD structure often (but not always) connects to the dust lanes along the leading edges of the bar at larger scales.

These correlations also validate the classification system itself, as they indicate that the classification bins are connected to physically relevant quantities and are not simply lost in the dust. However, these results were limited by the absence of bar strength measurements for most of the sample. Although Martini et al. (2003a) collected data on whether or not a galaxy was barred from the literature, these data varied substantially in quality and were deemed too heterogeneous to investigate correlations between the circumnuclear morphology and bar strength. In the next section, I apply this classification system to *HST* observations of a large sample of galaxies with measured bar strengths.

3. Bar Strength and Circumnuclear Structure

The new sample described here was compiled from all galaxies with published bar strengths (Buta & Block 2001; Laurikainen & Salo 2002; Block et al. 2004), visible-wavelength images with *HST*, and inclinations $R_{25} < 0.30$. Galaxies with low signal-to-noise (S/N), unfavorable placement on the WFPC2

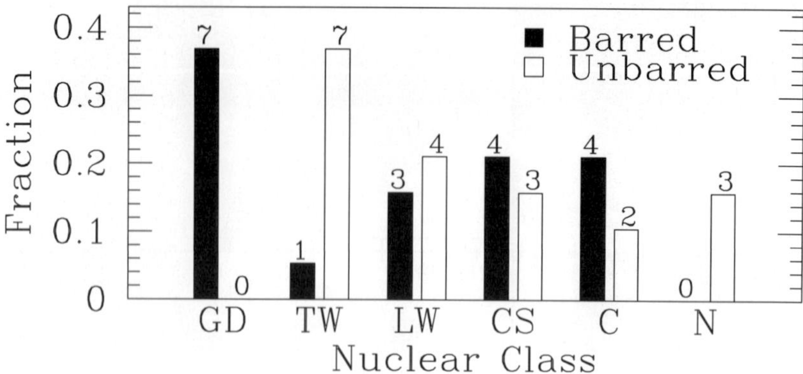

Figure 2. The frequency of the six circumnuclear classes in the sample of 19 barred and 19 unbarred galaxies studied by Martini et al. (2003b) and based on their Figure 2. GD structure is only found in barred galaxies, while TW structure appears to avoid barred galaxies.

detectors, or of type Scd ($T = 6$) or later were barred from inclusion, although Scd galaxies are discussed separately below. The final sample contains 48 galaxies of type S0 to Sc.

I used the structure map technique developed by Pogge & Martini (2002) to identify circumnuclear morphology. For this application, structure maps are superior to color maps because they can be applied to the entire (larger) field of view of the WFPC2 camera and many galaxies only have WFPC2 images. Mathematically, structure maps are:

$$S = \left[\frac{I}{I \otimes P}\right] \otimes P^t \tag{1}$$

where S is the structure map, I is the original image, P is the PSF, P^t the transpose of the PSF, and $\otimes$ is the convolution operator. Structure maps effectively emphasize structures on the scale of the PSF and deemphasize larger-scale spatial variations. In all of the structure maps shown here, dusty regions are dark and emission regions, such as star formation knots, are bright. Figure 3 shows images, color maps, and structure maps of four representative galaxies.

The galaxies were classified with the same system described in the previous section. The only modification is that instead of the fixed angular size of $19''$ employed by Martini et al. (2003a), I have chosen to classify the sample within a fixed 5% fraction of each galaxy's angular diameter D_{25}. This fractional size corresponds to a projected physical size range of $0.4 \rightarrow 2.4$kpc, while the projected physical size of the PSF is $2 \rightarrow 14$pc. Figure 4 presents structure maps of the central 5% of the 48 galaxies. There are fourteen galaxies common to this sample and Martini et al. (2003a) and as a check they were reclassified

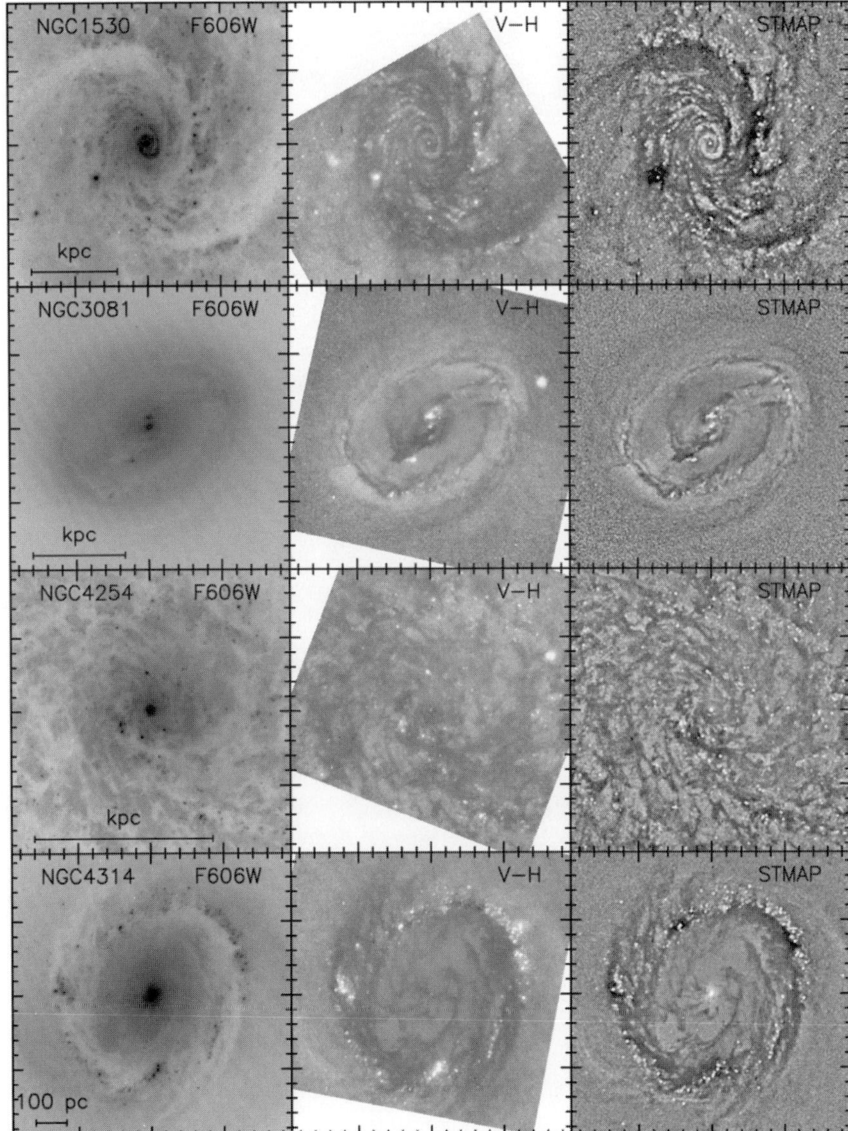

Figure 3. Comparison of V, $V - H$, and structure maps for NGC 1530, NGC 3081, NGC 4254, and NGC 4314. Both the color and structure maps effectively uncover dust features and star formation over a wider intensity range than the V image, even with the log intensity scaling shown. Dusty regions are dark, while emission line regions are light. The structure and color maps are similar, although the structure maps place greater emphasis on features near the resolution limit. Each panel is $19''$ on a side and displays north up and east to the left.

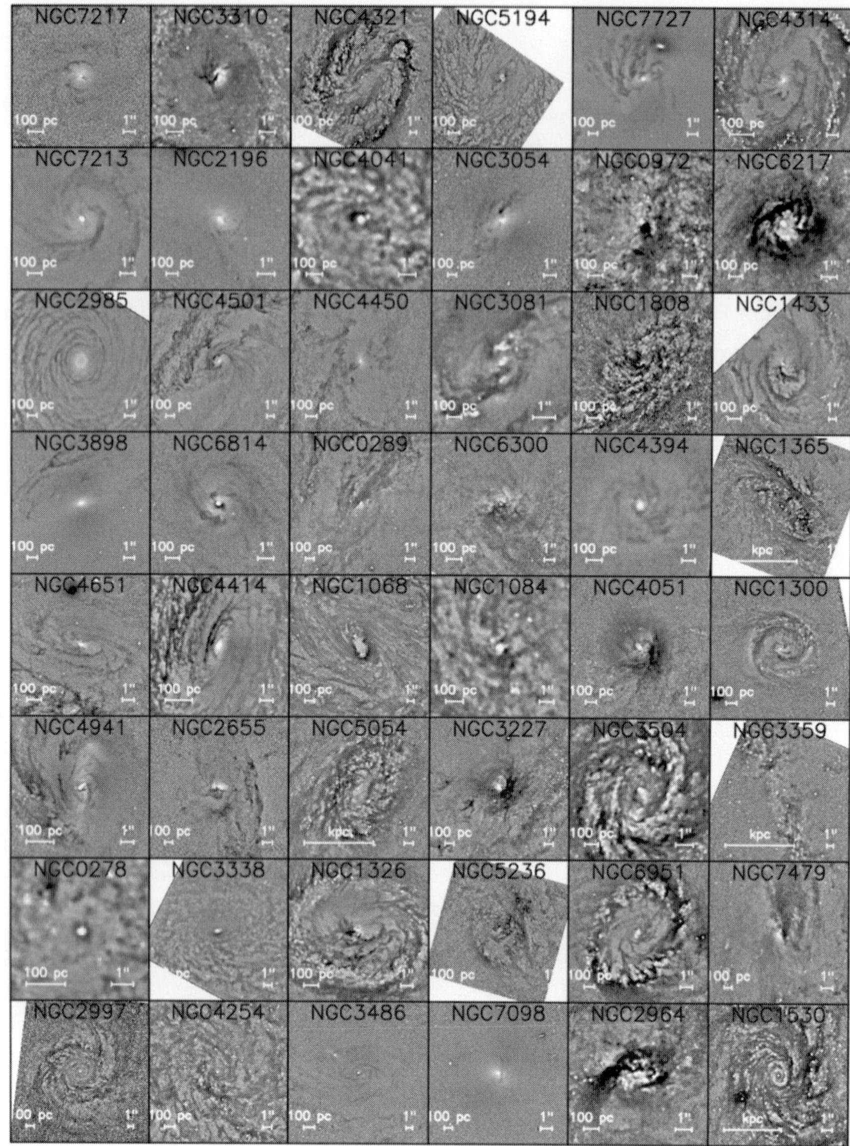

Figure 4. Structure maps for 48 galaxies with measured bar strengths ordered such that bar strength increases downward and to the right. Each panel shows the inner 5% of D_{25} from the RC3 catalog. The projected size of a kpc or 100pc is shown to the lower left, while a $1''$ scale bar is shown at lower right. North is up and east is to the left.

without reference to the prior classification. Nine received the same classification, three switched between the similar classes LW and CS, and only two (14%) changed significantly: NGC 6300 (C→GD) and NGC 4314 (LW→GD).

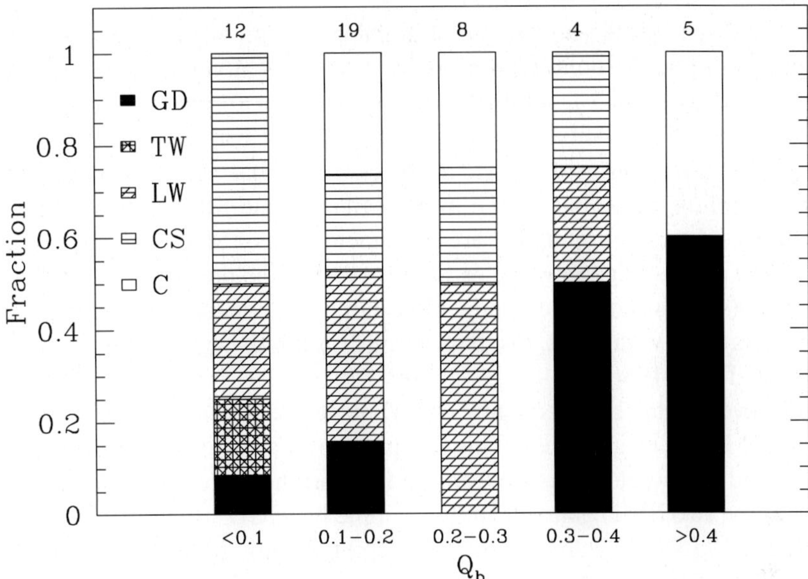

Figure 5. Fraction of each NC class in bins of Q_b. Only a small fraction of weakly barred galaxies have GD structure, while it is present in 5/9 galaxies with $Q_b > 0.3$. TW structure is only present in galaxies with $Q_b < 0.1$. The number of galaxies in each bin is shown above it.

Figure 5 shows the distribution of circumnuclear structure classes as a function of Q_b. The N class was not employed here because it was only populated by one galaxy (NGC 1398). Only four of 31 galaxies (13%) with $Q_b < 0.2$ have GD structure, while it is present in five of nine (56%) galaxies with $Q_b > 0.3$. More strongly barred galaxies are therefore more likely to have GD structure. The one GD galaxy with $Q_b < 0.1$ is NGC 6814, which was also classified GD in Martini et al. (2003a). This galaxy is listed as unbarred in the RC3, although it is classified as barred in the near-infrared.

There are also no galaxies with TW structure and $Q_b > 0.1$. Large pitch angle dust spirals are therefore not found in galaxies with a significant non-axisymmetric component. While only two galaxies were classified as TW, the probability that both would have $Q_b < 0.1$ is 4%. This result reinforces the suggestion of Martini et al. (2003b) that TW structure is only present in unbarred galaxies, and also supports recent simulation results (Maciejewski 2004).

Connection to larger-scale spirals: SB(s) and SB(r) galaxies

GD structure is preferentially found in galaxies with large Q_b and in many cases appears to be the continuation of the dust lanes along the leading edges of large scale bars, dust lanes that models show are formed by strong bars (Athanassoula 1992). Another historical measure of bar strength is whether the large-scale spiral arms originate at the ends of a bar SB(s), from an inner ring at the radius of the bar SB(r), or are intermediate SB(rs). Observations show that dust lanes along the bar are a characteristic of SB(s) galaxies, rather than SB(r) galaxies (e.g. Sandage & Bedke 1994). The presence of dust lanes suggests SB(s) bars should be strong, although hydrodynamic simulations find SB(s) spirals form with weak, fast bars and SB(r) spirals with strong, slow bars (Sanders & Tubbs 1980).

I have investigated the frequency of GD structure in all galaxies classified as SB in the RC3 (11 galaxies). This sample shows a correlation between GD structure and the connection between the bar and the large-scale spiral arms: Neither of the two SB(r) galaxies in this sample have GD structure, although it is present in $3/6$ SB(rs) galaxies and $2/3$ SB(s) galaxies. To test the connection between these classes and Q_b for a larger sample, I have computed the mean Q_b for all 45 galaxies classified as type SB(r): 0.28 (15), SB(rs): 0.35 (12), and SB(s) 0.43 (18). On average SB(s) is thus the most strongly barred type and SB(r) the weakest. However, the SB(s) sample does include more galaxies with late T type and large Q_b. If only galaxies with $T \leq 5$ are included (the range of the SB(r) sample), the mean value of Q_b for the SB(s) class decreases to 0.35 (11 galaxies). It would be valuable to revisit the SB(r)/SB(s) classification with near-infrared images.

Connection to global properties

I have also used this sample to investigate if circumnuclear structure depends on T type, luminosity, or distance. The sample was divided into early ($T \leq 1$; 10 galaxies), intermediate ($T = 2, 3$; 16), and late ($T = 4, 5$; 22) type bins. There is a gradual increase in the fraction of LW structure ($1/10 \rightarrow 4/16 \rightarrow 10/22$) and an approximately corresponding decline in the fraction of C structure ($4/10 \rightarrow 3/16 \rightarrow 3/22$). No change is observed in the fractional distribution of the remaining classes. The increase in the LW fraction at the expense of the C fraction suggests an increase in the fraction of galaxies that have circumnuclear gaseous disks, as a circumnuclear disk is required for spiral dust lanes to form (types GD, TW, LW, CS). There are no obvious trends with distance or luminosity, although this is not surprising because these galaxies span a relatively narrow range in luminosity and distance. This does indicate that the approximately factor of five range in the projected physical size of the kernel does not affect these results.

Very late-type galaxies

Figure 6 displays the nine Scd galaxies with measured Q_b in the *HST* archive, although only two meet the standards of the main sample. These two galaxies are both of type C, as are all but two of the other (lower S/N) galaxies. This may indicate that there is a dramatic increase in the fraction of galaxies with chaotic structure at very late T type, or it may only reflect the importance of high S/N for accurate classification of circumnuclear dust structure.

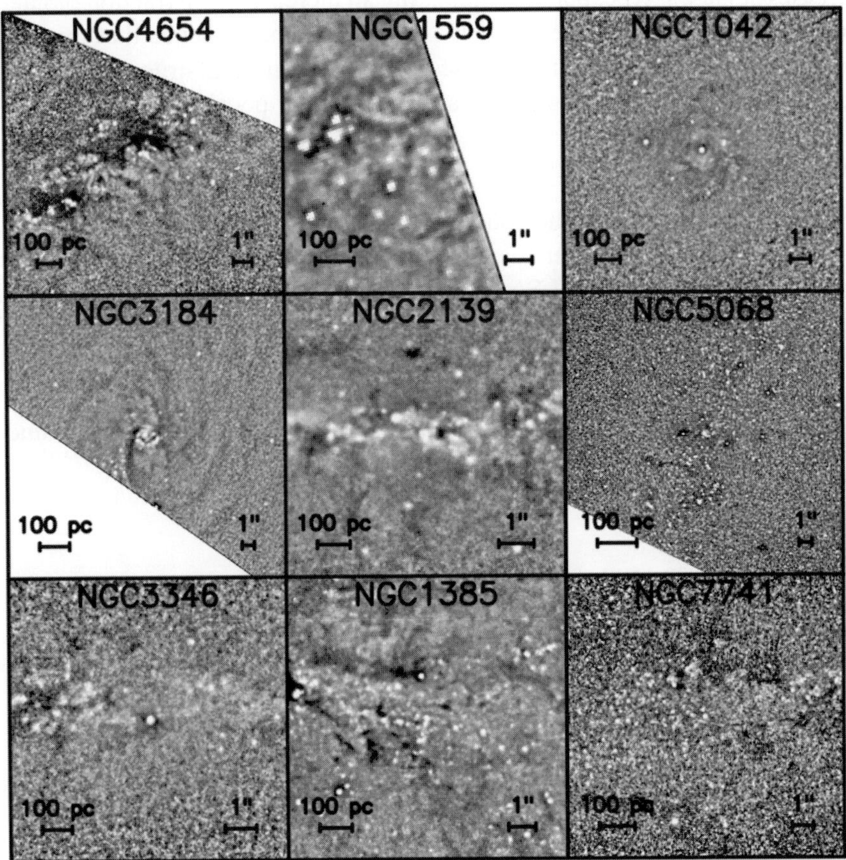

Figure 6. Same as Figure 4 for late-type galaxies ($T = 6$). Only NGC 2139 and NGC 3184 meet the same S/N and position requirements as the main sample. The structures in most panels are dominated by star forming regions, rather than dust.

4. Discussion and Summary

I have computed structure maps from *HST* data for 48 galaxies with measured bars strengths Q_b. These data clearly shown that the fraction of galaxies with GD structure increases sharply for stronger bars, while TW structure is only found in the most axisymmetric galaxies. GD structure is more common in galaxies classified as type SB(s), which are commonly observed to have dust lanes along the leading edges of their bars. Measurement of the mean Q_b for SB(s) and SB(r) galaxies shows that SB(s) galaxies generally have stronger bars. GD structure, SB(s) structure, and dust lanes along bars are therefore all correlated with bar strength. The fraction of galaxies with nuclear dust spirals increases at later T type, perhaps indicating an increase in the fraction of galaxies with circumnuclear gaseous disks. There is some evidence that this trend reverses at type Scd with an increase in the fraction of C structure, although the available data for these galaxies are significantly poorer quality.

Acknowledgments

I was supported during the course of this work by a Clay Fellowship at the Harvard-Smithsonian Center for Astrophysics. I acknowledge the support of the American Astronomical Society and the National Science Foundation in the form of an International Travel Grant, which enabled me to attend this conference. Support for this work was also provided by NASA through grant number AR-9547 from the Space Telescope Science Institute, which is operated by the Association of Universities for Research in Astronomy, Inc., under NASA contract NAS5-26555.

References

Athanassoula, E. 1992, MNRAS, 259, 345
Block, D.L. et al. 2004, AJ, *press*, (astro-ph/0405227)
Buta, R. & Block, D.L. 2001, ApJ, 550, 243
Buta, R., Laurikainen, E., & Salo, H. 2004, AJ, 127, 279
Combes, F. & Sanders, R.H. 1981, A&A, 96, 164
Kormendy, J. & Kennicutt, R.C. 2004, ARA&A, *in press*
Laurikainen, E., & Salo, H. 2002, MNRAS, 337, 1118
Maciejewski, W. 2004, in Carnegie Observatories Astrophysics Series, Vol. 1: Coevolution of Black Holes and Galaxies, ed. L. C. Ho (Pasadena: Carnegie Observatories, http://www.ociw.edu/ociw/symposia/series/symposium1/proceedings.html)
Martini, P., Regan, M.W., Mulchaey, J.S., & Pogge, R.W. 2003, ApJS, 146, 353
Martini, P., Regan, M.W., Mulchaey, J.S., & Pogge, R.W. 2003, ApJ, 589, 774
Pogge, R.W. & Martini, P. 2002, ApJ, 569, 624
Sandage, A., & Bedke, J. 1994, The Carnegie Atlas of Galaxies (Publ. 638; Washington, DC: Carnegie Inst. Washington)
Sanders, R.H. & Tubbs, A.D. 1980, ApJ, 235, 803

MORPHOLOGY OF BAR AND SPIRAL MODES: DO THEY RELATE?*

Preben Grosbøl
European Southern Observatory, Karl-Schwarzschild-Str. 2, D-85748 Garching, Germany

Abstract Deep surface photometry in the near-infrared K′-band of ~100 spiral galaxies was obtained with SOFI at the 3.5m NTT, La Silla, and used to analyze non-axisymmetric perturbations (e.g. bars and spiral structures) in disk galaxies. Synthetic rotation curves for the galaxies were constructed based on the axisymmetric light distribution in K and HI velocity width data.

Bars were detected in 85% of the galaxies in a subsample of ordinary SA spirals. Most bars are shorter than the scale length of the main disk and have pattern speeds in the range of 50-80 km/s/kpc assuming that they end at their co-rotation radius. A lack of tight, strong spirals is observed which may be due to non-linear damping of such structures.

The morphology of the bar-spiral interface region show several cases of phase offsets between end of bar and start of spiral. This is consistent with models in which bar and spiral have different pattern speeds.

Keywords: galaxies:spirals - galaxies:structure - infrared:galaxies

1. Introduction

Non-axisymmetric perturbations such as bars and spirals are very common in disk galaxies and may play an important role in the evolution of such systems by transferring angular momentum and enhancing star formation. It is now possible to acquired deep near-infrared (NIR) surface photometry of such systems and thereby study the detailed distribution of the old stellar disk population which constitute the major fraction of the stellar mass (Rix and Rieke 1993). Especially the K′ band at 2.1μ is well suited as attenuation by dust is insignificant.

The smooth spiral structures observed in NIR suggest that spiral density waves (Lin and Shu 1964) are important for the dynamics. The current paper

*Based on observations collected at the European Southern Observatory, La Silla, Chile.

D. Block et al. (eds.), Penetrating Bars through Masks of Cosmic Dust, 223–230.

extracts parameters for the spiral features in a sample of disk galaxies with the aim of testing theoretical predictions and quantify spiral structures.

2. Data and basic reductions

The sample consisted of 108 spiral galaxies all observed in the K′ filter with SOFI at the 3.5m NTT, La Silla. One part was taken from Grosbøl et al. (2004) while the remaining maps were obtained from a program designed to derive photometry of spiral galaxies in which SN Ia had been observed. The majority of maps reached a signal-to-noise ratio of 3 per □″ at ~20.5 mag/□″ whereas around 25 galaxies were observed 0.5-1 mag deeper. The average seeing of the final stacked images was of the order of 1″. The galaxies had intermediate inclination angles so that spiral structure in the disks could be studied. Although the sample does not have well defined statistical properties, being collected from several programs observed for different purposes, it represents a wide variety of spiral and bar types. The distribution of morphological types in the sample is 51 SA, 39 SX and 18 SB galaxies using the classification of de Vaucouleurs et al. (1991, hereafter RC3).

Standard reductions were applied including estimations of projection parameters by minimizing the m=2 constant phase term in the main disk as described by Grosbøl et al. (2004). The galaxies were decomposed in bulge and main exponential disk components where the former was allowed to consist of a spherical bulge following a modified Hubble law and a steep, central disk. Before perturbations in the disk were analyzed the spherical component was subtracted to avoid artificial bar-like feature due to de-projection effects. Residual in the central parts were seen in several cases possibly due to the bulge being non-spherical or not following the simple analytical form used. Without detailed kinematic information, it is not possible to derive an accurate estimate of the bulge geometric.

Non-axisymmetric perturbations in the disks were analyzed using 1D Fourier transforms of the azimuthal intensity variations in 1″ wide annuli in the plane of the disk. Sharp knots are often seen along the spiral arms on the K image suggesting a contribution of light from young object. Thus, the relative amplitudes derived will overestimate the perturbations of the old stellar disk population in the disk. Synthetic rotations curves for the galaxies were also constructed based on the decomposition of the K surface photometry (Kent 1985) and HI velocity width data available in RC3 using a Hubble constant of 75 km/s/Mpc. A maximum disk solution was used assuming that the main disk contributes 95% of the total circular velocity at its maximum while the halo accounts for the rest.

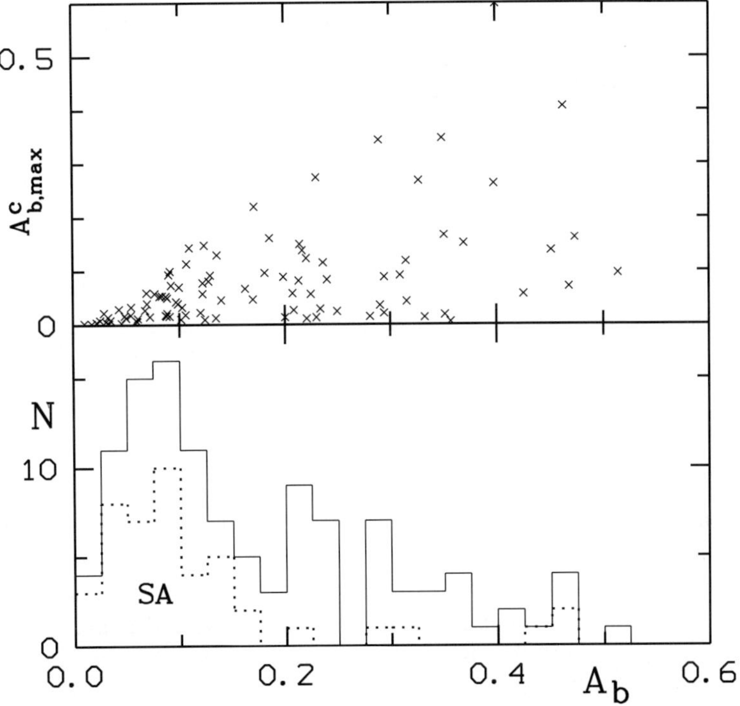

Figure 1. Histogram of the average bar amplitude A_b where the distribution of galaxies classified as SA is shown by the dotted line. The relation between A_b and maximum corrected bar amplitude $A^c_{b,max}$ is given in the top panel.

3. Strength and size of bars

Bars were identified on plots of the radial phase variation of the m=2 Fourier harmonic as features with almost constant phase. Strong dust lanes and star forming regions along bars can introduce so phase shifts. Thus, bars were defined as bisymmetric features with an absolute pitch angle $|i| > 60°$. It was also required that a bar was identified over a radius of at least $3''$. The average bar amplitude A_b was then estimated from the second harmonic excluding radii less than $2''$ to reduce effects due to unsatisfactory bulge subtractions. The histogram of A_b is shown in Fig. 1 where also the distribution of the ordinary SA galaxies is given. Even though the SA spirals show not bar structure in the visible, most of them have bars in the K band. Only 6 galaxies had no detectable bar perturbation at a 3% level corresponding ~15% of the SA galaxies or ~5% of all spirals. This estimate is a lower limit since incomplete removal of the bulge component may contribute to the detected systems.

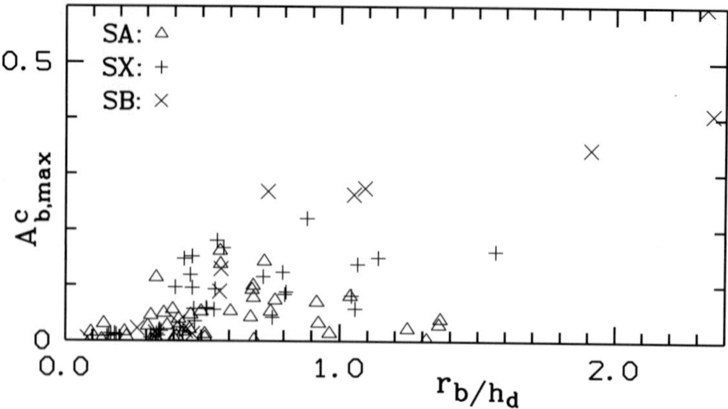

Figure 2. Distribution of maximum corrected bar amplitude $A^c_{b,max}$ as function of the bar length r_b divided by the exponential scale length h_d of the main disk.

The dynamic importance of a bar is better represented by its gravitational torque Q_t (Buta and Block 2001, Combes and Sanders 1981) than its amplitude A_b. However a simpler measure, A^c_b, was preferred and calculated by correcting the bar amplitude by the relative fraction of the surface brightness associated to the disk compared to the total as follows:

$$A^c_b = A_b v^2_{c_d} / v^2_c \tag{1}$$

where v_{c_d} and v_c are the circular velocities for the disk and the total galaxy, respectively. The maximum value of this corrected bar amplitude $A^c_{b,max}$ is shown in the upper panel of Fig. 1. It can be seen that $A^c_{b,max}$ mostly is smaller than the average bar amplitude due to the correction factor which in the bulge region is significantly smaller than unity. Observationally, the correction can be obtained by comparing the observed rotation curve of a galaxy with that estimated from fitting an exponential disk to its outer parts. It is implicitly assumed that the bar perturbation only affects the disk. The correction would be too high if a bar feature observed in the surface brightness distribution was due to a triaxial bulge.

The radial extents of the bars were measured manually using both the radial phase variation and direct maps. They are given in units of the exponential scale of the main disk as the abscissa in Fig. 2 while the corrected bar amplitude is shown as ordinate. Most bars in the sample are shorter than one scale length of the main disk with only a few SB galaxies being significantly longer such as NGC 1300, NGC 1365 and NGC 2487. Many of the bars found in SA galaxies are short. Since they are located in the bulge regions, their corrected amplitudes become very small.

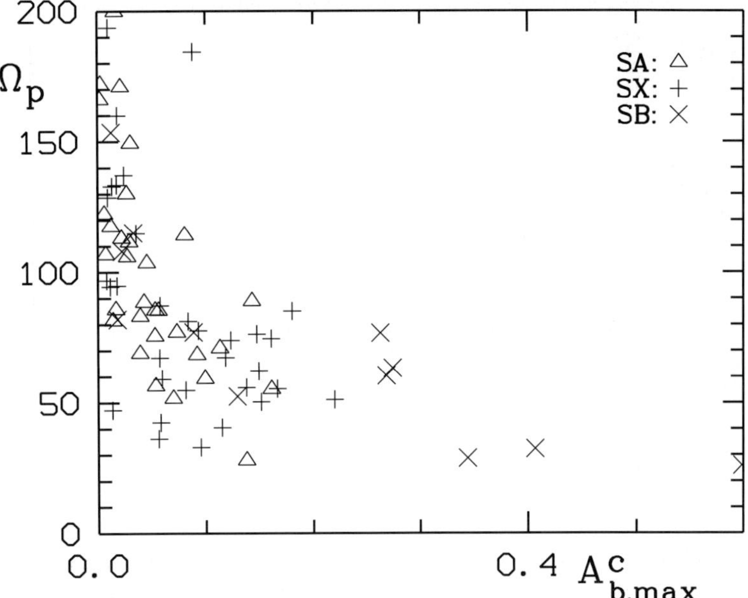

Figure 3. The pattern speed Ω_p (in km/s/kpc) of the bar, assuming it terminates at co-rotation, as function of its strength express by the maximum corrected bar amplitude $A^c_{b,max}$.

One can estimate the pattern speed of the bars by assuming that they terminate just inside their co-rotation (Contopoulos 1980) and using the synthetic rotation curves to derive the corresponding pattern speeds. The derived values are shown in Fig. 3 as function of their corrected amplitude using a Hubble constant of 75 km/s/kpc and systemic velocities from RC3.

The weak bars are often short and tend therefore to have high estimated values for Ω_p. These estimates have high uncertainties due to steeply rising rotation curves in the central parts of the galaxies. The group of bars with corrected maximum amplitudes in the range 0.05-0.30 have pattern speeds in the interval 50-80 km/s/kpc which corresponds well with values determined directly with the method of Tremaine and Weinberg (1984) (see, e.g., Merrifield and Kuijken 1995). A few galaxies with long, strong bars have estimated Ω_p values of the order of 30 km/s/kpc. There may be a weak trend for stronger bars to have lower pattern speeds but the current spread makes it insignificant.

4. Properties of spiral patterns

The radial region occupied by the spiral pattern was also determined visually from phase diagrams and relative intensity maps of the galaxies. The main spiral structure typically starts just outside the bar and consists of a symmetric

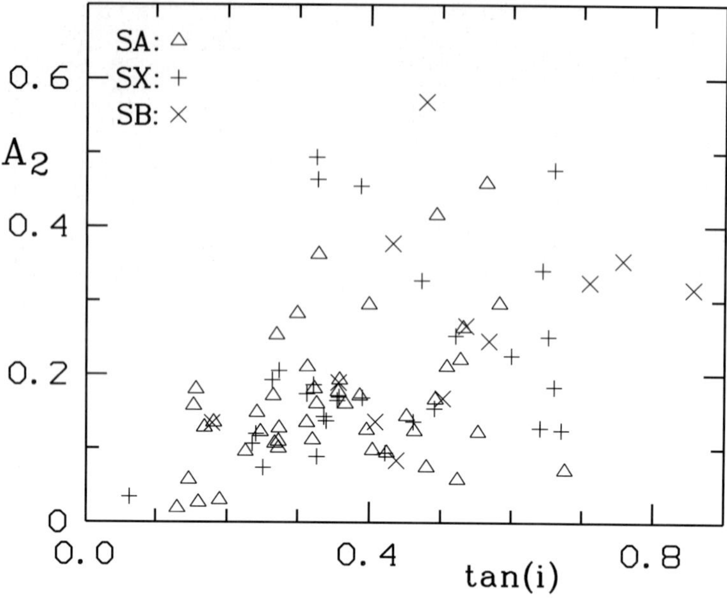

Figure 4. Distribution of the average amplitude A_2 of the main two-armed spiral pattern as function of its pitch angle i.

two-armed pattern. In the outer parts, this pattern often breaks up into multiple armes and changes its pitch angle. An increase of the m=4 harmonic relative to the m=2 term is frequently observed at this radius. Logarithmic spirals are fair approximations to the arms although they often show systematic departures such as being more open in the inner parts than further out.

The average pitch angle i and amplitude A_2 were measured for the main spiral pattern (see Fig. 4). Several of the galaxies with strong amplitudes A_2 may be overestimated compared to the true perturbation in the old stellar disk population due to strong star formation in their arms. The distribution covers the major part of the parameter space but tight, strong spirals are missing. Although it is difficult to detect spirals with $|i| < 5°$ due to their short inter-arm distance, the deficiency is also seen for more open pattern and is therefore likely not a selection effect. Spirals in this region would have high relative radial force perturbations which could lead to damping through non-linear effects. Studies of periodic orbits in realistic spiral potentials (Grosbøl 1993) indicate that such non-linear effects become important for relative radial force perturbations larger than 5% and tend to increase high Fourier harmonics. There may also be a deficiency of weak, open spiral patterns. This would suggest, if confirmed, that the amplification of spiral waves is efficient enough to move newly formed spiral pattern away from this region.

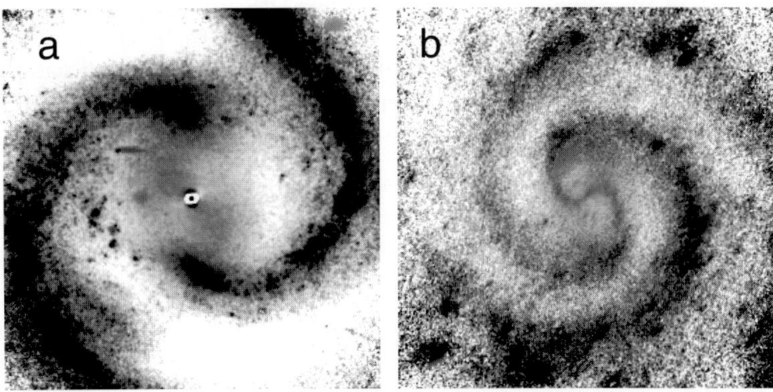

Figure 5. Face-on relative intensity maps in the K′ band of two spiral galaxies: a) NGC 1566 and b) NGC6118. Bulges were subtracted before the image were de-projected. A negative representation is used with black indicating excess intensity of more than 30% compared to their average radial profile of the galaxies.

5. Relations between bars and spiral arms

Comparison of simple properties of bars and spiral patterns (e.g. amplitude, extent and shape) did not shown any significant correlations. This is not surprising since the dispersion relation derived for galactic density waves (see e.g., Bertn et al. 1989) depends on several distribution functions such as the active disk mass and sound speed.

A direct interaction between bar and spiral pattern should be notable in their interface region. In many cases, there seems to be a physical or material connection between bar and spiral as observed in NGC 6118 (see Fig. 5b). On the other hand, significant phase offsets are seen in several systems such as in NGC 1566 (see Fig. 5a) where the bar is offset by more than 30° with respect to the start of the spiral pattern. The number of such offsets observed in this sample is consistent with the n-body models by Sellwood and Sparke (1988) in which the bar rotates faster then the spiral and interact through resonances.

6. Summary

The spiral structure including bars were studied in a sample of 108 disk galaxies. As subset consisting of ordinary SA galaxies was used to determine the distribution of weak bars or oval distortions. Bars were identified in ~85% of the SA systems down to the detection level of 3% relative intensity variation. The weak bars are often short compared to the scale length of the disk. The pattern speeds of the bars were estimated assuming that they terminate just inside their co-rotation. Excluding weak short bars, typical values for the pattern speed was found in the range of 50-80 km/s/kpc.

A majority of the galaxies had a symmetric, two-armed spiral pattern starting just outside the bar. In the outer parts, the pattern often breaks up into multiple arms or/and changes pitch angle. A lack of tight, strong patterns was observed. This indicates that such patterns have so strong relative, radial force perturbations that damping due to non-linear effects become significant.

References

Bertin, G., Lin, C. C., Lowe, S. A., & Thurstans, R. P. 1989, ApJ, 338, 78

Buta, R. & Block, D. L. 2001, ApJ, 550, 243

Combes, F. & Sanders, R. H. 1981, A&A, 96, 164

Contopoulos, G. 1980, A&A, 81, 198

de Vaucouleurs, G., de Vaucouleurs, A., Cowien, H., et al. 1991, Third reference catalogue of bright galaxies (New York: Springer)

Grosbøl, P. 1993, PASP, 105, 651

Grosbøl, P., Patsis, P. A., & Pompei, E. 2004, A&A, in press

Kent, S. M. 1985, ApJS, 59, 115

Lin, C. C. & Shu, F. H. 1964, ApJ, 140, 646

Merrifield, M. R. & Kuijken, K. 1995, MNRAS, 274, 933

Rix, H.-W. & Rieke, M. J. 1993, ApJ, 418, 123

Sellwood, J. A. & Sparke, L. S. 1988, MNRAS, 231, 25

Tremaine, S. & Weinberg, M. D. 1984, ApJ, 282, L5

BAR FORMATION BY GALAXY-GALAXY INTERACTIONS

Masafumi Noguchi
Tohoku University, Sendai, Japan

Abstract Bar formation process in interacting galaxies is reviewed on the basis of recent numerical simulations. I focus especially on the possible kinematical difference between tidally-induced bars and spontaneous bars: the former spans a wider range in the rotation rate, depending upon the parameters of the interaction as well as the internal structure of the host (i.e. perturbed) galaxy. I also discuss possible observational methods for discriminating between and weighing these two classes of galactic bars, including rotation rate measurements and environment statistics.

Keywords: barred galaxies, interactions, dynamics, environments

1. Introduction

Although the dynamical effects of galactic bars once formed have been investigated intensively by many studies, there are several interesting and important problems yet to be solved: when and how bars were created in disk galaxies, what made the gross variation of barred galaxies which are currently observed, and so on. Recent investigation of distant galaxy morphology is enabling a more direct approach to these issues than past studies. However, preliminary works have led to discrepant results concerning the bar abundance at different cosmological epochs (e.g., Abraham et al. 1999, Sheth et al. 2003), and a really fruitful achievement is yet to be made.

There have been proposed two major scenarios for the bar formation. One is the spontaneous bar model and the other is the tidally induced bar model. The former was discovered by numerical simulations of disk galaxies which reveal that relatively cool axisymmetric disks which maintain their flattened shapes mainly by rotational motions are violently unstable and prefer to form a bar within several disk rotation periods (e.g., Ostriker & Peebles 1973).

The systematic study on the creation of barred galaxies by galaxy-galaxy interactions was initiated by Noguchi (1987), followed by a series of numerical works (e.g., Noguchi 1988, Miwa & Noguchi 1998). Since the pioneering

D. Block et al. (eds.), Penetrating Bars through Masks of Cosmic Dust, 231–240.

work of Toomre & Toomre (1972), the interactions between galaxies such as collisions, mergers and close encounters have become one of the most interesting, exciting and important subjects in astrophysics (see Barnes & Hernquist 1992). These interactions make significantly peculiar morphologies called tails, bridges, shells, ripples, polar rings and so on. In most of early simulations on galaxy encounters, partaking disk galaxies were modelled as a point mass (or a rigid spherical potential) which supports a disk of test particles rotating around it. However, it has been revealed in middle 80's that the self-gravity of the galactic disks is an important ingredient of the dynamics of interacting galaxies even when the interaction does not lead to merger. Noguchi (1987) studied the close encounter between a disk galaxy and a point-mass companion by using N-body simulation, and had clarified that the generic form of strongly disturbed galactic disks is a long-lasting bar. This finding has raised a question about the origin of barred galaxies: namely, **are they intrinsic or extrinsic?** Here, we compare properties of spontaneous and tidal bars by numerical simulations with a hope to discriminate between these two classes of bars on the basis of observations. The content of this article is mainly based on Miwa & Noguchi (1998).

2. Numerical Models

In order to investigate the kinematic differences between tidally induced bars and spontaneously developed bars, we performed two different series of simulations. The first one deals with evolution of isolated disk galaxy models which suffer from the global bi-symmetric instability and make a bar spontaneously, and the second one simulates close encounter between a disk galaxy and a spherical perturbing galaxy.

The disk galaxy models used in our N-body simulation are composed of two components: an exponential disk and a spherical halo. Each component is constructed by 40000 collisionless particles interacting with each other gravitationally. The interstellar gaseous component is not considered for simplification. The gravitational potentials and forces are computed by using GRAPE-3AF system.

3. Spontaneous Bar Models

The mass of the galactic disk is the most important parameter to determine the stability against bar formation. Spontaneous models are characterized by relatively large disk masses. We ran two models which have a disk mass of 0.6 (Model A) and 0.35 (Model B) (Hereafter, the total mass of the disk galaxy is set to be unity). The disk mass in the former model is close to the upper limit in real galaxies which the observations of rotation curves suggest (e.g., Athanassoula et al. 1987). The latter represents a model which has the amount of disk mass near the threshold for bar instability. Toomre's parameter $Q = 1.5$ in these models.

4. Tidal Bar Models

In tidal bar models, we set the initial disk galaxy models such that the disk components are stable enough to prevent the global bi-symmetric instability. If the disk develops a bar spontaneously, analysis of the simulations becomes more complicated because of additional parameters such as bar phase at the time when the perturber passes the pericenter. Interaction between an already barred galaxy and a spherical galaxy was discussed by Gerin et al. (1990). Although the common occurrence of bars makes it unexceptional that already barred galaxies experience tidal interactions with other galaxies, we confine ourselves to axisymmetric precursors in this study.

There are two ways to stabilize the disk against bar instability. One is to reduce the disk mass fraction and another is to increase the random motions of disk stars. Correspondingly, we considered two different types of stable axisymmetric models: a) The light disk models (Models L, M, J, K, D, and G). The disk mass in these models ranges from 0.1 to 0.35. The Q parameter is set to be equal to 1.5. b) The hot disk models (Models E and F) . The disk masses are the same as in the spontaneous bar model, Model A. The main difference from the spontaneous case is that the initial Q value is set to be equal to 2.2 instead of 1.5. Motivation for running light disk models comes from the observation that several early-type (S0-S0a) disk galaxies may possess a disk component whose mass fraction with respect to the total galaxy mass is as small as ~ 0.1. This inference comes from the past optical and near-infrared photometry, which has led to the bulge-to-total luminosity ratio (B/T) ranging from ~ 0.1 to ~ 0.25 (e.g., Wainscoat et al 1990).

Perturbing galaxies were not constructed by self-gravitating particles but modelled by a rigid Plummer sphere with a scale length of 0.1 for simplicity. This simplification seems reasonable because the perturber passes well outside the target galaxy, and most particles in the target galaxy are not considered to be affected by the variation of the inner structure of the perturber. The mass of perturber was varied between 1 to 3. Parabolic prograde encounters with the pericenter distance of about twice the disk radius were simulated as the limiting cases which provide the maximum perturbations without resulting in mergers for a fixed perturber mass. The orbital plane of the perturber coincides with the disk plane of the target galaxy, and the perturber's motion is prograde.

5. Results

Figure 1 shows an example of tidal bars obtained in the numerical simulations. All the bars obtained in this study emerge after the passage of the perturbing galaxy. Once formed, those bars are usually a long-standing feature which survives to the end of the simulation like spontaneous bars.

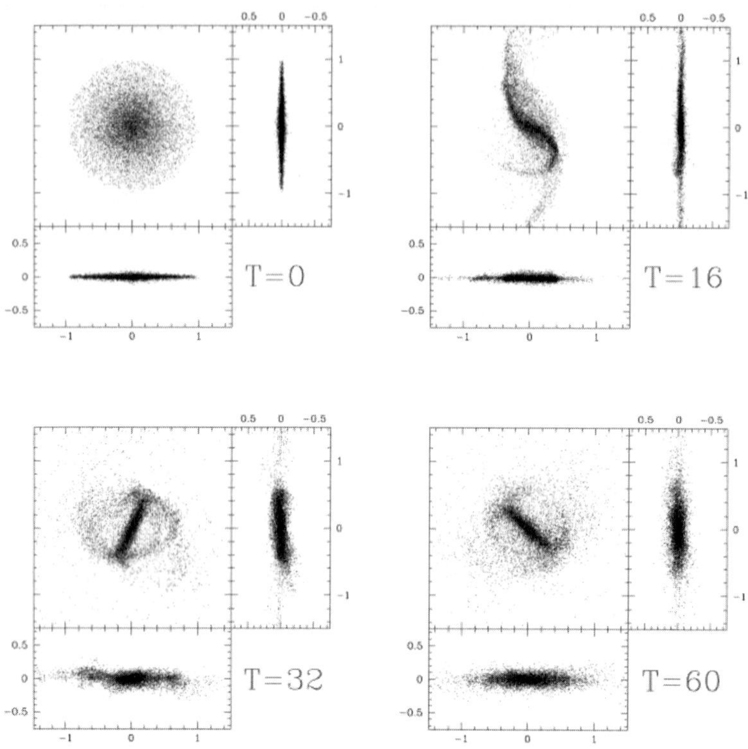

Figure 1. An example of tidally-induced bars (Model D). The face-on view and two edge-on views are shown. For a tipycal galaxy with a mass of $1.8 \times 10^{11} M_{\odot}$ and a radius of 20 kpc, the unit of time, T, becomes 0.99×10^{8}yr. The pericentric passage of the perturber occurs at $T = 12.6$.

The pattern speed is one of the most important quantities which characterize a galactic bar. It affects the global structure and dynamics of the host galaxy through creating various kinds of orbital resonances as well documented (e.g, Schwarz 1984, Combes & Gerin 1985). By this reason, we focus here upon how the pattern speed of the bar obtained in the simulation changes as a function of the model parameters. The main results are summarized in Figure 2.

Figure 2 shows that the mass of the disk component is the major factor to determine the bar pattern speed when the effect of the perturber is small (Models E, L, J, and D) or absent (Models A and B). Ω_p is a monotonically increasing function of M_d. Variation in random motions of disk stars (Q, compare Models A and E) or the degree of mass concentration (not indicated here)does not

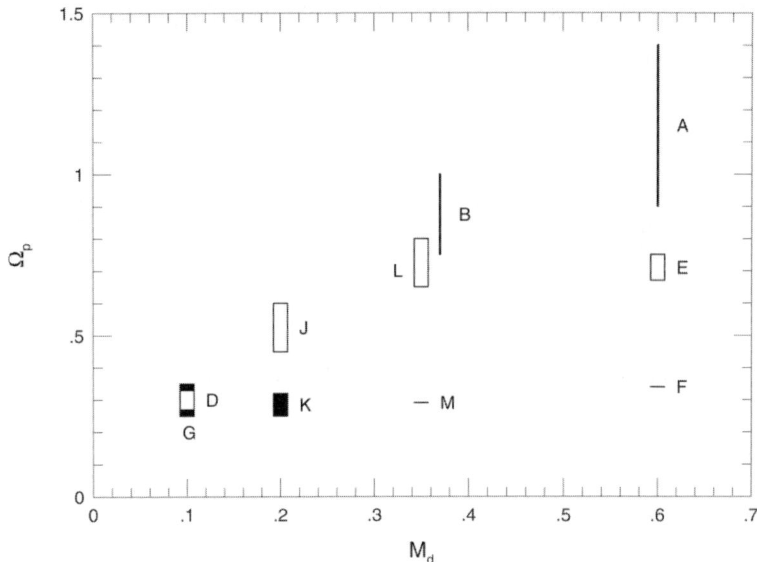

Figure 2. Pattern speed of the bar Ω_p plotted against the disk mass M_d. Models A and B are spontaneous models. Models E, L, J, and D are tidal models with the perturber mass, M_p, of 1, and Models F, M, K, and G are tidal models with $M_p = 3$. Models E and F have $Q = 2.2$, whereas all other models have $Q = 1.5$.

affect this correlation significantly. On the other hand, if the tidal perturbation is strong enough, the bar pattern speed depends hardly on the internal structure of the galaxy such as the disk mass. The constancy of Ω_p for the $M_p = 3$ model sequence (Models F, M, K, and G) is remarkable. The results summarized in Figure 2 illustrate clearly the change from one regime in which the bar structure is shaped largely by the internal property of the host galaxy to another regime in which strong tidal disturbancc erases intrinsic nature and imposes a universal feature determined by the tidal parameters. The first regime manifests itself by the continuity of Ω_p from spontaneous models (A and B) to tidal models (L, J, and D) as the disk mass decreases.

Here we discuss tidal bars in more details. As seen here, the tidal bars created in the hot disk models by a small perturber are the fast rotators, while the tidal bars created in the hot disk models by a large perturber rotates slowly. The influence of the perturber mass seems large when the disk mass in the target galaxy is large. The change of the pattern speed in these hot (and massive) disk models is caused primarily by the angular momentum transfer from the inner disk to the perturber at the time when the perturber passes the pericenter. Figure 3 shows the breakdown of the angular momentum loss suffered by the bar component in Models E and F having the same galactic structure but

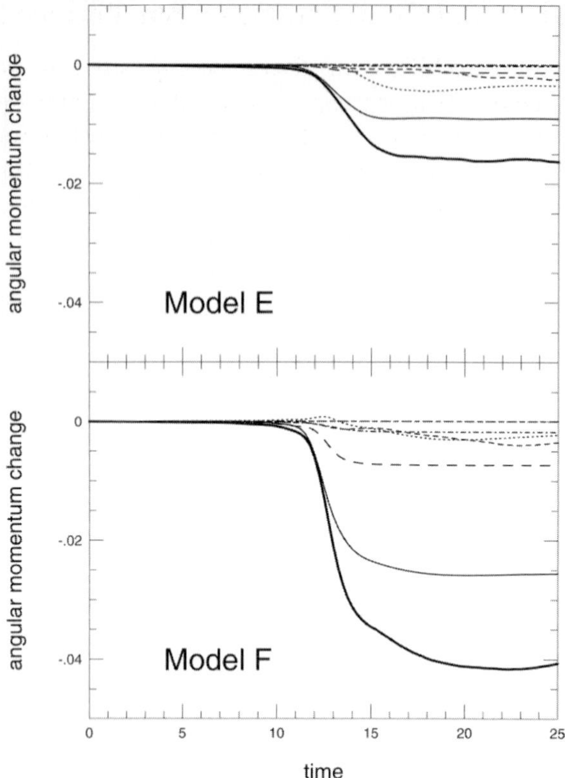

Figure 3. Total change of the angular momentum of the inner disk component (heavy solid line), where the bar is to form. The thin solid line indicates the portion absorbed by the perturbing galaxy, whereas other lines indicate the contribution of the outer disk, the halo and other components in decreasing the inner disk angular momentum.

different perturber masses. In both models, the perturber is the most efficient absorber of the bar angular momentum among various structural components, and a more massive perturber is more efficient, possibly leading to slower bar rotation. On the other hand, the perturber mass seems not to affect significantly the pattern speed of the bar in the light disk models. In this case, the small mass of the disk in which a tidal bar is created leads to a small pattern speed, regardless of the perturber mass.

As shown here, the pattern speed of tidal bars is generally smaller than that of spontaneous bars, and depends upon the characteristics of tidal interactions as well as the internal structure of the host galaxy.

6. Observational Confrontation : Bar Pattern Speed

Now we discuss possibilities of discriminating between spontaneous and tidal bars using the observational data. The result of the previous section may suggest detection of tidal bars using the pattern speed meaturements. The past attempts to derive the bar pattern speed are summarized as follows.

The most direct indicator of a bar pattern speed, Ω_p, is that proposed by Tremaine & Weinberg (1984), which utilize the colllisionless Boltzmann equation thought to reasonably describe the motion of the stellar component in a barred galaxy. This method has been applied to the SB0 galaxy NGC 936 and resulted in the ratio of the corotation radius to the bar length, $\mathcal{R} = 0.54 \sim 1.2$ in Kent (1987) and $1.0 \sim 1.8$ in Merrifield & Kuijken (1995).

A more common method for determining bar pattern speeds is to compare observations of gas velocities and densities with numerical models of gas flows. Since the first applications by Sanders & Tubbs (1980), who found the pattern speed in NGC 5383 by using published optical and HI velocity data and obtained $\mathcal{R} = 1.2$, this method has been applied to a dozen galaxies.

The relative position of corotation and other resonances can often be determined from optical features in the bar-spiral pattern, such as the rings and the shapes of dust lanes. If there is a rotation curve for a galaxy, then the location of any resonance feature can be used to determine the pattern speed, although some resonance features (OLR) may give the pattern speed more accurately than others (ILRs).

In general, studies of bar pattern speeds based on above methods indicate that the bars in early type galaxies end near 0.8 times the corotation radius on the average, giving a value of $\mathcal{R} \sim 1.2 \pm 0.2$. This is the region between the inner 4/1 resonance and the corotation resonance. These bars are called to rotate fast. The pattern speeds of bars in late type galaxies (later than Hubble type Sbc) are scarcely examined and therefore uncertain. It is sometimes argued that the late type bars contain no inner Lindblad resonance because they lack nuclear rings, ring-like starbursts, and offset dust lanes which are often associated with the presence of the ILRs. Provided that this is correct, it is not clear whether late-type bars end near inner Lindblad resonance or they end closer to corotation resonance and rotate faster than the $\Omega - \frac{\kappa}{2}$ curves.

What is the feasibility of pattern speed measurements in finding tidal bars? The measurements of bar pattern speed for candidate interacting galaxies are so far very meager. The result of Debattista et al. (2002) suggests a fast bar (like a spontaneous bar) in the post-interaction galaxy NGC 1023. On the other hand, the result of our numerical work suggests that the perturber must be sufficiently more massive (more than one magnitude in luminosity) to make a significant difference in the bar pattern speed between tidal and spontaneous bars (see Figure 2). A sample of interacting barred galaxies with a brighter companion should be obtained before this issue is thoroughly investigated.

7. Observational Confrontation : Environment Statistics

Another possible way to weigh the importance of tidal formation in the formation of barred galaxies is to make statistics on the dependence of bar incidence on galaxy environments. Because crowded environments such as clusters and groups are thought to be favorable to galaxy-galaxy interactions, we may see overabundance of bars in those regions if tidal formation is really operating. Unfortunately, there is no consensus on whether bar incidence or characteristics differ among various environments. Previous works in this area are summarized as follows.

van den Bergh (2002) has divided Northern 930 Shapley-Ames galaxies into the field, group, and cluster environments based mainly on inspection of Palomar Sky Survey prints. He used barred or non-barred classification made by two competent morphologists, using blue images. His result is that the bar fraction does not depend on the environments, and he argues that the bar formation is determined by instrinsic properties of the parent galaxy.

Thompson (1981) investigated Coma Cluster galaxies with diameter $>$ 16" and claimed overabundance of bars in the central region such that bars occur twice as often in Core ($r < 28'$) as in Annulus (mean $r = 50'$) or Outer Area (mean $r = 98'$), where r denotes the projected distance from the cluster center.

Elmegreen, Elmegreen & Bellin (1990, EEB) have carried out the first serious examination of bar occurrence in different environments. They compared three samples corresponding to different environments: the binary sample taken from Turner (1976)(87 galaxies) and Peterson (1979)(227 galaxies), the group sample taken from Geller & Huchra (1983) (460 galaxies) and Turner & Gott (1976)(221 galaxies), and the field sample taken from Turner & Gott (1976) (119 galaxies). All spiral galaxies in RC2 (de Vaucouleurs et al. 1976) were also used for comparison. Using morphology of the sample galaxies taken from RC2 and UGC (Nilson 1973), they reached the conclusions as follows.

1)Higher bar fraction in the binary sample is seen only for early Hubble types.
2) The smaller galaxy in a binary tends to be barred.
3) Excess of early-type galaxies is found in binary systems.

In order to explain these results in a consistent manner, they invoke the scenario that interactions turn perturbed galaxies into a **barred-and-earlier-type**.

Giuricin et al. (1993) have attempted to improve EEB's analysis by introducing objective environmental parameters. They are 1)ρ_σ, which is the measure for the number of nearby galaxies within the distance of σ, 2)C_R :

the number of galaxies with $M_B \leq -16$ and distance less than R, devided by $4\pi R^3/3$, and 3) distances $d_1(d_2, d_3, ...)$ of the first (second, third,...) nearest galaxy. The galaxy sample is taken from Nearby Galaxies Catalogue (Tully 1988) which is the combination of the magnitude-limited Revised Shapley-Ames Catalogue (Sandage & Tammann 1981) and the diameter-limited all-sky HI survey (Fisher & Tully 1981; Reif et al. 1982), and amounts to 2367 galaxies. Morphology of the sample galaxies was taken from RC3 (de Vaucouleurs et al. 1991). Their main results are
1) For early types (Sa-Sab), the bar fraction is larger in denser environments (in terms of $\rho_{0.25Mpc}$, $C_{0.5Mpc}$, and d_1).
2) For late types (Sb-Sm), the bar fraction is not dependent on environments.
3) Result 1) is due to less luminous ($M_B > -20$)early type galaxies.
4) For scales larger than 1Mpc, no dependence is observed for any morphological types.

They argue that these results agree with interaction scenario, because
1) Less massive galaxies tend to have a more massive companion.
2) Less massive galaxies tend to have a more gently rising rotation curve.
And these two characteristics are favorable to tidally-triggerred bar formation (Noguchi, 1987).

All the above studies have a serious limitation that the classification into barred or unbarred types is based on the published catalogue or made by eye-inspection and therefore is more or less subjective. Also the criteria for environment designation are different from one work to another. Apparently discrepant results mentioned above are likely to originate in these ambiguities.

8. Summary

The origin of galactic bars is among the most important problems in extragalactic astronomy requiring urgent investigation. It was found by numerical simulations that the bar characteristics, especially the pattern speed, vary widely among bars created by galaxy-galaxy interactions. Tidally-induced bars can rotate either fast or slowly, depending mainly upon the perturber mass. In order to find tidal bars utilizing their predicted slow rotation (compared with spontaneous bars) and assess the importance of tidal bar formation, we need a sample of interacting barred galaxies with a significantly more lunimous (by more than one magnitude) companion. We have also described past environmental statistics carried out in an attempt to clarify the origin of barred galaxies.

Acknowledgments

The author wishes to thank Erika Kamikawa for stimulating discussion on the various issues of barred galaxies.

References

Abraham, R.G., Merrifield, M.R., Ellis, R.S., Tanvir, N.R., & Brinchmann, J. 1999, *MNRAS*, 308, 569

Athanassoula, E., Bosma, A., & Papaioannou, S. 1987, *A&A*, 179, 23

Barnes, J.E., & Hernquist, L. 1992, *ARA&A*, 30, 705

Combes, F., Gerin, M. 1985, *A&A*, 150, 327

Debattista, V.P., Corsini, E.M., & Aguerri J.A. 2002, *MNRAS*, 332, 65

de Vaucouleurs, G., de Vaucouleurs, A., & Corwin, H.G. 1976, *Second reference catalogue of bright galaxies*, Austin: University of Texas Press (RC2)

de Vaucouleurs, G., de Vaucouleurs, A., & Corwin, H.G., Buta, R.J., Paturel, G., & Fouque, P. 1991, *Third reference catalogue of bright galaxies*, Springer-Verlag (RC3)

Elmegreen, D.M, Elmegreen, B.G., & Bellin, A.D. 1990, *ApJ*, 364, 415

Fisher, J. R., & Tully, R. B. 1981, *ApJS*, 47, 139

Geller, M.J., & Huchra, J.P. 1983, *ApJS*, 52, 61

Gerin, M., Combes, F., & Athanassoula, E. 1990, *A&A*, 230, 37

Giuricin, G., Mardirossian, F., Mezzetti, M., & Monaco, P. 1993, *ApJ*, 407, 22

Kent, S.M. 1987, *AJ*, 93, 1062

Merrifield, M.R. & Kuijken, K. 1995, *MNRAS*, 274, 933

Miwa, T., & Noguchi, M. 1998, *ApJ*, 499, 149

Nilson, P. 1973, *Uppsala General Catalogue of Galaxies* (UGC)

Noguchi, M. 1987, *MNRAS*, 228, 635

Noguchi, M. 1988, *A&A*, 203, 259

Ostriker, J.P., & Peebles, P.J.E. 1973, *ApJ*, 186, 467

Peterson, S.D. 1979, *ApJS*, 40, 527

Reif, K., Mebold, U., Goss, W. M., van Woerden, H., & Siegman, B. 1982, *A&AS*, 50, 451

Sanders, R.H., & Tubbs, A.D. 1980, *ApJ*, 235, 803

Sandage, A., & Tammann, G. A. 1981, *Revised Shapley-Ames Catalog of Bright Galaxies*

Schwarz, M. P. 1984, *MNRAS*, 209, 93

Sheth, K., Regan, M.W., Scoville, N.Z., & Strubbe, L.E. 2003, *ApJ*, 592, L13

Thompson, L.A. 1981, *ApJ*, 244, L43

Toomre, A., & Toomre, J. 1972, *ApJ*, 178, 623

Tully, R.B. 1988, *Nearby galaxies catalog*, Cambridge University Press

Turner, E.L. 1976, *ApJ*, 208, 20

Turner, E.L., & Gott, J.R., III 1976, *ApJS*, 32, 409

Tremaine, S., & Weinberg, M.D. 1984, *ApJ*, 282, L5

van den Bergh, S. 2002, *AJ*, 124, 782

Wainscoat, R.J., Hyland, A.R., & Freeman, K.C. 1990, *ApJ*, 348, 85

TRIGGERING AGNS - INTERACTIONS OR BARS?

Jeremy Lim[1], Cheng-Yu Kuo[1], Ya-Wen Tang[1], Jenny Greene[2], and Paul T. P. Ho[2]
[1]*Institute of Astronomy & Astrophysics, Academia Sinica, PO Box 23-141, Taiepei 106, Taiwan*
[2]*Harvard-Smithsonian Center for Astrophysics, 60 Garden Street, Cambridge, MA 02138, USA*

Abstract Studies in optical starlight have failed to defintively address whether galaxy interactions or galactic bars play an important role in triggering luminous active galactic nuclei (AGNs). Here, we present the first systematic imaging study of Seyfert (disk) galaxies in the 21-cm line of neutral atomic hydrogen (HI) gas. HI is the most sensitive and enduring tracer of galaxy interactions, and can reveal tidal features not otherwise visible in optical starlight. Our sample, selected from the Véron-Cetty & Véron (1998) catalog, comprises a volume-limited set of Quasi-Stellar Objects (QSOs) with absolute B-band nuclear magnitudes $M_{\rm B} \leq -23$ at redshifts $z < 0.07$ visible from the VLA, and all known galaxies with nuclear magnitudes $-19 \geq M_{\rm B} > -23$ (including also LINERS and HII galaxies) at $0.015 \leq z \leq 0.017$ in the northern hemisphere. We have detected most of the QSO hosts so far studied, and nearly all show strong HI distortions that can usually be traced to tidal interactions with a companion galaxy. Similarly, we have detected nearly all the AGN/LINER/HII galaxies observed, and find a much higher incidence of tidal interactions among the AGN hosts by comparison with our matched control sample. Those AGN hosts with uncertain or no clear tidal features show disturbed HI morphologies and/or kinematics, as well as HI companion galaxies, much more frequently than the control sample. Our study demonstrates that the undisturbed optical appearence of active galaxies can be deceptive, and implicates galaxy-galaxy interactions as a major factor in triggering luminous AGNs and low-luminosity QSOs. The majority of the active galaxies studied here appear to be at a relatively early stage of an encounter rather than late in a merger.

Keywords: Galaxies: active — galaxies: interactions — galaxies: Seyfert — quasars: general — radio lines: galaxies

1. Introduction

In the Local Universe, Seyfert galaxies favour early-type disk galaxies. Their luminous nuclear activity is generally believed to be initiated by a sudden inflow of gas from the host galaxy to fuel the central supermassive black hole. The two most popular competing mechanisms proposed to drive such gas in-

D. Block et al. (eds.), Penetrating Bars through Masks of Cosmic Dust, 241–250.

flows invoke torques generated internally by an axisymmetric potential such as a bar (e.g., beginning with Schwartz 1981; review by Shlosman et al. 1990), or those generated externally by gravitational interactions with a neighboring galaxy (e.g., beginning with Toomre & Toomre 1972; review by Barnes & Hernquist 1992). Identifying the actual mechanism(s) involved has become a focal problem in studies of galaxies hosting AGNs.

2. Optical Searches for Interactions and Bars

In optical starlight, only a small fraction of Seyfert galaxies are visibly interacting with neighboring galaxies. This also is true for those with AGNs sufficiently bright to be classified as QSOs (e.g., Bahcall et al. 1997; Boyce et al. 1998). Emphasizing the similar appearence (except for the bright nucleus) of active and inactive galaxies in the optical, de Robertis et al. (1998) searched for but were unable to find gross morphological differences between Seyfert galaxies and a matched control sample. Optical studies have therefore focussed on addressing whether there is an excess of close projected companion galaxies around Seyfert galaxies by comparison with matched control samples of inactive galaxies. For AGNs, the results are ambiguous: e.g., Dahari (1984) and Rafanelli et al. (1995) reported a positive excess of projected companion galaxies, but not Fuentes-Williams & Stocke (1988). Contrary to the simplest formulation of AGN unification models, some studies have even found an excess of projected companion galaxies around Seyfert 2 but not Seyfert 1 galaxies (e.g., Laurikainen & Salo 1995, Dultzin-Hacyan et al. 1999). The less extensive studies of low-redshift QSOs indicate an excess of projected companion galaxies compared with normal field galaxies (e.g., Bahcall et al. 1997), but here no control experiments have been made.

The above results have been interpreted in a number different ways. On the one hand, galaxy interactions may not be an important mechanism for triggering luminous AGNs. On the other hand, the active galaxy may have cannibalized a relatively low-mass companion galaxy in the recent past, and that this minor merger left no currently visible disturbances (e.g., de Robertis et al. 1998, Taniguchi 1999). To test this possibility, Corbin (2000) compared radial morphological asymmetries seen in both Seyfert and inactive galaxies, but found no significant differences in the level of asymmetry exhibited by the two samples. He therefore concluded that either such minor mergers occurred in the relatively distant past ($\gtrsim$ 1 Gyr ago) or not at all. Sanders et al. (1988) propose that QSOs are the immediate descendents of ultraluminous infrared galaxies, which in the vast majority of cases involve advanced (major) mergers between two massive disk galaxies. The QSO is triggered when sufficient gas reaches the central supermassive black hole, and becomes optically visible only after the surrounding gas and dust (which fed the starburst) has dispersed.

Optical studies that address whether there is an excess of bars in Seyfert galaxies by comparison with matched control samples of inactive galaxies have yielded similarly ambiguous results; e.g., the contrasting results of Arsenault (1989) and Moles et al. (1995). Not finding bars in the majority of the Seyfert galaxies in their sample even at the angular resolution of the HST, Regan and Mulchaey (1999) concluded that bars cannot be the primary mechanism for fuelling AGNs. In the latest study with the HST, Laine et al. (2002) found that $73\% \pm 6\%$ of the Seyfert galaxies in their sample exhibit bars compared with $50\% \pm 6\%$ of their control sample, a marginally significant difference. These results suggest that any real excess of bars in Seyfert galaxies must be relatively small.

3. Studies in Atomic Hydrogen Gas

Atomic hydrogen (HI) gas is well known to its practitioners as a sensitive and enduring tracer of gravitational interactions between galaxies. In normal disk galaxies, the HI gas disk usually extends about two times further out than the stellar optical disk; HI gas at the outskirts is more loosely bound and hence more easily perturbed by external gravitational forces, and being more widely dispersed also takes longer to relax. A particularly dramatic example of this can be seen in the M81 group of galaxies, which shows no visible evidence for large-scale disturbances in the optical, but shows interconnecting HI filaments between group members tracing tidal interactions (Yun et al. 1994).

Prior to the work described here, HI imaging studies of Seyfert galaxies have mostly concentrated on individual galaxies of particular interest. The available set of HI images therefore have different spatial resolutions and detection thresholds, and are not always sensitive to relatively diffuse large-scale structures characteristic of tidal features. Here, we present the first systematic HI imaging survey of Seyfert galaxies together with a matched control sample of inactive galaxies tailored for detecting tidal features.

4. Observations

Our observations were made in the 21-cm line of HI gas with the Very Large Array (VLA). Our first sample comprises all twenty (radio-quiet) Seyfert galaxies in the Véron-Cetty & Véron (1998) catalog with absolute B-band nuclear magnitudes $M_{\rm B} \leq -23$ (for $\rm H_o = 50$ Mpc km s^{-1}) at redshifts $z < 0.07$ and declinations $\delta > -35°$. This comprises a volume-limited sample of all known QSOs visible from the VLA, and contains roughly equal numbers of Seyfert 1 and 2 galaxies. The majority of these galaxies have been observed in the C-configuration of the VLA, providing an angular resolution of $\sim$15$''$ ($\sim$18 kpc at $z = 0.06$, for $\rm H_o = 70$ Mpc km s^{-1} henceforth when not explicitly stated) that is well matched to the optical sizes of massive disk galaxies and tailored towards detecting more extended HI tidal features. Very few of the host galaxies of these QSOs have been studied in sufficient detail in the optical to permit the selection of a matched control sample.

Our second sample comprises all twenty-eight disk galaxies in the Véron-Cetty & Véron (1998) catalog classified as candidate AGNs with nucelar magnitudes $-19 \geq M_{\rm B} > -23$ (for $\rm H_o = 50$ Mpc km $\rm s^{-1}$) at $0.015 \leq z \leq 0.017$ in the northern hemisphere. This is the nearest redshift range in which a significant number of Seyfert galaxies has been found with nuclear luminosities approaching those of QSOs. Four of the galaxies in our sample are classified as Seyfert 1, four are intermediate between Seyfert 1 and 2, eleven are Seyfert 2, five have Low-Ionization Narrow Emission-Line Regions (LINERs), two have nuclear HII regions, and the remaining two are unclassified. As a control experiment, we blindly selected twenty-seven inactive galaxies from the CfA Redshift Survey with the same overall distribution in Hubble type and (host) galaxy luminosity as our AGN sample, but at a factor of two lower redshift ($z \approx 0.008$) to reduce the required observing time to reach the same sensitivity. The majority of both the active and inactive galaxies have been observed in the D-configuration of the VLA, providing an angular resolution of $\sim 60''$ ($\sim$20 kpc at $z = 0.016$). For the control sample, we discarded data on the longer baselines to provide a spatial resolution and detection threshold that is matched to that of the AGN sample.

5. Results and Discussion

Data for about half the QSOs in our sample have been reduced. Some of the early results were presented by Lim & Ho (1999), and more up to date results by Lim et al. (2002). In brief, we have detected in HI most of the QSO host galaxies so far studied. All show HI tidal features or disturbed morphology and/or kinematics that can often be traced to interactions with a nearby companion galaxy. Illustrative examples are shown in Figure 1.

In this paper, we concentrate on our results for the more nearby AGN sample for which we have a matched control sample. For the AGN sample, we detected in HI twenty of the twenty-one galaxies so far observed, including all but one of the fifteen classified as Seyfert 1 or 2 galaxies. For the control sample, we detected twenty-one of the twenty-six galaxies so far observed. Although no selection was made based on their HI properties, we found that the overall distribution of the HI gas masses for the control sample is similar to that for the Seyfert sample.

Our results reveal a dramatic contrast between the HI spatial-kinematic structure of the control and Seyfert sample. Among the twenty-one inactive galaxies detected in HI, only one was found to exhibit HI tidal features tracing interactions with (two) nearby companion galaxies, even though no disturbances are visible in the optical as shown in Figure 2. The remaining twenty inactive galaxies show normal HI morphologies and kinematics, as well as apparently undisturbed optical disks, even though three have detectable HI companion galaxies within a projected separation of $\sim$50 kpc. Illustrative examples of these galaxies also are shown in Figure 2.

By contrast, at least six of the fourteen Seyfert galaxies detected exhibit HI tidal features, even though only two show visible optical disturbances. Two other Seyfert galaxies have very nearby optical companion galaxies that could not be spatially separated in our HI observations; nevertheless, the HI morphology and kinematics of these systems appear to be disturbed (in each case, the observed HI emission appears to be dominated or produced entirely by one galaxy). The remaining six Seyfert galaxies either have poorly resolved HI asymmetries that likely trace tidal features, or disturbed HI morphologies and/or kinematics. By comparison with the inactive galaxies, ten of the fourteen Seyfert galaxies have detectable HI companion galaxies within a projected separation of $\sim$50 kpc. Note that the (candidate) interacting companion galaxy can have either more, less, or comparable HI gas mass and optical luminosity to the Seyfert (host) galaxy. Illustrative examples of all the different cases described above for our AGN sample are shown in Figures 3 and 4.

6. Conclusions

In dramatic contrast with optical studies, our HI imaging study reveals that Seyfert (disk) galaxies hosting luminous AGNs and low-luminosity QSOs are frequently disturbed. In many cases, the disturbances can be traced directly to interactions with neighboring companion galaxies. The much higher frequency of tidal features and disturbances found in our AGN sample compared with our matched inactive control sample directly implicates, for the first time, galaxy interactions as an important if not primary mechanism for triggering luminous AGNs and low-luminosity QSOs in disk galaxies.

Our study also provides important clues on the nature of galaxy interactions that trigger optically-luminous AGNs. Many of the Seyfert galaxies in our sample appear to be experiencing a relatively early stage of an encounter, with the interacting companion galaxy still well separated optically from the Seyfert galaxy. These results contradict the suggestions that AGN activity is triggered primarily by either a major or minor merger in the recent past. Instead, the contrasting appearences of Seyfert galaxies in the optical and HI suggests that the nature of the interaction usually comprises an encounter that visibly perturb the gaseous but not stellar disk, or one in which AGN activity usually occurs only after the more tightly bound stellar disk have had sufficient time to relax.

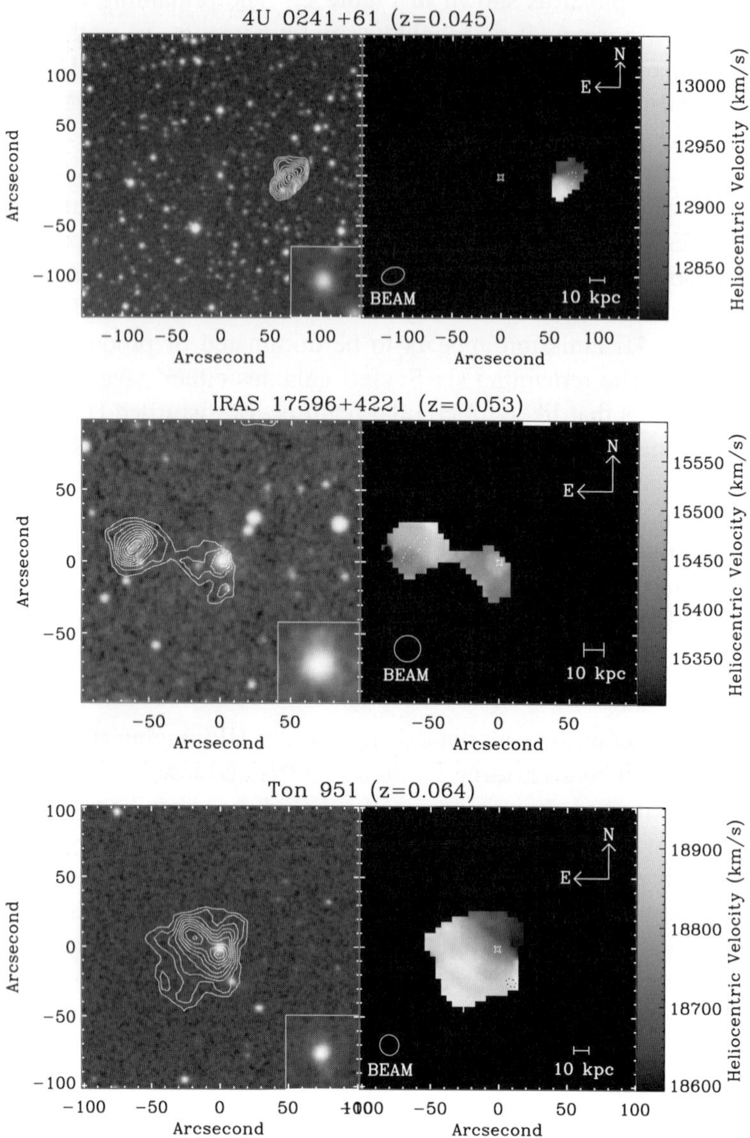

Figure 1. Left panels – contours of the integrated HI intensity overlaid on grayscale images of the optical fields (from the Digitized Sky Survey) centered on each QSO. The region immediately surrounding the QSO is shown magnified in the lower right inset of each panel. Right panel – grayscale-coded images of the intensity-weighted mean HI velocity, with the QSO position indicated by a star and the interacting companion galaxy as an ellipse. From top to bottom, the panels are arranged in order of decreasing projected separation between the QSO and and its interacting companion galaxy.

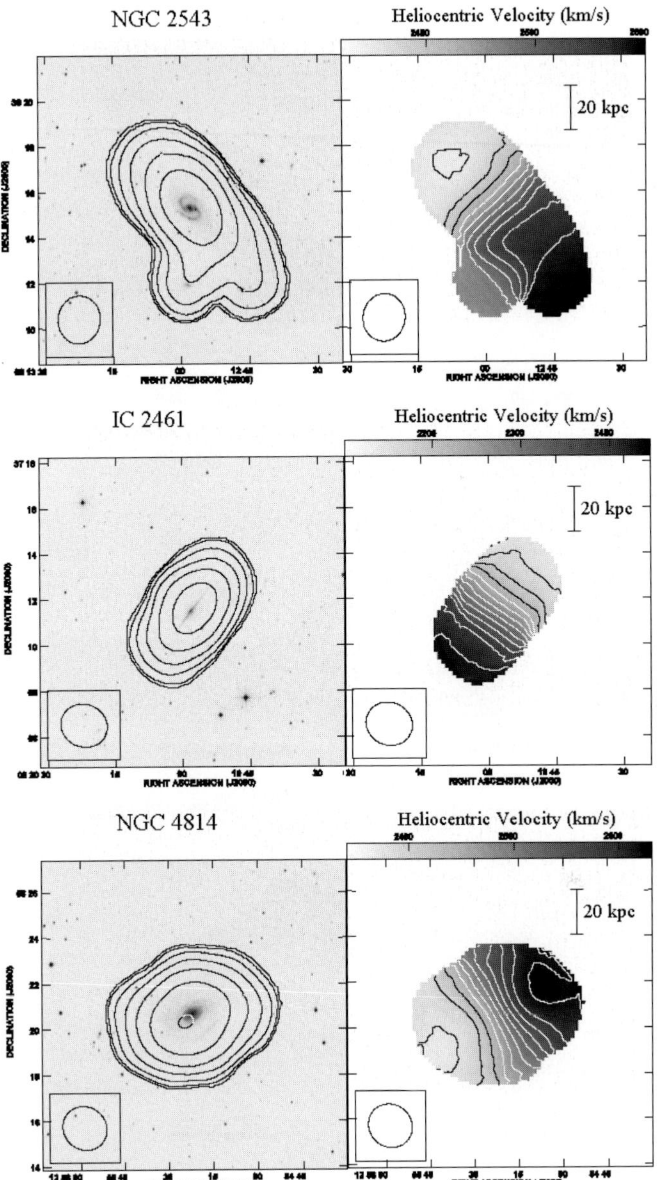

Figure 2. Same as in Figure 1, except that the objects here are from our control sample of inactive galaxies matched in host galaxy properties to our AGN sample. We found that only one galaxy in our control sample is interacting, with the interactions seen only in HI (top panel). The other galaxies appear to have normal rotating HI disks, as shown in the lower two panels for a highly inclined and a moderately inclined disk galaxy respectively.

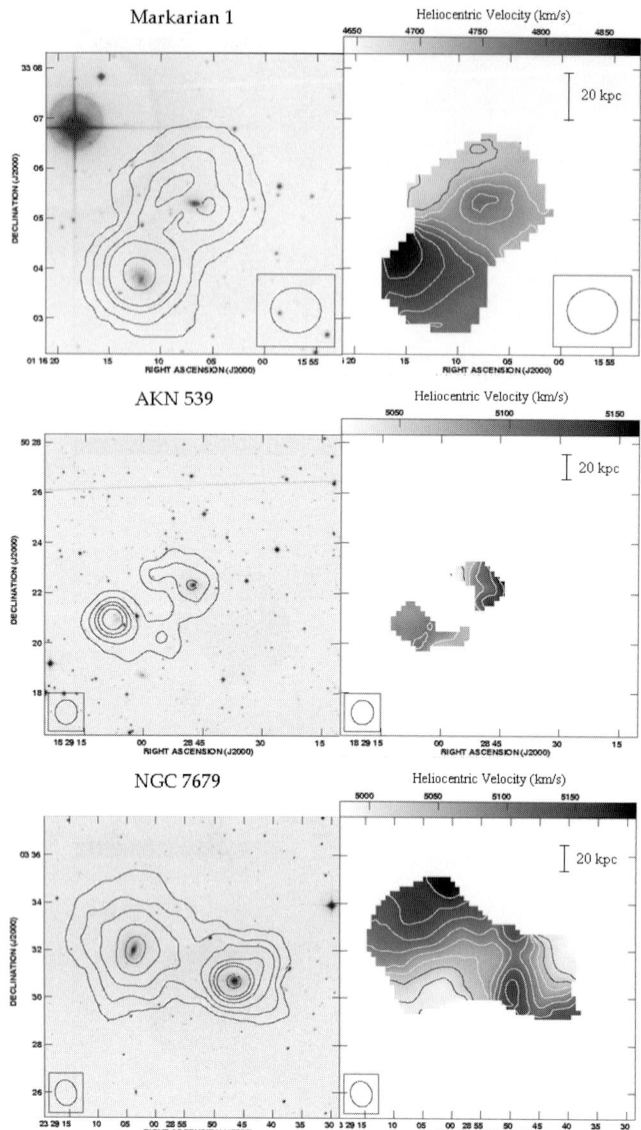

Figure 3. Same as in Figure 1, except that the objects here are from our AGN sample. In this figure, we show clear examples of interactions between the Seyfert galaxies and their neighboring companion galaxies. The tidal features are usually seen only or in HI and not in optical starlight. The bottom panel shows a comparatively rare example of two interacting Seyfert galaxies.

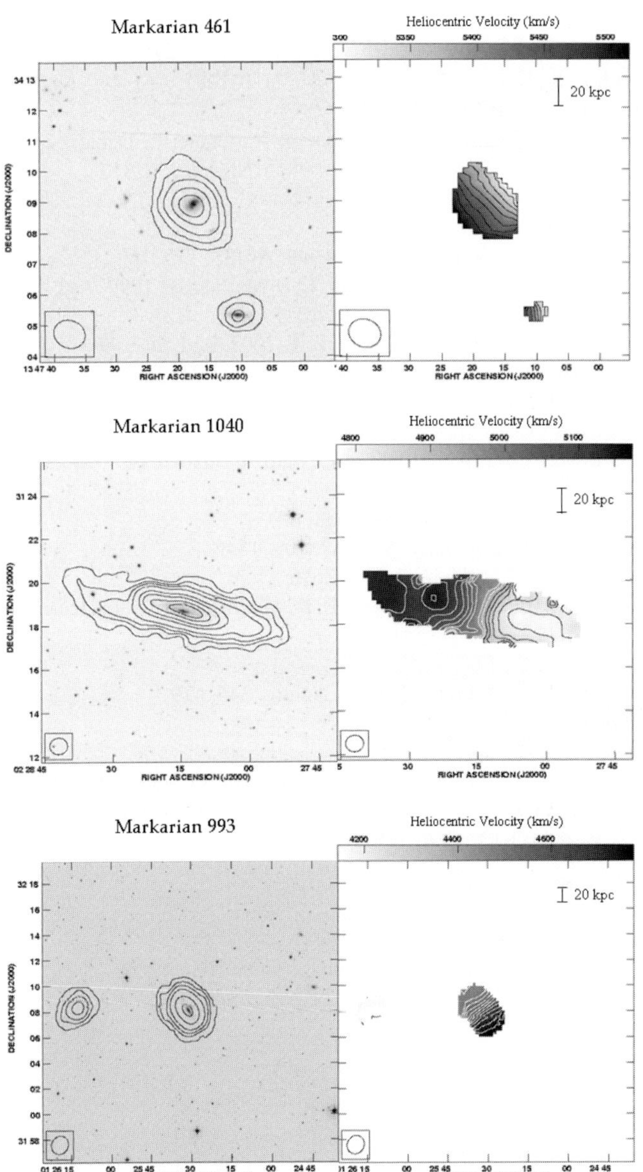

Figure 4. Same as Figure 3. The top panel shows a Seyfert galaxy that is likely to be interacting with its neighboring companion galaxy as seen in HI. The middle panel shows a Seyfert galaxy that may be interacting with its neighboring companion galaxy in the optical, but which cannot be properly separated in our HI observations. The bottom panel shows a Seyfert galaxy that appears to be relatively normal in HI (and in the optical), but which shows what may be a weak HI tidal feature on the side facing the neighboring companion galaxy.

References

Arsenault, R.1989, A&A, 217, 66

Bahcall, J. N., Kirkahos, S., Saxe, D. H., & Schneider, D. P. 1997, ApJ, 479, 642

Barnes, J. E. & Hernquist, L. ARA&A, 30, 705

Boyce, P. J., Disney, M. J., Blades, J. C., Boksenberg, A., Crane, P., Deharveng, J. M., Macchetto, F. D., Mackay, C. D., & Sparks, W. B. 1998, MNRAS, 298, 121

Corbin, M. R. 2000, 536, L73

Dahari, O. 1984, AJ, 89, 996

de Robertis, M. M., Kayhoe, K., & Yee, H. K. C. 1998, ApJSS, 115, 163

Dultzin-Hacyan, D., Krongold, Y., Fuentes-Guridi, I., & Marzian, P. 1999, ApJ, 513, L111

Fuentes-Williams, T. & Stocke, J. T. 1988, AJ, 96, 1235

Laine, S., Shlosman, I., Knapen, J. H., & Peletier, R. F. 2002, ApJ, 567, 97

Laurikainen, E. & Salo, H. 1995, A&A, 293, 683

Lim, J. & Ho, P. T. P. 1999, ApJ, 510, L7

Lim, J., Chuo, H., Wen, S., Liao, W., Ho, P. T. P. 2001, in QSO hosts and their environments, eds. I. Márquez, J. Masegosa, A. del Olmo, L. Lara, E. García, & J. Molina (Dordrecht: Kluwer), 191

Moles, M., Marquez, I., & Perez, E. 1995, ApJ, 438, 604

Rafanelli, P. Violato, M., & Buruffolo, A. 1995, AJ, 109, 1546

Regan, M. W. & Mulchaey, J. S. 1999, ApJ, 117, 2676

Sanders, D. B., Soifer, B. T., Elias, J. H., Madore, B. F., Matthews, K., Neugebauer, G., Scoville, N. Z. 1988, ApJ, 325, 74

Schwarz, M. P. 1981, ApJ, 247, 77

Shlosman, I., Begelman, M. C., & Frank, J. 1990, Nature, 345, 679

Taniguchi, Y. 1999, ApJ, 524, 65

Toomre, A. & Toomre, J. 1972, ApJ, 178, 623

Véron-Cetty, M. P. & Véron, P. 1998, ESO Scientific Rep. 18, A Catalog of Quasars and Active Nuclei (8th Ed.; Paris:ESO)

Yun, M. S., Ho, P. T. P., & Lo K. Y. 1996, Nature, 372, 530

TRIGGERED STAR FORMATION: FROM LARGE TO SMALL SCALES

Jan Palous [1], Pavel Jachym [1,2] and Sona Ehlerova [1]
[1] *Astronomical Institute, Academy of Sciences of the Czech Republic, Bocni II 1401, 141 31 Prague 4, Czech Republic,* [2] *LERMA, Observatoire de Paris, 61 avenue de l'Observatoire, 75014 Paris, France*

Abstract We give examples of star formation triggering. On the large scale, SF is triggered by collisions and interactions of galaxies, and by interactions of a galaxy with the environment in a cluster. On the intermediate scale, SF is triggered in barred galaxies by resonant rings. On the small scale, the star formation is self-regulated by the feedback from young, massive stars.

Keywords: galaxy interactions, star formation, feedback

1. Star formation triggering

The compression and collapse of the diffuse interstellar medium (ISM) to a stellar cluster may be triggered by a large scale or a small scale event. The formation places are giant molecular clouds (GMC) or clumps of gas accumulated in a turbulent ISM. The GMCs are supported by pressure and tidal fields against the gravitational collapse. The critical clump radius r_{crit} is given as

$$r_{crit} = \frac{\sigma_R}{\left[\frac{4}{3}G\pi\rho - (A-B)(3A+B)\right]^{1/2}}, \tag{1}$$

where σ_R is the radial component of the velocity dispersion, ρ is the average density of the GMC and $A = \frac{1}{2}\left[\frac{\Theta}{R} - \frac{d\Theta}{R}\right], B = -\frac{1}{2}\left[\frac{\Theta}{R} + \frac{d\Theta}{R}\right]$ are the Oort constants, $\Theta(R)$ is the linear galactic rotation curve and R is the galactocentric distance. A and B introduce into the formula (1) the effect of tidal forces.

Star formation is triggered when the critical radius r_{crit} decreases due to some reason below the radius of the clump r_{clump}: $r_{clump} > r_{crit}$. Such decrease happens when: 1) tides make $\frac{d\Theta}{dR} > 0$, in this case of rising rotation speed the tidal term decreases reducing the value r_{crit}; 2) σ_R decreases due to radiative cooling and dissipative shocks; 3) ρ increases due to gas inflow and accretion. Another type of triggering happens when the pressure of the ambient

D. Block et al. (eds.), Penetrating Bars through Masks of Cosmic Dust, 251–254.

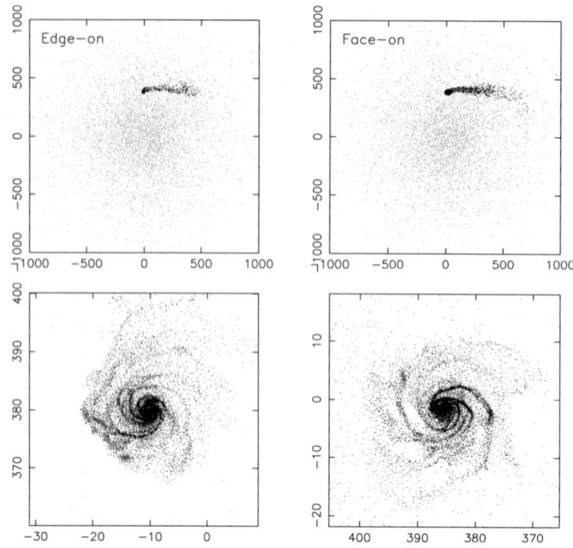

Figure 1. The motion of a spiral galaxy in a cluster of galaxies: numerical simulation of the gas stripping. Two different orientations of the galaxy disk relative to its orbit are shown: edge-on orientation - left panels, face-on orientation - right panels. Upper row shows the 2 x 2 Mpc area of the cluster, lower row shows the stripped galaxy disk in greater details.

medium increases. This may be due to some extragalactic pressure source like ram pressure from the interaction of the galaxy's ISM with the intracluster gas (IGM) or due to infall of a high velocity cloud into the galactic plane.

2. Reduced sheer and tidal fields

The galaxy encounters increase the rotation speed at some places of the tidal arms and bridges reducing locally the sheer and tidal fields triggering the star formation there. It may be the case of young blue super-clusters in the Antennae galaxies (Zhang et al. 2001). The test can be the velocity fields: if measured we could estimate if the local sheer is reduced. Galaxy pairs NGC 6621/2 and NGC 5752/4 are other examples of encounters (Keel & Born, 2003) reducing the tides in places, where the blue super-star clusters form. The barred spiral galaxy NGC 4314 in the constellation Coma Berenices gives an example of young stellar clusters concentrated in a nuclear star-forming ring. Star formation triggering is done at the inner inner Lindblad resonance in the bar-like galactic potential (Benedict et al. 2002).

3. Increased pressure in the ISM

The ram pressure of the hot X-ray IGM induced by a motion of a galaxy in a cluster increases the pressure in the ISM. It strips a fraction of gas, which streams away from its parent galaxy forming the gaseous tail. A large scale shock wave forms at the leading edge. The gas accumulates there and the compression triggers the star formation. An example of gas stripping is the case of C153 galaxy (Keel et al., 2003). In Fig. 1 we show numerical simulations of gas stripping using an SPH code (Jachym, 2004). The cluster of galaxies is approximated by a smooth gravitational potential and the hot diluted IGM. A spiral galaxy, which is composed of stars, dark matter and cold smoothly distributed gas in a disk + bulge + halo configuration, is exposed to the ram pressure as it moves in the cluster. As visible in Fig. 1, the gas is stripped to a long tail. Stripping is particularly effective when the galaxy approaches the center of the cluster, where it has the highest velocity and where the hot IGM gas concentrates. The orientation of the galactic disk relative to the orbit in the galaxy cluster plays an important role. The gas stripping is more intensive for the face-on orientation of the galaxy disk than for an edge-on orientation, since in the face-on case the effective cross section of the disk is larger. In the later orientation, the gas is collected and compressed at the leading edge of the disk, which triggers star formation there. With the face-on orientation, the gas stripping reduces the extent of the disk.

4. Increased gas density and cooling

Coherent expanding supershells are identified in Milky Way and other nearby galaxies. An example is shown in Fig. 2. This is the galactic supershell GS061+00+51 (Heiles, 1979) observed again with the Effelsberg radio-telescope by Ehlerova et al. (2001), who identified several other structures there, also visible in Fig. 2, e.g. the small spherical shell at $\sim (60^\circ, -1^\circ)$. The energy and the mass released from young and massive stars through stellar winds, fotoionizing radiation and supernova explosions modify or stop local collapses to stars. At the same time the same processes produce coherent shells, sheets and filaments, where the ISM accumulates, cools reducing its internal velocity dispersion, eventually triggering star formation. Star formation self-regulates through this feedback mechanism. An open question is how much of star formation is triggered by stellar feedback and what fraction is due to random compressions in the turbulent medium.

5. Summary

Triggering of star formation due to various events has been exemplified. Galaxy versus galaxy collisions, galaxy interactions with other galaxies and with galaxy environment have been labeled as the large scale, since the star formation is initiated simultaneously at different places of the galaxy. A small scale triggering is connected to expanding shells, sheets and filaments formed as a consequence of star formation are able under certain conditions compress

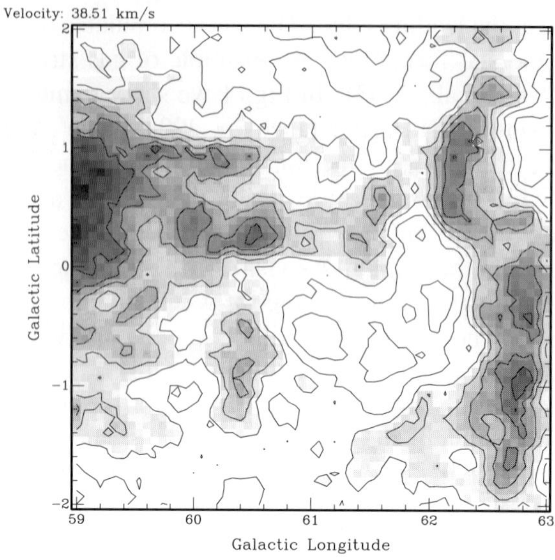

Figure 2. The HI supershell GS061+00+51 and its vicinity.

the ISM and trigger star formation at other places. Star formation triggering is present at various scales in the local universe and we assume that its ubiquity extends to high-z where galaxy versus galaxy collisions, environmental effects and star formation feedback appeared to be even more important than nowadays.

Acknowledgments

The authors gratefully acknowledge EARA Marie Curie doctoral fellowship programme, the cooperative project between CNRS and the Czech Republic "Gas-stars mass exchange and galaxy interactions" and the financial support by the projects of the Academy of Sciences of the Czech Republic No. K1048102 and AV0Z 1003909.

References

Benedict, C. F., Howell, D. A., Jorgensen I., Kenney, J. D. P., Smith B. J. (2002) AJ 123, 1411
Ehlerova, S., Palous, J., Huchtmeier, W. K. (2001) A&A 374, 682
Heiles, C. (1979) ApJ 229, 533
Jachym, P. (2004) PhD Thesis, Charles University, Prague, in preparation
Keel, W. C. & Born, K. D., 2003, AJ 126, 1257
Keel, W., Owen, F., Ledow, M., Wang, D. (2003). AAS...203, 4705K
Zhang, Q., Fall, M. S., Whitmore, B. C. (2001). ApJ 561, 727

INVESTIGATION OF AGE AND METALLICITY GRADIENTS IN SPIRAL GALAXIES

Barbara Cunow
Department of Mathematics, Applied Mathematics and Astronomy, University of South Africa, PO Box 392, Pretoria 0003, South Africa.
E-mail: cunowbhl@unisa.ac.za

Abstract Optical and near-infrared disc scalelengths are used to investigate age and metallicity gradients in the discs of spiral galaxies.The observed colour gradients are compared to model calculations, and the analysis indicates that the stars become younger and less metal-rich with increasing distance from the centre of the galaxy.

Keywords: galaxies: spiral – galaxies: stellar content – galaxies: structure

1. Introduction

It is known for a long time, that most spiral galaxies show colour gradients in their discs, in the sense that the outer regions appear bluer than the inner regions. There are three possibilities for creating such gradients: (i) age gradients in the stellar disc, (ii) metallicity gradients in the stellar disc, and (iii) dust extinction. Some studies indicate that the observed colour gradients are mainly caused by age gradients, whereas others find that metallicity gradients play an important role as well. As far as dust extinction is concerned, it is still not clear how much dust is found in spiral galaxies. Model calculations show that the observed colour gradients in disc galaxies can be severely affected by dust.

Peletier et al. (1994) and Byun, Freeman & Kylafis (1994) point out that studying the change of the observed colour gradients with inclination angle allows to separate affects due to age and metallicity gradients from those caused by dust. This can be achieved by measuring the colour gradients for a sample of galaxies which cover the whole range from face-on and edge-on view. This method is used here for studying age and metallicity gradients of the stellar populations in the discs of spiral galaxies.

In this work, the change of the colour gradients with inclination angle is investigated by studying the scalelength ratios $r_D(F_1)/r_D(F_2)$ of the sample

D. Block et al. (eds.), Penetrating Bars through Masks of Cosmic Dust, 255–259.

galaxies, where $r_D(F_1)$ and $r_D(F_2)$ are the disc scalelengths measured in two different wavelength regions, F_1 and F_2, as function of the apparent ellipticity ϵ. $r_D(F_1)/r_D(F_2)$ is a robust measure of the large-scale colour gradient in the disc, and ϵ is a measure of the inclination. In order to break the age-metallicity degeneracy, both optical and near-infrared data are used (see Bell & de Jong 2000 for details).

The observed scalelength ratios are compared to scalelength ratios obtained from model calculations. The simulations are done with different age and metallicity gradients in the stellar disc and different amounts of dust.

2. Data

The galaxy sample is the one of non-active galaxies studied by Cunow (2001). It consists of 32 spiral galaxies of type Sb, which cover the whole range from face-on to edge-on view.

For all objects, $BVRIJHK_S$ data were obtained using the 1.0 m telescope and the Infrared Survey Facility at the South African Astronomical Observatory, and from the Two Micron All Sky Survey. For each galaxy and filter, the disc scalelength was measured by fitting an exponential law to the surface brightness profiles. The apparent ellipticity was obtained from the shapes of the isophotes on the I image. Details are found in Cunow (2001, 2004).

3. Models

Using the method described in Cunow (2001), $BVRIJHK_S$ images of model galaxies were calculated for different colour gradients in the stellar disc and for various amounts of dust. The model galaxies consist of a luminous stellar disc and a dust disc. It is assumed that both the luminosity density of the stellar disc and the extinction coefficient of the dust disc decrease exponentially parallel to the plane of the galaxy and also perpendicular to it. In the following, the distance from the galactic centre measured parallel to the galactic plane is denoted by R, and the scalelength and the scaleheight of the luminous stellar disc by r_0 and z_0, respectively. The geometry chosen for the models is representative for an Sb galaxy.

The model images are calculated for different inclination angles and different amounts of dust. The values for the central optical depth in the B band for face-on view, τ_0^B, lie in the range between 0 and 10.

It is assumed that age and metallicity of the stellar populations in the stellar disc change with R, but do not vary perpendicular to the plane of the galaxy. It is further assumed that the average age $< A >$ and $\log(Z)$, where Z is the metallicity, change linearly with increasing R. The age gradient is denoted by a, the gradient of $\log Z$ by b.

For each R, the star formation history (SFH) is determined from the local value of $< A >$. It is assumed that the SFH has an exponential shape, and that the star formation time scale is determined by $< A >$ and the age of the oldest stars in the galaxy. The colour of the stellar disc at R is obtained from the SFH and from $BVRIJHK_S$ of a single stellar population with a metallicity which matches the local Z value. The values of r_0 for the different wavelength bands are determined from the colour profiles. Details are found in Cunow (2004).

The model simulations are done for different age and metallicity gradients. For simplicity purposes, it is assumed that a model galaxy has either an age or a metallicity gradient, but not both. The values for a lie in the range between -3.0 Gyr $r_0(K_S)^{-1}$ and 3.0 Gyr $r_0(K_S)^{-1}$, and those for b in the range between $-0.5\ r_0(K_S)^{-1}$ and $0.3\ r_0(K_S)^{-1}$. For each model image, r_D and ϵ are measured in the same way as for the real galaxies.

Bell & de Jong (2000) point out that such models allow the determination of relative ages and metallicites, but no information about the absolute ages and metallicities can be obtained. Hence the models used in this work are able to test age and metallicity gradients in the stellar disc, but they are not able to give absolute age and metallicity values.

4. Investigation of radial age and metallicity gradients

In order to test whether the observed colour gradients are dominated by age gradients or by metallicity gradients, the models with an age gradient and the ones with a metallicity gradient were fitted to the data separately. The scalelength ratios used for the analysis are $r_D(B)/r_D(I)$, $r_D(V)/r_D(J)$, $r_D(R)/r_D(H)$ and $r_D(I)/r_D(K_S)$.

For the models with an age gradient, the data are fitted best by the model with $a = -1.0$ Gyr $r_0(K_S)^{-1}$ and $\tau_0^B = 3$, giving a reduced χ^2 of 1.5. For the models with a metallicity gradient, the data are fitted best by the model with $b = -0.3\ r_0(K_S)^{-1}$ and $\tau_0^B = 3$, giving a reduced χ^2 of 1.2. Figure 1 shows $r_D(B)/r_D(I)$ and $r_D(I)/r_D(K_S)$ for the data and the best-fitting models. The $r_D(B)/r_D(I)$ values are represented equally well by both models, but $r_D(I)/r_D(K_S)$ is represented better by the model with a metallicity gradient than by the one with an age gradient.

These results indicate that the colour gradients in the stellar disc of the sample galaxies are dominated by metallicity gradients, but the presence of age gradients cannot be ruled out. The best-fitting models have negative age and metallicity gradients, indicating that the stars become younger and/or less metal-rich with increasing distance from the centre of the galaxy.

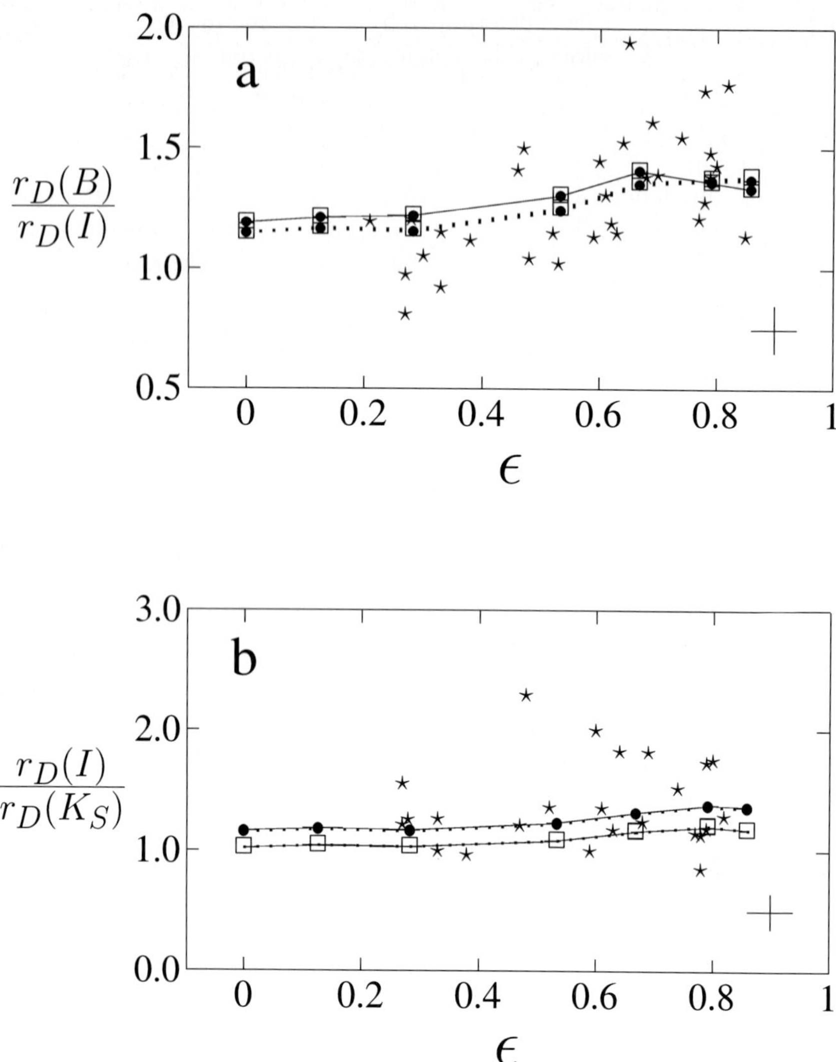

Figure 1. Disc scalelength ratios of the data and the best-fitting models plotted against apparent ellipticity ϵ. (a) shows the ratios between B and I, and (b) between I and K_S. The galaxy data are denoted by $\star$, $\square$ shows the models with a radial age gradient, and $\bullet$ the models with a radial metallicity gradient. The models with no vertical age gradient are denoted by solid lines, whereas the models with a vertical age gradient are denoted by dotted lines. The crosses in the lower right corner of each plot show the typical random error for an individual data point.

5. Vertical population gradients

In addition to radial age and metallicity gradients, vertical gradients might exist as well. Dalcanton & Bernstein (2002) find that their sample galaxies become systematically redder with increasing distance from the galactic plane, and they claim that these gradients are mainly caused by age gradients with the stars becoming systematically older with increasing distance from the galactic plane. The gradients measured by them lie between 0.3 Gyr h_z^{-1} and 1.2 Gyr h_z^{-1}, with h_z being the disc scaleheight.

In order to test whether vertical age gradients are present in the sample galaxies of this work, the models with $a = -1.0$ Gyr $r_0(K_S)^{-1}$, and those with $b = -0.3\ r_0(K_S)^{-1}$ were re-calculated with a vertical age gradient of 1.0 Gyr $z_0(K_S)^{-1}$. Due to this gradient, z_0 for the re-calculated models is different for different wavelength regions, it increases with increasing wavelength.

If these models are compared to the data, it is found that using the simulations with $a = -1.0$ Gyr $r_0(K_S)^{-1}$, the data are fitted best with $\tau_0^B = 3$ giving a reduced χ^2 of 1.4. For the models with $b = -0.3\ r_0(K_S)^{-1}$, the best fit is found for $\tau_0^B = 3$ giving a reduced χ^2 of 1.1. Hence the introduction of a vertical age gradient improves the model fits.

Figure 1 shows the data and the best-fitting models. For $r_D(I)/r_D(K_S)$, the predictions by the models with a vertical age gradient are almost identical to those without such a gradient. However, $r_D(B)/r_D(I)$ is represented better by the models with a vertical age gradient, even though the difference between the two sets of models is not very large. It can be concluded that a positive vertical age gradient might be present in the discs of the sample galaxies.

Acknowledgments

I thank the South African National Research Foundation NRF for financial support under the grant numbers 2050252 and 2053468. This work is based on observations collected at the South African Astronomical Observatory SAAO, Sutherland, South Africa, and has made use of the Two Micron All Sky Survey which is a joint project of the University of Massachusetts and the Infrared Processing and Analysis Center, California, USA.

References

Bell E.F., de Jong R.S., 2000, MNRAS, 312, 497
Byun Y.I., Freeman K.C., Kylafis N.D., 1994, ApJ, 432, 114
Dalcanton J.J., Bernstein R.A., 2002, AJ, 124, 1328
Cunow B., 2001, MNRAS, 323, 130
Cunow B., 2004, MNRAS, submitted
Peletier R.F., Valentijn E.A., Moorwood A.F.M., Freudling W., 1994, A&AS, 108, 621

SECULAR EVOLUTION AND THE GROWTH OF PSEUDOBULGES IN DISK GALAXIES

John Kormendy and Mark E. Cornell
Department of Astronomy, University of Texas, Austin, TX, USA

Abstract Galactic evolution is in transition from an early universe dominated by hierarchical clustering to a future dominated by secular processes. These result from interactions involving collective phenomena such as bars, oval disks, spiral structure, and triaxial dark halos. A detailed review is in Kormendy & Kennicutt (2004). This paper provides a summary illustrated in part with different galaxies.

Figure 2 summarizes how bars rearrange disk gas into outer rings, inner rings, and galactic centers, where high gas densities feed starbursts. Consistent with this picture, many barred and oval galaxies are observed to have dense central concentrations of gas and star formation. Measurements of star formation rates show that bulge-like stellar densities are constructed on timescales of a few billion years. We conclude that secular evolution builds dense central components in disk galaxies that look like classical – that is, merger-built – bulges but that were made slowly out of disk gas. We call these pseudobulges.

Many pseudobulges can be recognized because they have characteristics of disks – (1) flatter shapes than those of classical bulges, (2) correspondingly large ratios of ordered to random velocities, (3) small velocity dispersions σ with respect to the Faber-Jackson correlation between σ and bulge luminosity, (4) spiral structure or nuclear bars, (5) nearly exponential brightness profiles, and (6) starbursts. These structures occur preferentially in barred and oval galaxies in which secular evolution should be most rapid. Thus a variety of observational and theoretical results contribute to a new paradigm of secular evolution that complements hierarchical clustering.

Keywords: Galaxy dynamics, galaxy structure, galaxy evolution

1. Transition From Classical Bulges Built by Hierarchical Clustering to Pseudobulges Built by Secular Evolution

The relative importance of different processes of galactic evolution (Fig. 1) is changing as the universe expands. Rapid processes that happen in discrete events are giving way to slow, ongoing processes. Hierarchical clustering that builds classical bulges is giving way to the secular growth of pseudobulges.

D. Block et al. (eds.), Penetrating Bars through Masks of Cosmic Dust, 261–280.

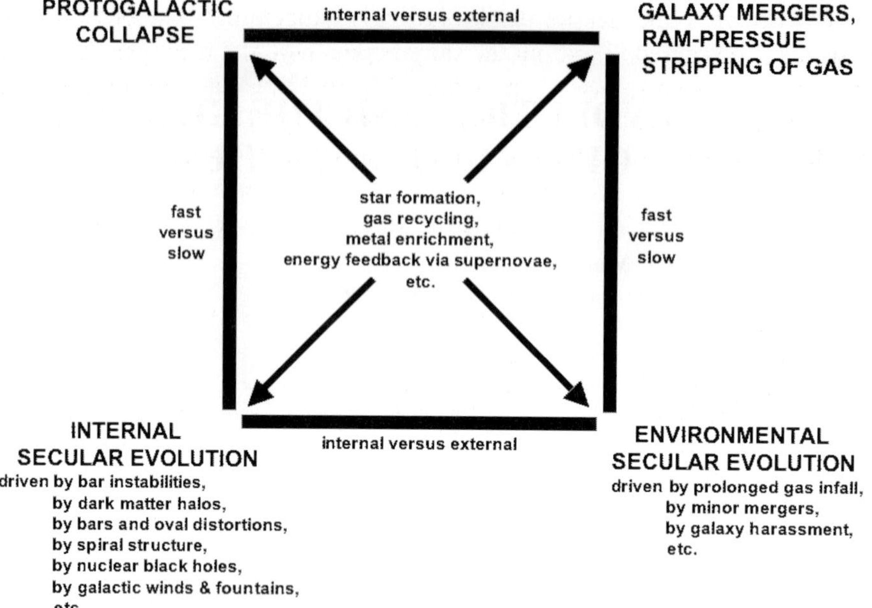

Figure 1. Morphological box (Zwicky 1957) of processes of galactic evolution. Updated from Kormendy (1982a), this figure is from Kormendy & Kennicutt (2004). Processes are divided vertically into fast (top) and slow (bottom). Fast evolution happens on a free-fall timescale, $t_{\rm ff} \sim (G\,\rho)^{-1/2}$; ρ is the density of the object produced and G is the gravitational constant. Slow means many galaxy rotation periods. Processes are divided horizontally into ones that happen internally in one galaxy (left) and ones that are driven by environmental effects such as galaxy interactions (right). The processes at center are aspects of all types of galaxy evolution. This paper reviews the internal and slow processes at lower-left.

Galactic evolution studies over the past 25 years show convincingly that hierarchical clustering (see White 1997 and Steinmetz 2001 for reviews) and mergers (Toomre 1977a, see Schweizer 1990 for a review) built and continue to build elliptical galaxies and elliptical-like classical bulges of disk galaxies. As the universe expands and as galaxy clusters virialize and acquire large velocity dispersions, mergers get less common (Toomre 1977a; Le Fevre et al. 2000; Conselice et al. 2003). Very flat disks in pure disk galaxies show that at least some galaxies have suffered no major merger violence since disk star formation began (see Freeman 2000 for a review). Therefore there has been time to reshape galaxies via the interactions of individual stars or gas clouds with collective phenomena such as bars, oval distortions, spiral structure, and triaxial dark matter halos. These secular processes are reviewed in Kormendy (1993) and in Kormendy & Kennicutt (2004). This paper provides a summary.

2. Secular Evolution of Barred Galaxies

Why do we think that secular evolution is happening? The observational evidence is discussed in § 4, but the story begins with the forty-year history of simulations of the response of gas to bars. Figure 2 illustrates this response and how well it accounts for barred galaxy morphology. The angular momentum transfer from bar to disk that makes the bar grow also rearranges disk gas into outer rings near outer Lindblad resonance (O in the figure at upper-left), inner rings near bar corotation (C), and dense concentrations of gas near the center.

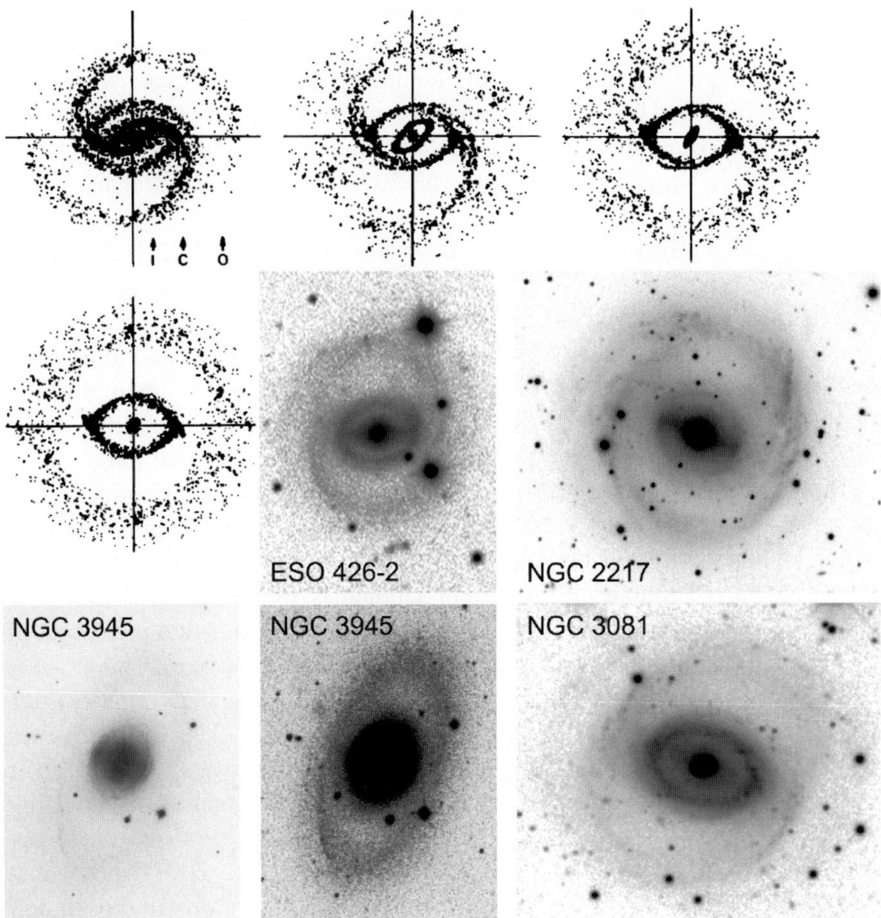

Figure 2. Evolution of gas in a rotating oval potential (Simkin, Su, & Schwarz 1980). The gas particles in this sticky-particle n-body model are shown after 2, 3, 5, and 7 bar rotations (top-left through center-left). Four SB0 or SB0/a galaxies are shown that have outer rings and a lens (NGC 3945) or an inner ring (most obvious in ESO 426-2 and in NGC 3081). Sources: NGC 3945 – Kormendy (1979); NGC 2217, NGC 3081 – Buta et al. (2004); ESO 426-2 – Buta & Crocker (1991). This figure is from Kormendy & Kennicutt (2004).

The essential features of Figure 2 are well confirmed by more recent, state-of-the-art simulations. In a particularly important paper, Athanassoula (1992) focuses on the gas shocks that are identified with dust lanes in bars. The shocks are a consequence of gravitational torques. Gas accelerates as it approaches and decelerates as it leaves the potential minimum of the bar. Therefore it piles up and shocks near the ridge line of the bar. Athanassoula finds that, if the mass distribution is centrally concentrated enough to result in an inner Lindblad resonance, then the shocks are offset in the forward (rotation) direction from the ridge line of the bar. That is, incoming gas overshoots the ridge line of the bar before it plows into the departing gas. The nearly radial dust lanes seen in bars are essentially always offset in the forward (rotation) direction. Compelling support for the identification of the shocks with these dust lanes is provided by the observation of large velocity jumps across the dust lanes (Pence & Blackman 1984; Lindblad, Lindblad, & Athanassoula 1996; Regan, Sheth, & Vogel 1999; Weiner et al. 2001; and especially Regan, Vogel, & Teuben 1997).

Shocks inevitably imply that gas flows toward the center. Because the shocks are nearly radial, the gas impacts them almost perpendicularly. Large amounts of dissipation make the gas sink rapidly. Athanassoula estimates that azimuthally averaged gas sinking rates are typically 1 km s^{-1} and in extreme cases up to $\sim$ 6 km s^{-1}. Because 1 km s^{-1} = 1 kpc (10^9 yr)$^{-1}$, the implication is that most gas in the inner part of the disk finds its way to the vicinity of the center over the course of several billion years, if the bar lives that long.

Crunching gas likes to make stars. Expectations from the Schmidt (1959) law are consistent with observations of enhanced star formation, often in substantial starbursts near the center. Examples are shown in Figure 3. Most of these are barred galaxies illustrated in Sandage & Bedke (1994). NGC 4736 is a prototypical unbarred oval galaxy. It is included to illustrate the theme of the next section that barred and oval galaxies evolve similarly.

Kormendy & Kennicutt (2004) compile gas density and star formation rate (SFR) measurements for 20 nuclear star-forming rings. The SFR densities are $1-3$ orders of magnitude higher than the SFR densities averaged over galactic disks. Gas densities are correspondingly high: nuclear star-forming rings lie on the extrapolation of the Schmidt law, SFR $\propto$ (gas density)$^{1.4}$. The BIMA Survey of Nearby Galaxies (SONG) (Regan et al. 2001) shows that *molecular gas densities follow stellar light densities, especially in barred and oval galaxies, even where the stellar densities rise toward the center above the inward extrapolation of an exponential fitted to the outer disk. Since star formation rates rise faster than linearly with gas density, this guarantees that the observed pseudobulges will grow in density faster than their associated disks. That is, pseudobulge-to-disk ratios increase with time.* Growth rates to reach the observed stellar densities are a few billion years.

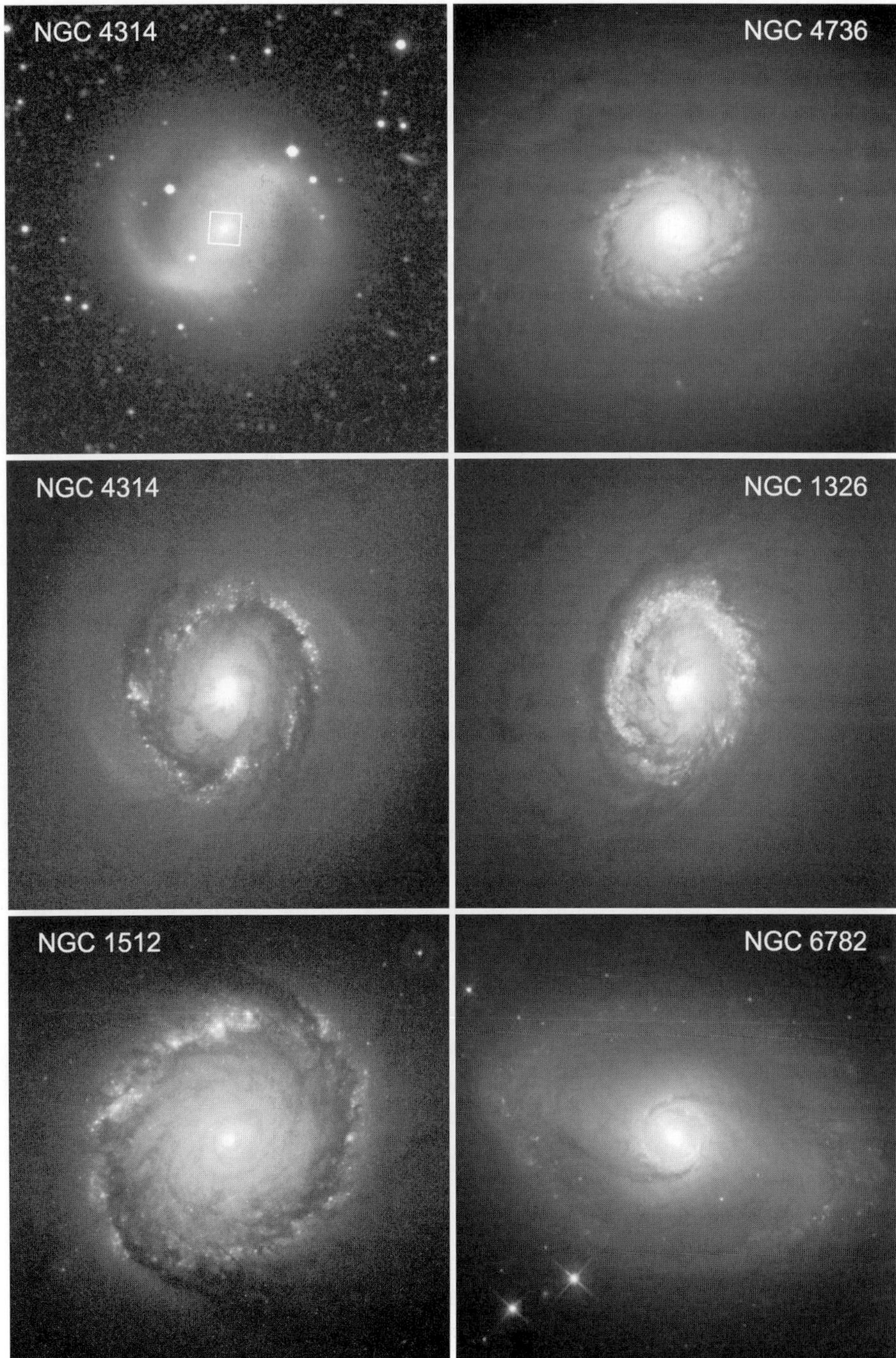

Figure 3. Nuclear star formation rings in barred and oval galaxies. Sources: NGC 4314 – Benedict et al. (2002); NGC 4736 – NOAO; NGC 1326 – Buta et al. (2000) and Zolt Levay (STScI); NGC 1512 – Maoz et al. (2001); NGC 6782 – Windhorst et al. (2002) and the Hubble Heritage Program. This figure is from Kormendy & Kennicutt (2004).

3. Secular Evolution of "Unbarred" Galaxies

How general are the results of Section 2? There are four reasons why we suggest that secular evolution and pseudobulge building are important in more than the $\sim 1/3$ of all disk galaxies that look barred at optical wavelengths:

1 As emphasized at this conference, near-infrared images penetrate dust absorption and are insensitive to the low-M/L frosting of young stars in galaxy disks. They show us the old stars that trace the mass distribution. They reveal that bars are hidden in many galaxies that look unbarred in the optical (Block & Wainscoat 1991; Spillar et al. 1992; Mulchaey & Regan 1997; Mulchaey et al. 1997; Seigar & James 1998; Knapen et al. 2000; Eskridge et al. 2000, 2002; Block et al. 2001; Laurikainen & Salo 2002; Whyte et al. 2002). About two-thirds of all spiral galaxies look barred in the infrared. Measures of bar strengths based on infrared images (Buta & Block 2001; Block et al. 2001; Laurikainen & Salo 2002) should help to tell us the consequences for secular evolution.

2 Many unbarred galaxies are globally oval. Ovals are less elongated than bars – typical axial ratios are ~ 0.85 compared with ~ 0.2 for bars – but more of the disk mass participates in the nonaxisymmetry. Strongly oval galaxies can be recognized independently by photometric criteria (Kormendy & Norman 1979; Kormendy 1982a) and by kinematic criteria (Bosma 1981a, b). Brightness distributions: The disk consists of two nested ovals, each with a shallow surface brightness gradient interior to a sharp outer edge. The inner oval is much brighter than the outer one. The two "shelves" in the brightness distribution have different axial ratios and position angles, so they must be oval if they are coplanar. But the flatness of edge-on galaxies shows that such disks are oval, not warped. Kinematics: Velocity fields in oval disks are symmetric and regular, but (1) the kinematic major axis twists with radius, (2) the optical and kinematic major axes are different, and (3) the kinematic major and minor axes are not perpendicular. Twists in the kinematic principal axes are also seen when disks warp, but Bosma (1981a, b) points out that warps happen at larger radii and lower surface brightnesses than ovals. Also, observations (2) and (3) imply ovals, not warps. Kormendy (1982a) shows that the photometric and kinematic criteria for recognizing ovals are in excellent agreement.

  Oval galaxies are expected to evolve similarly to barred galaxies. Many simulations of the response of gas to "bars" assumed that all of the potential is oval rather than that part of the potential is barred and the rest is not. NGC 4736 is a prototypical oval with strong evidence for secular evolution (Figures 3 and 6 here; Kormendy & Kennicutt 2004).

3 Bars commit suicide by transporting gas inward and building up the central mass concentration (Hasan & Norman 1990; Freidli & Pfenniger 1991; Friedli & Benz 1993; Hasan, Pfenniger, & Norman 1993; Norman, Sellwood, & Hasan 1996; Heller & Shlosman 1996; Berentzen et al. 1998; Sellwood & Moore 1999). Norman et al. (1996) grew a point mass at the center of an n-body disk that previously had formed a bar. As they turned on the point mass, the bar amplitude weakened. Central masses of 5 – 7 % of the disk mass dissolved the bar completely. Shen & Sellwood (2004) find that central masses with small radii, like supermassive black holes, destroy bars more easily than ones with radii of several hundred parsecs, like pseudobulges. A bar can tolerate a soft central mass of 10 % of the disk mass. Observations suggest that still higher central masses can be tolerated when the bar gets very nonlinear. The implication is this: Even if a disk galaxy does not currently have a bar, bar-driven secular evolution may have happened in the past.

4 Late-type, unbarred, but global-pattern spirals are expected to evolve like barred galaxies, only more slowly. Global spirals are density waves that propagate through the disk (Toomre 1977b). In general, stars and gas revolve around the center faster than the spiral arms, so they catch up to the arms from behind and pass through them. As in the bar case, the gas accelerates as it approaches the arms and decelerates as it leaves them. Again, the results are shocks where the gas piles up. This time the shocks have a spiral shape. They are identified with the dust lanes on the concave side of the spiral arms (Figure 4). Gas dissipates at the shocks, but it does so more weakly than in barred galaxies, because the gas meets the shock obliquely. Nevertheless, it sinks. In early-type spirals with big classical bulges, the spiral structure stops at an ILR at a large radius. The gas may form some stars there, but since the bulge is already large, the relative contribution of secular evolution is likely to be minor. In late-type galaxies, the spiral structure extends close to the center. Sinking gas reaches small radii and high densities. We suggest that star formation then contributes to the building of pseudobulges. Moreover, late-type galaxies have no classical bulges. So secular growth of pseudobulges can most easily contribute a noticeable part of the central mass concentration precisely in the galaxies where the evolution is most important.

  M 51 and NGC 4321 (Figure 4) are examples of nuclear star formation in unbarred galaxies. Their exceedingly regular spiral structure and associated dust lanes wind down close to the center, where both galaxies have bright regions of star formation (e. g., Knapen et al. 1995a, b; Sakamoto et al. 1995; Garcia-Burillo et al. 1998). They are examples of secular evolution in galaxies that do not show prominent bars or ovals.

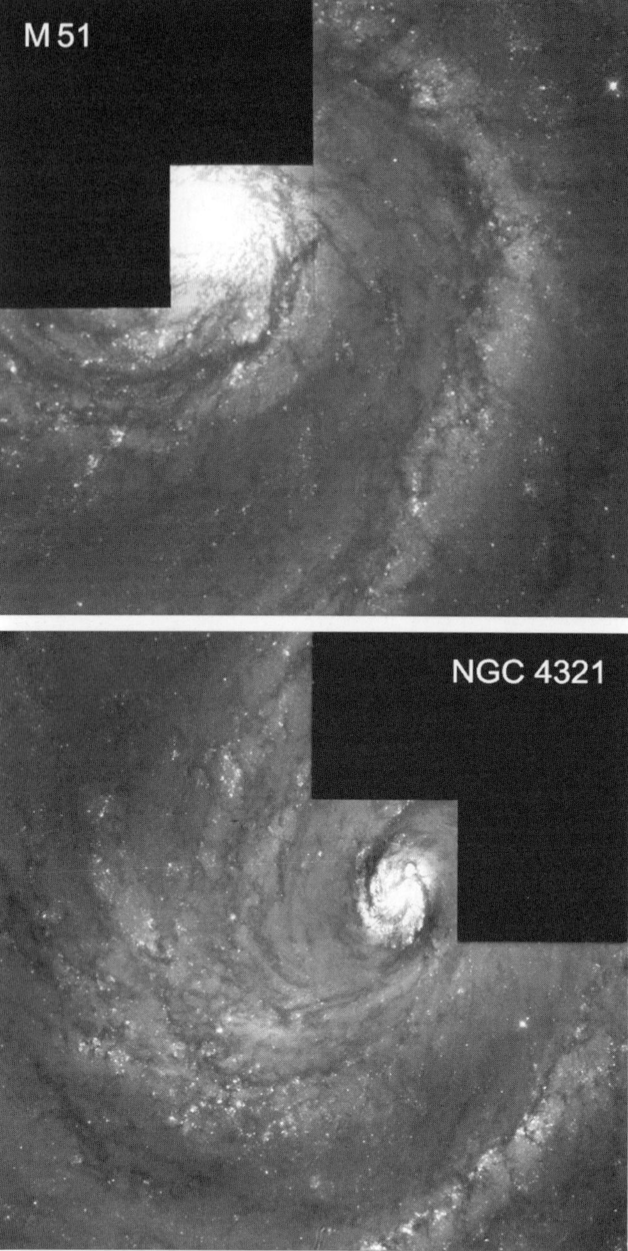

Figure 4. Nuclear star formation in the unbarred galaxies M 51 and NGC 4321 (M 100). Dust lanes on the trailing side of the global spiral arms reach in to small radii. As in barred spirals, they are are indicative of gas inflow. Both galaxies have concentrations of star formation near their centers that resemble those in Figure 3. These images are from the *Hubble Space Telescope* and are reproduced here courtesy of STScI.

4. The Observed Properties of Pseudobulges

Kormendy (1982a, b) suggested that what we now call pseudobulges were built by secular inward gas transport and star formation. Combes & Sanders (1981) suggested that boxy bulges formed from bars that heat themselves in the axial direction. Pfenniger & Norman (1990) discuss both processes. These themes – dissipational and dissipationless, secular pseudobulge building – have persisted in the literature ever since (see Kormendy & Kennicutt 2004).

How can we tell whether a "bulge" formed by these processes? Fortunately, pseudobulges retain enough memory of their disky origin so that the best examples are recognizable. Structural features that indicate a disky origin include nuclear bars, nuclear disks, nuclear spiral structure, boxy bulges, exponential bulges, and central star formation (Figures 3 and 4). We consider all of these to be features of pseudobulges, because the evidence is that all of them are built secularly out of disk material. Similarly, global spiral structure, flocculent spiral structure, and no spiral structure in S0 galaxies are all features of disks. In addition, pseudobulges are more dominated by rotation and less dominated by random motions than are classical bulges and ellipticals.

Spectacular progress in recent years has come from *HST* imaging surveys. We begin with these surveys. To make data available on more galaxies, we also provide a detailed discussion of two galaxies, NGC 4371 and NGC 3945, that are different from the ones discussed in Kormendy & Kennicutt (2004).

4.1 Embedded Disks: Spiral Structure, Star Formation

Renzini (1999) states the definition of a bulge: "It appears legitimate to look at bulges as ellipticals that happen to have a prominent disk around them [and] ellipticals as bulges that for some reason have missed the opportunity to acquire or maintain a prominent disk." Our paradigm of galaxy formation is that bulges and ellipticals both formed via galaxy mergers (e. g., Toomre 1977a; Steinmetz & Navarro 2002, 2003), a picture that is well supported by observations (see Schweizer 1990 for a review). But as observations improve, we discover more and more features that make it difficult to interpret every example of what we used to call a "bulge" as an elliptical living in the middle of a disk. Carollo and collaborators find many such galaxies in their *HST* snapshot survey of 75, S0 – Sc galaxies observed with WFPC2 in V band (Carollo et al. 1997, 1998; Carollo & Stiavelli 1998; Carollo 1999) and a complementary survey of 78 galaxies observed with NICMOS in H band (Carollo et al. 2001, 2002; Seigar et al. 2002). Figure 5 shows examples. These are Sa – Sbc galaxies, so they should contain bulges. Instead, their centers look like star-forming spiral galaxies. It is difficult to believe that, based on such images, anyone would define bulges as ellipticals living in the middle of a disk. Spiral structure happens only in a disk. Therefore these are examples of pseudobulges.

Figure 5. Sa – Sbc galaxies whose "bulges" have disk-like morphology. Each panel shows an $18'' \times 18''$ region centered on the galaxy nucleus and extracted from *HST* WFPC2 F606W images taken and kindly provided by Carollo et al. (1998). North is up and east is at left. Displayed intensity is proportional to the logarithm of the galaxy surface brightness.

4.2 Rotation-Dominated Pseudobulges

Figure 6, the $V_{\rm max}/\sigma - \epsilon$ diagram (Illingworth 1977; Binney 1978a, b), shows that pseudobulges (filled symbols) are more rotation-dominated than classical bulges (open symbols) which are more rotation-dominated than giant ellipticals (crosses). This is disky behavior, as follows. Seen edge-on, rotation-dominated disks have parameters that approximately satisfy the extrapolation of the oblate line to $\epsilon \geq 0.8$. Observed other than edge-on, they project well above the oblate line. In contrast, projection keeps $\epsilon \lesssim 0.6$ isotropic spheroids near the oblate line. The filled symbols therefore represent objects that contain rapidly rotating and hence disky central components. Of the most extreme cases, NGC 4736 is discussed in detail in Kormendy & Kennicutt (2004). Complementary photometric evidence for pseudobulges in NGC 3945 and NGC 4371 is discussed in the next subsection.

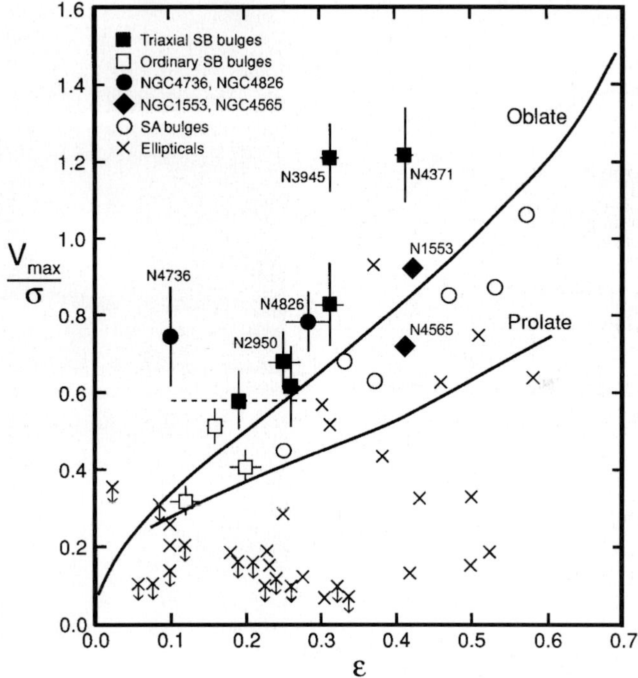

Figure 6. Relative importance of rotation and velocity dispersion: $V_{\rm max}/\sigma$ is the ratio of the maximum rotation velocity to the mean velocity dispersion interior to the half-light radius; $(V_{\rm max}/\sigma)^2$ measures the relative contribution of ordered and random motions to the total kinetic energy and hence, via the virial theorem, to the dynamical support that gives the system its ellipticity ϵ (Binney & Tremaine 1987). The "oblate" line describes oblate spheroids that have isotropic velocity dispersions and that are flattened only by rotation. The "prolate" line is one example of how prolate spheroids can rotate more slowly for a given ϵ because they are flattened partly by velocity dispersion anisotropy. This figure is from Kormendy& Kennicutt (2004).

4.3 Embedded Disks – II. Flat Pseudobulges

That some pseudobulges are essentially as flat as disks is inferred when we observe spiral structure (Figure 5), but it is observed directly in surface photometry of highly inclined galaxies (see Kormendy 1993 and Kormendy & Kennicutt 2004 for reviews). Figures 7 – 9 show examples.

The SB0 galaxy NGC 4371 contains one of the most rotation-dominated "bulges" in Figure 6. This result (Kormendy 1982b, 1993) already implies that NGC 4371 contains a pseudobulge.

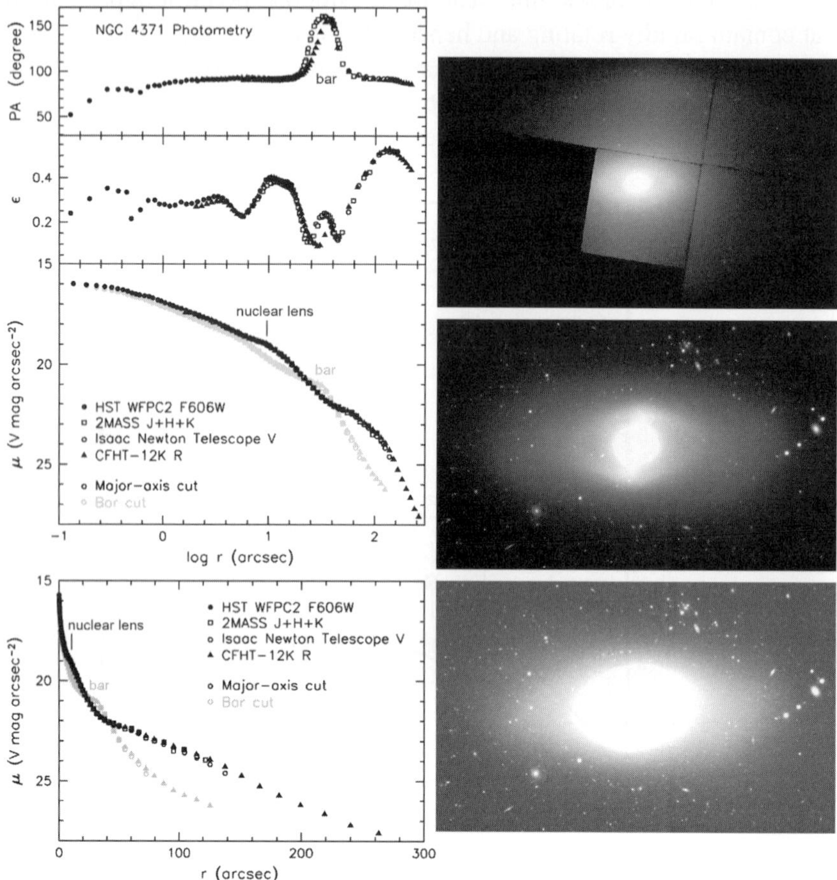

Figure 7. NGC 4371 pseudobulge – top image: 130″ × 80″ WFPC2 F606W image from the *HST* archive; middle and bottom: 406″ × 249″ CFHT 12K R-band images from Kormendy et al. (2004) at different logarithmic intensity stretches. North is up and east is at left. The plots show surface photometry, including brightness cuts along the major and bar axes (see the text) shifted to the V-band INT zeropoint derived using aperture photometry from Poulain (1988).

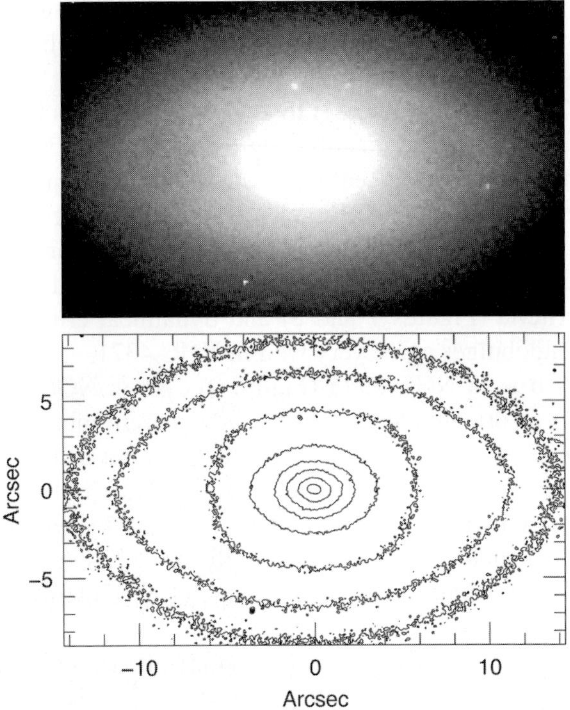

Figure 8. NGC 4371 nuclear lens. North is up and east is at left. The *HST* PC F606W image at the top is $28\overset{\prime\prime}{.}7 \times 17\overset{\prime\prime}{.}6$. The intensity stretch is logarithmic. The bottom panel shows isophotes chosen to distinguish the high-ellipticity lens from the rounder center. The contour levels are 19.6, 19.1, 18.6, 18.0, 17.6, 17.3, 16.8, and 16.3 V mag arcsec^{-2}. Compare the lens in NGC 1553 (Freeman 1975; Kormendy 1984).

Photometry strengthens the evidence that the "bulge" is disky. Kormendy (1979) concluded that "the spheroid is distorted into a secondary bar; i. e., is prolate". Wozniak et al. (1995) saw this, too, but noted that it could be a projection effect due to high inclination. Based on unsharp-masked images, Erwin & Sparke (1999) conclude that NGC 4371 contains a smooth nuclear ring and identify this – in effect, if not in name – as a pseudobulge.

Figure 7 shows our photometry. The ellipticities ϵ and position angles PA are based on ellipse fits to the isophotes. However, the isophotes are far from elliptical at some radii, so the bottom two panels show brightness cuts in 25° wedges along the major and bar axes. The bar is obvious as a shelf in surface brightness and as corresponding features in the ϵ and PA profiles. Interior to the outer exponential disk shown by the major-axis cut is a steep central rise in surface brightness that would conventionally be identified as the bulge. However, its properties are distinctly not bulge-like. It contains a shelf in surface brightness with radius $r \simeq 10''$; this is shown in more detail in Figure 8.

The outer rim of a shelf looks like a ring when an image is divided by a smoothed version of itself. The shelf has the brightness profile of a lens (cf. the prototype in NGC 1553: Freeman 1975; Kormendy 1984). We interpret it as a nuclear lens. The important point is this: The nuclear lens has essentially the same apparent flattening and position angle as the outermost disk. We cannot tell from Figure 7 whether it really is a disk or whether it is thicker than a disk and therefore prolate (a nuclear bar). Rapid rotation (Figure 6) makes the disk interpretation more likely. In either case, the nuclear lens is diagnostic of a pseudobulge (see Kormendy & Kennicutt 2004 for further discussion). Photometric criteria (Figures 7 and 8) and dynamical criteria (Figure 6) for identifying pseudobulges agree very well in NGC 4371.

The same is true in NGC 3945 (Figure 9). As in NGC 4371, the bar of this SB0 galaxy is oriented almost along the apparent minor axis. Therefore, when the "bulge" at $r \simeq 10''$ looks essentially as flat as the outermost disk, there is ambiguity about whether the inner structure is flat and circular or axially thick and a nuclear bar. It was interpreted as a nuclear bar in Kormendy (1979) and in Wozniak et al. (1995) and is illustrated as such in Kormendy & Kennicutt (2004). In contrast, Erwin & Sparke (1999) interpret it as "probably intrinsically round and flat – an inner disk". Erwin et al. (2003) reach the same conclusion in a detailed photometric study. Based on ϵ and PA profiles, they also identify an "inner bar" with radius $2''$. All of these features are well confirmed by our photometry (Figure 9). The main and "inner" bars are clear in the ϵ and PA profiles. The main bar also makes an obvious shelf in the bar-axis brightness cut, while the nuclear bar is so subtle that we regard it as an interpretation rather than a certainty. Our photometry is consistent with the interpretation either that the shelf at $r \simeq 10''$ in the major-axis cut is a nuclear bar (in which case the galaxy has three nested bars) or that this is a nuclear lens which is nearly circular and very flat. For the purposes of this paper, we do not have to decide between these alternatives. Either one is characteristic of a disk. Consistent with the conclusions of all of the above papers, either interpretation implies that the central rise in surface brightness above the galaxy's primary lens and outer ring is caused by a pseudobulge. This may have been added to a pre-existing classical bulge, but if so, the classical bulge dominates the light only in the central $1''.5$ (Figure 9; Erwin et al. 2003).

Present-day gas inflow and star-formation rates imply that dissipative secular evolution should be most important in Sbc – Sc galaxies (Kormendy & Kennicutt 2004). Classical bulges are the rule in Sas, but it is remarkable how easily one can find S0s with pseudobulges. We interpret this result as additional evidence for van den Bergh's (1976) "parallel sequence" classification, which recognizes that some S0s have smaller (pseudo)bulge-to-disk luminosity ratios than do Sa galaxies. The hint is that the secular evolution happened long ago, when the galaxies were gas-rich and before they were converted to S0s.

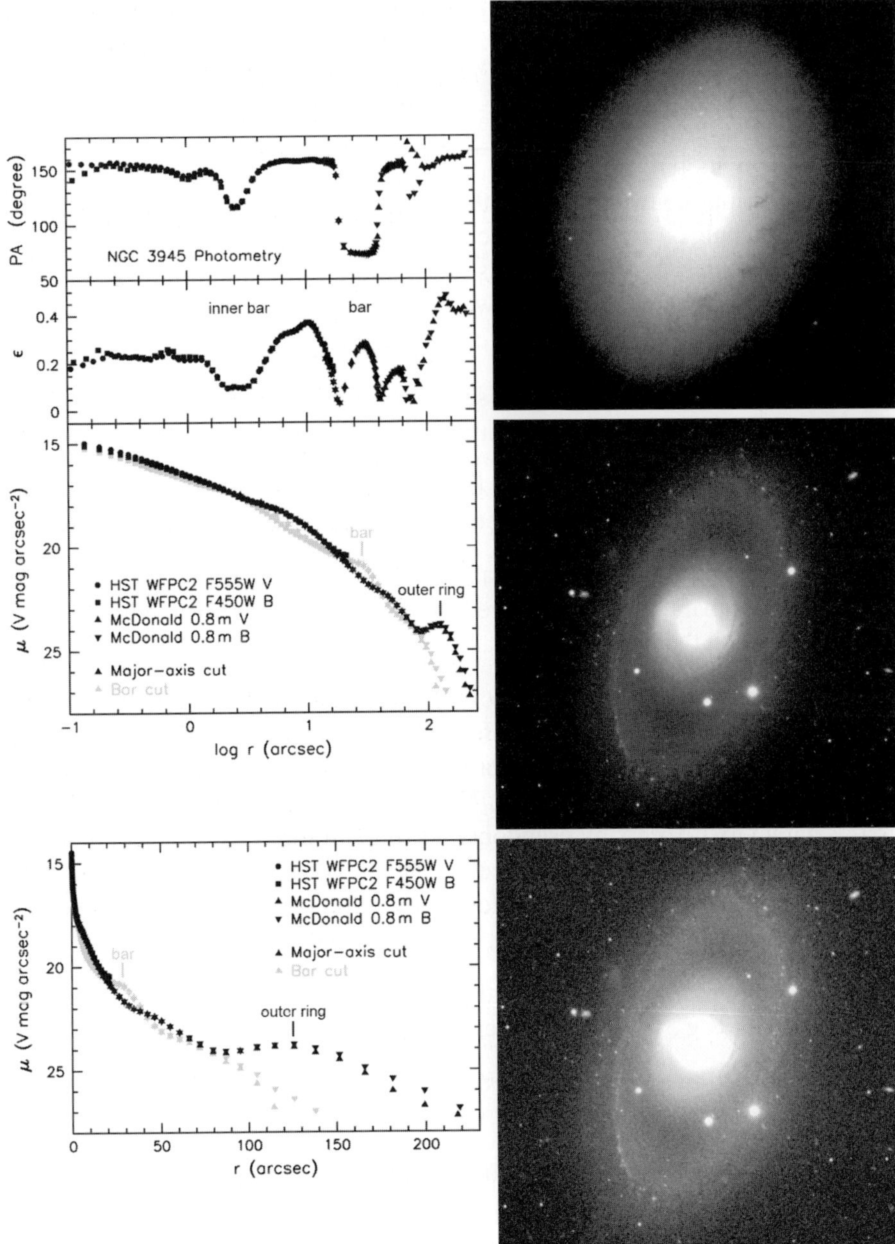

Figure 9. NGC 3945 pseudobulge – top image: $29'' \times 29''$ PC F450W image from the *HST* archive; middle and bottom: $6\farcm1 \times 6\farcm1$ WIYN B-band images from Buta et al. (2004) at different logarithmic intensity stretches. North is up and east is at left. The plots show ellipticity and position angle profiles from ellipse fits to the isophotes and major- and bar-axis cuts in 28° wedges. The McDonald Observatory 0.8 m telescope V-band images were zeropointed using aperture photometry from Burstein et al. (1987) and from Angione (1988). The other profiles are shifted to this zeropoint. Outer rings are usually elongated perpendicular to bars, so the apparent ellipticity of flat, circular isophotes is likely to be the one observed at the largest radii.

5. Preliminary Prescription for Recognizing Pseudobulges

We have space in this paper to review only a few features of pseudobulges. Kormendy (1993) and Kormendy & Kennicutt (2004) discuss others. In this section, we list these other features to provide a preliminary prescription for identifying pseudobulges.

Any prescription must recognize that we expect a continuum from classical, merger-built bulges through objects with some E-like and some disk-like characteristics to pseudobulges built completely by secular processes. Uncertainties are inevitable when we deal with transition objects. Keeping these in mind, a list of pseudobulge characteristics includes:

1 The candidate pseudobulge is seen to be a disk in images: it shows spiral structure or its apparent flattening is similar to that of the outer disk.

2 It is or it contains a nuclear bar (in face-on galaxies). Bars are disk phenomena; they are fundamentally different from triaxial ellipticals.

3 It is box-shaped (in edge-on galaxies). Box-shaped bulges are intimately related to bars; they are believed to be – or to be made by – edge-on bars that heated themselves in the axial direction.

4 It has $n \simeq 1$ to 2 in a Sersic (1968) function, $I(r) \propto e^{-K[(r/r_e)^{1/n}-1]}$, fit to the brightness profile. Here $n = 1$ for an exponential, $n = 4$ for an $r^{1/4}$ law, and $K(n)$ is chosen so that radius r_e contains half of the light in the Sersic component. Nearly exponential profiles prove to be a characteristic of many pseudobulges (e. g., Andredakis & Sanders 1994; Andredakis, Peletier, & Balcells 1995; Courteau, de Jong, & Broeils 1996; Carollo et al. 2002; Balcells et al. 2003; MacArthur, Courteau, & Holtzman 2003; Kormendy & Kennicutt 2004).

5 It is more rotation-dominated than are classical bulges in the $V_{\rm max}/\sigma - \epsilon$ diagram; e. g., $V_{\rm max}/\sigma$ is larger than the value on the oblate line.

6 It is a low-σ outlier in the Faber-Jackson (1976) correlation between (pseudo)bulge luminosity and velocity dispersion.

7 It is dominated by Population I material (young stars, gas, and dust), but there is no sign of a merger in progress.

If any of these characteristics are extreme or very well developed, it seems safe to identify the central component as a pseudobulge. The more of 1 – 7 apply, the safer the classification becomes.

Small bulge-to-total luminosity ratios B/T do not guarantee that a galaxy contains a pseudobulge, but if $B/T \gtrsim$ 1/3 to 1/2, it seems safe to conclude that the galaxy contains a classical bulge.

Based on these criteria, galaxies with classical bulges include M 31, M 81, NGC 2841, NGC 3115, and NGC 4594. Galaxies with prototypical pseudobulges include NGC 3885 (Figure 5), NGC 3945 (Figures 2, 6, 9), NGC 4314 (Figure 3), NGC 4321 (Figure 4), NGC 4371 (Figures 6, 7, 8), and NGC 4736 (Figures 3, 6). The classification of the bulge of our Galaxy is ambiguous; the observation that it is box shaped strongly favors a pseudobulge, but stellar population data are most easily understood if the bulge is classical.

6. Perspective

Internal secular evolution complements environmental processes such as hierarchical clustering and harrassment in shaping galaxies. Thirty years ago, Hubble classification was in active use, but we also knew of a long list of commonly observed features in disk galaxies, including lens components, boxy bulges, nuclear bars, and central starbursts, and also a list of unique peculiar galaxies (e.g., Arp 1966) that were unexplained and not included in morphological classification schemes. Almost all these features and peculiar galaxies now have candidate explanations within one of two paradigms of galaxy evolution that originated in the late 1970s. The peculiar objects have turned out mostly to be interacting and merging galaxies. And many previously unexplained features of disk galaxies are fundamental to our understanding that galaxies evolve secularly long after the spectacular fireworks of galaxy mergers, starbursts, and their attendant nuclear activity have died down.

Acknowledgments

It is a pleasure to thank Ron Buta and Marcella Carollo for providing many of the images used in the figures. David Fisher kindly took the images of NGC 3945 with the McDonald Observatory 0.8 m telescope, and Tom Jarrett provided the star-removed 2MASS images of NGC 3945. This paper is based partly on observations made with the NASA/ESA *Hubble Space Telescope*, obtained from the data archive at the Space Telescope Science Institute. STScI is operated by AURA, Inc. under NASA contract NAS 5-26555. We also used the NASA/IPAC Extragalactic Database (NED), which is operated by JPL and Caltech under contract with NASA.

References

Andredakis, Y. C., Peletier, R. F., & Balcells, M.: 1995. *MNRAS* **275**, 874
Andredakis, Y. C., & Sanders, R. H.: 1994. *MNRAS* **267**, 283
Angione, R. J.: 1988, *PASP* **100**, 469
Arp, H.: 1966, *Atlas of Peculiar Galaxies*, California Inst. of Technology, Pasadena
Athanassoula, E.: 1992. *MNRAS* **259**, 345
Balcells, M., Graham, A. W., Dom«nguez-Palmero, L., & Peletier, R. F.: 2003, *ApJ* **582**, L79
Benedict, G. F., et al.: 2002, *AJ* **123**, 1411

Berentzen, I., Heller, C. H., Shlosman, I., & Fricke, K. J.: 1998, *MNRAS* **300**, 49

Binney, J.: 1978a, *MNRAS* **183**, 501

Binney, J.: 1978b, *Comments Ap.* **8**, 27

Binney, J., & Tremaine, S.: 1987, *Galactic Dynamics*. Princeton Univ. Press, Princeton

Block, D. L., Puerari, I., Knapen, J. H. et al.: 2001, *A&A* **375**, 761

Block D. L., & Wainscoat, R. J.: 1991, *Nature* **353**, 48

Bosma, A.: 1981a, *AJ* **86**, 1791

Bosma, A.: 1981b, *AJ* **86**, 1825

Burstein, D., et al.: 1987, *ApJS* **64**, 601

Buta, R., & Block, D. L.: 2001, *ApJ* **550**, 243

Buta, R., Corwin, H. G., & Odewahn, S. C.: 2004, *The de Vaucouleurs Atlas of Galaxies*, Cambridge Univ. Press, Cambridge, in preparation

Buta, R., & Crocker, D. A.: 1991, *AJ* **102**, 1715

Buta, R., Treuthardt, P. M., Byrd, G. G., & Crocker, D. A.: 2000, *AJ* **120**, 1289

Carollo, C. M.: 1999, *ApJ* **523**, 566

Carollo, C. M., & Stiavelli, M.: 1998, *AJ* **115**, 2306

Carollo, C. M., Stiavelli, M., de Zeeuw, P. T., & Mack, J.: 1997, *AJ* **114**, 2366

Carollo, C. M., et al.: 2001, *ApJ* **546**, 216

Carollo, C. M., Stiavelli, M., & Mack, J.: 1998, *AJ* **116**, 68

Carollo, C. M., et al.: 2002, *AJ* **123**, 159

Combes, F., & Sanders, R. H.: 1981. *A&A* **96**, 164

Conselice, C. J., Bershady, M. A., Dickinson, M., & Papovich, C.: 2003, *AJ* **126**, 1183

Courteau, S., de Jong, R. S., & Broeils, A. H.: 1996, *ApJ* **457**, L73

Erwin, P., & Sparke, L. S.: 1999, *ApJ* **521**, L37

Erwin, P., Vega Beltran, J. C., Graham, A. W., & Beckman, J. E.: 2003, *ApJ* **597**, 929

Eskridge, P. B., et al.: 2002, *ApJS* **143**, 73

Eskridge, P. B., et al.: 2000, *AJ* **119**, 536

Faber, S. M., & Jackson, R. E.: 1976, *ApJ* **204**, 668

Freeman, K. C.: 1975, in *IAU Symposium 69, Dynamics of Stellar Systems*, ed. A. Hayli, Reidel, Dordrecht, 367

Freeman, K. C.: 2000, in *Toward a New Millennium in Galaxy Morphology*, ed. D. L. Block, I. Puerari, A. Stockton, & D. Ferreira, Kluwer, Dordrecht, 119

Friedli, D., & Benz, W.: 1993, *A&A* **268**, 65

Friedli, D., & Pfenniger, D.: 1991, in *IAU Symposium 146, Dynamics of Galaxies and Their Molecular Cloud Distributions*, ed. F. Combes & F. Casoli, Kluwer, Dordrecht, 362

Garcia-Burillo, S., Sempere, M. J., Combes, F., & Neri, R.: 1998, *A&A* **333**, 864

Hasan, H., & Norman, C.: 1990, *ApJ* **361**, 69

Hasan, H., Pfenniger, D., & Norman, C.: 1993, *ApJ* **409**, 91

Heller, C. H., & Shlosman, I.: 1996, *ApJ* **471**, 143

Illingworth, G.: 1977, *ApJ* **218**, L43

Knapen, J. H., et al.: 1995a, *ApJ*, **454**, 623

Knapen, J. H., et al.: 1995b, *ApJ*, **443**, L73

Knapen, J. H., Shlosman, I., & Peletier, R. F.: 2000, *ApJ* **529**, 93

Kormendy, J.: 1979, *ApJ* **227**, 714

Kormendy, J.: 1982a, in *Morphology and Dynamics of Galaxies, Twelfth Saas-Fee Course*, ed. L. Martinet & M. Mayor, Geneva Obs., Sauverny, 113

Kormendy, J.: 1982b, *ApJ* **257**, 75

Kormendy, J.: 1984, *ApJ* **286**, 116

Kormendy, J.: 1993, in *IAU Symposium 153, Galactic Bulges*, ed. H. Habing & H. Dejonghe, Kluwer, Dordrecht, 209

Kormendy, J., & Kennicutt, R. C.: 2004, *ARA&A* **42**, in press
Kormendy, J., & Norman, C. A.: 1979, *ApJ* **233**, 539
Kormendy, J., et al.: 2004, in preparation
Laurikainen, E., & Salo, H.: 2002, *MNRAS* **337**, 1118
Le Fevre, O., et al.: 2000, *MNRAS* **311**, 565
Lindblad, P. A. B., Lindblad, P. O., & Athanassoula, E.: 1996, *A&A* **313**, 65
MacArthur, L. A., Courteau, S., & Holtzman, J. A.: 2003, *ApJ* **582**, 689
Maoz, D., et al.: 2001, *AJ* **121**, 3048
Mulchaey, J. S., & Regan, M. W.: 1997, *ApJ* **482**, L135
Mulchaey, J. S., Regan, M. W., & Kundu, A.: 1997, *ApJS* **110**, 299
Norman, C. A., Sellwood, J. A., & Hasan, H.: 1996, *ApJ* **462**, 114
Pence, W. D., & Blackman, C. P.: 1984, *MNRAS* **207**, 9
Pfenniger, D., & Norman, C.: 1990, *ApJ* **363**, 391
Poulain, P.: 1988, *A&AS* **72**, 215
Regan, M. W., et al.: 2001, *ApJ* **561**, 218
Regan, M. W., Sheth, K., & Vogel, S. N.: 1999, *ApJ* **526**, 97
Regan M. W., Vogel, S. N., & Teuben, P. J.: 1997, *ApJ* **482**, L143
Renzini, A.: 1999, in *The Formation of Galactic Bulges*, ed. C. M. Carollo, H. C. Ferguson, & R. F. G. Wyse, Cambridge Univ. Press, Cambridge, 9
Sakamoto, K., et al.: 1995, *AJ* **110**, 2075
Sandage, A., & Bedke, J.: 1994, *The Carnegie Atlas of Galaxies*, Carnegie Inst. of Washington, Washington
Schmidt, M.: 1959, *ApJ* **129**, 243
Schweizer, F.: 1990, in *Dynamics and Interactions of Galaxies*, ed. R. Wielen, Springer, New York, 60
Seigar, M., et al.: 2002, *AJ* **123**, 184
Seigar, M. S., & James, P. A.: 1998, *MNRAS* **299**, 672
Sersic, J. L.: 1968, *Atlas de Galaxias Australes*. Obs. Astron. Univ. Nac. Cordoba, Cordoba
Sellwood, J. A., & Moore, E. M.: 1999, *ApJ* **510**, 125
Shen, J., & Sellwood, J. A.: 2004, in *Carnegie Observatories Astrophysics Series, Vol. 1: Coevolution of Black Holes and Galaxies*, ed. L. C. Ho, Cambridge Univ. Press, Cambridge, in press (astro-ph/0303130)
Simkin, S. M., Su, H. J., & Schwarz, M. P.: 1980. *ApJ* **237**, 404
Spillar, E. J., Oh, S. P., Johnson, P. E.,& Wenz, M.: 1992, *AJ* **103**, 793
Steinmetz, M.: 2001, in *Galaxy Disks and Disk Galaxies*, ed. J. G. Funes & E. M. Corsini, ASP, San Francisco, 633
Steinmetz, M., & Navarro, J. F.: 2002, *NewA*, **7**, 155
Steinmetz, M., & Navarro, J. F.: 2003, *NewA*, **8**, 557
Toomre A.: 1977a, in *The Evolution of Galaxies and Stellar Populations*, ed. B.M. Tinsley & R. B. Larson, Yale University Observatory, New Haven, 401
Toomre A.: 1977b, *ARA&A* **15**, 437
van den Bergh, S.: 1976, *ApJ* **206**, 883
Weiner, B. J., Williams, T. B., van Gorkom, J. H., & Sellwood, J. A.: 2001, *ApJ* **546**, 916
White, S. D. M.: 1997, in *The Evolution of the Universe: Report of the Dahlem Workshop on the Evolution of the Universe*, ed. G. Borner & S. Gottlober, Wiley, New York, 227
Whyte, L. F., et al.: 2002, *MNRAS* **336**, 1281
Windhorst, R. A., et al.: 2002, *ApJS* **143**, 113
Wozniak, H., et al.: 1995, *A&AS* **111**, 115
Zwicky F.: 1957, *Morphological Astronomy* Springer, Berlin

J. Navarro: If pseudobulges and classical bulges have radically different physical origins, how does one explain why there appears to be a continuity in the properties of their central supermassive black holes (when detected)?

J. Kormendy: This is a very important question. Half a dozen pseudobulges – including the one in our Galaxy – indeed satisfy the $M_{BH} - \sigma$ correlation. If larger samples confirm this result, it indicates that gas inflow always puts the same magic fraction ($\approx 0.12\%$) of the gas into the black hole independently of whether the inflow is rapid, as in a merger, or slow, as in secular evolution. It suggests a close connection between black hole feeding and (pseudo)bulge growth, i.e. with the microphysics of gas inflow and star formation. If even a few pseudobulges contain canonical 0.12% black holes, then this already argues against explanations of the black-hole – σ correlation that depend on dark matter or its initial density fluctuation spectrum.

There is a caveat. We have only weak constraints on how much classical bulge underlies the dominant pseudobulge in the above galaxies. I doubt that a 1/3 contribution by a classical bulge can securely be excluded in any of them. Since the black hole mass fraction in pseudobulges is slightly (not significantly) low, it is conceivable that black holes are correlating only with the classical bulge part of each galaxy. We need better measurements of the relative contributions of classical and pseudobulges in these and other galaxies.

BARS AND LENSES IN SPIRAL GALAXIES: CLUES FOR SECULAR EVOLUTION

Leslie K. Hunt[1], Carlo Giovanardi[2] and Matthew A. Malkan[3]
[1]*Istituto di Radioastronomia/Sezione Firenze, Italy,* [2]*INAF-Osservatorio Astrofisico di Arcetri, Firenze, Italy,* [3]*University of California, Los Angeles, CA, USA*

Abstract We analyze ~200 spiral galaxies in terms of elliptical isophotes, bar profile type, and non-axisymmetric perturbation amplitudes. More than half our sample is imaged in the H band, which enables us to derive gravitational potential and relative torque Q_g. We find that bar axial ratios increase significantly with decreasing Hubble type, both for exponential and flat bars; however, at a given Hubble type, exponential bars tend to have a larger axial ratio than flat ones. We have also identified four different torque morphologies, and find that one of these is prevalent in strong bars. These results are interpreted in the context of a secular evolution scenario, in which we speculate that early-type flat bars are the descendants of late-type exponential ones.

Keywords: Bars, spiral galaxies, near-infrared

1. Introduction

The two principal morphological components of spiral galaxies – bulges and disks – have been studied extensively, and 50 years of photometric analysis has quantified their shapes, sizes, and surface brightness distributions. Much less is known about the other important component in spiral morphology – bars – which only recently have come under intensive quantitative scrutiny.

The importance of bars and their potentially crucial role in galaxy evolution and dynamics is highlighted by the increasing number of studies devoted to them. Almost 20 years ago Elmegreen & Elmegreen (1985) discovered that bars come in basically two "flavors": "flat" bars with uniform intensities along their lengths, and prevalent in early-type spirals, and exponential bars whose intensity decreases roughly exponentially and which tend to reside in late types. Furthermore, early-type bars are usually longer relative to the galaxy size than late-type ones. Models showed that this could be attributed to the location of corotation which, because of the larger bulge-to-disk mass ratio in early-type disk galaxies, is closer in and enables a more efficient growth of the bar and angular momentum transfer (Combes & Elmegreen 1993).

D. Block et al. (eds.), Penetrating Bars through Masks of Cosmic Dust, 281–290.

Measuring bar strength is not an easy task, and a number of methods have been proposed. The initial visual definition of bars by Hubble was related to bar-interbar intensity contrast, which was quantified by Elmegreen & Elmegreen (1985). They also used Fourier amplitudes ($m = 2, 4$) to measure bar strength, but later it was found that even terms up to $m = 10$ can be significant in strong bars (Buta, Block, & Knapen 2003). Bar axial ratios $(b/a)_{\rm bar}$ were first studied for a large sample of galaxies by Martin (1995), and they also formed the basis for an objective morphological classification scheme developed by Abraham & Merrifield (2000). Bar strength may not be accurately measured through axial ratios, however, since it depends also on kinematics and orbital properties (Athanassoula 1992). Thus, to better assess the strength of non-axisymmetric perturbations in spiral disks, Buta & Block and collaborators (2001, 2001, 2002, 2003) followed by Laurikainen & Salo (2002, 2004) pioneered a technique to measure gravitational torques Q_g.

Nevertheless, even gravitational torques and non-axisymmetric perturbations may not be the final word on the strength of bars. Recent models by Regan & Teuben (2004) suggest that Q_g is degenerate with several bar characteristics that are involved with ring formation. Since gravitational potentials with vastly different bar orbit morphologies can have the same Q_g they conclude that it may not be a useful diagnostic of bar strength, and the effect a bar will have on galaxy evolution.

To gain further insight into the influence of bars on the evolution of a galaxy, we have combined many of the quantitative measures of bars mentioned above in an analysis of a large sample of near-infrared (NIR) and optical galaxy images. We assess bar axial ratios, sizes, and gravitational torques, and also derive isophotal properties of lenses for a subset of the sample.

2. The Sample

We have selected galaxies from the sample described in Moriondo et al. (1999) with well-defined bar classes (SA, SAB, SB) in the RC3 (de Vaucouleurs et al. 1991). The sample was imaged in the H band, and the data reduction and calibration is described in Moriondo et al. (1999). Two-dimensional bulge/disk decompositions have also been applied to the sample galaxies (Hunt et al. 2004); we will incorporate these structural parameters in our analysis.

To compare optical bar properties with those in the NIR, we also acquired the FITS images of our Moriondo et al. sample from the Digitized Sky Survey (DSS). The photographic density was converted to a roughly linear scale according to Lasker et al. (1990), after sky subtraction. Extensive tests showed that this procedure gives reasonably accurate isophotes and isophotal fits, at least in terms of ellipticity, position angle, and higher-order residuals (see below and Fig. 1).

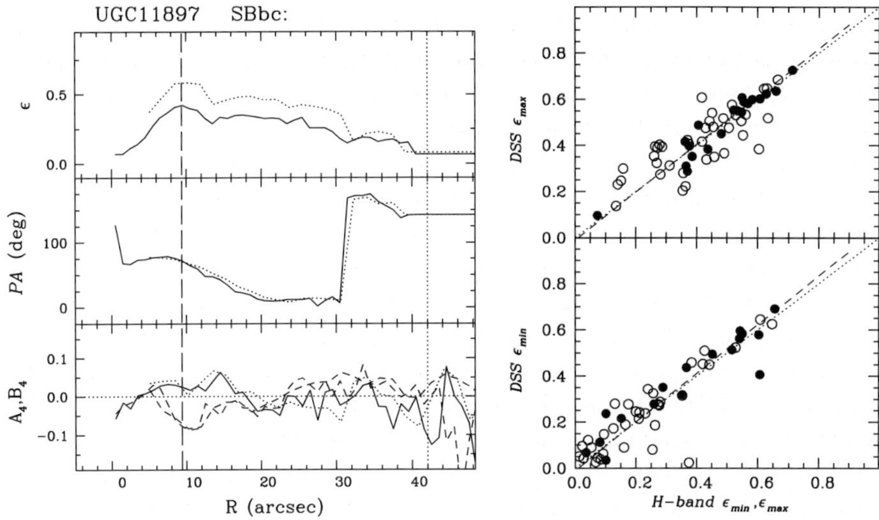

Figure 1. Left panel: Comparison of H-band and DSS radial profile for UGC 11897. H band is shown with a solid line, and DSS with a dotted one. The visually-measured bar length is shown as a vertical dashed line. Right panel: Comparison of$H$-band and DSS ellipticity extrema. Strong bars (SB) are shown as filled circles, and weak bars (SAB) as open ones.

Because the Moriondo et al. sample contains few early-type spirals, we selected a supplementary sample of 140 bright galaxies from RC3 with strong bars (SB), morphological type ranging from S0- to Sd, inclinations $i \leq 70°$, and B magnitude $\leq$ 12.5. DSS images of these galaxies were reduced as above. We preferred DSS images to 2MASS ones because of the better sensitivity of the former; the measurement of weak or low-contrast bars would not be guaranteed with 2MASS images, and our tests showed that the optical and NIR bar properties are indistinguishable.

3. The Analysis

Visually Measured Bar Lengths and Orientations

We measured the ends of the bar from a visual greyscale display of both the H-band and DSS images, and derived bar radius and mean position angle. Included in our definition of bar length are any structures (such as ansae) which occur along the bar, right up to the start of the spiral arms. In galaxies with known lenses (Kormendy 1979), lens sizes were measured in the same way.

Because of potential differences in bar frequencies between the H band and the DSS, we have taken particular care to inspect both sets of images in each bar class for *visual* differences in bar detectability in the two wavelength regimes. While we find some discrepancies relative to the assigned bar class, they are the same in both H and DSS. Contrary to what is usually claimed, we found that bars are often less discernible in the H band than in the DSS images, a phenomenon that appears to result mostly from decreased *contrast* in the NIR. While most bars stand out in the DSS images either because of enhanced surface brightness or because of delineation by dust lanes, most bars in H are not distinguished by high surface brightness (and are obviously much less affected by dust).

Axial Ratios and Sizes

The brightness contours of the H-band and DSS images have been fit to ellipses with the IRAF/STSDAS routine `ellipse`. The left panel of Fig. 1 shows an example of this procedure, with the DSS profile shown as a dotted line, and the H-band as a solid one. We investigated potential differences between H and the DSS by comparing the values of ϵ_{max} and ϵ_{min} for each, and plotting them against the visually measured bar length R_{bar}; these are shown in the right panel of Fig. 1. The small scatter of the regressions implies that isophotal properties of these galaxies are similar in H and in the optical to the 5% level, thus that bar characteristics themselves are virtually indistinguishable. We can therefore safely combine the RC3 sample (DSS) and the Perseus one (H band). Our data also confirm that bar lengths are well reproduced by ellipticity *maxima* (e.g., Athanassoula et al. 1990) (correlations not shown).

Fig. 2 shows that lens lengths are well correlated with ellipticity *minima*. From the correlations shown in Figures 2 and among R_{bar}, ϵ_{max}, and ϵ_{min}, we conclude that lenses are, on average, 1.4 times larger than bars. Moreover, properties of the bars themselves may determine the size of a lens, whether or not the lens itself happens to be present.

Since the bar axial ratio varies along the bar, a single value must be chosen; we have taken $(b/a)_{bar} = 1 - \epsilon_{max}$. Since bar axial ratios (and bar lengths themselves) suffer from projection effects, we have attempted to rectify them assuming that bars are two-dimensional systems, coplanar with the disk. Only three of our galaxies are inclined more than 60°, so the corrections are typically 30% or less. However, when these corrections are applied to the observed $(b/a)_{bar}$, the scatter increases and the trends described in the next section become noisier. Martin (1995) also found that rectification introduces additional scatter in his data. We have therefore adopted the observed values for R_{bar} and for $(b/a)_{bar}$, rather than the rectified ones. All the trends reported here are present in the corrected data, but are less significant because of the larger scatter.

Flat and Exponential Bar Profiles

Using the bar angle derived from the visual measurements, we have extracted major- and minor-axes surface brightness profiles of the bars from the H-band images, averaged over 2 pixels perpendicular to the run of the profile. Two representative results are shown in in Fig. 3, where the solid line denotes

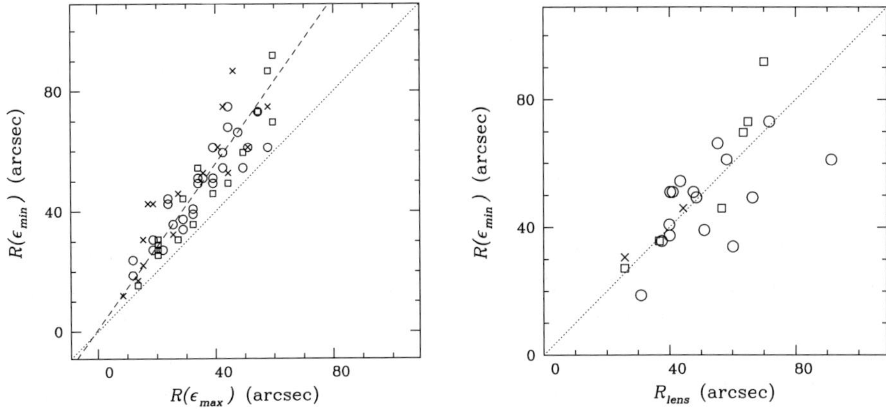

Figure 2. **Left panel:** Comparison of lens and bar lengths. Radius of minimum ellipticity $R(\epsilon_{min})$ plotted against $R(\epsilon_{max})$. **Right panel:** $R(\epsilon_{min})$ plotted against visually-measured lens radius. Early type spirals (T≤1) are shown as open circles, intermediate types ($1 < T \leq 4$) as open squares, and late types ($T \geq 5$) as ×s. In all panels, the dotted line denotes equality, and a dashed one the best-fit regression.

the profile along the bar, and the dotted line perpendicular to the bar. The galaxies have not been deprojected (c.f., Elmegreen et al. 1996b). Also noted in Fig. 3 is the bar profile type ("flavor"), which we have defined according to Elmegreen & Elmegreen (1985). Essentially, a bar is said to have a "flat" profile when its brightness remains relatively constant or decreases slowly along its length, and an "exponential" profile when the surface brightness falls off with roughly the same slope along both the major and minor axes.

Gravitational torques: non-axisymmetric structure

Following Buta & Block (2001), we have quantified the gravitational forcing of the bar by measuring the nonaxisymmetric force field inferred from our NIR images, assuming constant M/L. The relative strength Q_g of the nonaxisymmetric perturbation is the tangential force $F_T = \frac{1}{r}\partial\Phi/\partial\phi$ divided by the azimuthally averaged radial force $F_R = \partial\Phi/\partial r$ (see also Laurikainen & Salo 2002). To calculate the potential, we first remove foreground stars and deproject the images, then convolve them with an exponential vertical disk structure having a scale height 20% of the radial scale lengths from our B/D decompositions. Since the convolution function depends very little on the exact form of the model used for the vertical distribution (Laurikainen & Salo 2002), the exponential is convenient because the convolution integral can be calculated analytically. We then decompose the potential into polar Fourier components,

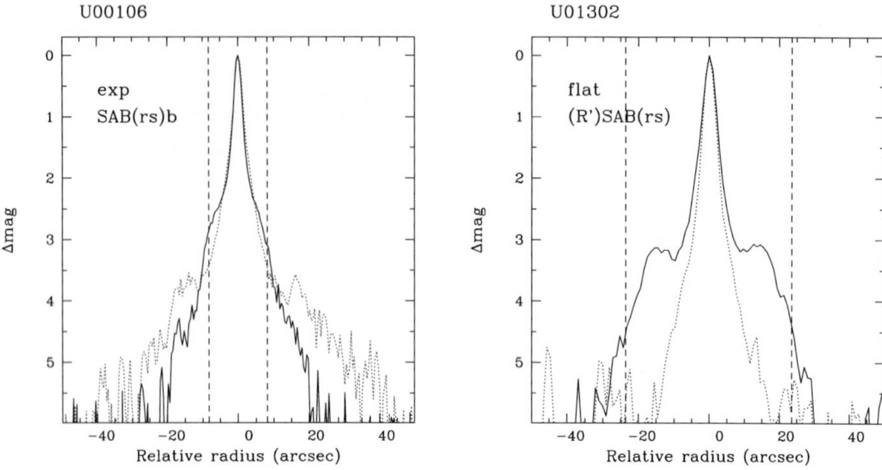

Figure 3. Examples of the H-band bar surface brightness profiles: the major-axis profile (along the bar) is shown with a solid line, and the minor axis (perpendicular to the bar) with a dotted one. Our determination of bar profile type is given in the upper-left corner; R_{bar} is shown as a vertical dashed line.

and calculate F_T, F_R, and their ratio, Q_g. Representative examples of the results of this operation are shown in Fig. 4.

Visual inspection of the Q_g images shows that they can be divided into roughly four morphology classes as shown in Fig. 5: symmetric 2-lobed butterfly (S2B) diagrams in which the "wing" of the butterfly contains a symmetric 2-lobed sub-structure; asymmetric 2-lobed butterfly (A2B) diagrams in which the two lobes are not of equal intensity; 1-lobed butterflies (1B) in which there is no substructure in the "wing"; and irregular (Irr) structure in which there is no clearly-defined butterfly. These morphologies appear to be part of a continuous trend, and we hypothesis (but have not yet proved) that the S2B's result from a strong $m = 4$ (or higher order) term(s), relative to $m = 2$. The S2B morphology seems to signify a two-component bar structure, since axisymmetric bulges would not give the two strong gradients as the perpendicular distance from the bar increases. Examples of all Q_g morphologies can be found in both exponential and flat bars.

4. Results and Discussion

• When $(b/a)_{bar}$ is plotted against morphological type T (see Fig. 6), there is a clear trend for bar axial ratios to decrease with increasing Hubble type; **earlier Hubble types tend to have "fatter" bars with a significance of 4σ.** This trend for bar axial ratios is similar in kind to the trend found by Buta et al. (2004) for maximum Q_g as a function of Hubble type; bars tend to be

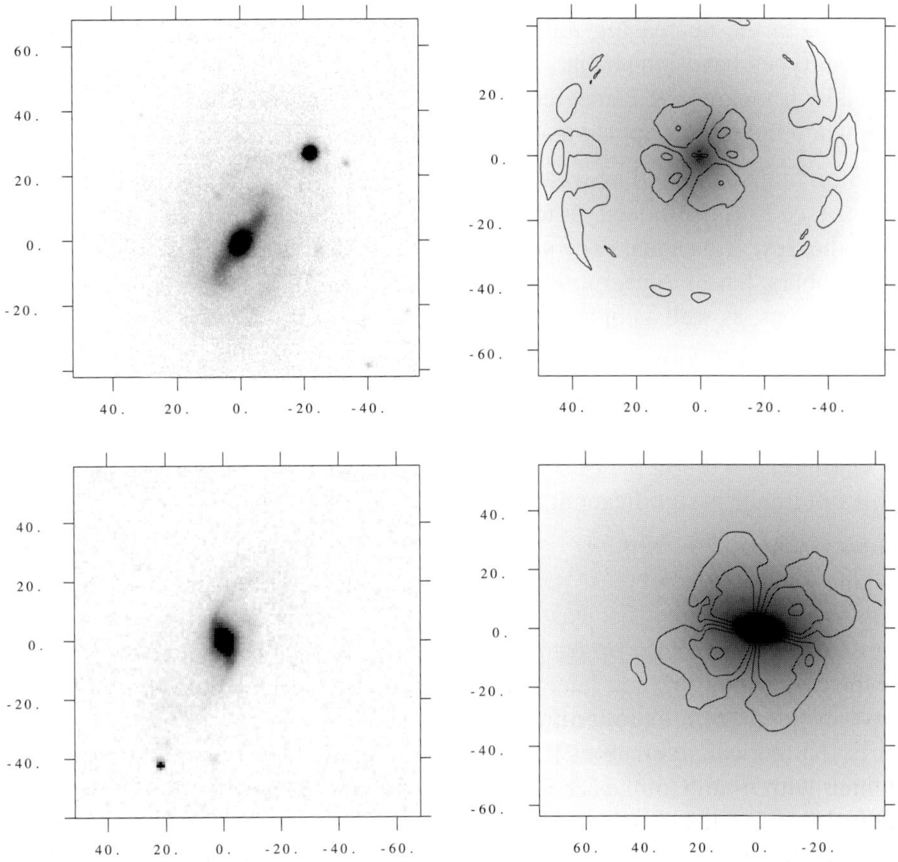

Figure 4. **Left panels:** H band images of U 380 (top) and U 12486 (bottom) before rotation, deprojection. **Right panels:** gravitation potential as grey scale, with contours of Q_g superimposed. U 380 has a flat bar, and U 12486 an exponential one. Horizontal and vertical axes are in arcsec, and orientation is N up, E left.

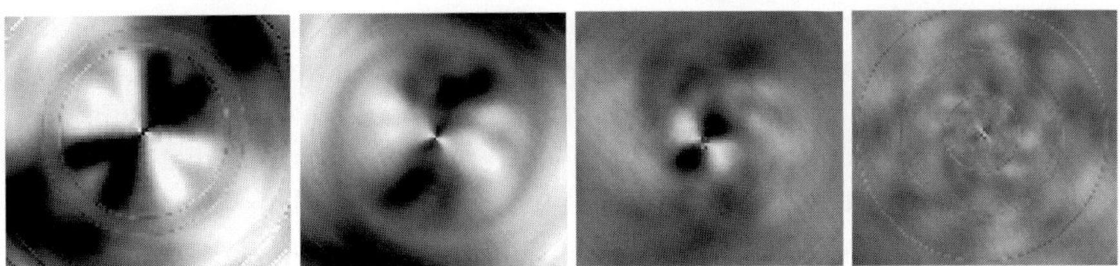

Figure 5. Examples of the Q_g torque morphology classes. From left to right: S2B (UGC 1111); A2B (UGC 89); 1B (UGC 1688); Irr (UGC 1437).

stronger [thinner and larger Q_g(max)] with increasing Hubble type, an effect also due to the decreasing influence of the bulge. We also find that earlier Hubble types have longer bars than early types (not shown), a trend already noted by Elmegreen & Elmegreen (1985). Thinner bars tend to be classified as SB, while fatter ones appear as SAB, or indeed not classified (SA).

- Structures classified as weak bars have axial ratios that approach those of lenses (see Fig. 6). Indeed, in terms of their isophotal properties, the **thinnest and shortest lenses are very similar to the fattest and longest bars**. Similarly to bars, lenses appear to increase in size going from late to early morphological types, as well as increasing their axial ratio.

- **At a given T, exponential bars have larger $(b/a)_{bar}$ and A2B, 1B, and Irr Q_g morphology, while flat bars tend to have smaller $(b/a)_{bar}$ and S2B morphology** (see Figs. 6, 7). For Sb's, all bars with $Q_g(\text{max}) > 0.4$ are S2B's: $< Q_g(\text{max}) >= 0.48 \pm 0.17$ for S2B, $< Q_g(\text{max}) >= 0.27 \pm 0.07$ for A2B+1B. This difference is significant at the 4σ level.

Clues for Secular Evolution?

One way to interpret these results is in the context of secular evolution as the origin of Hubble type. If at least some fraction of early type spirals results from bar-induced bulge build-up from later types, early-type bars themselves could be the descendants of their later type counterparts. We **speculate that flat bars have evolved from exponential bars**, as galaxies migrate from late types to early. Numerical simulations show that bars tend to get longer, stronger, and thinner with time (Combes & Elmegreen 1993; Athanassoula 2003). We find observationally that flat bars tend to be thinner and longer than exponential ones, and that strong flat bars have a S2B torque morphology. Such a bar has a flat profile because of resonances developed at the tips which appear as ansae, perhaps the beginning of inner rings; this process takes time, and requires a relatively large central concentration. In practice, the $(b/a)_{bar}$ of a flat bar is smaller for a given T because a_{bar} is longer as result of the ansae. Moreover, the predominantly S2B Q_g morphology in strong flat bars implies a more complex, morphologically evolved, bar structure.

Lenses can also be interpreted in this framework, as the result of bar evolution. Their isophotal properties are very similar to inner rings (Buta 1995), as well as to the fattest and longest bars. It is probable that a lens results from a bar-induced resonance (Buta & Combes 1996), and possible that once a bar starts to dissolve and become triaxial, it first appears as a lens. Such a process, because of the resonances required, can only occur in a galaxy with enhanced central concentration, i.e., an early type.

References

Abraham, R.G., & Merrifield, M.R. 2000, ApJ, 120, 2835

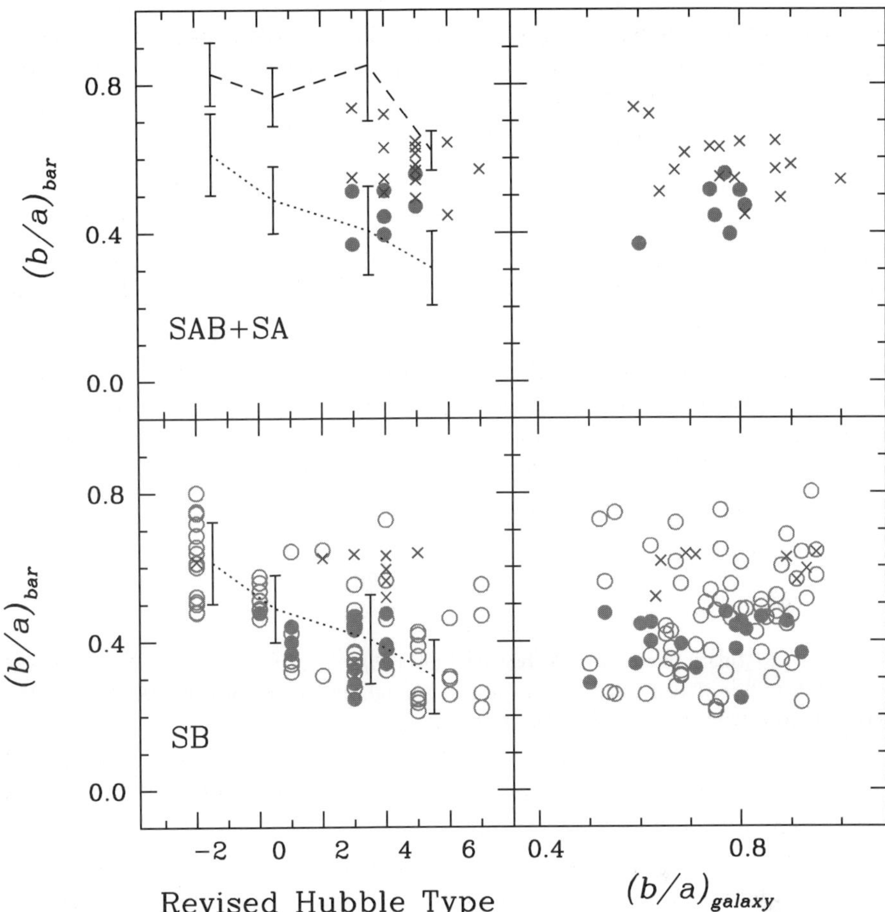

Figure 6. The bar axial ratio plotted against Hubble index T. The upper panel shows the weak bars (formal class SAB, SA), and the lower panel strong ones (SB). Exponential-profiled bars are shown as $\times$'s, intermediate-type profiles as open squares, and flat-profiled bars as filled circles; DSS bars are shown as open circles. The dotted (dashed) line and error bars show the median trends with morphological type for bars (lenses). Right panels show the run of bar axial ratio with galaxy inclination; the random distribution of the points lends support to our non-rectification of the observed ratios.

Athanassoula, E., Morin, S., Wozniak, H., Puy, D., Pierce, M.J., Lombard, J., & Bosma, A. 1990, MNRAS, 245, 130

Athanassoula, E. 1992, MNRAS, 259, 328

Athanassoula, E. 2003, MNRAS, 341, 1179

Block, D.L., Bournaud, F., Combes, F., Puerari, I., & Buta, R. 2002, A&A, 394, L35

Block, D.L., Puerari, I., Knapen,J.H., Elmegreen, B.G., Buta, R., Stedman, S., & Elmegreen, D.M. 2001, A&A, 375, 761

Buta, R. 1995, ApJS, 96, 39

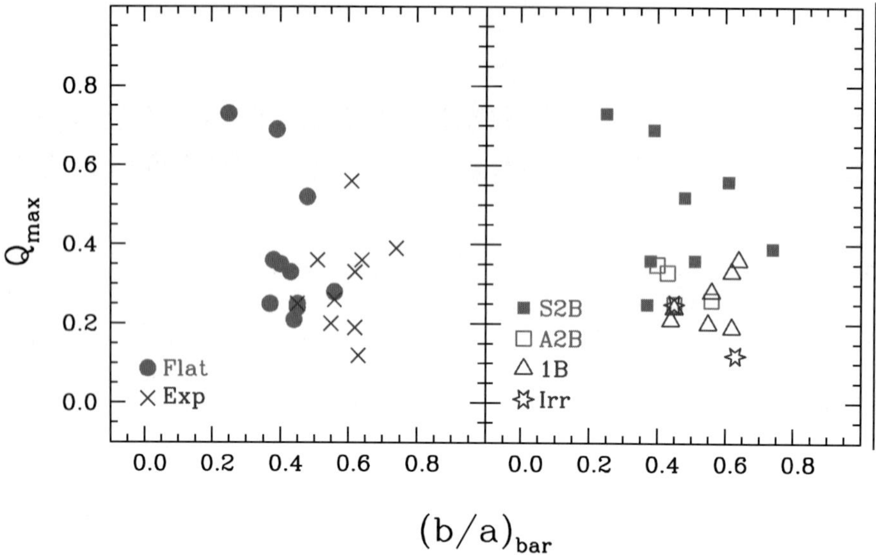

Figure 7. Q_g(max) plotted against bar axial ratio $(b/a)_{bar}$. In the left panel, galaxies are coded by bar "flavor" (exponential, flat), and in the right by Q_g morphology (S2B, A2B, 1B, Irr). Here we have selected a subset of galaxies with roughly the same Hubble type: $< T > \approx$Sb.

Buta, R., & Block, D.L. 2001, ApJ, 550, 243

Buta, R., Block, D.L., & Knapen, J.H. 2003, ApJ, 126, 1148

Buta, R., & Combes, F. 1996, Fundamentals of Cosmic Physics, 17, 95

Buta, R., Laurikainen, E., & Salo, H. 2004, AJ, 127, 279

Combes, F., & Elmegreen, B.G. 1993, A&A, 271, 391

de Vaucouleurs, G., de Vaucouleurs, A., Corwin, H. G., Buta, R., Paturel, G., & Fouque, P. 1991, Third Reference Catalogue of Bright Galaxies (Springer, New York) (RC3)

Elmegreen, D.M., Elmegreen, B.G., Chromey, F.R., Hasselbacher, D.A., & Bissell, B.A. 1996, AJ, 111, 1880

Elmegreen, B.G., Elmegreen, D.M., Chromey, F.R., Hasselbacher, D.A., & Bissell, B.A. 1996, AJ, 111, 2233

Elmegreen, B.G. & Elmegreen, D.M. 1985, ApJ, 288, 438

Hunt, L.K., Pierini, D., & Giovanardi, C. 2004, A&A, 414, 905

Kormendy, J. 1979, ApJ, 227, 714

Lasker, B.M., Sturch, C.R., McLean, B.J., Russell, J.L., Jenkner, H., & Shara, M.M. 1990, AJ, 99, 2019

Laurikainen, E., & Salo, H. 2002, MNRAS, 337, 1118

Martin, P. 1995, AJ, 109, 2428

Moriondo, G., et al. 1999, A&AS, 137, 101

Regan, M.W., & Teuben, P.J. 2004, ApJ, 600, 595

EVOLUTION AND IMPACT OF BARS OVER THE LAST NINE GYR: EARLY RESULTS FROM GEMS

Shardha Jogee[1], Fabio D. Barazza[1], Hans-Walter Rix[2], James Davies[1], Inge Heyer[1], Marco Barden[2], Steven V.W. Beckwith[1], Eric F. Bell[2], Andrea Borch[2], John A. R. Caldwell[1], Christopher Conselice[3], Boris Haussler[2], Catherine Heymans[2], Knud Jahnke[4], Johan H. Knapen[5], Seppo Laine[6], Gabriel M. Lubell[7], Bahram Mobasher[1], Daniel H. McIntosh[8] Klaus Meisenheimer[2], Chien Y. Peng[9], Swara Ravindranath[1], Sebastian F. Sanchez[4], Isaac Shlosman[10], Rachel S. Somerville[1], Lutz Wisotzki[4] and Christian Wolf[11]

[1] *Space Telescope Science Institute, 3700 San Martin Drive, Baltimore MD 21218, U.S.A,* [2] *Max-Planck Institute for Astronomy, D-69117 Heidelberg, Germany,* [3] *Caltech, Dept. of Astronomy, Pasadena, CA 91125, U.S.A,* [4] *Astrophysikalisches Institut Potsdam, D-14482 Potsdam, Germany,* [5] *University of Hertfordshire, Hatfield, Herts AL10 9AB, UK* [6] *Spitzer Science Center, Caltech, Pasadena, CA 91125, U.S.A,* [7] *Vasser College, Dept. of Physics and Astronomy, Poughkeepsie, NY 12604, U.S.A,* [8] *University of Massachusetts, Amherst, MA 01003, U.S.A,* [9] *University of Arizona, Tucson, AZ 85721, U.S.A,* [10] *University of Kentucky, Lexington, KY 40506-0055, U.S.A,* [11] *University of Oxford Astrophysics, Oxford OX1 3RH, UK*

Abstract Bars drive the dynamical evolution of disk galaxies by redistributing mass and angular momentum, and they are ubiquitous in present-day spirals. Early studies of the Hubble Deep Field reported a dramatic decline in the rest-frame optical bar fraction $f_{\rm opt}$ to below 5% at redshifts $z > 0.7$, implying that disks at these epochs are fundamentally different from present-day spirals. The GEMS bar project, based on $\sim$ 10,000 galaxies with HST-based morphologies and accurate redshifts over the range 0.2–1.3, aims at constraining the evolution and impact of bars over the last 9 Gyr. We present early results indicating that $f_{\rm opt}$ remains $\sim$ constant with a lower limit of $\sim$ 30% over $z \sim 0.2$–1.3, corresponding to lookback times of $\sim$ 2.5–9 Gyr. The bars detected at $z > 0.6$ are primarily strong with ellipticities of 0.4–0.8. Remarkably, both the bar fraction and range of bar sizes observed at $z > 0.6$, appear to be comparable to the values measured in the local Universe for bars of corresponding strengths. Implications for bar evolution models are discussed.

Keywords: Galaxy Evolution,, Bars, Dynamics, Galaxy Surveys

D. Block et al. (eds.), Penetrating Bars through Masks of Cosmic Dust, 291–300.

1. Open issues in bar-driven galaxy evolution

Large-scale bars are ubiquitous in present-day spirals, with reported optical bar fractions of $\sim$ 30 % for both strong and weak bars (e.g., Eskridge *et al.* 2002aa, hereafter ES02; § 3). Such bars can drive the dynamical evolution of disk galaxies by redistributing mass and angular momentum. In particular, they are believed to efficiently drive gas from the outer disk of galaxies into the inner kpc via gravitational torques. There is mounting observational evidence for bar-driven gas inflow. Observations of cold or ionized gas velocity fields show evidence for shocks and non-circular motions along the large-scale stellar bar (e.g., Regan, Vogel, & Teuben 1997). Barred galaxies show shallower metallicity gradients across their galactic disks than unbarred ones (Martin & Roy 1994). CO(J=1–0) interferometric surveys show that, on average, the molecular gas central concentration in the inner kpc is higher in barred than in unbarred galaxies (Sakamoto *et al.* 1999). Once gas reaches the central regions of bars and builds up large densities comparable to the Toomre critical density for the onset of gravitational instabilities, intense starbursts are triggered (Jogee *et al.* 2004a). The frequency of large-scale stellar bars is in fact significantly higher in starburst galaxies than in normal galaxies (Hawarden *et al.* 96; Hunt & Malkan 99), while the question of whether Seyferts have an excess of large-scale bars remains under investigation (Regan *et al.* 1997; Knapen *et al.* 2000; Laurikainen *et al.* 2004). Although bars in the local Universe are fairly well studied, several fundamental aspects of bar-driven galaxy evolution remain open:

- When and how did bars form? Are they a recent ($z<0.7$) phenomenon or were they abundant 10 Gyr ago? Two conflicting results have been reported to date. Abraham *et al.* (1999) claim a dramatic decline in the rest-frame optical bar fraction ($f_{\rm opt}$) at $z > 0.7$, based on WFPC2 images of a small sample of $\sim$ 50 moderately inclined spirals in the Hubble Deep Field (HDF). They find $f_{\rm opt}$ drops from $\sim$ 24% at $z \sim 0.2$–0.6 to below 5% at $z > 0.7$. A second study (Sheth *et al.* 2003), based on NICMOS images of the HDF, report an *observed* bar fraction of 5% at $z \sim 0.7$–1.1. Out of 95 galaxies in this redshift interval, they find four large bars with a mean semi-major axis a of 12 kpc (1.4″). Since NICMOS has a large effective PSF, this study fails to detect smaller bars which, at least locally, constitutes the majority of bars (see § 3). The authors argue that their observed bar fraction of 5% at $z > 0.7$ is at least comparable to the local fraction of large (12 kpc) bars, and thus, by extrapolation, the overall optical bar fraction over all bar sizes (1.5-15 kpc) probably does not decline at $z > 0.7$. This conclusion suffers from small number statistics and relies heavily on the extrapolation over bar sizes.

- Are bars long-lived or do they dissolve and reform over a Hubble time? Early studies (e.g., Hasan & Norman 1990; Norman, Sellwood, & Hasan 1996) proposed that once a large central mass concentration (CMC) builds up in the inner 100 pc of a galaxy, for instance via bar-driven gas inflow, it will destroy or weaken the bar by inducing chaotic orbits and reducing bar-supporting orbits. Subsequently, Athanassoula (2002) found that in simulations with live halos, bars are more difficult to destroy due to resonant angular momentum exchange. Shen & Sellwood (2004) find that in purely stellar N-body simulations, the bar is quite robust to CMCs. Furthermore, after a bar is destroyed by a CMC, the disk left behind is dynamically hot, and does not readily reform a new bar unless it is cooled significantly. Recently, Bournaud & Combes (2002) proposed that bars are destroyed primarily due to the reciprocal torques of gas on the stars in the bar (rather than by the CMC *per se*). In their models, bars can dissolve and reform recurrently over timescales $\leq$ 3–4 Gyr, provided the galaxy accretes sufficient cold gas over a Hubble time.

- Bar-driven evolutionary scenarios purport that bars may transform late-type spirals into intermediate types (e.g., Scd to Sbc) via gas inflows which build pseudo-bulges, and via bending instabilities (e.g., Sellwood 1993) or vertical ILRs (e.g., Combes *et al.* 1990) which drive stars to large scale height into a bulge-like configuration. At intermediate redshifts, Katz & Weinberg (2002) suggest bars can solve the dark matter halo cusp-core controversy. While there is a wealth of circumstantial evidence (see review by Kormendy & Kennicutt 2004) consistent with bar-driven secular evolution, no systematic tests of this picture have been made either at local or intermediate redshifts.

Using $\sim$ 10,000 galaxies with superb HST images and accurate redshifts over $z \sim$ 0.2–1.3 (corresponding to lookback times $T_{\rm back}$ of $\sim$ 2.5–9 Gyr) from the **G**alaxy **E**volution from **M**orphologies and **S**EDs (GEMS;Rix *et al.* 2004; § 2) survey, we have started a comprehensive study of the properties (fractions, sizes, strengths) of bars as a function of redshift, and host galaxy properties. The GEMS bar project will constrain how bars evolve and impact the structural evolution of galaxies over the last 9 Gyr, with the eventual goal of resolving some of the above open issues. We describe here our methodology, early results, and upcoming highlights. These findings are described in more detail in Jogee *et al.* 2004b. We assume a flat cosmology with $\Omega_M = 1 - \Omega_\Lambda = 0.3$ and a Hubble constant H_0 = 70 km s^{-1} Mpc^{-1} throughout this paper.

2. Imaging and Characterization Bars out to $z \sim 1$

GEMS (Rix *et al.* 2004) is a two-color (F606W and F850LP) imaging survey with the HST Advanced Camera for Surveys (ACS) of a large-area (800

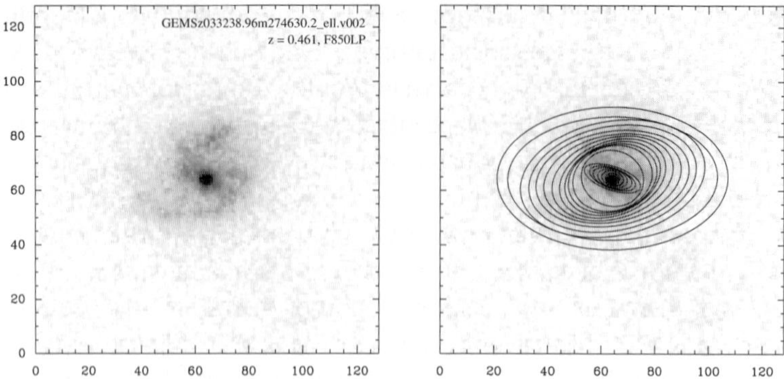

Figure 1a. **Identification and characterization of bars at intermediate redshifts** – The left panel shows the GEMS image of a galaxy at $z \sim 0.5$ (lookback time $T_{\rm back} \sim 5$ Gyr) with a bar, prominent spiral arms, and a disk. The right panel show the same image with an overlay of the fitted isophotes. The latter clearly trace the bar, spirals, and outer disk.

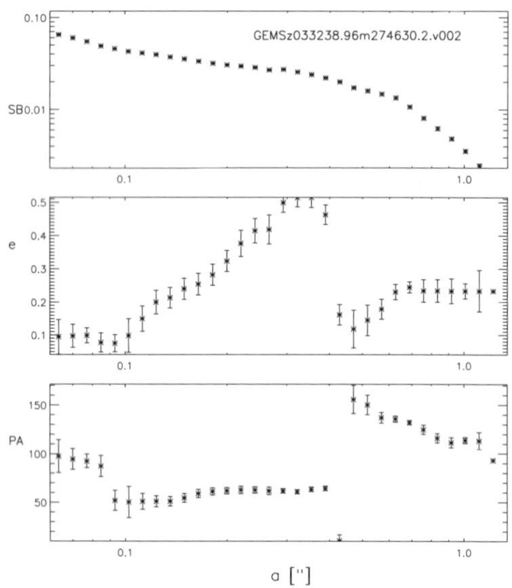

Figure 1b. **Identification and characterization of bars at intermediate redshifts** – The radial plots of the surface brightness (top), ellipticity e (middle), and PA (bottom) generated by the isophotal fits in the previous figure are shown. The bar with a semi-major axis $a \sim 0.31''$ causes e to rise smoothly to a maximum, while the PA has a plateau over the region dominated by the x_1 orbits. At larger a, beyond the bar end, e drops sharply. The spiral arms lead to a twist in PA and varying e, and subsequently, the disk dominates.

arcmin^2 or $\sim$ 28′ × 28′) field centered on the *Chandra* Deep Field South (CDF-S). The limiting 5σ depth for compact sources is m$_{AB}$(F606W)=28.3 and m$_{AB}$(F850LP)=27.1. ACS imaging from the GOODS project (Giavalisco et al. 2004) is used in the central quarter field. GEMS provides morphologies for $\sim 10,000$ galaxies in the redshift range $z \sim 0.2$–1.3 where SEDs and accurate redshifts ($\sigma_z/(1+z) \sim 0.02$) down to R_{AB}= 24 exist from the COMBO-17 project (Wolf *et al.* 2003). The dataset are further enhanced by panchromatic $Chandra$, $Spitzer$, and ground-based observations covering the X-rays to the far-IR and radio.

The GEMS ACS dataset offer many advantages for studying bars, compared to earlier WFPC2/NIC3 imaging surveys. The dataset provide a factor of 100 improvement in number statistics, an effective ACS PSF which is 2–3 times better than the WFCP2 and NIC3 PSFs, and redder wavelength coverage and sensitivity via the ACS F850LP filter/CCD combination. The superior PSF allows us to probe bars down to small sizes ($0.12''$ or 1 kpc; see § 3). Finally, the large area ($\sim 120\times$ HDF area) coverage reduces the effect of cosmic variance.

We identify and characterize bars out to $z \sim 1.3$ by performing isophotal analyses of the F606W (V-band) and F850LP (z-band) images using an automated iterative version of of the IRAF STSDAS "ellipse" routine. Isophotal fits are a good guide to the underlying orbital structure of a galaxy and provide a robust way of identifying and characterizing bars (e.g., Wozniak *et al.* 1995; Jogee *et al.* 2002). A bar is identified by its characteristic signature in the radial profiles of ellipticity (e), surface brightness, and position angle (PA), as illustrated in Figs. 1a and 1b. The ellipticity is required to rise smoothly to a *global* maximum above 0.25 while the PA has a plateau over the region dominated by the x_1 orbits, followed by a drop ≥ 0.1 in e as the bar-to-disk transition occurs. The ellipticity, PA, and semi-major axis for both the bar and outer disk can be identified from these profiles, thereby allowing the profiles to be subsequently deprojected to derive the intrinsic bar strength and size. We also use the disk ellipticity and inferred inclination i to reject all galaxies with $i < 60°$, since morphological classification is unreliable in highly inclined galaxies.

3. Early Results and Future Highlights

We use two approaches to characterize the optical bar fraction $f_{\rm opt}$. The first is to identify bars in the reddest observed band (F850LP) at all redshifts, with the idea that redder wavelengths are less impacted by extinction, and consequently better trace the old stellar potential. The disadvantage of this method is that different, increasingly bluer rest-frame bands are then traced in the two redshift bins of interest (0.24$< z \leq$0.65 and 0.65$< z \leq$1.3) as illustrated in Table 1a. A complementary approach, illustrated in Table 1b, is to use both F606W and F850LP images in order to work in a fixed rest-frame band (*B*/*V*) and avoid significant bandpass shifting.

We present early results based on a subsample of 600 galaxies distributed in the overlap area of GEMS and GOODS. Figure 2 illustrates examples of

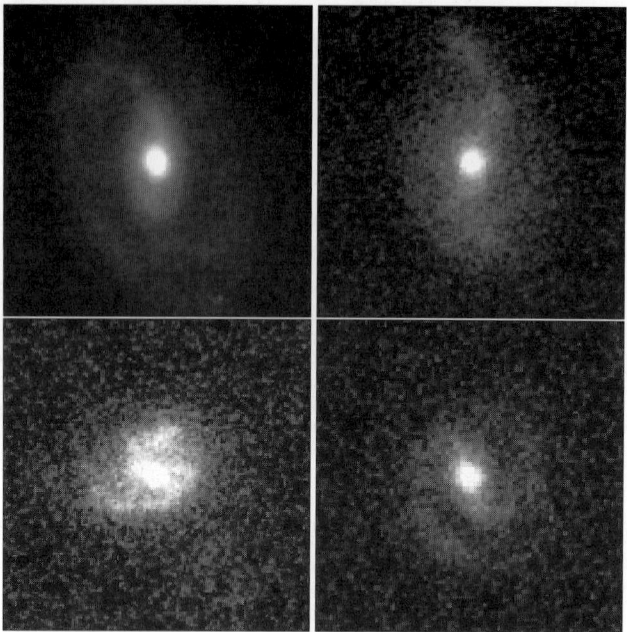

Figure 2. Examples of bars and spiral arms in GEMS galaxies at $z \sim 0.3$ and 0.4. (top panel:left to right), and at $z \sim 0.5$ and 0.9 (bottom panel:left to right)

intermediate redshift galaxies with bars and spiral arms in this sample. While being only a small fraction of the total GEMS dataset, this sample is 3–6 times larger than previous WFPC2 and NIC3 samples used to study bars in the HDF. An inclination cutoff of $i < 60°$ and a magnitude cutoff of $R_{\rm AB} < 24$ mag are imposed in order to ensure accurate photometric redshifts, spectral typing, and reliable morphological analyses. Galaxies were assigned E/S0 and (spiral+ starburst) types based on comparing their rest frame $U-V$ colors and absolute $M_{\rm V}$ magnitudes with template SEDs from Kinney et al (1996) and Coleman, Wu, & Weedman (1980). We imposed an absolute magnitude cutoff of $M_{\rm V} <$ -19. The rest-frame optical bar fraction $f_{\rm opt}$ in a given redshift bin was taken as ($N_{\rm bar}$/$N_{\rm tot}$), where $N_{\rm bar}$ is the number of bars detected, and $N_{\rm tot}$ is the number of spirals and starbursts. An alternative approach involves using the single fit Sersic index n to identify disk galaxies as the main contributor to $N_{\rm tot}$. Our exciting result, based on the first method, are highlighted below:

- *We find that the rest-frame optical bar fraction $f_{\rm opt}$ remains $\sim$ constant at 30-35% over $z \sim$ 0.2–1.3 or lookback times of $\sim$ 2.5–9 Gyr. In particular, we do not find evidence of a dramatic decline in $f_{\rm opt}$ to below 5% at $z >$* 0.7. This result holds whether we choose to use the reddest observed band (F850LP) in both redshift bins (Table 1a), or choose to work in the same fixed rest-frame band (Table 1b). To further test the robustness of the results, we repeated the analyses treating the spiral and starburst types separately, and found no change. We also performed an extra-consistency check by using the F606W images over the $0.65 < z \leq 1.3$ interval to estimate the bar fraction (f_{uv}) in the rest-frame UV. Strong bars are not expected to be delineated by recent massive SF as the strong shocks on their leading edges are not conducive to SF. As expected, the estimated f_{uv} is much lower (5%) than f_{opt}. The results on f_{opt}, therefore, seem quite robust. Analyses of a smaller sample of ACS images of the tadpole field is leading to similar conclusions (Elmegreen *et al.* 2004)

Table 1a. Optical bar fraction f_{opt} in the reddest available filter (F850LP)

Redshift	$0.24 < z \leq 0.65$	$0.65 < z \leq 1.3$
Filter	F850LP	F850LP
Rest-frame	I/V	V/B
Bar Fraction	33%	35%

Table 1b. Optical bar fraction f_{opt} in a fixed rest-frame (B/V)

Redshift	$0.24 < z \leq 0.65$	$0.65 < z \leq 1.3$
Filter	F606W	F850LP
Rest-frame	B/V	V/B
Bar Fraction	25%	35%

- What kind of bars are we detecting at $z > 0.6$? The bars at $z \sim 0.6$–1.3 have semi-major axes $a \sim 0.15''$–$1.8''$ and 1.6–14 kpc, and it is noteworthy that the majority of these bars have a below $0.5''$ and 5 kpc, and would consequently be hard to detect without the superb resolution of ACS. The bars have rest-frame B-band ellipticities e_B in the range 0.4–0.8 and the majority have $e_B > 0.5$ (Fig. 3). *The detected bars are thus strong bars.* The fact that we not identify weak ($0.25 < e_B < 0.4$) bars at $z \sim 0.6$–1.3 (Fig. 3) could lend itself to several interpretations. One possibility is that bars at $z \sim 1$ were on average stronger than local ones (e.g., Bournaud & Combes 2002). Another possibility is that cosmological dimming, loss of spatial resolution, and the potentially heavy impact of dust and SF at increasing redshifts make it harder to detect such weak bars. In order to quantify these effects we are artificially redshifting local galaxies out to $z \sim 1.3$ in rest-frame B-band, assuming the current cosmology *du jour,* and folding in the ACS and GEMS survey parameters. Weak short bars are indeed hard to detect, but the exact effect of their non-detectability on the overall fraction is not yet fully quantified.
- How do the bars at $z > 0.6$ compare to local ones? Since carefully defined, large volume-limited samples of barred galaxies for comparison

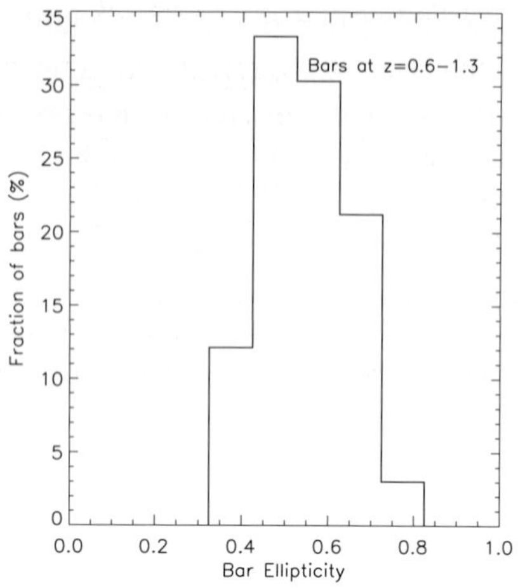

Figure 3. **Bars strengths –** The distribution of bar strengths (ellipticities e_{B}) in the rest-frame B/V is shown for bars at $z \sim 0.6$–1.3, over lookback times of 6–9 Gyr.

at $z \sim 0$ do not yet exist, we choose to compare the $z > 0.6$ bars to the local Ohio State University (OSU) sample of bright galaxies (Table 2; ES02) *It is remarkable to note that strong (e_{B} =0.4–0.8) bars which were in place 6–9 Gyr ago, and those at current epochs, have a similar rest-frame optical bar fraction* f_{opt} (Table 1b and 2) *and a similar bar size distribution* (Fig. 4). The simplest naive interpretation of these results is that bars are long-lived over timescales ≥ 6 Gyr, and possibly over a Hubble time. However, we withhold any definite conclusions until we have compared the normalized bar sizes (i.e., the ratio of bar size to disk size) at these two epochs, and until we have increased our sample sufficiently so that the data can be binned over 1 Gyr interval over the last 9 Gyr.

- *Concluding remarks:* In recent years, a fundamental puzzle was whether the bar fraction shows a dramatic decline at $z > 0.6$. Mounting evidence (Jogee *et al.* 2004b, Elmegreen *et al.* 2004, Sheth *et al.* 2003) now suggests this is not the case. The emphasis should now be to push investigations beyond mere estimates of bar fractions. The next challenge is to distinguish between the two main bar evolution scenarios: long-lived bars over a Hubble time versus models of recurrent destruction/reformation of bars (e.g., Bournaud & Combes 2002). In order to attack this problem, we plan a systematic comparison of the properties of bars (strength, normalized sizes) and host galaxies (central mass concentration, B/D, masses), over each Gyr interval out to lookback times

Table 2. Local bar fractions in the optical ($f_{\rm opt}$) and near-IR ($f_{\rm NIR}$) based on the OSU sample

	Strong Bars	Weak Bars	Strong+Weak Bars	No Bars
f in B-band using RC3 bar strength [a]	**37%**	34%	67%	33%
f in B-band using bar ellipticity $e_{\rm B}$ [b]	**33%**	28%	61%	39%
f in H-band using bar ellipticity$e_{\rm H}$ [c]	49%	21%	70%	30%

Notes to table — The local bar fractions in this table are based on the OSU sample, after excluding all galaxies with inclination $i > 60\,°$. a. The bar fractions shown here are from ES02 who use the RC3 classes of SB, SAB, and SA for "strong bars" , "weak bars", and "no bars"; b. The bar fractions shown here assume a B-band bar ellipticity $e_{\rm B} \geq 0.4$ to define strong bars, and $0.25 \leq e_{\rm B} < 0.4$ for weak bars; c. The bar fractions shown here assume an H-band bar ellipticity $e_{\rm H} \geq 0.4$ to define strong bars.

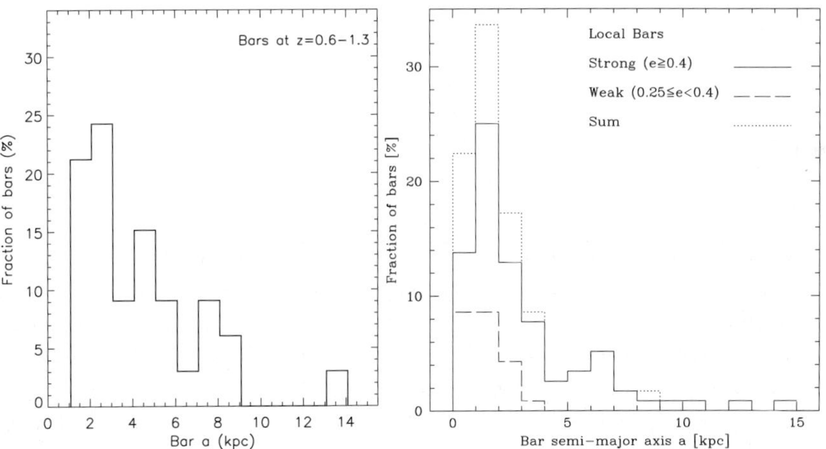

Figure 4. **Distribution of bars sizes 6–9 Gyr ago compared to the present epoch –** $Left$: The distribution of bar semi-major axes a in rest-frame B/V is shown for bars at $z \sim$ 0.6–1.3 (lookback times $T_{\rm back}$ of $\sim$ 6–9 Gyr) in the GEMS survey. Note that most bars have a < 5 kpc (0.6″ at $z \sim$ 0.9) and would be difficult to detect without the exquisite resolution of ACS. $Right$: Ditto, but for local bars in the OSU sample after excluding inclined galaxies with $i > 60°$. The distribution is shown for strong ($e_{\rm B} \geq 0.4$; solid line) and weak ($0.25 \geq e_{\rm B} < 0.4$; hashed line) local bars. The strong local bars are directly comparable to the bars detected at $z \sim$ 0.6–1.3, as the latter have ellipticities $e_{\rm B} \geq 0.4$. There is a striking similarity in the size distribution of strong bars at the present epoch and 6–9 Gyr ago.

of 9 Gyr. Another important area to explore is the relationship of bars to starburst and AGN activity out to $z \sim 1$, since the cosmic density of both types of activity show a dramatic increase from z=0 to 1. The combination of GEMS ACS imaging, $Chandra$ observations, and inflowing $Spitzer$ data makes these timely issues to address, and hold the promise of exciting science ahead.

Acknowledgments

SJ acknowledges support from the National Aeronautics and Space Administration (NASA) under LTSA Grant NAG5-13063 issued through the Office of Space Science, and thanks Paul Eskridge for kindly providing images and parameters for galaxies in the OSU survey.

References

Abraham, R. G., Merrifield, M. R., Ellis, R. S., Tanvir, N. R., & Brinchmann, J. 1999, MNRAS, 308, 569

Athanassoula, E. 2002, ApJL, 569, L83

Bournaud, F. & Combes, F. 2002, A&A, 392, 83

Coleman, G. D., Wu, C.-C., Weedman, D. W. 1980, ApJS, 43, 393

Elmegreen, B. G., et al. 2004, ApJ, submitted

Eskridge, P. B., et al. 2002, ApJS, 143, 73

Giavalisco, M., et al. 2004, ApJL, 600, L93

Hasan, H. & Norman, C. 1990, ApJ, 361, 69

Hawarden, T. G., et al. 1986, MNRAS, 221, 41

Hunt, L. K. & Malkan, M. A. 1999, ApJ, 516, 660

Jogee, S., Knapen, J. H., Laine, S., Shlosman, I., Scoville, N. Z., & Englmaier, P. 2002, ApJL, 570, L55

Jogee, S., Scoville, N. Z., & Kenney, J. 2004a, ApJ, in press

Jogee, S., Barazza, F., Rix, H.-W., Davies, J., Heyer, I., Barden, M., Beckwith, S. V. W., Bell, E. F., Borch, A., Caldwell, J. A. R., Conselice, C., Haussler, B., Heymans, C., Jahnke, K. , Knapen, J. H., Laine, S., Lubell, G., Mobasher, B., McIntosh, D. H., Meisenheimer, K., Peng, C. Y., Ravindranath, S., Sanchez, S. F., Shlosman, I., Somerville, R. S., Wisotski, L., & Wolf, C. 2004b, ApJL, submitted

Kinney, A.L., Calzetti, D., Bohlin, R.C., McQuade, K., Storchi-Bergmann, T., Schmitt, H.R. 1996, ApJ, 467, 38

Knapen, J. H., Shlosman, I., & Peletier, R. F. 2000, ApJ, 529, 93

Kormendy, J. & Kennicutt, R. C. 2004, ARAA, submitted.

Kinney, A.L., Calzetti, D., Bohlin, R.C., McQuade, K., Storchi-Bergmann, T., Schmitt, H.R. 1996, ApJ, 467, 38

Laurikainen, E., Salo, H., & Rautiainen, P. 2002, MNRAS, 331, 880

Martin, P. & Roy, J. 1994, ApJ, 424, 599

Norman, C. A., Sellwood, J. A., & Hasan, H. 1996, ApJ, 462, 114

Regan, M. W., Vogel, S. N., & Teuben, P. J. 1997, ApJL, 482, L143

Rix, H., et al. 2004, ApJS, 152, 163

Sakamoto, K., Okumura, S. K., Ishizuki, S., & Scoville, N. Z. 1999, ApJ, 525, 691

Shen, J. & Sellwood, J. A. 2004, ApJ, 604, 614

Sheth, K., Regan, M. W., Scoville, N. Z., & Strubbe, L. E. 2003, ApJL, 592, L13

Weinberg, M. D. & Katz, N. 2002, ApJ, 580, 627

Wolf, C., Meisenheimer, K., Rix, H.-W., Borch, A., Dye, S., & Kleinheinrich, M. 2003a, A&A, 401, 73

Wozniak, H., Friedli, D., Martinet, L., Martin, P., & Bratschi, P. 1995, A&A Suppl., 111, 115

FIRST PHYLOGENETIC ANALYSES OF GALAXY EVOLUTION

Didier Fraix-Burnet
Laboratoire d'Astrophysique de Grenoble BP 53, F-38041, Grenoble cedex 9 (France)

Abstract The Hubble tuning fork diagram, based on morphology, has always been the preferred scheme for classification of galaxies and is still the only one originally built from historical/evolutionary relationships. At the opposite, biologists have long taken into account the parenthood links of living entities for classification purposes. Assuming branching evolution of galaxies as a 'descent with modification', we show that the concepts and tools of phylogenetic systematics widely used in biology can be heuristically transposed to the case of galaxies. This approach that we call "astrocladistics" has been first applied to Dwarf Galaxies of the Local Group and provides the first evolutionary galaxy tree. The cladogram is sufficiently solid to support the existence of a hierarchical organization in the diversity of galaxies, making it possible to track ancestral types of galaxies. We also find that morphology is a summary of more fundamental properties. Astrocladistics applied to cosmology simulated galaxies can, unsurprisingly, reconstruct the correct "genealogy". It reveals evolutionary lineages, divergences from common ancestors, character evolution behaviours and shows how mergers organize galaxy diversity. Application to real normal galaxies is in progress. Astrocladistics opens a new way to analyse galaxy evolution and a path towards a new systematics of galaxies.

Keywords: Galaxies: fundamental parameters – Galaxies: evolution – Galaxies: formation

1. Introduction

Galaxies have been found by Hubble to be isolated systems of stars (Hubble 1922, 1926). We know they also contain gas and dust, while black holes are certainly rather common. The Hubble (1936) diagram was originally devised to introduce evolution as a link between the different morphological types. This unification scheme has lost its meaning, but the Hubble tuning fork remains a reference in depicting galaxy diversity. However, numerous classifications, attached to particular observational criteria, have appeared, rendering difficult a visualisation of evolution of galaxies globally.

The biological world is also made of complex objects in evolution. The hierarchical organization of diversity on a tree-like structure is caused by evo-

D. Block et al. (eds.), Penetrating Bars through Masks of Cosmic Dust, 301–305.

lution in a "descent with modification" scheme (Darwin 1859). A long history of classification of living organisms has lead to the establishment of powerful methodologies to reveal the evolutionary relationships between the different species. One of the most widely used nowadays is cladistics (Hennig 1965). It does not compare objects on appearance or global similarity, but rather on the evolutionary states of their characters (descriptors).

An attempt to apply such a methodology to astrophysics was proposed by Fraix-Burnet et al (2003). Astrocladistics has thus been developed through some conceptualization and formalization of galaxy classification, formation, evolution and diversification. Several samples have now been analysed. In this talk, I present a summary of the present status of astrocladistics and its achievements. More details are to be found elsewhere (Fraix-Burnet et al 2004a, 2004b, 2004c, Fraix-Burnet and Davoust 2004).

2. Some basics of astrocladistics

The detailed presentation of astrocladistics with its formalism, concepts and tools, is done in Fraix-Burnet et al (2004b, 2004c). Only a very brief overview is given in this section.

A galaxy is fully described and characterized by its basic constituents (stars, gas, dust) and all their properties. A galaxy is thus a complex object. Its evolution is the evolution of its constituents, so that their properties can be seen as characters with different evolutionary states. Hence a galaxy is represented by a vector having as many components as characters or descriptors allowed by observations.

The formation of a galaxy should be understood as the formation of the galaxy as we see it, that is with all the character states as observed. Hence any physical or chemical process that affects any of its properties may result in the formation of a new galaxy of a different class or species. We have identified five such formation processes (Fraix-Burnet et al 2004c): assembling, secular evolution, interaction, accretion-merging, ejection-sweeping. In each case, material (basic constituents) of the progenitors is transmitted to the descendant while being modified. This is fundamentally the darwinian "descent with modification" mechanism. A big difference is that in the case of galaxies the modification is much more brutal because a galaxy can give birth to a new galaxy of a different species.

Evolution of galaxies is governed by these formation processes which occur several times during the history of the Universe, always in different conditions. Diversity is thus generated, and is organized in a hierarchical way. Note that this has nothing to do with the hierarchical scenario of galaxy formation that considers only the mass. The astrocladistic analysis, using the matrix made with the galaxies and their characters, tries to arrange the objects on a tree-like structure, grouping them from the derived (evolved, that is inherited from a common ancestor) character states they share. It is then possible to define species or classes, and thus build a phylogenetic classification.

3. Analyses of two samples of simulated galaxies

In order to illustrate the validity of our approach and to better understand the application of cladistics to astrophysics, two samples of simulated galaxies have been analysed (Fraix-Burnet et al 2004b, 2004c). They were extracted from the GALICS (Galaxies In Cosmological Simulations) database, which results from a hybrid model for hierarchical galaxy formation studies, combining the outputs of large cosmological N-body simulations with simple, semi-analytic recipes to describe the fate of the baryons within dark matter haloes (Hatton et al 2003). The main advantage is that the entire history is known for each galaxy with all the formation processes and progenitors. Note that cladistics does not aim at reconstructing genealogies (which in reality would require the impossible identification of all parents), but phylogenetic trees (which reveal the evolutionary relationships). Galaxies in GALICS are characterized principally by photometric properties (about 100).

The first sample (50 galaxies) is made with galaxies that have never known any merging, so that their diversification is due simply to secular evolution and gas accretion (no interaction are considered in the simulations). We show (Fraix-Burnet et al 2004b) that astrocladistics correctly establishes the right chronology. Diversity in this case is organized in a hierarchical pattern, as expected.

Among the five formation processes, merging of galaxies is somewhat peculiar. First, it clearly illustrates how galaxies disappear to form a new one, an important concept in astrocladistics, whatever the formation process. Second, the transmitted material of the merging galaxies is mixed together. This might look like biological hybridization which is known to affect the tree-like structure of diversity. However we show that this is not the case and confirm our conclusion using a sample with 43 galaxies having underwent different numbers of mergers in their history (Fraix-Burnet et al 2004c).

4. Origin of the morphological dichotomy in the Dwarf galaxies of the Local Group

The first application of astrocladistics to real galaxies has been performed on the Dwarf galaxies of the Local Group (Fraix-Burnet et al 2004a). The 36 objects are described by 24 characters (photometry, masses, line fluxes, metallicity, kinematics) plus morphology (spheroidal, irregular or intermediate).

Since this last parameter is the only one to be qualitative and subjective, it has not been considered in the analysis but merely projected on the result tree.

A robust tree is found, showing for the first time that diversity among a sample of real galaxies is organized hierarchically. Not only does this validates our approach, but the power of a cladistic analysis can now be exploited for galaxies. One very important result is that the morphological segregation is found *and* explained: irregular galaxies are all gathered on a branch which separated from spheroidal galaxies sometimes in the evolution because of a strong accretion of hydrogen. This is an indication that morphology is kind of a gross summary of more fundamental properties. However, the tree brings much more information in the form of consistent evolutionary interpretations. For instance, some objects are known to be exceptionally bright for spheroidal dwarf galaxies (as bright as irregulars). Our result yields a simple explanation: luminosity increases during evolution and these objects happen to be more evolved than other spheroidals and as much as irregulars.

5. Virgo galaxies: first step toward a new classification

Two samples of galaxies belonging to the Virgo cluster are currently being analysed (Fraix-Burnet and Davoust 2004). Results are still preliminary, but a robust tree has been found with a sub-sample of 21 galaxies described by 35 characters (large band and line photometry, some kinematical data and line ratios). This shows that the hierarchical organization of galaxy diversity exists among real galaxies. Equally important, our study reveals a large consistency among evolution of most characters. For instance blue luminosity, B-V and U-B colours, line widths, maximum velocity dispersion, do all evolve quite regularly along the tree. It is thus possible to define different classes that are evolutionary related. More work has to be done for a complete interpretation.

In the analysis, like in others, qualitative characters such as morphological type and presence/absence of a bar were not used. But their projection afterwards onto the result tree is very informative. First, the morphological type evolves quite regularly along the tree, again demonstrating that it is kind of a gross summary of more fundamental and quantitative properties of galaxies. Second, the character 'presence or absence of a bar' is erratic on the tree, being apparently not relevant for the evolutionary state of a galaxy.

6. Conclusions

Astrocladistics is now producing more and more results while the concepts and tools have been largely clarified in the frame of galaxy diversity and evolution. The two fundamental ingredients (a galaxy is described by evolutive characters, diversity is caused by evolution and organized in a hierarchy) have been validated.

The application of cladistics to galaxy evolution requires a precise formalization of some definitions and concepts. In particular, it makes a difference between the definition, the description and the nomenclature of galaxies. This leads to the obvious result that the evolution of a galaxy is the evolution of its fundamental constituents. Their properties are the true descriptors of galaxies and characterize their evolution. This has been successfully checked on simulated and real galaxies.

The hierarchical organization of galaxy diversity, expected from the formation processes that generate new galaxy species, is found in all samples studied so far, either simulated or real. This is an extremely important result because if the existence of a tree-like structure remains true for all classes of galaxies at all epochs, it then will be possible to track the ancestral galaxy species back to the very first objects of the Universe.

Astrocladistics is a methodology to understand galaxy diversity via evolutionary relationships. It groups objects according to their history and using evolutionary states of their characters. It naturally builds a classification, but this can be done only after several analyses of different kinds of galaxies are achieved. This will be the next step of our work and this new galaxy classification will be based on solid taxonomical rules.

References

Darwin, C. (1859). *The Origin of Species*. Penguin, London.

Fraix-Burnet, D., Choler, P., and Douzery, E.J.P. (2003). Ap&SS, 284, 535 (astro-ph/0303410).

Fraix-Burnet, D., Choler, P., and Douzery, E.J.P. (2004a). Submitted to Proc. Natl. Acad. Sci.

Fraix-Burnet, D., Choler, P., Douzery, E.J.P., and Verhamme, A. (2004b). Submitted to A&A.

Fraix-Burnet, D., Douzery E.J.P., Choler, P., and Verhamme, A. (2004c). Submitted to A&A.

Fraix-Burnet, D., and Davoust, E. (2004). In preparation.

Hatton, S., Devriendt, J.E.G., Ninin, S., Bouchet, F.R., Guiderdoni, B., and Vibert, D. (2003). MNRAS, 343, 75 (astro-ph/0309186).

Hennig, W. (1965). Annual Review of Entomology 10, 97.

Hubble, E.P. (1922). ApJ, 56, 162

Hubble, E.P. (1926). ApJ, 64, 321

Hubble, E.P. (1936). *The Realm of Nebulae*. New Haven:Yale Univ. Press.

A UNIFIED PICTURE OF DISK GALAXIES WHERE BARS, SPIRALS AND WARPS RESULT FROM THE SAME FUNDAMENTAL CAUSES

Daniel Pfenniger and Yves Revaz
Geneva Observatory, University of Geneva, CH-1290 Sauverny, Switzerland

Abstract Bars and spiral arms have played an important role as constraints on the dynamics and on the distribution of dark matter in the optical parts of disk galaxies. Dynamics linked to the dissipative nature of gas, and its transformation into stars provide clues that spiral galaxies are driven by dissipation close to a state of *marginal stability* with respect to the dynamics in the galaxy plane. Here we present numerical evidences that warps play a similar role but in the transverse direction. N-body simulations show that typical galactic disks are also marginally stable with respect to a bending instability leading to typical observed warps. The frequent occurrence of warps and asymmetries in the outer galactic disks give therefore, like bars in the inner disks, new constraints on the dark matter, but this time in the outer disks. If disks are marginally stable with respect to bending instabilities, our models suggest that the mass within the HI disks must be a multiple of the detected HI and stars, i.e., disks must be heavier than seen. But the models do not rule out a traditional thick halo with a mass within the HI disk radius similar to the total disk mass.

Keywords: Bars, spiral arms, warps, dark matter, galaxy dynamics

1. Introduction

Over the years we discover progressively the true nature of galaxies by confronting observations to theory and vice versa. A simple tool to rank the importance of the various physical ingredients at play in galaxies is a more complete form of the virial theorem than usually discussed:

$$\tfrac{1}{2}\ddot{I} = \underbrace{2E_{\rm kin}}_{>0} + \underbrace{E_{\rm grav}}_{<0} + \underbrace{3P_{\rm int}V}_{>0} \underbrace{-\,3P_{\rm ext}V}_{<0} + \underbrace{E_{\rm mag}}_{>0} \ldots \approx 0\,. \quad (1)$$

It allows to rank the main energies (bulk kinetic, gravitational, internal and external pressures, magnetic, etc.) determining the system equilibrium measured by its moment of inertia I in the volume V. Clearly the *magnitudes of the interacting energies* (mainly bulk kinetic against gravitational energy in galax-

D. Block et al. (eds.), Penetrating Bars through Masks of Cosmic Dust, 307–316.

ies) rank the importance of each factor. A Taylor expansion of this equation in time also shows that an evolution along a sequence of quasi-steady equilibria is determined to first order by the *magnitude of the interacting powers* (mainly gas cooling against mechanical heating power).

The main point emphasized here is that the same rules found to well explain to first order the horizontal properties of spiral galaxies, i.e., gravitational physics supplemented by energy dissipation, are also able to explain the ubiquitous warps. By comparing numerical simulations of thin self-gravitating disks to the observed properties of spirals, in particular to their frequent warps, a coherent dynamical and evolutive picture of spiral galaxies emerges, with the suggestive hint that two types of dark matter are involved: 1) the classical extended dark halos much thicker than the disks, and 2) a dark component coeval with the HI disks similar to the one proposed in Pfenniger & Combes (1994).

Nowadays the need of at least two types of dark matter is actually well motivated. From the big-bang predicted baryogenesis most of the baryons remain to be found, and on the other hand, also from cosmology non-baryonic dark matter is required to obtain a coherent description of large scale structure formation and the Universe cosmological parameters. Since baryons are known to be strongly dissipative and sometimes collisional, contrary to the expected non-baryonic dark matter, we have no grounds to expect that in galaxies their respective spatial distributions should coincide.

2. The role of bars and spirals

The bar instability has been used in the 70's for supporting the idea that an extended dark matter halo must exist to prevent bar formation (Ostriker & Peebles, 1973). Indeed the early N-body simulations showed that bars result spontaneously from a dynamical plane instability in a collisionless disk with an initial flat core. But because theoreticians were wrongly assuming that bars were exceptional, they imagined a hot and massive collisionless component coexisting in the optical disks as a solution to prevent the quick formation of bars, despite the awareness by skilled observers such as de Vaucouleurs that bars are frequent. Subsequent higher resolution and infrared observations revealed that bars are in fact even more frequent and found in a majority of spirals, as reminded in several papers at this conference.

The reverse problem was thus discussed many times: how bars and hot dark halos can coexist, since dark halos were then no longer viewed as hypothetical. It was also understood over the years that barless disks can exist when the central density profile is too centrally concentrated.

Spiral density waves are a more general but less robust version of the bar phenomenon. The main reason for spiral formation is now well understood as resulting from the non-linear growth of a spontaneous gravitational instability in the disk plane with the same origin as for bars: a kinematically too cold disk is gravitationally unstable, and the non-linear result is typically a bar in the initial flat core and spiral arms in the outer differentially rotating region. Without invoking more than Newtonian physics it was also found that the typical double exponential disks profiles are also a natural asymptotic state for a collisionless disk passing through a bar instability (Pfenniger & Friedli, 1991).

The non-linear structures resulting from the gravitational instability are never strictly stationary, but evolve secularly (over several rotational periods). In pure collisionless disks they tend to destroy the spiral arms and later bars, so obviously something must regenerate them in real galaxies.

The theory of bars and spirals is presently incomplete because the full self-consistent problem is very non-linear. Thus no analytical theory able yet to *predict* the full development of strongly rotating collisionless self-gravitating disks. Only the brute force N-body simulations are able to do it.

Since the Newtonian physics involved in these N-body experiments is very well understood, and that the numerical codes can to a large extent be trusted because different versions implementing different techniques have been developed and checked over several decades by many groups, we can use these N-body techniques to predict and explain the behaviours of galaxies. The results of such N-body simulations should be taken as seriously as analytical developments in celestial mechanics. As example of success of N-body techniques is the prediction that bars may evolve into peanut-shaped structures. This was first empirically found in N-body experiments (Combes & Sanders, 1981; Combes et al., 1990), understood theoretically (Pfenniger & Friedli, 1991), and later confirmed by observations (e.g., Bureau & Freeman, 1999).

Coupled to this well understood underlying physics, numerous independent studies of the mass to light ratio in the optical parts of spirals have determined that a substantial fraction of the gravitating mass there is well explained by the detected baryons (e.g. Sancisi, 2003). This ensures that we have at least a basic physical understanding of thc inner parts of galaxies.

3. The role of gas

Since disk galaxies as star producing systems must contain also a lot of gas, its effects must be considered on the long run. First, dust polluted gas is very efficient to lose its thermal energy by infrared radiation, so galaxies must be seen, besides in first approximation as rotating self-gravitating objects, in second approximation also as energy dissipating structures. Gravitationally bound rotating structures slowly losing energy tend to rapidly converge towards thin disks because then angular momentum is a quantity much harder to dissipate away than energy.

So the frequent occurrence of spiral arms (after all disk galaxies are called spirals!) follows directly from the constant competition between the effect of cooling driving disks towards the gravitational Safronov-Toomre instability threshold. Any further cooling leads to strong reaction from the disk by dynamical heating. As long as gas cooling continues to be efficient the natural long term state of spiral galaxies is therefore to stay close to the marginal stability threshold.

Numerous studies show that galactic disks, including the Milky Way, have disks close to a marginal stability state with a Safronov-Toomre parameter Q close to unity. Because of this marginal state, spiral galaxies do react strongly to other perturbations such as galaxy interactions by amplifying spiral arms.

Many studies have been trying to determine whether spirals and bars result either from galaxy interactions or from a proper disk instability. The more fundamental cause of spirals and bars is actually the internal marginal stability state making galactic disks very reactive to various perturbations. Even small satellite interactions trigger large responses from a disk in the form of grand design spirals. The name of "spirals" for disk galaxies is in the end an excellent way to characterize their close to marginal stability state, showing both that dissipation acts and dynamics reacts.

A corollary of such a marginally stable state is that a steady state is unlikely. Instead evolution is to be expected as long as the marginal stability state is maintained by the competing factors, gas cooling against dynamical heating.

As by-product the large scale dynamical instability of galactic disks leads to local interstellar gas compression, shocks and turbulence, cascading down to smaller scale gas instabilities (e.g., Fleck, 1981; Elmegreen, 2004). At the bottom of the cascade the most visible effect of the gas "turbulence" is star formation (Klessen, 2004), which implies gas consumption. The long term effects of large scale instabilities is to transform progressively the dissipative component into a collisionless stellar component. By consuming gas the cooling agent becomes rarer, and by forming stars the dynamical heating more effective in counter-balancing gas cooling. In addition the mechanical energy output produced especially by massive stars provides a second important source of heating competing gas cooling.

4. Constraints on dark matter forms

So the slow transformation of matter from gas rich, but also dark matter rich disks to gas poor, star rich and dark matter poor structures already indicates that the above picture is broadly consistent. Gas poor disk galaxies (S0's, Sa's) have namely typically less prominent and open spiral arms in more symmetric disk, while gas rich spirals (Sd's, Sc's) have large open spirals in irregular disks. The fact that along the spiral sequence the visible gas represents always a minor fraction of the mass indicates that some of the dark mass must be gas in order to be able to form subsequently all the stars that are seen in S0's and Sa's (Pfenniger et al., 1994).

However the fact that S0's and Sa's still contain a fraction of dark mass while showing very little star formation indicates also that some of the dark mass is in a form that cannot easily form stars. Around 40% of the total mass within the HI disk radius might be in a dark collisionless form. Therefore the above considerations show already that we can have a consistent dynamical picture of disk galaxies including the gas and star formation aspects provided that *two* forms of dark matter exist: one, close to the visible gaseous form for explaining the properties of the spiral sequences as an evolutive sequence of dissipative gravitating disks, and one non-gaseous form for explaining the remaining "indestructible" dark mass in the evolved part of the sequence, the S0's and Sa's.

5. The role of warps

All these considerations have been made considering the plane dynamics of spiral galaxies, except for the bulge growth via vertical instabilities in the inner stellar disk.

But what about the dynamical effects transverse to the disks in the outer regions? Namely, a notorious puzzle in spiral galaxies is the ubiquitous warp phenomenon which has eluded a clear explanation up to now. For instance warps are unlikely to result from resonant normal modes, because the soft edges of galaxy disks damp discrete modes (Hunter & Toomre, 1969). Normal modes generated by massive inclined dark halo (Dekel & Shlosman, 1983; Sparke, 1984; Sparke & Casertano, 1988) are ruled out by dynamical friction that damp the warp in a few dynamical times (Dubinski & Kuijken, 1995). Only particular triaxial halos can produce a torque that leads to a warp with a straight line of node and negligible back reaction (Petrou, 1980). Interactions are efficient in warping disks (Hernquist, 1991; Huang & Carlberg, 1997), however they cannot be invoked in isolated warped galaxies.

Warps are especially obvious in the HI outer disks, but to a lesser amplitude the stellar disks are also warped. Statistics of warps in HI (Bosma, 1991; Richter & Sancisi, 1994; Garcia-Ruiz et al., 1998) and in the optical band (Reshetnikov & Combes, 1998, 1999; Sanchez-Saavedra et al., 1990; Sanchez-Saavedra et al., 2003) reveal that more than half the spiral galaxies are warped and asymmetric. Warps are also linked to large scale disk horizontal asymmetries, both signatures showing that the outer spiral disks are not as well virialized as the inner optical disks.

To answer the question about the general cause of warps, we have first tried to well understand the dynamics of ideal isolated and purely self-gravitating

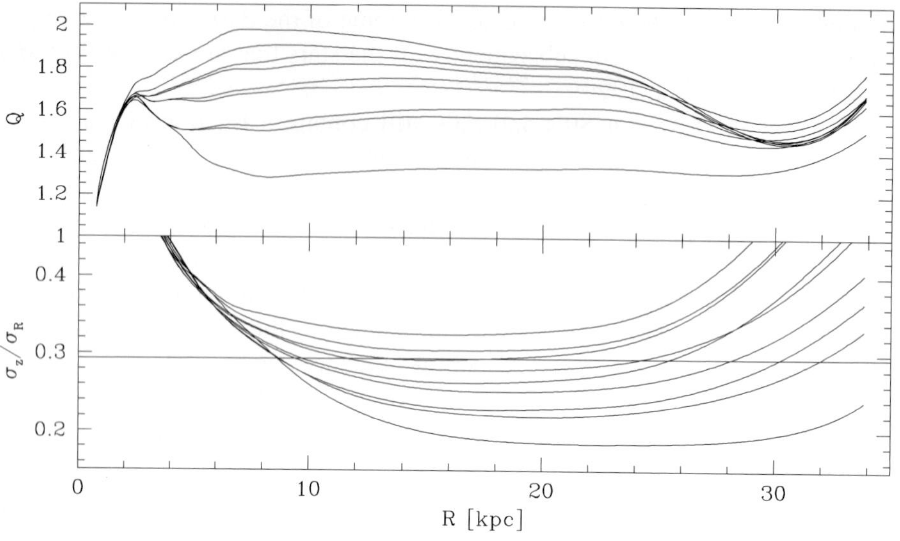

Figure 1. The radial stability parameter Q (top) and the ratio σ_z/σ_R (bottom) as a function of the galactic radius for different models. In both graphics, the curves correspond, from bottom to top, in the interval $R = 10 - 20$, to the models of increasing thickness, respectively.

disks of collisionless particles. In a second step we will introduce energy dissipation, since disks form for the single reason that the energy dissipation rate is much faster than the angular momentum transport rate. Therefore energy dissipation must be taken as the second most important factor in understanding galaxies, after the pure gravitational dynamics of collisionless matter.

Consequently we have undertaken first to study in detail massive self-gravitating disks with various degrees of flattening by means of N-body simulations (Revaz & Pfenniger, 2004). The simulated disks are made of a stellar bulge and an exponential disk components, and a collisionless heavy disk component proportional to the HI disk, including a density depression in the optical disk, and a flaring thickness almost proportional to radius. The Milky Way is the template galaxy for guiding the choice of the various mass ratios and scale lengths. The mass components and profiles are also such that an almost flat rotation curve are obtained for any thickness of the heavy disk component. By solving the Jeans equations separately for each mass component, we can start simulations with an almost equilibrium model, but with various degrees of velocity dispersion ratios $\sigma_r/\sigma_z(R)$, while keeping an initial Toomre parameter Q well above 1 on almost the full radial range (see Fig. 1).

The main result is that conformally to predictions made long ago by Toomre (1966) and Araki (1985), too flat disks are unstable with respect to bending instabilities. The instability in thin sheets is just related to the velocity dispersion

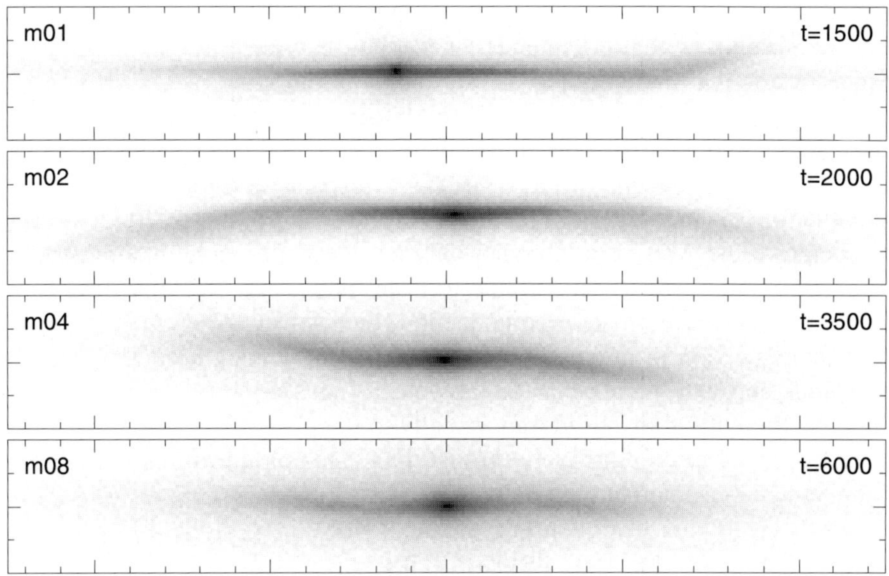

Figure 2. Edge-on projections of different heavy disk models having developed spontaneously a warp. The box dimensions are $100 \times 16\,\mathrm{kpc}^2$

ratios between the vertical and radial velocity dispersions σ_z and σ_R. When $\sigma_z/\sigma_R < 0.293$ the sheet bends spontaneously with growth rates of order of Gyr. In thin disks this translates to first S-shaped warp growing modes for slightly unstable disks, and secondly for strongly unstable disks to U-shaped warp modes persisting for Gyrs (see Fig. 2).

If, like bars and spiral arms, warps results of internal disk instabilities, not only we obtain a unified picture of galaxies, but also several new clues about the dark matter nature and its distribution. For the same fundamental cause, the marginal stability of self-gravitating disks subject to an energy dissipation, disk produce spontaneously bars, spirals and warps that counteract dissipation by mechanical heating.

In order to be in such a warped state self-gravitating disks must first obviously have a dissipative component. Dusty gas can be identified as the primary cause of energy losses. Second, the mass distribution must be sufficiently self-gravitating and thin in order to reach the Araki stability threshold. This provides an interesting constraint for the gravitating mass.

The Araki criterion immediately tells us that a disk made of classical smooth gas would never be transversally unstable because the gas pressure would be isotropic. But a classical gaseous disk dissipating its heat decreases correspondingly its pressure, and inevitably after some time reaches the Safronov-Toomre radial instability threshold, at which point the subsequent non-linear evolution depends on the detailed microscopic physics of the gas. In the galaxy disk case we know that the interstellar medium is widely non-homogeneous, which means that the isotropic pressure assumption is not necessarily valid. We also know well the most visible result of the gas instability, star formation. Once star form they evolve as collisionless matter, and disks with such fluids are well known to evolve towards anisotropic dispersions with σ_z/σ_R ratios of the order of 0.5 (Pfenniger & Friedli, 1991; Huber & Pfenniger, 2001).

Therefore the Araki criterion can only be met with a combination of collisionless matter property to be able to have rather large velocity dispersion anisotropy, and a dissipative component, which lowers over time faster the velocity dispersion in z than in R. The dynamical heating produced radially by spirals and bars increases also the velocity anisotropy if the radial heating is inefficiently transferred transversally to the plane. The radial dynamics of disks indeed heats effectively through bars and spiral arms essentially the radial kinematics, maintaining it above $Q \sim 1$.

6. The two types of dark matter

So the ubiquitous existence of warps in disk galaxies is a strong hint that they are sufficiently massive and thin to be transversally unstable to warp modes, preferentially S-shaped modes. U-shaped modes are also possible if a disk is driven sufficiently deep below Araki's threshold. If this is the case then we must have a substantial mass component almost as thin as HI-disks that behaves as collisionless matter for several rotational periods. In order to regenerate warps, dissipation is essential, without it a too anisotropic disk heats dynamically until a stable thicker state is reached.

If disks react to bending instabilities by kinematical heating transverse to the disk, then one must expect that warped disks are maintained close to the marginal state balancing gas dissipation with dynamical heating.

Then the question is whether such warps may constrain the traditional thick and hot dark halos made of collisionless matter. By adding a corresponding potentials to the initial N-body models, we have calculated up to which halo mass with given flattening a marginal state to bending would be kept (Revaz & Pfenniger, 2004). It turns out that the effective disk thickness provides a strong constraint on the dark halo mass, but almost no constraint on its flattening. In all studied cases the exact halo flattening is very little constrained by the marginal stability state above a density flattening around $0.3 - 0.5$, which is anyway the range usually considered in cosmological simulations. In contrast, the relative mass of a hot dark halo within a radius comparable to the HI disk radius is directly related to the precise massive disk thickness: the thicker the marginally unstable disk is, the lesser mass can be contained in hot spheroidal thick halo assuming that the warp results from a bending instability. For models parameters fitted to the Milky Way, the halo is at most as heavy as the disk.

Since we can estimate the HI disk thickness, we can give a constraint of the dark halo mass if the disk dark matter has a thickness similar to the HI. The Milky Way has a known HI thickness and a known warp, therefore if this warp results from a disk marginal stability state then the dark halo mass within a radius similar to the HI disk radius ($\sim 30\,\mathrm{kpc}$) is constrained to be below 0.4. This value is similar to the dark matter fraction found in evolved Sa's, S0's.

7. Conclusions

By realizing that bars, spirals and warps are different effects resulting from that same fundamental causes, a slightly energy dissipative gas component acting secularly on a mostly collisionless rotating self-gravitating system, we obtain new clues about dark matter.

First confirming several studies about the horizontal dynamics of galaxies, we arrive at the conclusion that spiral disks are to first order self-gravitating and collisionless, they must contain more matter than seen. This matter must be weakly collisional in order to develop anisotropic velocity dispersions in R and z. It is clear that the radial velocity dispersion is rapidly regulated by the radial dynamical instabilities, so the anisotropy increases in z through a dissipative component, identified with the dusty gas.

Dissipation must be sufficiently effective in order to maintain the disks close to instability, as witnessed by the spirals and warps, but also most of the mass cannot be strongly collisional, otherwise the velocity dispersion would be isotropic and no bending instability would occur. The physical solution that we are investigating is close to the clumpuscule model in Pfenniger & Combes (1994), where much of the mass is condensed in the form of cold, dense planet-mass molecular hydrogen clumps, stabilized in their core by a solid or liquid phase of molecular hydrogen (Pfenniger, 2004ab; Revaz & Pfenniger, 2004b).

The warped N-body models do not rule out traditional massive thick halos with extended core. However the mass of the massive halo can hardly exceed the massive disk mass if this disk possesses a warp produced by a bending instability. The models do not constrain well the dark halo axis ratio, provided it is above ~ 0.3.

Thus we arrive at a bi-modal solution for dark matter in galaxies that seems to satisfy all the known constraints, from observational to galaxy dynamics and evolution constraints, and to cosmological constraints. Of course the nature of the dark matters remain to be better understood and discovered by observations. Encouragingly, tiny clumps of molecular hydrogen have been recently detected (Heithausen, 2004), but the nature of the hot, non cuspy dark halos remain to be found.

Acknowledgments

This work has been supported by the Swiss National Science Foundation.

References

Araki, S. 1985, Ph.D. Thesis, Massachusetts Inst. Technology

Bosma, A. 1991, in : Warped disks and inclined rings around galaxies, Casertano S., Sackett P., Briggs F.H. (eds.), Cambridge University Press, 181

Bureau, M., Freeman, K.C. 1999, AJ, 118, 126

Combes F., Debbasch F., Friedli D., Pfenniger D. 1990, A&A 233, 82

Combes F., Sanders R.H. 1981, A&A 96, 164

Dekel A., Shlosman I. 1983, in : Internal Kinematics and Dynamics of Galaxies, E. Athanassoula (ed.), IAU 100, 177

Dubinski, J., Kuijken, K. 1995, ApJ, 442, 492

Elmegreen, B. 2004, astro-ph/0405555

Fleck R.C. 1981, APJ 246, L151

Garcia-Ruiz I., Kuijken K., Dubinski K. 1998, in : Galactic Halos: A UC Santa Cruz Workshop, D. Zaritsky (ed.), ASP Conference Series, 136, 385

Hernquist L. 1991, in : Warped disks and inclined rings around galaxies, Casertano S., Sackett P., Briggs F.H. (eds.), Cambridge University Press

Heithausen, A. 2004, ApJ, 606, L13

Huang, S., Carlberg, R.G. 1997, ApJ, 480, 503

Huber, D., Pfenniger, D. 2001, A&A, 374, 465

Hunter, C., Toomre, A. 1969, ApJ, 155, 747

Klessen, R.S. 2004, astro-ph/0402673

Ostriker, J.P., Peebles, P.J.E. 1973, ApJ 186, 467

Petrou, M. 1980, MNRAS, 191, 767

Pfenniger, D. 2004a, in preparation

Pfenniger, D. 2004b, in The Dense Interstellar Medium in Galaxies, S. Pfalzner et al. (eds.) Springer, p. 409

Pfenniger, D., Combes, F. 1994, A&A, 285, 91

Pfenniger, D., Combes, F., Martinet, L. 1994, A&A, 285, 79

Pfenniger, D., Friedli, D. 1991, A&A, 252, 7

Reshetnikov, V., Combes, F. 1998, A&A, 337, 9

Reshetnikov, V., Combes, F. 1999, A&AS, 138, 101

Revaz, Y., Pfenniger, D. 2004a, A&A in press

Revaz, Y., Pfenniger, D. 2004b, A&A in preparation

Richter, O.G., Sancisi, R. 1994, A&A 290, L9-L12

Sanchez-Saavedra, M.L., Battaner, E., Florido, E. 1990, MNRAS, 246, 458

Sanchez-Saavedra, M.L., Battaner, E., Guijarro, A., et. al. 2003, A&A, 399, 457

Sancisi, R. 2003, at IAU Symposium 220 "Dark Matter in Galaxies"

Sparke, L.S. 1984, MNRAS, 211, 911

Sparke, L., Casertano, S. 1988, MNRAS, 234, 873

Toomre, A. 1966, Geophys. Fluid Dyn., 46, 111

ON THE GENERATION OF THE HUBBLE SEQUENCE THROUGH AN INTERNAL SECULAR DYNAMICAL PROCESS

Xiaolei Zhang
US Naval Research Laboratory, 4555 Overlook Ave. SW, Washington, DC 20375, USA

Abstract The secular evolution process, which slowly transforms the morphology of a galaxy over its lifetime, could naturally account for observed properties of the great majority of physical galaxies if both stellar and gaseous accretion processes are taken into account. As an emerging paradigm for galaxy evolution, its dynamical foundation had been established in the past few years, and its observational consequences are yet to be fully explored. The secular evolution picture provides a coherent framework for understanding the extraordinary regularity and the systematic variation of galaxy properties along the Hubble sequence.

Keywords: Generation of the Hubble Sequence, Secular evolution of galaxies

1. Introduction

The dominant view over the past few decades has been that the structural properties of galaxies remain largely unchanged unless galaxies were perturbed by violent events such as mergers (Toomre & Toomre 1972). This view is particularly favored by the currently popular hierarchical clustering/cold dark matter (CDM) paradigm of structure formation and evolution (Peebles 1993 and the references therein). In the early 1980s, photometric and kinematic evidence in the buldges of late-type galaxies, and hints from N-body simulations of barred galaxies which incorporated a dissipative gas component, prompted several investigators to speculate that a fraction of late-type bulges might be formed from gas accretion under the bar potential (Kormendy 1982; Combes & Sander 1981). These early observations and speculations have since been further developed into one version of the secular evolution scenario, which emphasizes the role of dissipative gas accretion in the formation of the so-called "pseudo bulges" in late type galaxies (Kormendy & Kennicutt 2004 and the references therein).

Even though the role of mergers during the early phases of galaxy assembly is still to be assessed, growing evidence has shown that at least since $z \sim 1$

D. Block et al. (eds.), Penetrating Bars through Masks of Cosmic Dust, 317–327.

the rate of merger appears to have been significantly reduced (Conselice et al. 2003), and during subsequent time the significance of merger in transforming galaxy morphology is likely to be overshadowed by the slower secular evolution process (Kormendy & Kennicutt 2004).

However, secular evolution involving gas alone under barred potential leads to some apparent paradoxes. First of all, as emphasized by Andredakis and coworkers (Andredakis, Peletier, & Balcells 1995), the continuity of the galaxy properties across the entire Hubble sequence, highlighted for example by the continued variation of Sersic index n in fitting the bulge surface density profile, indicates that there is not an apparent break in the formation mechanism between the late-type disk galaxies and mid-to-early-type disk galaxies. However, due to the paucity of gas compared to stars in most galaxies, manifesting as a gaseous-to-stellar mass ratio of 1/10 or less (which is true even for high-redshift late-type disk galaxies, presumably a result of rapid star formation following the accretion shocks of dissipative disk formation, D. Sanders 2002 private communication; F. Combes 2004 private communication), there simply is not enough of a reservoir of gas for use to build up the bulges of intermediate Hubble type galaxies such as our own, which has a bulge mass comparable to the disk mass, not to say for galaxies of even earlier Hubble types. Secondly, bulges of intermediate to early type galaxies, including our own, consists mostly of stars of very old age (Jablonka, Gorgas & Goudfrooij 2002), and could not have been built up by the secular accretion of gas over a Hubble time which subsequently formed stars. Even though Kormendy and Kennicutt (2004) argued that some of these bulges may be termed "classical", i.e., they could have formed out of the dissipative collapse at an early stage of galaxy formation, they nonetheless pointed out that a good fraction of these apparently old bulges have stellar kinematics which were rotation-dominated, and are related to the kinematics of their disks, hinting at a secular evolution origin for their formation.

These and other apparent paradoxes, such as the existence in some galaxies of dense core in the central region which appears to be formed by radial gas accretion and yet is kinematically decoupled from the bulge stars, can be naturally resolved if we allow also the possibility of stellar accretion. Such a possibility was not explored in the past decades mainly due to a theoretical barrier, i.e., the well-known result that the stellar motion in a galaxy containing quasi-stationary non-axisymmetric patterns conserves the Jacobi integral in the rotating frame of the pattern (Binney & Tremaine 1987), and the orbital motion of stars under such potentials generally will not exhibit secular delay or increase, since there is no wave and disk star interaction except at the wave-particle resonances which are usually localized for quasi-steady wave patterns (Lynden-Bell & Kalnajs 1972).

2. Physical Mechanisms for Producing the Secular Morphological Evolution of Galaxies

It was first demonstrated in Zhang (1996, 1998, 1999) that secular orbital changes of stars across the entire galaxy disk are in fact possible due to the collective instabilities induced by the unstable density wave modes such spirals, bars, as well as by the skewed 3D density distributions reflected in the twisted isophotes of many high-redshift galaxies. These skewed global patterns were shown to lead to a secular energy and angular momentum exchange process between the disk matter and the wave pattern, mediated by a local gravitational instability, or a collisionless shock, at the potential minimum of the pattern (Zhang 1996). The integral manifestation of this process is an azimuthal phase shift between the potential and density spirals, which results in a secular torque action between the wave pattern and the underlying disk matter. As a result of the torquing of the wave on the disk matter, the matter inside the corotation radius loses angular momentum to the wave secularly and sinks inward. The wave carries the angular momentum it receives from the inner disk matter to the outer disk and deposits it there, causing the matter in the outer disk to drift further out.

Closed form evolution rates can be derived for this process. In the quasi steady state, the rate of angular momentum exchange between a skewed density wave pattern (spiral, bar, etc.) and the basic state of the disk, per unit area, is given by (Zhang 1996, 1998)

$$\overline{\frac{dL}{dt}}(r) = -\frac{1}{2\pi}\int_0^{2\pi} \Sigma_1(r,\phi)\frac{\partial \mathcal{V}_1(r,\phi)}{\partial \phi}d\phi, \tag{1}$$

which, for two sinusoidal waveforms, is given by

$$\overline{\frac{dL}{dt}}(r) = (m/2)A_\Sigma A_\mathcal{V} sin(m\phi_0), \tag{2}$$

where A_Σ and $A_\mathcal{V}$ are the amplitudes of the density and potential waves, respectively, and ϕ_0 is the phase shift between these two waveforms.

The orbital decay rate it induced on a disk star (or a gas clump) can be derived from equation (2) as

$$\frac{dr}{dt} = -\frac{1}{2}F^2 v_0 \tan(i)\sin(m\phi_0), \tag{3}$$

where F is the fractional wave amplitude (which is the geometric average of the fractional density and potential wave amplitude), v_0 is the circular velocity of the star, i is the pitch angle of the spiral. m is the number of spiral arms, respectively.

From the above expressions we can easily see why the skewness of the pattern is needed: It is the skewness of the mass distribution which leads to the potential/density phaseshift through the Poisson equation (Zhang 1996). Without phaseshift, there will be no secular angular momentum exchange between the wave and the disk matter at the quasi-steady state, and thus no secular mass redistribution. A perfect oval or triaxial mass distribution without any skewness will not be the direct driver for secular evolution.

For our own Galaxy, if we assume that over the past Hubble time the Galactic spiral pattern has an average 20% amplitude and 20 degree pitch angle, the orbital delay rate can be calculated using equation (3) to be about 2 kpc of orbital decay per Hubble time, Therefore, a star in the Sun's orbit will not make it all the way in to the inner Galaxy in a Hubble time. However, the corresponding mass accretion rate across any Galactic radius inside corotation is about $6 \times 10^9 M_\odot$ per Hubble time. A substantial fraction of the Galactic Bulge can thus be built up in a Hubble time.

Another important consequence of spiral-induced wave-basic state interaction is the secular heating of the disk stars. Since a spiral density wave can only gain energy and angular momentum in proportion to the pattern speed Ω_p, and a disk star loses its orbital energy and angular momentum in proportion to its circular speed Ω, an average star cannot lose the orbital energy entirely to the wave, the excess energy serves to heat the star. The resulting orbital velocity diffusion rate is given by

$$D = (\Omega - \Omega_p) F^2 v_c^2 \tan(i) \sin(m\phi_0). \tag{4}$$

Using parameters appropriate for the Galaxy, this can quantitatively explain the observed age-velocity dispersion relation of the solar neighborhood stars (Zhang 1999; Wielen 1977). This secular heating process, since it originates from a local gravitational instability at the arms, serves to increase all three space velocities of stars. The vertical component of the velocity dispersion allows the stars to gradually drift out of the galactic plane as they spiral inward to become bulge stars. The corresponding energy injection into the interstellar medium, if used as the top-level energy input for turbulence cascade, can quantitatively account for the size-linewidth relation of the Galactic molecular clouds (Zhang et al. 2001; Zhang 2002).

The secular morphological evolution process leads to the Hubble type of an average galaxy to evolve from the late to the early (Zhang 1999). The secular evolution speed is expected to be faster for cluster galaxies than for isolated field galaxies, which were observed to evolve slowly, because the rate of secular evolution is proportional to the wave amplitude squared and pattern pitch angle squared (see, e.g., equations 3, 4), and cluster galaxies are found to have large amplitude and open spiral patterns excited through the tidal interactions with neighboring galaxies and with the cluster potential as a whole. This naturally accounts for the morphological Butcher-Oemler effect.

The secular evolution of galaxies through the internal process mediated by skewed non-axisymmetric patterns is a rigorous consequence of the Newtonian dynamics and the assumption of global self-consistency. The analytical equations for the evolution rates (equations 3 and 4) are quantitatively confirmed in N-body simulations (Zhang 1998). Detailed comparison of the predictions of the secular evolution theory with the observed galaxy properties and with other theories of galaxy evolution can be found in (Zhang 2003). In what follows, we will highlight the role of secular evolution in generating the structural properties and scaling relations of galaxies.

3. Generation of the Structural Properties of Galaxies Along the Hubble Sequence through Secular Evolution

The properties of disk galaxies vary systematically along the Hubble sequence from the late to the early Hubble types. An important descriptor of the structural characteristics of disk galaxies is the rotation curve. After assembling hundreds of observed rotation curves for nearby disk galaxies, Persic, Salucci, & Stel (1996) found that these rotation curves fall onto a two-dimensional surface in the three-parameter space of normalized galactic radius, velocity, and the absolute magnitude. Since for the nearby galaxies the variation of the absolute magnitude corresponds to the variation of Hubble type (i.e., bigger and brighter galaxies generally have earlier Hubble types), this means that the rotation curve shapes also vary systematically for galaxies of varying Hubble types. In Figure 1, we plot three typical rotation curves generated using the data from Persic et al. (1996).

Using dynamical relations for equilibrium axisymmetric/spherically symmetric mass distributions, i.e. $V(R)^2 = \frac{GM_{dyn}(R)}{R}$, as well as $M_{dyn}(R) = \int_0^R \Sigma_{proj}(r) 2\pi r dr = \int_0^R \rho(r) 4\pi r^2 dr$, where $M_{dyn}(R)$ is the dynamical or total mass within a galactic radius R (including the contributions from both the luminous and the dark components), and $\Sigma_{proj}(R)$ is the total projected surface density (disk plus spheroidal) at radius R, we can infer the underlying mass and surface density distributions corresponding to the different regimes for each type of rotation curves (c.f. Figure 1).

(I) Steeply rising solid body rotation curve, as observed for the inner regions of early type galaxies.

This regime is characterized by solid body rotation ($\Omega = \Omega_0$ = constant, or a nearly linear rise of rotation velocity $V(R) = \Omega_0 R$). Using $V(R)^2 = \Omega_0^2 R^2 = GM_{dyn}/R$, we obtain $M_{dyn} = \frac{\Omega_0 R^3}{G}$, which implies a constant volume density $\rho = \rho_0 = \frac{\Omega_0}{G}\frac{3}{4\pi}$. Therefore the projected surface density can be obtained by equating $M_{dyn}(R) = \int_0^R \Sigma_{proj}(r) 2\pi r dr$ and $M_{dyn}(R) = \frac{3}{4}\pi R^3 \rho_0$, which gives $\Sigma_{proj}(R) = 2\rho_0 R$, i.e. the solid-body regime corresponds to a constant column density ρ_0 and linearly rising projected surface density $\Sigma_{proj}(R) \propto R$.

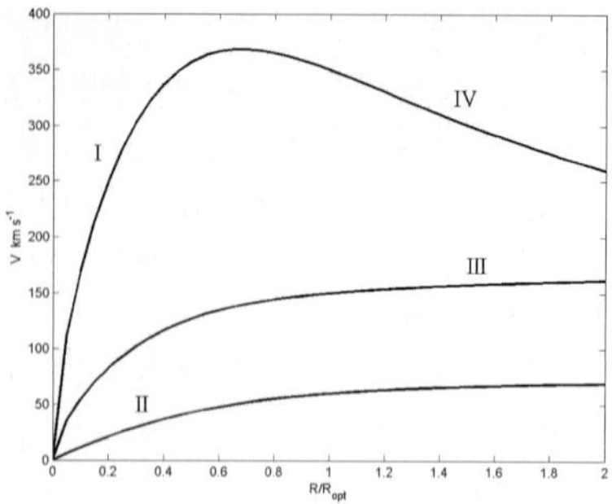

Figure 1. Typical rotation curves for disk galaxies of varying absolute magnitudes, as well as four characteristic regimes which are analyzed in the main text. Top: M=18; middle: M=21; bottom: M=24, where M is the absolute magnitude of the representative galaxy (data from Persic et al [1996]).

(II) Gently rising rotation curve, as representative of the very late type disks, including most of the dwarf galaxies and low-surface-brightness (LSB) disk galaxies.

This regime can be shown to correspond to approximately constant projected surface density, as found to be the case for many LSB disks. Assume $\Sigma_{proj}(R) = \Sigma_0$ = constant, we obtain $M_{dyn}(R) = \int_0^R \Sigma_0 2\pi r dr = \Sigma_0 \pi R^2$, which we equate to $V(R)^2 R/G$ to obtain $V(R)^2 = \sqrt{\Sigma_0 \pi R G}$, which indeed corresponds to a gently rising rotation curve.

Instead of constant projected surface density, many LSBs are found to have instead constant volume density (de Blok 2003). These two views (constant projected surface density, or constant mass volume density) can be consistent if the scale height in the central regions of these LSBs are nearly constant, as for a pancake-type proto galaxy cloud collapse.

(III) Flat rotation curve, as is representative of the outer part of most Sb/Sc galaxies.

In this regime, the projected surface density profile should be close to $1/R$, as shown below. Assume $\Sigma_{proj}(R) = \frac{\Sigma_0 R_0}{R}$, where $\Sigma_0 = \Sigma_{proj}(R = R_0)$, we have $M_{dyn}(R) = \int_0^R \Sigma_{proj}(r) 2\pi r dr = \Sigma_0 R_0 2\pi R$. Using again $V(R)^2 = GM_{dyn}/R$, we obtain $V(R)^2 = G\Sigma_0 R_0 2\pi R/R$, or $V(R) = \sqrt{G\Sigma_0 R_0 2\pi}$ = constant. Therefore the flat rotation curve regime corresponds to 1/R projected surface density distribution.

(IV) The falling rotation curve regime, as representative of the outer part of the early type galaxies.

In this regime the projected surface density $\Sigma_{proj}(R)$ falls off faster than 1/R, and can be up to $\Sigma_{proj} = 0$ (Keplerian). For the extreme case of Keplerian rotation curve due to a concentrated central mass distribution, $\Sigma_{proj}(R_{outer}) \approx 0$, and $V(R)^2 = \frac{GM_{dyn}}{R}$, or $V(R) = \sqrt{\frac{GM_{dyn}}{R}}$, which is indeed falling.

In the secular evolution scenario a typical galaxy starts its life with properties similar to that of an LSB galaxy. The high-z correspondence of the local LSBs appears to be the damped L_α system (Wolfe 2001), which also possesses disk-like morphology with large scale length, and a constituting mass component which is gas rich and metal poor.

As the proto galaxy disk contracts and condenses, the global spiral instability develops, which begins the mass-redistribution process that transforms the flat surface density distribution to a more and more centrally concentrated mass distribution. The intermediate stage of this evolution resembles Freeman's type II disks (Freeman 1970), and the later stage with a a 1/R projected surface density resembles a Freeman's type I disk.

As the evolution progresses, the inward-accreted disk stars rise out of the disk plane to become bulge stars. The thermalized three-dimensional bulge mass will have nearly constant volume density and undergoes solid-body rotation. The outer part of the visible disk starts to be drained of matter, producing the falling rotation curve.

This entire evolution sequence therefore corresponds to a galaxy evoling from the lower rotation curve to the upper rotation curve in Figure 1.

The twisted isophotes often found for the Hubble Deep Field galaxies are expected to evolve into the observed variation of major axis with radius for elliptical galaxies (Faber & Jackson 1976), possibly bypassing the disk formation phase, at least for some galaxies. Like the spiral structure, the twisted isophotes are results of dissipation and differential rotation, and just as in the case of spiral structure, these three-dimensional skewed structures will also lead to a phaseshift between the potential and density distributions, resulting in accelerated secular morphological changes.

4. Generation and Evolution of the Galaxy Scaling Relations Along the Hubble Sequence

Galaxy scaling relations (Faber & Jackson 1976; Tully & Fisher 1977; Djorgovski & Davis 1987; Dressler et al. 1987) own their existence first and foremost to the fact that galaxies are equilibrium configurations and satisfy the Virial theorem relation (see, e.g., Binney & Tremaine 1987). This relation dictates that during any quasi-equilibrium evolution process of a gravitational system, $-E \sim T \sim -V/2$ where T is the kinetic energy, V is the potential energy, and $E = T + V$ is the total energy of the system. The fact that spiral galaxies can remain on the Tully-Fisher relation during secular morphological evolution despite the fact that stars can't radiate about 1/2 of the converted potential energy away, is due to the fact that spiral density wave actually carries about 1/2 of the dissipated orbital energy (equal to potential energy) away to the outer disk to be deposited there (Zhang 1999). The existence of a collective dissipation process at the spiral arms also made possible the regulated conversion of orbital velocity to random velocity, so a single kinetic energy (the half that is not being carried away) can enter the Virial relation no matter how it is apportioned between the circular and the random motions.

According to the Virial theorem (we set G=1 in the following derivations of scaling relations) $M_{dyn} = V_e^2 R_e$, where M_{dyn} is the dynamical mass of a galaxy, V_e is the effective velocity spread which corresponds to the total velocity dispersion for an elliptical galaxy or the maximum rotation velocity for a spiral galaxy, and R_e is an effective radius which makes the Virial relation hold.

Define the mean surface brightness of a galaxy SB as $SB \equiv \frac{L}{R_e^2}$ where L is the luminosity of the galaxy, and the dynamical mass-to-light ratio M_{dyn}/L can be written as $M_{dyn} \equiv (M_{dyn}/L) * L$. Therefore it follows that

$$(M_{dyn}/L)^2 * L^2 = M_{dyn}^2 = V_e^4 R^2 = V_e^4 \frac{L}{SB}, \tag{5}$$

or

$$(M_{dyn}/L)^2 L = \frac{V_e^4}{SB}. \tag{6}$$

Therefore we finally have from

$$L = V_e^4 \frac{1}{SB} \frac{1}{(M_{dyn}/L)^2}. \tag{7}$$

We thus see that in order to obtain the classical Tully-Fisher relation, which has $L \propto V_e^4$, we must have $SB \cdot (M_{dyn}/L) \sim$ constant. This can be accomplished in two ways: to have SB and M_{dyn}/L each being constant; or, to have the variations of the two factors offset each other during an evolution process.

Traditionally, the former was assumed to be the case (see, e.g. the discussions and references in Shu 1982). It is now known that both these quantities vary considerably for galaxies along the Hubble sequence. The surface brightness is found to be higher for earlier type galaxies (McGaugh & de Blok 1998 and the references therein), whereas the dynamical mass-to-light is found to be lower for the earlier Hubble types (Zwaan et al. 1995; Bell & de Jong 2001). The observed opposing trends of variation of surface brightness and dynamical mass-to-light ratio is naturally explained as the outcome of the secular baryonic mass accretion and the increase fraction of luminous baryon mass, especially towards the central region of a galaxy as its Huuble type evolve from late to early. The Tully-Fisher relation can be maintained as long as the increase in surface brightness is accompanied by a corresponding decrease in the dynamical mass to light ratio during the secular evolution process.

In the secular evolution scenario, since most of the elliptical galaxies (i.e. those lower luminosity, so-called disky ellipticals) are the end results of evolution from the initial condition of disks, we speculate that spiral galaxies may satisfy similar kind of fundamental plane relation just as ellipticals. This has been found to be so (see, e.g., Pharasyn, Simien, & Heraudeau 1997). The fundamental plane relation for spirals can likewise be derived from the Virial theorem. Starting from

$$V_e^2 = G\frac{M_{dyn}}{R_e}, \tag{8}$$

we have

$$V_e^2 = G\frac{M_{dyn}}{L}\frac{L}{R_e} = L^{1/2}(\frac{L}{R_e^2})^{1/2}\frac{M_{dyn}}{L}, \tag{9}$$

which leads to

$$10\log V_e = -(1+2\beta)M_t - \mu_e + constant, \tag{10}$$

where M_t is the absolute magnitude and μ_e is the face-on average surface brightness in mag/arcsec^2, and where we have assumed M$_{dyn}$/L $\propto L^\beta$. Pharasyn et al. (1997) found that fitting I band and K band data of a sample of spiral galaxies to this relation resulted in $\beta \approx -0.15$, i.e. the mass-to-light ratio slightly decreases with increasing luminosity, which is consistent with what we would expect from a secular evolution picture, i.e. as the secular evolution advances, L increases and M$_{dyn}$/L decreases.

The fundamental plane relation for elliptical galaxies and bulges were first obtained by Faber et al. (1987), Djorgovski & Davis (1987), and Dressler et al. (1987). One type of the fundamental plane relation for ellipticals can be written as (c.f. equation 2a and 2b of Djorgovski & Davis 1987)

$$L \sim V_e^{3.45} SB^{-0.86}. \tag{11}$$

This relation can be re-written into the form

$$M_t(R_e) = -8.62(\log V_e + 0.1\mu_e) + constant, \tag{12}$$

which can be further written as we arrive at

$$10\log V_e = -1.25M(R_e) - \mu_e + constant. \tag{13}$$

Comparing equations (10) and (13), we see that apart from a possible difference in the "constant" term (accounting for the change in total luminosity), the only difference between the spiral and elliptical fundamental plane relations is in the different sign of the mass-to-light ratio exponent β: $\beta = -0.15$ for spirals and $\beta = 0.13$ for ellipticals and bulges, where β is defined through

$M_{dyn}/L = L^{\beta}$. This difference is apparently brought about by the fact that spiral galaxies still have varying reserves of baryonic dark matter to form stars, therefore as the secular evolution proceeds (and therefore L increases) the mass-to-light ratio decrease; whereas elliptical galaxies have essentially exhausted their central baryonic dark matter supply, thus the ellipticals in more advanced stage of evolution (which also generally have larger L) will experience greater degree of dimming, which is reflected in the increase of mass-to-light ratio with L.

This research was supported in part by funding from the Office of Naval Research.

References

Andredakis, Y.C., Peletier, R.F., & Balcells, M. 1995, MNRAS, 275, 874
Binney, J., & Tremaine, S. 1987, Galactic Dynamics (Princeton:Princeton Univ. Press)
Bell, E.F. & de Jong, R.S. 2001, **ApJ**, 550, 212.
Combes, F., & Sanders, R.H. 1981, A&A, 96, 164
Conselice, C.J., Kershady, M.A., Dickinson, M., & Papovich, C. 2003, AJ, 126, 1183
de Blok, W.J.G. 2003, presented in Dark Matter in Galaxies, IAUS 220
Djorgovski, S. & Davis, M. 1987, **ApJ**, 313, 59
Dressler, A. et al. 1987, **ApJ**, 313, 42
Faber, S.M., et al. in Nearly Normal Galaxies, ed. S.M. Faber (New York: Springer), 175
Faber, S.M., & Jackson, R.E. 1976, **ApJ**, 204, 668
Freeman, K.C. 1970, **ApJ**, 160, 811
Hamabe, M., & Kormendy, J. 1987, in Structure and Dynamics of Elliptical Galaxies, ed T. de Zeeuw (IAU), 379
Jablonka, P., Gorgas, J., & Goudfrooij, P. 2002, Ap&SS, 281, 367
Kormendy, J. 1982, in 12th Advanced Course of the SSAA, eds. L. Martinet, & M. Mayor (Geneva Observatory: Geneva), 115
Kormendy, J., & Kennicutt, R. 2004, ARA&A, in press
Lynden-Bell, D., & Kalnajs, A.J., 1972, **MNRAS**, 157, 1
McGaugh, S.S., & de Blok, W.J.G. 1998, **ApJ**, 499, 41
Peebles, J. 1983, Physical Cosmology (Princeton: Princeton Univ. Press)
Persic, M., Salucci, P., & Stel, F. 1996, **MNRAS**, 281, 27
Pharasyn, A., Simien, F., & Heraudeau, Ph. 1997, in Dark and Visible Matter in Galaxies, ASP Conf. Series 117, eds M. Persic & P Salucci (San Francisco: ASP), 180
Shu, F.S., 1982, The Physical Universe: An Introduction to Astronomy (Mill Valley: University Science Books)
Toomre, A., & Toomre, J. 1972, ApJ, 178, 623
Tully, R.B., & Fisher, J.R. 1977, **A&A**, 54, 661
Wielen, R. 1977, **A&A**, 60, 263
Wolfe, A.M., 2001, in Galaxy Disks and Disk Galaxies, eds J.G. Funes, S.J. & E.M. Corsini (San Francisco: ASP), 619
Zhang, X. 1996, **ApJ**, 457, 125; 1998, **ApJ**, 499, 93; 1999, **ApJ**, 518, 613
Zhang, X., 2002, Ap&SS, 281, 281
Zhang, X. 2003, JKAS, 36, 223
Zhang, X., Lee, Y., Bolatto, A., & Stark, A.A., 2001, ApJ, 553, 274
Zwaan, M.A., van der Hulst, J.M., de Bolk, W.J.G., & McGaugh, S.S. 1995, **MNRAS**, 273, L35

J. Kormendy: The essential idea from which all of your results follow is the phase shift between the ridge line of the spiral density wave and the potential felt by the disk stars. Can you give us a simple heuristic understanding of why this phase shift happens and of how big it is in degrees?

X. Zhang: The phase shift exists for all density wave patterns that are skewed, including spirals and bars (but wouldn't be there if a bar is perfectly straight). It can be understood from the Poisson equation side because gravity is long-range, and potential as the integral of density does not have to look like a density.

For the particular type of density falloff of a spiral mode, the phase shift is such that potential lags density inside co-rotation, and vice versa outside co-rotation.

The angular momentum flux associated with a skewed pattern which has a phase shift of the above signs on both sides of co-rotation allows the negative angular momentum to be dumped onto each annular ring inside co-rotation and positive angular momentum dumped outside co-rotation. This leads to the spontaneous growth of the spiral mode in the linear regime, and to the evolution of basic state mass distinctions at the quasi-steady state. For a typical Sb galaxy, the phase shift is of the order of a few tens of degrees.

THE ANGULAR MOMENTUM PROBLEM AND THE FORMATION OF BULGELESS GALAXIES

Elena D'Onghia[1,2] & Andreas Burkert[2]
[1] *Max-Planck-Institut für extraterrestrische Physik, 85748 Garching, Germany*
[2] *University Observatory Munich, Scheinerstrasse 1 81679 Munich, Germany*

Abstract The specific angular momentum of Cold Dark Matter (CDM) halos in a ΛCDM universe is investigated. Their dimensionless specific angular mometum $\lambda' = \frac{j}{\sqrt{2}V_{vir}Rvir}$ with V_{vir} and R_{vir} the virial velocity and virial radius, respectively depends strongly on their merging histories. We investigate a set of simulations of the Cold Dark Matter models (ΛCDM) to explore the specific angular momentum content of halos formed through various merging histories. We show that halos with a quiet merging history, dominated by minor mergers and accretion until the present epoch, acquire by tidal torques, on average, only 2% per cent of the angular momentum required for their rotational support ($\lambda' = 0.02$), whereas observational data for a sample of late-type bulgeless galaxies indicates that those galaxies reside in dark halos with exceptionally high values of $\lambda' \approx 0.06 - 0.07$. By minor mergers and accretion the specific angular momentum of dark halos is preserved or increases slowly with time, but not enough to spin up to the observed values for late-type dwarf galaxies. Feedback processes have been invoked to solve the problem that gas from looses a large fraction of its specific angular momentum during infall. Our results indicate that cosmological models of bulgeless galaxy formation have an even more severe problem as even without any angular momentum loss the specific angular momentum gained through smooth merging and accretion will be a factor of 3 smaller than observed.

Keywords: cosmology: theory – galaxies: formation

1. Introduction

In the current paradigm for structure formation, dark matter is assumed to be cold and collisionless and luminous galaxies form by gas cooling into dark matter halos, which grow by gravitational accretion and merging in a hierarchical fashion (White & Reese 1978). In this standard picture, the model proposed by Fall & Efstathiou (1980, hereafter FE80) for disk formation links the sizes of galactic disks to the angular momentum of their parent dark matter

D. Block et al. (eds.), Penetrating Bars through Masks of Cosmic Dust, 329–333.

halos. This theory is able to produce disks with sizes that are in agreement with observations if the gas initially had the same specific angular momentum as dark matter halos show today and if the gas preserved its specific angular momentum during the protogalactic collapse phase. The angular momentum of galaxies results from torques due to tidal interactions with neighbouring structures, acquired early, before the halo decoupled from the Hubble expansion.

Many models for the formation of galactic disks have been proposed based on this picture by FE80. Most of them incorporate the mass accretion history of halos and are able to reproduce many properties of observed disk galaxies (e.g. Mo, Mao & White 1998). However, in these models, the angular momentum of the dark matter halos is assigned without accounting for the merging history of halos. Recent results from numerical N-body simulations have pointed out that the effect of major mergers is to increase the mean angular momentum content of the halos (Gardner 2001; Vitvitska et al. 2002). This is explained by the orbital angular momentum of the merging halos which dominates the final net angular momentum of the remnant.

However, numerical simulations that incorporate gas dynamics have difficulties to make realistic disk galaxies in the current cosmological paradigm of CDM. In simulations, the disks are smaller, denser and have much lower angular momenta then observed disk systems (e.g. Navarro & Steinmetz 2000).

Previous work has focused on angular momentum properties of halos that had at least one dominant major merger during their evolution. However, major merger events tend to destroy disks, producing spheroidal stellar systems, like bulges or early-type galaxies (see e.g. review by Burkert & Naab 2003). Here we explore the angular momentum properties of halos that host pure disk galaxies, or bulgeless galaxies and that never experienced a major merger.

2. Bulge to Disk ratio vs. Spin parameter

We performed three simulations within a ΛCDM cosmological universe with $\Omega_0 = 0.3$, Ω_Λ=0.7, h=H$_0/70$ km s^{-1} Mpc^{-1} and $\sigma_8 = 0.9$. The simulated volume was 15 h^{-1} Mpc box size in all runs. Each simulation was performed using the publicly available version of the smoothed particle hydrodynamics (SPH) code GADGET (Springel, Yoshida & White 2001). The simulations began from a spatially uniform grid of 128^3 equal-mass particles with Plummer–equivalent softening comoving length of $10h^{-1}$ kpc. The particle masses was $1.34 \times 10^8\ h^{-1}$ M$_\odot$. We identify halos with the classic friend-of-friend (FOF) method using a linking $b = 0.15$. All halos at z=0 with masses between M= $10^{11} - 10^{12}\ h^{-1}$ M$_\odot$, containing at least 1000 particles are analysed.

The angular momentum of a galaxy, J, is commonly expressed in terms of the dimensionless spin parameter $\lambda = J\sqrt{|E|}/\mathrm{G}M^{5/3}$where E is the total energy, and M is the total mass. The value of the spin parameter represents the ratio between the actual angular velocity of the system, divided by that needed to support the system purely by rotation. The halo spin parameter distribution in simulations is found to be well approximated by the lognormal function. In practise, it is more convenient to use the modified spin parameter (Bullock et al. 2001):

$$\lambda' = \frac{j}{\sqrt{2}V_{vir}R_{vir}} \tag{1}$$

where $j = J/M$ is the specific angular momentum. For ΛCDM cosmology the lognormal parameters are found to be $< \lambda' >= 0.035 \pm 0.006$ and $\sigma_{\lambda'}$=0.5-0.6 (Bullock et al. 2001).

We trace each identified halo backward in time, following the mass of the most massive progenitor as a function of redshift during $0 < z < 3$. We start when the mass of the most massive progenitor is 30% of the z=0 mass. Assuming instead 50% of the halo mass at present time does not change results. To identify mergers, we denote a halo as a major merger remnant if at some time during $0 < z < 3$, its major progenitor was classified as a single group in one output, but two separate groups with a mass ratio$\leq 4:1$ in the preceding output.

For each halo we identify the time of the last major merger. We assume that accretion events and minor mergers encrease the mass of the disk component whereas major mergers convert all the available gas into stars and destroy disks, forming a stellar spheroid (for a recent review, see Burkert & Naab 2003). Neglecting galactic winds and assuming a universal baryon fraction for all infalling substructures, the final ratio between bulge mass B and the sum of bulge and disk mass (B+D) is given by the ratio of the dark halo mass at the time of the last major merger, divided by the dark halo mass at $z = 0$.

Fig.1 shows the B/(B+D) ratio at z=0 of halos analysed in one of our simulations as a function of their spin parameter λ'. For better visualisation we just plot the results of one run as the other two runs show similar trends. Different symbols are used to denote halos that have their last major merger at different epochs (see legend in Fig.1). The plot shows that the halo spin parameter is a function of the ratio between bulge mass and total baryonic mass. Many halos with high B/(B+D) ratios also show higher λ' values than the average $< \lambda' >\approx 0.035$ in the lognormal distribution (filled symbols in Fig.1). Most of these halos experienced a recent major merger in the redshift range $0 < z < 1$. Halos that have not experienced any major merger from $0< z < 3$, show a lognormal distribution significantly lower than the average referred to all halos.

In the general picture of FE80 disks should have the same distribution of total specific angular momentum as the dark matter halos. Hence they should have the same value for the spin parameter, $\lambda'_{disk} \approx \lambda'$. This is expected because all the material experiences the same external torques the early expansion phase before separating into two distinct components as a result of dissipative processes during the collapse phase.

We find that disk-dominated late-type galaxies inhabiting halos that have not experienced a major merger have a distribution of λ'_{disk} that peaks around the value of 0.023 which is substantially smaller than expected from observed rotation curves. We use the results of Van den Bosch, Burkert & Swaters (2001, BBS hereafter) who determined spin parameters for galactic disks of 14 late-type bulgeless dwarf galaxies. Fig.1 shows the spin parameters of their galaxies, assuming a mass-to-light ratio of unity in the R band. The result is surprising: none of our models lies in the region of the diagram covered by the observational data.

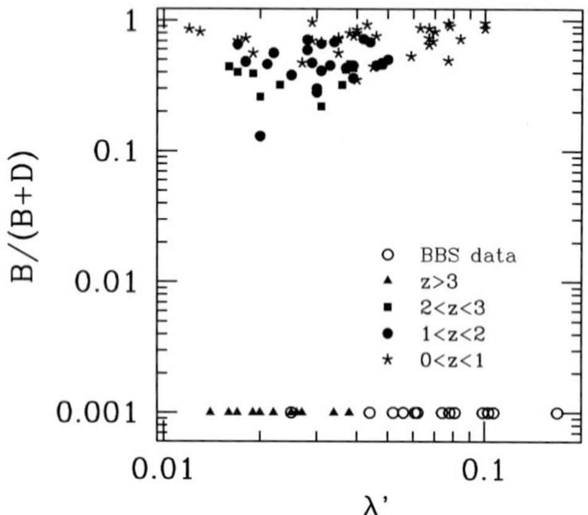

Figure 1. The bulge mass (B) compared to the total baryonic mass (B+D) is plotted for each halo against its spin parameter computed at z=0. Different symbols mark halos that had the last major merger at different epochs. The spin parameter values measured for 14 late-type dwarf galaxies by van den Bosch, Burkert & Swaters (2001)(BBS) are also plotted.

3. Summary

The net result seems to be that tidal torques generate halos with typical spin parameters of $\lambda' = 0.02$ whereas data for bulgeless galaxies indicate halos with values of $\lambda' = 0.06 - 0.07$. At the moment it is not at all clear how one can reconcile the observations and theory. It is known that the spin parameter distribution for the collapsed objetcs is insensitive to the shape of the initial power spectrum of density fluctuations, to the environment and the adopted cosmological model (Lemson & Kauffmann 1999).

A modification of the nature of dark matter does not seem to solve the problem either. Recent works show that warm dark matter halos have systematically smaller spins than their counterparts in ΛCDM (e.g. Knebe et al. 2002), although in this scenario the presence of pancakes could provide more efficient torques on the protohalos.

Feedback was invoked as a mechanism to prevent the process of drastic angular momentum loss of infalling gas. However, we have shown here that the dark halos that experienced no major mergers have already too low an angular momentum to produce the observed disks and no feedback process is know that would increase the the specific angular momentum of the gas. The origin of extended bulgeless disk galaxies remains a puzzle.

Acknowledgments

The authors wish to thank Giuseppe Murante for providing his initial conditions for one of the simulations and for his help with the merging tree.

References

Bullock, J.S., Dekel, A., Kolatt, T.S., Kravtsov, A.V., Klypin, A.A., Porciani, C., & Primack, J.R. 2001, ApJ, 555, 240

Burkert, A. & Naab, T. 2003, in "Galaxies and Chaos", eds. G. Contopoulos & N. Voglis, Springer, p. 327

Fall, S.M., Efsthatiou, G. 1980, MNRAS, 193, 189 (FF80)

Gardner, J.P. 2001, ApJ, 557, 616

Knebe, A., Devriendt, J., Mahmood, A., & Silk, J. 2002, MNRAS, 329, 813

Lemson, G., & Kauffmann, G. 1999, MNRAS, 302, 111

Mo, H.J., Mao, S., White, S.D.M. 1998, MNRAS, 295, 319

Navarro J.F., Steinmetz, M. 2000, ApJ, 538, 477

Springel, V., Yoshida, N., & White, S.D.M. 2001, NewA, 6, 79

Vitvitska, M., Klypin, A.A., Kravtsov, A.V., Bullock, J.S., Primack, J.R., Wechsler, R.H. 2002, ApJ, 581, 799

White, S.D.M., Reese, M.J. 1978, MNRAS, 183, 341

Bakubung

DISKS EVOLUTION IN A COSMOLOGICAL FRAMEWORK

Anna Curir[1], Paola Mazzei[2] and Giuseppe Murante[1]
[1] *INAF - Astronomical Observatory of Torino (Italy),* [2] *INAF -Astronomical Observatory of Padova (Italy)*

Abstract We investigate the bar instability in stellar disks embedded inside a dark matter halo evolving in a fully cosmological framework. We present cosmological simulations in which the role of the disk-halo mass ratio and the Q parameter are emphasized. Disks expected to be stable according to classical criteria form indeed weak bars. This result is due to dynamical properties of tha cosmological halo, which is far from stability and isotropy.

Keywords: Galaxies; Spirals; Spiral arms and bars; Galactic disks; Galactic halos

1. Introduction

A possible role of the Dark Matter (DM) halo in enhancing the bar instability in stellar disks has been recently explored (Curir and Mazzei, 1999; Athanassoula, 2003), and a progressive effort for improving the models for the haloes including the disks has been made in the last years (Mazzei and Curir, 2001; Athanassoula and Misiriotis, 2002), taking account also the informations about shapes, densities and concentration coming from the cosmological hierarchical clustering scenario.
The new idea here is to imbed a disk inside a fully cosmological halo selected in a suitable slice of Universe. We want to explore how the role of that halo can be different with respect to the classical scenario. In particular we want to investigate, besides the role played by the halo–disk mass ratio, a possible role of the halo substructure and evolution.

2. Methods

The galaxy model consists of a truncated exponential disk, self-consistently embedded in a dark matter halo extracted from a cosmological simulation. To obtain the DM halo, we performed a low-resolution (128^3 particles) simulation of a "concordance" ΛCDM cosmological model: Ω_m =0.3, Ω_Λ=0.7, H_0=100

D. Block et al. (eds.), Penetrating Bars through Masks of Cosmic Dust, 335–339.

h km s^{-1} Mpc^{-1}, h=0.7, where Ω_m is the total matter content of the Universe, Ω_Λ the cosmological constant component, and H_0 the value of the Hubble constant. The initial redshift is z=20. We selected one suitable DM halo with mass M$\sim 10^{11}h^{-1}$ M$_\odot$, and resampled it with a multi-mass tecnique.
The high-resolution DM halo has been followed to z=0; no significant merger involved the halo after a redshift of 5. The virial mass of the halo at redshift z=0, is $M_{halo} = 1.14 \cdot 10^{11}h^{-1}$ M$_\odot$, and the corresponding virial radius is $R_{vir} \sim 125h^{-1}$kpc . The galactic disk has been embedded in the halo at a redshift z=2, and z=1.
The spatial distribution of the star particles in the disk model follows the exponential surface density law: $\rho_{stars} = \rho_0 \exp -r/r_0$ where r_0 is the disk scale legth, $r_0 = 4h^{-1}$kpc, and ρ_0 is the surface central density. The disk is truncated at five scale lenghts, $R_{disk} = 20h^{-1}$kpc . Circular velocities are assigned analytically to disk particles through the global (disk+halo) potential, obtained using Jeans equations. Velocity dispersions are monitored through a Toomre-like parameter Q. We investigated two possibilities for the value of Q: the calssical stable disk having Q=1.5, and a more cold situation in which Q=0.5.

Our model of galaxy is, by choice, very semplified. Our aim is to study the *gravitational* effect of the halo on the disk and to have hints on the *gravitational* feedback of the disk itself on the halo.

3. Simulations

Cosmological cases

The main parameters and the initial properties of this set of simulations are listed in Table 1. We verify that the inclusion of the disk does not result in significant changes in the accretion history of the DM halo. Simulations 1, 2 and 3 in Table 1 refer to a *cold* disk ($Q = 0.5$). In simulation 1, at the final time (i.e. z=0) the baryon fraction inside R_{vir}, $f_b = M_{disk}/M_{disk+DM} \sim 0.34$, is 44% less than its initial value, 0.53. The final baryon's fraction of simulation 2 is $\simeq 0.16$, compared with its initial value, 0.28. Simulation 3 provides $f_b \simeq 0.05$ at z=0, 50% less than its initial value.
Simulations 4 and 5 provide the same final baryon's fractions as the corresponding simulations with the lower Toomre's parameter.
Simulation 6 and 7, which explore the role of the initial redshift on the bar instability, provide quite the same final values of the baryon's fraction as the corresponding simulations 2 and 3. Thus Neither the Toomre parameter nor the initial redshift affect the evolution of this ratio which is driven by the mass of the stellar disk. While the baryonic fraction of simulation 1 is too high to be consistent with the cosmological value, simulations 3, 5 and 7 give baryonic

fractions within 2σ to such a value. Simulations 2, 4 and 6 are in between these two extremes. Both the disk parameters (i.e. Q and M_{disks}) and the redshift of disk immersion sightly affect the halo configuration at z=0 as shown in Table 2.

Table 1. Initial properties of the disks

N	Q	M_{disk}	z	M_{DM}	R_{DM}
1	0.5	1	2	0.64	0.44
2	0.5	0.33	2	1.94	1.31
3	0.5	0.1	2	6.4	4.38
4	1.5	0.33	2	1.94	1.31
5	1.5	0.1	2	6.4	4.38
6	0.5	0.33	1	2.0	0.97
7	0.5	0.1	1	6.7	6.64

M_{disk}=mass of the disk in $5.9 \times M_{\odot}$
M_{DM}=mass of dark matetr inside the disk radius
R_{DM}= halo-to-disk mass inside the disk radius

We point out that some of the initial R_{DM} ratios (see col. 6 of Table 1) are below the classical threshold instability value (2.2, Athanassolua et al., 1987), while the final R_{DM} ratio (see col. 3 of Table 2) is greater than such value for all our simulations, with the exception of simulation 1; nevertheless a bar is always noticed (see col.s 4 and 5).

Table 2. Results at z=0

N	M_{DM}	R_{DM}	S_m	a_{max}	$pseudo-bulge$	$bars\,in\,bars$
1	0.64	0.41	0.42	7	y	n
2	0.77	2.39	0.33	8	y	n
3	0.73	7.41	0.8	3.8	n	y
4	0.78	2.40	0.48	5	weak	n
5	0.73	7.41	0.70	6.5	n	y
6	0.79	2.43	0.35	6.8	weak	n
7	0.77	7.73	0.58	5.0	n	y

S_m= strength of the bar
a_{max}=major axis correponding to the maximum strength

Isolated cases

We also performed several simulations of the isolated disk+halo system using the halo as extracted from our cosmological simulations at z=2. By comparing results of this set of simulations with the previous ones we aim to disentangle the effect of cosmology on the system evolution. We examined the effect of different number of disk particles, i.e. 56000, 18800, 9400, to look at the effect of particle resolution. We carried out simulations with different softening lengths to deepen the role of this important parameter too. Moreover, we performed one run with disk configuration kept *frozen* for $T = 0.5$ Gyr (longer than t_{dyn} of the halo, $\sim$ 0.29Gyr) during the halo evolution, to focus on the impact of the halo reaction to the disk immersion on the onset and growth of the bar instability inside the disk. We also performed isolated runs using an analytical isothermal halo having the same mass as our cosmological one. By comparing these simulations to the cosmological ones we strongly enhance the role of the spin, of the anysotropy and of the more realistic density distribution which chrachterize our approach.
By comparing Fig. 1 with Fig. 2 one can realize that the cosmological environment triggers a strong bar instability in the ligth disk having mass one tenth of the halo mass, whereas the same mass ratio does not produce any bar inside an analytical halo.

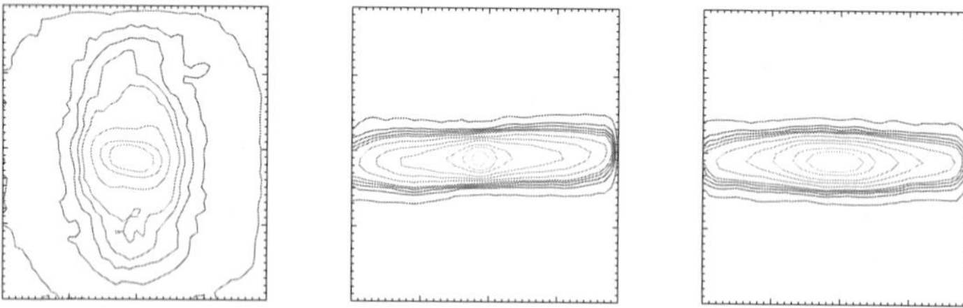

Figure 1. Face-on , edge-on and side on morphology of the stellar disk in simulation 5 at z=0

4. Results

Our results can be summarized as follows:

- stellar disks of different properties, i.e. mass and Q parameter, embedded in the same halo and evolving in a fully consistent cosmological scenario, develop long living bars lasting from the time of the immersion, up to redshift 0;

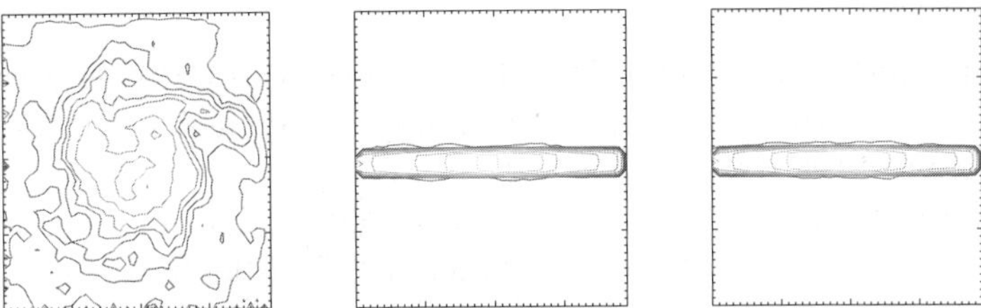

Figure 2. As in fig. 1 for the same disk in an analytical halo followed for the same time span

- the strength of the bar at z=0 is weakly depending on the Q parameter, for a given disk mass. However for the same disk-to-halo mass ratio, colder disks show stronger and longer bars. Thus the less massive *warm* disks entail the weakest bars. Moreover bar in bar is a common feature in their face-on morphology:
- the classical criterion to account for bar instability, i.e the halo-to-disk mass ratio inside the disk radius, R_{DM}, cannot be validated in a cosmological framework where a bar always develops, due the halo evolution. Thus a bar develops also for a R_{DM} ratio classically forbidden;
- simulations performed embedding different disks in the same halo, extracted at z=2 from the cosmological framework, show that the effects of the large scale structures are negligible only in the less massive, halo dominated disks, moreover by comparing our results (Curir et al., 2004) with Curir and Mazzei (1999) we deduce that live unrelaxed halos are the more suitable to mimic cosmological halos as far as the bar instability is concerned.

Large-scale effects are uninfluent in triggering the bar instability but the anisotropy, the substructure and the continuous collapse of a cosmological halo have a crucial effect in enhancing and fuelling such a instability also in the cases where halo *ad hoc* models provided stability predictions.

References

Athanassoula, E. (2003) MNRAS 341,1179

Athanassoula, E., Bosma, A., Papaioannou ,S. (1987) A&A, 179,23

Athanassoula, A., Misiriotis, A. (2002) MNRAS, 330, 35

Curir, A. and Mazzei, P.(1999) A&A, 352, 103

Mazzei, P. and Curir, A. (2001) A&A, 372, 803

Curir, A., Mazzei, P. and Murante G., New Astronomy, submitted

GALAXY FORMATION AND THE COSMOLOGICAL ANGULAR MOMENTUM PROBLEM

Andreas M. Burkert[1] and Elena D'Onghia[1,2]
[1] *University Observatory Munich, Scheinerstr. 1, D-81679 Munich, Germany;*
[2] *Max-Planck-Institut fur Extraterrestrische Physik, Karl-Schwarzschild-Str. 1, D-85741 Garching, Germany*

Abstract The importance of angular momentum in regulating the sizes of galactic disks and by this their star formation history is highlighted. Tidal torques and accretion of satellites in principle provide enough angular momentum to form disks with sizes that are in agreement with observations. However three major problems have been identified that challenge cold dark matter theory and affect models of galaxy evolution: (1) too much angular momentum is transferred from the gas to the dark halos during infall, leading to disks with scale lengths that are too small, (2) bulgeless disks require more specific angular momentum than is generated cosmologically even if gas would not lose angular momentum during infall, (3) gravitational torques and hierarchical merging produce a specific angular momentum distribution that does not match the distribution required to form exponential disks naturally; some gas has exceptionally high angular momentum, leading to extended outer disks while another large gas fraction will contain very little specific angular momentum and is expected to fall into the galactic center, forming a massive and dominant bulge component. Any self-consistent theory of galaxy formation will require to provide solutions to these questions. Selective mass loss of low-angular-momentum gas in an early phase of galaxy evolution currently seems to be the most promising scenario. Such a process would have a strong affect on the early protogalactic evolution phase, the origin and evolution of galactic morphologies and link central properties of galaxies like the origin of central massive black holes with their global structure.

Keywords: galaxies: disks, formation, kinematics and dynamics, structure - dark matter

1. Introduction

The origin of the distribution of mass and angular momentum in disk galaxies is yet an unsolved astrophysical puzzle. Eggen, Lynden-Bell and Sandage (1962) argued that spiral galaxies like the Milky Way formed by rapid infall of an initially uniform sphere of gas into a centrifugally supported disk. It was soon realized that these disks would have characteristic exponential surface

D. Block et al. (eds.), Penetrating Bars through Masks of Cosmic Dust, 341–356.

density distributions if the initial gas sphere would be in solid body rotation and if the gas would preserve its initial specific angular momentum distribution $M(< j)$ during infall (Mestel 1963, Crampin & Hoyle 1964, Innanen 1966, Freeman 1970). Here, $M(< j)$ is the cumulative mass of gas with angular momentum less or equal to j.

Since these first pioneering studies our insight into galaxy formation has substantially changed and improved. Current cosmological models consider a dissipationless cold dark matter (CDM) component to dominate structure formation in the Universe (Blumenthal et al. 1984). Initially small dark matter density perturbations in the early Universe decouple from the Hubble flow, collapse into virialized dark matter halos and merge into larger and larger structures. Gas accumulates within the extended dark halos, dissipates its kinetic energy and settles into the equatorial plane as soon as centrifugal equilibrium is being reached, forming fast rotating disks that subsequently turn into stars (White & Rees 1978). Fall & Efstathiou (1980) argued that gas and dark matter should initially be well mixed. In this case, the specific angular momentum distribution of the gas should be equal to that of the dark halo. If $M(< j)$ would be preserved during infall into the equatorial plane, the exponential disk scale length R_d would be directly related to the specific angular momentum λ of the dark halo. Here λ is the dimensionless spin parameter (e.g. Peebles 1969)

$$\lambda = \frac{J|E|^{1/2}}{GM_{vir}^{5/2}} \tag{1}$$

where J, E, and M_{vir} are the total angular momentum, energy and virial mass of the halo, respectively and G is Newton's constant. This shifted the focus to a more detailed investigation of the spin properties of dark halos and a determination of λ. Peebles (1969) suggested that halos would acquire non-negligible specific angular momentum by the gravitational interaction of their building blocks with neighboring structures. Subsequent cosmological N-body simulations (Barnes & Efstathiou 1987, Efstathiou & Barnes 1983, Zeldovich & Novikov 1983, Cole & Lacey 1996) confirmed this mechanism. They showed that prior to collapse the angular momentum of a dark fluctuation grows roughly linearly with time, as predicted by linear tidal-torque theory (White 1984) until it decouples from the Hubble flow, collapses and virializes. The λ distribution of virialized halos turned out to be well described by a log-normal (Steinmetz & Bartelmann 1995, Cole & Lacey 1996, Gardner 2001, Bullock et al. 2001)

$$p(\lambda)d\lambda = \frac{1}{\sigma_\lambda\sqrt{2\pi}} \exp\left(-\frac{ln^2(\lambda/\lambda_0)}{2\sigma_\lambda^2}\right) dln\lambda \tag{2}$$

with median value of $\lambda_0 = 0.042 \pm 0.006$ and dispersion $\sigma_\lambda = 0.5 \pm 0.04$. A more partical spin parameter λ' was proposed by Bullock et al. (2001):

$$\lambda' = \frac{J}{\sqrt{2} M_{vir} V_{vir} R_{vir}} \tag{3}$$

with R_{vir} the halo virial radius and $V_{vir}^2 = GM_{vir}/R_{vir}$ its virial velocity. The spin parameters λ and λ' turn out to be very similar due to the fact that dark halos are well described by a universal density distribution (Navarro, Frenk & White, 1997, NFW) and the λ' distribution also follows a log-normal with best fit values $\lambda'_0 = 0.035 \pm 0.005$ and $\sigma_{\lambda'} = 0.5 \pm 0.03$ (Bullock et al. 2001).

Given λ', and assuming that the disk gas will have the same specific angular momentum as the dark halo, the exponential disk scale length R_d can be easily determined. Adopting a flat rotation curve with velocity v_c, the disk's specific angular momentum is (Mo et al. 1998)

$$j_d = 2R_d v_c = \sqrt{2}\lambda' V_{vir} R_{vir}. \tag{4}$$

For typical NFW halos with concentrations $c = 10$, the peak velocity of the dark matter rotation curve which is in general of order the observed peak rotation velocity is $v_{peak} = 1.2V_{vir} \approx v_c$. For a flat ΛCDM-cosmology with cosmological parameters $\Omega_m = 0.3$, $\Omega_\Lambda = 0.7$ and a Hubble constant of 70 km s^{-1} Mpc^{-1} the virial parameters of dark halos are coupled by the relations (Navarro & Steinmetz 2000)

$$R_{vir} = 270 \times \left(\frac{M_{vir}}{10^{12} M_\odot}\right)^{1/3} kpc = 215 \times \left(\frac{V_{vir}}{100 km/s}\right) kpc. \tag{5}$$

Combining these questions leads to a relationship between the exponential disk scale length R_d and the disk's rotational velocity v_c:

$$R_d = 8.5 \left(\frac{\lambda'}{0.035}\right)\left(\frac{v_c}{200 km/s}\right) kpc. \tag{6}$$

A more detailed investigation which takes into account adiabatic contraction of the dark halo (Jesseit et al. 2002) and the combined gravitational force of the disk and dark halo shows that equation 6 overestimates the disk scale length by a small amount of order 10% to 20%.

Figure 1 compares the observations (Courteau 1997) with the prediction of equation 6, adopting a typical value of $\lambda' = 0.035$. Note that the slope reflects directly the spin parameter λ'. The theoretical predictions lead to a correlation that is steeper than observed. The dashed line shows a best fit model which requires a value of $\lambda' = 0.025$. If the rotation speeds of galactic disks are approximately the same as the maximum circular velocities of their

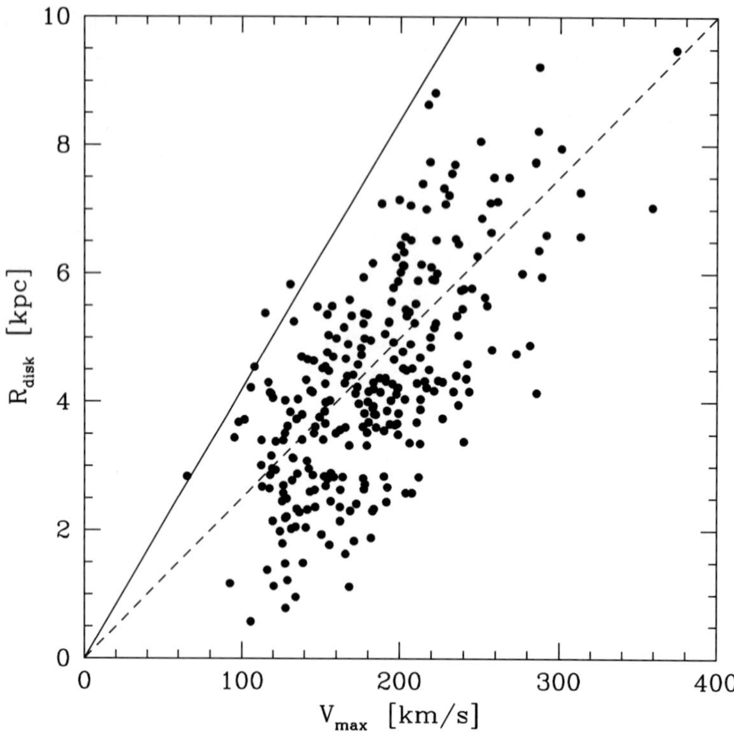

Figure 1. The observed disk scale lengths versus their maximum rotation velocities are shown for the Courteau (1997) sample. The solid line shows the theoretically predicted correlation for $\lambda' = 0.035$. The dashed curve corresponds to $\lambda' = 0.025$

dark halos, and if the gas would have the same specific angular momentum as the dark halo, cosmological models predict disk scale lengths that are about a factor of 1.4 larger than observed. Good agreement could be achieved if the gas retained only 70% of the available angular momentum during infall. A similar conclusion was reached by Navarro & Steinmetz (2000) and Mo, Mao & White (1998) who investigated the structural properties of galactic disks within characteristic NFW-type dark halos in greater details.

2. Angular Momentum Loss during Galactic Disk Formation

Numerical simulations have become one of the most powerful tools for exploring galaxy formation. Hydrodynamical simulations of disc formation were initiated by Navarro & Benz (1991), who included radiative cooling by hydrogen and helium, and attempted to account for star formation and feedback processes. In contrast to the conclusion drawn in the previous chapter, the simulated galaxies however failed to reproduce their observed counterparts: the disks were found to be too small and were more centrally concentrated than actual galaxies (Navarro & White 1993). In addition, star formation in the models was overly efficient, converting gas into stars too fast. Many of the shortcomings of the early modeling could be accounted for by the limited resolution of the simulations and the way in which feedback was treated. As star formation was very efficient in low mass halos at high redshifts, a large number of dense, compact stellar systems formed that lateron were collected in the innermost regions of larger galaxies that formed through the merging of smaller objects. Angular momentum loss by dynamical friction drove these star clusters to the center where they formed large, dense bulges instead of extended disks.

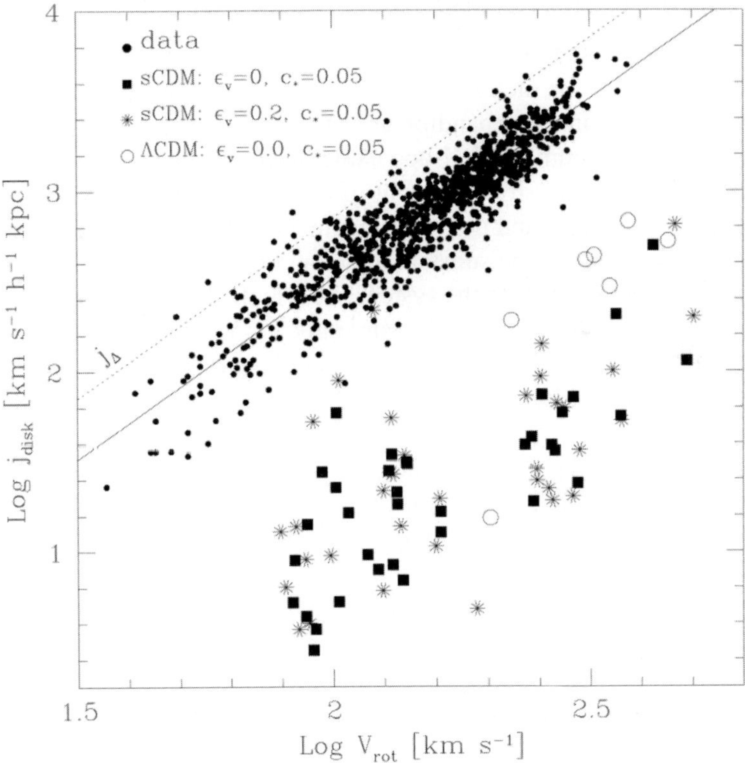

Figure 2. The specific angular momentum $j_{disk} = R_d \times v_{rot}$ of model disks with scale length R_d and rotational velocity v_{rot} is compared with observational data (filled points). Figure adapted from Navarro & Steinmetz (2000).

The cosmological angular momentum problem was reinvestigated lateron in greater details by Navarro & Steinmetz (2000) with high-resolution N-body/gas-dynamical simulations that included star formation and feedback to examine the origin of the I-band Tully-Fisher relation for different comologies. Although the slope and the scatter could be well reproduced in the simulations, the models failed to match the zero point (Fig. 2). Again, the galaxies were too compact with respect to the observations even with realistic feedback formulations that were calibrated to reproduce the empirical correlations, connecting the local star formation rate with the gas surface density (Kennicutt 1998).

A possible solution is suppression of early cooling of the gas by strong feedback from supernovae which prevents drastic angular momentum loss and produces better fits to the observations (Sommer-Larsen et al. 2003, Abadi et al. 2003). However, even in these simulations the disk systems typically contained denser and more massive bulges than observed late-type galaxies, indicating that the specific angular momentum problem is not completely solved.

3. Spin Parameter and Halo Merging History

In the Fall & Efstathiou (1980) model, the scale size of a galaxy is determined by its angular momentum, which is acquired by tidal torques from neighboring objects in the expanding universe, prior to the collapse of the halo. Recent results from numerical N-body simulations have suggested that major mergers might be especially important to increase the mean angular momentum content of the halos and their spin parameters (Gardner 2001; Vitvitska et al. 2002). This is due to the substantial amount of orbital angular momentum which a major merger adds to the system and which dominates the final net angular momentum of the remnant (Gardner 2001). Minor mergers, in contrast, have little effects on the angular momentum budget of galaxies.

Semi-analytical models of galaxy evolution (Kauffmann et al. 1993) cannot investigate the angular momentum properties of a halo self-consistently. They can however predict their merging history and by this their spin parameters, if the spin is coupled with the history of minor and major mergers. This question has been investigated in details by Vitvitska et al. (2002). They followed the evolution of the cumulative mass and the spin parameter of the major progenitors for three high-resolution halos and found that the spin parameters clearly change with time. Instead of a gradual increase, λ is increasing with every major merger and subsequently decreasing as a result of minor merging. They proposed that the net angular momentum of dark matter halos originates from statistically random merging with preferentially the last major merger dominating. In this picture, the evolution of the halo angular momentum is quite different from the standard tidal torques scenario, in which the angular momentum grows steadily at early times and its growth flattens lateron.

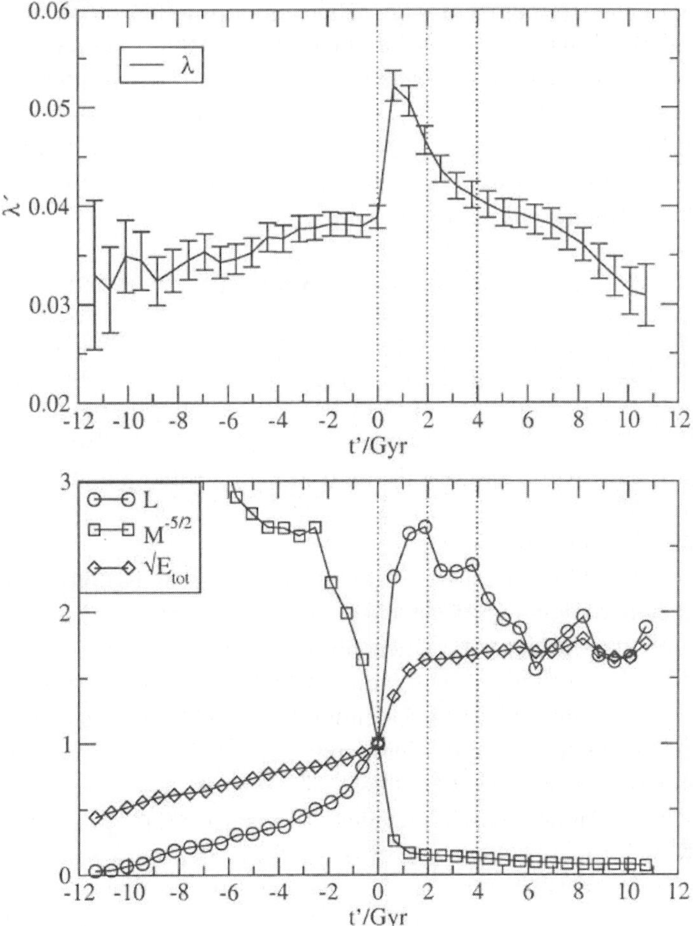

Figure 3. The dimensionless spin parameter λ is shown in the upper panel for dark halos which experienced a major merger. The lower panel shows the evolution of the total angular momentum L, the total mass M and the total energy E_{tot} with time. The time axis t' is centered on the epoch of major merging.

This conclusion has recently been questioned by Hetznecker & Burkert (in preparation). Figure 3 shows the evolution of the mean spin parameter λ', averaged over a large set of dark halos that experienced at least one major merger in their lifetime. The time axis is centered at the epoch of the major merging. Note that indeed a substantial increase in λ' is visible when the merger occurs. However, afterwards, within the next 2 Gyrs, λ' decreases again, approaching the same low value that had been achieved prior to the merging event. Hetznecker & Burkert argue that the temporary increase in λ' is a result of the fact that dark halos are out of virial equilibrium during this epoch. Dark particles oscillate into and out of the virial radius and the most bound particle, with respect to which the angular momentum is typically defined, is not a good measure of the center of mass. Neglecting unrelaxed dark halos, the authors find no difference between major and minor mergers and an average λ'-distribution that is on average shifted to somewhat lower values compared with previous studies that include unrelaxed halos.

4. A Test Case: Bulgeless Galaxies

Most work has up to now focused on angular momentum properties of halos that had at least one dominant major merger during their evolution. However, major merger events tend to destroy disks, producing spheroidal stellar systems, like bulges or early-type galaxies (see e.g. review by Burkert & Naab 2003). Not much work has been devoted to explore the angular momentum properties of halos that host pure disk galaxies, or bulgeless galaxies and that never experienced a major merger.

Recently D'Onghia & Burkert (2004) performed three N-body simulations in a ΛCDM cosmological universe and explored the angular momentum properties of halos that did not experience any major merger from redshift 3 until the present time and that are in principle good candidates to host bulgeless galaxies. The authors traced each identified halo backward in time, following the mass of the most massive progenitor as a function of redshift during $0 < z < 3$ (for details, see D'Onghia & Burkert, these proceedings). They found that disk-dominated late-type galaxies inhabiting halos that have not experienced a major merger have a distribution of λ' that peaks around a value of 0.023 which is substantially smaller than expected from observed rotation curves of bulgeless disks. To demonstrate this they compared their results with the sample of van den Bosch, Burkert & Swaters (2001) who determined spin parameters for 14 late-type bulgeless disk galaxies.

Fig.4 shows the probability distribution of the spin parameter of halos that did not experience any major mergers since z=3 (filled region) and compares it with the normalized probability distribution of λ'_{disk} for the sample of galaxies measured by van den Bosch, Burkert & Swaters (2001) assuming a mass-to-light ratio of unity in the R band. The galaxies show a distribution that follows a lognormal distribution with an average value for $\lambda'_{disk} \approx 0.067$, and a dispersion of $\sigma_{\lambda'} \approx 0.31$. This is a factor of 3 larger than predicted by the numerical simulations. Again, galactic disks forming in these halos would be too small. However, this time, the problem cannot be solved by feedback processes which prevent the loss of specific angular momentum of clumpy infalling gas.

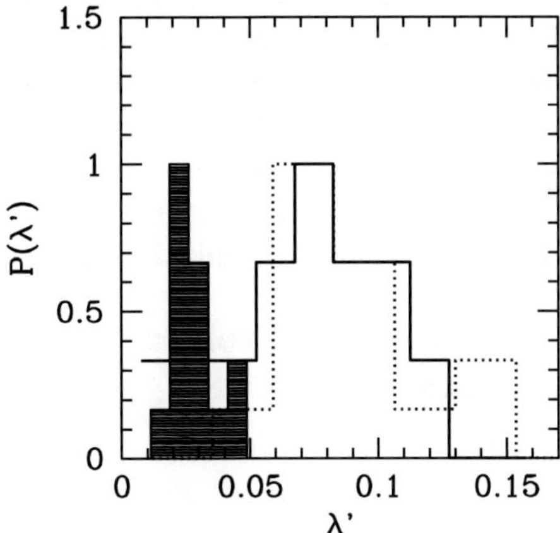

Figure 4. The distribution of the spin parameters λ' of dark halos that experienced no major merger is shown by the filled histogram and compared to the spin distribution of observed bulgeless disk galaxies (solid line). The dotted curve shows the λ' distribution if the spins of dark halos without major mergers would be multiplied by a factor of 3.15.

It is remarkable that multiplying the λ' values of simulated halos by a factor 3.15 (dotted histogram in Fig. 4) reproduces the peak and dispersion of the observed λ'_{disk} distribution quite well. Halos without major mergers acquire their specific angular momentum through tidal torques in the early epochs of evolution (Barnes & Efsthatiou 1987), when the density contrasts were small, in accordance with the prediction of the linear theory. The net result are halos with typical spin parameters of $\lambda' = 0.02$ whereas data for bulgeless galaxies indicate halos with values of $\lambda' = 0.06 - 0.07$, pointing out a new angular momentum problem, especially for bulgeless galaxies.

It's not clear how to overcome this problem. It is known that the spin parameter distribution for the collapsed objects is insensitive to the shape of the initial power spectrum of density fluctuations, to the environment and the adopted cosmological model (Lemson & Kauffmann 1999).

In the Fall & Efsthatiou (1980) model it is assumed that the gas and the dark matter in a protogalaxy have the same distribution of specific angular momentum. However the two components undergo different relaxation mechanisms: the dark matter experiences collisionless violent relaxation and the gas shocks and dissipates its kinetic energy and settles into the central regions. Van den Bosch et al. (2002) used numerical simulations in a cold dark matter cosmology to compare the angular momentum distributions of dark matter and non-radiative gas. They showed that gas and dark matter have identical angular momentum distributions after the protogalactic collapse phase if, and only if, cooling is ignored, in agreement with the standard assumptions. In addition, they found that 5% to 50% of the mass has negative specific angular momentum. Realistic disks do not typically contain counterrotating material. This suggests that during the cooling process the gas with negative specific angular momentum collides with material with positive specific angular momentum to build a bulge component. Thus, even without substructures and consequent dynamical friction a bulge would be likely to form in all galaxies due to this process, making the formation of bulgeless galaxies even more difficult.

5. The Angular Momentum Problem of Low-Mass Elliptical Galaxies

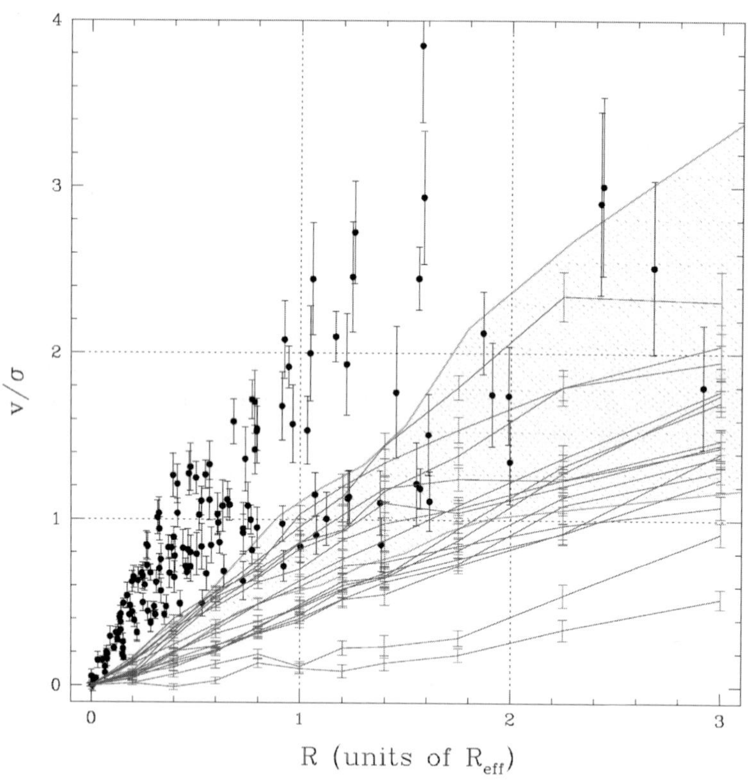

Figure 5. Comparison of observed v/σ (circles with error bars) and v/σ for edge-on numerical merger remnants resulting from collisions of disk galaxies. The shaded area corresponds to the range occupied by Bendo & Barnes (2000). Both distributions are not compatible with the observed one, even in this edge-on case.

As outlined previously, numerical simulations indicate an angular momentum problem for disk galaxy formation that results from the fact that gas loses a substantial fraction of its angular momentum while settling into the equatorial plane. Cretton et al. (2001) however also found a related problem for fast rotating low-mass elliptical galaxies. Figure 5 shows the observed radial dependence of v/σ of their early-type galaxy sample with v the observed line-of-sight rotational velocity of the stellar spheroid along the major axis and σ the local line-of-sight velocity dispersion. The shaded region and the solid lines show the results of numerical simulations of unequal and equal-mass spiral galaxy mergers which reproduce many of the global properties of ellipticals remarkably well (Naab & Burkert 2003, Burkert & Naab 2003). Note that none of these models leads to curves that rise as steeply as observed for many cases of low-mass ellipticals which show values of $v/\sigma \approx 2$ within 1-2 effective radii. Theoretical models instead predict values of $v/\sigma \approx 1$.

Like in the case of galactic disk formation, tidal torques and dynamical friction during the major merger event efficiently remove specific angular momentum from the baryonic stellar component, leading at the end to stellar spheroids with rotational velocities that do not exceed much the velocity dispersion.

6. The Specific Angular Momentum Distribution Problem

Early estimates assumed that the detailed specific angular momentum distribution of dark halos is well described by a hypothetical uniform sphere in solid body rotation. This question was reinvestigated again by Bullock et al. (2001) using a large statistical sample of halos that was drawn from a high-resolution ΛCDM simulation. They found that the angular momentum distribution of dark halos has indeed a universal form which however strongly deviates from the previously expected uniformly rotating sphere. The angular momentum distribution can be well fitted by (Fig. 6)

$$M(<j) = M_{vir} \frac{\mu j}{j_0 + j} \tag{7}$$

where $M(<j)$ is the cumulative total mass of dark matter with angular momentum less than j and $\mu > 1$ is a free shape parameter. The characteristic specific angular momentum j_0 is determined by

$$j_0 = \sqrt{2} V_{vir} R_{vir} \lambda' / b(\mu) \tag{8}$$

with

$$b(\mu) = -\mu \ln(1 - \mu^{-1}) - 1. \tag{9}$$

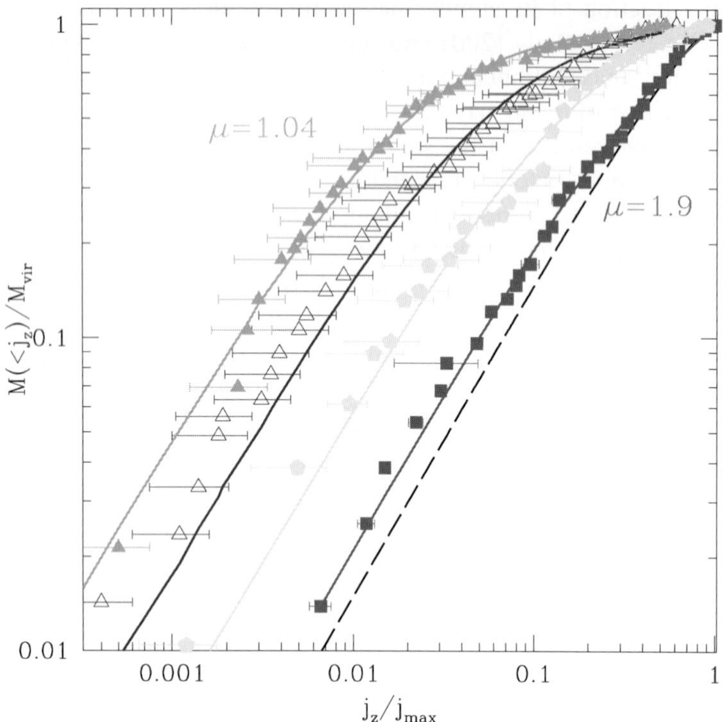

Figure 6. The total mass $M(< j)$ of dark matter with angular momentum less than j is fitted by equation (7). Figure adapted from Bullock et al. (2001).

Note, that this formula ignores any material with negative angular momentum (Chen et al. 2003) which makes the problem of low-angular momentum gas even worse, as discussed in chapter 5.

Following Bullock et al. (2001) we explore the surface density distribution of galactic disks, adopting $\lambda' = 0.04$, a NFW halo with concentration $c = 14$ and a disk baryon fraction of $f_{disk} = M_{disk}/M_{vir} = 0.03$. We keep μ as a free parameter within the measured range of $1.01 \leq \mu \leq 2$ and study its effect on the structure of the disk. Assuming that the dark halo contracts adiabatically due to the gravitational force of the infalling gas, one can calculate the radius r_* in the equatorial plane where gas of a given angular momentum j_* reaches centrifugal equilibrium. r_* is given by the implicit equation

$$j_* = Gr_*(M_{NFW}(r_{cont}) + M_{gas}(< j_*)) \tag{10}$$

where r_{cont} is the initial radius of the dark halo mass shell that after adiabatic contraction ends up at r_* (Jesseit et al. 2002; Blumenthal et al. 1984):

$$r_{cont} = r_* \frac{M_{NFW}(r_{cont}) + M_{gas}(< j_*)}{M_{NFW}(r_{cont})}. \tag{11}$$

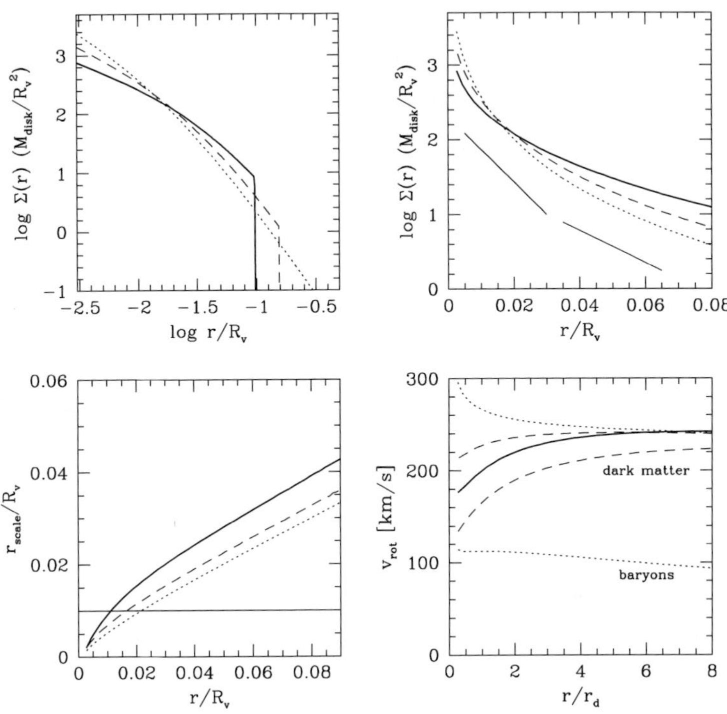

Figure 7. The predicted disk surface density distribution is shown for disks that form from gas with a specific angular momentum profile as predicted by equation 7. The dotted, dashed and solid lines show cases with $\mu = 1.06$, $\mu = 1.25$ and $\mu = 2.0$, respectively. The upper left panel shows a log-log representation. The upper right panel shows the corresponding log-linear plot and compares it with an exponential disk of scale length $0.01R_v$ and $0.02R_v$, where R_v is the dark halo virial radius. The lower left panel shows the local disk scale length, normalized to the virial radius. Typical observed scale lengths are $0.01R_v$. The lower right panel shows the corresponding rotation curves adopting a dark halo mass of $10^{12}M_\odot$, which is characteristic for the Milky Way.

The upper left panel of figure 7 shows the normalized disk surface density profiles for $\mu = 1.06$, $\mu = 1.25$ and $\mu = 2.0$. The profiles agree with Bullock et al. (2001, see their Fig. 20). In this representation it is difficult to estimate which profile, if any, would fit an exponential disk. Therefore the upper right panel shows again the same profiles, however now in a log-linear representation and focussing on the observed exponential disk regime of $r/R_d \leq 4$, where a disk scale length of $R_d \approx 0.01 - 0.02R_{vir}$ has been assumed. An exponential disk with scale length of $0.01R_{vir}$ or $0.02R_{vir}$ is shown by the straight thin solid lines. Even for large values of $\mu \geq 2$ the profiles are not well fitted by an exponential. All profiles instead rise above the exponential in the inner and outer regions. The situation becomes even more clear in the lower left planel which shows the local exponential scale length defined as $r_{scale} = (dln\Sigma/dr)^{-1}$ for all three values of μ. In all cases, r_{scale} is continuously and steeply increasing with radius, with no sign of a plateau at a characteristic scale length of order $0.01 - 0.02R_{vir}$. If one assumes that all material with small scale length $r_{scale} < 0.01R_{vir}$ forms a bulge component one can calculate the predicted bulge-to-disk ratio as function of μ. We find that for all reasonable values of μ, galactic disks should harbour large bulges with bulge-to-disk mass ratios of more than 20% which is not consistent with the population of late-type disk galaxies.

Viscous effects and secular evolution might change the surface density profiles of galactic disks and redistribute their specific angular momentum distribution. It has indeed been shown e.g. by Slyz et al. (2002) that exponential stellar disks would arise naturally from rather arbitrary initial conditions if the star formation timescale is equal to the viscous timescale. It is however unlikely that viscous effects would increase the specific angular momentum of the gas in the innermost regions, by this reducing the bulge mass fraction.

7. Summary

Although, on average, cosmological models of structure formation in a ΛCDM universe generate enough spin to explain the origin of extended galactic disks several problems still remain to be solved. These include the loss of angular momentum during gas infall, the specific angular momentum distribution and exponential disk formation and the origin of bulge-less galaxies. The last problem might be even more puzzling if many bulges formed by secular processes as summarized recently in Kormendy & Kennicutt (2004). Feedback has been invoked as a mechanism to prevent the process of drastic angular momentum loss of infalling gas. (van den Bosch, Burkert & Swaters 2002; Maller & Dekel 2002; Maller, Dekel & Somerville 2002). However, D'Onghia & Burkert (2004) have shown that the dark halos that experienced no major mergers have already too low an angular momentum to produce the observed disks and it is not clear which feedback process would increase the specific angular momentum of the gas beyond that of the dark component in order to explain this result. Selective outflow of especially low-angular momentum gas could provide another solution. Again it seems difficult to understand bulgeless galaxies in this context as in general a bulge component would be required to generate the kinetic energy to drive galactic winds. Another interesting scenario was proposed by Katz et al. (2003) and Birnboim and Dekel (2003) who suggested that gas in high-redshift disk galaxies is accreted cold and without virial shocks directly from filaments. More work along these lines is clearly required to understand the origin of galaxies and their angular momentum in greater details.

Acknowledgments

The authors wish to thank David Block and Ken Freeman for organising a conference on Penetrating Bars through Masks of Cosmic Dust.

References

Abadi, M.G., Navarro, J.F., Steinmetz, M., Eke, V.R. 2003, ApJ, 591, 499
Barnes, J.E., & Efsthatiou, G. 1987, ApJ, 319, 575
Bendo, G. & Barnes, J. 2000, MNRAS, 316, 315
Birnboim, Y., Dekel, A. 2003, MNRAS, 345, 349
Blumenthal, G.R., Faber, S.M., Primack, J.R. & Rees, M.J. 1984, Nature, 311, 527
Bullock, J.S., Dekel, A., Kolatt, T.S., Kravtsov, A.V., Klypin, A.A., Porciani, C., & Primack, J.R. 2001, ApJ, 555, 240
Burkert, A. & Naab, T. 2003, in Galaxies and Chaos, eds. G. Contopoulos & N. Voglis, Springer, p. 327
Chen, D. N., Jing, Y.P., Yoshikaw, K. 2003, ApJ, 597, 35
Cole, S. & Lacey, C. 1996, MNRAS, 281, 716
Courteau, S. 1997, AJ, 114, 2402
Crampin, D.J, & Hoyle, F. 1964, ApJ, 140, 99
Cretton, N., Naab, T., Rix, H.-W. & Burkert, A. 2001, ApJ, 554, 291
de Jong, R.S., Lacey, C. 2000, ApJ, 545, 781
D'Onghia, E. & Burkert, A. 2004, ApJL submitted, (astro-ph/0402504)
D'Onghia, E. & Burkert, A. 2004, this procceding.
Efstathiou, G. & Barnes, J. 1983, in Proc. 3rd Moriond Astrophys. Meeting, Formation and Evolution of Galaxies and Large Structures in the Universe, ed. J. Audouze & J. Tran Thanh Van (Dordrecht: Reidel), p361
Eggen, O.J., Lynden-Bell, D. & Sandage, A.R. 1962, ApJ, 136, 748
Fall, S.M., Efsthatiou, G. 1980, MNRAS, 193, 189
Freeman, K.C. 1970, ApJ, 160, 811
Gardner, J.P. 2001, ApJ, 557, 616
Jesseit, R., Naab, T. & Burkert, A. 2002, ApJ, 571, L89
Innanen, K.A. 1966, AJ, 71, 64
Katz, N., Keres, D., Dave, R., Weinberg, D.H. 2003, in "The IGM/Galaxy Connection: The Distribution of Baryons at z=0", Vol. 281., eds. Jessica L. Rosenberg and Mary E. Putman
Kauffmann, G., White, S.D.M. & Guiderdoni, B. 1993, MNRAS, 264, 201
Kennicutt, R.C. 1998, ARAA, 36, 189
Kormendy, J. & Kennicutt, R.C. 2004, ARAA, in press
Lemson, G., & Kauffmann, G. 1999, MNRAS, 302, 111

Maller, A.H., & Dekel, A. 2002, MNRAS, 335, 487
Maller, A.H., & Dekel, A., Somerville, R. 2002, MNRAS, 329, 423
Mestel, L. 1963, MNRAS, 126, 553
Mo, H.J., Mao, S. & White, S.D.M. 1998, MNRAS, 295, 319
Naab, T., & Burkert, A. 2003, ApJ, 597, 893
Navarro, J. & Benz, W. 1991, ApJ, 380, 320
Navarro, J., Frenk, C.S. & White, S.D.M. 1997, ApJ, 490, 493
Navarro, J.F., Steinmetz, M. 2000, ApJ, 538, 477
Navarro, J. & White, S.D.M. 1993, MNRAS, 265, 271
Peebles, P.J.E. 1969, ApJ, 155, 393
Slyz, A.D., Devriendt, J.E.G-, Silk, J. & Burkert, A. 2002, MNRAS, 333, 894
Sommer-Larsen, J., Gøtz, M., Portinari, L. 2003, ApJ, 596, 47
Steinmetz, R.S. & Bartelmann, M. 1995, MNRAS, 272, 570
Van den Bosch, F.C., Burkert, A., & Swaters, R.A. 2001, MNRAS, 326, 1205
Van den Bosch, F., Abel, T., Croft, R.A.C., Hernquist, L. & White, S.D.M. 2002, ApJ, 576, 21
Vitvitska, M., Klypin, A.A., Kravtsov, A.V., Bullock, J.S., Primack, J.R., Wechsler, R.H. 2002, ApJ, 581, 799
White, S.D.M. 1984, MNRAS, 286, 38
White, S.D.M. & Rees, M.J. 1978, MNRAS, 183, 341
Zeldovich, Y.B. & Novikov, I.D. 1983, in Relativistic Astrophysics, ed. G. Steigman (Chicago: Univ. Chicago Press), p384

THE PROBLEMS WITH GALAXY FORMATION

George Lake
Washington State University, Pullman, WA

Abstract We review the problems with galaxy formation in the current ΛCDM model. The first difficulty is that pure disk galaxies seem to be impossible to build, a problem that is related to both the "angular momentum catastrophe" and the "small scale structure problem". The latter problem is that one sees thousands of satellites in simulations of galaxy halos, while real galaxies appear to have very few. These two problems become intertwined as the lumps cause problems in the formation of disks. They stir things up and transfer angular momentum. Retrograde moving lumps collide with material that might have made a disk and force it all into something that isn't flattened. Finally, more distant satellites might ruin a disk at late times. The last of the small scale problems is the high central densities of dark matter that are predicted but also not seen. We will review the situation and how it started with something that we thought we understood in the 1980's and has gotten progressively murky over the last several years.

Keywords: Galaxy formation, cold dark matter, angular momentum, galaxy morphology

1. Introduction

A theme of this meeting is the "Hubble tuning fork". Nearly 70 years ago, the Hubble (1936) sequence gave a consistent schema to the properties of galaxies. This clearly deserves an explanation in any theory of galaxy formation. The sequence ranges from purely spheroidal systems to pure disks. The full classification scheme includes descriptions of bulge-to-disk ratios, "bars", "lenses", pitch angles of spiral arms and a variety of more detailed features (Sandage 1961). The relative importance of these features is weighted differently in different classification schemes (deVaucouleurs 1959; Morgan 1970; van den Bergh 1960a,b,1976). Our host for this conference has emphasized that many of these features may change if one looks in the infrared versus the optical (Block and Puerari 1999).

It's remarkable how much we understand about the morphology of galaxies, the formation of bars and spirals and the relationship of pitch angles of spirals to luminosity are a few of the details that are reasonably well understood. On the large scale, we have a terrific theory in Cold Dark Matter (CDM) that

D. Block et al. (eds.), Penetrating Bars through Masks of Cosmic Dust, 359–375.

has transformed to Lambda CDM (ΛCDM) with the discovery of dark energy. This theory has passed some remarkable tests and enabled a new precision cosmology with measurements from WMAP (Spergel *et al.* 2003).

For now, our emphasis is on applying the theory of large scale structure to galaxy formation. We will be interested in the crude parameter of the bulge-to-disk ratio with some key questions being: “how are disks made? how are ellipticals made? how are the composite systems made?"

There is no shortage of theories for the origin of the Hubble sequence. Jeans (1938) noted that the flattest spheroids (E7) had the maximum flattening of the Maclaurin sequence of fluid ellipsoids. Since they are fluids, they are rotationally flattened and become bar unstable if spun any faster. He proposed that angular momentum was the control parameter and the dynamical instability of rapidly rotating systems separated disks from spheroids. While it's been nearly 30 years since the discovery that ellipticals aren't rotationally flattened (Bertola and Capaccioli 1975), Jeans' idea still has some compelling elements. Inside key resonances, bars push material inward and they move material outward in the outer parts. This is a nice mechanism to create two component systems. The ellipticals do indeed all rotate too slowly to expect the formation of a bar by dynamical instability. However, the bulges that live with disks are all very flat and all rotate so quickly that they would be bar unstable if they were completely self-gravitating (Lake 1983). Hence, Jeans' idea might yet prove to have some role in separating disks from bulges during formation.

As an interesting aside, Jeans original depiction of the Hubble tuning fork was motivated as much by the behavior of fluid ellipsoids as it was by observations of galaxies. This means that the Hubble sequence doesn't fit Sandage's model for the "best use" of morphological classification. For years, many SO galaxies such as NGC 3115 were classified as E7 because the Hubble tuning fork was so strongly motivated by an equivalent diagram for Maclaurin, Reimann and Dedekind ellipsoids.

A variant of Jeans' idea was to link angular momentum to “overdensity" and hence to the clustering environment (Blumenthal *et al.* 1984). Unfortunately, the range of spins expected from tidal torques is too narrow and poorly correlated with overdensity (Barnes and Efstathiou 1987, Gardner 2001). Since ellipticals are dense and spin slowly while spirals are diffuse and rapidly rotating, the evolution of the angular momentum is at the heart of the matter.

For a long time, the standard model for forming spheroids and disks invoked two distinct epochs of formation. Spheroids were imagined to have formed at higher redshift with Compton cooling off the microwave background while disks were the product of radiative cooling at late times (Hoyle 1945; Gott 1977; Gunn 1982). This reflected the idea that disks were easy to make. The angular momentum distribution and cooling rate was just right if tidally torqued gas collapsed thru a dark halo (Ostriker and Rees 1977, Fall and Efstathiou 1980). It was more difficult to imagine how one made the dense and slowly rotating elliptical. Since the really old globular clusters were associated with the spheroidal component, a much earlier epoch of formation seemed the best solution (*c.f.* Gunn 1982).

Over time, it also became clear that spheroids could be made by merging (Toomre 1977) or "noisy collapses" (Lake and Carlberg 1988a,b). Merging has had it's up and downs. There is a lot of phenomenology of elliptical galaxies that seems at odds with the idea of pure mergers of disks (Ostriker 1980) and this has gotten more pronounced as so many ellipticals show only passive evolution from an early redshift of formation, but some do show signs of "youth" at $z \sim 1$ (van der Wel *et al.* 2004). Galaxies do merge and the remnants have very high mean and central densities (Hibbard and Yin 1999) and their velocity dispersions fit with ellipticals (Lake and Dressler 1986) which seemed like the worst of the early objections to the scheme. The product also has low angular momentum. Many of these features owe to braking of the luminous lumps by dynamical friction within the dark halo (Lake and Dressler 1986)

Toomre's proposal was motivated by the objects in Arp's (1966) Atlas of Peculiar Galaxies that looked like they were cold disks in the midst of merging. What's become clear over the years is that noisy, violent collapses are very different from quiet ones, an issue discussed in §4.

In trying to understand galaxy morphology in a cosmological context, we have focused on the origin of the most basic parameter of the bulge-to-disk ratio. At different times in the history of understanding the origin of morphologies, disks have been considered to be "given" or easy to form with the emphasis being on how one made spheroids. We will show that the puzzle is now how disks are made.

In the last millenium, it was believed that realistic disks would form if the gas retained its angular momentum gained by torques from nearby structures and cooled within dark matter halos (White & Rees 1978, Fall & Efstathiou 1980, Fall 1983, Mo, Mao & White, 1998). First simulations of galaxy formation that included star formation (Lake & Carlberg 1988, Katz 1992) provided strong evidence that hierarchical models do create rotationally supported stellar systems. These showed that the disks also needed quiet collapse environments. To date, simulations of galaxy formation in a full cosmological context have not yet been able to form pure disk galaxies. High resolution simulations of dark matter halos in CDM models have far more substructure ("satellites") than observations. These lumps create several problems: dynamical friction suffered by the dense gaseous lumps and subsequent angular momentum transfer inhibits the formation of disks (Lake and Carlberg 1988a). Since these lumps occur in all CDM collapses, the typical disk scale lengths are much smaller than those observed (Navarro & White 1994). In addition, many of the lumps are retrograde since the angular momentum owing to tidal toques is a small bias on the overall large velocity dispersion of material in the halo. These retrograde lumps hit material that is directly rotating and preventing material in either lumps from being a part of the final disk. Finally, the "satellites" that are lumps in the outer part of the halo are abundant and their collision with the disk would destroy the stellar disk component of spiral galaxies at late times (Moore et al. 1999).

These well-known problems go by the name "the small scale crisis" and the clearly related "disk angular momentum problem". We need to add one last one, the theory predicts very high central densities for the dark matter in contrast to what we see (Moore *et al.* 1996ab). In the end, it's as though CDM has put a sign up in the Universe that says "no disk galaxies allowed", but they are there anyway. Why?

2. The Problems with Galaxy Formation Part 1a: The Lumps that Aren't Seen

The formation of structure in the universe by the hierarchical clustering is an elegant and well-defined theory that explains observations of the universe on large scales (Blumenthal *et al.* 1984). In early simulations, each level of merging appeared to erase the previous generation of substructure, so the model was too efficient to be consistent with the observed hierarchy of structures on the scale of galaxy groups and clusters (White and Rees 1978) . It was not at all clear why the merging would produce smooth galaxies and then stop at larger mass scales. The theorists predicted *"overmerging"* and this effect was reproduced in simulations for several years (White et al. 1987, Frenk et al. 1988). While overmerging was clearly a virtue on the scale of galaxies, it was a severe problem for rich clusters of galaxies. Solutions focused on the role of gas dynamics in making lumps within rich clusters of galaxies (Katz & White 1993). Eventually, Moore, Katz and Lake (1996) showed that numerical heating dominated over physical mechanisms unless simulations had nearly 10^6 particles within the virial radius of a cluster. Simulations with this resolution reversed the picture, overmerging disappeared and halos the size of the Milky Way are predicted to have nearly the same scaled distribution of substructure as the Virgo cluster (Moore et al. 1999, hereafter M99; Klypin et al 1999).

This strong prediction can be tested observationally. A Milky Way sized halo should have $\sim$500 satellites within 500 kpc, with circular velocities greater than 5% of the parent halo's velocity, *i.e.* $V_{cir}/V_{parent} > 0.05$, in contrast to a scant 11 that areobserved (Klypin et al. 1999; M99).

In this branch of astrophysics, when there is a problem, there always seems to be a feedback mechanism to solve it. We refer to this as "an appeal to the gas fairies", which will seem appropriate as you begin to understand the complicated flying dances that they are sometimes asked to perform. In this case, the gas fairies are invoked to give small galaxies a particular structure. The stellar components of the Milky Way satellites must accumulate in the core regions of their dark halos where the characteristic velocities σ are smaller than the asymptotic value of V_{cir}. The observed velocities of Milky Way's satellites would be re-mapped to much higher peak values than expected, shifting the objects plotted in Fig.1 (left panel) to the right until they match the theoretical prediction (Hayashi et al. 2003). There are still many satellites missing at lower peak velocities compared to CDM predictions, but these are declared to have gone dark owing to the ejection of gas from systems with low escape velocities of only 20 - 60 km s^{-1}. These objects are also deficient in the field (*c.f.* Kauffmann, White & Guiderdoni 1993) where the same processes could keep them from being observed.

D'Onghia and Lake (2004) found that the situation is far more vexing. ROSAT found a new class of objects that are called fossil groups (Ponman *et al.* 1994). Fossil groups are defined as having a dominating giant elliptical galaxy with the next brightest object being 2 magnitudes fainter, embedded in a X-ray halo with a luminosity 10-60 % of the Virgo cluster (Vikhlinin *et al.* 1999, Jones *et al.* 2003). RXJ1340.6+4018 at redshift 0.171 is the archetype with a bright isolated elliptical galaxy $M_R = -22.7$, surrounded by dark matter and a hot gaseous halo. The spatial extent of the X-ray emission, $\sim$ 500 kpc, the total mass, $\sim 6 \cdot 10^{13}$ $M_{\odot}$, and the mass of the hot gas correspond to a galaxy cluster $\sim$ 40% as massive as Virgo, and the optical luminosity of the central galaxy is comparable to that of cluster cD galaxies (Jones *et al.* 2000). Five additional fossil groups have been confirmed spectroscopically. For one of them, RXJ1416.4. the X-ray temperature is estimated to be $\sim$ 1.5 keV (Jones *et al.* 2003).

Fossil groups are the most massive overmerged systems that have been seen. We note that there has been some confusion about overmerging on the group scale and even some confusion about what to call groups. Zabludoff and Mulchaey (2000) have pointed to the Local Group as an archetypal loose group. We would refer to the Local Group as a "future group" since it is still collapsing and reserve the term "group" for bound and virialized objects. The reason that we belabor this is that if we mistakenly treated the Local Group as a bound and virialized group, then the Milky Way would be the second brightest galaxy and this would appear to be anomalously massive compared with other substructure functions. There is also a tendency to ask for too much in the defininition "overmerging". Fossil groups have one dominant galaxy that sits at the center of the X-ray isophotes and is at the kinematic center. Since there are a few other faint galaxies in these systems, some have asked: "why haven't they overmerged". So, we should be clearer about this term.

Galaxies like the Milky Way are the archtypes of overmerged structures, while the Virgo cluster serves the same role for systems that have preserved their substructure. So, the cleanest way to talk about overmerging is to compare a substructure function to these two systems. The substructure function is defined as the number of objects with a circular velocity greater than a fraction of the parent's circular velocity. There is a weak dependence of substructure function on mass within the ΛCDM model. In the left panel of figure 1 (adopted from D'Onghia and Lake 2004), we show substructure functions for the fossil group, a few clusters, the Local Group and Centaurus. The right panel shows composite functions derived from an average luminosity function (Hunsberger, Charlton and Zaritsky 1998) for 39 Hickson (1982) compact groups (HCGs) and 5 loose groups in Zabludoff & Mulchaey (2000),

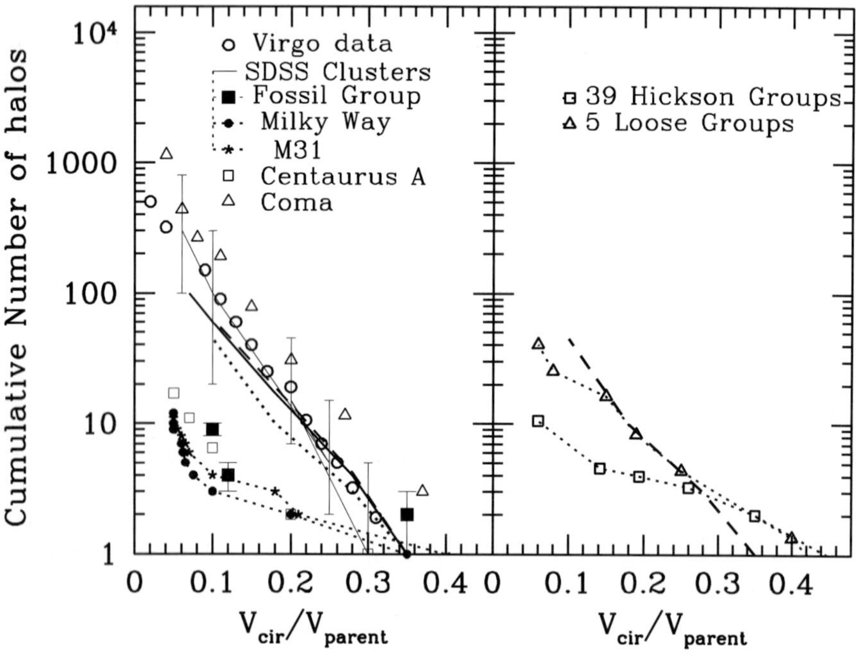

Figure 1. The substructure functions are shown for a variety of systems. In the left panel, there are the clusters like Virgo, Coma and a composite from the SDSS that look like the predictions of the ΛCDM simulations. There are then the "overmerged systems" that include typical galaxies and a Fossil Group that is roughly 40% of the mass of Virgo. The right panel shows some Hickson groups and loose groups that have somewhat different behavior. They have more substructure than expected at the highest masses and then flatten out.

The substructure functions show that the fossil group RXJ1340.6+4018 is in an overmerged state similar to the Milky Way. The HCGs look more complicated. These are selected to have at least 4 galaxies within 3 magnitudes of the brightest. As a result, the substructure function starts off fairly high at the bright end, but then it quickly flattens out. A few of the HCG look like relaxed groups with observed X-ray emission (Heldson and Ponman 2000, Horner, Muchotsky and Scharf 1999), but even the richest HCGs with X-ray emission often break up into multiple components (Zabludoff and Mulchaey 2000). HCGs often appear to have undisturbed spirals which is impossible if they were really in bound virialized groups (Moore *et al.* 1999ab). Hernquist, Katz and Weinberg (1995) have suggested that many of the HCGs are projections of filaments along the line of sight. Such a projection projection would produce their odd substructure functions.

The objects that are missing around galaxies like the Milky Way and M31 are missing in the field as well. This has been used to argue for a mass cutoff in the substructure function. One could either propose a model where there was a roll off in the number of small mass objects (*c.f.* Governato *et al.* 2004) or appeal to their low binding energies to blow out gas and make them invisible (*c.f.* Hayashi *et al.* 2003). The fossil groups are extremely interesting insofar as they run counter to this argument. The fossil groups are missing objects that are clearly present in the Virgo cluster and the field. Indeed, they are missing objects that are nearly as bright as L* galaxies.

3. The Problems with Galaxy Formation Part 1b: Are the Lumps Felt but not Seen?

Moore *et al.* (1998) noted that the substructure that was predicted by CDM simulations could just be dark. There were concerns about their consequences, especially the problem of heating up the disks of thin galaxies. However, if they were there and dark, they should be detected by lensing using the brightness ratios of multiple images. With this technique, Dalal and Kochanek (2002) used a sample of isolated ellipticals that are likely the centers of fossil groups. They find that a few percent of the mass in halos is in clumps with masses in the range of $10^6 - 10^9 M_{\odot}$. This is still a factor of a few below what is seen in simulations (Moore *et al.* 1999; Klypin et al. 1999; Ghigna et al. 2000). It could be lower still as the expectations were normalized to the properties of an isolated galaxy while the sample of galaxies was characteristic of the galaxies that live in the middle of much more massive fossil groups. On the other hand, Zentner and Bullock (2003) found that Dalal and Kochanek's model underestimated the substructure since they placed substructures uniformly in the halo rather than allowing for stripping and destruction in the central regions where they had the greatest sensitivity. Most of the substructure that is causing anomalous lensing will have to be projected within the Einstein radius of the lens rather than be within that radius of the center within the parent halo. All of this debate over normalization may be irrelevant. Evans and Witt (2003) argue that the case for lumps has been overstated as they fit cases of anomalous brightness ratios with a smooth lens,

There should be a lot of lumps that are more massive than 10^9 $M_\odot$ in fossil groups and they have very different consequences. When the lumps are this large, the positions of images will shift betraying individual lumps rather than just altering the statistical properties of brightness ratios. This was not seen by Dalal and Kochanek (2002). If there were dark lumps as large as L*, these would cause shifts in the center of mass that could be seen with weak lensing maps. Here, one would compare the centers defined by the brightest galaxy, the X-ray emission and the lensing map. These will disagree if there is significant clumping. When the X-ray isophotes are smooth and well-determined, the center of the X-ray emission generally agrees with the location of the brightest galaxy (Mulchaey 2000), but the lensing map should be a more sensitive test.

We can now move on to what are likely bigger problems with the lumps expected in ΛCDM .

4. The Problems with Galaxy Formation Part IIa: No Pure Disk Galaxies Allowed

In the early 1980's, Carlberg and I started a project to put galaxy formation and morphology into a cosmological context. At the time, it seemed that the main parameters would be the initial angular momentum and the cooling timescale. There were other parameters that we considered such as the ratio of dark to luminous matter, but it seemed likely that these were fixed within a narrow range. We had unique access to the Cray computers at AT&T Bell Laboratories when nearly every other Cray was behind a large fence in a defense laboratory. At Universities, you couldn't justify a computer unless it was oversubscribed. At AT&T, an oversubscribed computer meant that human resources were being wasted, so the load level was targeted to an average of 30% to enable important tasks such as the rapid design of new generation switches. I made a deal with the head of the computing center so that we could run galaxies at low priority and disappear when the paying customers wanted the machine. As a result, we were able to explore a wide range of parameters as well as looking at galaxy formation in "cosmological context". Our notion of this seems extremely dated today. We started the simulations as isolated top hats in expansion with maximum expansion being just a factor of two from the starting radius.

We ran a lot of simulations before we understood that most of the parameters weren't significant. We revised the code and ignored a lot of the final parameters in the final runs. We quickly realized that we were making nice disks that were thin and had spiral arms and we also had ellipticals that had modest flattening and were slowly rotating. The thing that it took us a while to understand (this work predated graphics workstations) was that the slowly rotating ellipticals were much denser than the spirals even though they started with the same density and angular momentum. The difference was the degree of substructure in the collapses. The quiet collapses behaved as they were expected to and formed disks. An aside is that we were able to make nice thin disks, but the object that we showed in our paper was the worst, puffiest thing that we still called a disk. We did this to make it clear how we had classified the final objects, but it caused some confusion to those who ran simulations and reproduced our results (Katz 1992; Steinmetz and Muller 1995).

A simple way to characterize the specific angular momentum of galaxies is their rotation velocity times their half mass radius, $v_{rot}r_{half}$. This number is nearly a factor of 10 lower for ellipticals than it is for spirals. We found this difference in our simulations even though our initial states had identical specific angular momentum. The difference for the ellipticals was in three factors, each contributing about a factor of two. The first was that the lumpy collapses transferred angular momentum from the inner parts where there was lots of gas and lumps to the outer parts where it was mostly dark matter. The chaotic collapses also threw a lot of the highest angular momentum material to large radius where it never cooled, reducing the specific angular momentum of the material that made stars and matching the phenomenology of ellipticals having hot gas in their halos. Finally, while an elliptical is more dense than a spiral in the inner parts, the $e^{-r^{1/4}}$ compared to an exponential disk means that there is also a lot more material at large radii. Hence, an elliptical has twice as much angular momentum relative to a disk for the same value of $v_{rot}r_{half}$.

In recent years, even the quiet collapse models for disks have run into trouble. If angular momentum is conserved in detail, the disks are too centrally concentrated (van den Bosch 2001). Tidal torques are such a weak bias that much of the gas in a protogalaxy is counter rotating, making it impossible to form a pure disk with detailed conservation of angular momentum (van den Bosch *et al.* 2002). We are back to: "when your back is to the wall, invoke feedback", or "here come the gas fairies". Even in the quiet collapse model, there is a push to add preheating to solve this problem of the general angular momentum distribution and the negative angular momentum gas. However, pre-heating tends to unbind the distant gas and leaves behind an even larger fraction of negative angular momentum gas (van den Bosch *et al.* 2003).

5. The Problems with Galaxy Formation Part IIb: The Latest Failures to Make Disks

If the abundant subhalos were the only problem, any mechanism that reduced the lumpiness of galaxy assembly would lead to realistic disks. There were also some resolution problems in the early simulations of "the angular momentum castastrophe". It was telling that the final state was always an object in solid body rotation that was the size of the viscous scale. Simulations with N $\sim 10^4$ suffer from excessive artificial viscosity that would overestimate the angular momentum transfer, as was early recognized in previous works (e.g Sommer–Larsen, Gelato & Vedel 1999). A poorly resolved disk will also be damaged by two–body effects that continue to transfer angular momentum from the disk even after its fully formed (Mayer et al. 2001).

Together, the failure to build galactic disks similar to those observed and the overabundance of substructure pose a formidable challenge to the ΛCDM paradigm, suggesting that a fundamental ingredient is missing from our understanding of the formation of galactic and subgalactic structures. Let's see what the gas fairies can do for us. There could be strong feeback from SN (e.g Thacker & Couchman 2000, 2001) or an external UV background produced by QSOs and massive stars (Quinn, Katz, & Efstathiou 1996, Benson et al. 2002). However, as we mentioned at the end of the last section, some things seem to get worse with feedback (van den Bosch *et al.* 2003).

The problems seem firmly attached to the global model so alternatives to ΛCDM (*c.f.* Spergel & Steinhardt 2000) have been considered. Warm Dark Matter (WDM) (Pagels & Primack 1982) lacks power on small scales which will certainly reduce the number of subhalos and allow a smoother accretion of gas, forming larger disks (Bode, Ostriker & Turok 2001, Colin, Avila-Reese, & Valenzuela 2000). Some of these solutions have been explored in recent studies (e.g. Thacker and Couchman 2001; Sommer–Larsen & Dolgov 2001; Sommer–Larsen, Gotz & Portinari 2003; Abadi *et al.* 2003).

These most recent works have formed a rotationally supported stellar component, but there is a massive stellar spheroid in each case as well. In simulations with just a few thousand particles representing the virialized part of a galaxy dark matter halo (e.g Eke, Navarro & Steinmetz 2001), a clear disk can not be seen. Thacker and Couchman (2001) used $\sim 1.5\ 10^4$ dark matter (DM) particles with the halo virial radius; their "bulge" component is still almost twice as massive as the thin disk. The bulge component decreases to only 50% of the stellar disk in the simulation presented by Abadi et al (2003) that had $3.6\ 10^4$ DM particles within the virial radius of their simulated galaxy. While these results are encouraging, previous simulations do not clarify if stronger feedback was necessary, or just sufficient to avoid systematic loss of angular momentum during the formation of galactic disks. Runs with larger particle numbers also had larger disks, suggesting that resolution effects still play a major role. Important differences in the results might also arise from the different merging histories of each individual halo simulated, the spin of their halos (e.g Thacker and Couchman [2001] simulated a halo with a larger than average spin which in turn makes it easier to form larger disks) and the different SN feedback recipes used in various works.

Governato *et al.* (2004) ran a new set of smoothed particle hydrodynamic (SPH) simulations that used over a million dark matter particles and nearly a quarter of a million gas particles. These draw a galaxy from a cosmological volume that was 100 Mpc across. A candidate halo for a "typical spiral galaxy" is selected based the halo's environment (relatively isolated), mass and internal velocities. This approach gives true meaning to the phrase "cosmological context". The simulations had cooling, star formation, supernova feedback and a UV background. They were able to form disk dominated galaxies in a 2 different (ΛCDM and ΛWDM) flat cosmologies. The resources required for such a simulation are not conducive to parameter studies.

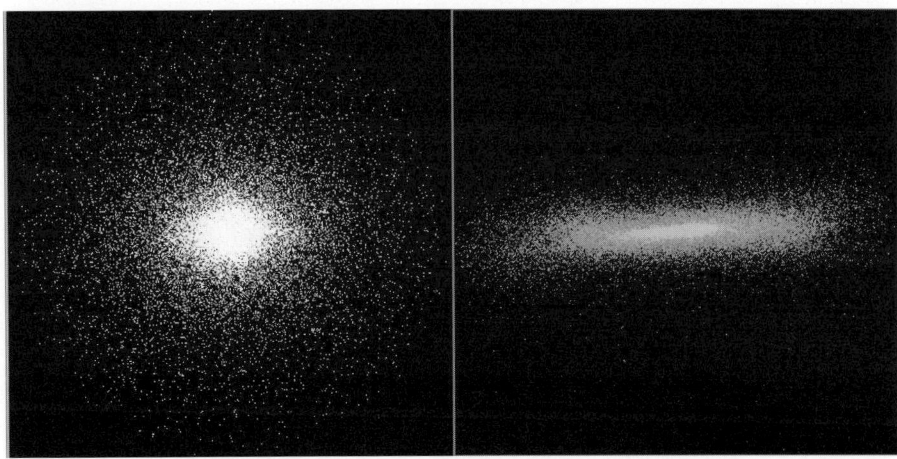

Figure 2. The bulge and the disk of the ΛCDM simulation. The panels are 40 kpc across. The left panel shows all the stars that formed before a redshift of 6 while the right panel shows those that formed after a z of 1.5. The disk shows a thin and thick component that largely owes to the heating by the larger satellites.

The final galaxies in both simulations were disk dominated with characteristic rotation velocities of 250-275 $km\ s^{-1}$, the ratio of bulge-to-disk was 1:2.8 in ΛCDM and 1:4 in ΛWDM). The stars and gas lost 60% of their angular momentum in ΛCDM run versus 10% in the ΛWDM run. The loss in the ΛWDM run likely owed to numerical effects such as artificial viscosity. The ΛWDM galaxy was more disk dominated at higher redshift, but it was bar unstable resulting in some transfer of material to the inner parts enhancing the bulge component of the fit at the expense of the disk component. The disk was nearly at the threshold for the bar instability and the growth and transfer of material was likely boosted a bit by numerical effects. The continued flux of satellites heated the disk of the ΛCDM galaxy to a radial dispersion of over 100 $km\ s^{-1}$ while the ΛWDM galaxy had a very reasonable value of 40-50 $km\ s^{-1}$.

The final result is very encouraging, but it's still not clear that a "pure disk" can be produced in this cosmology. In their resolution studies, Governato *et al.* (2004) found that they had not reached results that had converged. There are obvious problems with viscous transport of angular momentum, but more subtle ones are likely as well. For example, we noted that there is problem with retrograde material which is made even worse when it's in lumps. The ΛWDM run smooths out the lumps, but it could be that SPH with feedback is too effective at eliminating the retrograde material. This will become clearer as resolution increases with both the power of computers and the increasing efficiency of algorithms.

6. The Hubble Sequence of Masses

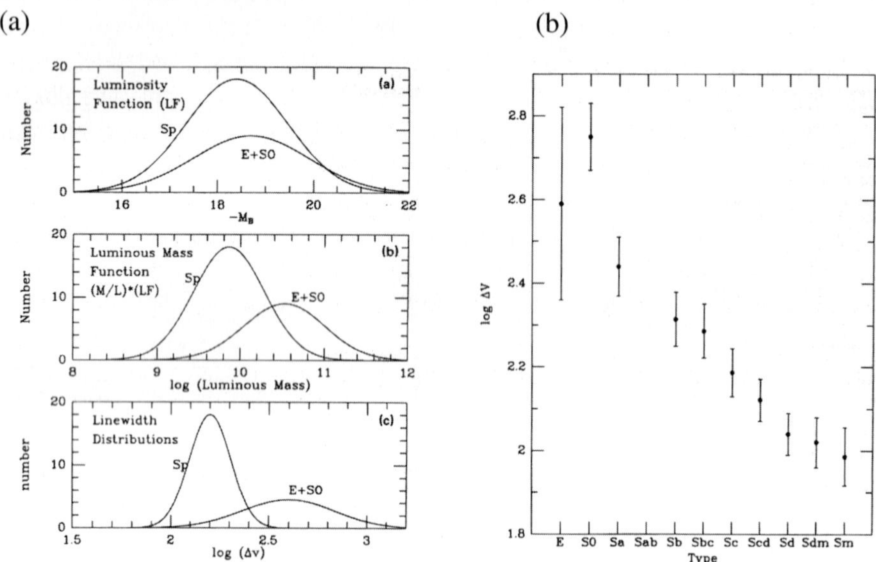

Figure 3. These show the tight correlation between Hubble type and mass or linewidth. The left figure shows that the correlation is masked somewhat by the difference in M/L ratios for the two types of galaxies. As a result, the luminosity functions for spirals and ellipticals (taken from Binggeli, Sandage and Tammann [1985] is more similar that the luminous mass functions or the linewidth distributions. The panel on the right shows that the correlation between linewidth and type is a clear progression across the Hubble sequence. The data for types Sb-Sm was compiled by Meisels with Sa's added from the sample of Rubin *et al.* (1985) and S0's from Dressler (1987). Both panels are from Lake and Carlberg (1988). Together with the Fisher-Tully relation, this is a powerful constraint on the role of secular evolution in the current Universe. Very few early type galaxies could be the result of the secular evolution of late types.

One of the most interesting features of the Hubble Sequence is the strong correlation of Hubble type with mass or linewidth. If you know the linewidth of a galaxy, you also know it's Hubble type to $\pm$ half a type (where Sab is half a type from Sb) with more than 80% probability (Meisels 1983, Lake and Carlberg 1988). Lake and Carlberg (1988) used this feature together with the idea that noisy collapses make ellipticals and quiet collapses make disks to propose that there is a mass cutoff to lumps. The observed lump distribution fits this. Making the lumps dark takes us back to the problem of "no pure disks allowed".

7. The Problems with Galaxy Formation Part III: Where are the cuspy centers?

One of the first CDM problems on small scales was the central density profiles. The theory predicts very cuspy density profiles while the observations show that the central density profiles are shallow with a fairly large core radius (Moore 1994; Moore *et al.* 1999). As the simulations have gotten better, the cores have gotten steeper (Ghigna *et al.* 2000). Even in the papers that work hard to stretch the observations of galaxies to meet the ΛCDM predictions (Swaters *et al.* 2003), the best that can be done is to throw out several galaxies and match simulations with poor resolution that produce lower concentration halos with less singular cores. Lensing studies of clusters also find flat central density profiles (Tyson, Kochanski., and dell'Antonio 1998).

The first thought that comes to mind is that lumps might be able to stir up the center and produce a core. But, it's easy to see why this will fail. A lump can only get into the center without disruption if it has a higher density than the central region. If it has a higher density and a smaller mass, it will have an even higher phase density than the material that is already in the core. Hence, the effect of lumps will be to build up an even higher density core. Ma and Boylan-Lolchin (2004) tried to use lumps to stir up the center of a halo and lower the density. From reading their paper, you might think this worked. However, they showed just what is expected. The particles that were originally in the center are moved out, but these are displaced by even higher density material from the lumps creating a steeper final profile.

Another proposal to soften the central density and create a dark matter core uses resonant interactions of a bar to move material out of the central region (Weinberg and Katz 2003, Holley-Bockelmann, Weinberg, and Katz 2003). Cores are seen in clusters of galaxies as well as disk galaxies, so presumably one would have to use resonant heating of lumps that live outside the core in the case of clusters of galaxies. If lumps can transform the halo, it's reasonable to ask why the simulations resolve lumps but the profiles just get progressively steeper with increasing resolution. Since the interaction involves resonances, advocates claim that simulations with fewer than 10 or even 100 Million particles are too noisy to see the resonant interactions. This assertion is problematic. In simulations with fewer than 100,000 particles, scattering out of resonances is dominated by the particle noise. However, once one reaches a million particles, the scattering out of resonances is dominated by the lumps rather than the particles (Moore, Katz and Lake 1996). As the particle number is increased beyond a million particles, the particle noise decreases but the lumps are better preserved (Borgani *et al.* 2002; Taylor, Silk and Babul 2003). The ratio of the heating from lumps versus particles goes as $\eta N_{lump} \lambda$, where η is in the fraction of mass in lumps with N_{lump} particles and λ is a factor of order unity that takes into account the fact that the lumps are "softer" than the particles (this parameter is related to the difference in the Coulomb logarithm for a softened particle versus a lump that is "softened" by it's half mass radius). As a result, simulations with $\sim 10^6$ particles are about as quiet as ΛCDM simulations will get, so the resonant heating mechanism will not work within the standard ΛCDM model. If most of the lumps were eliminated and the largest ones acted resonantly like a bar, then one might suppose that ΛWDM would work. But, ΛWDM wouldn't cut enough power to reduce lumps that dominate the scattering in clusters and would cut too much power to work in dwarf galaxies (where one could appeal to the bar again though). It would only work in the mass range where the cores are so poorly determined that there isn't a known problem. To rephrase this argument, the lumps dominate the scattering of particles out of resonance in both simulated and real clusters of galaxies. So, it is difficult to see how resonant interactions could create a core in the dark matter distribution of a cluster of galaxies. A greater problem with this solution is evident in the paper by Holly-Brockelmann in this volume. While the bar creates a core in the dark matter, the transferred material in the disk makes the final rotation curve STEEPER. The shallow rotation curves of late type galaxies is the fundamental problem that must be solved.

8. Summary

There are several interrelated problems on small scales in the current model:

- Galaxies and clusters seem to have cores while the simulations produce cuspy central profiles
- There is too little structure on small scales compared to the ΛCDM model
  - The problem with too little structure on small scales creeps upward in fossil groups, so it may not to be a simple mass cutoff
  - If the lumps are there but dark, there is a problem with making pure disk galaxies. Since the fossil groups are purely elliptical systems, making the lumps dark there would not have such a problem
- Disks are hard to make. It would get easier without structure on small scales, but retrograde material might remain a problem.

Cutting power below a specific mass scale helps some of these problems, but it doesn't do anything for the problem of cores. There is the odd problem with fossil groups, but maybe it's reaching far enough from the basic set of problems that it should be solved independently. Finally, there is the problem with making pure disks given the amount of retrograde material that is expected.

There are some interesting "peculiarities" to galaxy formation as well. While most of these will likely fit into the current cosmological context, they are clearly worth looking at. Most ellipticals are really old and fit models of passive evolution with redshifts of formation greater than 3 (Holden *et al.* 2004). These galaxies are clearly more massive than the disk galaxies that appear to have formed more recently. Why is there such a strong correlation of formation epoch with mass? This correlation is opposite to the naive expectation that massive objects take longer to form. Is this just a correlation of the star formation time rather than the dynamical assembly? In the noisy collapses, ellipticals might make all their stars promptly whereas disks have to wait until the disk has built up to be unstable to the formation of giant star forming clouds. There is an odd inverse age-metalicity relationship on extragalactic scales. The oldest populations are in ellipticals and are metal-rich while the youngest are in metal-poor dwarfs. This is normally attributed to binding energies, but there might be more to this.

Here, we've looked at the problems on the smallest scales where one expects a cosmological model like ΛCDM to apply. There has been a tendency to "overdefend" these models. There were talks for decades that tried to convince the world that $\Omega = 1$ when there wasn't any evidence that it might be close. While a flat Universe fits very well, the original problem with $\Omega \neq 1$ was that it put us in a special epoch. It was an intolerable cosmic coincidence that we should live in that special epoch. Now, we have a lot of very strange coincidences where we are in the epoch where all of the ratios between dark energy, matter, radiation and normal matter are neither fantastically large nor small. A sinking suspicion might be that dark energy and matter and two parts of a complicated equation of state. This could well manifest itself by changing the power spectrum on small scales and imprinting of cores in the dark matter distribution.

Acknowledgments

I wish to thank the many fabulous collaborators that I've had on these topics: Ray Carlberg, Elena D'Onghia, Fabio Governato, Ben Moore, Joachim Stadel, Sebastiano Ghigna, Jeff Gardner, Tom Quinn and James Wadsley. This work has been supported by the NSF and NASA. I still owe an enormous debt to Nils-Peter Nilson for huge amounts of Cray time while at AT&T Bell Labs.

References

Abadi, M.G., Navarro, J.F., Steinmetz, M. and Eke, V. 2003 *Ap. J.*, 591, 499

Benson, A. J., Bower, R. G., Frenk, C. S., and White, S. D. M. 2000, *MNRAS*, 314, 557

Bertola, F. and Capaccioli, M. 1975, *Ap. J.*, 200, 439

Borgani, S., Governato, F., Wadsley, J., Menci, N., Tozzi, P., Quinn, T., Stadel, J. and Lake, G. 2002, *MNRAS*, 336, 409

Colin, P., Avila-Reese, V., and Valenzuela, O. 2000, *Ap. J.*, 542, 622

Dalal, N. and Kochanek, C. S. 2002, *Ap. J.*, 572, 25

D'Onghia. E. and Lake, G. 2004, preprint, astro-ph/0309735

Dressler, A. 1987, *Ap. J.*, 317, 1

Eke, V. R., Navarro, J. F., and Steinmetz, M. 2001, *Ap. J.*, 554, 114

Evans, N. W. and Witt, H. J. 2003, *MNRAS*, 345, 1351

Frenk, C.S., White, S.D.M., Davis, M., Efstathiou, G. 1988, *Ap. J.*, 327, 507

Gardner, J. P. 2001, *Ap. J.*, 557, 616

Ghigna, S., Moore, B., Governato, F., Lake, G., Quinn, T., Stadel, J. 2000, *Ap. J.*, 544, 616

Gott, J. R. 1977, *ARA&A*, 15, 235

Governato, F, Mayer, L, Wadsley, J., Gardner, J.P., Willman, B., Hayashi, E., Quinn, T., Stadel, J. and Lake, G. 2004, preprint, astro-ph/0207044, *Ap. J.*, in press.

Gunn, J. E. 1983, IAU Symp. 100: Internal Kinematics and Dynamics of Galaxies, 100, 379

Hayashi, E., Navarro, J. F., Taylor, J. E., Stadel, J., Quinn, T. 2003, *Ap. J.*, 584, 541

Helsdon, S.F., Ponman, T.J. 2000, *MNRAS*, 319, 933

Hernquist, L. Katz, N. and Weinberg, D. H. 1995, *Ap. J.*, 442, 57

Hibbard, J. E. and Yun, M. S. 1999, *ApJ*, 522, L93

Hickson, P. 1982, *Ap. J.*, 255, 382

Holden, B. P., Stanford, S. A., Eisenhardt, P., and Dickinson, M. 2004, *AJ*, 127, 2484

Holley-Bockelmann, K., Weinberg, M. D. and Katz, N. 2003, preprint, astro-ph/0306374

Horner, D. J., Mushotzky, R. F., Scharf, C. A. 1999, *Ap. J.*, 520, 78

Hoyle, F. 1945, *MNRAS*, 105, 287

Hubble, E. P. 1936, *Realm of the Nebulae*, Yale University Press,

Hunsberger, S. D., Charlton, J. C., Zaritsky, D. 1998 *Ap. J.*, 505, 536

Jeans, J.. 1938, *Cosmology and Cosmogony*, Cambridge University Press,

Jones, L. R., Ponman, T. J., Forbes, D. A. 2000, *MNRAS*, 312, 139 (JPF00)

Jones, L.R., Ponman, T.J., Horton, A., Babul, A., Ebeling, H., Burke, D.J. 2003, *MNRAS*, 343, 627

Katz, N., Whitr, S.D.M. 1993, *Ap. J.*, 412, 455

Kauffmann, G., White, S. D. M., Guiderdoni, B. 1993, *MNRAS*, 264, 201

Klypin, A., Kravtsov, A., Valenzuela, O., Prada, F. 1999, *Ap. J.*, 522, 82

Lake, G. and Dressler, A. 1986, *Ap. J.*, 310, 605

Lake, G. and Carlberg, R. G. 1988, *AJ*, 96, 1581

Lake, G. and Carlberg, R. G. 1988, *AJ*, 96, 1587

Ma, C-P. and Boylan-Kolchin, M. 20034 preprint, astro-ph/0403102.

Mayer, L., Governato, F., Colpi, M., Moore, B., Quinn, T., Wadsley, J., Stadel, J., and Lake, G. 2001, *Ap. J.*, 559, 754

Meisels, A. 1983, *A&A*, 118, 21

Mo, H . J., Mao, S., and White, S. D. M., 1998, MNRAS, 295, 319

Moore, B., Katz, N., Lake, G. 1996, *Ap. J.*, 457, 455

Moore, B., Katz, N., Lake, G., Dressler, A., and Oemler, A. 1996, *Nature*, 379, 613

Moore, B. et al. 1999, *ApJ*, 524, L19 (M99)

Morgan, W. W. 1970, IAU Symp. 38: The Spiral Structure of our Galaxy, 38, 9
Mulchaey, J. 2000, ARA&A, 38, 289
Ostriker, J. P. 1980, *Comm. Astrophys..*, 8, 177
Pagels H. and Primack, J.R., 1982, *Phys. Rev. Lett..*, 48, 223
Ponman, T.J. et al. 1994, *Nature*, 369, 462
Quinn, T., Katz, N., and Efstathiou, G. 1996, *MNRAS*, 278, L49 (Washington, D.C.: Carnegie Institution of Washington)
Sommer-Larson, J., and Dolgov, A., astro-ph/9912166, *Ap. J.*in press.
Sommer-Larsen, J., Gelato, S., and Vedel, H. 1999, *Ap. J.*, 519, 501
Sommer-Larson, J., Gotz, M. and Portinari, L. 2003 *Ap. J.*, 596, 47
Spergel, D. N. and Steinhardt, P. J. 2000, *Phys. Rev. Lett.*, 84, 3760
Spergel, D. N., et al. 2003, *ApJS*, 148, 175
Steinmetz, M. and Muller, E. 1995, *MNRAS*, 276, 549
Swaters, R. A., Madore, B. F., van den Bosch, F. C., and Balcells, M. 2003, *Ap. J.*, 583, 732
Taylor, R. E., Silk, J. and Babul, A. 2003, in IAU Symposium 220, "Dark Matter in Galaxies", ed. S. Ryder, D.J. Pisano, M. Walker and K. Freeman (San Francisco: ASP).
Thacker, R. J., and Couchman, H. M. P., 2000, ApJ, 545, 728
Thacker, R. J. and Couchman, H. M. P. 2001, *ApJ*, 555, L17
Toomre, A. 1977, in *Evolution of Galaxies and Stellar Populations*, ed. B. M. Tinsley and R. B. Larson (New Haven: Yale University Observatory), p. 401
Tyson, J. A., Kochanski, G. P., and dell'Antonio, I. P. 1998, *ApJ*, 498, L107
van den Bergh, S. 1960a, *Ap. J.*, 131, 215
van den Bergh, S. 1960b, *Ap. J.*, 131, 558
van den Bergh, S. 1976, *Ap. J.*, 206, 883
van den Bosch, F. C. 2001, *MNRAS*, 327, 1334
van den Bosch, F. C., Abel, T., Croft, R. A. C., Hernquist, L., and White, S. D. M. 2002, *Ap. J.*, 576, 21
van den Bosch, F. C., Abel, T., and Hernquist, L. 2003, *MNRAS*, 346, 177
van der Wel, A., Franx, M., van Dokkum, P. G., and Rix, H.-W. 2004, *ApJ*, 601, L5
Vikhlinin, B.R. et al. 1999, *Ap. J.*, 520, L1
Weinberg, M. D. and Katz, N. 2002, *Ap. J.*, 580, 627
Zabludoff, A. and Mulchaey, J. 2000, *Ap. J.*, 539, 136
Zentner, A.R., Bullock, J.S., preprint, astro-ph/0304292
White, S. D. M., and Reese, M. J. 1978, *MNRAS*, 183, 341
White, S.D.M., Davis, M., Efstathiou, G., Frenk, C.S. 1987, *Nature*, 330, 451

THE INTERPLAY BETWEEN BARS AND DARK MATTER HALOS

Kelly Holley-Bockelmann
University of Massachusetts, USA

Abstract The evolution of a stellar bar transforms both the galaxy disk and its host dark matter halo. We present high resolution, fully self-consistent N-body simulations that clearly demonstrate that dark matter halo central density cusps flatten as the bar torques the halo. This effect is independent of the bar formation mode and occurs even for rather short bars. The halo and bar evolution is mediated by resonant interactions between orbits in the halo and the bar pattern speed, as predicted by linear perturbation theory. The bar lengthens and slows as it loses angular momentum, a process that occurs even in rather warm disks.

Keywords: galaxies: spiral, galaxies: kinematics and dynamics, galaxies: structure, methods: n-body simulations

1. Introduction

It is widely accepted that a galactic bar will rearrange the stellar and gaseous disk, but the bar's effect on the dark halo is much more controversial. Once thought to be merely a rigid, stabilizing mass that inhibits bar growth, a dark matter halo can communicate with a galaxy disk, goading bar formation and responding to the bar in a subtle interplay that is dictated, in part, by linear perturbation theory. In general, linear perturbation theory suggests that angular momentum transfer drives long-term galaxy evolution, and this transfer is mediated by orbits that are in resonance with quasi-periodic perturbers. In the case of a barred galaxy, the bar provides a huge, organized source of angular momentum from which resonant orbits in the rest of the system can draw; substantial angular momentum is transferred from the bar to the halo via resonances between the bar pattern speed and the orbits of dark matter particles in the inner halo (Lynden-Bell & Kalnajs 1972; Tremaine & Weinberg 1984; Athanassoula 2002). This orbit-based picture of galaxy evolution starts from a fundamentally different perspective than the 'global response' paradigm, and leads to new insights about what drives secular galaxy evolution.

Getting the physics right on the 'microscopic' level of an individual orbit response can have effects on a galactic scale with cosmological implications.

D. Block et al. (eds.), Penetrating Bars through Masks of Cosmic Dust, 377–386.

Cold dark matter (CDM) halos in every mass regime are thought to form with a characteristic density profile, expressed as $\rho(r) \propto r^{-\gamma}(1+r/r_s)^{\gamma-3}$ (Navarro et al 1997, hereafter NFW). While disagreements remain on the precise value of the central slope, γ, which range between -1 and -1.5 (Moore et al 1998; Power et al 2003), there is a consensus that primordial dark matter halos are universally cuspy. This prediction is testable using rotation curve decomposition to constrain both luminous and dark matter density profiles. Though this technique suffers from degeneracies in interpretation, many real dark matter halos appear to be much less cuspy than those predicted by standard CDM (de Blok et. al. 2001; McGaugh 2000; Swaters et. al. 2003), and some are even consistent with a flat central profile, or 'core'.

If the resonant dynamic regime outlined above applies to real galaxies, and if bars are a natural phase of early galaxy evolution, the cusp–core controversy could be reconciled through bar-halo interactions. In this scenario, primordial disks form in a cuspy dark matter halo. Such young galaxies are repeatedly perturbed by satellites (Tóth & Ostriker 1992; Steinmetz & Navarro 2002), which triggers a primordial bar (Binney & Tremaine 1987; Walker, Mihos & Hernquist 1996). The length of these first-generation bars can be much larger than those formed through internal disk instabilities, perhaps encompassing the entire disk. For such a large and massive bar, the reservoir of resonant halo orbits stretches from deep inside the halo cusp to well outside the radius of the disk, allowing a broad range of halo radii to accept angular momentum via resonant exchange. In idealized calculations, the torque from a primordial bar could destroy an NFW cusp out to half the bar radius, or 10 kpc in Milky Way units (Weinberg & Katz, 2002).

2. A Bar's Impact on a Cuspy Dark Matter Halo

We conducted a suite of fully self-consistent, multi-million particle simulations of barred galaxy evolution in cuspy halos using a state-of-the art code that minimizes Poisson noise (see Holley-Bockelmann, Weinberg & Katz 2004 for details). We constructed an equilibrium exponential disk embedded in an NFW halo (see Table 1 for simulation parameters of one of these experiments), and either allowed the bar to form naturally through disk instablities, or triggered it using an external potential which mimiced the 'natural' bar. The evolution of the externally and self-triggered systems were virtually identical. Figure 1 shows the face-on and edge-on view of a bar 0.18 time units (0.4 Gyr) after formation, as well as the response induced in the dark matter halo. The bar's behavior throughout the simulation is unmistakable: it lengthens, strengthens, and dramatically slows its rotation (Figures 2 and 3).

In a bulk sense, the halo clearly siphons angular momentum from the disk. Figure 4 plots the angular momentum evolution of the halo, disk, and bar.

Figure 1. Surface density maps of the model after bar formation. The brightness corresponds to the logarithm of the density with white (black) being the most (least) dense and black over 5 orders of magnitude. The horizontal scale for each panel is 10 R_d. Left: Face on view of the stellar component. Center: The disk edge on. Right: Face on view of the dark matter particles, with a superimposed contour plot of the $m = 2$ component of the halo potential to accentuate the bar wake.

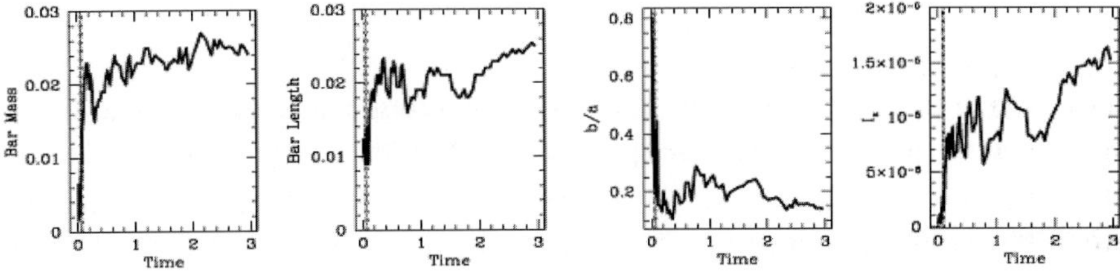

Figure 2. The bar mass (far left panel), bar length (left center), minor to major axis ratio (right center), and the z-component of the moment of inertia (far right) as a function of time. The vertical line is the time when the triggering stops.

Table 1. Simulation Parameters. [a]

Component	*Particle Number*	*Mass*	R_d	z_0	Q	R_{vir}	c
Disk	500000	8.7 x 10^{10} $M_\odot$	3 kpc	0.3 kpc	1.2		
Halo	5000000	1.3 x 10^{12} $M_\odot$				300 kpc	15

[a] Scaled from a $M_{\rm vir} = R_{\rm vir} = 1$ system to Milky Way units. In the text, we will present data first in system, then in Milky Way units.

Overall, the halo gains 16 % of the initial total angular momentum and the disk loses 16.5 %. The bar mediates this angular momentum transfer, both from the bar to the outer disk and from the disk to the halo. The bar itself loses 30 % over the course of the simulation, with about half transferred to the rest of the disk and half to the halo.

The real evidence that this evolution is governed by resonant interactions can be found by looking at *which* halo orbits gain angular momentum. Linear

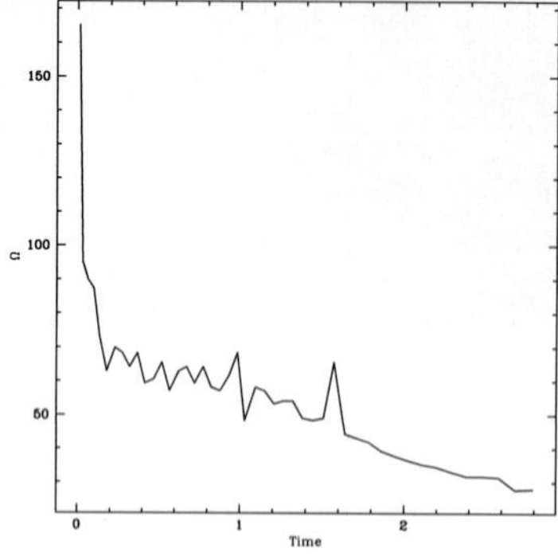

Figure 3. The pattern speed of the bar as a function of time.

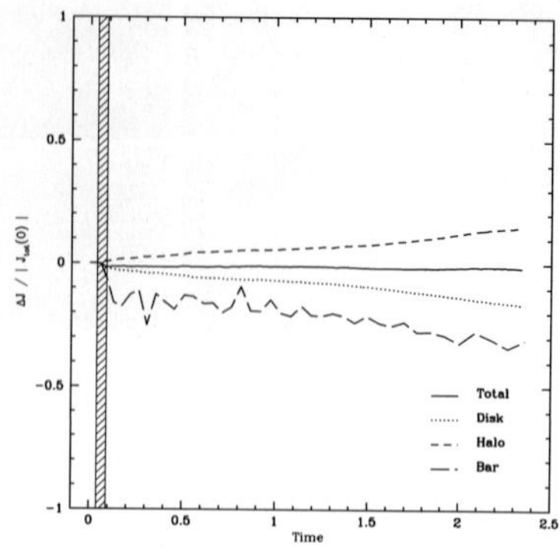

Figure 4. The change of angular momentum in each component, normalized by the initial total angular momentum.

perturbation theory predicts that angular momentum exchange will take place *only* at confined islands in phase space, the regions that correspond to halo-bar resonances. The planar resonances for a bar rotating with frequency $\Omega_{\rm bar}$ take the form:

$$l_r\Omega_r + l_\phi\Omega_\phi = m\Omega_{\rm bar}, \tag{1}$$

where Ω_r and Ω_ϕ are the radial and azimuthal orbital frequencies, respectively, l_r and l_ϕ are integers, and m is the azimuthal multipole index (m=2 for a bar). Ignoring phase, there are two non-degenerate conserved quantities for a spherical halo, e.g. total energy E and angular momentum J. A resonance will describe a curve in this E–J plane, and angular momentum changes should only take place along these curves. Typically, these islands of resonant angular momentum exchange occur in positive and negative pairs, as orbits either gain or lose J depending on their direction just before resonance in the bar's rotating frame.

Figure 5 shows the change in angular momentum for halo particles between two times, T_1 and T_2. The halo phase-space distribution is plotted in the E, $\kappa = J/J_{\rm circ}$ plane. (κ measures an orbit's eccentricity, and $\kappa = 0(1)$ corresponds to radial (circular) orbits). The contours depict the change in the z-component of the angular momentum, ΔJ_z. Over plotted and labelled are the loci of resonances described by equation 1 and calculated from the N-body phase space at T_1. For our models, these are typically near vertical lines.

In Figure 5, we plot the evolution between $T_1 = 0.5$ (1.1 Gyr) and $T_2 = 0.8$ (1.75 Gyr), after the bar has completed approximately 10 rotations. At this point, the bar has a stable figure and pattern speed and the resonance signature is unambiguous, with nearly all the angular momentum exchange in the halo

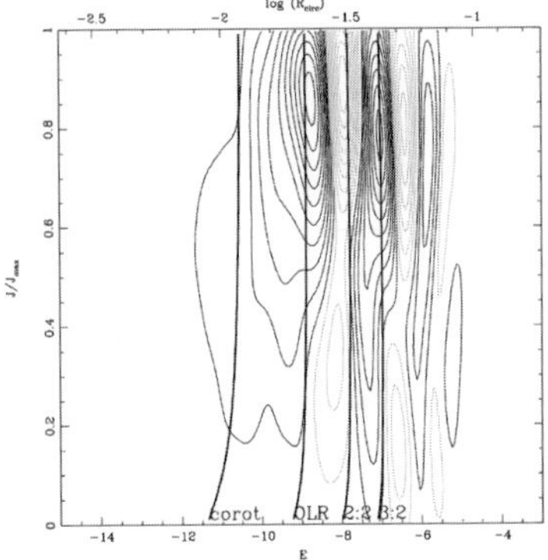

Figure 5. The angular momentum, ΔJ_z, exchanged between $T_1 = 0.5$ (1.1 Gyr) and $T_2 = 0.8$ (1.75 Gyr) for the halo phase-space distribution. See text for axis details. Angular momentum gain (loss) is represented by black (grey) contours. The nearly vertical lines are the positions of major resonances at time T_1, and each are labelled.

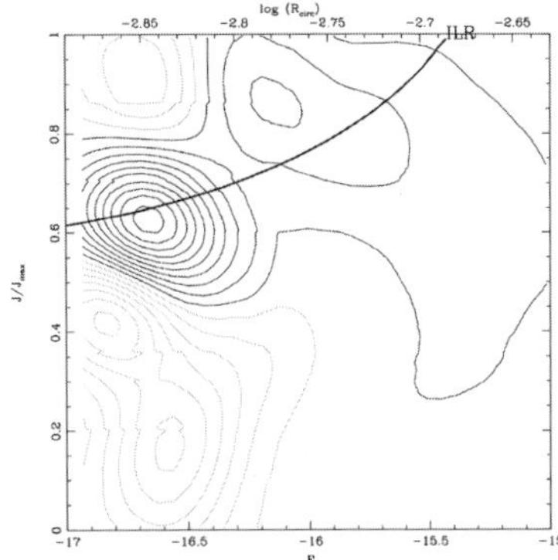

Figure 6. The relative angular momentum, $\Delta J_z/J_{\rm tot}$, exchanged between $T_1 = 0.5$ (1.1 Gyr) and $T_2 = 0.8$ (1.75 Gyr) for the inner halo. Only the inner region of the halo is plotted; outside this inner region, the relative angular momentum exchange is negligible.

restricted to easily identifiable low-order resonances with the bar's rotation. The main participants in this exchange are the corotation resonance ($0:2:2$), the Outer Lindblad Resonance ($1:2:2$), and resonances at $2:2:2$ and $3:2:2$.

Figure 5 shows the phase-space locations that dominate the angular momentum lost by the bar and act to slow the bar. However, since halo orbits have different amounts of angular momentum, the ΔJ_z contours shown are not useful for gauging the effect on the structure of the inner halo. In Figure 6, we plot the relative change in angular momentum ($\Delta J_z/J_{\rm tot}$, where $J_{\rm tot}$ is the initial total angular momentum of an orbit) over the same time period. The largest relative change in angular momentum takes place within the central region, and nearly all of it can be attributed to the Inner Lindblad Resonance ($l_r : l_\phi : m = -1:2:2$).

With so much of the halo cusp gaining angular momentum, it must evolve. Figure 7 shows that the halo density profile changes dramatically. After $t = 1$ (2.2 Gyr), the halo profile begins to deviate from an NFW profile at $R = 1.67\times 10^{-3}$ virial units (500 pc), quickly flattening to a $\gamma = 0$ cusp. At $R = 1\times 10^{-3}$ (300 pc), the central halo density has decreased to half its original value. The bar-induced flattening continues to the end of the simulation at $t = 2.25$ (4.95 Gyr), producing a core of about $R = 3\times 10^{-3}$ (900 pc).

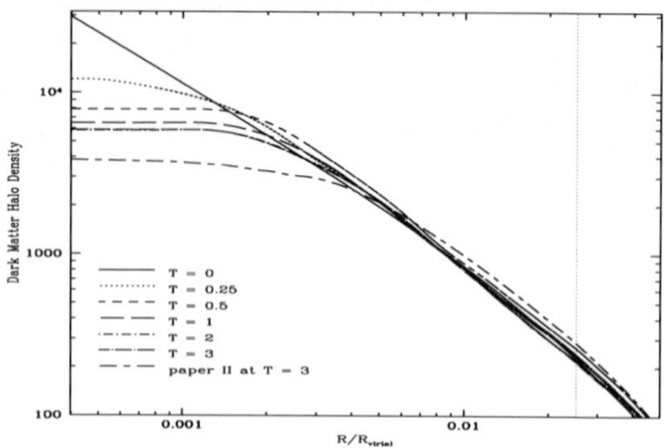

Figure 7. Initial and final halo density profiles. The bottom dashed line shows the final density profile predicted by linear perturbation theory for a perturber with a mass, radius and elongation of the average bar parameters for our experiment. The vertical dashed line shows the final bar radius.

3. Summary

We find that bars can drive the evolution of a fully self-consistent disk and dark matter halo. A disk scale length sized bar can flatten an NFW cusp, as it pumps angular momentum into orbits in the central regions of the halo. Furthermore, we find that resonant interactions are indeed responsible for transfering this angular momentum from the bar to the halo; angular momentum is deposited at discrete regions in halo phase space that correspond to low-order resonances with the bar pattern speed.

Do these experiments alone solve the cusp-core controversy? In a word: no. Our N-body experiments demonstrate that an initially corotating, scale-length sized bar will generate a core in the density profile that extends out as far as $R = 0.003$ (900 pc). Current rotation curve decompositions of high surface brightness spirals (Salucci & Burkert 2000) suggest halo core radii of at least $R = 0.04$ (12 kpc). The dynamics of the gas and stellar components may overwhelm the dark matter signal in these systems, however. Low surface brightness galaxies (LSBs) are thought to be dark matter dominated and can, therefore, be used to obtain a cleaner dark matter signal from the rotation curve. Typical halo core radii for LSBs range from 0.001 (300 pc) to 0.02 (6 kpc). It appears that scale-length-sized bars, like those studied here, do not generate cores this large in cuspy ΛCDM halos.

However, the lengths of *tidally-triggered* bars can be over 10 times larger than those in our simulations (see Figure 21 in Holley-Bockelmann, Weinberg, & Katz 2004). Since linear perturbation theory suggests that the destruction of the halo cusp scales with the size of the perturbing quadrupole, we expect that larger bars will generate proportionately larger cores. This scenerio ties the epoch of cusp disruption to the epoch of mergers/satellite encounters, early in the life of a galaxy. One can envision multiple epochs of bar formation and destruction during the galaxy's hierarchical assembly, characterized by relatively quiescent periods punctuated by mergers. It is plausible that a new disk is rebuilt after each major merger and that the quadrupole of the merger remnant, or else a minor merger, will trigger a large bar without disrupting the disk. A larger core at the end of each merger-punctuated epoch facilitates the formation of a larger bar and, in turn, a larger core. Eventually, a 'late-generation' bar will form entirely within the confines of a dark matter core, and (even if the galaxy is dark matter-dominated), the bar will only do minimal work on the halo and will barely slow (Figures 8 and 9).

Taken more generally, our bar–halo mechanism predicts an intrinsic dispersion in galaxy properties due to differences in evolutionary history: the present-day morphology will depend on the bar's formation mechanism; its relative size, strength and age; the galaxy's overall merger history; and its gas content. This can lead to galaxies with various degrees of cusp flattening, disk and bar evolution.

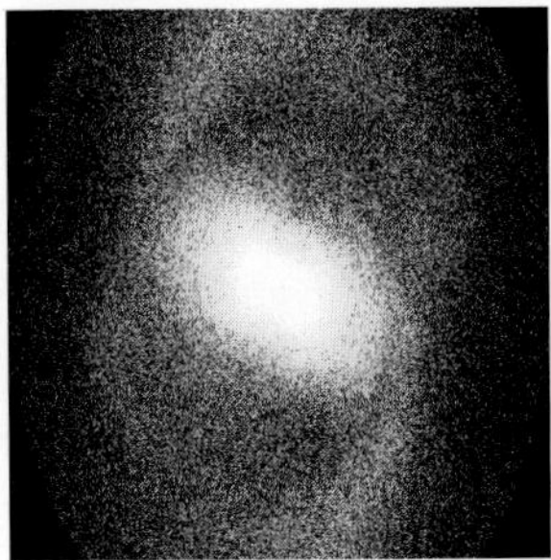

Figure 8. Face on view of a 'late-generation' bar formed in an evolved NFW model with a flat inner core within $r = 0.5R_{vir}$. The horizonatal scale is $10R_d$.

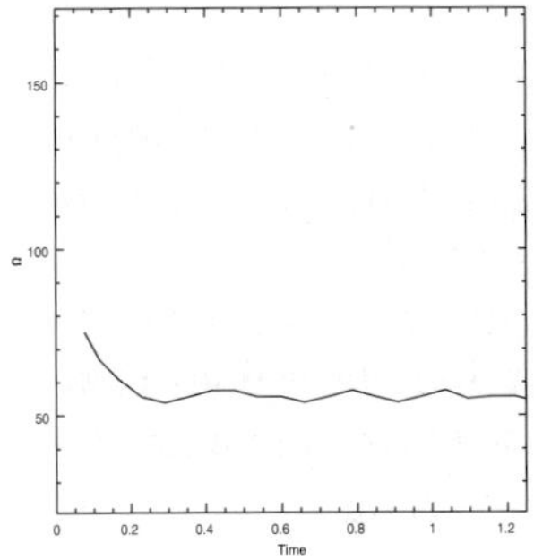

Figure 9. Pattern speed of the bar as a function of time. The bar slows mildly between t=0 and t=0.2 due to a transient m=2 spiral pattern (left) that extends beyond the core.

Most of the evidence against dark matter cusps in galaxies concerns LSB dwarfs. The same analysis used to indicate the lack of a central dark matter cusp also shows that LSB dwarfs are three or four times more baryon deficient than normal galaxies (Van den Bosch & Swaters 2001). Our bar mechanism not only provides a natural explanation for the existence of dark matter cores, but for the existence of LSB dwarfs as well. If a strong bar forms in a gas rich dwarf galaxy disk, the gas loses substantial angular momentum, is driven towards the center where it generates a strong starburst and, due to the shallow potential well of the dwarf galaxy, is expelled as a supernova-driven wind

(Dekel & Silk 1986). The work done on the galaxy during this process will further expand the core. The remaining galaxy, therefore, would be an LSB with a dark matter core; those dwarf galaxies that had the strongest bars will have the largest cores and the lowest surface brightnesses.

This presupposes that LSB dwarfs begin as baryon-rich, barred, HSB galaxies. Since there are so few barred galaxies in the present, baryon-deficient LSB population, it has long been suspected that bar formation is suppressed in LSBs, even when 'triggered' by satellites (Mayer & Wadsley, 2003). We constructed a wide range of warm, Toomre stable LSB disks in NFW halos, and found that there was always a way to form a bar; more centrally disk-dominated models formed small bars via disk instabilies (Figures 10 and 11), more centrally halo-dominated models formed large bars from tidal interactions (Figure 12 and 13). We have found that Poisson noise from models with too few particles can counteract the effect of a tidal trigger and artificially suppress bar formation in N-body experiments of Toomre stable LSBs. There is no reason to believe, then, that either the current or primordial LSB population are necessarily exempt from bar-halo evolution.

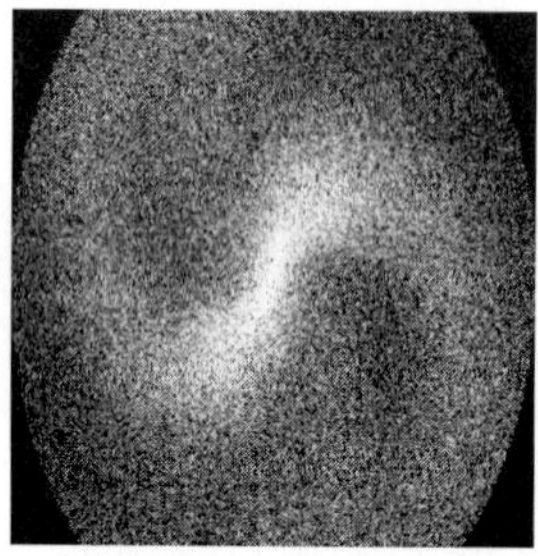

Figure 10. Face on view of an instability-driven bar in a low surface brightness galaxy. The horizontal scale is about $4R_d$.

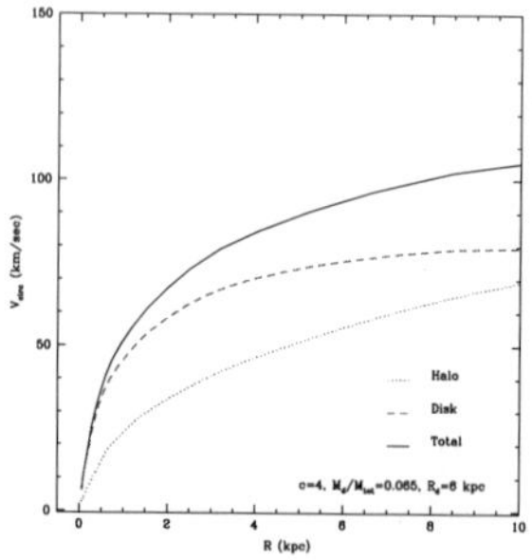

Figure 11. Initial rotation curve for this model. For $M_{vir} = 4 \text{ x } 10^{11} M_{\odot}$, $R_{vir} =$ 200 kpc, and $M/L = 2$, the central surface brightness μ_B is 22.67.

Our experiments set a lower limit to the size of the core generated by early bars. This lower limit already has implications that are astrophysically relevant. For example, Gondolo & Silk (1999) have argued that dark matter halo cusps may produce a neutrino signal from particle dark matter annihilation, though this requires that the cusp continues inward to 1 pc. These neutrino signatures will not exist for any galaxy that has gone through a barred phase (unless a supermassive black hole has subsequently induced a cusp (Ullio et al 2002)).

Bars are ubiquitously produced in galaxy simulations either through local instabilities or tidal interactions (Barnes & Hernquist 1992; Noguchi 1996; Steinmetz & Navarro 2002). Recent cosmological simulations designed to

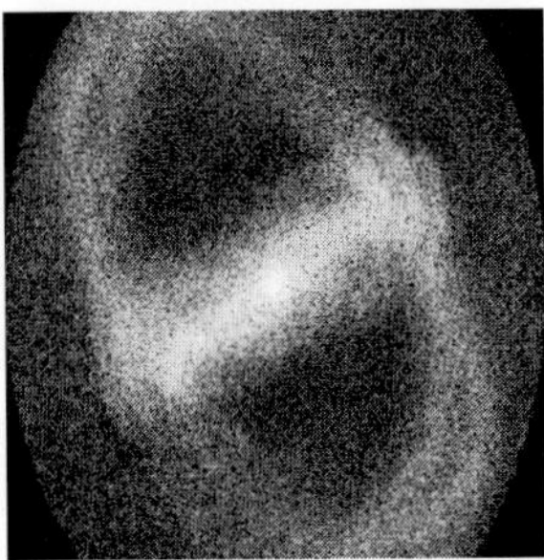

Figure 12. Face on view of a tidally-triggered bar in a low surface brightness galaxy. The strong bar was excited by a quadrupole that mimiced a satellite with $M_{sat} = 0.1$ at $R_{flyby} = 0.13$.

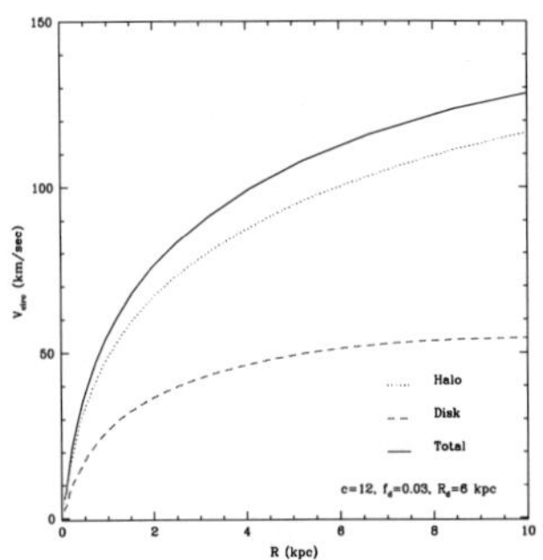

Figure 13. Initial rotation curve for this model. For the same physical units as Figure 11, $\mu_B = 23.5$

track morphological galaxy evolution predict that the bar phase is a natural byproduct of galaxy evolution (Steinmetz & Navarro 2002). An $L_\star$ Sb galaxy at $z = 0$ should have experienced a large bar by $z = 1.5$, and this should be observable with NGST. Both the higher galaxy gas fraction (Somerville, Primack, & Faber 2001) and the higher interaction rate (Le Fèvre et al 2000; Kolatt et. al. 2000) should enhance high redshift bar formation. Despite all this theoretical prejudice, the high redshift bar fraction of Sb galaxies is claimed to be smaller than the local bar fraction: 5% for $0.6 < z < 0.8$ versus 30% in the local Universe (van den Bergh et al. 2002; Abraham et al. 1999). This might be a selection effect. The identification of high redshift bars might be hampered by low sensitivity (van den Bergh et al. 2002), for example. Moreover, the classification technique used to identify high redshift bars may misclassify ones undergoing strong starbursts (Jogee et al. 2002), a characteristic event for a newly-formed bar in a gas rich environment (Friedli & Benz 1993; Sheth et al. 2002). In fact, recent NICMOS data have revealed that there is no significant evidence for a decrease in the fraction of barred spirals out to $z \sim 0.7$ (Sheth et al 2003). Answers to this mystery will help define the epoch of bar formation, constrain the "duty cycle" of bars and reveal the importance of bars in driving galaxy evolution.

Acknowledgments

This work was supported in part by NSF AST-0205969 and AST-9988146, and by NASA ATP NAG5-12038 and LTSA NAG5-13102.

References

Abraham et al. 1999, MNRAS, 308, 569
Athanassoula, E. 2002, ApJ, 569, 83
Barnes, J., & Hernquist, L. 1992, ARAA, 30, 705
van den Bergh et al. 2002, AJ, 123, 2913
Binney, J. & Tremaine, S. 1987, Galactic Dynamics, Princeton University Press
de Blok et al. 2001, ApJ, 552, L23
van den Bosch, F. & Swaters, R. 2001, MNRAS, 325, 101
Dekel, A. & Silk, J. 1986, ApJ, 303, 39
Friedli, D., & Benz, W. 1993, AAP, 268, 65
Gondolo, P., & Silk, J. 1999, PRD, 83, 9, L1719
Holley-Bockelmann, K., Weinberg, M., & Katz, N. 2004, MNRAS, in press. [astro-ph/0306374]
Jogee et al. 2002, ApJ, 570, L55
Kolatt et al. 2000, astro-ph/0010222
Le Fèvre et al. 2000, MNRAS, 311, 565
Lynden-Bell, D. & Kalnajs, A. 1972, MNRAS, 157, 1
Mayer, L. & Wadsley, J. 2004, MNRAS, 347, 277
McGaugh, S. 2002, Proceedings of the Yale Workshop " The Shapes of Galaxies and Their Dark Halos", Edited by Priyamvada Natarajan, 186
Moore et al. 1999, MNRAS, 310, 1147
Navarro, J., Frenk, C., & White, S. D. M. 1996, ApJ, 462, 563
Noguchi, M. 1996, *Barred Galaxies*, IAU Circular, 157, 339
Power et al. 2003, MNRAS, 338, 14
Salucci, P., & Burkert, A. 2000, ApJ, 537, L9
Sheth et al. 2002, AJ, 124, 2581
Sheth et al. 2003, ApJL, in press [astro-ph/0305589]
Somerville et al. 2001, MNRAS, 320, 504
Steinmetz, M., & Navarro, J. 2002, New Astronomy, 7, 155
Swaters et al. 2003, ApJ, 583, 732
Tóth, G., & Ostriker, J. P. 1992, ApJ, 389, 5
Ullio et al. 2002, PRD , 66, 12, 123502
Tremaine, S., & Weinberg, M. 1984, MNRAS, 209, 729
Walker, I., Mihos, C., & Hernquist, L. 1996, ApJ, 460, 121
Weinberg, M., & Katz, N. 2002, ApJ, 580, 627

RECENT RESULTS FROM THE SPITZER SPACE TELESCOPE: A NEW VIEW OF GALAXY MORPHOLOGY AND CLASSIFICATION

Giovanni G. Fazio, Michael A. Pahre, Steven P. Willner, and Matthew L. N. Ashby
Harvard-Smithsonian Center for Astrophysics, 60 Garden Street, Cambridge, MA 02138, USA

Abstract The IRAC instrument on the Spitzer Space Telescope has imaged two dozen nearby galaxies. The images reveal the reddening-free spatial distributions of the stellar photospheric emission and of the warm dust in the ISM. These two components provide a new framework for galaxy morphological classification, in which the presence of spiral arms and their emission strength relative to the starlight can be measured directly and with high contrast. Four mid-infrared classification methods are explored, three of which are based on quantitative global parameters (colors, bulge-to-disk ratio) similar to those used in the past for optical studies. In this limited sample, all correlate well with traditional B-band classification.

Two lenticular galaxies exhibit warm dust emission emanating from spiral arms—an IR feature that is not apparent even in an HST color map. Spiral arms of dust in lenticular galaxies may provide evidence for an evolutionary link between them and early-type spiral galaxies.

Warm dust emission in M 81 can be traced all the way to the nucleus, unlike the near-ultraviolet emission. Finally, the presence of an AGN in M 81 can be inferred both from nuclear colors redder than the bulge and the presence of a point source at the center.

Keywords: infrared – galaxy classification – interstellar dust

1. Introduction

The morphological classification scheme introduced by Hubble (1926, 1936), based on blue-light images, has been modified periodically over the years (e.g., Sandage 1961; de Vaucouleurs 1959; Kormendy 1979; Buta 1995) but remains a fundamental method by which astronomers continue to sort and compare galaxies. While the method has proven very successful over much of the last century, its dependence on fundamental, physical properties intrinsic to the galaxies is indirect due to the complicated emission processes sampled in the B-band.

D. Block et al. (eds.), Penetrating Bars through Masks of Cosmic Dust, 389–404.

In this paper, we revisit the morphological classification scheme using images with resolution and sensitivities similar to those used for traditional optical classifications—but at mid-infrared wavelengths. While some nearby galaxies could be resolved by previous infrared space missions operating at these wavelengths, and equal or better spatial resolution can be obtained from the ground, the combined wide-field coverage and sensitive infrared imaging detectors of the IRAC instrument on the Spitzer Space Telescope permits us to explore an entirely new region of parameter space for imaging nearby galaxies. These images demonstrate a new approach to galaxy classification based on the properties of the galaxy *interstellar medium* relative to its starlight.

2. Spitzer/IRAC In-Flight Instrument Performance

The Spitzer Space Telescope, NASA's Great Observatory for infrared exploration, was launched on 2003 August 25, into a heliocentric orbit, trailing the Earth. The telescope consists of an 85 cm cryogenically-cooled mirror and three focal-plane instruments, which provide background-limited imaging and spectroscopy covering the spectral region from 3 to 180μm wavelength. Incorporating large-format infrared detector arrays, with the intrinsic sensitivity of a cryogenic telescope, and the high observing efficiency of a solar orbit, Spitzer has already demonstrated dramatic improvements in capability over previous infrared space missions. Following 60 days of In-Orbit-Checkout and 30 days of Science Verification, normal operations began on 2003 December 1. To date, all spacecraft systems continue to operate extremely well (Werner et al. 2004). The Telescope has an expected lifetime of 5 to 6 years. More than 75% of the observing time will be available for General Observers.

Two of the Spitzer Space Telescope instruments, the Infrared Array Camera (IRAC; Fazio et al. 2004), and the Multiband Infrared Photometer for Spitzer (MIPS; Rieke et al. 2004) are designed as imaging instruments, although the MIPS also has a spectral energy distribution (SED) channel for very low-resolution spectroscopy. IRAC covers the wavelength region between 3.2 and 9.4 μm, while MIPS covers the region between 21.5 and 175 μm. The third instrument, the Infrared Spectrograph (IRS; Houck et al. 2004) provides low and moderate-resolution spectroscopic capabilities (5 to 40 μm), although it has two small imaging peak-up apertures that can also be used for imaging.

The Infrared Array Camera (IRAC) is a simple four-channel camera that obtains simultaneous broad-band images at 3.6, 4.5, 5.8, and 8.0 μm. Two nearly adjacent 5.2×5.2 arcmin^2 fields of view in the focal plane are viewed by the four channels in pairs (3.6 and 5.8 μm; 4.5 and 8 μm). All four detector arrays in the camera are 256×256 pixels in size (1.22 arcsec pixels), with the two shorter wavelength channels using InSb and the two longer wavelength channels using Si:As IBC detectors. IRAC is a general-purpose, wide-field camera that can be used for a large range of astronomical investigations. In-flight observations with IRAC have already demonstrated that IRAC's sensitivity, pixel size, field of view, and filter selection are excellent for studying galaxy structure and morphology. IRAC was built by the NASA Goddard Space Flight Center (NASA/GSFC) with the Smithsonian Astrophysical Observatory (SAO) having management and scientific responsibility.

Additional information on the Spitzer Space Telescope and its instruments can be found at the Spitzer Science Center (SSC) web site.[1]

3. Sample

The galaxies presented here are a small but representative subset of a sample of about 100 scheduled to be observed with Spitzer. These galaxies were drawn from a complete sample from Ho, Filippenko, & Sargent (1997), which contains nearly 500 galaxies in the northern sky at $B_T < 12.5$ mag. A subsample of approximately 100 galaxies was drawn from it under the constraints that: (1) the galaxies should have small sizes (less than 7 arcmin), so that they would fit in the instantaneous instrument field-of-view; and (2) the two-parameter space of luminosity and morphological type is fully spanned. The galaxies were randomly drawn within these constraints. Several low surface brightness galaxies were added to the subsample, as the parent sample was judged to be missing such galaxies. The sample properties in luminosity and morphological type are shown in Figure 1. The sample will be more fully described in a future paper. It is fortunate that the sample observed to date with Spitzer nearly fully spans the classical morphological sequence (only Sa is missing).

Two additional galaxies (NGC 3031=M81, Willner et al. 2004; and NGC 300, Helou et al. 2004) were observed for Early Release Observations; the first one is in the parent sample, while the latter is southern and is not. One more (NGC 4038/40399, the "Antennae") was drawn from observations described in Wang et al. (2004).

We adopt the RC3 optical classifications (de Vaucouleurs et al. 1991) as provided by NED throughout this paper for comparison purposes with the exception of NGC 5363: it is mis-classified as I0? but is most likely intended to be the type S0pec. We have adopted E/S0pec for it.

4. Observations

The data were taken with the InfraRed Array Camera (IRAC; Fazio et al. 2004) on the *Spitzer Space Telescope* during the first five IRAC campaigns of normal operations (2003 December – 2004 April). Quantitative data in this paper are given in instrumental magnitudes relative to Vega. The calibration of IRAC is based on circular aperture photometry of stellar point sources. We have adopted an extended photometry calibration using the elliptical galaxy

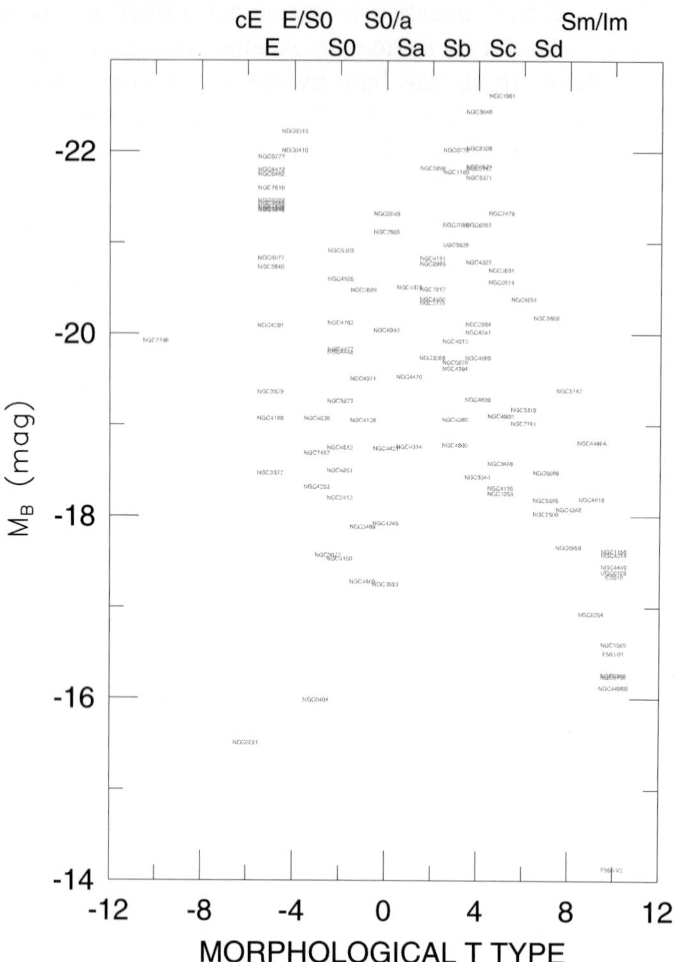

Figure 1. Optical morphologies and luminosities for the Spitzer sample. This sample of $\sim$ 100 galaxies was selected from the $\sim$ 500 galaxy parent sample of Ho, Filippenko, & Sargent (1997) in a manner that fully spans the space of luminosity and morphological type. The galaxies were also constrained mostly to have diameters $<$ 5 arcmin, so that they would fit within the IRAC field-of-view.

NGC 777, which shows the least evidence for any dust emission, and we assume a pure stellar spectrum matching that of an M0 III star. If NGC 777 does have dust emission after all, then the other galaxies will have larger emission than derived here. Pahre et al. (2004a) give some more details of the calibration.

For each galaxy, the $[3.6] - [4.5]$ color is nearly constant with radius and is consistent with stellar photosphere emission. Dust emission is likely to be important at longer wavelengths. The stellar emission was subtracted from the 5.8 and 8.0 μm images using the 3.6 and 4.5 μm images suitably scaled to match the theoretical colors of M0 III stars: $[3.6] - [4.5] = -0.15, [4.5] - [5.8] = +0.11, [5.8] - [8.0] = +0.04$ mag (M. Cohen and T. Megeath, private comm.). The resultant images are referred to as "non-stellar emission."

Surface photometry was measured on the galaxies using the IRAF task ELLIPSE, following the methodology of Pahre (1999). Circular aperture magnitudes were also measured for each galaxy. One-dimensional models of $r^{1/4}$ bulge plus exponential disk were fit to the isophotal surface brightness profiles and the circular aperture magnitude curves of growth. The B/D values from the two approaches were averaged, and half the difference is taken as the uncertainty.

The general infrared SED properties of various galaxy components are shown in Figure 2, along with the IRAC filter bandpasses. Starlight is blue in the IRAC bandpasses, because the mid-infrared wavelengths are longer than the peak of the blackbody radiation even for cool M giant stars. (Note that M giants are *bluer* than Vega in the $[3.6] - [4.5]$ color due to CO absorption in the 4.5μm bandpass.) Warm dust emission appears in lines of polycyclic aromatic hydrocarbon (PAH) molecules in the 5.8 and 8.0μm bandpasses. Emission from an AGN is *redder* than starlight in the IR bandpasses.

5. Infrared Morphologies of Galaxies

Images of the galaxies at 3.6 and 8.0 μm are shown in Figures 3 and 4, organized by optical morphological type. A color representation of the images at 3.6, 4.5, and 8.0 μm is shown at the end of the volume. The 3.6μm flux samples the unreddened stellar light distribution, while the 8.0μm flux samples that same starlight (dimmed by a factor of more than four since these wavelengths are on the Rayleigh-Jeans tail) plus emission lines of PAH tracing out warm dust.

Comparing Figures 3 and 4, early-type galaxies and the bulges of spiral galaxies are dominated by starlight emission. The disks of late-type galaxies, on the other hand, are dominated by warm dust emission.

Galaxy morphology is more clearly delineated in the dust emission (Figure 4) than it is in the traditional, optical blue light images. The warm dust provides a clean tracer of the reddening-free interstellar medium, which has high contrast and can trace spiral arms all the way to the center of a galaxy (see M 81 below). Furthermore, there is a changing ratio of starlight to warm dust emission along the galaxy sequence. This is the basis of a new method of classifying galaxies in the infrared at $3.2 < \lambda < 9.4\mu$m using Spitzer im-

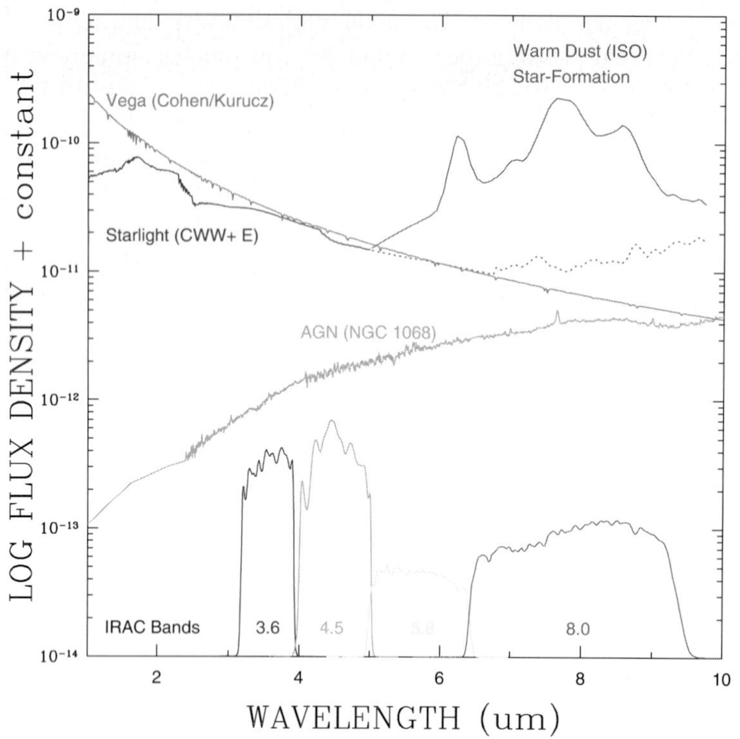

Figure 2. The spectral energy distributions of various galaxy components (starlight, warm dust, and active nuclei) at the IRAC wavelengths. The IRAC bandpasses are shown at the bottom of the figure. Warm dust emitting in the PAH lines, and active nuclei, both appear red in the IRAC bandpasses, while stellar light appears blue. The presence of CO absorption in the 4.5μm bandpass also results in cool, late-type stars having a bluer color $[3.6] - [4.5]$ than earlier-type stars like Vega.

ages. This is shown graphically in the color figure appearing at the end of the volume, where the starlight is color-coded as blue and the warm dust as red.

Other IR quantities which correlate with optical morphological type are shown in Figure 5. Two colors correlate well: the stellar $[3.6] - [4.5]$ color and the stellar vs. warm dust color $[3.6] - [8.0]$. The color $[3.6] - [4.5]$ is redder for late-type galaxies because they have a young stellar population that partially masks the the CO absorption, characteristic of late-type giant stars, in the 4.5 μm bandpass. (See Figure 2). The color $[3.6] - [8.0]$ is redder for late-type galaxies due to the warm dust emission in the PAH lines in the 8.0 μm bandpass. The IR bulge-to-disk ratio also shows a good correlation. Any of

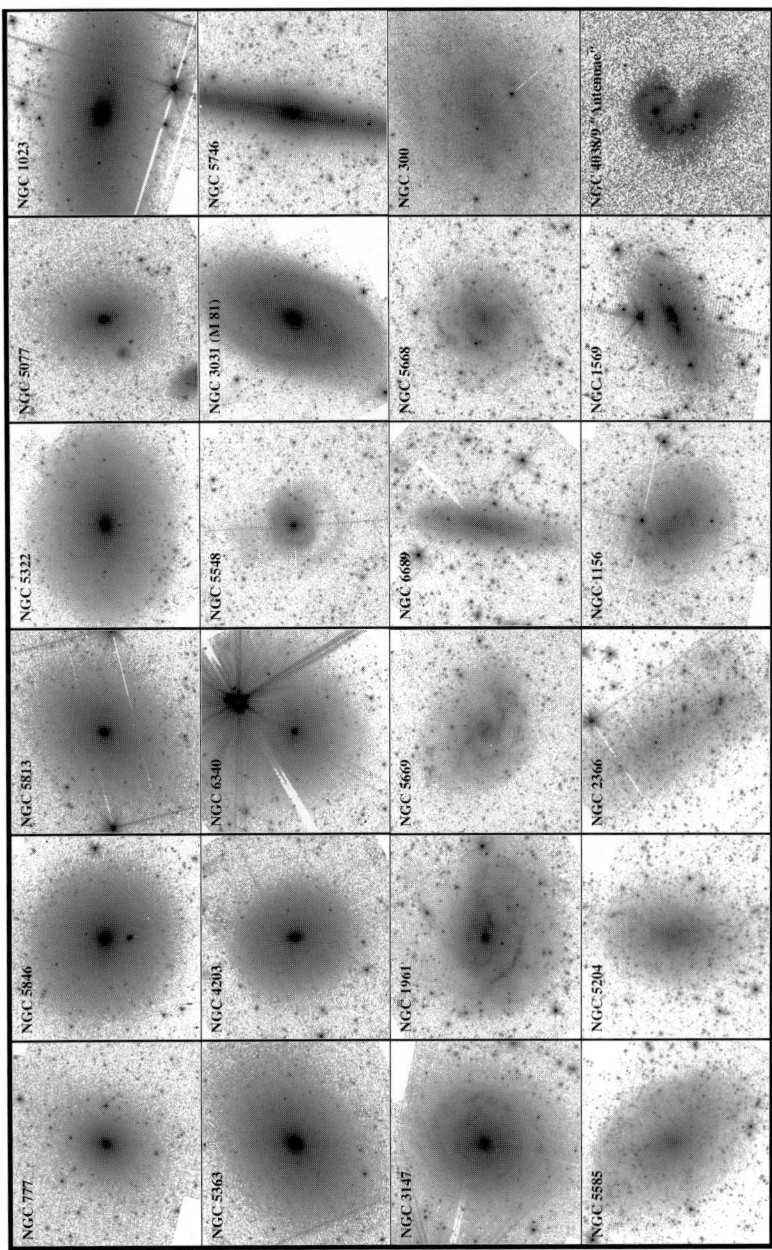

Figure 3. Images of the galaxies at $\lambda = 3.6\mu$m. The mosaic is organized from early- to late-types from left to right.

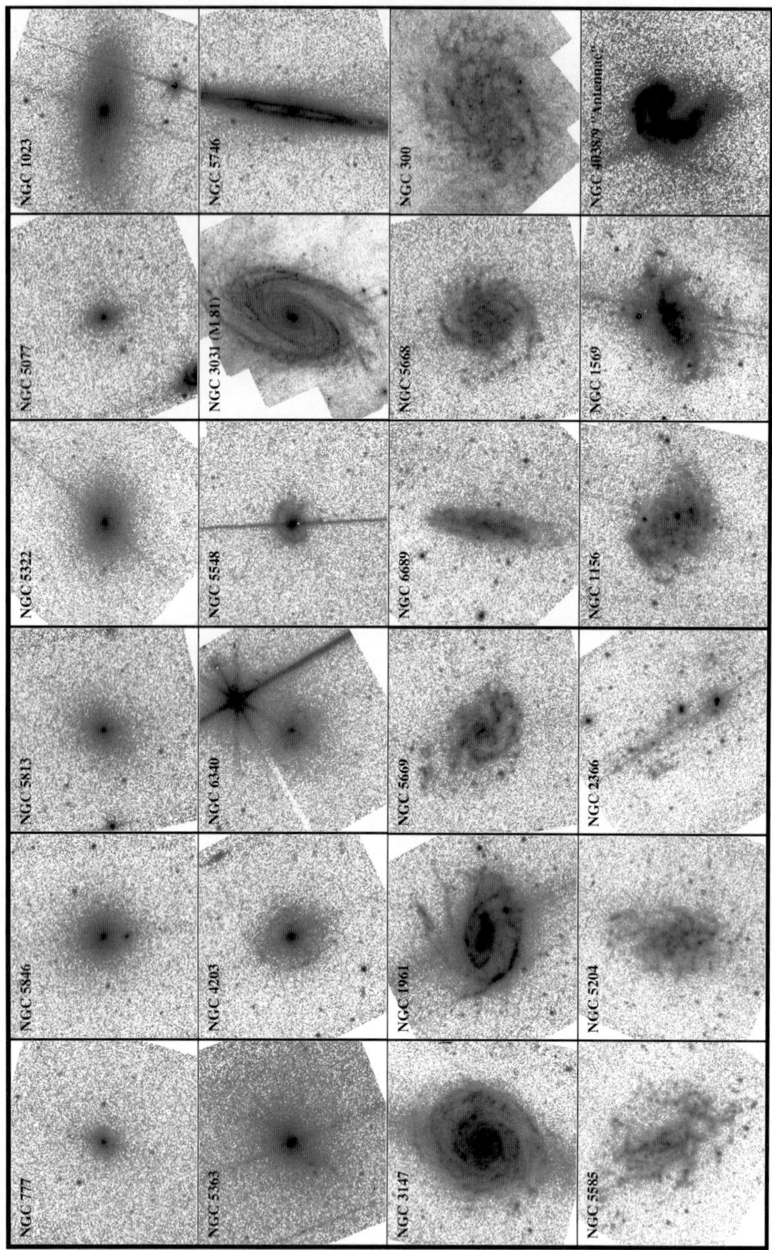

Figure 4. Images of the galaxies at $\lambda = 8.0\mu$m.

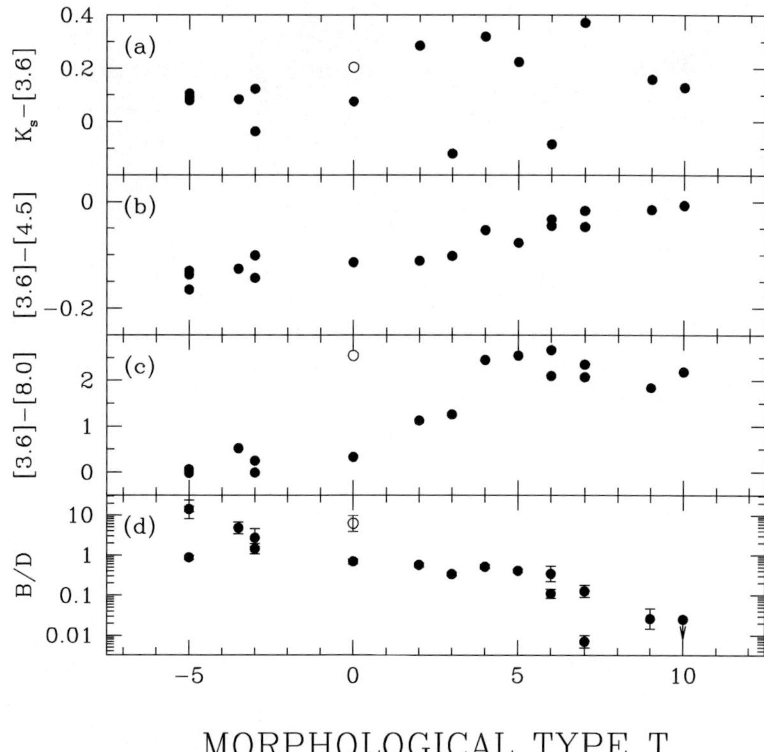

Figure 5. Correlation between various colors and the 3.6μm bulge-to-disk-ratio with optical morphological type. The $[3.6]-[4.5]$ color shows that the starlight changes from blue for early-types to red for late-types. A similar trend is found for $[3.6]-[8.0]$, which samples the ratio of starlight to warm dust emitting at the longer wavelengths.

these quantities, or all of them together, are suitable stand-ins for galaxy morphological type.

Why do infrared morphological classification and these other three quantities measured at infrared wavelengths work so well? The 3.6 and 4.5μm light is a good tracer of stellar mass (for a wide range of metallicity and age) free of dust obscuration, hence B/D measured at these wavelengths samples a fundamental galaxy property. The dust emission is a high contrast tracer of the interstellar medium, thereby making the morphology easier to discern. In galaxies of various types, the 8.0μm bandpass can exhibit anything from $\sim$ 100% starlight (elliptical galaxies) to $\sim$ 95% dust emission (NGC 1961, 3147, 4038/4039), and thus the $[3.6]-[8.0]$ color provides a direct measure of increasing ISM content along the galaxy sequence.

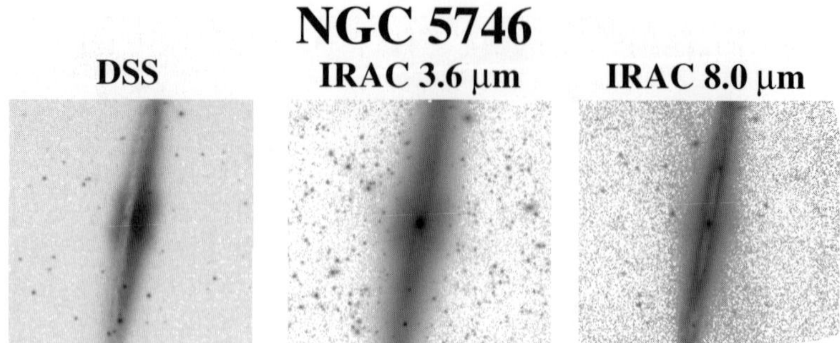

Figure 6. Edge-on spiral galaxy NGC 5746. At optical wavelengths (left), the galaxy shows a prominent dust lane. The IRAC image at 3.6μm (middle) is free of dust obscuration and shows a hint of a spiral arm. The 8.0 μm image (right) shows the ring prominently as well as several arms outside of it, demonstrating that this is a Sab(r) galaxy. The ring and arms are prominent in the PAH emission lines in the 8.0 μm band.

6. NGC 5746 at IR Wavelengths

A good example of the power of infrared galaxy classification is NGC 5746 (Figure 6), an edge-on spiral that cannot be reliably classified at optical wavelengths. Even the near-infrared emission from 2MASS shows a little bit of obscuration, but the near-IR light is otherwise mostly smooth. (NGC 5746 is among the 100 largest galaxies in the 2MASS XSC.) At IR wavelengths, however, a bright ring appears along with several arms outside of it. We suggest Sab(r) as the galaxy's morphological type.

7. Spatial Distribution of Warm Dust in Lenticular Galaxies

The warm gas phase was detected in early-type galaxies via its dust emission at far-IR wavelengths with IRAS (Jura et al. 1987; Knapp et al. 1989) and at optical wavelengths from atomic gas emission (Caldwell 1984; Phillips et al. 1986). Later, ISO detected polycyclic aromatic hydrocarbon (PAH) lines in the mid-IR (Lu et al. 2003; Xilouris et al. 2004). Of order half or more early-type galaxies exhibit either far-IR dust or optical ionized gas emission, although the detection rate in the infrared could be lower if the effects of an AGN are removed (Bregman et al. 1998). Current knowledge of the distribution of warm dust emission within early-type galaxies indicates that it mostly follows the light distributions (e.g., Athey et al. 2002) and hence is thought to arise from processes other than star formation such as AGN mass loss. The exceptions to this statement are a few galaxies with somewhat extended structures in 15μm ISO images that have been interpreted as dust lanes (Xilouris et al. 2004).

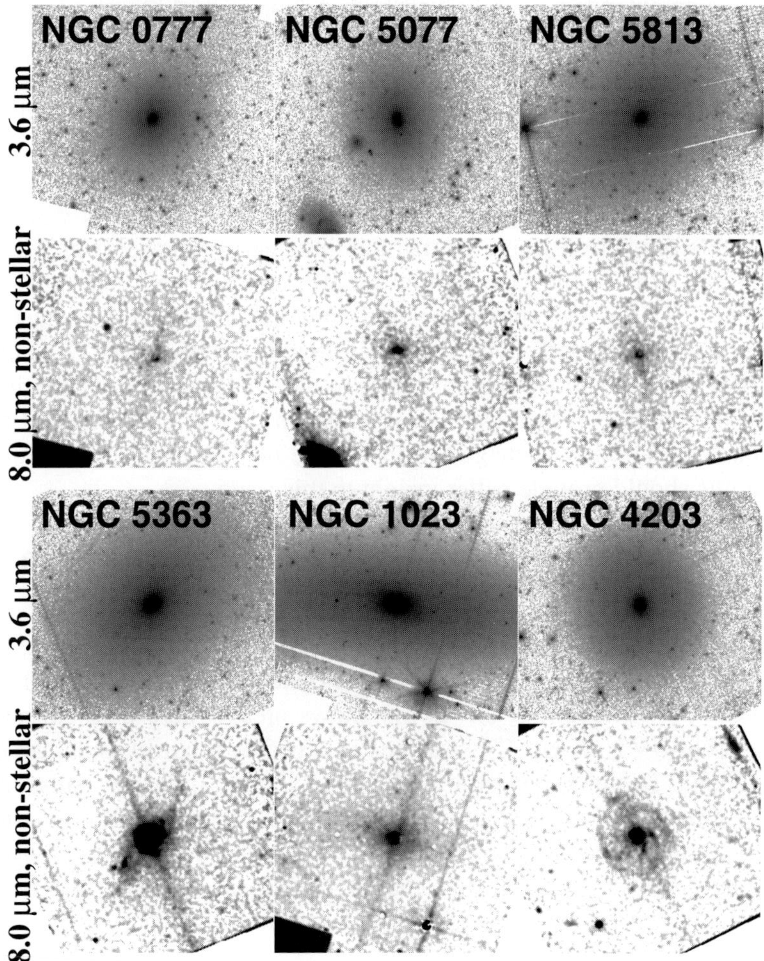

Figure 7. Warm dust emission in early-type galaxies. Six galaxies are shown, each with separate images in stellar emission (3.6μm) and warm dust emission (8.0μm "non-stellar"). The top three galaxies are classified as ellipticals and show no sign of warm dust emission at 8.0 μm. The bottom three galaxies are classified as lenticulars and show resolved warm dust. NGC 4203 appears face-on with several spiral arms emanating out from its nucleus, NGC 5363 appears somewhat inclined and shows two arms, and NGC 1023 shows a smooth distribution of dust.

The mid-IR imaging capability of the Spitzer improves the spatial resolution over previous missions by a factor of a few and the sensitivity by more than an order of magnitude. These new images (Figure 7; Pahre et al. 2004b) show that several early-type galaxies have substantial structure in the distribution of their warm dust as traced by PAH emission bands at 6.2–8.6μm. In particular, two lenticular galaxies show warm dust organized into spiral arms, and a third shows an extended, smooth distribution. For NGC 4203, this warm dust emission is not seen as obscuration at optical wavelengths with the Hubble Space Telescope (Erwin & Sparke 2003), further demonstrating the power of the infrared to highlight new galaxy morphological features that are invisible at optical wavelengths.

Galaxies with optical brightness similar to these have been studied with IRAS in order to detect (or place strong limits on) the presence of warm dust. A reasonable prediction would be that the three early-type galaxies with dust emission at 8.0μm should also exhibit far-IR emission from that dust, while those without 8.0μm dust emission would show only upper limits in the far-IR. All but one of the six early-type galaxies follow the expectations. The exception is NGC 1023, whose stringent limits in the far-IR seem to contradict the presence of extended 8.0μm warm dust emission. The larger sample of early-type galaxies that will be imaged in the future both at mid- and far-infrared wavelengths should provide better insight to this issue.

8. Comparison of Warm Dust and UV Continuum in M 81

An image of the non-stellar emission in M 81 is shown in Figure 8. Spiral arms of warm dust can be seen all the way into the nuclear region. Comparison to the near-UV (Marcum et al. 2001) shows areas where dust emission is prominent but UV is absent. This can be explained by relatively small amounts of extinction. The inner spiral arms especially show little UV emission but plenty of dust emission. It is an open question whether the PAH emission in the inner spiral arms continues to act as a tracer of star formation activity, and extinction removes the UV, or if instead PAH exists in a non-star-forming ISM illuminated by the general UV radiation field from the evolved stars.

9. Identifying the AGN in M 81

The presence of an AGN in M 81 can be inferred in two ways: (1) nuclear colors significantly redder than the bulge, and (2) point source residual after fitting and subtracting two-dimensional models of the bulge and disk light. The first one is shown in Figure 9, and both are described by Willner et al. (2004). The nucleus of M 81 is more than 0.4 mag redder than the inner bulge, which cannot be explained by any reasonable stellar photosphere model.

10. Summary

Images of two dozen galaxies taken at $3.2 < \lambda < 9.4\mu$m with the IRAC instrument on the Spitzer Space Telescope show that:

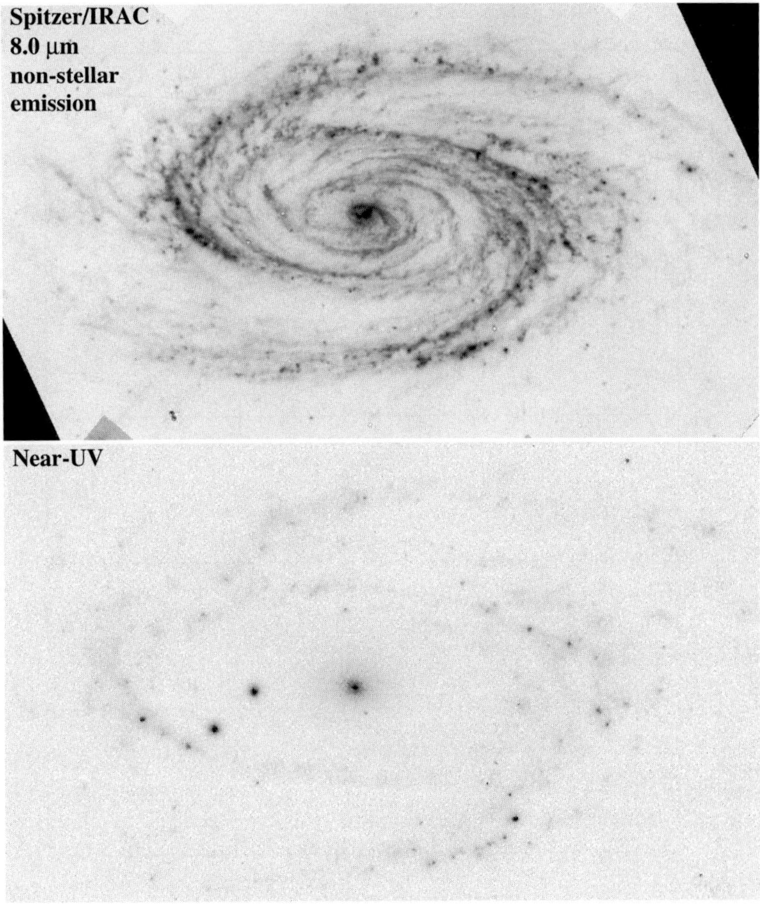

Figure 8. M 81 in warm dust emission and UV continuum. (top) IRAC 8.0μm image of warm dust emission; the stellar emission has been subtracted using models based on shorter wavelength images. (bottom) Near-UV image from Marcum et al. (2001).

1 The mid-IR cleanly separates emission from interstellar matter and starlight.

2 The mid-IR dust emission, particularly the PAH feature at $\lambda = 7.7$ μm, is a clear tracer of the presence of interstellar matter. The emission shows high contrast against stellar emission at the same wavelength.

3 The mid-IR light provides an entirely new scheme by which to classify galaxies–primarily based on the ratio of their ISM to starlight emission.

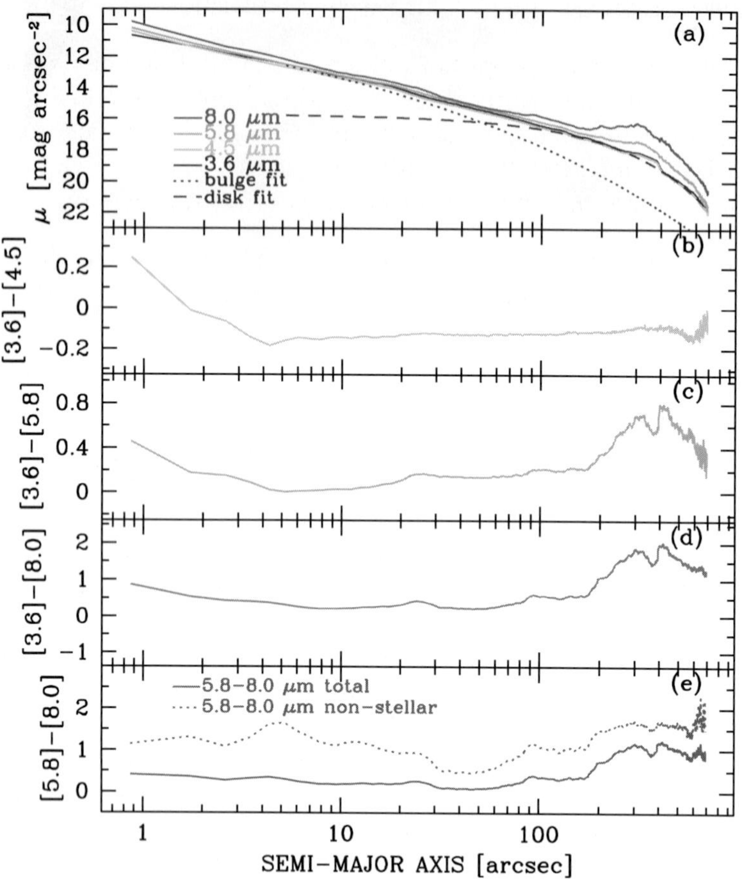

Figure 9. Surface photometry and color profiles for M 81. (a) Surface photometry in the IRAC bands along with bulge plus disk model fit to the 3.6μm data. (b) Color profile in $[3.6] - [4.5]$ which shows $+0.4$ mag redder colors in the nucleus, indicative of an AGN. (c) and (d) Color profile in $[3.6] - [5.8]$ and $[3.6] - [8.0]$, which are both redder than starlight alone due to PAH emission at the long wavelengths. (e) Color profiles in $[5.8] - [8.0]$ for both the combined starlight and warm dust and the warm dust alone. The color of the warm dust is roughly consistent with PAH color $[5.8] - [8.0] = 2.06$ mag predicted by Li & Draine (2001).

4 The colors of stellar photospheres in the mid-IR vary only a small amount with population age or mass function, and hence the stellar emission is a direct tracer of stellar mass. The bulge-to-disk-ratios measured at 3.6 and 4.5μm therefore sample the mass ratio of the stellar content, not a mixture of stellar content and recent massive star formation activity.

5 Two colors ([3.6] – [4.5] and [3.6] – [8.0]), as well as the 3.6 μm bulge-to-disk-ratio, correlate well with traditional morphological type and hence can be regarded as a stand-in for galaxy classification.

6 Three of six early-type galaxies observed exhibit dust emission that is organized into spiral arm or inner disk-like structures, The [5.8] – [8.0] color of the dust emission matches that for dust in actively star-forming galaxies and theoretical models of PAH emission, Two out of three galaxies that show 8.0μm dust emission also show far-IR emission.

7 Active galactic nuclei can be readily identified both via their mid-infrared colors and as point source residuals in two-dimensional modelling.

Acknowledgments

This work is based on observations made with the Spitzer Space Telescope, which is operated by the Jet Propulsion Laboratory, California Institute of Technology under NASA contract 1407. Support for the IRAC instrument was provided by NASA through Contract Number 960541 issued by JPL. The IRAC GTO program is supported by JPL Contract # 1256790. M.A.P. acknowledges NASA/LTSA grant # NAG5-10777. IRAF is distributed by the National Optical Astronomy Observatories, which are operated by the Association of Universities for Research in Astronomy, Inc., under cooperative agreement with the National Science Foundation. This publication makes use of data products from the Two Micron All Sky Survey, which is a joint project of the University of Massachusetts and the Infrared Processing and Analysis Center/California Institute of Technology, funded by the National Aeronautics and Space Administration and the National Science Foundation. This research has made use of the NASA/IPAC Extragalactic Database (NED) which is operated by the Jet Propulsion Laboratory, California Institute of Technology, under contract with the National Aeronautics and Space Administration. The Digitized Sky Surveys were produced at the Space Telescope Science Institute under U.S. Government grant NAG W-2166.

Notes

1. http://www.spitzer.caltech.edu

References

Athey, A., Bregman, J., Temi, P., & Sauvage, M. 2002, ApJ, 571, 272

Bregman, J. N., Snider, B. A., Grego, L., & Cox, C. V. 1998, ApJ, 499,

Buta, R. 1995, ApJS, 96, 39

Caldwell, N. 1984, PASP, 96, 287

Erwin, P. & Sparke, L. S. 2003, ApJS, 146, 299

Fazio, G. G., et al. 2004, ApJS, in press (astro-ph/0405616)
Helou, G. X., et al. 2004, ApJS, in press
Ho, L. C., Filippenko, A. V., & Sargent, W. L. W. 1997, ApJS, 112, 315
Hubble, E. P. 1926, ApJ, 64, 321
Hubble, E. P. 1936, Realm of the Nebulae (New Haven: Yale University Press)
Jura, M., Kim, D. W., Knapp, G. R., & Guhathakurta, P. 1987, ApJ, 312, L11
Knapp, G. R., Guhathakurta, P., Kim, D., & Jura, M. A. 1989, ApJS, 70, 329
Kormendy, J. 1979, ApJ, 227, 714
Li, A. & Draine, B. T. 2001, ApJ, 554, 778
Lu, N., et al. 2003, ApJ, 588, 199
Marcum, P. M., et al. 2001, ApJS, 132, 129
Pahre, M. A., Ashby, M. L. N., Fazio, G. G., & Willner, S. P. 2004a, ApJS, in press (astro-ph/0405594)
Pahre, M. A., Ashby, M. L. N., Fazio, G. G., & Willner, S. P. 2004b, ApJS, submitted
Phillips, M. M., Jenkins, C. R., Dopita, M. A., Sadler, E. M., & Binette, L. 1986, AJ, 91, 1062
Sandage, A. 1961, The Hubble Atlas of Galaxies (Washington: Carnegie Institution)
de Vaucouleurs, G. 1959, Handbuch der Physik, 53, 275
de Vaucouleurs, G., de Vaucouleurs, A., Corwin, H. G., Buta, R. J., Paturel, G., & Fouque, P. 1991, Third Reference Catalogue of Bright Galaxies (Springer-Verlag: New York)
Wang, Z., et al. 2004, ApJS, in press
Werner, M. W., et al. 2004, ApJS, this issue
Willner, S. P., et al. 2004, ApJS, in press (astro-ph/0405626)
Xilouris, E. M., Madden, S. C., Galliano, F., Vigroux, L., & Sauvage, M. 2004, A&A, 416, 41

USING BARS AS SIGNPOSTS OF GALAXY EVOLUTION AT HIGH AND LOW REDSHIFTS

Kartik Sheth[1], Karin Menendez-Delmestre[1], Nick Scoville[1], Tom Jarrett[1], Linda Strubbe [1,2],Michael W. Regan[3], Eva Schinnerer[4], and David L. Block[5]
[1]*California Institute of Technology, Pasadena, CA,* [2]*University of California - Berkeley,* [3]*Space Telescope Science Institute,* [4]*National Radio Astronomy Observatory,* [5]*University of the Witwatersrand*

Abstract An analysis of the NICMOS Deep Field shows that there is no evidence of a decline in the bar fraction beyond z$\sim$0.7, as previously claimed; both bandshifting and spatial resolution must be taken into account when evaluating the evolution of the bar fraction. Two main caveats of this study were a lack of a proper comparison sample at low redshifts and a larger number of galaxies at high redshifts. We address these caveats using two new studies. For a proper local sample, we have analyzed 134 spirals in the near-infrared using 2MASS (main results presented by Menendez-Delmestre in this volume) which serves as an ideal anchor for the low-redshift Universe. In addition to measuring the mean bar properties, we find that bar size is correlated with galaxy size and brightness, but the bar ellipticity is not correlated with these galaxy properties. The bar length is not correlated with the bar ellipticity. For larger high redshift samples we analyze the bar fraction from the 2-square degree COSMOS ACS survey. We find that the bar fraction at z$\sim$0.7 is $\sim$50%, consistent with our earlier finding of no decline in bar fraction at high redshifts.

1. Bars: Signposts of Galaxy Evolution

Bars are ubiquitous in local disk galaxies (e.g., RC3, Eskridge et al. 2000, Menendez-Delmestre et al. 2004). They play an important role in evolving galaxies by transporting vast amounts of gas to the center (Sakamoto et al. 1997; Sheth et al. 2004), igniting circumnuclear starburst activity (Ho et al. 1997 and references therein), reducing the chemical abundance gradient (Martin & Roy 1997), and perhaps feeding black holes (Shlosman et al. 1988, see Knapen and Laurikainen in this volume). Bars also form bulges, and perhaps evolve galaxies along the Hubble sequence (e.g., Norman, Sellwood & Hasan 1996; see also Kormendy, Bournaud, and Combes in these proceedings). At high redshifts bars were expected to be fairly common because of dynamically colder disks and increased merging activity. So it was a surprise when the

D. Block et al. (eds.), Penetrating Bars through Masks of Cosmic Dust, 405–414.

first studies of high redshift galaxy morphology found an apparent paucity of barred spirals beyond z∼0.5–0.7 (van den Bergh et al. 1996; Abraham et al. 1999; van den Bergh et al. 2000). If true these results set strict constraints on cosmological simulations of galaxy formation by requiring that disks were not sufficiently massive, or were dynamically too hot until only ∼6 Gyr ago. These constraints would be in stark contrast to conclusions from studies of the cosmic star formation history (e.g., Steidel et al. 1999 and references therein) and thinness of local disks (e.g., Toth & Ostriker 1992) which argue that massive disks were already in place by z∼1.

Bunker (1999) used a near-infrared 1.6μm NICMOS GTO image to show that in at least one galaxy, a bar was missed by the early studies because they observed the high redshift galaxies in the rest-frame blue / ultraviolet light where bars are difficult to identify. The increasingly better visibility of bars at longer wavelengths is a well-known effect (see for example Figure 1, and Menendez-Delmestre et al and references therein in these proceedings). We decided to investigate whether the apparent decline in the bar fraction at high redshifts could be due to such a selection effect? This question was the focal point of Sheth et al. (2003) and whose results we briefly summarize here.

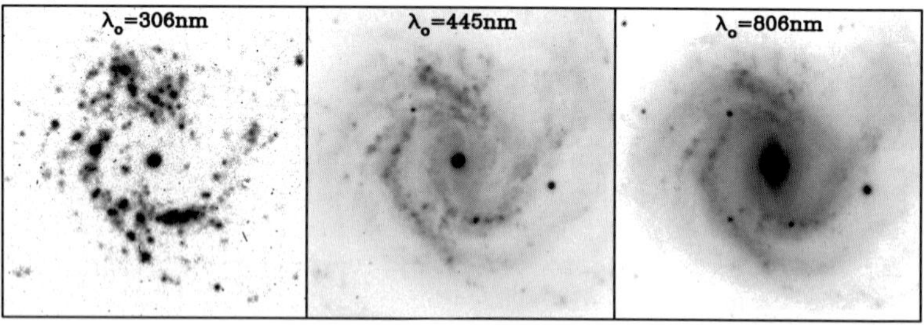

Figure 1. Left panel: UV appearance of NGC 4303, simulated using a continuum-subtracted Hα image. Note how the bar is completely invisible. Middle panel: Blue-band image. The bar is faint and difficult to identify. Right panel: I-band image. Only now does the bar become visible.

2. The Bar Fraction in the NICMOS Deep Field

We analyzed the bar fraction and bar properties in the WFPC2 (V, and I-band) and NICMOS (H-band data) for the northern Hubble Deep Field(Williams et al. 1996; Dickinson et al. 2000). The NICMOS data are ideal for studying galaxies in the redshift range 0.7,z<1 because these data observe the galaxies in rest-frame V through I-band light.

The Bar Identification Technique

In contrast to previous studies which identified bars by eye, by ellipticity breaks, or a change in isophote position angles between an inner and outermost isophote (e.g., van den Bergh 1996; Abraham et al. 1999;), we applied a more stringent two signature criteria to identify bars. We demanded that a) the ellipticity increase monotonically and then drop with a sharp change of at least $\Delta\epsilon > 0.1$, and b) the position angle remain constant over the bar region and change by Δ PA $> 10^\circ$ after the bar region. We require that the bar be symmetric by fixing the center for the galaxy fitting algorithm. Our algorithm misses bars if the galaxy is highly inclined, if the bar position angle is the same as the galactic disk, if the underlying galactic disk is too faint to be adequately imaged, or if the data have inadequate resolution to resolve bars. Our method, though perhaps overly strict, has the advantage of ensuring a firm lower limit to the bar fraction.

As discussed in Sheth et al. (2003), at $z < 0.7$, we identify five barred spirals, and two candidate barred spirals, consistent with the prior analysis of the HDFN by Abraham et al. (1999). At $z > 0.7$ we identify four barred spirals and five candidate barred spirals, including two possible candidates at z=1.66 and z=2.37. The four barred spirals are shown in Figure 2. For comparison, in the previous WFPC2 HDFN studies, van den Bergh et al. (1996) found no barred spirals, and Abraham et al. (1999) found two barred galaxies beyond $z{\sim}0.5$. If we use the Abraham et al. (1999) magnitude cutoff of I(AB)=23.7, the total number of disk-like galaxies drops to 31; amongst these we identify three barred spirals. Though we detect a few more bars, the total number of barred spirals is still small. Does this reflect a true decline in the bar fraction at $z > 0.7$? We argue that when the spatial resolution of the observations are considered in the context of bar visibility, there is no evidence of a decline in bars at $z>0.7$.

Spatial Resolution & the Visibility of Bars

In Figure 3, we show the apparent angular size of various galactic structures as a function of redshift. Overlaid are detection limits for various telescopes adopting a five PSF detection threshold; five PSFs is an appropriate choice (Menendez-Delmestre et al. 2004, also in this volume). The figure shows that at 0.8μm WFPC2 data is only marginally capable of detecting a 5 kpc structure beyond $z{\sim}0.7$. The NICMOS data have even coarser resolution (longer λ and larger pixels), and even though these data are not affected by bandshifting until $z > 2$–3, they only detect structures with sizes $\gtrsim$ 10 kpc. The average size of the four bars identified at $z > 0.7$ is 12 kpc.

The most important point of note from Figure 3 is that a measurement of the bar fraction must take into account the size of bars and the available spatial res-

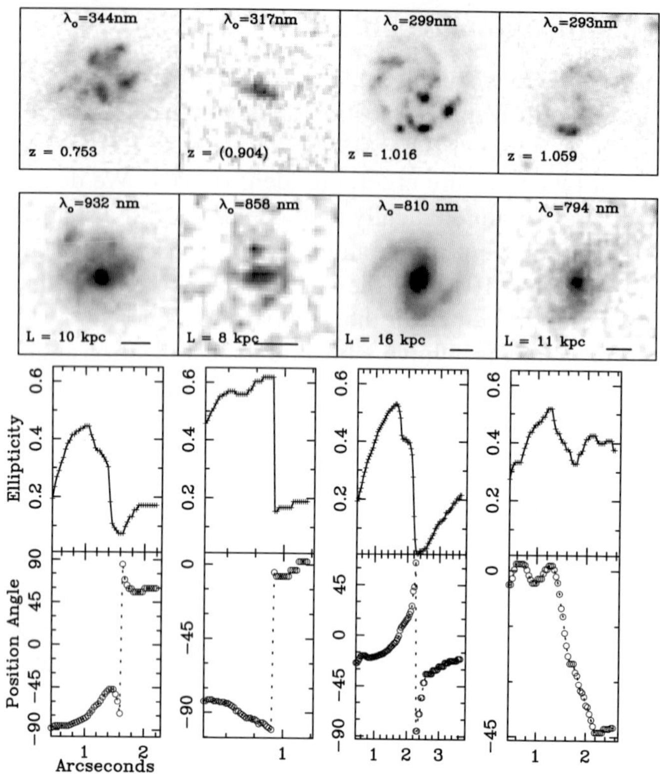

Figure 2. Barred spirals at z > 0.7, arranged by redshift. The top row shows the optical (F606W, V-band) WFPC2 images and the second row shows the near-infrared (1.6μm, H-band) NICMOS images. 0.5" scale is shown with a horizontal segment in the lower right of each panel. The rest-frame wavelength for each galaxy is listed inside the top of each panel.

olution of the data. Although we only detect three or four bars in the NICMOS data one must note that these data are biased towards detecting only the largest bars. Therefore, when we compare the bar fraction at different redshifts, we must compare the fraction for bars of equal sizes. For a representative sample of local galaxies (SONG, Regan et al. 2001, Sheth et al. 2004), we find that bars with sizes > 12 kpc are rare; only one out of 44 galaxies in SONG has a bar larger than 12 kpc. Thus the fraction of bars we detected in the NICMOS Deep Field (4/95 galaxies for the entire sample, or 3/31 galaxies using an I-magnitude cutoff) is similar, and perhaps even, larger than the bar fraction seen in SONG. Thus we concluded that there was no evidence for a decline in the bar fraction at z > 0.7, as previously claimed.

Our results are hampered by small number statistics (we discovered only a handful of bars at z > 0.7 and our local comparison sample contained forty four galaxies), and difficulty in defining comparable samples at high and low

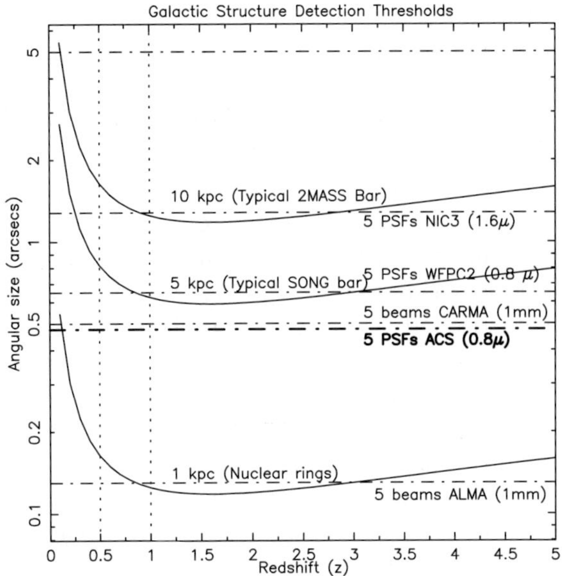

Figure 3. The detection threshold of various galactic structures as a function of redshift for different telescopes and instruments is shown. The horizontal dotted-dashed lines are an arbitrary 5 PSF or 5 beam limit. Note that at 0.8μm, even the WFPC2 data is only marginally capable of detecting a typical 5 kpc bar at z > 0.7; ACS z-band is only slightly better. The NICMOS data can only detect the large bars at z > 0.5. Also shown are capabilities of two new millimeter arrays, CARMA and ALMA which will be ideal for probing the gas kinematics in high redshift systems.

redshifts (see Sheth et al. 2003). From the existing analysis, it is difficult to conclusively state how the bar fraction varies as a function of redshift, and how the bar properties vary. In the next two sections we outline new results from two separate on-going studies aimed at overcoming these two main caveats of our NICMOS study.

3. Defining a Proper Local Sample: The 2MASS Local Galaxy Atlas

How well do we know the fraction of nearby bars? What are their properties (size, strength)? How do these properties depend on the host galaxy properties? Answers to these questions are of fundamental importance before any comparison is made to the high redshift Universe.

Since bars are best studied in the infrared, we decided to address all of these questions using the 2MASS Large Galaxy Sample. Results of this study were presented in a poster at this conference by Menendez-Delmestre. We chose all spirals of type Sa-Sd, with $i < 65^0$ for a sample of 134 galaxies in J+H+K. We ran the same ellipse fitting algorithm on these galaxies as we did previously on the high redshift sample and identified bars using the two signature criteria described above.

As shown in Figure 2 of Menendez-Delmestre et al. (this volume), the fraction of barred spirals in the 2MASS sample is 58% (see Figure 1 in Menendez-Delmestre et al.). Another 21% of the galaxies are identified as candidate bars where usually there is an ellipticity signature but no corresponding position angle change. We examined each of the candidate bars by eye and found that about two-thirds of the candidates were in fact barred. So the total fraction of barred spirals in the infrared ∼72%, similar to the fraction found by Eskridge et al. (2000). The bar fraction changes slightly with the T-type with the highest fraction (80%) in T=3 but there is no significant trend in the fraction with the Hubble type. Unfortunately the 2MASS sample has low signal to noise and we were unable to quantify the bar fraction in galaxies of type later than Sd. Note that the overall bar fraction in RC3 is lower, but not significantly different than the 2MASS fraction. This suggests that overall the change in the bar fraction between B-band (RC3) and the near-infrared is not large but also note that the effect may become severe at wavelengths shorter than the B-band, as shown by Figure 1.

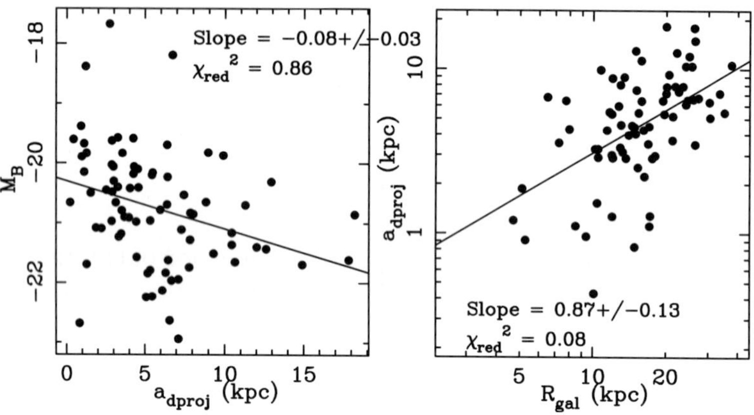

Figure 4. Left Panel: Absolute B-magnitude vs. deprojected bar semi-major axis. Right Panel: Deprojected bar semi-major axis vs. galaxy size. There are weak trends of larger bars in brighter and larger galaxies.

For the 78 barred spirals in 2MASS, we derive a median deprojected bar semi-major axis of 5.1 kpc. This result is different from the SONG sample we considered in Sheth et al. (2003). The reason for the difference is that 2MASS surveys a larger volume but is shallower. As a result we observe larger and brighter galaxies which tend to have longer bars (see below). We measure a mean ellipticity of 0.45 but the most notable result from the distribution of ellipticities is the scarcity of thin bars ($\epsilon > 0.7$). Previously similar results have been interpreted as evidence for a second or later generation of bars (e.g., Block et al. 2002, see also Bournaud, Combes in this volume) because in numerical simulations the first generation of bars is expected to be strong (thin) whereas subsequent generations are expected to be weaker (fatter). However note that the measurement of ellipticity can be affected by bulge light. Ellipse fitting does not automatically account for bulge light and as a consequence gives a higher ellipticity than the actual value. When we compare the bar ellipticity and bar length we find no correlation indicating that larger bars are preferentially fatter or thinner. We find that bars come in all shapes and sizes.

How do the bar properties depend on the host galaxy? From the 2MASS data, we find a trend of larger bars in larger and brighter galaxies (Figures 4). This is expected given that the bar instability is correlated with the mass of the disk.

Correlations of the bar strength (we use ellipticity as a proxy for strength) and the host galaxy parameters are very weak (Figures 5). Unlike the bar length, there is no correlation between the bar ellipticity and galaxy size or brightness.

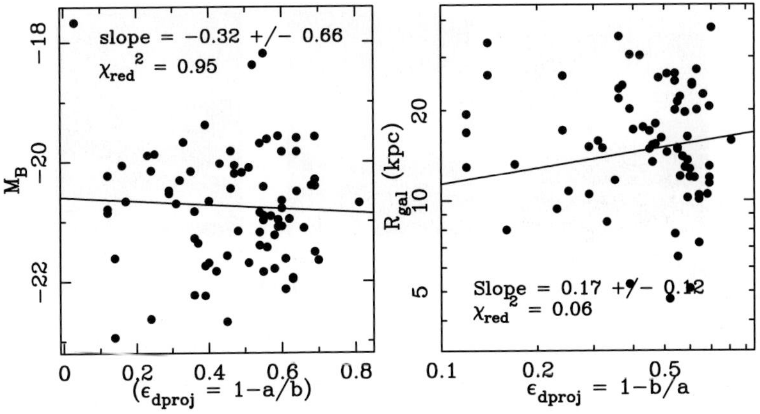

Figure 5. Left Panel: Absolute B-magnitude vs. deprojected bar ellipticity. Right Panel: Size of galaxy vs. deprojected bar ellipticity. There is no correlation between the bar ellipticity and the host galaxy size or brightness.

Note that these correlations are not immune to biases in the analysis. For instance the ellipse fitting technique always underestimates the bar ellipticity (Sheth et al. 2000, 2004) because the bulge light affects the galaxy isophotes to make the fitted ellipses fatter/thicker than the real underlying bar. Hence the lack of "strong" bars in early-Hubble type galaxies is not necessarily a real effect. Another bias is related to resolution. Small bars are difficult to identify especially in galaxies with bright bulges and nuclei. So any trend of larger bars in earlier type galaxies should also be viewed with caution. Nevertheless the trends of larger bars in larger galaxies and brighter galaxies are relatively robust.

4. COSMOS: Overcoming Small Number Statistics

The 2MASS analysis by Menendez-Delmestre et al. (2004) offers an excellent analysis of the local sample of bars and their host galaxies. We can now extend these results to higher redshifts to understand how the bar fraction and bars evolved over time.

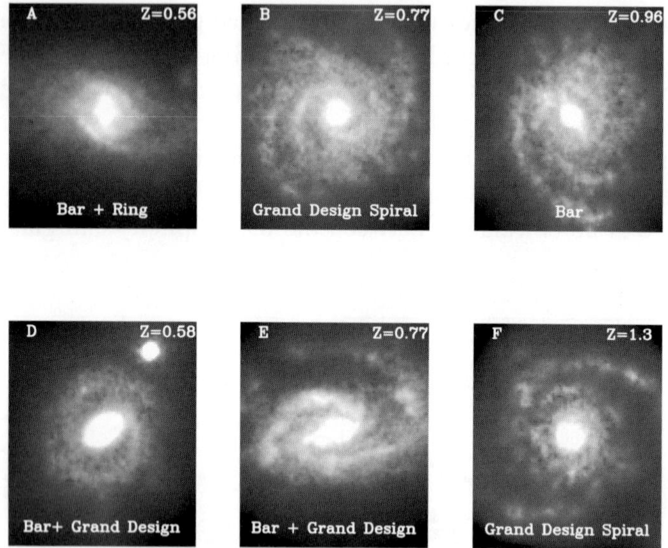

Figure 6. Typical barred and spiral galaxies from COSMOS data. Galaxy labeled A is virtually a twin of NGC 4303 shown in Figure 1. It shows a strong bar and a ring encircling the bar. The redshifts are photometrically derived from ground-based BVRIz Subaru data and are reliable to z 0.8. So galaxies C and F may in fact be at lower redshifts than indicated here.

Until recently, the Hubble Deep Fields were the standard windows into the high redshift Universe. These data already indicated that defining a comparative sample at high redshifts is difficult. For example the fraction of irregular objects increases to nearly 30% to z$\sim$1 (e.g., Griffiths et al. 1994, Abraham et al. 1996). Lilly et al. (1996) concluded that disks evolve passively in luminosity whereas Shude et al. (1998) argue in favor of significant evolution. Certainly some of these discrepancies are a result of cosmic variance and scarcity of high quality data. These gaps, however, are now being filled with surveys like GOODS, UDF, DEEP2, and GEMS.

The most ambitious and revolutionary new survey is COSMOS (http://www.astro.caltech.edu/cosmos) which is observing a contiguous 2-square degree equatorial field using the Advanced Camera for Survey on the Hubble Space Telescope, and multi-wavelength data from X-rays to radio. The size of the field is important because it overcomes cosmic variance and provides a truly representative view of the high redshift Universe. COSMOS is an ideal treasury program with a vast range of applications. In particular, for the study of galaxy morphology COSMOS is a gold mine (or more appropriately for this conference, a diamond mine) of opportunity. It will provide spectroscopic redshifts for over 100,000 galaxies with 10,000 galaxies in each of five bins of from z=0.5 to z=2. The spatial resolution of ACS data is exquisite. As shown in Figure 3, one can easily identify structures with sizes as small as 3 kpc in diameter. If Figure 6 we show some typical examples of disk galaxies at three photometrically determined redshifts.

The Bar Fraction in COSMOS

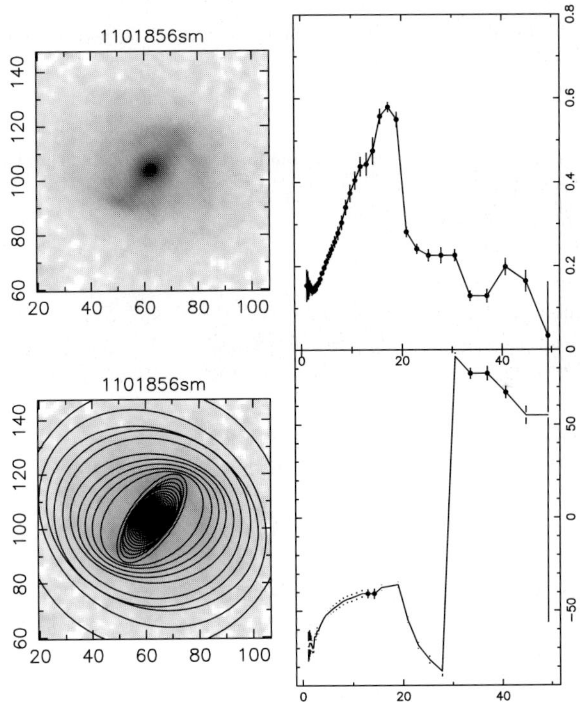

Figure 7. Example of ellipse fitting for a barred spiral in COSMOS.

In order to image such a large field with adequate sensitivity (COSMOS is within half a magnitude of the GOODS data) we chose to use the broader I-band filter over the z-band filter for COSMOS. Hence the analysis of the bar fraction can be done with full confidence to z$\sim$0.7–0.8 and perhaps to even higher redshifts. To date approximately 200 fields out of 600 have been observed. For these fields we first identified a sample of spiral galaxies from a photometric catalog derived from BVRIz data from observations on the Subaru teelscope for the entire COSMOS field. We only chose those galaxies that had photometric redshifts derived with the highest confidence (>95%). We ran the same ellipse fitting algorithm on these galaxies. An example of such a fit is shown in Figure 7. We have done a preliminary analysis for the redshift bin $0.6 < z < 0.8$ for $\sim$ 700 galaxies. We find that the fraction of barred spirals is $\sim$50%, consistent with the local fraction. Although further analysis is necessary with better redshifts and the complete sample, the results already indicate that the bar fraction is not declining significantly, at least out to z$\sim$0.8.

References

Abraham, R. G., Tanvir, N. R., Santiag o, B. X., Ellis, R. S., Glazebrook, K., van den Bergh, S. 1996, MNRAS, 279, L47

Abraham, R. G., Merrifield, M. R., Ell is, R. S., Tanvir, N. R., & Brinchmann, J. 1999, MNRAS, 308, 569

Block, D.L., Bournard, F., Combes, F., Puerari, I., & Buta, R. 2002, A&A, 394, 35

Bunker, A,J. 1999, in "Photometric Redshifts and the Detection of High Redshift t Galaxies", ASP Conference Series, Vol. 191, Eds. Weymann, R., Storrie-Lombard i, L., Sawicki, M., & Brunner, R.

Dickinson, et al. 2000, ApJ, 531, 624

Eskridge, P., et al. 2000, AJ, 119, 536

Griffiths, R. E. et. al. 1994, ApJL, 435, 19

Ho, L. C., Filippenko, A. V., & Sargent, W. L. W. 1997, ApJ, 487, 591

Lilly, S., Lefevre, O., Hammer, F., Cramp ton, D., Schade, D. J., Hudon, J. D., Tresse, L. 1996, IAUS, 171, 209

Martin, P., & Roy, J. 1994, ApJ, 424, 599

Norman, C. A., Sellwood, J. A., & Hasan, H. 1996, ApJ, 462, 114

Regan, M. W., Helfer, T. T., Thornley, M. D., Sheth, K., Wong, T., Vogel, S. N., Blitz, L., & Bock, D. C.-J. 2001, ApJ, 561, 218

Sakamoto, K., Okumura, S. K., Ishizuki, S., Scoville, N. Z. 1999, ApJ, 525, 691

Sheth, K., Regan, M.W., Vogel, S.N., & Teuben, P.J. 2000, ApJ, 532, 221

Sheth, K., Regan, M.W., Scoville, N.Z., & Strubbe, L.E. 2003, ApJL, 592, 13

Sheth, K., Vogel, S. N., Regan, M. W., Teuben, P.J., Harris, A. I., Thornley, M. D., & Helfer, T.T. 2004, ApJ, submitted

Shude, M., Mao, H. J. & White, S. D. M. 1998, MNRAS, 297, L71

Steidel, C.C. 1999, PNAS, 96, 4232

Toth, G., & Ostriker, J.P. 1992, ApJ, 389, 5

van den Bergh, S., Abraham, R. G ., Ellis, R.S., Tanvir, N. R., Santiago, B., & Glazebrook, K. G. 1996, AJ, 112, 359

van den Bergh, S., Cohen, J., Ho gg, D. W., & Blandford, R. 2000, AJ, 120, 2190

van den Bergh, S., Abraham, R.G ., Whyte, L.F., S. K., Ishizuki, S., Scoville, N. Z. 1999, ApJ, 525, 691

THE VIMOS VLT DEEP SURVEY: REDSHIFT DISTRIBUTION OF A $I_{AB} \leq 24$ SAMPLE, AND THE EFFECT OF ENVIRONMENT ON GALAXY EVOLUTION

Olivier Le Fevre[1], O. Ilbert[1], C. Marinoni[2], S. Paltani[1], G. Vettolani[2], L. Tresse[1], E. Zucca[10] C. Adami[1], M. Arnaboldi[7], S. Arnouts[1] S. Bardelli[10], M. Bolzonella[11], M. Bondi[2], A. Bongiorno[11], D. Bottini[3], G. Busarello[7], A. Cappi[10], S. Charlot[5], P. Ciliegi[10] , T. Contini[4], S. Foucaud[3], P. Franzetti[3], B. Garilli[3], I. Gavignaud[4], L. Gregorini[2], L. Guzzo[6], A. Iovino[6], V. Le Brun[1], D. Maccagni[3], D. Mancini[7], B. Marano[11], C. Marinoni[1], G. Mathez[4], A. Mazure[1], H.J.McCracken[8], Y.Mellier[8], B. Meneux[1], P. Merluzzi[7], R. Merighi[10], R. Pello[4], J.P. Picat[4], A. Pollo[6], L. Pozzetti[10], M. Radovich[7], V. Ripepi[7], D. Rizzo[4], R. Scaramella[2], M. Scodeggio[3], G. Zamorani[10], A. Zanichelli[2],

[1]*Laboratoire d'Astrophysique de Marseille, UMR 6110 CNRS-Universite de Provence, Traverse du Siphon-Les trois Lucs, 13012 Marseille, France, email: olivier.lefevre@oamp.fr,* [2]*Istituto di Radio-Astronomia - CNR, Bologna, Italy,* [3]*IASF - INAF, Milano, Italy,* [4]*Laboratoire d'Astrophysique - Observatoire Midi-Pyrenees, Toulouse, France,* [5]*Max Planck Institut fur Astrophysik, 85741 Garching, Germany,* [6]*Osservatorio Astronomico di Brera - INAF, via Brera, Milan, Italy,* [7]*Osservatorio Astronomico di Capodimonte - INAF, via Moiariello 16, 80131 Napoli, Italy,* [8]*Institut d'Astrophysique de Paris, UMR 7095, 98 bis Bvd Arago, 75014 Paris, France,* [9]*Observatoire de Paris, LERMA, UMR 8112, 61 Av. de l'Observatoire, 75014 Paris, France,* [10]*Osservatorio Astronomico di Bologna - INAF, via Ranzani 1, 40127 Bologna, Italy,* [11]*Universita di Bologna, Departimento di Astronomia, via Ranzani 1, 40127 Bologna, Italy*

Abstract The knowledge of the general framework of galaxy evolution from direct observation is necessary to understand the morphologies and substructures like bars which are frequently observed in nearby galaxies. We report here on the progress of the VIMOS VLT Deep Survey (VVDS). This survey is aimed to study the evolution of the main population of galaxies from $z \sim 5$ to present from the observation of an unprecedented large sample of galaxies with measured redshifts using VIMOS on the ESO-VLT. The first epoch sample of $\sim$ 10000 objects observed with $I_{AB} \leq 24$ yields 8626 galaxy redshifts, and 64 QSOs. We are showing here first results on the evolution of the Luminosity Function (LF) of galaxies in under dense vs. over dense environments: the LF is steeper in low redshift environements at all redshifts up to $z \sim 1$, and the luminosity evolution seems to have the same amplitude for either type of environment over this red-

D. Block et al. (eds.), Penetrating Bars through Masks of Cosmic Dust, 415–422.

shift range. We quantify for the first time the evolution of the bias between the distribution of galaxies and dark matter haloes, with a $\sim 40\%$ evolution of the bias between $z = 0.7$ and $z = 1.4$ probably indicating that merging is playing a major role in driving the evolution of galaxies.

Keywords: Deep redshift surveys, galaxy evolution, large scale structure evolution

1. Introduction

The general evolution process leading to the morphological distribution of galaxies observed today, and sub-structures like bars in many of them, is yet to be completely understood. A careful census of the galaxy population all along the lifetime of the universe is necessary to understand the evolution of galaxies. This requires dedicated instrumentation on the largest ground and space telescopes, to conduct large and deep galaxy surveys.

The properties of some specific galaxy populations at high redshifts now appear quite well established. The identification of galaxies with the lyman-break technique at redshifts 3 to 4, is now extended down to redshifts 2 (Steidel et al., 2004). The population of extremely red galaxies (EROs) is being probed by K-selected surveys (Cimatti et al., 2002; Abraham, et al., 2004). Each of these surveys provide a view of high redshift galaxies with its own selection highlight, and relating ther relative significance to the complete evolution scenario is the subject of intense debate.

Making a complete census of the galaxy population up to $z \sim 5$, above a given luminosity threshold is now becoming possible on large samples. Although evidently biased by the band of selection, magnitude selected samples provide the advantage to identify within the same survey all the galaxies above a fixed luminosity in a given volume of the universe. This allows to derive global quantities describing the main galaxy population, with a relatively easy to control bias, and perhaps more importantly, attempt to relate these quantities measured at different epochs in a global evolution scenario.

We are describing here the progress of the VIMOS VLT Deep Survey (VVDS), a magnitude selected survey going down to $I_{AB} = 24$ for the first time with a high completness level. In particular we focus onto the evolution of galaxies related to the local environment, and present the first measurements of the evolution of the bias between the galaxies and the underlying dark matter at $z \sim 1$.

2. The VVDS: status

The redshift measurement of the first epoch VVDS-Deep observations performed in the fall of 2002 is now complete (Le Fevre et al., 2003). To $I_{AB} \leq 24$, we have a secure measurement of 7174 galaxies, 58 QSOs and 559 stars in the VVDS-02h, and 1452 galaxies, 8 QSOs and 139 stars in the CDFS (

Le Fevre et al., 2004b). The completeness in redshift measurement is $\sim 85\%$, with a sampling of the total population down to this magnitude of $\sim 25\%$. The final check of the VVDS-Wide measurements for $I_{AB} \leq 22.5$ is being completed.

Survey observations are continuing. A total of $\sim$ 900 objects for the VVDS-Deep and $\sim$20000 objects for the VVDS-Wide, have been observed in the period december 2003 - April 2004. A total of 14 nights of observations are planned from June to September 2004. We expect the processing of these new observations to proceed fast as all the data processing and redshift measurement tools have been significantly improved from the experience gained with the first epoch data, together with the expertise of the team in measuring redshifts.

3. Redshift distribution of a $I_{AB} \leq 24$ sample

Measuring redshifts

To measure redshifts of a complete sample of galaxies with $0 < z \leq 5$ is not an easy task. The number of possible redshift solutions to consider when measuring the spectrum of a faint object with only a few significant spectral features in the observed bandpass can be large, especially for objects with a flat spectrum. We have concentrated our effort into a computer aided tool for the astronomers to measure redshifts. KBRED is a correlation engine with increasing levels of complexity from a simple cross correlation between the observed spectrum and reference templates, up to a more sophisticated correlation approach including the decomposition of the spectrum using a PCA analysis (Scaramella et al., 2004). The key elements in the development of this tool into a reliable tool have been the building of galaxy templates from the VVDS observations going as far as possible into the rest-frame UV, ultimately shortward of Lyα, and the display to the user of all possible redshift solutions and their associated best template and correlation peak quality in a single summary panel.

Redshift distribution

The VVDS is the first attempt to secure a large sample of purely magnitude selected galaxies and AGNs as faint as $I_{AB} = 24$. Rather than putting emphasis into a specific galaxy population in a narrow redshift range, the VVDS offers a complete census of the galaxy population from $z \sim 0$ to $z \sim 5$. The redshift distribution of 8626 galaxies with $I_{AB} \leq 24$, observed in the VVDS-CDFD and the VVDS-02h field, is shown in Figure 1.

The combination of the wavelength coverage of the VIMOS LRRED grism, and of the strong OH emission from the sky, seriously affects our capability in measuring redshifts in the range $2.2 < z < 2.8$ as is immediately evident in Figure 1, with a dip in the redshift distribution in this redshift range. This has been simulated extensively by Paltani et al. (2004). Extrapolating from the redshift distribution through this redshift range, we estimate that $\sim 3\%$ of this redshift population have escaped our measurement, and contributes significantly to our incompleteness.

The redshift distribution observed in the the combined VVDS-2hr and VVDS-CDFD fields has a median $< z >= 0.76$, and decreases above $z \sim 1$ with a high redshift tail going up to $z \sim 5$, the highest secure redshift measured being $z = 5.1$. There are 476 galaxies with $1.4 \leq z \leq 2.2$, and 334 with $2.8 \leq z \leq 5.1$ measured in this field.

The I-band selection of the VVDS means we are looking at the evolved stellar population above $4000A$ rest frame up to $z \sim 1.2$, while above this redshift, it is increasingly aiming at the UV continuum from young stars. Although the observed redshift distribution of galaxies is the result of several evolution processes, it has been proposed (Kauffmann & Charlot, 1998) to use this distribution to discriminate between a purely passive evolution scenario with ageing of the stellar populations, and a hierarchical formation model with merging playing a major role. A pure luminosity evolution model (PLE) seems to over-predict the number of galaxies by a factor 1.5 at $z \sim 1.5$, 2 at $z \sim 2$ and 3 at $z \sim 4$. A more detailed analysis is in preparation (Le Fevre et al., 2004a).

4. The evolution of the galaxy population vs. the local environment

The influence of the environment on the evolution of galaxies is well demonstrated (e.g. Dressler, 1980). Among several others, two main processes are identified, mergers of sub-units to form more massive galaxies in the hierachical growth picture, with more mergers in higher density environments, and accretion of cold gas from the large scale structure environment in which galaxies are embeded (Combes, 2004). Directly observing the properties of galaxies from low density to high density environments is necessary to evaluate the relative contributions of these processes.

We have derived the local galaxy density in the VVDS-02h field using a top hat smoothing and Wiener filtering techniques, applied on $\sim 5-15$Mpc scales (Marinoni et al., 2004). While the survey is in progress, the currently sampled volume in the VVDS-02hr is $30 \times 30 \times 6000$ Mpc^3. Using the local galaxy density estimation δ_g, the VVDS sample is split in two samples at each redshift, one with galaxies in a "low density" environment with $\delta_g \leq 0$, the other with galaxies in a "high density" environment with $\delta_g > 0$. The luminosity function for each sub-sample is then computed as shown in Figure 2. There

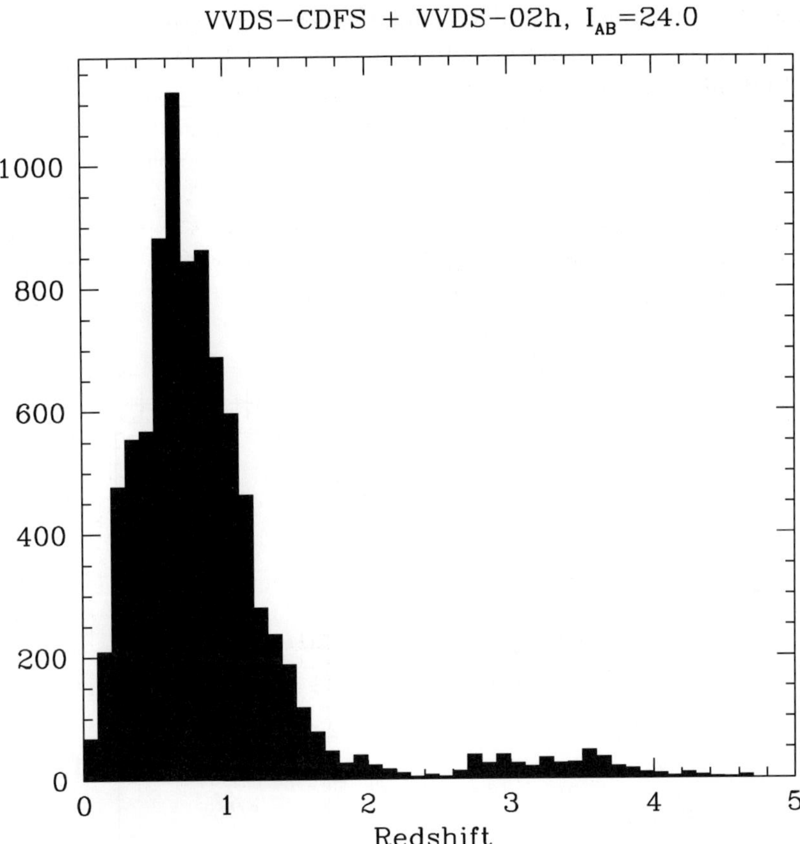

Figure 1. Redshift distribution of 8626 galaxies in the VVDS-02hr and VVDS-CDFS fields, from a magnitude selected sample with $I_{AB} \leq 24$. The incompleteness is $\sim 85\%$. In the redshift domain $2.2 \leq z \leq 2.8$, the efficiency in measuring redshifts is reduced due to the combination of the wavelength domain and the strong sky OH emission, producing the observed dip (see text).

are two main conclusions coming from this analysis: (1) the slope of the LF of galaxies in under-dense regions is significantly steeper than in over-dense regions and this difference is already present at $z \sim 1$, (2) the evolution in luminosity is comparable in under-dense and over-dense regions. The first point can be interpreted as a variant of the morphology-density relation observed in clusters of galaxies (Dressler, 1980): denser environments are populated by earlier galaxy types than in more sparse environments. and this difference is already in place at $z \sim 1$. The second point seems to give an indication that the physical phenomena driving luminosity evolution are dominated by processes internal to galaxies rather than related to the environment. This will be investigated in more details in Ilbert et al. (2004).

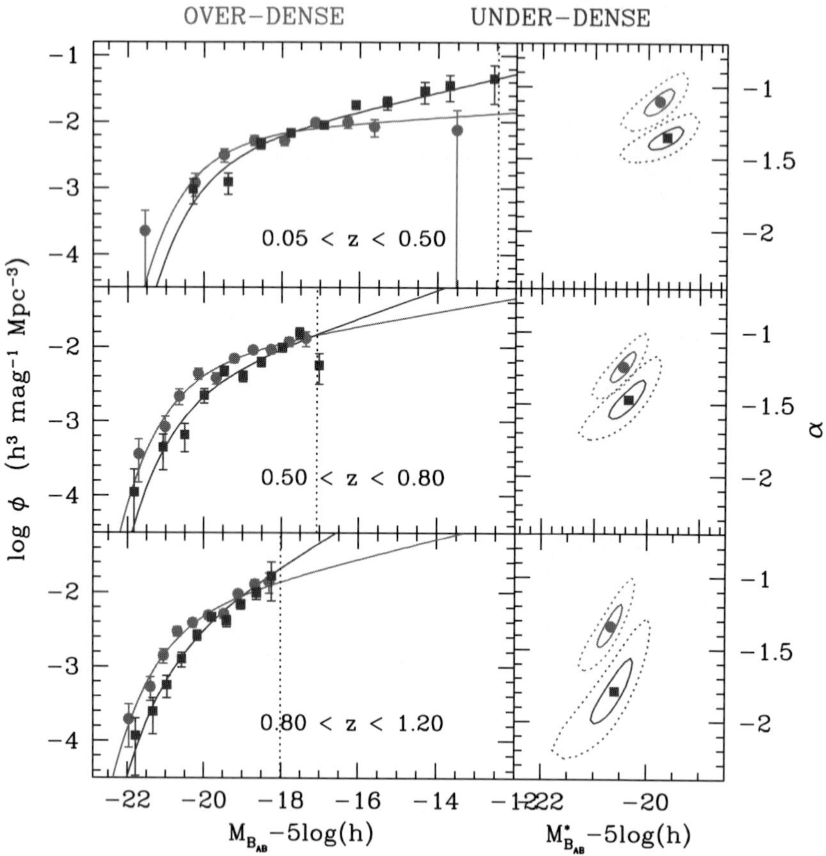

Figure 2. Evolution of the luminosity function of galaxies in the VVDS-02hr field vs. the local environment. At each redshift, the luminosity function of galaxies in "high density" environments is compared to the LF of galaxies in "low density" environments (see text)

5. Measuring the bias between galaxies and the underlying dark matter

The bias is computed from the analysis of the probability distribution function (PDF) of galaxies, compared to the PDF of dark matter haloes as derived from the ΛCDM predictions, as a function of redshift (see e.g. Dekel & Lahav, 1999). With the current geometry of the observations available in the VVDS-02h field, the bias analysis can be conducted in the redshift range

$0.7 \leq z \leq 1.4$. Figure 3 presents the evolution of the bias between the galaxies observed in the VVDS and the dark matter, compared to the models of Fry (1996), Matarrese et al. (1997). The bias evolves by more than 40% from z=0.7 to z=1.4. This evolution seems to be consistent with an evolution of the galaxies driven by merging (Marinoni et al., 2004).

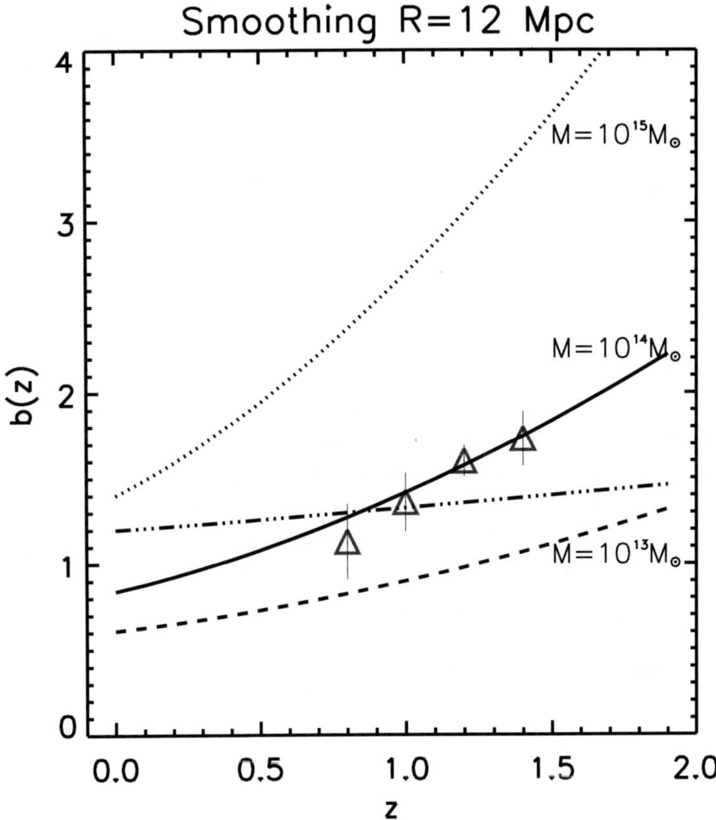

Figure 3. Evolution of the bias between the luminous galaxies and dark matter haloes. The full line is a model from Matarrese et al. (1997) with evolution mostly driven by mergers, which seems to fit best the observed bias evolution (see text)

6. Summary

The VVDS is an on-going deep redshift survey of the distant universe, pushing the observations of the main population of galaxies up to $z \sim 5$. The redshift distribution N(z) of the "deep" I_{AB} selected sample is peaked at $z \sim 0.8$, with a flat high redshift tail extending to $z \sim 5$. We are able to quantify the evolution of the luminosity function of galaxies vs. environment, showing that the LF of galaxies in low vs. high density environments is already differentiated at $z \sim 1$. We have computed for the first time the evolution of the bias between galaxies and dark matter haloes: the bias rises by 40% between $z = 0.7$ and $z = 1.4$, consistent with a merger-driven evolution. More insight into the connection between the environement and galaxy evolution will come with detailed morphology observations with HST-ACS in the frame of the HST-COSMOS survey, in connection with the knowledge of the underlying cosmic web from the deep spectroscopic survey with VIMOS.

References

Abraham, R., et al., 2004, AJ, in press, astro-ph/0402436
Cimatti, A., et al., 2002, A&A, 381, L68
Combes, F., 2004, these proceedings
Dekel, A., Lahav, O., 1999, Ap.J., 520, 24
Dressler, A., 1980, Ap.J.,
Fry, 1996, Ap.J., 461, L65
Ilbert, O., and the VVDS team, 2004, A&A, in preparation
Kauffmann, G., and Charlot, S., 1998, MNRAS, 297, L23
Le Fevre, O., Vettolani, G., and the VVDS team, astro-ph/0402203
Le Fevre, O., and the VVDS team, 2004a, in preparation
Le Fevre, O., Vettolani, G., and the VVDS team, 2004b, A&A, in press, astro-ph/0403628
Marinoni, C., and the VVDS team, 2004, in preparation
Matarrese, S., et al., MNRAS, 1997, 286, 115
Paltani, S., and the VVDS team, 2004, in preparation
Scaramella, R., and the VVDS team, 2004, in preparation
Steidel, C.C., et al., 2004, Ap.J., in press, astro-ph/0401439

AN *HST* ACS/WFC Hα IMAGING SURVEY OF NEARBY GALAXIES

Rolf A. Jansen
Dept. of Physics and Astronomy, Arizona State University, Box 871504, Tempe, AZ 85287-1504
Email: Rolf.Jansen@asu.edu

Abstract Previous studies of nearby galaxies show large discrepancies between different star formation (SF) indicators on large ($\gtrsim$100 pc, or even global) scales: the strikingly complex interplay of young stars, dust and ionized gas are the primary cause of this variance. The few galaxies in the *HST* Archive with both Hα and mid-UV images show this complex geometry to extend down to pc scales.
To provide a benchmark for understanding SF processes at spatial resolutions that are unattainable from the ground in normal and star-bursting, barred and non-barred galaxies, I am conducting an Hα SNAPshot survey with ACS/WFC of $\sim$24 galaxies of all Hubble types, that are nearby but beyond the Local Group and that were previously imaged with *HST* in the mid-UV and red.
These data can be used to address a wide range of astrophysical problems, but the immediate goals are to: (1) spatially resolve the dust clouds and filaments which strongly affect mid-UV and Hα derived SF rates, (2) test how the large-scale correlation between Hα and mid-UV flux breaks down on pc scales, and (3) model the propagation of star formation by comparing the SF over time scales of $\sim$100 Myr (via mid-UV) and $\sim$5 Myr (via Hα). This will (4) significantly improve our insight into, and calibration of SF in UV-bright galaxies at high z, and into the cosmic SF history.

Keywords: Galaxies: star formation — galaxies: ISM — galaxies: fundamental parameters — galaxies: nearby

Quantitative Star Formation Tracers

Accurate measurements of star formation (SF) in external galaxies based on tracers like the mid-UV continuum, Hα recombination line, [O II] emission line, and radio continuum emission are hampered by systematic uncertainties due to the presence of interstellar dust[26,27][9][3,4,5]. This dust is intermixed with interstellar gas and distributed in highly complex geometries down to scales that have only been resolved from the ground in the very nearest Local Group galaxies[4] but are readilly apparent in *HST* images[8],[12]. In their ground-based spectrophotometric study, Jansen et al. [13−16] showed

D. Block et al. (eds.), Penetrating Bars through Masks of Cosmic Dust, 423–427.

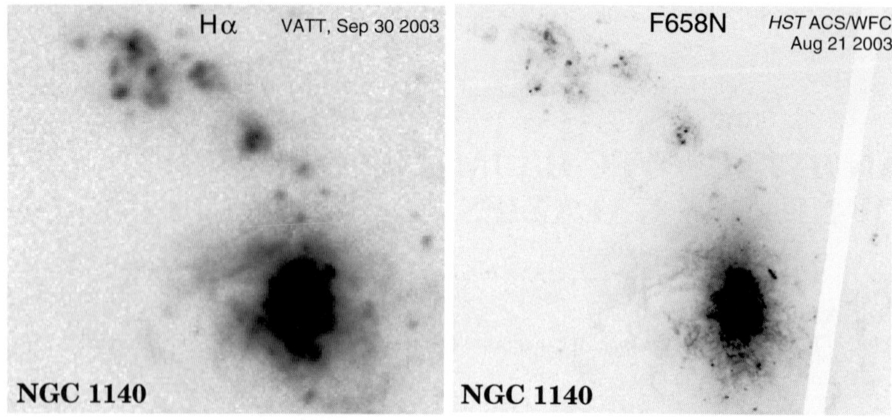

Figure 1. Hα images of star-bursting dwarf (IBm) galaxy NGC 1140 obtained from the ground under good conditions (VATT, Mt.Graham; σ=0.″9) and obtained as part of the ACS/WFC SNAPshot survey. The large uncertainties and discrepancies in global SFR calibrations can only be resolved by studying the tracers most sensitive to dust in a diverse nearby galaxy sample *at HST resolution*.

strong correlations between a galaxy's Hubble type, wavelength-dependent UBR morphology, luminosity and global emission line strengths of [O II] and Hα (hence the global star formation rate [SFR]), demonstrating the need for a relatively large and varied galaxy sample. However, ground-based studies are significantly limited by the seeing ($\gtrsim$0.″4), both in imaging and spectroscopy, and are therefore not suited for the study of individual SF sites beyond the Local Group. Since the emission emerging from a galaxy, especially in the UV, is dominated by regions of low obscuration[7] and may itself depend on the SFR[11], the resulting large uncertainties and discrepancies in global SFR calibrations can only be resolved by studying the tracers most sensitive to dust in a diverse nearby galaxy sample *at HST resolution*[8],[21],[5].

An ACS/WFC Hα SNAPshot survey of Nearby Galaxies

In Cycles 9+10, Windhorst et al. imaged a sample of 89 nearby galaxies with WFPC2 in the mid-UV F300W and red F814W filters[28,17] supplementing the 17 galaxies in the *HST* Archive for which such images were already available. The combined sample spans the Hubble sequence in morphological type, and is well represented in late-type/irregular, merging and interacting galaxies which are more common at higher redshifts.

Jansen et al. are currently conducting an ACS/WFC Hα SNAPshot survey (PID #9892) of nearby galaxies selected from this sample. 48 nearby galaxies were selected with $cz \lesssim 2500$ km s^{-1}, sampling different SF environments, for which sufficient S/N could be obtained in the F658N Hα filter in a partial *HST* orbit, and without prior WFPC2 Hα images of acceptable quality in the Archive. I chose to aim for the longer holes in the *HST* schedule (~32–55 min) to ensure well-exposed Hα images and sensitivity to faint extended emission, even though fewer such opportunities are available. It allowed fitting three WFC exposures within a SNAP visit and therefore to also obtain a short F625W (r') exposure for continuum subtraction and to aid in selecting low-extinction sightlines (see below). As of this writing (May 2004), 19 galaxies were imaged and by the end of Cycle 12 this should be a random subset of ~50% of the pool of 48 targets. The ACS/WFC Hα SNAPshots provide both exquisite spatial resolution and spectral information in one of the most important atomic emission lines, that has not been mapped yet in a significant number of galaxies with *HST* images in both Hα and the mid-UV.

Science goals for the Hα survey

(1) *Measure the intrinsic correlation of Hα and mid-UV flux as tracers of current high mass star formation, by distinguishing low and high extinction sight-lines at $\lesssim$10 pc spatial resolution.* We will model the Lyman continuum luminosity and corresponding high-mass SFR[18,20] for the the sight-lines with the very lowest extinction (as determined from comparison of the F300W, F625W, and F814W images) —taking into account possible changes in recent SF history or spatial propagation of the SF— and so constrain the ratio of Hα to mid-UV flux to be used in SF calibration at high z. We will study whether that ratio varies as a function of a galaxy's Hubble type, or whether it depends on (a) the distance from the galaxy center (e.g., metallicity gradients), (b) the size or luminosity of a H II-region, (c) local morphological structures, like bars, spiral arms, ridges, bubbles or filaments, and (d) on the proximity to other H II-regions.

(2) *Measure the effect of averaging over sight-lines with varying dust column densities on the derived star formation rates, calibrate lower resolution measurements, and model the SFR calibrations at higher redshifts.* The effects of dust, age and metallicity are extremely difficult to disentangle[2]. The high spatial resolution of ACS/WFC, however, helps significantly (its $0\farcs05$ pixels equal to $\lesssim$10 pc out to 37 Mpc). For the more extinguished sight-lines, we will test the interstellar extinction law[10],[7],[22] with simple assumptions on the dust geometry, for a much larger and more varied sample of galaxies than entered these studies, using the intrinsic Hα–mid-UV flux-ratios measured along the most transparent sightlines as reference. We will then measure the impact of averaging over varying dust column densities on the derived SFRs and properly model the SFR calibration at high z, taking into account the metal production history. Studies with this kind of detail have so far been possible only in Local Group galaxies[25],[4]. This study will complement the latter (which compared

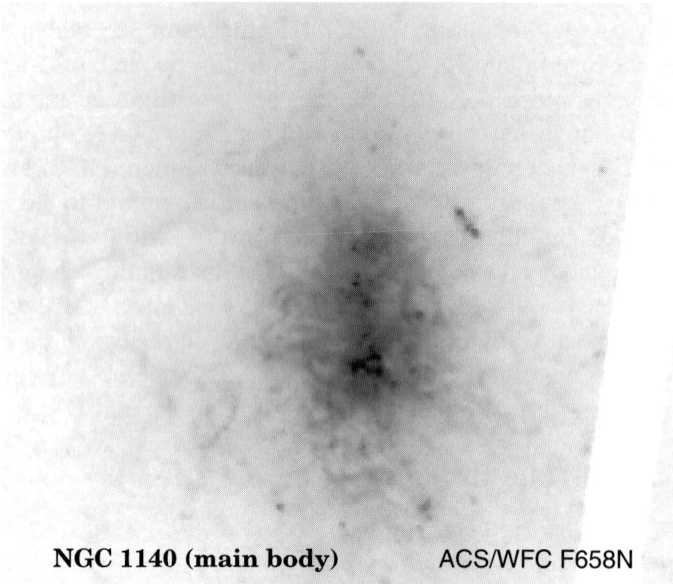

Figure 2. Detail of the full ACS/WFC Hα image of NGC 1140 centered on the high surface brightness "main body" or "bar" portion of the galaxy. The filamentary nature of the distribution of the Hα emission is apparent down to the smallest scales ($\sim$5 pc).

UIT mid- and far-UV imagery with ground-based Hα at $\sim 3''$ resolution) by spatially resolving individual SF sites.

(3) *Investigate whether the stellar populations responsible for the mid-UV emission coincide spatially with those responsible for exciting the H I gas.* Available (lower resolution) images of nearby galaxies show distinctive spatial distributions for Hα and mid-UV light. The high spatial resolution attainable with ACS/WFC allows us to examine to what extent this trend continues down to pc scales: because of different timescales probed by mid-UV and Hα light, this gives powerful insight into the mechanisms by which SF propagates within galaxies on time-scales $\lesssim$100 Myr. It also gives us the opportunity to study in detail the fraction of Hα emitted by diffuse ionized gas[24,23], and to study outflows, bubbles and superbubbles driven by SNe that have recently exploded in or near H II-regions (as seen in the Hα image of NGC 5253 [8]).

(4) *Probe the faint end of the H II-region LF — do size and luminosity depend on galaxy type?.* We will measure whether the faint-end slope of the H II-region LF depends on galaxy morphology —and if so, how— and, by comparing diameters and luminosities, what fraction of H II-regions is density bounded. For example, Kennicutt et al. [19] and Caldwell et al. [6] find systematic changes in the shape of the H II-region LF with Hubble type, as would result from differences in the intrinsic molecular cloud mass spectrum. The high end

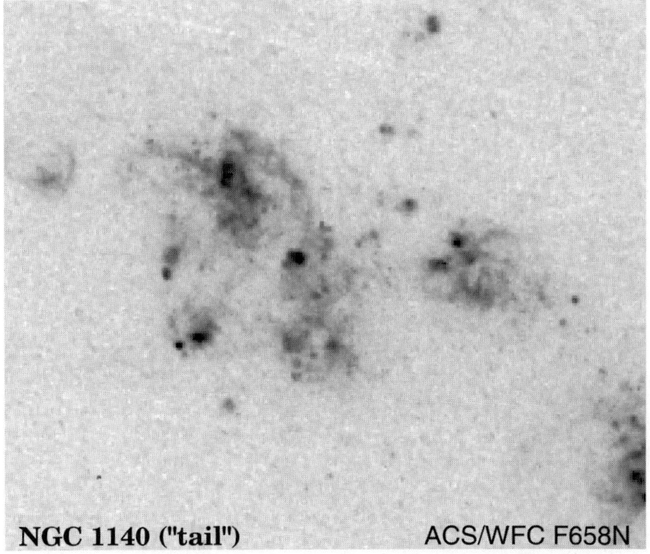

Figure 3. Detail of the full ACS/WFC Hα image of NGC 1140 centered on the "tail" of the galaxy. The knots of star formation visible in the ground based image (Fig. 1a) are resolved into the brightest stars and/or compact young clusters.

of the mass spectrum is sensitive to shear due to differential rotation, and so a dependence on Hubble type could naturally emerge. Beckman et al. [1] find that the H II-region LF in several disk galaxies with Cepheid distances shows "glitches" near an apparently invariant luminosity of $L^*_{\rm H\,II} \sim 10^{38.6}$. We will investigate to what extent these results are due to the limited spatial resolution in these studies.

Acknowledgements

The author gratefully acknowledges support from NASA grant GO-9892 and thanks the team of co-Is of this program.

References

[1] Beckman et al. 2000, AJ 119, 2728
[2] Bell & de Jong 2000, MNRAS 312, 497
[3] Bell & Kennicutt 2001, ApJ 548, 681
[4] Bell et al. 2002, ApJ 565, 994
[5] Bell 2002, ApJ 577, 150
[6] Caldwell et al. 1991, ApJ 370, 526
[7] Calzetti et al. 1994, ApJ 429, 582
[8] Calzetti et al. 1997, AJ 114, 1834
[9] Calzetti 2001, PASP 113, 1449
[10] Cardelli et al. 1989, ApJ 345, 245
[11] Clarke & Oey 2002, MNRAS 337, 1299
[12] de Grijs et al. 2001, AJ 121, 768
[13] Jansen 2000, Ph.D. Thesis (Univ. of Groningen)
[14] Jansen et al. 2000a, ApJS 126, 271
[15] Jansen et al. 2000b, ApJS 126, 331
[16] Jansen et al. 2001, ApJ 551, 825
[17] Jansen et al. 2004, ApJS (in prep.)
[18] Kennicutt 1983, ApJ 272, 54
[19] Kennicutt et al. 1989, ApJ 337, 761
[20] Kennicutt 1998, ARA&A 36, 189
[21] Leitherer 2001, in: "The Central kpc of Starbursts and AGN", p.457
[22] Macri et al. 2001, ApJ 549, 721
[23] Martin 1997, ApJ 491, 561
[24] Martin & Kennicutt 1997, ApJ 483, 698
[25] Oey & Kennicutt 1997, MNRAS 291, 827
[26] Sullivan et al. 2000, MNRAS 312, 442
[27] Sullivan et al. 2001, ApJ 558, 72
[28] Windhorst et al. 2002, ApJS 143, 113

HST MID-UV IMAGING OF NEARBY GALAXIES

Rogier A. Windhorst, Violet A. Taylor, and Rolf A. Jansen
Department of Physics and Astronomy, Arizona State University, Box 871504, Tempe, AZ 85287

Abstract We summarize Hubble Space Telescope projects that imaged a significant sample of nearby galaxies of all types and inclinations in the mid-UV with WFPC2 and some in the near-IR with NICMOS. The sample emphasizes high-surface brightness late-type, irregular, peculiar and merging galaxies, since these objects are the most prevalent at high redshifts. The sample is used to establish a local benchmark for quantitative classifications of galaxies seen by *HST* in their rest-frame UV, and in the next decade also by the James Webb Space Telescope.

We review how nearby galaxies change their rest-frame structure and morphology from the rest-frame UV to the optical as a function of galaxy type and inclination, and other studies that are being done with this *HST* sample. When observed in the rest-frame mid-UV — early- to mid-type galaxies are more likely misclassified as later types than late-type galaxies are misclassified as earlier types. This is because later-type galaxies are dominated by the same young and hot stars in all filters from the mid-UV to the red, and so have a smaller "morphological K-correction" than true earlier type galaxies. This K_{morph} can explain a small part, but certainly not all of the excess faint blue late-type galaxies seen in deep *HST* fields. Contrary to early–mid-type galaxies, the majority of late-type galaxies become redder outward. While young UV/blue-bright stellar populations dominate their inner morphology, most late-type galaxies are inferred to have a significant halo or (thick) disk of older stars. We discuss these results in a cosmological context.

Keywords: galaxies: elliptical — spiral — irregular — mergers — peculiar — ultraviolet

1. The Need to Reliably Classify Faint Optical Samples

Since the mid 1990's, faint galaxies have been observed systematically with the *Hubble Space Telescope* (*HST*). The best statistics, spatial sampling, and areal coverage are achieved in the red, and good *HST* structural information is available today for $\gtrsim 10^5$ galaxies with I$\lesssim$26–27 mag. These galaxies come from the two Hubble Deep Fields ("HDF"s; Williams et al. 1996, 1998), the UltraDeep Field (UDF; Beckwith et al. 2004), their flanking fields, the *HST* Medium-Deep Survey, and other *HST* parallel surveys (e.g., Griffiths et al. 1994; Driver et al. 1995a; Abraham et al. 1996, 1999; Roche et al. 1997). The median half-light radius of faint field galaxies with I$\sim$26 mag is $r_e \simeq$ 0.2–

D. Block et al. (eds.), Penetrating Bars through Masks of Cosmic Dust, 429–440.

0.3″ (Odewahn et al. 1996, Cohen et al. 2003). *HST*'s 2.4 m aperture limits the useful wavelength for proper galaxy classification to $4000 \lesssim \lambda \lesssim 8000$ Å, because of the high spatial resolution and good sampling required. *HST* can determine morphology and structure with relative ease for a very large number of galaxies, and so explore a part of parameter space that constrains galaxy formation and evolution, which is complementary to spatially resolved spectroscopy and kinematics (see Koo 2004 and Le Fèvre 2004; this Vol.).

The most dramatic result from these *HST* studies was that at faint fluxes, late-type/irregular galaxies dominate the faint blue galaxy counts (Driver et al. 1995a; Glazebrook et al. 1995; Odewahn et al. 1996; Windhorst et al. 1997, 2000). These are believed to be star-forming galaxies that underwent substantial evolution since $z \lesssim 3$–4 (Ellis et al. 1996; Pascarelle et al. 1996; Abraham et al. 1999). Ellipticals and early-type spirals evolved much less since $z \lesssim 1$ (Driver et al. 1995a; Lilly et al. 1998; Cohen et al. 2003). Driver et al. (1998) suggested differential evolution of galaxies as a function of type, implying a gradual formation of the Hubble sequence with cosmic time.

2. The Morphological K-correction and How to Address it

In the deepest *HST* fields, where reliable classifications are achievable to $I \lesssim 26$ mag (e.g., Odewahn et al. 1996; Driver et al. 1998), the sampled redshift range is $z \simeq 1$–3, while the bulk of the galaxies is at $z \lesssim 2$. Faint high redshift galaxies observed in the *I*-band are therefore primarily seen in the rest-frame mid-ultraviolet (mid-UV), or 2000–3200 Å, which primarily traces high surface brightness ("SB") regions populated by high densities of young ($\lesssim 1$ Gyr) hot stellar populations. The reliability of results derived from faint *HST* galaxy morphology is therefore fundamentally limited by the uncertain rest-frame mid-UV morphology of nearby galaxies. Although these faint, late-type/irregular objects resemble some classes of nearby late-type and peculiar galaxies (e.g., Hibbard & Vacca 1997), they need not be physically late-type objects. Instead, they may be earlier-type galaxies that look dramatically different in the rest-frame UV. Because of the wavelength dependence of nearby galaxy morphology (especially towards the UV), faint galaxy classifications will depend on the rest-frame wavelength sampled. This "morphological K-correction" can be quite significant, and must be quantified in order to distinguish genuine evolutionary effects from simple band-pass shifting.

Morphological K-corrections were derived from *UIT* images by Bohlin et al. (1991) and Giavalisco et al. (1996), showing that nearby galaxy morphology can change quite dramatically below 3600–4000 Å, where the hot (young) stellar population — located mainly in spiral arms and H II-regions — dominates the spectral energy distribution (SED) and where bulges essentially disappear (Burgarella et al. 2001; Marcum et al. 2001; Kuchinski et al. 2001). Galaxies therefore often appear to be of later Hubble type the further one looks into the rest-frame UV. Qualitatively, this is easy to understand: in the optical/near-IR, we see the accumulated luminous phases of long-lived ($>$1 Gyr) stars, which emit most of their energy at longer wavelengths, whereas the mid-UV samples the star-formation rate (SFR) averaged over the past Gyr or less. The mid-UV includes the longest wavelengths where young stars can dominate the integrated galaxy light, and traces primarily presently active star forming regions, or those regions where star-formation (SF) has only recently shut down. A number of authors have explored the effects of band-pass shifting using multicolor optical images to extrapolate to the rest-frame UV on a pixel-by-pixel basis using both ground-based data (e.g., Hibbard & Vacca, 1997; Brinchmann et al. 1998), mid-UV FOCA balloon images (Burgarella et al. 2001), *UIT* far-UV images (Kuchinski et al. 2000), and *HST* images of galaxies at moderate redshifts (Abraham et al. 1999; Bouwens, Broadhurst, & Silk 1998). The galaxy shape and size distributions measured in deep *HST* images appear to considerably exceed the effects of band-pass shifting alone.

To address these issues unambiguously, we present a systematic *HST* imaging project with the Wide Field and Planetary Camera 2 ("WFPC2") in the mid-UV and red for a representative sample of $\sim$100 nearby galaxies. A subset was also imaged with the Near-Infrared Camera and Multi-Object Spectrograph ("NICMOS") in H-band to study the older stellar populations, and with the Advanced Camera for Surveys ("ACS") in H-α to trace their recent SF and correlate this with the mid-UV images. This survey will help us better understand the *physical* drivers of the rest-frame mid-UV emission, i.e., the relation between SF and the global physical characteristics of galaxies, their recent SF history, and the role of dust absorption/scattering. Details are given in Windhorst et al. (2002), Odewahn et al. (2002; & this Vol.), Eskridge et al. (2003), de Grijs et al. (2003), Jansen (2004; this Vol.), and Taylor et al. (2004).

3. Strategy, Sample Selection, and HST Observations

With the *HST* WFPC2, we imaged $\sim$100 nearby galaxies through one or, whenever possible, two wide-band mid-UV filters below the atmospheric cut-off: F300W ($\lambda_{cent} \simeq$ 2930 Å; $\Delta\lambda \simeq$ 740 Å FWHM) and F255W ($\lambda_{cent} \simeq$ 2550 Å; $\Delta\lambda \simeq$ 395 Å FWHM). To this we add short red F814W exposures to provide red-leak removal and study the older stellar population. The F255W, F300W filters are logarithmically about equally spaced in λ from Johnson U, B, and so add significantly to the ground-based optical–near-IR color baseline.

Over the last decade, several groups did systematic ground-based imaging of a large number of nearby galaxies covering all Hubble types and inclinations. Most of these were observed in $UBVR$, but a subset also in the near-IR $IJHK$ filters (Eskridge et al. 2002, see also this volume). We selected our HST sam-

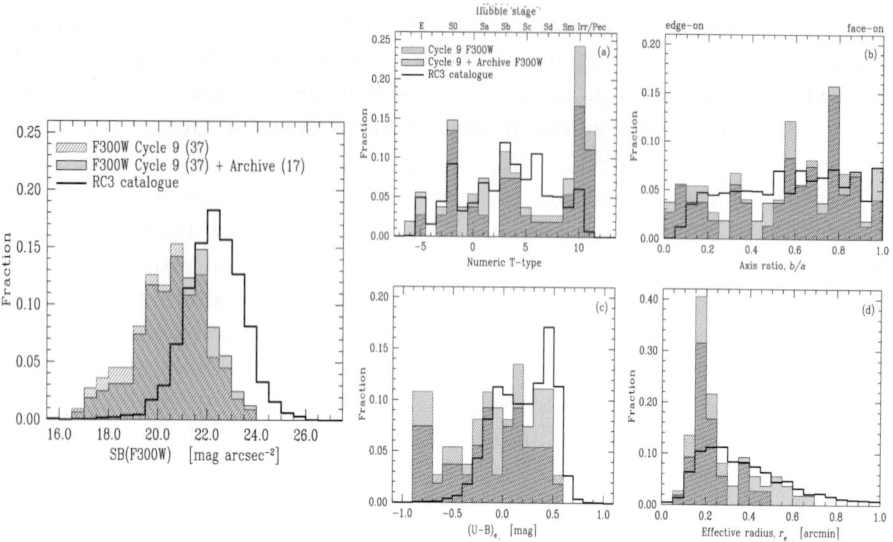

Fig. 1 (LEFT PANEL): The distribution of predicted average mid-UV surface brightness (SB) out to r_e for the galaxies observed in *HST* Cycle 9 (*hashed histogram*) and for the full galaxy sample (*solid histogram*), which includes galaxies with mid-UV images in the *HST* Archive. For comparison, we show the SB-distribution for the 3009 galaxies in the RC3 with measured B_T, $(U - B)$ and r_e (*open histogram*). **Fig. 2a–2d (RIGHT FOUR PANELS):** Other properties of the selected *HST* mid-UV galaxy sample. As in Fig. 1, we plot the Cycle 9 galaxies (*hashed*), the full sample (*solid*), and the RC3 (*open*) versus (*a*) morphological type; (*b*) apparent axis ratio, b/a; (*c*) average $(U - B)$ color out to r_e; and (*d*) effective radius r_e. Our sample approximates the RC3 distribution, except that for comparison with high redshift samples we emphasize the later types, bluest galaxy colors, and smallest sizes.

ple of ∼100 galaxies from these ground-based samples, which include: (1) 86 face-on spirals (de Jong & van der Kruit 1994); (2) 220 spirals from the OSU survey (Eskridge et al. 2000, 2002); (3) 113 galaxies from Frei et al. (1996); (4) 100 galaxies with 1500 Å and 40 with 2500 Å images from the *Astro*/*UIT* mission (Kuchinski et al. 2001; Marcum et al. 2001); (5) 48 edge-on galaxies (de Grijs et al. 1997); (6) 150 UGC galaxies that are morphologically irregular, peculiar, or mergers (Hibbard & Vacca 1997); and (7) 49 late-type dwarf spiral galaxies and compact, high-SB luminous blue galaxies (Matthews & Gallagher 1997). The last two samples provide likely local counterparts of the peculiar and irregular galaxies seen with *HST* in large numbers at high redshifts.

The selection criteria for inclusion in the *HST* sample are: (a) B-band half-light radius 0.1' $\lesssim r_e \lesssim$ 1.0', so that the object fits inside a WFPC2 or NIC3 mosaic; (b) predicted average mid-UV SB out to r$\simeq r_e$: 18 $\lesssim \mu_{F300W} \lesssim$ 22.5–23.0 mag arcsec^{-2}(Fig. 1), so that a galaxy can be detected out to r$\simeq$2–3r_e in one orbit with sufficiently high S/N to allow structural features to be recognized; (c) distribution of Hubble types and (d) the distribution over apparent axis ratio (inclination) should both be representative for nearby galaxy samples (Fig. 2).

Fig. 1 shows that the galaxies observed with *HST* have higher predicted mid-UV SB than the median in the RC3, which is $\mu_{F300W} \simeq 22.3$ mag arcsec^{-2}. This is not an overriding concern, since high-redshift samples are similarly biased (or more so) in favor of high-SB galaxies due to the severe $(1+z)^4$ cosmological SB-dimming. Based on Fig. 1, one could apply weights to each observed *HST* galaxy to better represent the true local galaxy distribution when complete samples are in order. Figures 2a–2d show the distributions of the *HST* mid-UV sample over morphological type, axis ratio b/a, (U–B) color within r_e, and effective radius r_e. Our selected mid-UV sample deliberately overemphasizes late-types, bluer, and smaller galaxies, since these are the dominant galaxy population at high redshifts. Fig. 2b shows that our selected mid-UV galaxies have a similar b/a distribution as the RC3. Since there is no significant trend in the galaxy b/a distribution from the RC3 level (B$\lesssim$15 mag) to the HDF limit (B$\lesssim$28 mag; Odewahn et al. 1997), our nearby *HST* comparison sample is thus a fair one in terms of galaxy ellipticities at all redshifts.

In total, $\sim$100 galaxies were observed with WFPC2 in Cycle 9–10, and about 12 with NICMOS/NIC3 in H-band and 20 with ACS in *H*-α in Cycle 12 (Jansen 2004; this Vol.). Typically, we exposed 2$\times$800–1000 sec in F300W and 2$\times$100–160 sec in F814W, while some exposures came from the HST Archive. Exposures times were made flexible to optimally use the full *HST* orbit allocated per galaxy. The resulting 1-orbit SB-sensitivity matches that achieved in deep *HST* I-band images for faint galaxies, accounting for SB-dimming. The resulting point source sensitivity resolves many galaxies into their brightest star-forming regions (OB associations, young star clusters), and resolves the brightest stars in the nearest objects.

4. Discussion of the HST Images

In this section, we give a high level discussion of the *HST*/WFPC2 (Fig. 3) and some of the recent NIC3 images obtained thus far (Fig. 4). The *HST* images have in general a much higher dynamic range than the ground-based images. This sometimes leads to a different galaxy classification than in the ground-based images or the RC3, and also makes it difficult to properly display the full range of structures seen within a galaxy. The best display of the *HST* images is given in the electronic version of Windhorst et al. (2002).

• **Early-type Galaxies:** The regular early-type galaxies (elliptical and S0's) in our sample show in general a significant increase in SB from the mid-UV to the red, reflecting an overall lack of a young stellar population, as also seen in the far-UV *UIT* images of Marcum et al. (2001) and Kuchinski et al. (2000). Out of 7 early-type galaxies imaged, two have small blue nuclear features. Two galaxies are merger remnants and have significant dusty disks in the mid-UV, but will likely soon evolve into early-type galaxies. Three early-type galaxies

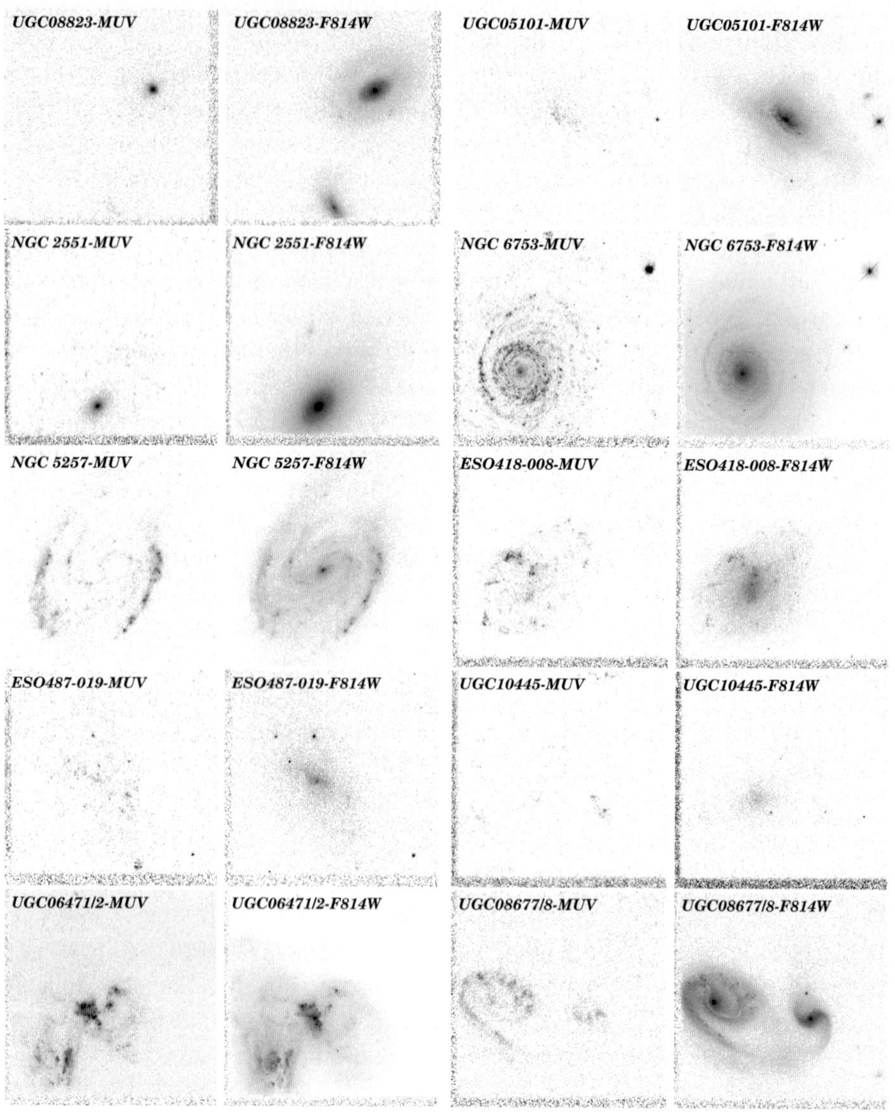

Fig. 3: Summary of our *HST*/WFPC2-images (75"×75" FOV) of nearby galaxies in F300W ("mid-UV"; left panels) and F814W (right panels). **(Row 3.1)** The Seyfert UGC08823 with an old stellar population in F814W, and the merger remnant UGC05101 which is dusty in the mid-UV; **(Row 3.2)** The early-type spiral NGC2551 and mid-type grand-design spiral NGC6753; **(Row 3.3)** The late-type barred spiral NGC5257 and the small spiral ESO418-008; **(Row 3.4)** The late-types ESO487-019 and UGC10445, each with significant K_{morph} (which is an exception rather than the rule for the late-type galaxies observed in our sample); **(Row 3.5)** The mergers UGC06471-2 and UGC08677-8 with various wavelength dependent structures. The best full-size displays are given in the electronic version of Windhorst et al. (2002).

become dominated by point sources in the mid-UV, indicating weak optical-UV AGN, Seyfert, or LINER nuclei (Fig. 3.1). While these are small number statistics, their presence in our sample is due to our pre-selection of galaxies with high predicted average SB in the mid-UV within r_e, resulting in a number of early-type galaxies that are dominated in the mid-UV by AGN.

If AGN generally reside in bulge-dominated galaxies (e.g., Magorrian et al. 1998), then the red and old stellar population of the underlying early-type galaxies will be generally faint in the UV, but the presence of a (weak) AGN will result in blue (U-B) colors and therefore a high SB the mid–near UV, and so inclusion into UV-selected samples. Similarly, ground-based U- or B-band selected surveys result in significant numbers of point-like AGN in early-type galaxies at moderate redshifts (z$\gtrsim$0.3), which are thus classified as QSO's. However, had these objects been selected from the ground at much longer wavelengths (i.e., in the I-band), then they would have shown up as early-type galaxies in good seeing. This leads us to wonder to what extent the (strong) cosmological evolution of AGN selected through their UV/blue excess at modest redshifts (Koo & Kron 1988; Boyle et al. 2000) could in part be due to a "morphological K-correction" of early-type galaxies with weak AGN — the strong SB-dimming of their UV-faint stellar population will render the early-type AGN host galaxies invisible at intermediate to higher redshifts.

• **Mid-type Spirals:** In general, spiral arms are more pronounced in the mid-UV, as *UIT* showed to be generally true in the far-UV (Bohlin et al. 1991, Hill et al. 1992, Kuchinski et al. 2001, Marcum et al. 2001). However, mid-type spirals and star-forming galaxies appear in general more similar from the mid-UV to the optical than early-type galaxies (Fig. 3.2 & 3.3). Some appear as later types in the mid-UV, and a few show drastic changes in type from the optical to the mid-UV, equivalent to a change in T-type of ΔT$\gtrsim$3 (Fig. 3.4b). One galaxy shows a spectacular resonance ring of hot stars, while the remainder disk becomes essentially invisible in the mid-UV (Eskridge et al. 2003).

We observe a variation in color of the galaxy bulges/centers in spirals, and a considerable range in scale and SB of the individual star-forming regions. Dust features in mid-type spirals can be well traced by comparing the F300W to the F814W images, and are visible in lanes or patches (possibly trailing spiral density wave patterns?), in pockets, and/or bubbles. A curious feature is that almost without exception, the mid-type spirals in our sample have their small nuclear bulges bisected by a dust-lane, which is often connected to the inner spiral arm structure, as seen by comparing the F814W and mid-UV images. This is interesting in the context of the finding that all bulge-dominated systems have a central black-hole with an average mass $M_{bh}{\simeq}0.005{\times}M_{bulge}$ (Magorrian et al. 1998). The small nuclear dust-lane in most mid–late-type spirals may be involved in feeding the inner accretion disk, if present. Some edge-on galaxies are very faint in F300W when compared to F814W, others emit/transmit significantly more in F300W, with some intermediate cases. All show a F300W/F814W flux ratio increasing from the inside out, as expected for a decreasing dust content from the inside out, and/or a strong radial gradient in the stellar population, if we assume that the dust and stars are well mixed (i.e., the stars are not preferentially located in front of most of the dust; see e.g., Jansen et al. 1994, 2000; Witt & Gordon 1996; Kuchinski et al. 1998).

• **Late-type Galaxies and Irregulars:** The late-type and irregular galaxies imaged are a heterogeneous mixture. The majority of these galaxies show a F300W morphology that is similar to F814W. Important differences are seen, however, due to recognizable dust-lanes blocking out the F300W light. Dust is visible in pockets, holes or bubbles, perhaps due to supernova-induced outflows, or outflows fueled by bright star-forming regions, such as seen in M82 at *HST* resolution (de Grijs et al. 2001). Some late-type galaxies are physically smaller galaxies with what appears to be the beginning of spiral structure. Others are Magellanic Irregulars with various regions of stochastic SF. Star-formation "ridges" are seen in late-type galaxies, as well as hot stars or star-clusters that are particularly conspicuous in the mid-UV (F300W and/or F255W). A few late-type galaxies would be classified significantly differently when observed in the mid-UV than in the F814W passband, especially when observed under less than perfect atmospheric seeing conditions (see Fig. 3.4). A quantitative discussion of the classification changes as a function of rest-frame wavelength will be given by Taylor et al. (2004).

• **Peculiars and Mergers:** The majority of the peculiars and interacting galaxies show a F300W morphology that is similar to that in F814W. Several mergers have spectacular dust-lanes that thread along with the spiral arms, which are tidally distorted during the interaction (Fig. 3.5), while most others have absorption in apparently random locations associated with a star-burst.

5. Galaxy Outskirts from a Cosmological Perspective

The combined *HST*+ground-based UBVR light profiles show that early–mid type galaxies have only small radial color-gradients in the outer parts, where disks tend to get somewhat bluer at larger radii (Fig. 4 here; de Jong & van der Kruit 1994; de Jong 1996; de Grijs et al. 1997; Jansen et al. 2000). Corrected for reddening by dust, this behavior indicates a (thin) disk that grows with cosmic time through the infall of gas and minor mergers that did not recently upset the disk. Remarkably, the majority of the *late-type* galaxies in our sample shows the opposite behavior, becoming significantly *redder* at larger radii. Fig. 4a and 4b compare an example of each trend, and in Taylor et al. (2004) we quantify the color-gradients for the total sample. In general, color-gradients

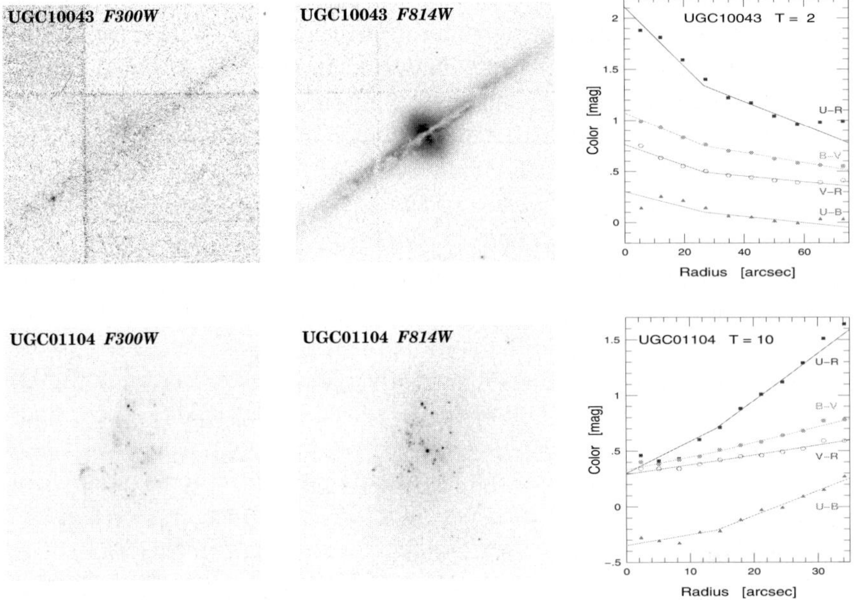

Fig. 4: WFPC2-images in F300W and F814W and color-gradients in (U–B), (B–V), (V–R), and (U–R) for **(a)** UGC10043 [*top*] and **(b)** UGC01104 [*bottom*]. UGC10043 shows the usual behavior for early–mid-type spirals of getting *bluer* with increasing radius. Irregular galaxy UGC01104, on the other hand, becomes progressively *redder* toward its outer regions. This is the more common behavior for late-type/irregular galaxies in our *HST* samples, and suggest the existence of an old halo or (thick) disk.

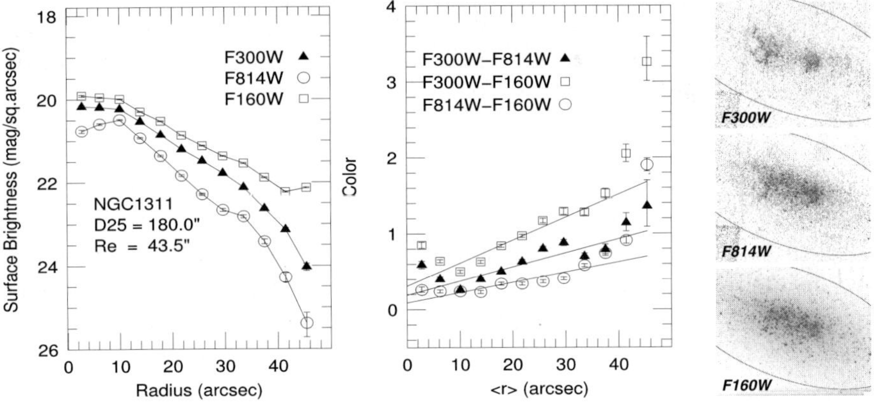

Fig. 5: Profiles of surface brightness [*left*] and color [*middle*] measured from our HST/WFPC2 F300W, F814W and NIC3 F160W images [*right*]. The ellipses in each image correspond to the outermost points used in the profiles. NGC1311 is an Sm galaxy that is well-resolved with HST into its stellar populations from the mid-UV to the near-IR and that *becomes redder* with increasing distance from its center, a trend seen in the majority of late-type galaxies in our sample. Significantly more stars are seen in H-band (through most of the dust), and a dimmer unresolved, older stellar population is apparent.

get redder outwards for the majority of late-type galaxies, and this is in general more often seen for galaxies of lower luminosity, smaller half-light radius, and lower average SB (Tully et al. 1996; Jansen et al. 2000; Taylor et al. 2004). This trend appears in all filters, thereby ruling out sky-subtraction errors are the major cause (Fig. 5), and appears somewhat more pronounced for rounder galaxies than for flatter galaxies. *While young UV/blue-bright stellar populations clearly dominate their inner morphology, most late-type galaxies thus appear to have a significant old halo or (thick) disk population.*

The idea of old Population II envelopes in dwarf irregular galaxies was first proposed by Baade (1958). Tully et al. (1996) found a significant number of nearby dwarf galaxies with red outer envelopes, and Jansen et al. (2000) found a similar result in the NFGS Survey. Recently, Dalcanton & Bernstein (2002) suggested red envelopes in a sample of late-type edge-on galaxies. These findings constrain hierarchical models of galaxy formation (e.g.Katz et al. 1999; Kauffmann et al. 1999). Dalcanton & Bernstein (2002) favor a scenario where late-type galaxies formed in an early epoch of significant mergers ($\gtrsim$6 Gyr ago or $z \gtrsim 1$), resulting in *red* outer colors, which they explain as an old (thick) disk. These late-type galaxies would have remained largely undisturbed since $z \simeq 1$, and recently undergone only more centrally concentrated SF.

Important related clues to the nature of late-types can be found from the galaxy counts as a function of type, which are now available over the entire flux range $15 \lesssim B \lesssim 27$ mag (Cohen et al. 2003), so they can be properly normalized at the bright end. At the faint end ($B \gtrsim 24$ mag or $z \gtrsim 1$), the type-dependent counts not only show the well-known excess of late-type galaxies, but also show a significant excess of early–mid types at $z \gtrsim 1$. Cohen et al. (2003) explain this as the Cosmological Constant itself gradually turning off the epoch-dependent merger rate, when it started to significantly affect the expansion at $z \lesssim 1$. Today's red outskirts in late-type galaxies could thus be the smoking gun of the cosmic accelleration induced by Λ. As a consequence, most (major) mergers occurred early-on in galaxies of all types and luminosities, so halos and (thick) disks were formed early-on and are still seen as such today. Disks of giant galaxies are still growing today, because of infalling surrounding gas and minor mergers, but are otherwise evolving passively. Late-type low-luminosity systems are mostly left with stochastic star-formation in their inner parts today, which is no longer fed by their environment in a major way.

6. Discussion and Conclusions

In summary, our *HST* mid-UV imaging data base shows that in galaxies where star-formation is sufficiently pronounced, it can dominate the morphology from the mid-UV though the optical, resulting in very little change in morphology from the UV to the red. However, when stellar populations older than about 1

Gyr produce most of the optical light, we see changes in morphology between the optical — where these stars tend to produce relatively regular structures due to the effects of orbital mixing within the galaxies — and the mid-UV — where younger stars whose locations still reflect the distribution of their birthplaces. An additional complication is introduced by the presence of dust obscuration. Dust lanes or dust clouds that are nearly transparent in the visible can be opaque in the mid-UV, thereby changing the apparent morphology. In summary, the qualitative results from our study are:

• (1) High-SB early-type galaxies in the optical show a variety of morphologies in the mid-UV that *can* lead to a different morphological classification, although not necessarily always as later–type. The often rather peculiar mid-UV morphology of early-type galaxies is generally quite different than that of the real late-type galaxies as seen in the mid-UV.

• (2) About half of the mid-type spirals in the optical appear as later morphological types in the mid-UV, but not all mid-type spiral galaxies do look dramatically different in the mid-UV. Their mid-UV images show a considerable range in the scale and SB of individual star-forming regions. A comparison of F300W to F814W images yields good sensitivity to dust features.

• (3) The majority of the heterogeneous subset of late-type, irregular, peculiar and merging galaxies display mid-UV morphologies that are similar to those seen in the red, but with important differences due to recognizable dust-features absorbing the bluer light, and due to hot stars, star-clusters, and star-formation "ridges" that are bright in the mid-UV. Less than one third of the galaxies classified as late-type in the optical appears sufficiently different in the mid-UV to result in a different classification.

Our *HST* mid-UV survey of nearby galaxies shows that — when observed in the rest-frame mid-UV — early- to mid-type galaxies are more likely misclassified as later types than late-type galaxies are misclassified as earlier types. This is because later types are dominated by the same young and hot stars in all filters from the mid-UV to the red, and so have a smaller "morphological K-correction" than true earlier types. The morphological K-correction can thus explain part, but certainly not all of the excess faint blue late-type galaxies seen in deep *HST* fields. Classification of faint galaxies in the rest-frame mid-UV will likely result in some fraction of early–mid type galaxies being misclassified as later-types, likely a larger fraction than vice versa.

In conclusion, it is unlikely that the morphological K-correction can explain all of the faint blue galaxy excess as misclassified earlier-type galaxies. And the morphological K-correction cannot explain the slight excess of *early–mid type* galaxies at faint magnitudes ($B \gtrsim 24$ mag) with respect to passively evolving models, as found by Odewahn et al. (1996) and Cohen et al. (2003), since the main misclassification error goes in the opposite direction, as discussed above. Instead, our mid-UV survey seems to support the conclusion of Cohen et al. (2003) that the number of faint galaxies is larger than the non-evolving predictions *for all galaxy types*, but more significantly so for the later types. They give a possible explanation of this finding in terms of hierarchical formation models, where the Cosmological Constant itself was responsible for gradually turning off the strongly epoch-dependent merger rate at $z \lesssim 1$, since in WMAP cosmology the universe became dominated by Λ and started expanding exponentially for $z \lesssim 0.4$. Hence, Λ may be responsible for the gradual transition from the $z \gtrsim 1$ Universe, where galaxy formation and evolution was driven by major mergers, to the Universe at $z \lesssim 0.4$, where at best minor mergers or infall affected the more passive, secular evolution of giant galaxies.

Acknowledgments

We acknowledge support from NASA grants GO-8645.*, GO-9124.*, GO-9824.*, GO-9892.*, and AR-8765.*, awarded by STScI, which is operated by AURA for NASA under contract NAS 5-26555.

References

Abraham, R. G., et al. 1996, MNRAS, 279, L47
Abraham, R. G., et al. 1999, MNRAS, 308, 569
Baade 1958, AR 5, 303
Beckwith, S. V. W., et al. 2004, in prep.
Bohlin, R. C., et al. 1991, ApJ, 368, 12
Bouwens, R. J., et al. J. 1998, ApJ, 506, 579
Boyle, B. J., et al. 2000, MNRAS, 317, 1014
Burgarella, D., et al. 2001, A&A, 369, 421
Cohen, S., et al. 2003, AJ, 125, 1762
Conselice, C. J., et al. 2000, AJ, 119, 79
Cowie, L. L., et al. 1995, *Nature*, 377, 603
Dalcanton & Bernstein 2002, AJ 124, 1328
de Grijs, R., & Peletier, R. F. 1997, A&A, 320, L21
de Grijs R., et al. 2001, AJ, 121, 768
de Grijs R., et al. 2003, MNRAS, 342, 259
de Jong, R. S., et al. 1994, A&ApS, 106, 451
de Vaucouleurs, G., et al. 1991, 3rd Reference Catalogue of Bright Galaxies (New York: Springer)
Driver, S. P., et al. 1995a, ApJ, 449, L23
Driver, S. P., et al. 1995b, ApJ, 453, 48
Driver, S. P., et al. 1998, ApJ, 496, L93
Ellis, R.S., et al. 1996, MNRAS, 280, 235
Eskridge, P. B., et al. 2003, ApJ586, 923
Eskridge, P. B., et al. 2002, ApJS 143, 73
Giavalisco, M., et al. 1996, AJ, 112, 369
Glazebrook, K., et al. 1995, MNRAS, 275, L19
Hibbard, J. E., et al. 1997, AJ, 114, 1741
Hill, J. K. et al. 1992, ApJ, 395, L37
Jansen, R. A., et al. 1994, MNRAS, 270, 373
Jansen, R. A., et al. 2000a, ApJS, 126, 271
Katz et al. 1999, ApJ 523, 463
Kauffmann et al. 1999, MNRAS 303, 188
Koo, D. C., & Kron, R. G. 1988, ApJ, 325, 92
Kuchinski, L. E., et al. 1998, AJ, 115, 1438
Kuchinski, L. E., et al. 2001, AJ, 122, 729
Lilly, S. J., et al. 1998, ApJ, 500, 75
Marcum, P. M., et al. 2001, ApJS, 132, 129
Matthews, L. D. et al. 1997, AJ, 114, 1899
Odewahn, S. C., et al. 1996, ApJ, 472, L013
Odewahn, S. C., et al. 1997, AJ, 114, 2219
Odewahn, S. C., et al. 2002, ApJ, 568, 539
Pascarelle, S. M., et al. 1996b, *Nature*, 383, 45
Roche, N., et al. 1997, MNRAS, 288, 220
Taylor, V. A., et al. 2004, AJ, in preparation
Williams, R. E., et al. 1996, AJ, 112, 1335
Windhorst, R. A., et al. 1997, in "The Ultraviolet Universe at Low and High Redshift", Ed. W. H. Waller, et al. (New York: AIP Press), Vol. 408, 242
Windhorst, R. A., et al. 2000, in "Toward a New Millennium in Galaxy Morphology", Ed. D. Block et al. Astroph. and Space Sc., Vol. 269–270, 243
Windhorst, R. A., Taylor, V. A., Jansen, R. A., Odewahn, S. C., Chiarenza, C. A., Conselice, C. J., de Grijs, R., de Jong, R. S., MacKenty, J. W., Eskridge, P. B., Frogel, J. A., Gallagher III, J. S., Hibbard, J. E., Matthews, L. D., & O'Connell, R. W. 2002, ApJS, 143, 113 ("the HST mid-UV team")
Witt, A. N., & Gordon, K. D. 1996, ApJ, 463, 681

BULGES, DISKS, AND KINEMATICS OF GALAXIES AT Z ~ 1

David C. Koo
UC Observatories/Lick Observatory & Department of Astronomy and Astrophysics
University of California, Santa Cruz, CA, USA

Abstract DEEP is a major spectral survey of faint field galaxies with the Keck Telescopes. The goals include exploring galaxy formation and evolution, mapping distant large scale structures, and constraining cosmology. After giving an overview of DEEP, we highlight prior and current DEEP work on the properties of bulges and disks at redshifts $z \sim 1$, including a comparison of the colors of distant field bulges and those within a rich cluster at $z \sim 0.83$. Kinematics are used to explore the evolution of the Tully-Fisher relation for disks, evolution of the Fundamental Plane for bulge-dominated galaxies, and the masses and nature of luminous blue compact galaxies.

Keywords: bulges, spheroids, disks, kinematics, field galaxies, cluster galaxies, photometry, evolution

1. Overview of DEEP

DEEP (Deep Extragalactic Evolutionary Probe) was initiated over 10 years ago to use the Keck 10 m Telescopes for a major spectral survey of faint field galaxies. The use of DEIMOS (DEep Imaging Multi-Object Spectrograph) divides DEEP into two phases (see Table 1). The first (DEEP1) is comprised of a suite of pilot surveys of 10's to 100's of galaxies using pre-DEIMOS spectrographs on Keck. DEEP1 was designed to establish the technical feasibility and scientific scope of the second phase (DEEP2) that does use DEIMOS.

DEEP2 is distinguished from prior redshift surveys, including DEEP1, by its large sample of 50,000 galaxies to $R_{AB} \sim 24$. The survey uses BRI two-color diagrams to preselect galaxies near $z \sim 1$ for three of its four fields. The exception is the EGS (see Table 1). A competitive survey currently underway is the VMOS-VLT Deep Survey (see Le Fevre in these proceedings).

DEEP2 is also distinguished by using spectral resolution high enough to yield rotation curves and linewidths. Such internal kinematics of galaxies provide a powerful new dimension related to masses of galaxies. Such masses are

D. Block et al. (eds.), Penetrating Bars through Masks of Cosmic Dust, 441–451.

Table 1. DEEP Survey Characteristics

-	DEEP1	DEEP2
Telescopes	Keck I & II; HST	Keck I & II
Instruments [a]	LRIS, ESI, HIRES, NIRSPEC	DEIMOS, LRIS-B
Survey Period	1995-2001 (30 Nights)	2002-2005 (120 Nights)
Fields [b] (FOV)	HDF-N & FF ($8' \times 8'$)	0230+00 ($30' \times 120'$)
-	GSS 1417+52 ($4' \times 42'$)	1417+52 ($16' \times 120'$) EGS
-	SA68 0017+15 (five $4' \times 7'$)	1652+35 ($30' \times 120'$)
-	-	2330+00 ($30' \times 120'$)
-	-	GOODS-N & S
No. Galaxies	1000 to $I \sim 23.5$ (1-4h)	1HS: 50,000 to $R_{AB} \sim 24$ (1h)
& Depth (exp)	-	3HS: few 1000 to $I \sim 24$ (3-10h)
Photometry	KPNO ($UBRI$); HST(VI)	UH (BRI); HST(part EGS:VI)
Science	Spheroid/Bulge Evol.	1HS: non-HST DEEP1 Science &
-	Disk Surface Brightness Evol.	Clustering Evol. at z $\sim$ 1
-	Compact & High z Galaxy Evol.	Lum. Funct. (z, color, etc.)
-	Tully Fisher & Fund. Plane Evol.	Volume Test (Dark Energy)
-	Red & Blue Galaxy Evol.	3HS: HST DEEP1 Science &
-	AGN/Variability/Lum. Funct.	"desert $z \sim 1.4 - 2.5$" galx Evol.
-	Star Form. & Metallicity Evol.	Red Galaxy Age & Metallicity
[a] ESI:	Echelle Spectrograph Imager	
LRIS:	Low Resolution Imaging Spectrog.	B - Blue side (UV sensitive)
HIRES:	High Resolution Echelle Spectrog.	
NIRSPEC:	Near Infrared Spectrog.	
DEIMOS:	DEep Imag. Multi-Obj. Spectrog.	
[b] HDF:	Hubble Deep Field;	N-North & FF-Flanking Fields
GOODS:	Great Obs. Origins Deep Surveys	N-North & S-South
GSS & EGS :	Groth Strip Survey	Extended Groth Strip
SA68:	Selected Area 68	-

intimately tied to the dark matter halo masses which in turn are the fundamental components best modeled by theoretical simulations of galaxy formation.

The key to DEEP2's success is DEIMOS. This spectrograph allows about 120 targets per mask and uses a huge (8K x 8K) red sensitive CCD mosaic. Together with other design advances, DEIMOS achieves roughly a 7 fold increase in efficiency compared to LRIS. The reader can check papers by Davis *et al.* (2003) and Faber *et al.* (2003), as well as to URL: http://www.ucolick.org/~loen/Deimos/deimos.html, for more details on the reduction of data from and performance of DEIMOS. For more information about the DEEP, DEEP1, and DEEP2 projects, the reader is referred to URL: http://deep.ucolick.org/ and http://deep.berkeley.edu.

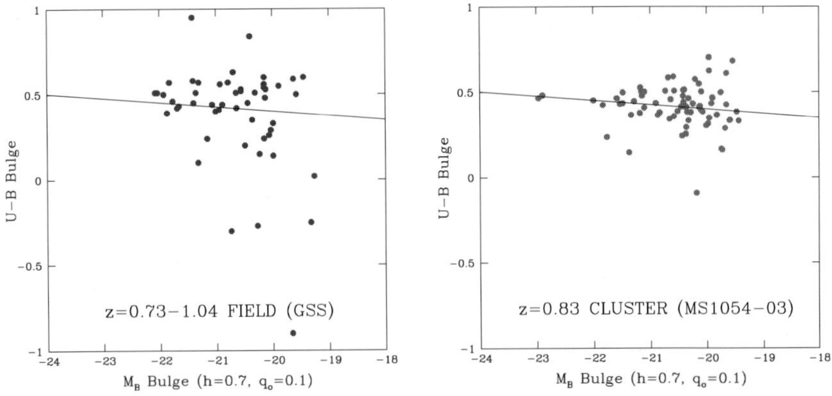

Figure 1. $U - B$ vs M_B for bulges within galaxies in the rich redshift z ∼0.83 cluster MS1054-03 and in the field at redshifts z from 0.73 to 1.04. Local Sbc galaxies have $U - B$ colors about 0 and E/S0 about 0.5. The solid line is from a fit to the color-magnitude relation of galaxies in cluster MS1054-03 (van Dokkum et al. 2000). The field spheroids are on average *redder*, i.e., above this line, while the cluster spheroids roughly track the line.

In the following sections, we will highlight some of the DEEP programs aimed at understanding the evolution of bulges and disks.

2. Evolution of Bulges and Spheroidals

We have undertaken several approaches to explore the evolution of spheroids and luminous early-type galaxies, several of which exploit kinematics. In one study (Im *et al.* 2002), the luminosity function of over 100 early-type galaxies (E/S0) was derived, where the early-type class was selected on the basis of B/T being larger than 0.4; low levels of asymmetries; and high levels of smoothness using the GIM2D software for structural parameter extractions (Simard *et al.* 2002). Keck redshifts were complemented by photometric redshifts. The main result is that there is evidence for roughly 1 magnitude of luminosity brightening back to redshifts $z \sim 1$, but otherwise no evidence for any dramatic drop in number density (Im *et al.* 2002). This result supports hierarchical models of galaxy formation in an open or accelerating universe. In another work (Im *et al.* 2001), a more detailed study was made of the blue E/S0s. Such blue galaxies are particularly interesting as candidates for ellipticals in formation at relatively low redshifts. Kinematic information, including some very high resolution velocity widths as measured with an Echelle system (ESI: Sheinis *et al.* 2000), were used and we found nearly all to be of low dynamical mass, rather than being bonafide massive E/S0 galaxies undergoing intense star formation.

In the third study (Gebhardt *et al.* 2003), we explored the fundamental plane (luminosity: size: velocity dispersions) of 35 galaxies using LRIS data and

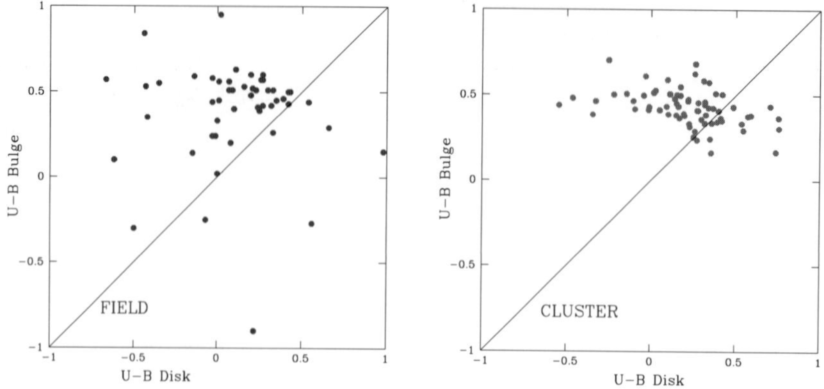

Figure 2. $U - B$ bulges vs $U - B$ disks for cluster and field galaxies as in previous two figures. Note that any correlations of colors between bulges and disks are very weak for these luminous spheroids, so any secular processes (which predict high correlations) are likely to be minor factors for these distant luminous bulges.

found evidence for nearly 2 mag of luminosity brightening back to redshifts $z \sim 1$. This amount would nominally suggest recent formation, if one adopts only passive evolution after a single burst of star formation, but oddly, the colors of these galaxies were moderately red.

This result was supported by a purely photometric HST study of the luminosities and colors of luminous bulge subcomponents rather than of the entire galaxy (Koo *et al.* 2004). While the size-luminosity relation indicated nearly the same amount of luminosity evolution ($\sim$ 1.5mag), virtually all bulges were nevertheless found to be very red (restframe $U - B \sim 0.5$). In a followup study to check whether field bulges were as red as those in clusters (Koo *et al.* 2004, in preparation) , the galaxies in cluster MS1054-03 at a redshift of $z \sim 0.8$ were analyzed with the same procedues and quality of data as used for those in the field. We find that in both environments, almost all luminous spheroids are very red (see Fig. 1), and thus presumably old, a result appearing to be independent of B/T, disk colors (see Fig. 2), and disk inclinations. We also find that the spheroid colors are almost always the same or redder than the disk for both cluster and field galaxies (see Fig.2). This result is expected in either the monolithic or merger models for bulge formation.

Blue spheroids that were found in the field are a small fraction ($< 15\%$) of the total sample and have properties that do not match well to those expected for a young, massive spheroid (classical bulge) in formation. Their luminosities, surface brightnesses, and kinematics are all too low to match local luminous bulges after fading. While we find no good evidence for a signficant population of *genuine* blue spheroids at redshifts $z < 1$, some may, however, qualify as pseudobulges formed from secular evolutionary processes (Kormendy & Kennicutt 2004).

The color-mag distributions of both cluster and field spheroids (Fig. 1) show a shallow slope, low scatter, and a red color surprisingly close to the colors of spheroids today. This lack of significant color evolution along with signficant luminosity evolution is difficult to explain by simple passive evolution after an initial brief burst. A plausible explanation includes a "drizzling" of post-burst star formation after an early burst of metal-rich star formation. This proposed additional, but mild amount of later star formation is supported from numerous other observations that show emission lines among distant E-S0 galaxies (e.g., Schade *et al.* 1999, Willis *et al.* 2002, van Dokkum & Ellis 2003).

3. Evolution of Spiral Disks

DEEP has also undertaken a variety of studies of distant disks, mostly in the form of exploring the evolution of the Tully-Fisher relation using Keck rotation curves and of the disk surface brightness using structure decomposition from HST images. Such studies of the Tully-Fisher relation (velocity vs. luminosity), as well as that of other scaling relations when one adds size or surface brightness, are critical to test whether various theories of disk formation are correct (Faber *et al.* 2001).

As seen in Fig.3, emission-line rotation curves of likely spirals can be well measured with Keck's low resolution spectrograph (LRIS: Oke *et al.* 1995) to redshifts near $z \sim 1$ for galaxies as faint as $I \sim 22$ with 1–2 hour exposures (Vogt *et al.* 1996, 1997). Our more complete sample of about 100 rotation

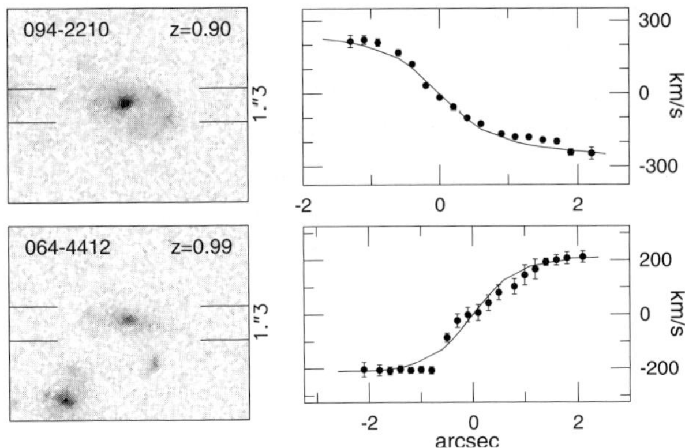

Figure 3. Examples of the rotation curves measured for two high redshift galaxies (Vogt et al. 1996), the upper with total $I \sim 21.4$ and the lower with $I \sim 22.4$. Besides the ID and redshift at the top, the images show the orientation and width of the slit.

curves (Vogt *et al.* 1999, Vogt *et al.* 2004) support the original conclusion that the optical Tully-Fisher relation for spirals near redshifts $z \sim 1$ show only modest ($<$ 0.6mag) changes relative to that seen locally (Vogt *et al.* 1996, 1997). These results appear on the surface to disagree with the older claims for more extensive evolution of 1.5 to 2.0 mag (Rix *et al.* 1997, Simard *et al.* 1998, Mallen-Ornelas *et al.* 1999), but the differences may reflect the selection criteria adopted. While our high-quality Tully-Fisher sample included galaxies that were slightly elongated and resolved along the slit, i.e. larger disk systems, the other samples were generally limited to very blue or strong emission line targets, some of which may be very compact (see below). The small amount of luminosity evolution is supported by our study of the surface brightness of distant disks (Simard *et al.* 1999). We find, after careful accounting for selection effects, no significant evolution.

Several other groups have also been successful in measuring the rotation curves of distant field galaxies and come to slightly different conclusions regarding disk evolution. For example, Ziegler *et al.* (2002) and Boehm *et al.* (2003) have an extensive survey on the VLT that has now reached over 110 ro-

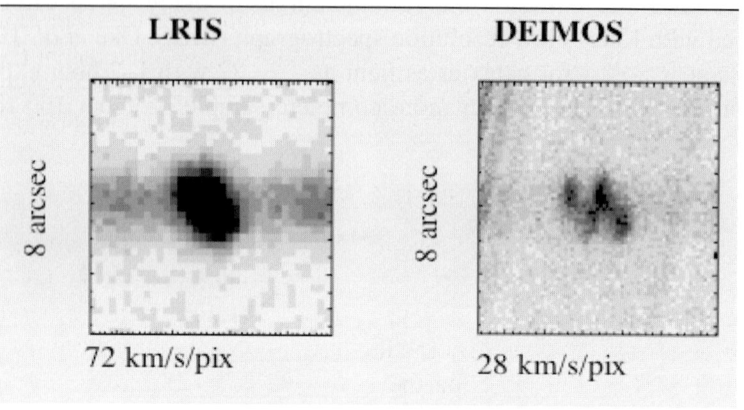

Figure 4. Example of the improved measurements of internal kinematics of galaxies using the higher spectral resolution of the second generation Keck spectrograph, DEIMOS, as compared to that from the first generation faint object spectrograph, LRIS. The large 8Kx8K CCD detector of DEIMOS allows the use of such high resolutions for a given spectral range. The DEIMOS spectrum shows a well-defined rotation curve as seen via the [OII] 3727A doublet, which has an intrinsic separation of its two lines of 220 km/s.

tation curves in the redshift range from 0.1-1. They find that the Tully-Fisher relation shows a mass-dependent luminosity evolution such that the lower mass galaxies exhibit 1-2 mag of luminosity brightening back in time while the most massive ones show little evolution. Barden *et al.* (2003) have a program to study the rotation curves of redshift $z \sim 1$ galaxies using the near-infrared spectrograph (ISAAC) on the VLT. They find brightening by more than 1 magnitude in restframe B, which they say is the result of a combination of increasing surface brightness, slightly smaller disk sizes, and decreasing rotation speeds. Milvang-Jensen *et al.* (2003) find significant evolution at the 0.8 mag level by redshift $z \sim 0.5$, but their sample was only 19 field galaxies.

There is clearly a strong need to increase the sample sizes, redshift, faintness, and spectral quality. DEEP2 will provide gains in all these directions. In numbers, DEEP2 will provide many 1000's of spatially resolved rotation curves; in depth, DEEP2 is easily reaching well beyond a redshift of $z = 1$ (see Fig. 6). Instead of working with a single $z = 1.34$ massive disk (van Dokkum & Stanford 2001), from which the authors conclude any luminosity evolution in the Tully-Fisher relation is less than 0.7 mag in restframe V, DEEP2 will have perhaps 100's in the same redshift range. In faintness, the 1 hour DEEP2 sample is not pushing new ground, but several efforts are underway to take much longer exposures for subsamples of the full survey. Finally, the superior quality of the DEIMOS resolution will improve the measurements (see Fig. 4). The key stumbling block has been the lack of supporting HST color imaging data. This is presently less of an issue with the recent Cycle 13 allocation of 126 orbits to cover a 10' x 60' strip within the EGS field and with community access to the HST images of the GOODS and GEMS fields.

4. Evolution of Luminous Compact Galaxies

We follow the discussion of distant bulges and disks with another important class at intermediate redshifts: luminous blue compact galaxies (Koo *et al.* 1995; Guzman *et al.* 1996). These galaxies with active star formation have luminosities around L^* and become increasingly more common at intermediate redshifts. They may also be the lower redshift counterparts to the "Lyman Break Galaxies" (Steidel *et al.* 1996). The class does not easily fit either the classical early-type galaxies (though some have morphologies that appear to be similar) or the disk galaxies (though most have average light profiles that more resemble exponential disks than $r^{1/4}$) and highlights well the importance of kinematics. Like most faint galaxies, these compact galaxies are too small to yield more than linewidths for kinematic data. Assuming linewidths are reliable measures of the true gravitational potential, we can then add HST sizes estimate dynamical masses. To match HI or Hα values of kinematics, the linewidths need an upward correction of 40% (Rix *et al.* 1997, Telles & Terlevich 1993).

The dynamical masses of distant luminous blue compact galaxies have been found to be especially interesting. For some, we find that luminosity can be a poor gauge of their estimated dynamical masses. Though many have the typical luminosities (L^*) of massive galaxies, their velocity widths (σ) may be smaller than 30 km-s^{-1} as seen in the line profiles shown in Fig.5 (Koo *et al.* 1995). These small velocity widths were possible to measure using the high resolution echelle spectrograph on Keck (HIRES: Vogt *et al.* 1994) . The resultant dynamical masses yield M/L ratios that span a wide range (Guzman *et al.* 1996, Guzman *et al.* 1997, Phillips *et al.* 1997). A more recent study of the stellar masses using near-infrared photometry confirms that the stellar masses are proportionately low as well (Guzman *et al.* 2003). Our results suggest that some luminous blue compacts may be the progenitors of quiescent spheroidals today or perhaps the bulges of local spirals (Hammer *et al.* 2001) or even the precursors of pseudobulges (Kormendy and Kennicutt 2004); that the downsizing scenario (Cowie *et al.* 1996) may apply to these galaxies (Guzman *et al.* 1997); and that such galaxies seen at $z < 1$ may be lower-redshift counterparts to the Lyman-drop galaxies seen at higher redshifts $z \sim 3$ (Steidel *et al.* 1996). A key point is that optical luminosities and dynamical mass are seen to be

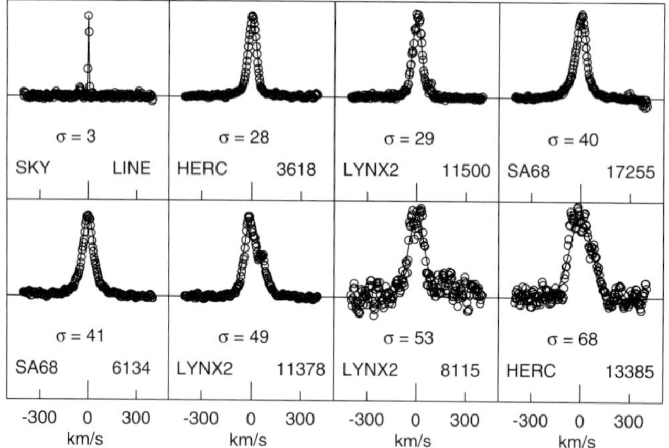

Figure 5. Panel of Keck echelle spectra (HIRES) showing emission line profiles from a sample of luminous (L^*) blue compact galaxies at redshifts z ~0.1–0.7 (Guzman et al. 1996). The σ are the FWHM/2.35 velocity widths in km s^{-1}. The dynamic range of the M/L in rest-frame B for these compact galaxies spans a factor of 45. The HIRES instrumental profile is shown in the upper-left panel.

poorly correlated for this sample, i.e. stable and constant M/L may be a poor assumption at least for some classes of galaxies. This result clearly demonstrates the necessity, usefulness, and promise of kinematics as an important new dimension to discern the evolution of different galaxy populations.

5. Summary

The main theme that arises from our various DEEP programs is that galaxy evolution is a complex problem. Galaxies are diverse in size, luminosity, structure, etc.; are composed of subcomponents which may experience different star formation and dynamical histories and evolution; and reside in a wide range of environments involving different physical mechanisms for their evolution. We have established that kinematics are both feasible with 8-10 m class telescopes and valuable for understanding distant galaxies. For example, we find relatively little evolution in the Tully-Fisher relation or disk surface brightness to redshifts $z \sim 1$, as well as little evidence for evolution in the fundamental plane, volume density, or luminosity beyond that expected from passive evolution for early-type galaxies to $z \sim 1$. The colors of bulges are, however, redder than expected and thus a puzzle. On the other hand, luminous blue compact galaxies appear, whether at intermediate redshifts $z < 1$ or at high redshifts $z \sim 3$, to have low dynamical masses and are suggested to be possible progenitors of quiescent low-mass spheroidals or the bulges of spirals today or the building blocks of larger galaxies rather than dynamically massive ellipticals undergoing formation via monolithic collapse.

The lessons from our DEEP phase-one programs indicate great promise for our current main survey DEEP2 of 50,000 galaxies. Such large numbers are vital for analysis after subdivision of the full sample by a wide range in luminosity, size, M/L, structure, redshift, and environment. We are optimistic that the kinematic data will yield new studies of evolution that rely on mass, including mass functions; M/L functions; Tully-Fisher, Fundamental Plane, and other scaling laws; merger rates; and dark matter distributions (halo vs disk; large scale structure vs mass). Precision measurements of cosmology will also be possible with cosmological volume tests of samples that are kinematically selected, such as galaxies with sufficiently high internal velocities or galaxy groups above a threshold velocity dispersion.

Acknowledgments

I thank the DEEP team, especially many of its junior members, for providing the vast bulk of the preliminary data shown here. The cluster bulge work is a collaborative effort with Susmita Datta and Vy Tran. DEEP and DEIMOS were initiated through the NSF Science and Technology Center for Particle Astrophysics (CfPA) with additional funding support from other NSF programs, NASA, CARA, UCO/Lick, Sun, and Quantum.

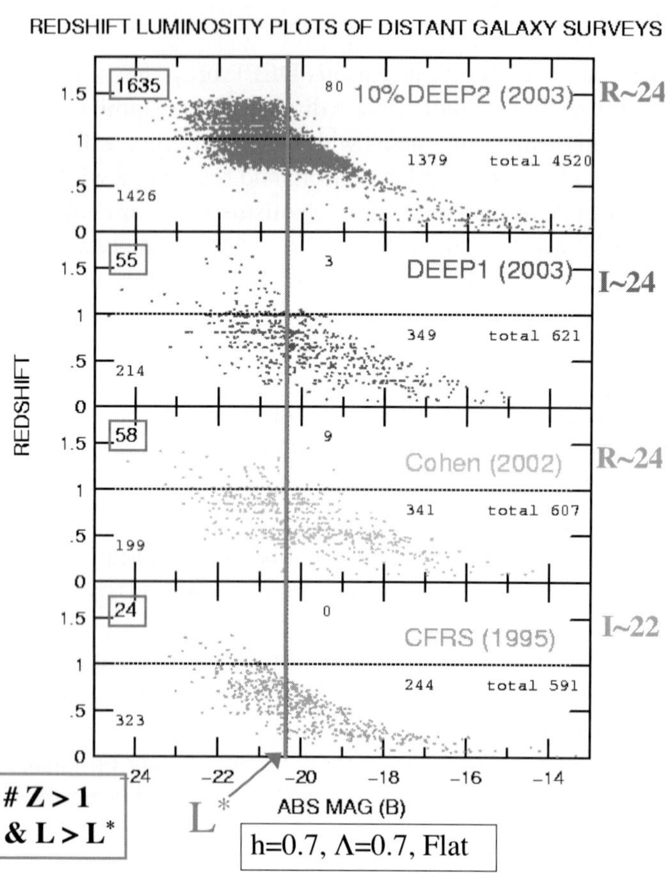

Figure 6. Redshift vs luminosity (M_B assuming indicated cosmology) for four distant field galaxy surveys as indicated. CFRS: Canada France Redshift Survey (Lilly et al. 1995); Cohen: Caltech Faint Galaxy Redshift Survey (Cohen et al. 2000). The numbers on the right-hand side are the magnitude limits in the indicated passbands. Each of the panels is divided into quadrants by a vertical line at L^* of the local luminosity function of field galaxies and a horizontal line at redshift $z = 1$; the number of galaxies from each survey in each quadrant is noted, with the luminous, $z > 1$ sample size specified in a box at the upper left of each subpanel. Note that DEEP2, with only 10% of the survey completed, already has 1635 such galaxies, more than 10× the sum of the other three prior surveys (137).

References

Barden, M., *et al.* 2003, astro-ph/0302392
Boehm, A., *et al.* 2003, A&A, in press, astro-ph/0309263
Cohen, J. G., *et al.* 2000, ApJ, 538, 29
Cowie, L. L., et al. 1996, *AJ*, **112**, 839
Davis, M., *et al.* 2003, SPIE, 4834, 161
Faber, S. M., et al. 2001, *ASP Conf. Ser.*,**230**, 517-526
Faber, S. M., *et al.* 2003, SPIE, 4841, 1657
Gebhardt, K., *et al.* 2003, ApJ, 597, 239
Guzman, R. et al. 1996, *ApJ*, **460**, L5
Guzman, R. et al. 1997, *ApJ*, **489**, 559
Guzman, R. *et al.* 2003, ApJ, 586, L45
Hammer, F. *et al.* 2001, ApJ, 550, 570
Im, M., et al. 2001, *AJ*, **122**, 750
Im, M., *et al.* 2002, ApJ, 571, 136
Koo, D. C. et al. 1995, *ApJ*, **440**, L49
Koo, D. C., *et al.* 2004, ApJ, submitted
Kormendy, J., & Kennicutt, R. C. 2004, ARA&A, in press
Lilly, S. J., *et al.* 1995, ApJ, 455, 50
Mallen-Ornelas, G. et al. 1999, *ApJ*, **518**, L83
Milvang-Jensen, B., *et al.* 2003, MNRAS, 339, L1
Oke, J. B. et al. 1995, *PASP*, **107**, 375
Phillips, A. C. et al. 1997, *ApJ*, **489**, 543
Rix, H.-W., et al. 1997, *MNRAS*, **285**, 779
Schade, D., *et al.* 1999, ApJ, 525, 31
Sheinis, A. I., et al. 2000, *Proc. SPIE*, **4008**, 522
Simard, L., *et al.* 1998, ApJ, 505, 96
Simard, L., et al. 1999, *ApJ*, **519**, 563
Simard, L., *et al.* 2002, ApJS, 142, 1
Steidel, C. C., et al. 1996, *AJ*, **112**, 352
Telles, E., & Terlevich, R. 1993, *Ap&SS*, **205**, 49
van Dokkum, P. G., *et al.* 2000, ApJ, 541, 95
van Dokkum, P. G., & Stanford, S. A. 2001, ApJ, 562, L35
van Dokkum, P. G., & Ellis, R. S. 2003, ApJ, 592, L53
Vogt, N. P. et al. 1996, *ApJ*, **465**, L15
Vogt, N. P. et al. 1997, *ApJ*, **479**, L121
Vogt, N. P. 1999, *ASP Conf. Ser.*,**193**, 145
Vogt, N. P. et al. 2004, *ApJ*, in preparation
Vogt, S., et al. 1994, *Proc. SPIE*, **2198**, 362
Willis, J. P., *et al.* 2002, MNRAS, 337, 953
Ziegler, B. L., *et al.* 2002, ApJ, 564, L69

References

FOURIER DECOMPOSITIONS OF GALAXIES

S. C. Odewahn

The University of Texas, Hobby-Eberly Telescope, HC75 Box1337-10, Ft. Davis, TX, 79734, USA

Abstract Automated surface photometry and pattern classification techniques are combined to morphologically classify galaxies. The two-dimensional light distribution of a galaxy is reconstructed using Fourier series fits to azimuthal profiles computed in concentric elliptical annuli centered on the galaxy. Both the phase and amplitude of each Fourier component have been studied as a function of radial bin number for a large collection of galaxy images using principal component analysis. Up to 90% of the variance in many of these Fourier profiles may be characterized in as few as 3 principal components and their use substantially reduces the dimensionality of the classification problem. Supervised learning methods in the form of artificial neural networks are used to train galaxy classifiers that detect morphological bars at the 85-90% confidence level and can identify the Hubble type with a 1σ scatter of 1.5 steps on the 16-step stage axis of the revised Hubble system (RHS).

Keywords: galaxies: automated morphological classification

1. Introduction

A useful classification system must relate members of different classes to well understood general properties of the objects being systematically classified. It has long been known (de Vaucouleurs 1977, Buta et al. 1994) that the stage axis of the Hubble sequence produces smooth, strong correlations with well known global properties: color, surface brightness, maximum rotational velocity, and gas content. These measured quantities are linked directly to very important physical properties like stellar population fraction and star formation rate. Buta and Combes (1995) review how RHS family and variety convey information about the specific dynamical properties of disk systems such as the degree of differential rotation or the presence of certain families of resonant orbits among the stellar component of the disk. Hence, a quantitative system describing the morphological properties of galaxies is clearly desirable. Here I review a method presented in Odewahn et al. 2002 which constitutes a truly morphological approach to galaxy classification based on the Fourier recon-

D. Block et al. (eds.), Penetrating Bars through Masks of Cosmic Dust, 453–457.

struction of galaxy images and the subsequent pattern analysis based on the amplitude and phase angle of the Fourier components used.

2. The Method

The two-dimensional luminosity distribution observed in most galaxies most often displays a high degree of azimuthal symmetry. Local departures from this overall radial symmetry in the form of high spatial frequency components (arms, bars, rings) form the basis for additional criteria to visual galaxy classification systems like the RHS. A method which quantitatively describes departures from azimuthal symmetry in galaxy images would seem to be a rich starting place for developing a machine automated classification system based on supervised learning. In Odewahn et al. 2002 we adapted a moments-based image analysis technique to fit Fourier components to fixed-grid azimuthal profiles in elliptical annuli to reconstruct galaxy images. One may think of this step as an optimized data compression technique for reducing the dimensionality of our ultimate pattern classification problem: the recognition and characterization of morphological features in galaxies. Pattern classification is performed using an artificial neural network (Odewahn et al. 1992).

3. Image Reconstruction and Information Compression

The radial surface brightness profile of a galaxy is computed in elliptical annuli centered on the galaxy center. The rational here, valid in most cases, is that we are isolating the disk component of the galaxy and hence establishing a way of defining the equatorial plane of the system. Within each annulus, we compute the run of flux density with position angle, θ, in the equatorial plane of the galaxy, to form the azimuthal surface brightness profile. To describe each azimuthal profile, and reduce the dimensionality of our classification problem, each azimuthal profile is modeled with a Fourier series. The moment equations used to fit this series, and the relations used to derive amplitude and phase terms for each Fourier coefficient are given in Odewahn et al. 2002.

A practical application of this method to a barred spiral galaxy imaged with HST is shown in Figure 1. Through experimentation with both HST- and ground-based images, it was determined that using 17 elliptical annuli and up to the $m = 5$ Fourier term consistently reproduces the basic morphological features of most galaxy images. The Fourier-reconstructed image fits are comprised of radially-dependent sets of Fourier amplitudes and phase angles. Using 17 annuli, 6 Fourier amplitudes (including the $m = 0$ term) and the corresponding phase angles to parameterize each image, we thus describe each galaxy with a 221 element classification vector. An input classification vector with over 100 elements is rather large for most pattern classifiers trained in the presence of noise. To distill our image parameterization further, we have chosen to characterize the radial properties of each Fourier component.

The mean radial trends in both the amplitude and phase of the coefficients were studied in a sample of 196 nearby and distant galaxies. A clear trend exists among the 2θ and 4θ amplitudes: barred systems have significant power in the inner rings, AB systems have systematically lower power, and A galaxies have the lowest amount of 2θ and 4θ power in even the inner elliptical annuli. In the inner region of the reconstructed images, the mean Fourier profiles clearly delineate the presence of a bar. Phase terms are important in distinguishing barred galaxies from purely spiral systems. In the case of a barred galaxy, the 2θ and 4θ phase angles remain fixed with ring number (radius), whereas with a two-armed spiral pattern we expect to see a systematic variation of 2θ and 4θ phase angles with radius since the flux density peaks are changing their position angle smoothly as we progress outward in the radial annuli.

A principal component analysis (PCA) was made for each set of Fourier profiles from galaxies having reliable stage and family classifications. For most Fourier coefficients it was determined that 85% to 90% of the profile variance could be described with only 2 or 3 principal components. Hence, using the eigenvectors determined from this analysis, we can characterize the Fourier profile of a random galaxy using only 2 or 3 parameters. These principal component (PC) spaces show strong segregation by morphological properties (see Figure 2), and are used to form 8 to 10 element input vectors to the ANN classifier. Such an image vector constitutes a very reasonable size for training a supervised classifier with the 100-200 patterns (galaxies) available in this work.

4. Results

A set of 262 images were collected using a sample of galaxies having $b/a > 0.4$ and weighted mean stage estimates from at least 3 visual classifiers (including RC3 when available). Morphologically-dependent Fourier image parameters were computed for each galaxy using the LMORPHO package (Odewahn et al. 2002). Backpropagation ANNs were trained following the procedures outlined in Odewahn (1997) using 3 different types of input vector sets and 6 different network architectures. In practice, a final type was assigned for each input vector using the weighted mean value of the output node pattern. This allows one to assign not only the highest weight classification, but the rms scatter in this node-weighted mean value can be used to assign a classification confidence.

The Fourier-based stage classifiers developed in this experiment clearly perform as well as a human classifier (an rms scatter about 1.5 steps in the 16-step RHS scale). This method of parameterizing a galaxy image preserves the in-

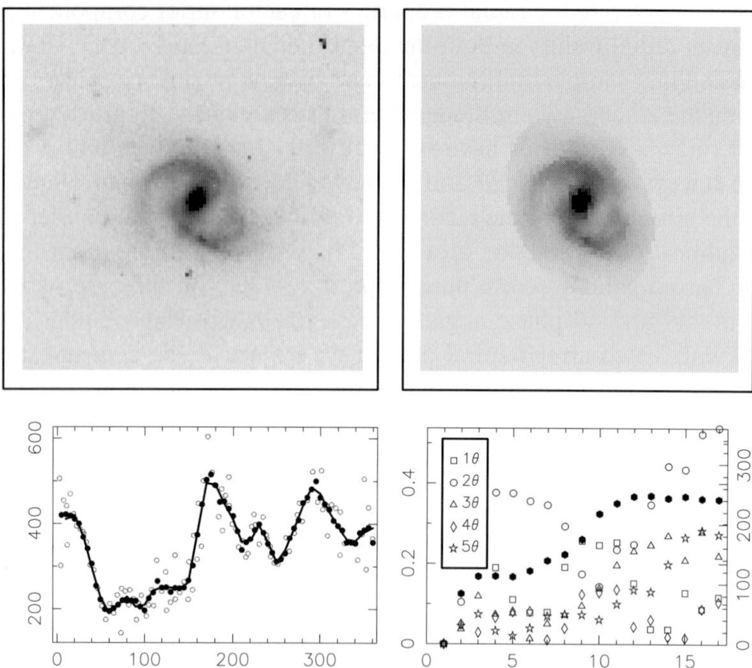

Figure 1. The Fourier image-modeling method described in the text applied to a galaxy observed with HST. The original I image is shown in the upper-left panel, and the Fourier model image is shown in the upper-right. In the lower-left panel we plot the azimuthal profile for the annulus over-plotted in the galaxy image. Open circles are individual pixel values, solid points represent the smooth points, and the solid line represents the Fourier series fitted to this profile. In the lower-right panel we plot the five Fourier amplitudes as a function of radial annulus number using open symbols (left numeric scale). Finally, the solid points represent the phase angle, measured in degrees in the equatorial plane, of the 2θ component (right numeric scale).

formation content needed to emulate the process used by a human classifier. ANN-based bar classifiers were derived in similar fashion using 71 A galaxies and 61 B galaxies, achieving a success rate performance of 92% among training patterns and 85% among test patterns. In a second experiment the DBNN method of Philip et al. (2000) was applied to the identical training and testing patterns as for the backpropagation ANN. After boosting, the training set was found to produce a 94.1% successful classification rate and the test set was found to produce a success rate of 87.5%. The performance of the DBNN classifier was found to be marginally better than the backpropagation-trained network, however a large sample of training data will be needed to assess if this improvement is significant. Nevertheless, both methods produced automated

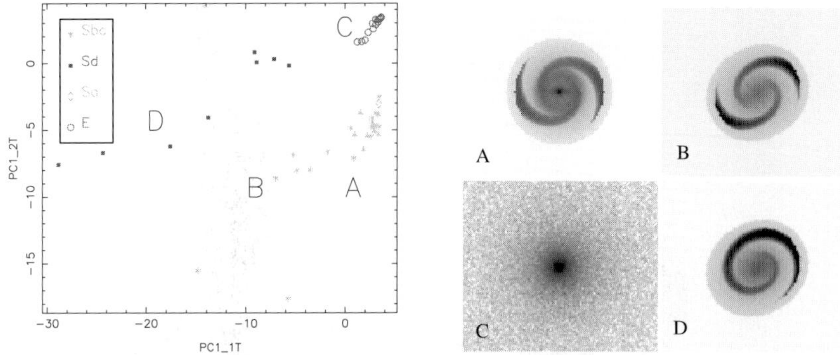

Figure 2. A parameter space formed from the first principal component of the 1θ Fourier amplitude profile, PC_1T, and the first principal component of the 2θ Fourier amplitude profile, PC_2T. The large letter labels correspond to the images in the right panel. (b) [Right] Sample model images produced by LMORPHO that were used to compute the parameter space in the left-hand panel. Objects in the top row (A,B) are idealized mid-type spirals (with B inclined assuming an optically thin disk model), object C (lower left) is a typical elliptical model, and object D (lower right) is the one-armed spiral morphology often seen in compact groups (see compact group 16 in the catalog of Hickson 1993). We have used model galaxy images in this figure to clearly demonstrate the power of the method, but tests with real galaxy images have been found similarly effective.

family classifiers that are able to discriminate bar presence with a roughly 90% probability of success.

Acknowledgments

I wish to thank my collaborators in this work: R. Windhorst, S. Cohen, R. Jansen, V. Taylor, and S. Philip. I am thankful for travel support from McDonald Observatory.

References

Buta, R., Mitra, S., de Vaucouleurs, G., & Corwin, H.G. 1994, AJ, 107, 118.

Buta, R. & Combes, F. 1996, Fund. Cosmic Physics, 17, 95

Philip, N. S, Joseph, K. B., Kembhavi, A., & Wadadekar, Y 2000, Proceedings of Automated Data Analysis in Astronomy, Narosa Publishing, New Delhi, 125

Odewahn, S. C., Cohen, S., Windhorst, R. A. & Philip, N. S. 2002 ApJ 568, 539

Odewahn, S.C. 1997, Nonlinear Signal and Image Analysis, Annals of the New York Academy of Sciences, 188, 184

Odewahn,S.C., Stockwell, E.B., Pennington, R.M., Humphreys, R.M., and Zumach, W.A. 1992, AJ, 103, 318

de Vaucouleurs, G. 1977, *The Evolution of Galaxies and Stellar Populations*, ed. B. Tinsley and R. Larson, 43, (New Haven, USA:Yale University Press)

THE EVOLUTIONARY STATUS OF CLUSTERS OF GALAXIES AT Z $\sim$ 1

Holland Ford[1], Marc Postman[2], J. P. Blakeslee[1], R. Demarco[1], M. J. Jee[1], P. Rosati[4], B. P. Holden[3], N. Homeier[1], G. Illingworth[3], R. L. White[2]
[1]*Johns Hopkins University, Baltimore, MD, USA* [2]*Space Telescope Science Institute, Baltimore, MD, USA* [3]*Lick Observatory, University of California, Santa Cruz, CA 95064* [4]*Karl-Schwarzschild-Strasse 2, D-85748 Garching, Germany*

Abstract Combined HST, X-ray, and ground-based optical studies show that clusters of galaxies are largely "in place" by $z \sim 1$, an epoch when the Universe was less than half its present age. High resolution images show that elliptical, S0, and spiral galaxies are present in clusters at redshifts up to $z \sim 1.3$. Analysis of the CMDs suggest that the cluster ellipticals formed their stars several Gyr earlier, at $z \gtrsim 3$. The morphology–density relation is well established at $z \sim 1$, with star-forming spirals and irregulars residing mostly in the outer parts of the clusters and E/S0s concentrated in dense clumps. The intracluster medium has already reached the metallicity of present-day clusters. The distributions of the hot gas and early-type galaxies are similar in $z \sim 1$ clusters, indicating both have largely virialized in the deepest potentials wells.

In spite of the many similarities between $z \sim 1$ and present-day clusters, there are significant differences. The morphologies revealed by the hot gas, and particularly the early-type galaxies, are elongated rather than spherical. We appear to be observing the clusters at an epoch when the sub-clusters and groups are still assembling into a single regular cluster. Support for this picture comes from CL0152 where the gas appears to be lagging behind the luminous and dark mass in two merging sub-components. Moreover, the luminosity difference between the first and second brightest cluster galaxies at $z \sim 1$ is smaller than in 93% of present-day Abell clusters, which suggests that considerable luminosity evolution through merging has occurred since that epoch. Evolution is also seen in the bolometric X-ray luminosity function.

Keywords: Clusters of Galaxies, Cluster Evolution, Galaxy Evolution, High Redshift

1. Introduction

The Advanced Camera for Surveys (ACS) IDT is using the new capabilities of the ACS (Ford et al. 2002) to answer fundamental questions about clusters and cluster galaxies at redshifts $z \sim 1$, an epoch when they are approximately half the age of the Universe. Our goals include constraining the formation ages

D. Block et al. (eds.), Penetrating Bars through Masks of Cosmic Dust, 459–476.

and the SF history of early-type galaxies, measuring the fundamental properties of cluster galaxies (e.g. structure, morphology, and luminosity) and their relationships within the clusters, measuring the evolution of cluster and galaxy characteristics from $z \sim 1$ to the present, and investigating the assembly of the brightest cluster galaxies. We also aim to establish links between clusters at $z \sim 1$ and proto-clusters at $z \sim 2$ to 5 (Miley et al. 2004), though this will not be discussed here. In this paper we describe the results to date from our ongoing study of eight clusters at $z \sim 1$. In section 3 we discuss CL1252 and CL0152 in detail, the two clusters where our analysis has progressed furthest. In subsequent sections we discuss and compare properties of the entire sample, and then end with a discussion of the evolutionary status of the clusters.

2. Cluster Selection and Cluster Properties

Five of the clusters in our program were initially identified from the ROSAT Deep Cluster Survey (Rosati et al. 1998; RDCS), one from the Einstein Extended Medium Sensitivity Survey (Gioia & Luppino 1994; MS1054), and two from a Palomar deep near-infrared photographic survey (CL1604+4304 & CL1604 +4321; Gunn et al. 1986). The reality of the clusters has been confirmed by extensive spectroscopy with ground based telescopes. The properties of the clusters and their ACS observations are summarized in Table 1. The number of spectroscopically confirmed galaxies in each cluster is in parentheses in column 2. The velocity dispersions in column 3 are in the clusters' rest frames. The age of the Universe T_z at redshift z assumes $h = 0.7$, $\Omega_m = 0.30$, $\Omega_\Lambda = 0.70$, giving $T_0 =$ 13.47 Gyr today, the cosmology we use throughout this paper unless stated otherwise.

Table 1. Properties of High Redshift Clusters Observed with ACS.

Cluster	*Redshift*[a] *T_z(Gyr)*	*Rest Frame Vel. Dispersion*	*X-ray Lum.* *(10^{44} ergs s^{-1})*	*Filters*[b]	*Total Orbits*
MS1054	0.831(143) 6.5	1112	23.3	V,i,z	24
CL0152	0.837 (102) 6.5	1632[c]	7.8	r,i,z	24
CL1604+4304	0.897 (22) 6.2	1226	2.0	V,I	4
CL1604+4321	0.924 (44) 6.1	935	<1.2	V,I	4
CL0910	1.101 (10) 5.4	N/A	1.5	i,z	8
CL1252	1.237 (36) 4.9	760[d]	2.5	i,z	32
CL0848-A,B	1.265 (40) 4.8	640 (A)	1.5 (A) $\sim$1 (B)	i,z	24

[a]The number of spectroscopically confirmed members is in parentheses.
[b]Capital letters are Johnson Filters and small case letters are Sloan filters.
[c]Demarco et al. 2004
[d]Girardi et al. 2004

3. The Clusters CL1252 and CL0152

CL1252

Figure 1 shows the center of one of our most distant clusters, CL1252 at $z = 1.237$. Spectroscopically confirmed members are marked with circles (passive galaxies) and squares (emission line galaxies). Figure 2 shows the confirmed cluster members in a combined i, z mosaic image of four overlapping ACS pointings. The two figures show several facts. The cluster is large; the most distant confirmed members are $\sim 170''$ (~ 1.4 Mpc) from the "center" of the cluster. Like other clusters at this redshift, the projected image of CL1252 is elongated. Finally, the passive, primarily early-type galaxies, are more strongly concentrated to the center and to the axis of the cluster than are the later type galaxies with [OII]λ3727 emission, i.e. the star-forming galaxies. The X-ray and lensing results discussed below show that the early-type galaxies are primarily in the deepest part of the cluster's potential.

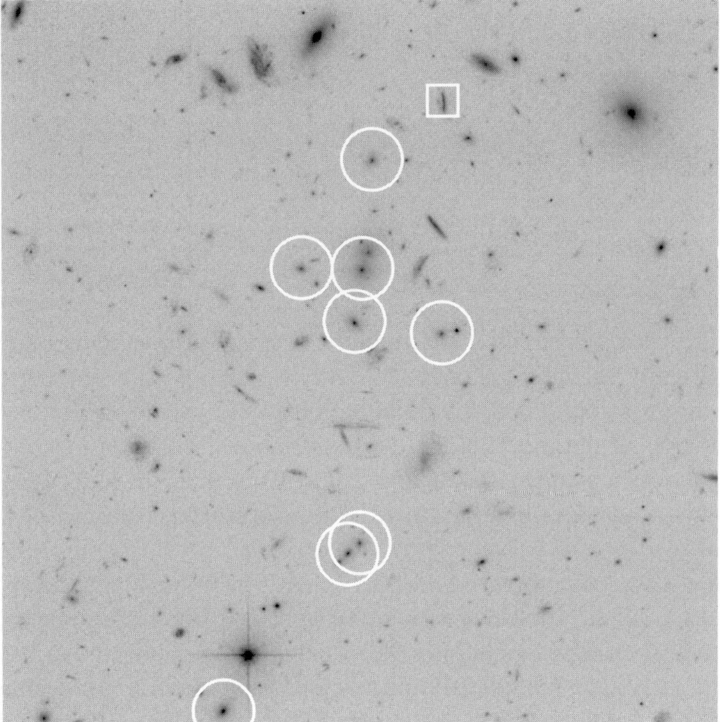

Figure 1. An ACS composite i,z image of the center of the cluster CL1252 at $z = 1.237$. The field is $70''$ square (~ 580 kpc in the restframe). Spectroscopically confirmed members are circled; emission line galaxies are boxed. The circles are $6''$ (50 kpc) in diameter. The "red" early-type cluster members are very conspicuous in a composite ACS i,z/VLT-K image.

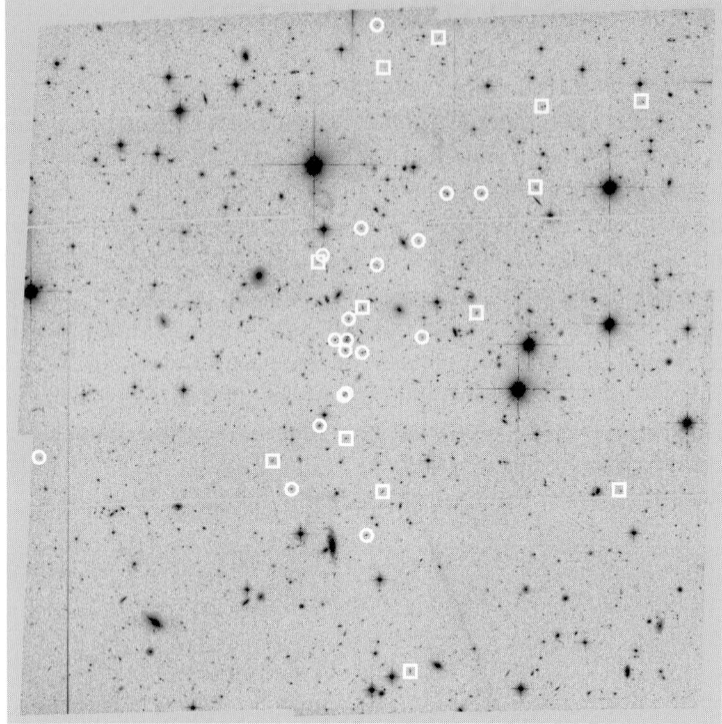

Figure 2. An ACS composite i,z mosaic image of four overlapping fields centered on CL1252 (symbols as in Figure 1). The field is $\sim 350''$ on a side (~ 2.9 Mpc in the restframe).

Figure 3 shows the spatial distribution of spectroscopically confirmed and photometrically selected galaxies in CL1252, along with a number of other clusters in our sample, coded by Hubble type. The figure, which is a visual representation of the morphology-density relationship (MDR), shows that E and S0 galaxies are concentrated along an axis, whereas the latter type glaxies have a much wider distribution. The MDR for six clusters is discussed at length in Section 4.

Figure 4 shows the observed ACS F775-F850LP color-magnitude (CM) relation for CL1252. The relation is quite tight and implies an intrinsic color scatter in these bandpasses of only 0.024 mag for the ellipticals, or 0.030 mag for all the early-type galaxies (ellipticals and S0s). Such a low scatter at this redshift when the universe was less than 5 Gyr old implies either a very high degree of synchronization in the formation of the stars in different galaxies (an unlikely scenario given the stochastic nature of hierarchical assembly), or that the galaxies are already advanced in age so that the fractional age differences are small. Given an assumed form of the star formation history, it is possible to derive the mean age of the galaxy population.

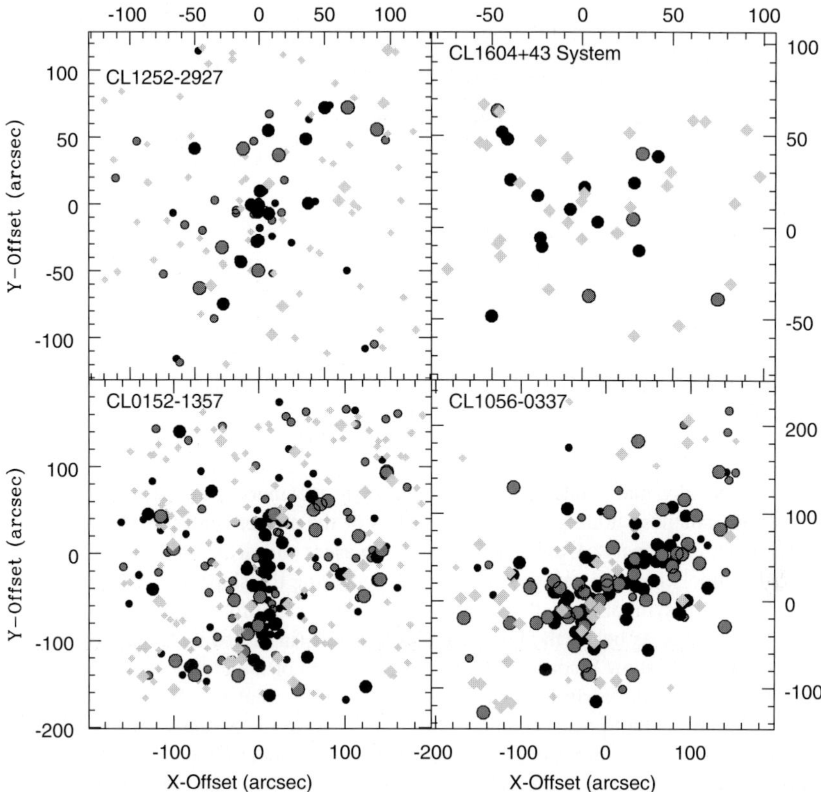

Figure 3. The spatial distribution of galaxy types in five clusters. Light grey diamonds are spirals, dark grey circles are S0s, and black circles are ellipticals. Large symbols are spectroscopically confirmed members and small symbols are candidate cluster members based on Bayesian photometric redshifts. Data for the two 16-hr clusters are combined into one plot.

We simulated the evolution in the observed galaxy colors using two simple star formation histories (Blakeslee et al. 2003: B03). In the first, the galaxies form in single bursts randomly distributed over the interval between the epoch of recombination and some ending time prior to the epoch at which the cluster is observed. In the second, the galaxies form stars at constant rates between randomly selected times prior to the epoch at which they are observed. The true star formation history is likely somewhere between the extremes of these single burst and constant formation models. The two models imply mean ages of 2.6 to 3.3 Gyr for the elliptical galaxies with a scatter in age of about 35%. This means that the stars in these galaxies formed over a period of about 1 Gyr, centered near a redshift of $z \approx 3$. The same models give a mean age of 1–2

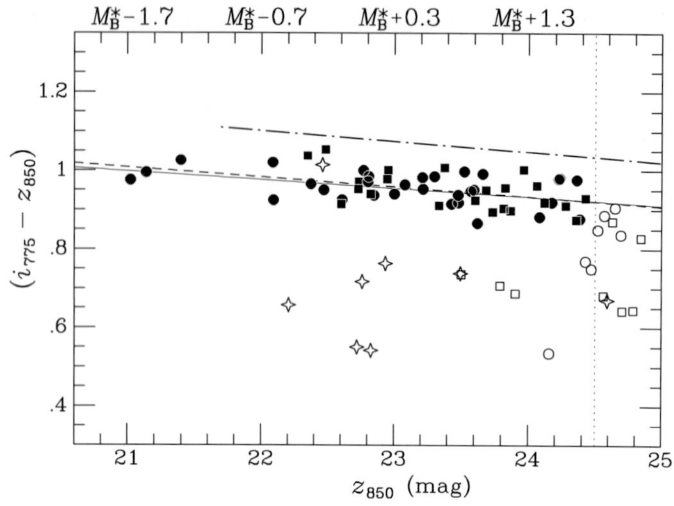

Figure 4. ACS color-magnitude diagram for confirmed members of the CL1252 cluster, and other early-type galaxies within a 2 arcminute radius of the cluster center (excluding spectroscopically known interlopers). Circles and squares represent elliptical and S0 galaxies, respectively. Solid symbols indicate galaxies that we use for fitting the slope and scatter of the CM relation, while open symbols (all of which lack spectroscopic information) were rejected as probable interlopers or as below the faint magnitude cutoff (indicated by the dotted line). Finally, the star symbols show 8 confirmed late-type members of the cluster, most of which are significantly bluer than the early-type CM relation. Two representative linear fits are shown: a fit to the 15 confirmed elliptical members (solid line) and to the 52 early-type red-sequence galaxies (including probable but unconfirmed members). The labels at top give the approximate luminosity conversion, assuming the WMAP cosmology and -1.4 mag of luminosity evolution, such that $M_B^* = -21.7$ (AB). The relation for the Coma cluster, transformed to these bandpasses at $z = 1.24$ (no evolution correction), is indicated by the dot-dashed line.

Gyr for the S0 population, with a scatter of $\sim$ 50%, and of course the blue colors and large scatter of the late-type galaxies imply ongoing star formation. These results are consistent with the features observed in the galaxy spectra. Thus, galaxies in protoclusters at $z \approx 3$ must have experienced very high rates of star formation, which declined sharply to the modest levels observed near $z \sim 1$. ACS observations of protocluster candidates at these high redshifts appear to be consistent with this view (Miley et al. 2004).

Lidman et al. (2004) come to very similar conclusions from their analysis of a CMD measured from deep VLT J,K$_s$ images of CL1252. Using instantaneous single-burst solar-metallicity models,they conclude that the average age of galaxies in the center of CL1252 is 2.7 Gyrs.

Rosati et al. (2004a: R04) combined Chandra and XMM-Newton observations of CL1252 to measure the temperature, gas mass, metallicity, and bolometric luminosity of CL1252 within a 60$''$ radius (500 kpc). Their results are summarized in Table 2. The total mass in the last column is within a radius of 536 ± 40 kpc.

Table 2. X-ray Properties of CL1252.

$L_{[0.5-2.0]}$ 10^{44} *erg s*$^{-1}$	$L_{[Bol]}$ 10^{44} *erg s*$^{-1}$	T_X *keV*	Z_{gas} $Z_{\odot}$	M_{gas} $10^{13} M_{\odot}$	M_{tot} $10^{14} M_{\odot}$
$1.9^{+0.3}_{-0.3}$	$6.6^{+1.1}_{-1.1}$	$6.0^{+0.7}_{-0.5}$	$0.36^{+0.12}_{-0.10}$	$1.8^{+0.3}_{-0.3}$	$1.9^{+0.3}_{-0.3}$

The left hand panel of Figure 5 shows the projected distribution of the total mass in CL1252 derived by Lombardi et al. (2004:L04) from an analysis of the weak lensing in the ACS images. The middle panel shows adaptively smoothed X-ray contours from R04's Chandra observations of the cluster. The right hand panel shows the smoothed VLT K-band light distribution of photometrically selected cluster members (Toft et al. 2004: T04). The centroids of the distributions of X-ray gas and galaxy light are very close to one another. The adaptively smoothed X-ray image shows an edge brightening on one side that suggests the gas and the brightest concentration of galaxies are moving in a direction parallel to the long axis of the cluster defined by its early-type galaxies (cf Figures 1, 2, and 3). If this interpretation is correct, the hot gas trapped in the deepest potential is interacting with a lower density gas associated with galaxies further down the axis of the cluster. However, the projected distribution of total matter is lagging rather than leading the compressed edge of the hot gas. Unless the mass distribution is being significantly affected by the mass associated with an obvious foreground cluster, this fact argues against a scenario wherein collisionless cold dark matter is leading hot gas that is retarded by pressure forces.

Despite its age of less than 5 Gyr, R04 conclude that CL1252 is well thermalized, with thermodynamical properties, as well as metallicity, very similar to those of clusters of the same mass at low redshift. Nonetheless, the elongation and large angular extent of the cluster, as well as the leading edge of the X-ray gas, suggest that it is still collapsing, with mergers and gas stripping of many of the galaxies yet in the future. The relatively high value of the metallicity is consistent with a scenario wherein the major episode of metal enrichment and gas preheating by supernovae occurred at $z \sim 3$.

CL0152

Figure 6 shows a composite ACS i,z image of the richest of two prominent subclusters in CL0152. Spectroscopically confirmed galaxies are circled. Figure 7 shows an overlay of Chandra X-ray contours on the entire ACS field. The hot gas is confined to two components that coincide with two concen-

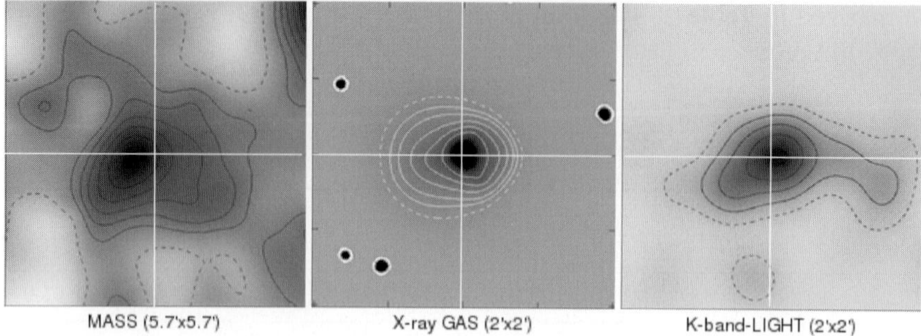

Figure 5. Mass, hot gas, and luminous matter in CL1252. The left panel is the projected mass distribution derived from L04's analysis of weak lensing in ACS images. The middle panel is the X-ray contours from adaptive smoothing of R04's Chadra observations of the cluster. The right panel is T04's smoothed K-band light distribution of photometrically selected cluster members. The images are rotated 90 degrees counter clock wise with respect to the images in Figures 1 and 2.

trations of early-type galaxies. As discussed below, the X-ray properties and mass distribution derived from our weak lensing analysis suggest that the two mass components are merging. Confirmed members with no emission lines are circled, and star-forming galaxies with [OII]λ3727 emission are marked by "stars" (Homeier et al. 2004). The cluster is larger than the $\sim 350''$ (~ 2.7 Mpc) field of the four overlapping ACS exposures. The confinement of the star-forming late-type galaxies to a ring or shell around the two subclusters is very striking. Two bright X-ray sources in the field coincide with two galaxies that are confirmed members. One of the two galaxies is an isolated (barred) spiral. The other is a very "disturbed" spiral in a compact group of five galaxies, and appears to have had a recent close interaction with another galaxy. Images and spectra of the two galaxies are shown in 8. Both galaxies have broad MgII 2800 lines, showing that they are Seyferts (Demarco et al. 2004).

Jee et al. (2004: J04) used ACS images of CL0152 to measure the gravitationally induced shear in the weekly distorted background galaxies that fill the field around the cluster. The PSF of ACS has a complicated shape, which also varies across the field. J04 constructed the PSF model of ACS from an extensive investigation of 47 Tuc stars in sufficiently uncrowded regions. They verified that the model PSF accurately describes the *actual* PSF variation pattern in the cluster observations after a slight adjustment of ellipticity is applied.

Figure 9a shows the mass reconstruction created from the shear field with a maximum-likelihood algorithm overlaid on a smoothed luminosity distribution that is based on the spectroscopically confirmed cluster members. The mass map is dominated by the dark matter within the cluster. The correspondence between the mass map and the luminosity distribution is quite good. Figure 9b

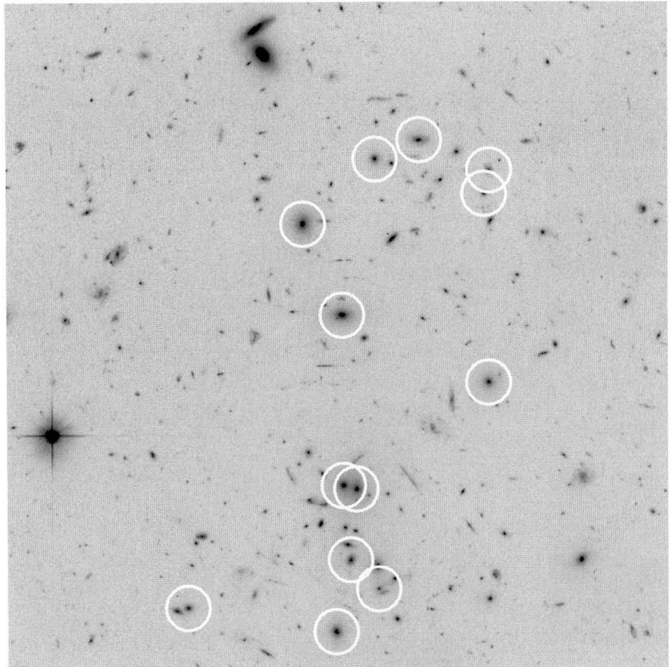

Figure 6. A composite ACS i,z image of the brighter of the two most conspicuous sub-clusters in CL0152. The field size is $90''$ ($\sim$ 690 kpc in the restframe) Spectroscopically confirmed members are circled ($6''$ diameter; $\sim$ 50 kpc in the restframe). There are several thin arcs from lensed background galaxies that are much bluer than the early-type galaxies in the cluster. The lensed galaxy just below and to the left of the two overlapping circles has two components that are mirror images, indicating that the galaxy is very close to a caustic.

shows the smoothed X-ray contours derived from J04's reanalysis of archival Chandra observations. The figure shows that the peaks of the X-ray emission in the two brightest subclusters are lagging behind the peaks in the luminous matter and the dark matter. The displacement of the southern X-ray peak relative to the galaxies was noted by Maughan et al. (2003: M03). They suggested that the relatively collisionless galaxies are moving ahead of the gas. J04's (dark) mass and luminosity maps strengthen this suggestion. The two subclusters appear to be merging due to their mutual gravitational attraction. If the mass is cold dark matter, it is collisionless, whereas the merger of the hot gas in the two subclusters will be slowed by pressure forces. Further support for this picture comes from M03's suggestive evidence that there is a faint ridge of higher temperature gas midway between the two components and perpendicular to the long axis of the cluster.

In a standard Λ CDM cosmology, J04 find a mass of $1.92 \pm 0.3 \times 10^{14} M_{\odot}$ within a $50''$ radius (380kpc). This value agrees with M03's $2.4^{+0.4}_{-0.3} \times 10^{14} M_{\odot}$

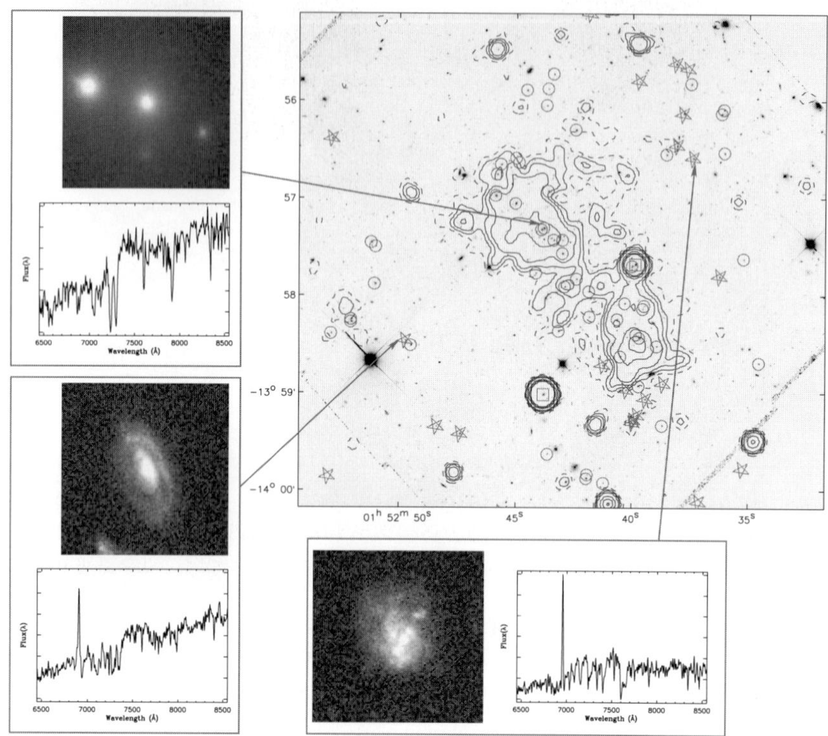

Figure 7. Star-forming galaxies in CL0152 and morphology of the X-ray emitting gas. Spectroscopically confirmed passive galaxies are circled and galaxies with star formation, as indicated by [OII]λ3737 emission, are shown with a "star". The latter are typically spirals and late-type galaxies, while the former are mostly E/S0s. The insets show typical morphologies and spectra. The spatial segregation of the star forming later type galaxies is very striking. The Chandra X-ray isophotes (Demarco et al. 2004; Maughan et al. 2003) are 3, 5, 7, 10, 20 and 30 sigma above the background.

within the same radius derived from the Chandra X-ray observations. Transforming to $\Omega_M = 0.3$, $\Omega_\Lambda = 0.7$, $H_0 = 100$ for comparison with Joy et al.'s (2001: J01) Sunyaev-Zeldovich mass of $2.1 \pm 0.7 \times 10^{14} M_\odot$ within $65''$, J04 find $1.7 \pm 0.2 \times 10^{14} M_\odot$ for $r \leq 65''$, a value encompassed by J01's larger error bars.

In summary, the relatively high spatial resolution of J04's weak lensing mass distribution reveals several concentrations of mass that coincide with luminous substructure within the cluster. The offsets between the peaks in the X-ray emitting gas and the peaks in the dark matter provide evidence that the cluster

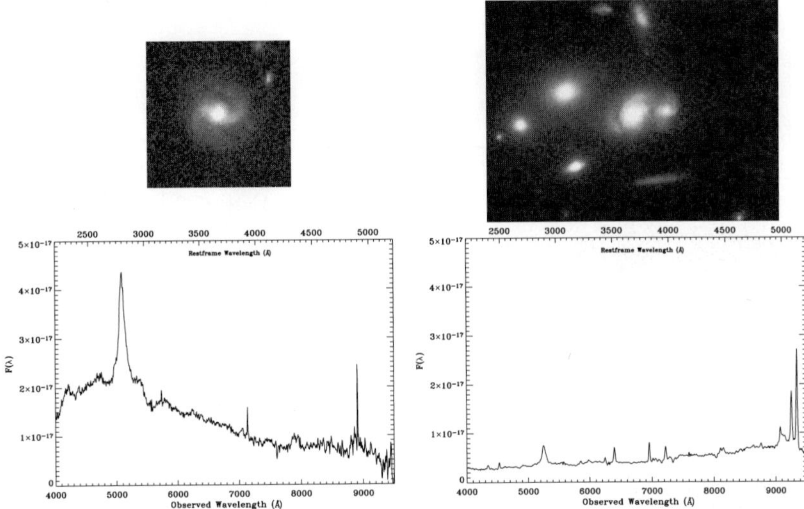

Figure 8. Two X-ray bright Seyfert galaxies in CL0152. The respective sizes of the left and right panels are $5'' \times 5''$ (38 × 38 kpc) and $10'' \times 7.5''$ (76 × 57 kpc). The AGN in the right hand panel is the brightest spiral and appears to be interacting with a fainter, smaller spiral. The projected separation of the two spirals is $1.15''$ (8.8 kpc). The strong broad emission line in each spectrum is MgII2800.

components are merging. The mass derived from the weak lensing agrees with masses derived from the SZ effect and from the X-ray emission.

4. Global Properties of the Clusters

Cluster Morphology-Density Relation

The correlation between morphology and density is a fundamental characteristic of the local universe (Dressler 1980: D80; Postman & Geller 1984: PG83; Goto et al. 2003). The morphology-density relation (MDR) and its evolution are therefore essential predictions of any viable large-scale structure formation scenario. Dressler et al. (1997: D97) measured the MDR at $z \sim 0.4$ and concluded that the fraction of S0 galaxies has increased significantly over the past 4 Gyr, suggesting that these galaxies are relatively recent structures formed from later-type systems infalling into clusters. Smith et al. (2004: Sm04) used WFPC2 images of 6 clusters in the range $0.76 \leq z \leq 1.27$ (many of which are in common with our ACS survey) to obtain the first measurement of the MDR at look back times up to 8.8 Gyr. Sm04 find that the form of the MDR has undergone significant evolution particularly at the high density end where the early-type (E+S0) fraction has increased from 0.7 ± 0.1 at $\sim 10^3$ galaxies Mpc^{-2} at $z \sim 1$ to ≥ 0.9 at the present epoch. At low densities, very

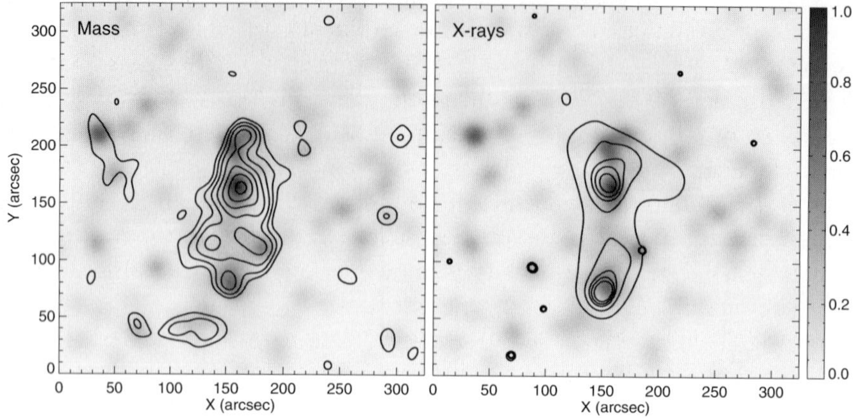

Figure 9. The left hand panel shows J04's mass contours overlaid on the smoothed luminosity distribution derived from the i_{775} image. The map resolves five mass concentrations within the elongated main body of the cluster. There are an additional four mass concentrations outside the main body that are associated with confirmed cluster members. The right hand panel shows smoothed Chandra X-ray contours overlaid on the luminosity distribution. The hot gas is lagging behind the luminous and dark matter along and perpendicular to the axis of the cluster.

little evolution is detected. They propose a series of simple models to explain this trend and nearly all result in an early-type population at $z \sim 1$ in which S0 galaxies are quite rare (typically $\leq 10\%$ of the population).

We visually classified the morphologies of all galaxies in each of our ACS images regardless of position or color. The morphological classification was performed on the full sample of $\sim$3,500 galaxies brighter than 24 mag by one of us (MP) but 3 other team members classified a subset of 20% of these to provide an estimate of the uncertainty in the classifications. The classifications were done in the ACS band that samples at least part of the rest-frame B-band in each cluster. Unanimous or majority agreement between all 4 classifiers in the overlap sample was achieved for 75% of the objects brighter than $i_{775} = 23.5$. There is no significant systematic offset between the mean classification for the 3 independent classifiers and the classification by MP giving confidence that the full sample was classified in a consistent manner. MP also classified all galaxies from our ACS exposure of MS1358 ($z = 0.33$) that were in common with the extensive study performed by Fabricant et al. (2000). Agreement between the MP classifications and those from Fabricant et al. was achieved $\sim$ 80% of the time with no systematic bias seen in the discrepant classifications.

We compute a projected density using the same prescription as D80, D97, and Sm04. For CL0152 and MS1054, we have a sufficient number of spectroscopic redshifts that we can compute the MDR just using confirmed members. We correct the measured density for incompleteness in our redshift survey and to match the same fiducial luminosity limit used by D80 - although we allow for evolution of the characteristic galaxy luminosity (e.g., Postman, Lubin, & Oke 2001). For CL1252 at $z = 1.24$, we derive the MDR using a photo-z selected sample. We confirm that the photo-z derived MDR is not significantly different from the spectroscopic redshift result. We have corrected the densities for CL1252 for incompleteness and contamination due to the fact the scatter in the photometric redshifts is significantly larger than the cluster velocity dispersion. The MDR expressed as the early-type fraction (E+S0) as a function of local projected density is shown in Figure 9. We also show the MDR's derived for the current epoch (D80) and at $z \sim 1$ by Sm04. We find that our ACS-based MDR exhibits less evolution than the Sm04 result but is still significantly different from the current epoch MDR. The excellent angular resolution of ACS allows us to make a direct measurement of the S0 fraction. We find S0 fractions of 0.31 ± 0.10 and 0.19 ± 0.20 at $z = 0.83$ and $z = 1.24$, respectively. Both values are higher than the extrapolations made by Smith et al. based on the observed S0 fractions at $z = 0.5$. However, the difference is at best a 2 sigma result. Furthermore, the ACS data suggest a weak dependence of the S0 fraction on projected density that is similar to what is seen today. These S0 fractions are comparable with what D97 find at $z = 0.4$ but are less than the local S0 fraction, which is approaching $0.5 - 0.6$ in cluster cores (D80, PG83, Poggianti 2001). If these results hold up as we extend our analysis to the entire cluster sample it would suggest an early ($z > 1.2$) formation of a 20–30% population fraction of lenticular systems that undergoes a doubling in population only in the past 4 Gyrs. One caveat is that the 3 clusters used in the analysis here are all relatively X-ray luminous systems – it remains to be seen if there is a significant correlation between X-ray luminosity and the rate at which the MDR evolves. A more thorough description of our MDR measurement will be discussed by Postman et al. (2004).

Brightest Cluster Galaxies

Studying the brightest cluster galaxies (BCGs) in our clusters can reveal essential clues to the timescales for their assembly process. The first significant difference is that 3 of the 6 BCGs are S0 or later. It is rare in current epoch clusters to find such a high fraction of BCGs with disks. In the clusters with later type BCGs, the process of galactic cannibalism may have either not yet begun or may only just be getting underway. We fitted the 2D surface brightness distributions of each BCG and used the best-fit model to measure the metric luminosity within a radius of 14.5 kpc ($h = 0.7$, $\Omega_m = 0.3$, $\Omega_\Lambda = 0.7$). The photometric measurements were transformed to the rest-frame AB B-band so that we could compare the data with Bruzual-Charlot predictions for a passively evolving early-type SED. We find that the BCGs in our clusters are, on average, less luminous than a passively evolving BCG normalized to match

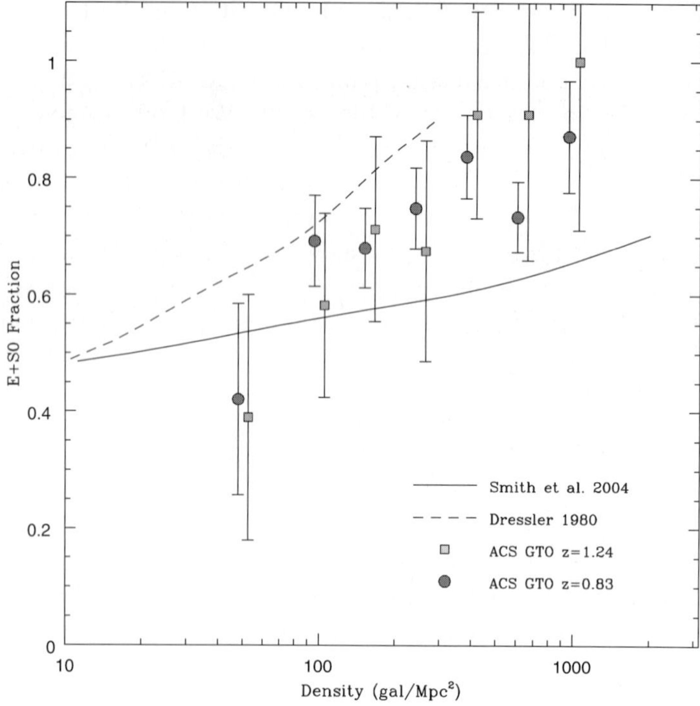

Figure 10. The morphology-density relation for two epochs: z=0.83 (7 Gyr ago) and z=1.24 (8.6 Gyr ago). The early-type fraction includes all galaxies classified as E or S0. Results from previous studies are shown for comparison. The Smith et al. (2004) result (based on WFPC2 imaging) covers a similar range of redshifts. We find a significantly higher fraction of S0 galaxies at both epochs compared to their extrapolation from z=0.50. We also see less evolution in the MDR than they report but we are not inconsistent with their result.

the mean current epoch BCG metric luminosity. The exception is the BCG for MS- 1054, which is quite consistent with these predictions. This suggests that many of these BCGs will still undergo significant merger events. For example, the residuals after our best fit model for the BCG and 2nd-ranked in CL1252-29 are subtracted show substantial structure consistent with tidally stripped stars. These two galaxies are close together and thus appear to be in the process of merging. As the 2nd-ranked galaxy is nearly as luminous as the BCG, the luminosity of the final merger will be nearly double, making it comparable to the luminosity of a current epoch BCG. A further indication that BCGs at $z \sim 1$ are still assembling is that the difference between the 1st and 2nd-ranked luminosity within the above metric radius is on average only 0.09 mag (again the MS-1054 M2-M1 value is the outlier at 0.30 mag). Only 8%

of current epoch BCGs have M2-M1 as small as 0.09 mag (Postman & Lauer 1995).

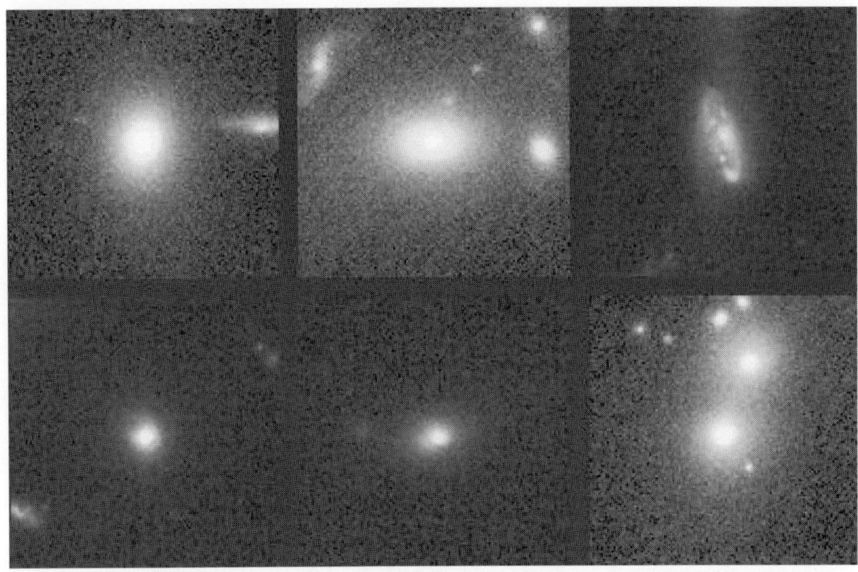

Figure 11. BCGs identified in 6 of our intermediate redshift clusters. From left to right in the top row: CL0152-13, MS1054-03, CL1604+4304. From left to right in the bottom row: CL1604+4321, CL0910+54, CL1252-29.

Table 3. Properties of the Brightest Cluster Galaxies in Six Intermediate Redshift Clusters.

Cluster	*RA(J2000)* *Obs Color*	*DEC(J2000)* ABB_{rest}	z_{obs} $(U\text{-}B)_{rest}$	*Type* $(M2 - M1)$	*Obs Mag*
MS1054	10:57:00.0	-03:37:36	0.8313	E	20.14 (i)
	1.58 (V-i)	-22.877	1.30	0.301 (i)	
CL0152	01:52:46.0	-13:57:00	0.8342	E	20.73 (i)
	1.22 (r-i)	-22.516	1.26	0.088 (i)	
CL1604+4304	16:04:25.0	+43:03:21	0.8966	Sb/c	20.89 (I)
	1.65 (V-I)	-22.497	1.15	0.027 (I)	
CL1604+4321	16:04:36.7	+43:21:41	0.9222	S0/a	21.31 (I)
	1.15 (V-I)	-22.236	0.73	0.009 (I)	
CL0910	09:10:45.7	+54:41:25	n/a	Sa?	21.52 (z)
	1.03 (i-z)	-22.337	1.18	0.174 (z)	
CL1252	12:52:54.4	-29:27:18	1.2343	E	21.30 (z)
	0.96 (i-z)	-23.046	1.22	0.121 (z)	

Color-Magnitude Diagrams

Figure 12 shows that a well defined "red sequence" appears to be a characteristic of all clusters at $z \sim 1$. Although we do not yet have final constraints on the ages based on the scatter in color for all these systems, the striking definition of the sequences suggests that, as with CL1252, we are observing the ellipticals in these clusters at an epoch when they are $\sim$ 3–4 Gyr old.

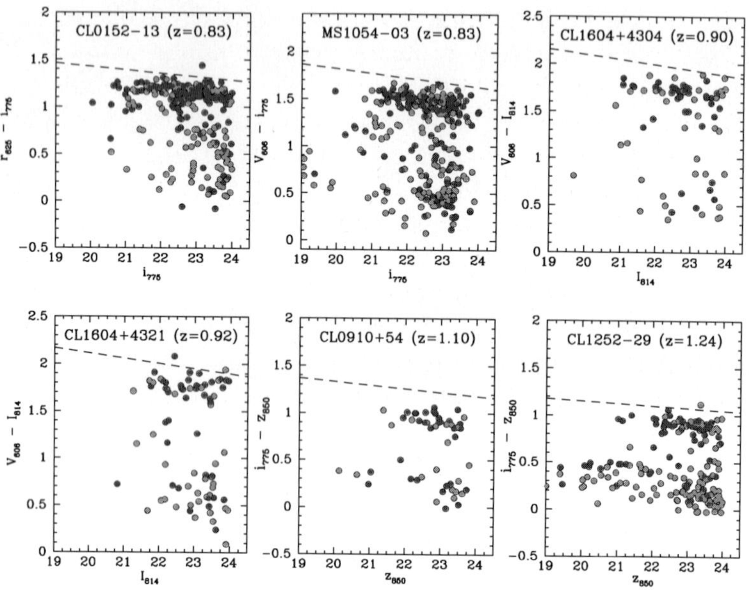

Figure 12. The CMDs for E and S0 galaxies along the line of sight to six $z \sim 1$ clusters. Ellipticals are dark grey and the S0s are light grey. The dashed line is where the Coma Cluster CMD would lie if Coma were simply redshifted to the relevant epoch and observed in the indicated passbands.

5. Discussion and Summary

Deep ACS and NICMOS images (this paper; van Dokkum et al. 2001) show that elliptical, S0, and spiral galaxies are present in clusters at redshifts up to $z \sim 1.3$. Analysis of the CMDs in clusters at $z \sim 1$ (B03, L04, van Dokkum & Stanford 2003, Holden et al. 2004) suggest that the cluster ellipticals underwent the bulk of their star formation 2.6 to 3.3 Gyrs earlier ($z \sim 3$). The morphology-density relation is well established at $z \sim 1$, with star forming spirals segregated in the outer parts of clusters and E/S0s concentrated in dense clumps. Thus, the properties of the majority of the luminous elliptical galaxies are well established. The one exception is that the magnitude difference,

$M2 - M1$, between the second and first brightest galaxies in the clusters at $z \sim 1$ is smaller than in 93% of present-day Abell clusters.

In contrast, the spatial distribution of early-type galaxies in the $z \sim 1$ clusters is primarily elongated, often with two or more dense concentrations of galaxies (sub-clusters and groups). In general, the X-ray morphology follows the galaxy concentrations, and, with the exception of MS1054, the X-ray emission is not spherical (Rosati 2004b: R04b). In some clusters the X-ray iso-contours suggest interactions between sub-clusters (e.g. CL0152). There is evidence that the cluster bolometric X-ray luminosity L_X evolves from high to low redshift, but there is little evolution in the co-moving density (R04b). Finally, there appears to be mild evolution of the L_X vs T_X relation (Ettori et al. 2004), in disagreement with simple models. The combination of the two results suggests significant non-gravitational heating at earlier epochs.

When all of the facts are taken together, the following picture emerges. Clusters of galaxies are largely "in place" by $z \sim 1$, an epoch when the Universe was $\sim$40% its present age. The early-type galaxies in these clusters, primarily found in high density regions, had largely ceased star formation by the time the Universe was 2.7 to 3.8 Gyrs old. The intracluster medium had already reached the metallicity of present-day clusters by $z \sim 1$ (Tozzi et al. 2003). The injection of metals from supernovae was likely one of the mechanisms that heated the gas in addition to adiabatic compression as the clusters collapse. This injection of heat "puffs up" the gas, which counteracts the expected cosmological evolution of the L_X vs T_X relation, yielding the mild evolution observed. The distribution of the hot gas and early-type galaxies is very similar in $z \sim 1$ clusters, suggesting that both have virialized in the center of the deepest potentials in the clusters.

In spite of the many similarities between $z \sim 1$ and present-day clusters, there are many significant differences. The distributions of the hot gas and the early-type galaxies are irregular and elongated rather than spherical. Thus, we appear to be observing the clusters at an epoch when they were still assembling from groups and sub-clusters. Support for this picture comes from the leading edge of the hot gas in CL1252 (R04) and from CL0152 (J04) where the gas appears to be lagging behind the luminous and dark mass in two merging sub-components. The BCGs appear to have considerable evolution ahead of them via mergers, for instance the two central galaxies in CL1252 appear ready to merge in the near future. Because spiral BCGs are rare in nearby clusters, the one in CL1604+4304 is likely to undergo morphological transformation or fade to lesser prominence as its gas is expended, The overall merging process will also result in the bolometric X-ray luminosities increasing as the $z \sim 1$ clusters evolve. In spite of a strong selection bias for the most luminous clusters, only one of our $z \sim 1$ clusters (MS1054) is brighter than R04b's fiducial $L_X^* \sim 3 \times 10^{44}$ ergs sec^{-1}. Using R04b's value for the evolution in luminosity, $L^* = L_0^*(1+z)^B$, with $B \sim -2.25$, the clusters will brighten by approximately a factor of 5 by the present epoch.

Many tasks remain for the future. We and others must fully characterize all of the clusters that are being studied with ACS, Chandra, XMM-Newton, and large ground based telescopes. Finally, we need to find clusters at earlier stages of evolution between the present limit $z \sim 1.3$ and the apparent proto-clusters being studied at redshifts $z \geq 2$ (Miley et al. 2004).

Acknowledgments

ACS was developed under NASA contract NAS5-32865, and this research was supported by NASA grant NAG5-7697.

References

Blakeslee, J., Franx, M., Postman, M., Rosati, P., Holden, B. P., et al. 2003, ApJ, 96, 143 (B03)
Cole, S., Lacey, C., Baugh, C., & Frenk, C. 2000, MNRAS, 319, 168
Demarco, R. et al. 2004, in preparation (D04)
Dressler, A. 1980, ApJ, 236, 351
Dressler, A., Oemler, A., Couch, W., et al. 1997, ApJ, 490, 577
Ettori, S., Tozzi, P., Borgani, S., & Rosati, P. 2004, A&A, 417, 13
Fabricant, D., Franx, M., & van Dokkum, P. 2000, ApJ, 539, 577
Ford, H.C. et al. 2002, Proc. SPIE, 4854, 81
Gioia, I. & Luppino, G. 1994, ApKS, 94, 583
Girardi et al. 2004, in preparation
Gunn, J.E., Hoessel, J.G., & Oke, J.B. 1986, ApJ, 306, 30
Holden, B., Stanford, S. A., Eisenhardt, P. & Dickinson, M. 2004, AJ, 127, 2484
Homeier, N. et al. 2004 in preparation
Jee, J. et al. 2004, in press
Joy, M., et al. 2001, ApJ, 551, L1
Lidman, C., Rosati, P., Demarco, R., et al. 2004, A&A, 416, 829 (L04)
Lombardi et al. 2004, in preparation
Maughan, B. J., Jones, L. R., Ebeling, H., Perlman, E., Rosati, P., et al. 2003, ApJ, 587, 589
Miley et al. 2004, this conference proceedings
Poggianti, B. 2001, MmSAI, 72, 801
Postman, M., & Geller, M. 1983, ApJ, 281, 95
Postman, M., Lubin, L., & Oke, J. B. 2001, AJ, 122, 1125
Postman, M., et al. 2004, in prep
Postman, M., & Lauer, T. 1995, ApJ, 440, 28
Rosati, P., Della Ceca, R., Norman, C., & Giacconi, R. 1998, ApJ, 492, L21
Rosati, P., Tozzi, P., Ettori, et al. 2004a, AJ,127, 230
Rosati, P. 2004b, Carnegie Obs. Astro. Ser. 3: Clusters of Galaxies: Probes of Cosmological Structure and Galaxy Evolution, ed. J. Mulchaey et al. (Cambridge: Cambridge Univ. Press)
Smith, G., Treu, T., Ellis, R., Moran, S., & Dressler, A. 2004, astro-ph/0403455
Toft et al. 2004, A&A, in press, astro-ph/0404474
Tozzi, P. et al. 2003, ApJ, 593, 705
van Dokkum, P. G., Stanford, S. A., Holden, B. P., et al. 2001, ApJ, 552, L101
van Dokkum, P. G. & Stanford, S. A. 2003, ApJ, 585, 79

DISTANT $z > 2$ PROTOCLUSTERS AND THEIR GALAXIES

George Miley[1], Roderik Overzier[1,2], Bram Venemans[1], Andrew Zirm [1], Huub Rottgering[1] and Holland Ford[2]
[1]*Sterrewacht Leiden, Postbus 9513, 2300RA Leiden, The Netherlands,* [2]*Department of Physics and Astronomy, The John Hopkins University, Baltimore, Maryland 21218, USA*

Abstract Observations from the VLT and the HST show that the most luminous radio galaxies are excellent tracers of protoclusters at $z > 2$. Several on-going projects to find and study distant protoclusters are reviewed. After summarising the macro-properties of the identified clusters, the detection of various different galaxy populations are described. Finally, preliminary results on the properties of the protocluster members are presented and remarks are made on the galaxy sizes and morphologies. The Lyα - emitting protocluster galaxies are smaller than field Lyman break galaxies at the same redshift.

1. Introduction

In this talk I shall present results from two projects designed to investigate distant protoclusters and their constituents. The first project involves observations in Lyα with the VLT and is a Ph.D. project for Bram Venemans. The second project is a part of the focused study of clusters with the Advanced Camera for Surveys (ACS) carried out by the ACS Science Team, and described by Holland Ford in the previous talk. It is a Ph.D. project for Roderik Overzier. Because of its high spatial resolution, large field, and excellent sensitivity, the ACS is well suited for studying the morphologies of galaxies in the protoclusters and for finding additional galaxies on the basis of the Lyman break feature in their spectra.

These two projects form part of a large programme involving ground-based and space facilities, designed to provide unique information on the evolution of galaxies and clusters between redshifts of ~ 5 and ~ 1.

2. Need to Study Distant Protoclusters

Within standard Cold Dark Matter (CDM) scenarios the first stars and stellar systems should form through gravitational infall of primordial gas in large

D. Block et al. (eds.), Penetrating Bars through Masks of Cosmic Dust, 477–487.

CDM halos (e.g. White & Rees 1978). Numerical simulations suggest that as these halos merge they form web-like networks of young galaxies and re–ionized gas (e.g. Baugh et al. 1998). The most massive galaxies, and the richest clusters emerge from regions with the largest overdensities. The red-shift "spikes" found to occur in the distribution of $z > 3$ Lyman break galaxies (e.g. Steidel et al. 1998) indicate that the early Universe is indeed highly clumped. To understand galaxy evolution properly it is therefore essential to investigate effects dependant on environment. $z > 2$ ancestors of nearby rich clusters (henceforth termed "protoclusters") are amongst the most extreme and obvious targets for such an investigation.

3. Finding $z > 2$ Protoclusters

There have been several reports of galaxy overdensities in the early Universe (e.g. Pascarelle et al. 1996; Keel et al. 1999; Moller and Fynbo 2001; Steidel et al. 1998,2000; Shimasaku et al. 2003). Besides a galaxy overdensity or redshift "spike", an additional characteristic signature of a distant protocluster is the presence of a bright large clumpy galaxy that can be associated with the progenitor of a massive dominant cluster galaxy.

Radio galaxies have the properties expected of these progenitors. They are (i) the IR-brightest and presumably most massive galaxies at $z > 2$ (De Breuck et al. 2000), (ii) have extremely clumpy continuum structures, strikingly similar to simulations of forming, dominant cluster galaxies (Pentericci et al. 1998, Kurk et al. 2003) and (iii) are surrounded by giant clumpy Lyman α halos whose 100 kpc-scale sizes are comparable with the envelopes of cD galaxies (e.g. van Ojik et al. 1996, 1997).

For this reason we conducted a large VLT program to find $z > 2$ protoclusters around distant luminous radio galaxies. Deep narrow and broad band imaging were used to locate candidate galaxies having excess Lyman α emission in surrounding cluster-sized regions. Followup spectra with the VLT and Keck confirmed that these candidates have similar redshifts to the target radio galaxies. Table 1 shows a list of our targets, with some results. All 5 radio galaxies studied with the VLT to sufficient depth have > 20 spectroscopically-confirmed Lyman α and/or Hα companions, with galaxy overdensities (redshift spikes) of $\sim 5 - 15$, structure sizes of > 3 Mpc and velocity dispersions of $\sim 300 - 1000$ km/s, centred at the velocity of the radio galaxies (see Fig. 2; Pentericci et al. 2000; Kurk 2000, 2003; Venemans et al. 2002, 2004).

Venemans et al. (2002, 2004) estimate that the masses of these structures are $\gtrsim 10^{14-15}$ M$_{\odot}$, determined from the volume occupied by the overdensity, the mean density of the Universe at the protocluster redshift, the measured galaxy overdensity, and the bias parameter (see e.g. Steidel et al. 1998). These masses are comparable to the mass of the Coma Cluster. Regions around powerful

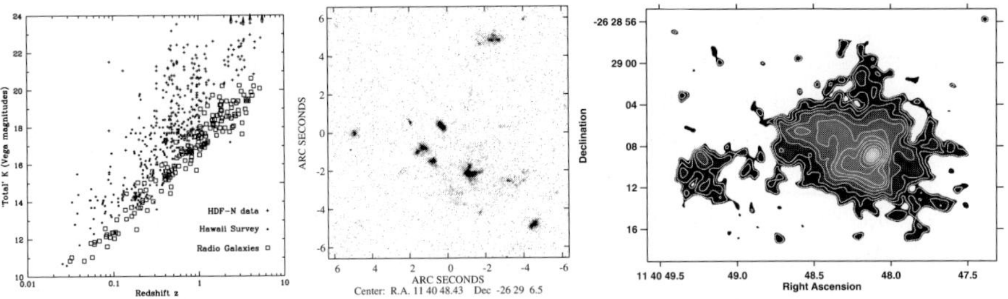

Figure 1. Evidence that distant luminous radio galaxies are amongst the most massive forming central cluster galaxies in the early Universe. **a)** Composite Hubble K-z diagram of radio-loud and radio-quiet galaxies. The radio-loud galaxies trace the upper envelope of K luminosity (De Breuck et al. 2000). **b)** A 7-orbit WFPC2 image of 1138–26, showing an extremely clumpy morphology that is strikingly similar to theoretical simulations of forming massive central cluster galaxies (Pentericci et al. 1998, Kurk et al. 2003a,b). The sizes, profiles and luminosities of the clumps are similar to those of Lyman break galaxies (e.g. Steidel et al. 1996). **c)** A narrow–band image of the Lyα gas halo of 1138–26, which extends by $\sim 20''$ (160 kpc) and from which the massive galaxy is presumably forming.

Table 1. **Protoclusters discovered with the VLT.**
Shown are the radio galaxy, its redshift, the number of candidate Lyα emitters in the imaging ('IMG'), the number of spectroscopically confirmed Lyα emitters (excluding the radio galaxy), ('SPC') and the velocity dispersion of the confirmed cluster members. For details see Pentericci et al. (2000), Kurk et al. (2000,2003), Venemans et al. (2002, 2004), and Miley et al. (2004).

Radio Galaxy	z	IMG	SPC	Δv (km s^{-1})	Notes
PKS 1138–262	2.16	70	14	~ 1000	Hα[1]
MRC 0052–241	2.86	73	37	~ 900	
MRC 0943–242	2.92	~ 70	29	~ 800	
MRC 0316–257	3.13	85	31	~ 625	[OIII][2], Lyα[3]
TN J1338–1942	4.10	50	32	~ 350	
TN J0924–2201	5.19	~ 20	6		

[1] Hα in ISAAC narrow-band filter. [2] [OIII] $\lambda 5007A$ in ISAAC narrow-band filter. [3] Lyα in ACS narrow-band filter (F502N).

distant radio galaxies therefore exhibit properties expected of protoclusters: (i) measured galaxy overdensities, (ii) inferred masses comparable with rich clusters, (iii) bright clumpy radio galaxies surrounded by cD-sized Lyman α halos, consistent with the progenitors of dominant cluster galaxies. Because **all these ingredients are present together**, the observed structures around radio galaxies are the most plausible ancestors of rich clusters (see also e.g. Ivison et al. 2000, Smail et al. 2003, Stevens et al. 2003).

Are radio-selected protoclusters special? There are strong statistical arguments that they are not. Because the radio lifetimes (few $\times 10^7$ yr) are small

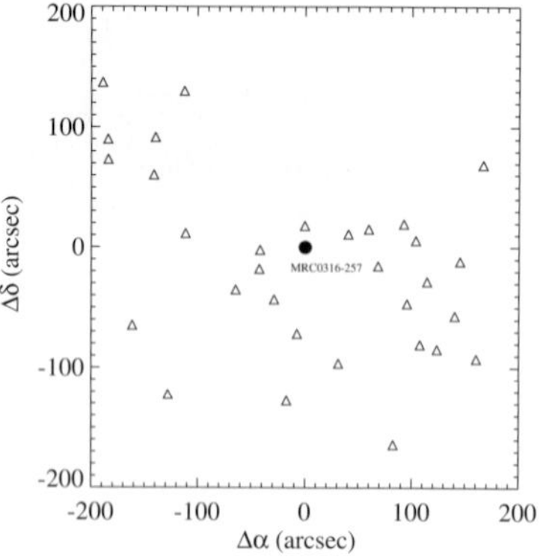

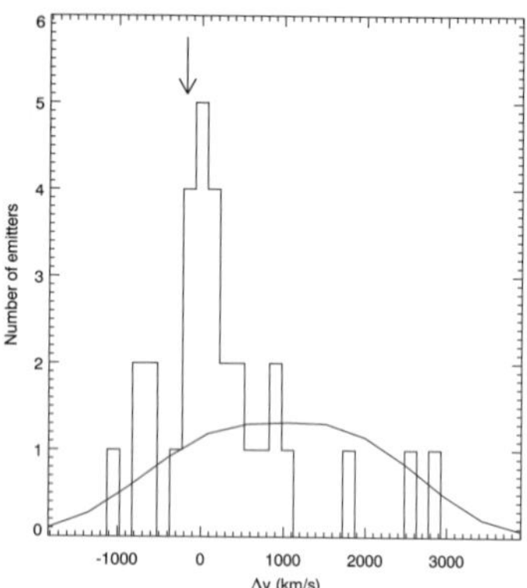

Figure 2. Protocluster 0316–26 at $z = 3.1$. **a)** Angular distribution of confirmed Lyα emitters (triangles). **b)** Velocity distribution of Lyα emitters. Evidence that the overdensity in Lyα emitters is associated with the radio galaxy includes (i) the coincidence of the peak in the distribution with the mean velocity of the radio galaxy, indicated by the arrow, and (ii) the distribution (FWHM 1500 km s^{-1}) is a factor of ~ 2 narrower than the width of the narrow-band filter used for the candidate detection. Similar distributions were observed for 5/5 $z > 2$ radio galaxy targets observed in our VLT program.

relative to cosmic evolution, there must be hundreds more protoclusters hosting "dead" radio sources than those having "living" ones. Indeed, the space densities of luminous radio sources is consistent with every protocluster having harbored a luminous radio–loud galaxy or quasar at some time during the AGN era $2 < z < 5$ (Venemans et al. 2002). When the protoclusters evolve to $z \sim 1$, the radio sources will be extinguished (following evolution in the space-density of AGN). Hence nearby rich clusters would not generally be expected to harbor active luminous radio sources, consistent with observation.

4. Probing Protocluster Galaxy Populations

Techniques used to detect distant galaxies are all biased. Detections based on continuum features (e.g. Ly break, 4000 A break) or emission lines (e.g. Lyα or Hα) select galaxies with properties tuned to the filters used. Studies of galaxy evolution must therefore sample galaxies using as many different techniques as possible.

Only $\sim 25\%$ of (field) Lyman break galaxies have substantial Lyα fluxes (Steidel et al. 2000), so the Lyα emitters found in our initial VLT detections must represent only a fraction of protocluster galaxies. We have therefore augmented our Lyα searches with additional galaxy detection techniques, including Lyman and 4000 A breaks ('extremely red objects' (EROs)) and excess

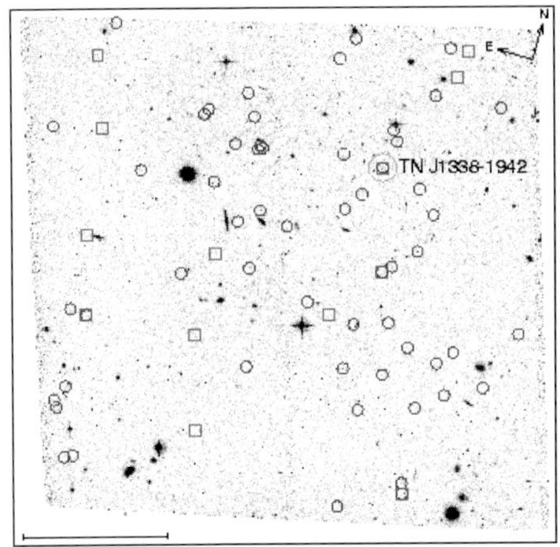

Figure 3. Evidence for a Lyman-break population in the 1338–19 protocluster at $z = 4.1$ from g, r, i, ACS observations (Miley et al. 2004; Overzier et al. 2004 in prep). Objects having Lyman breaks at the protocluster redshift would be observed as g-dropouts. An excess of g-dropouts (circles) is observed and these are concentrated towards the radio galaxy TN J1338–1942. Lyα galaxies are indicated by squares. Scale bar is $1'$ ($\sim$ 500 kpc at $z = 4$).

H α and [OIII] emission (see Table 1). Comparisons of galaxy properties (e.g. line fluxes, morphologies, SEDs, spatial distributions) between the different galaxy populations are being used to constrain galaxy evolution models.

These comparative population studies are already yielding fruitful results. Additional galaxy populations have been detected in our protocluster targets at $z = 2.2$ and $z = 4.1$. For 1138–26 at $z = 2.2$, VLT observations identified distinct Hα and 4000 A break populations in addition to the Lyα one (Kurk et al. 2003). The Hα and ERO populations are systematically more concentrated towards the radio galaxy than the Lyα one, consistent with the Lyα emitters being younger and less massive. For 1338–19 at $z \sim 4.1$, the most distant of our protoclusters, a Lyman break population has been detected with the ACS, additional to the VLT Lyα emitters (Fig. 3; Miley et al. 2004). The population of LBGs is spatially concentrated in the vicinity of the radio galaxy.

5. PROCESS - The Evolution of Protocluster Galaxies

Radio-selected protoclusters are unique laboratories for studying galaxy evolution. First, they provide substantial numbers of galaxies at specific redshifts, allowing the **optimum rest-frame filters** to be chosen for a systematic study of galaxy evolution. Secondly, comparing **the most extreme overdense regions** with field galaxies (from e.g. GOODS) is an efficient way of investigating changes in galaxy properties as a function of environment.

Motivated by the observations described above, we have therefore begun a large long–term programme (PROCESS: PROtoCluster Evolution Systematic Study) to obtain snapshots of the evolutionary state of massive clusters at 4 strategically chosen redshifts between z $\sim$ 4 and z $\sim$ 1. Galaxies, clusters and AGN all undergo dramatic evolution over this redshift range. Observations of SEDs and morphologies of galaxies in the PROCESS protoclusters are being used to **disentangle the history of galaxy assembly and star formation** and its dependance on environment. PROCESS will make use of the HST, Spitzer and ground-based telescopes (VLT & Keck).

The targets in PROCESS include 3 objects from Table 1: 1338–19 at z = 4.1, 0316–26 at z = 3.1 and 1138–26 at z = 2.2. Each of these targets has more than 20 spectroscopically identified companions. The prime "low-redshift" cluster chosen for comparison is Cl 1252–29 at z = 1.2. This is one of the most distant known X-ray clusters selected from the ROSAT Deep Cluster Survey (Rosati et al. 1998, 1999). Because it has been well studied in the ACS/GTO program, it is an excellent well-evolved cluster to compare with the high redshift protoclusters.

6. Sizes and Morphologies of Ly α Emitters

The ACS is the prime instrument for studying the sizes and morphologies of protocluster galaxies. We have recently obtained preliminary information on the sizes and morphologies of Lyα emitters in two of our richest VLT protoclusters, both PROCESS targets.

0316–26 at z = 3.1. This object has 32 spectroscopically confirmed Lyα members (Venemans et al., in prep.). It is an outstanding PROCESS target for three reasons. Firstly, its redshift corresponds to an epoch when clusters, galaxies and AGN were all evolving rapidly and for which several distinctive galaxy selection techniques can be used to detect different galaxy populations. Secondly, [OIII]λ5007 is moved into a VLT/ISAAC narrow-band filter, so that [OIII] excess galaxies can be detected and unique information obtained on [OIII] equivalent widths. Thirdly, Lyα is fortuitously redshifted into the ACS F502N narrow-band filter, permitting deep, wide-field Lyα studies at high resolution. 40 orbits on 0316–26 are scheduled with the ACS during the next year (2005), that should enable a definitive study of this key object to be made.

Here we present results from a 3-orbit HST/ACS observation, obtained in July 2002 with the F814W filter, centred at 8333A. Of the 32 confirmed Lyα emitting sources, 19 (including the radio galaxy) were located in the area imaged by the ACS. Three continuum morphologies of Lyα emitters in 0316–26 are shown in Fig. 4 (bottom row).

To quantify the size of these objects, the half light radius (r_h) of each emitter was measured using the program SExtractor and empirically corrected for

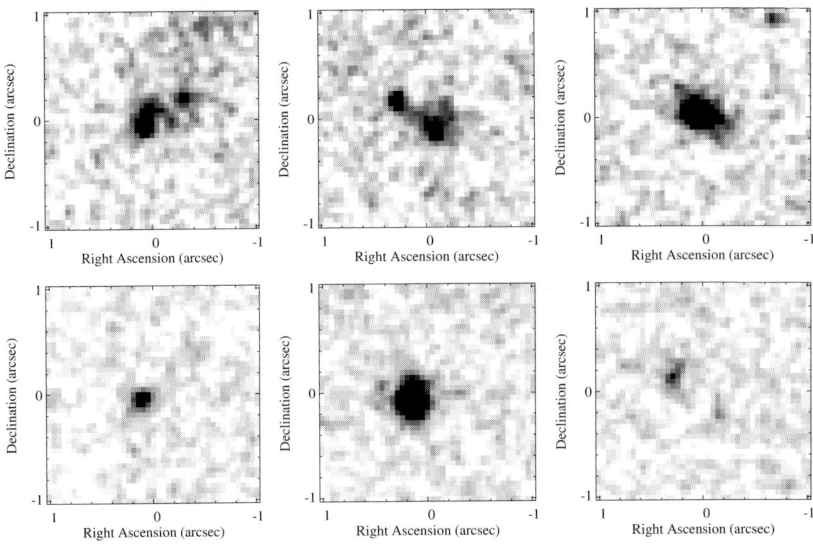

Figure 4. Examples of the continuum morphologies of Lyα emitters from HST/ACS observations. Top row shows Lyα emitters in 1338–19, observed through the $F775W$ filter ($\sim 1500A$ rest-frame). Bottom row shows Lyα emitters in 0316–26, observed through the $F814W$ filter ($\sim 2000A$ rest-frame).

an underestimate at faint fluxes (Venemans et al., in prep.). The corrected half light radii of the emitters range from $0\farcs06$ to $0\farcs18$. Translating the sizes directly to physical sizes, the measured half light radii correspond to 0.5–1.5 kpc, with a median size of $\sim$1 kpc.

The sizes of the confirmed Lyα emitters were compared with those of non-protocluster Lyman break galaxies ('LBGs') at similar redshifts taken from the Great Observatories Origins Deep Survey. (GOODS). The average size of GOODS "field" LBGs at $z \sim 3$ is $0\farcs28$ or $\sim$ 2.3 kpc (Ferguson et al. 2003). Lyα emitters in our 0316–26 protocluster are significantly smaller than (non Lyα selected) LBGs at the same redshift.

1338–19 at $z = 4.1$. The redshift of the rich protocluster associated with TN J1338–1942 corresponds to an age of only $\sim$ 1.4 billion years after the Big Bang. The ACS observations were carried out in May–June 2003, as part of the GTO high-redshift cluster programme. 18 orbits were split over 4 broad-band filters, *g, r, i* and *z*, thereby bracketing redshifted Lyα at 6214A. The ACS field covers 13 of the 32 spectroscopically confirmed Lyα emitters. The Lyα emitters are faint in continuum with i-band ($\sim 1500A$ rest-frame) magnitudes of 25.3–28.1. Examples of their morphologies are shown in Fig. 4 (top row).

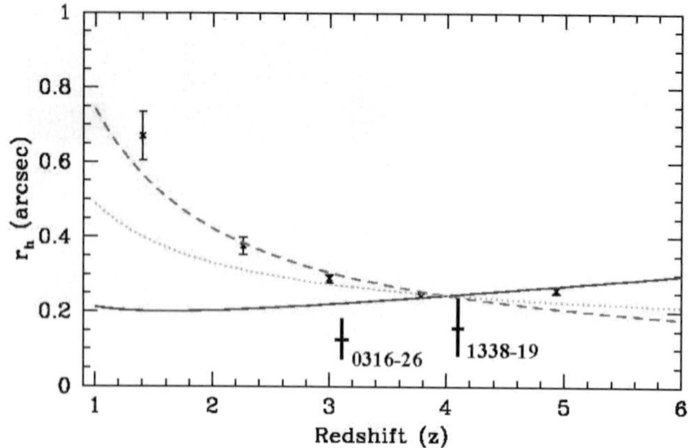

Figure 5. The redshift-size evolution of LBGs from Ferguson et al. (2004). Average sizes of the Lyα emitting protocluster galaxies in 0316–26 at $z = 3.1$ and 1338–19 at $z = 4.1$ are indicated, with the vertical bars indicating the range of measured sizes.

Preliminary (uncorrected) measurements from SExtractor give half-light radii of $0.07'' - 0.24''$ Excluding point-like objects, the mean half-light radius is $\sim 0.15''$. This is $\sim 2\times$ smaller than that measured for $z \sim 4$ LBGs in GOODS.

Taking our results at $z \sim 3$ and $z \sim 4$ together, there appears to be a trend for the mean sizes of Lyα-selected galaxies in protoclusters to be smaller than the mean sizes for Lyman break field galaxy at similar redshifts. This is shown in Figure 5. Lyα galaxies are generally fainter than the faintest of the LBGs measured by Ferguson et al. (2004) and It is not clear whether their smaller sizes is connected with (i) an overdense protocluster environment, (ii) their larger Lyα equivalent widths or (iii) their fainter continuum luminosities. A more detailed study is underway to disentangle the various effects (Overzier et al. 2004, in prep.). This will include a comparison between the continuum morphologies of the Lyα emitters and equally faint (non-Lyα) LBGs selected from our own protocluster dataset.

The protocluster Lyα emitters are undergoing significant star formation. Star formation rates (SFR) of the Lyα emitters computed separately from the Lyα luminosities and the 1500 A UV continuum fluxes give approximately the same result, namely 0.5–20 $M_{\odot}$ yr^{-1}, with a mean of a few $M_{\odot}$ yr^{-1}. More detailed analysis of the SEDs should enable us to constrain the amount of dust extinction in the protocluster galaxies.

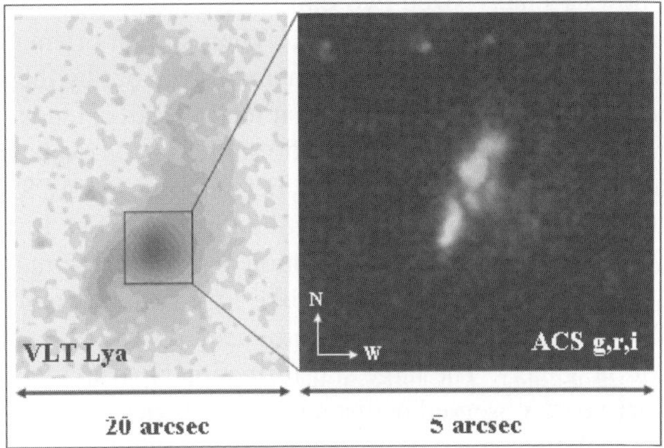

Figure 6. The radio galaxy TN J1338–1942, showing the large-scale Lyα halo observed with VLT (left), and high-resolution details of the (mostly) Lyα emission observed with HST/ACS (right).

The Pecular Radio Galaxy TN J1338–1942

The host galaxies of the luminous radio sources in our protoclusters are of considerable interest, because they are presumed to be forming massive galaxies that will develop into the dominant galaxies of evolved clusters. The ACS image of 1338–19 reveals intriguing new information about the Lyα and continuum morphologies of this radio galaxy.

Venemans et al. (2002) drew attention to the peculiar nature of the Lyα emission from 1338–1942. Although giant Lyα halos are a common feature of distant luminous radio galaxies, they are usually distributed symmetrically around the radio galaxy. In the case of TN J1338–1942, the halo is highly asymmetric, with a tail that extends for $\sim$15" (110 kpc) to the north-west. This Lyα tail points along the radio source axis and away from the centre of the protocluster as defined by the distribution of Lyα emitting galaxies. The agreement in orientation is intriguing and may indicate that the evolution of the radio galaxy is intimately linked to the evolution of the protocluster.

The ACS image of the Lyα emission from the inner region of the radio galaxy is shown in Figure 6 (right) together with the VLT Lyα image showing the large Lyα tail (left). Because the equivalent width of Lyα is large ($\sim$500 A) and its spatial extent is much larger than that of the continuum emission, the ACS image is dominated by Lyα. An important question concerns whether the Lyα emitting gas is primeval material falling into the forming radio galaxy, or processed gas flowing outwards (e.g. a superwind as described by Heckman et al. 1990).

The ACS shows a galaxy-sized Lyα component elongated in the direction of the radio emission and the large-scale halo. Their similar orientation indicates that at least some of the gas is entrained in and driven out by the radio jet as it propagates outwards. However, the fact that the Lyα tail extends more than 70 kpc beyond the edge of the observable radio source needs explanation.

Of particular interest is the peculiar Lyα feature in the ACS image pointing towards the southwest and the dark zone directly above it, suggestive of absorption. Although the effect of absorption and scattering may well distort the observed morphology, such a wedge-like Lyα feature would be expected from an outflowing enriched superwind produced by violent starbursts in the inner region of the galaxy. The large star formation rate (up to a few thousand solar masses per year), observed in distant radio galaxies (e.g. Dey et al. 1997) could generate sufficient pressure to power such a wind. From the Lyα line profile, the velocity of the outward flowing gas must be a few hundred km/s, implying a time needed for the gas to reach the end of the tail of $> 10^8$ y. Such a star formation rate would be sufficient to build up the entire stellar population of a massive galaxy if it persists for $\sim 10^8$ yr, the dynamical lifetime of the Lyα tail. Spectroscopy and spectrapolarimetry could test the validity of such a scenario.

7. Conclusions

Distant luminous radio galaxies are excellent signposts for pinpointing the most distant protoclusters. They provide unique laboratories for studying massive galaxies and clusters during a key period in the development of the Universe. This is the time when clusters become virialised and dramatic evolution occurs both in the space density of active galaxies and in star formation.

VLT observations have revealed 5 protoclusters, with a total of more than 120 Lyα emitting galaxies in overdense regions of the Universe. We have found strong evidence for additional galaxy populations in these protoclusters, using Lyman break and Hα techniques. Preliminary data suggests that the sizes of Lyα emitters in distant protoclusters are significantly smaller than the sizes of field Lyman break emitters at similar redshifts. A large project is underway to investigate the dependance on redshift and environment of the various galaxy populations in 4 key protoclusters. The ultimate goal of this project is to trace and disentangle the histories of star formation and structure assembly of galaxies in clusters.

Acknowledgments

The results that presented here were obtained under the auspices of several large collaborations. We particularly thank Wil van Breugel, Jaron Kurk, Carlos De Breuck and all the members of the ACS Science Team. Harry Ferguson kindly allowed us to use superimpose our data on the diagram reproduced in Figure 5. The ACS was developed under NASA contract NAS 5-32865. This research has been supported by NASA grant NAG5-7697 and an equipment grant from Sun Microsystems, Inc. The Space Telescope Science Institute is operated by AURA Inc., under NASA contract NAS5-26555. GKM acknowledges grants from the Royal Netherlands Academy for Arts and Sciences, the Netherlands Organisation for Scientific Research (NWO), the Netherlands Space Organization (SRON) and the Kerkhoven-Bosscha Fund.

References

Baugh, C.M. et al. (1998). ApJ 498, 504
Brand, K. et al. (2003). astro-ph/0302170
De Breuck, C. et al. (2000). A&AS, 143, 303
Dey, A. et al. (1997). Ap. J. ,490, 698. (1997)
Ferguson, H. et al. (2003). astro-ph/0309058
Francis, P. et al. (2001). ApJ, 554, 1001
Heckman, T. M. et al. (1990). Ap J. Suppl. 74, 833
Ivison, R. J. et al. (2000). ApJ, 542, 271
Keel, W. et al. (1999). AJ, 118, 2547
Kurk, J. D. et al. (2000). A&A, 358, L1
Kurk, J. D. et al. (2003). Ph.D. Thesis, University of Leiden
Miley, G. K.et al. (2004). Nature, 427, 47
Moller, P. and Fynbo, J. U. (2001). A&A,372, L57
van Ojik, R. et al. (1996). A&A, 313, 25
van Ojik, R. et al. (1997). A&A, 317, 358
Overzier, R. et al. (2004). in prep.
Pascarelle, S. et al. (1996). AJ, 456, L21
Pentericci, L. et al. (1998). ApJ, 504, 139
Reuland, M. et al. (2003). in prep.
Rosati, P. et al. (1998). ApJL, 492, 21
Rosati, P. et al. (1999). ApJ, 118, 76
Shapley, A. et al. (2003). ApJ, in press (astro-ph/0301230)
Smail, I. et al. (2003). Ap.J. 583, 551
Shimasaku, K. et al. (2003). ApJ, 586, L111
Steidel, C. C. et al. (1998). ApJ, 492, 428
Steidel, C. C. et al. (2000). ApJ, 532, 170
Stevens, J. A. et al. (2003). Nature, 425,264S
Venemans, B. et al. (2002). ApJL, 569, 11
Venemans, B. et al. (2003). submitted to A&A
Windhorst, R. A. et al. (1998). Ap. J, 494, L27
White, S. et al. (1978). MNRAS, 183, 341

THE GALAXY STRUCTURE-REDSHIFT RELATIONSHIP

Christopher J. Conselice
California Institute of Technology, Pasadena, CA USA

Abstract There exists a gradual, but persistent, evolutionary effect in the galaxy population such that galaxy structure and morphology change with redshift. This galaxy structure-redshift relationship is such that an increasingly large fraction of all bright and massive galaxies at redshifts $2 < z < 3$ are morphologically peculiar at all wavelengths from rest-frame ultraviolet to rest-frame optical. There are however many examples of morphologically selected spirals and ellipticals at these redshifts. At lower redshifts, the bright galaxy population smoothly transforms into normal ellipticals and spirals. The rate of this transformation strongly depends on redshift, with the swiftest evolution occurring between $1 < z < 2$. This review characterizes the galaxy structure-redshift relationship, discusses its various physical causes, and how these are revealing the mechanisms responsible for galaxy formation.

Keywords: Audio quality measurements, perceptual measurement techniques

1. Introduction

The structures and morphologies of galaxies change with time. Determining the history and cause of this galaxy structure-redshift relationship, including the origin of modern galaxy morphologies (i.e., ellipticals, disks) is perhaps the missing, and until now overlooked, link in understanding galaxy formation. While we currently have a good understanding of global galaxy formation and evolution, such as the star formation and mass assembly history (e.g., Madau et al. 1996; Dickinson et al. 2003), we are only beginning to understand *how* galaxy formation occurs as opposed to simply when. During the last few years it has become clear with the advent of wide-field imaging surveys from the ground, and from space using the Hubble Space Telescope, that galaxy structure evolves (e.g., Driver et al. 1995; Glazebrook et al. 1995; Abraham et al. 1996; Brichmann & Ellis 2000; Conselice et al. 2004b). There is a clear galaxy structure (or morphology)-redshift relationship such that galaxies in the more distant universe are peculiar while those in the local universe are more regular

D. Block et al. (eds.), Penetrating Bars through Masks of Cosmic Dust, 489–508.

or normal[1] (van den Bergh et al. 2001). Determining the physics behind the morphology-redshift relationship is critical for any ultimate understanding of galaxies, and the physical causes of galaxy structure and its evolution.

The morphology-redshift relationship can furthermore potentially be used as a key test of galaxy formation models. Theories behind galaxy formation can be divided into two main ideas - the monolithic collapse of material early in the universe to form stars and galaxies within a very short time (e.g., Larson 1975; Tinsley & Gunn 1976) and the hierarchical formation scenario (e.g., White & Rees 1978; Blumenthal et al. 1984; White & Frenk 1992; Cole et al. 2000). Observationally, we know that galaxies do not appear to form rapidly in the early universe, but have an extended star formation history that does not decline significantly until the universe was about half its current age. Likewise, about half of all stellar mass in the universe formed between $z \sim 1$ (8 Gyrs ago) and today (Dickinson et al. 2003). The fact that star formation occurs over time, and not quickly at very high redshift, largely rules out rapid collapses as the primary method for forming all galaxies. High redshift galaxies also tend to be small with likely small stellar masses (Ferguson et al. 2004; Papovich et al. 2001). Therefore a large fraction of all galaxies must have formed gradually throughout time. Understanding this process, that is what is causing mass to build up in galaxies, requires studying their internal properties.

There are several ways to measure the physical processes responsible for forming galaxies which can potentially explain the observed morphology-redshift relationship (§3). One method, and by far the most common is to study global galaxy properties, such as the evolution of stellar mass and star formation, and to compare these with predictive models (e.g., Somerville et al. 2001). Other methods, which are now just being explored, involve probing the internal features of high-z galaxies either through spectroscopy or high resolution imaging. While integral field spectroscopy for high-z galaxies is still in its infancy, understanding the internal structural features of high redshift galaxies in now in a golden age, utilizing new techniques (e.g., Conselice et al. 2000a; Peng et al. 2002) with high resolution Hubble Space Telescope imaging (e.g., Giavalisco et al. 2004; Rix et al. 2004).

The idea that the structures of galaxies hold clues towards understanding their current and past formation histories is a new, and perhaps still controversial, idea. There is however increasing amounts of evidence that suggests galaxy structure reveals fundamental past and present properties of galaxies (see Conselice 2003 and references within and §2). Utilizing these tools, we

[1] Throughout this review, I refer to 'regular' or 'normal' galaxies to denote systems that are on the present day Hubble sequence; namely ellipticals and spirals. Peculiar galaxies and/or mergers are not normal galaxies according to this criteria. This differs from the usual meaning of normal which refers to galaxies without the presence of an active galactic nuclei (AGN).

can begin to determine the origin of the galaxy morphology-redshift relationship. Understanding this relationship will also likely help us determine the physical formation mechanisms, and the history of galaxy assembly. Furthermore, it is now possible to compare observations of the galaxy morphology-redshift relationship with theoretical models based on cosmological and dark matter assumptions, connecting the universe as a whole to its constitute galaxies. I argue in this review that the galaxy morphology-redshift relationship is a pillar for understanding galaxy formation as well as possibly the relationship between the baryonic content of galaxies and their dark matter halos, the evolution of galaxies and their black holes, as well as the relationship between cosmological parameters and the evolution of galaxy structure. In summary, I will address in detail the following issues:
(i) What is the galaxy structure-redshift relationship and how does it evolve?
(ii) What is the physical causes for the formation and evolution of the galaxy structure-redshift relationship?
(iii) What does the evolution of the galaxy-structure redshift relationship tell us about galaxy formation?

I do not address some morphological evolution problems, such as the formation and evolution of bars, rings or other internal structures, as these are dealt with in other contributions (e.g., Jogee, Sheth). Throughout I will assume a cosmology with $H_0 = 70$ km s^{-1} Mpc^{-1} and relative densities of $\Omega_\lambda = 0.7$ and $\Omega_m = 0.3$.

2. The Physical Basis of Galaxy Structure

Before describing the galaxy morphology-redshift relationship, and what it implies for understanding galaxy formation, I will review our current understanding of how galaxy structure correlates with physical properties of galaxies. It has been known for decades that, broadly speaking, galaxy morphology correlates with the physical properties of galaxies such as luminosities, sizes, gas content, colors, masses, and mass to light ratios (e.g., Roberts & Haynes 1994). Generally, early-type galaxies (ellipticals) are larger, more massive, contain older stellar populations, and are found in denser areas. Later type galaxies, such as spirals, are bluer, contain younger stellar populations, more gas, and are less massive overall than the ellipticals. The correlation between these physical properties and Hubble/de Vaucouleur classifications are however not as strong. While in the mean these properties change with Hubble type, there is significant overlap in any given property across the Hubble sequence[2].

[2]Room prohibits a detailed discussion of all the problems with Hubble classifications. The fact that physical properties only correlate in the mean for a given Hubble type is only one of many issues. For a detailed discussion of this see Appendix A from Conselice (2003). It is still useful to separate in the broadest sense, ellipticals from spirals, as I do here, as these are fundamentally different galaxy types.

Table 1. The co-moving densities of bright ($M_B < -20$) galaxies as a function of morphological type in units of log (Gpc^{-3})[a]

Redshift	Ellipticals	Spirals	Peculiars
0.0	6.4±4.6	6.9±4.91	5.3±4.2
0.5	6.3±6.0	6.6±6.1	5.7±5.5
0.8	6.3±5.9	6.2±5.9	5.4±5.4
1.0	6.4±6.0	6.5±6.1	6.1±5.7
1.2	6.2±5.8	6.1±5.9	5.7±5.7
1.3	6.0±5.8	5.9±5.6	6.3±6.0
1.5	5.7±5.5	...	5.4±5.4
1.6	...	5.4±5.4	5.7±5.6
1.7	5.4±5.4	5.7±5.5	6.1±5.9
1.9	5.7±5.5	5.4±5.4	5.7±5.7
2.0	5.9±5.7	6.1±5.9	6.4±5.9

[a]Galaxies at $z = 0$ are taken from the "Third Reference Catalogue of Bright Galaxies" (de Vaucouleurs et al. 1991). The other galaxy densities at $z > 0$ are an average between the densities found for each type in the Hubble Deep Fields North and South.

The quantitative structures of galaxies, on the other hand, correlate strongly with physical properties, including: the current star formation rate, the stellar mass, galaxy radius, central black hole mass, and merging properties. It is impossible to describe all but a few of these correlations here. The first, and oldest, is that the light concentration of an evolved stellar population correlates with its luminosity, stellar mass, and scale (e.g., Caon, Capaccioli & D'Onofrio 1993; Graham et al. 1996; Bershady et al. 2000; Conselice 2003). This was first noticed by the failure of the de Vaucouleurs $r^{1/4}$ surface brightness profile to fit the surface brightness distributions of elliptical galaxies brighter than or fainter than $M_B \sim -20$ (e.g., Schombert 1986). It was realized later that the shape of the surface brightness profile for early types correlates strongly with its absolute blue magnitude (Binggeli & Cameron 1991; Caon et al. 1993). It was later shown that the general Sersic profile, with its concentration parameter n, gives a much better fit than the de Vaucouluer's profile for all early types (Graham et al. 1996). The central light concentration, as measured through the Sersic n index, or a non-parametric concentration index (C) (Bershady et al. 2000), also correlates with the mass of central black holes (Graham et al. 2001).

While the concentration of a galaxy's light profile correlates with the stellar mass, magnitude, and size of a galaxy, in a sense revealing the past formation history of a galaxy, there are many indicators in the structures of galaxies for ongoing galaxy formation. For example, recently it has been shown that the clumpiness of a galaxy's light distribution correlates and traces the location of star formation (e.g., Takamiya 1999; Conselice 2003). The clumpiness is measured by quantifying the fraction of a galaxy's light in the rest-frame B-band in high spatial frequency structures. The ratio between the amount of light in these high spatial frequency structures and the total light gives a measure of the clumpiness. This can be demonstrated by the strong correlation between the clumpiness index S (Conselice 2003) and Hα equivalent widths and colors of star forming galaxies. There is also a strong relationship between the dynamical state of a galaxy and the presence of a merger. Generally, merging galaxies are asymmetric, while non-mergers are not (Conselice et al. 2000a,b; Conselice 2003). This has been shown in numerous ways, including empirical methods (Conselice 2003) and the correlation of internal HI dynamics and asymmetries of stellar distributions (Conselice et al. 2000b).

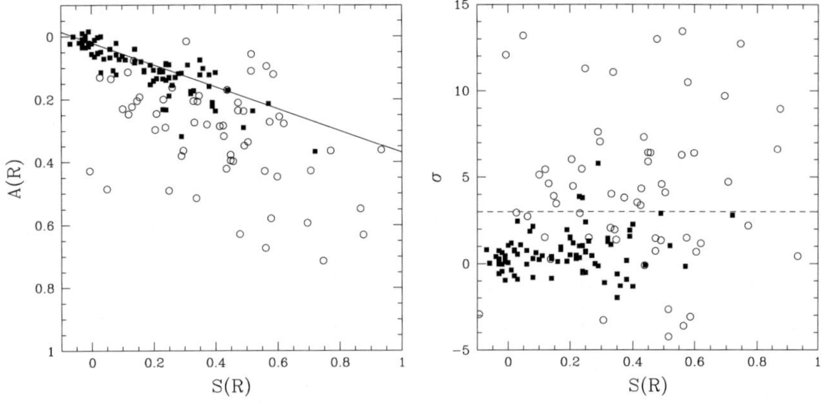

Figure 1. The relationship between the asymmetry index (A) and the clumpiness index (S). For normal galaxies (black squares, left panel) there is a strong correlation between A and S such that $A = (0.35 \pm 0.03) \times S + (0.02 \pm 0.01)$. The galaxies which deviate from this relationship are the ongoing major mergers, shown in the left panel as open circles. The right panel shows the deviation from the $A - S$ relationship in units of the scatter of the asymmetry values of the normal galaxies (σ). Generally, only the mergers deviate from this relationship by more than 3σ. For a physical reasoning behind these correlations see the text and Conselice (2003).

As shown in Conselice et al. (2000a) asymmetric light distributions can also be caused by star formation. However, decomposing light and kinematic structures in galaxies show that primary asymmetries are not the result of star formation, which forms in clumps or clumpy light (e.g., Elmegreen 2002), but from large scale lopsidedness of the size of the galaxy itself (Andersen et al. 2001). Likewise, there is a strong correlation between the asymmetry parameter and the clumpiness parameter for normal star forming galaxies (Conselice 2003; Figure 1). Galaxies with high clumpiness values (S), which correlates with high amounts of star formation, have correspondingly higher asymmetry values. However, this correlation deviates for systems involved in major mergers, such as nearby ultraluminous infrared galaxies (Figure 1). The nature of this deviation is such that a galaxy undergoing a merger has too high an asymmetry for its clumpiness, demonstrating that large asymmetries are produced in large scale features and not in clumpy, star formation like regions (Conselice 2003; Mobasher et al. 2004; Figure 1)[3].

3. The Galaxy Morphology-Redshift Relationship

3.1 Summary

The summary figure for understanding the galaxy structure-redshift relationship is shown in Figure 2. This relationship can be summarized simply as: *At higher redshifts (early times) the fraction of bright galaxies that are peculiar in structure and morphology increases gradually at the expense of both spirals and ellipticals.*

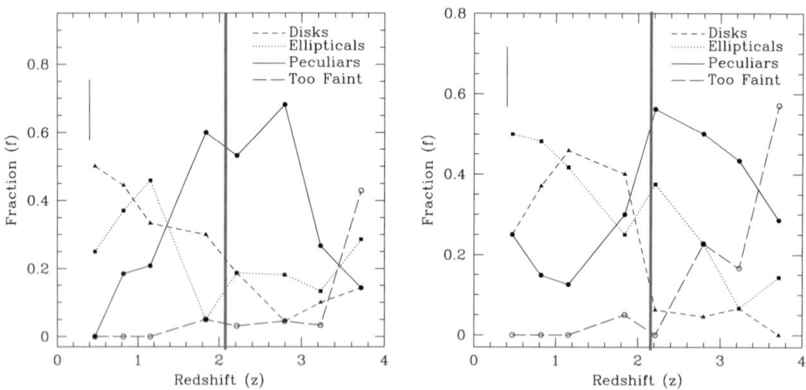

Figure 2. **The galaxy structure-redshift relationship.** This figure shows the evolution in relative fractions of different galaxy types as a function of redshift for classifications in the I_{814} (left panel) and H_{160} (right panel) band images of the HDF-N. The short vertical solid line on each plots gives the average error for these fractions. The long vertical line is the redshift limit for detecting spirals and ellipticals with $M_B < -20$. Types plotted on this are: disks (short dashed), ellipticals (dotted), peculiars (solid) and galaxies which are too faint for a classification (long dashed).

[3]The concentration index (C), asymmetry index (A), and clumpiness index (S) form the CAS morphology system described in detail in Conselice (2003). With these three parameters the major classes of nearby galaxies can be distinguished. Although this idea is not explicitly described in this review, it is used in papers described here, such as Conselice et al. (2004b) and Mobasher et al. (2004) to determine galaxy types.

The final state of galaxy evolution surrounds us and the modern universe is dominated by galaxies that can be classified on the Hubble sequence. A large fraction of all modern massive and bright galaxies are either ellipticals or spirals; only roughly 1-2% of all bright galaxies with $M_B < -20$ can be classified as peculiars (e.g., Marzke et al. 1998) (see Table 1). This changes gradually with redshift up to $z \sim 1$ and then more rapidly between $1 < z < 2$; at $z \sim 1$ most galaxies have relaxed morphologies while at $z \sim 2$ most galaxies are peculiar (Conselice et al. 2004b; Table 1).

3.2 Galaxy Structures at Low Redshift $z < 1$

At redshifts $z < 1$ most of the bright ($M_B < -20$) and massive galaxies ($M_* > 10^{10}$) are normal galaxies, that is ellipticals and spirals (Table 1; Figure 2). This relative fraction remains largely similar out to $z \sim 1$, with some important exceptions. In general, the co-moving density of elliptical and disk galaxies remains constant, to within a factor of 2, out to $z \sim 1$ with a slight decline (Figure 3; Brinchmann & Ellis 2000; Conselice et al. 2004b).

There is a more pronounced change in other features of normal galaxies from $z \sim 1$ to $z \sim 0$. These properties include co-moving B-band luminosity densities (ρ_{L_B}), and stellar mass densities (ρ_*). While the number density evolution of Hubble types is the physical manifestation of the galaxy structure-redshift relationship, the evolution of these other properties can reveal important clues for how this relationship is put into place, and how it might be evolving. The rest-frame B-band luminosity luminosity density evolution for galaxies of known morphology is also shown in Figure 3. There is a clear decline in luminosity densities at $z < 1$ for all galaxies, including ellipticals and spirals. This peak in the B-band luminosity density, which is produced in normal galaxies, must be due to recent star formation, as the stellar mass density for normal galaxies grows with redshift (Figure 3). The stellar mass density for ellipticals is half of its modern value at $z \sim 1$ in the Hubble Deep Field North (HDF-N). There is however perhaps an over density of ellipticals at $z \sim 1$ in the HDF-N, and cosmic variance is an issue. Although a lower density of early types would only enhance the evolution in stellar mass for these systems. This effect is also seen in studies considering galaxies on the 'red sequence', defined by the tight correlation between magnitude and color. The stellar mass in these red sequence galaxies increases by a factor of two from $z = 1$ to $z = 0$ (Bell et al. 2004), exactly the increase found when considering morphologically selected early types. Because of the large amount of co-moving luminosity in normal galaxies at $z \sim 1$, star formation must be occurring in early type galaxies during this time (see also Stanford et al. 2004). Luminosity and stellar mass functions suggests that this evolution is occurring for the lower mass and lower luminosity systems (Conselice et al. 2004b), while the higher mass systems are perhaps largely formed by $z \sim 1$, or even earlier (e.g., Glazebrook et al. 2004).

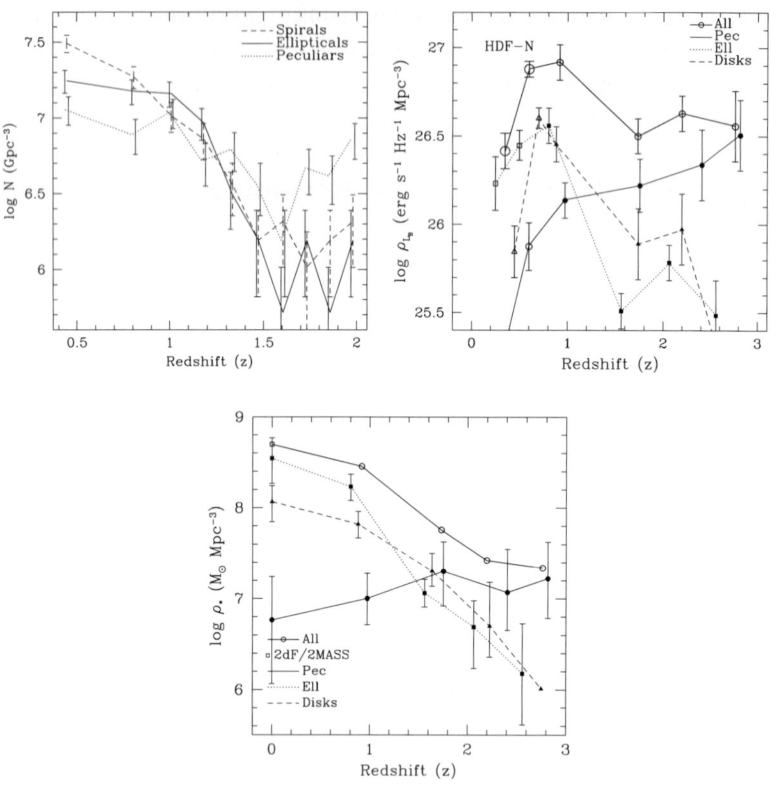

Figure 3. The relative co-moving number (N), rest-frame B-band luminosity (ρ_{L_B}), and stellar mass density (ρ_*) of galaxies as a function of redshift from deep NICMOS imaging of the Hubble Deep Field North (Conselice et al. 2004b). Points at redshifts $z < 0.5$ are taken from Brinchmann & Ellis (2000), Fukugita et al. (1998) and the 2dF/2MASS surveys (Cole et al. 2001).

3.3 Galaxy Structures at Medium Redshift $1 < z < 2$

Morphological counts of galaxies from early Hubble Space Telescope imaging found a large increase in the number of peculiar/irregular galaxies at fainter magnitudes (e.g., Driver et al. 1995; Glazebrook et al. 1995). It was however unknown during these early Hubble observations what the redshifts, and therefore, the characteristics of these peculiar galaxies were. When redshifts for these faint galaxies became available, it was argued that Hubble types appeared in abundance by $z \sim 1$, and evolve only slightly down to lower redshifts (van den Bergh et al. 2001; Kajisawa & Yamada 2001).

Ultimately what is desired is a determination of Hubble types as a function of redshift for galaxies of different luminosities and stellar masses. This was performed for bright galaxies in the Hubble Deep Field North and South by Conselice et al. (2004b). The results of this are shown in Figure 3 for galaxies brighter than I = 27. As described in §3.2 there is a rapid decline in the number of normal galaxies between $z \sim 1$ and $z \sim 1.5$, such that the co-moving density increases by 8.3 $\times 10^3$ Gpc^{-3} Gyr^{-1} for ellipticals and 5.7 $\times 10^3$ Gpc^{-3} Gyr^{-1} for spirals during this 1.6 Gyr period. As discussed briefly in §3.5 this change in morphology is not caused by the so-called morphological k-corrections in which galaxies appear different at different wavelengths. It can however be partially produced by selection effects, although a strong drop is also found when considering galaxies at a fixed absolute magnitude (Figure 2). Both spirals and ellipticals with $M_B < -20$ should be found in the Hubble Deep Fields up to $z \sim 2$, and at even higher redshifts if passive evolution is considered (e.g., Conselice et al. 2004b).

The redshift range $1 < z < 2$ is obviously critical for understanding the final onset and production of the Hubble sequence and the origin of the galaxy structure-redshift relationship. It is also the epoch (with a short 2.5 Gyrs!) where the star formation rate, AGN activity and stellar mass assembly is at its highest. Understanding how the galaxy structure-redshift relationship is evolving in this epoch is critical for understanding the causes behind galaxy formation. It is therefore worth spending some time discussing what is found morphologically and structurally in the galaxy population between $1 < z < 2$.

Figure 4 shows Advanced Camera for Surveys (ACS) images of the brightest galaxies in the rest-frame optical found within the Chandra Deep Field South Great Observatory Origins Survey (GOODS) imaging. Clearly, there is a rich morphological mix at this redshift, with many galaxies appearing similar to modern ellipticals and spirals, but with important structural differences that make them fundamentally different from modern normal galaxies. Another way to investigate this population is to study systems that have spectral energy distributions that likely place them at $1 < z < 2$, such as the extremely red objects, discussed in §4.1. The images of these galaxies reveal that these almost normal galaxies have outer shell like features, and what appears to be large star forming complexes. Understanding the physical causes behind these features will help reveal the formation mechanisms of galaxies.

3.4 Galaxy Structures at High Redshift $z > 2$

The structures and morphologies of galaxies at $z > 2$ are just now being studied in detail. Early work in this area suggested that galaxies selected by

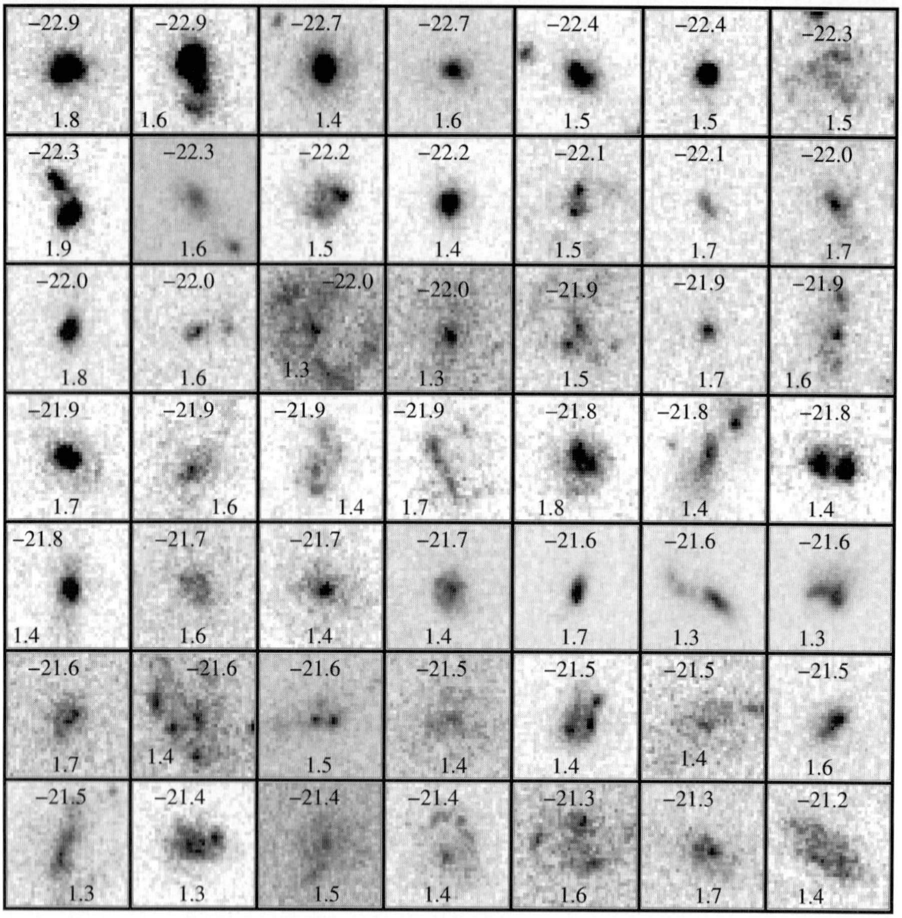

Figure 4. The brightest galaxies in ACS GOODS images whose photometric redshifts place them at $1 < z < 2$. These are ordered from brightest to faintest down to $M_B = -21$. The upper number is the M_B of each galaxy and the lower number is its redshift. There is a large diversity of properties, from systems that appear very peculiar to those that look similar to normal galaxies. Scale of these images is $\sim 2''$ on each side, corresponding to ~17 kpc at these redshifts.

the Lyman-break technique have compact structures with outer envelopes (Giavalisco et al. 1996). These compact structures have half-light radii a few kpc in size, similar to the bulges of modern spirals or moderate luminosity spheroids. Many of these compact galaxies have steep light profiles and asymmetrically distributed outer nebulosity . The depth of this early imaging was quickly superseded by analyses of the Hubble Deep Field North which showed a rich diversity of galaxy structures (Figure 5; Ferguson, Dickinson & Williams

2000). The morphologies of these galaxies is still however a largely unexplored area of parameter space.

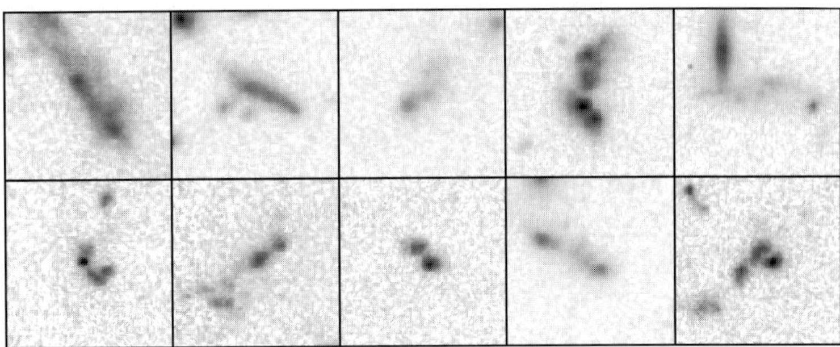

Figure 5. The morphologies of the brightest galaxies at $z > 2$ in the Hubble Deep Field North showing the peculiar and non-compact structures of these galaxies, many with several bright central regions or knots.

One of the reasons the morphologies of $z > 2$ galaxies have not been studied in detail is that describing their structures is not a trivial problem, as very few galaxies at high redshift can be identified as objects that would fit on the Hubble Sequence (e.g., Giavalisco et al. 1996; Conselice et al. 2003a; Lotz et al. 2003). One way to approach this problem is to use a purely descriptive approach described above, while another is to use quantitative techniques to characterize these structures. The has been done in Conselice et al. (2003a) and Conselice et al. (2004b) for galaxies at $z > 2$. The CAS systems shows that at the highest redshifts there is a real dichotomy in the galaxy population such that the most luminous and most massive galaxies are consistent with undergoing a major merger based on CAS indices, particularly the asymmetry index (Conselice et al. 2003a). What is found is that half of all bright, $M_B < -21$ or massive $M_* > 10^{11}$ $M_\odot$ galaxies are actively undergoing a major merger. The fainter and lower mass galaxies have peak merger fractions at $z \sim 2.5$ that are only 20% or lower. The relative fraction of mergers declines at lower redshifts very quickly for these massive and luminous galaxies with a power law index $(1+z)^{3-5}$ (Conselice et al. 2003). A comparison to hierarchical assembly models of galaxies is shown in Figure 6 using GALFORM simulations (e.g., Benson et al. 2002). These models over predict the number of major mergers occurring for the brightest galaxies at $z < 1$, consistent with the fact that there are too many bright K-band selected galaxies at $z \sim 1 - 1.5$ than predicted in these Cold Dark Matter based models (Somerville et al. 2004; see also §4.3).

3.5 Morphological K-corrections

A primary problem in understanding the galaxy structure-redshift relationship is to constrain the effects of the morphological k-correction, whereby galaxy structure changes as a function of wavelength. The main problem is that most deep high resolution imaging is done in the optical, probing up to $\lambda = 1\mu$m, allowing for a sampling of the rest-frame optical, < 4000 A, only up to $z \sim 1.5$. At redshifts higher than this we begin to sample the rest-frame ultraviolet light from galaxies. For galaxies at $z \sim 3$, for example, the z-band filter (F850L) (the reddest GOODS and ACS Hubble Ultra Deep Field filter) samples $\sim$ 2500 A, the near ultraviolet, where only young stars with ages < 100 Myrs are sampled.

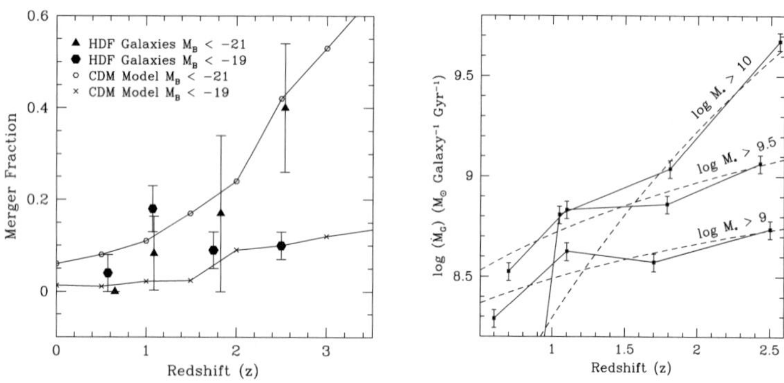

Figure 6. Left panel: major merger fractions to $z \sim 3$ at magnitude limits $M_B = -21$ and -19. Semi-analytical model predictions are also shown. Right panel: Stellar mass accretion history from major mergers as a function of initial mass (see Conselice et al. 2003a).

This is potentially a problem because both qualitatively, and quantitatively, galaxies have very different structures in their rest-frame optical and UV light (e.g., Bohlin et al. 1991; Kuchinski et al. 2001; Windhorst et al. 2002; Papovich et al. 2003). The largest differences are found for galaxies that are composed of old and young stellar components which are not spatially mixed, such as early type spirals or Hubble classifications Sa and Sb (e.g., Windhorst et al. 2002). The one type of galaxy that looks nearly identical at UV and optical wavelengths are starbursting galaxies, or galaxies whose structures are dominated by star formation (e.g., Conselice et al. 2000c; Windhorst et al. 2002).

Similarly, when examining galaxies at different redshifts, there are different types of morphological k-corrections, depending on the redshift and galaxy type. At $z < 1.5$ morphological k-corrections are not an issue as we are able to sample rest-frame optical light. At higher redshifts the NICMOS camera on HST allows us to determine the rest-frame optical structures and morphologies, although only limited field coverage exists (e.g., Dickinson et al. 2000). The result of this imaging is that galaxies that look irregular and distorted in the rest-frame UV also appear distorted in the rest-frame optical (Teplitz et al. 1998; Thompson et al. 1999), with some possible and important exceptions (e.g., Giavalisco 2002; Labbe et al. 2003; Conselice et al. 2004a,b). While these are rare systems, they do exist, and examples of morphologically selected ellipticals are found out to $z > 2$. These normal galaxies will likely become more common as we probe deeper in the infrared with high resolution wide field imaging.

4. The Nature of the Peculiar High-z Galaxy Population

4.1 Extremely Red Objects

Extremely red objects (EROs), usually defined by an optical-infrared color limit, usually using the criteria (R-K) $> 5 - 6$, or (I-K) $> 4 - 5$ (e.g., Daddi et al. 2000) were traditionally thought to be early type or dusty galaxies at $z > 0.8$. Extremely red galaxies are red because of a large 4000 A break produced by aged, or dusty, stellar populations. EROs are now also found in the near infrared with a (J-K) limit that locates objects at redshifts $z > 2$. The ERO population is therefore a good one for determining the basic properties of evolved galaxies at high redshifts.

The morphologies of EROs are mixed, with a strong redshift dependence (Figure 7). Systems at $z < 1.2$ are typically evolved galaxies, while those at $z > 1.2$ are more irregular or disky like systems (Moustakas et al. 2004). The spectra of EROs are also mixed, with about half showing signs of evolved stellar populations, and the other half show emission lines (Cimatti et al. 2002). Morphological studies of EROs demonstrate that a large fraction, perhaps the majority of the K < 20 objects, are disks (Yan & Thompson 2003). It thus appears likely that the ERO definition, far from finding only specific galaxy populations, is sampling all morphological types, and is a good sample for studying galaxies evolving onto the Hubble sequence. This has been effectively done by Moustakas et al. (2003) who studied the redshift distributions of EROs in the GOODS-South field as a function of morphology. Moustakas et al. found that the lower redshift EROs are dominated by early and late-type galaxies, while higher redshift samples at $z > 1.5$ are dominated by galaxies that cannot be place on the Hubble sequence (Figure 7). Since the ERO definition should find the most evolved galaxies, it is unlikely that there are a significant number of normal bright ellipticals at $z > 1.5$ which are being missed.

Recently, there have been discovery claims for a new population of red galaxies at $z > 2$. These are similar to the lower redshift ERO population in that they are identified by a color cut that isolates galaxies based on the

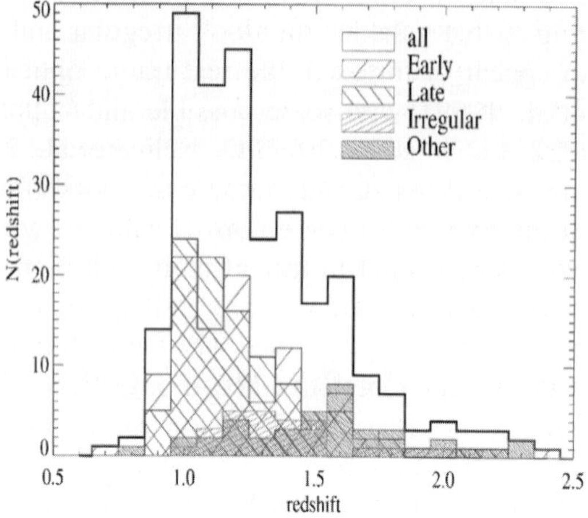

Figure 7. The redshift distributions for extremely red galaxies from the GOODS-South field (Moustakas et al. 2004) showing how the morphological distribution for this popualtion shifts to the 'other' or peculiar types at higher redshifts.

Balmer break, but uses the color between two infrared filters, normally the J and K bands (e.g., (J-K) > 2.3). These systems are typically at $z > 2$ and may be quite distinct from the Lyman-break galaxy population (Franx et al. 2003; van Dokkum et al. 2004). The morphologies of these systems have not been studied in detail, partially because there are so few examples, yet bright K-band selected galaxies with K < 20 often have irregular UV morphologies (Labbe et al. 2003; Daddi et al. 2004) indicating star formation. There are however some hints for spiral structures and more regular compact near infrared morphologies in some of these systems systems (Labbe et al. 2003; Daddi et al. 2004). These infrared EROs have large derived stellar masses and are generally consistent, based on clustering analyses and stellar population arguments, to be the most evolved systems at high redshift, and are the best candidates for being the progenitors of evolved galaxies found in dense regions today.

4.2 Luminous Diffuse Objects and Chain Galaxies

In Conselice et al. (2004) a new galaxy type, found abundantly between $1 < z < 2$, is described and characterized. These galaxies, which have no local counterparts, were discovered based on their low light concentrations and high luminosities, and are called Luminous Diffuse Objects (LDOs). These objects are fairly common with surface densities 1.8 arcmin^{-2}, and co-moving number densities of 5×10^5 Gpc^{-3} within the GOODS South field (Conselice et al. 2004). These objects were independently discovered by Elmegreen et al.

(2004a) and are likely face on counterparts of the 'chain galaxies' discussed in Cowie et al. (1995) (Elmegreen et al. 2004b).

Elmegreen et al. (2004b) compared the colors of the star forming knots in their sample of LDOs with the knots found in chain galaxies, finding a very similar color distribution. This suggests that chain galaxies are the edge-on versions of LDOs. Chain galaxies were known even at discovery to be large complexes of star forming regions (Cowie et al. 1995). Formation scenarios for these systems are discussed in Elmegreen et al. (2004b), and are consistent with large amounts of star formation occurring after gas in initial disks fragment to produce several large clumps. In models, these clumps are predicted to form through energy dissipation and later merge together to form bulges (Immeli et al. 2004). The fact that there are no obvious bulge components in LDOs is a clue that bulge formation may occur *after* disk formation, not before, as is generally assumed in hierarchical models.

LDOs are likely in a phase where a large fraction of their stellar mass is being assembled. The star formation rate in these systems is on average 4 $M_{\odot}$ year^{-1} before correcting for dust, and these systems have starburst spectral energy distributions (Figure 8). These systems account for up to 50% of all the star formation occurring between $1 < z < 2$, where a large fraction of the stellar mass in galaxies formed (Dickinson et al. 2003). Since these are likely disks in formation (Conselice et al. 2004; Elmegreen et al. 2004b; Figure 8), they are unlikely to dominate the growth of stellar mass between $1 < z < 2$ since most stellar mass at $z \sim 0$ is locked up in ellipticals (e.g,. Fukugita et al. 1998). This remains somewhat of a mystery, unless some of these LDOs merger to form early type galaxies at $z < 1.5$, which is unlikely (Conselice et al. 2003a).

4.3 Galaxy Mergers

A major, but still largely under-studied aspect of galaxy formation, is the role of galaxy mergers in the formation of galaxies. The merging of galaxies to form larger systems is the cornerstone of the idea behind the modern galaxy formation model, dark matter, and cosmology (Cole et al. 2000). Constraining this process observationally is just now being done. The first problem is identifying, confidently, which galaxies at high redshift are undergoing a major merger (defined as a merger mass ratio 3:1 or higher). There are a few methods for finding galaxies which are merging, or which soon will. The traditional method for identifying mergers is to identify galaxies in kinematic pairs that are close (< 20 kpc), and at nearly the same radial velocity (with < 500 km s^{-1}). Identifying pairs at high redshift in this manner is difficult due to the inability to determine radial velocities for complete samples, and it is thus largely applicable only at redshifts $z < 1$.

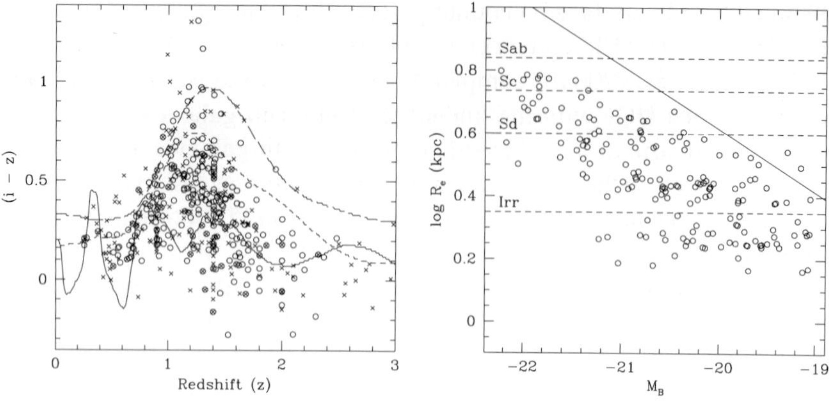

Figure 8. Left panel: The distribution of $(i - z)$ colors for LDOs (circles) as a function of redshift with two Coleman, Wu and Weedman spectral energy distributions and a Kinney et al. starburst model plotted (see Conselice et al. 2004a). These are from bluest to reddest - starburst (solid line), Scd (dashed), Sbc (long dashed). Right panel: Absolute magnitude-effective radius relationship for LDOs. The solid line is the canonical Freeman disk relationship at $z \sim 0$. The dashed horizontal lines show the effective radii of different nearby galaxy types.

A promising method explored in Conselice (2003) is identifying galaxy mergers which are in progress, that is systems that have already merged and are undergoing dynamical relaxation. One way to identify these systems is through their chaotic kinematic structures revealed through integral field spectroscopy or velocity curve work (e.g., Erb et al. 2004). Alternatively, and more affordable, but perhaps less accurate, is to use the stellar structures of galaxies. In §2 the methods and reasoning behind using the asymmetry index to find mergers are explained, and the results of these techniques applied to high redshift galaxies are described in §3.4.

Since the modern paradigm for forming galaxies implicitly assumes massive galaxies form by merging, it is important to test these models. The agreement between Cold Dark Matter based models and the data, shown in Figure 6, is good at high redshift, but fails by a significant amount to reproduce the merger fractions for bright galaxies at lower redshifts. This is likely because massive galaxies are forming earlier than low mass galaxies, which may or may not be an effect of environment - massive galaxies are also more likely to be found in dense environments (Dressler 1980). We can examine this in more detail to determine how the masses of high redshift galaxies are built up through mergers and star formation induced by merging. The amount of stellar masses added to a galaxy during observed major merger, assuming a mass ratio of 1:1, can be calculated from the star forming properties of galaxies found at high redshift (e.g., Papovich et al. 2001; Figure 6). Based on the merger rates calculated in Conselice et al. (2003a), a typical Lyman-break galaxy at $z \sim 3$ will undergo $\sim$ 5 major mergers, with accompanying star formation, before $z \sim 1$ and is unlikely to have a major merger at $z < 1$. The mass added by each of these mergers, plus the likely amount of new stars produced in star formation, is enough to create a $> M_*$ (Cole et al. 2001) galaxy by $z = 1$. This also suggests that the galaxy structure-redshift relationship can be described as a cooling of the galaxy population from an era of rapid mergers that has been steadily declining since $z \sim 3$.

Minor mergers are harder to constrain, yet are likely a major method for adding material to normal galaxies at $z < 1$ (e.g., Patton et al. 2002; Bundy et al. 2004). This method, or secular evolution, are the most likely possibilities for driving evolution in the galaxy population at $z < 1$, when up to 50% of all stellar mass formed.

4.4 High Redshift Luminous Infrared Galaxies

A significant galaxy population at high redshift are the luminous infrared galaxies detected by redshifted mid-infrared emission from dust grains at observed wavelengths 10 - 850 μm. The existence of these sources is suggested by the large far-IR and sub-millimeter backgrounds (Puget et al. 1996). These luminous infrared sources are very common at $z > 2$, over an order of magnitude as common as their local counterparts (Chapman et al. 2003a). With the launch of the *Spitzer Space Telescope* the study of these galaxies is very much in progress, although important features of these galaxies are already known. Luminous infrared galaxies in the nearby universe are found to be heated by a mixture of AGN activity and star formation (Sanders & Mirabel 1996), although the most luminous sources with $L > 10^{12}$ $L_\odot$ are nearly all starburst induced major mergers. The relative contributions at high redshift is also a mixture of these two types, although preliminary *Spitzer* observations show that the spectral energy distributions of high redshift luminous infrared galaxies are largely composed of star forming systems (Frayer et al. 2004).

The rest-frame UV (or observed optical) properties of sub-mm sources observed with the Hubble Space Telescope reveal that these galaxies would be included in magnitude limited optical studies (such as the HDF) and that their structures are very peculiar in appearance (e.g., Chapman et al. 2003b). There are two pieces of evidence that suggests these morphologically peculiar sub-mm/radio selected luminous infrared galaxies at $z > 2$ are undergoing major galaxy mergers. The first is the very large line widths found in CO mapping (Genzel et al. 2004). Galaxies with similar large velocity widths in the nearby universe are undergoing major mergers (Conselice et al. 2000b). The other evidence is that the morphologies of sub-mm sources are generally peculiar (e.g., Chapman et al. 2003b) and are quantitatively consistent with undergoing major mergers (Conselice et al. 2003b). The merger fraction for sub-mm detected galaxies is in fact higher than for the most massive Lyman-break galaxies (Conselice et al. 2003b). This suggests that the most massive galaxies at high redshift, which are probably these sub-mm sources (Tecza et al. 2004), are forming by mergers.

5. Relationship to the Dark Universe

The structure of galaxies, and its evolution, potentially relates directly with the existence of dark matter, dark energy and black holes. I only briefly discuss this here as much of this is work for the future. One example is that galaxy mergers might not occur in the abundance we see in the absence of dynamical friction produced by dark matter halos (e.g., Sellwood 2004). Detailed modeling of this however has not yet been done. As discussed, black holes are directly traceable with the concentration of galaxy light as discussed in §2. Dark energy is also likely imprinting its effects, and is perhaps the fundamental cause, of the morphology-redshift relationship.

The relationship between the velocity dispersion of spheroids and the mass of their central black holes (Gebhardt et al. 2000; Ferrarese & Merritt 2000) is a fundamental property of spheroids. This relationship, which is effectively between the scale of a spheroid system and its central black hole, is also projected in the concentration index-black hole mass relationship. This relationship appears to hold up to $z \sim 1.2$ based on the correlation between galaxies with X-ray emission and the CAS concentration index (Grogin et al. 2003). X-ray sources up to $z \sim 1.2$ are found in galaxies with the highest light concentrations, suggesting that the bulge/central black hole relationship is in place by these redshifts. Understanding this relationship at higher redshift, as well as how dark matter condenses and evolves with galaxy structure, are topics for future investigations with 20-30 meter ground based telescopes and the James Webb Space Telescope.

6. Acknowledgments

Its a pleasure to thank my collaborators and colleagues, especially Mark Dickinson, Richard Ellis, Kevin Bundy, and Casey Papovich for helping shape my evolving understanding of this material. I also thank David Block and the organizing committees for inviting me to present this contribution and their patience in receiving this review.

References

Abraham, R.G., et al. 1996, ApJS, 107, 1
Andersen, D.R., et al. 2001, ApJ, 551, 131L
Bell, E.F., et al. 2004, ApJ, 608, 752

Benson, A.J., Lacey, C.G., Baugh, C.M., Cole, S., & Frenk, C. 2002, MNRAS, 333, 156
Bershady, M.A., Jangren, A., & Conselice, C.J. 2000, AJ, 119, 2645
Binggeli, B., & Cameron, L.M. 1991, A&A, 252, 27
Blumenthal, G.R., et al. 1984, Nature, 311, 517
Brinchmann, J., & Ellis, R.S. 2000, ApJ, 536, 77L
Bundy, K. Fukugita, R.S., Ellis, R.S., Kodama, T., & Conselice, C.J. 2004, ApJ, 600, 123L
Caon, N., Capaccioli, M., & D'Onofrio, M. 1993, MNRAS, 265, 1013
Chapman, S.C., Blain, A.W., Ivison, R.J., & Smail, I.R. 2003, Nature, 422, 695
Chapman, S.C., Windhorst, R., Odewahn, S., Yah, H., Conselice, C. 2003b, ApJ, 599, 92
Cimatti, A., et al. 2002, A&A, 381, L68
Cole, S., Lacey, C.G., Baugh, C.M., & Frenk, C.S. 2000, MNRAS, 319, 168
Cole, S., et al. 2001, MNRAS, 326, 255
Conselice, C.J., Bershady, M.A., & Jangren, A. 2000a, ApJ, 529, 886
Conselice, C.J., Bershady, M.A., & Gallagher, J.S. 2000b, A&A, 354, 21L
Conselice, C.J., et al. 2000c, AJ, 119, 79
Conselice, C.J. 2003, ApJS, 147, 1
Conselice, C.J., Bershady, M.A., Dickinson, M., & Papovich, C. 2003a, AJ, 126, 1183
Conselice, C.J., Chapman, S.C., Windhorst, R.A. 2003b, ApJ, 596, 5L
Conselice, C.J., et al. 2004a, ApJ, 600, L139
Conselice, C.J., Blackburne, J., & Papovich, C. 2004b, astro-ph/0405001
Cowie, L.L., Hu.E.M., & Songaila, A. 1995, AJ, 110, 1576
Daddi, E., et al. 2004, ApJ, 600, 127L
Daddi, E., Cimatti, A., & Renzini, A. 2000, A&A, 362, 45L
de Vaucouleurs, G., et al. 1991, "Third Reference Catalogue of Bright Galaxies"
Dickinson, M., Papovich, C., Ferguson, H.C., & Budavari, T. 2003, ApJ, 587, 25
Dickinson, M., et al. 2000, ApJ, 531, 624
Dressler, A. 1980, ApJ, 236, 351
Driver, S.P., Windhorst, R.A., & Griffiths, R. 1995, ApJ, 453, 48
Elmegreen, B.G. 2002, ApJ, 577, 206
Elmegreen, D.M., Elmegreen, B.G., & Hirst, A.C. 2004a, ApJ, 604, 21L
Elmegreen, D.M., Elmegreen, B.G., & Sheets, C.M. 2004b, ApJ, 603, 74
Erb, D.K., Steidel, C.C., Shapley, A.E., Pettini, M., & Adelberger, K.L. 2004, astro-ph/0404235
Ferguson, H.C., Dickinson, M., & Williams, R. 2000, ARA&A, 38, 667
Ferguson, H.C., et al. 2004, ApJ, 600, 107L
Ferrarese, L., & Merritt, D. 2000, ApJ, 539, 9L
Franx, M. et al. 2003, ApJ, 587, 79L
Frayer, D.T., et al. 2004, astro-ph/0406351
Fukugita, M., Hogan, C.J., & Peebles, P.J.E. 1998, ApJ, 503, 518
Gebhardt, K., et al. 2000, ApJ, 539, 13L
Genzel,R., et al. 2004, astro-ph/0403183
Giavalisco, M., Steidel, C., Macchetto, F.D. 1996, ApJ, 470, 189
Giavalisco, M. 2002, ARA&A, 40, 579
Giavalisco, M. et al. 2004, ApJ, 600, L93
Glazebrook, K., et al. 2004, astro-ph/0401037
Glazebrook, K., et al. 1995, MNRAS, 275, 19L
Graham, A., Lauer, T.R., Colless, M., & Postman, M. 1996, ApJ, 465, 534
Graham, A., Erwin, P., Caon, N., & Trujillo, I. 2001, ApJ, 563, 11L
Grogin, N., et al. 2003, ApJ, 595, 684
Immeli, A., Samland, M., Westera, P., & Gerhard, O. 2004, astro-ph/0406135
Kajisawa, M., & Yamada, T. 2001, PASJ, 53, 833

Kuchinski, L.E., Madore, B.F., Freedman, W.L., & Trewhella, M. 2001, AJ, 122, 729
Larson, R.B. 1975, MNRAS, 173, 671L
Labbe, I., et al. 2003, ApJ, 591, 95L
Lotz, J.M., Primack, J., & Madau, P. 2003, astro-ph/0311352
Marzke, R.O., et al. 1998, ApJ, 503, 617
Madau, P., et al. 1996, MNRAS, 283, 1388
Mobasher, B., et al. 2004, ApJ, 600, 143L
Moustakas, L.A., Casertano, S., Conselice, C.J., et al. 2004, ApJ, 600, 131L
Papovich, C., et al. 2003, ApJ, 598, 827
Papovich, C., Dickinson, M., & Ferguson, H.C. 2001, ApJ, 559, 620
Patton, D.R. et al. 2002, ApJ, 565, 208
Peng, C.Y., Ho, L.C., Impey, C.D., & Rix, H.-W. 2002, AJ, 124, 266
Puget, J.-L., et al. 1996, A&A, 308, 5
Rix, H.-W., et al. 2004, ApJS, 152, 163
Roberts, M.S., & Haynes, M.P. 1994, ARA&A, 32, 115
Sanders, D.B., & Mirabel, I.F. 1996, ARA&A, 34, 749
Schombert, J.M. 1986, ApJS, 60, 603
Sellwood, J.A. 2004, astro-ph/0401398
Somerville, R.S., Primack, J.R., & Faber, S.M. 2001, MNRAS, 320, 504
Somerville, R.S., et al. 2004, ApJ, 600, 135L
Stanford, A., et al. 2004, AJ, 127, 131
Takamiya, M. 1999, ApJS, 122, 109
Tecza, M., et al. 2004, ApJ, 605, 109L
Teplitz, H., Gardner, J., Malmuth, E., & Heap, S. 1998, ApJL, 507, 17L
Thompson, R.I., et al. 1999, AJ, 117, 17
Tinsley, B.M., & Gunn, J.E. 1976, ApJ, 203, 52
van den Bergh, S., Cohen, J., Crabbe, C. 2001, AJ, 122, 611
van Dokkum, P. et al. 2004, astro-ph/0404471
White, S.D.M., & Rees, M.J. 1978, MNRAS, 183, 341
White, S.D.M., & Frenk, C.S. 1991, ApJ, 379, 52
Windhorst, R.A., et al. 2002, ApJS, 143, 113
Yan, L., & Thompson, D. 2003, ApJ, 586, 765

THE COSMIC BACKGROUND: EVOLUTION OF INFRARED GALAXIES AND DUST PROPERTIES. A LECTURE DEDICATED TO THE MEMORY OF MAYO GREENBERG

Jean-Loup Puget, Guilaine Lagache and Herve Dole
Institut d'Astrophysique Spatiale, Universite Paris Sud, F-91405 Orsay, France

Abstract In recent years it became clear that a population of galaxies radiating most of their power in the far-infrared contribute an important part of the whole star formation activity in the universe. These galaxies emit up to 99% of their energy output in the infrared by dust. The optical properties of dust in galaxies are thus very important to understand this population. Mayo Greenberg who has brought many of the ideas underlying our understanding of the physics of interstellar dust has been a pioneer in predicting the important role of organic solid material formed on dust grains in molecular clouds and of potential importance of transient heating of very small particles. It appeared with the *ISO* data on galaxies and even more today with the first *Spitzer* data that these mechanisms are important globally for the observations of infrared galaxies at significant redshifts. The understanding of their evolution is one of the keys to the understanding of galaxy built up and evolution.

Keywords: infrared galaxies, cosmic background, galaxy formation

1. Ultra Luminous Infrared Galaxies

A fraction of the stellar radiation produced in galaxies is absorbed by dust and re-radiated in the mid and far-infrared. In our Galaxy, this concerns only about a third of the total luminosity, much less in elliptical galaxies. The effect of the "dust mask" has been identified for a long time as an obvious nuisance for optical observations and implies a "correction" to account for the total energy output of galaxies. Now that galaxies have been found for which most of the radiation is coming out at long wavelengths, the infrared part of the spectrum cannot be treated as a "correction" to optical observations. Although a few very luminous galaxies were observed in the seventies, it is really with the IRAS survey that a proper census of the infrared emission of galaxies at low redshift was properly carried out. The luminosity function at 60 and 100 μm

D. Block et al. (eds.), Penetrating Bars through Masks of Cosmic Dust, 511–521.

is dominated by $L_\star$ spiral galaxies as could be expected, but a high luminosity power law tail of luminous galaxies was found. This population, although not dominant, was carrying a substantial of the infrared energy production in the local universe which was only one third of the optical one. It was also found that these luminous galaxies were often associated with interacting of merging galaxies. Some were clearly starbursts and others AGNs.

With *ISO*, the sensitivity allowed to investigate the more distant universe up to a redshift of about one and to investigate through spectroscopic studies whether AGN activity or starburst activity was powering Ultra Luminous InfraRed Galaxies (ULIRG).

The simplest expectation on the one hand was that heavy elements, and thus dust, should have decreasing abundances when going to high redshifts and thus the ratio of infrared to optical should decrease.The association with mergers, on the other hand, could mean an increase of the fraction of infrared galaxies in the past due to the higher number density.

We'll refer to "infrared galaxies" and to "optical galaxies" for short to mean galaxies in which the infrared emission, respectively optical, emission dominates. The two populations are rather well separated (not necessarily meaning that one galaxy does not go from one class to the other).

2. Properties of interstellar dust and the spectral energy distribution of infrared galaxies

For a population where most of the stellar radiation is absorbed and reradiated by dust, dust properties and the physics of the absorption and emission are essential as they determine the Spectral Energy Distribution (SED) of the galaxy.

Interstellar dust is the topic to which Mayo Greenberg brought a major contribution. In fact he was really visionary on some topics of the interstellar dust physics. Dust grains which contribute most of the optical extinction must have a size between 0.01 and 0.1 μm to account for the fact that the albedo drops significantly for wavelengths larger than about 1 μm. Considering the energy density of the radiation in a galactic disc like ours, the temperature of a dust grain is rather low: 15 to 25 K. The emission is peaked at typically 100 μm: the wavelength of the emitted radiation is much larger than the grain size when the absorbed radiation has a wavelength comparable to the grain size. This has an important consequence. The emissivity in the far-infrared will decrease roughly like the square of the wavelength which in turn makes the temperature dependence on the radiation energy density u very weak ($T \simeq u^{1/6}$). For a galaxy like ours the infrared part of the SED peaks at 170 μm when for a ULIRG it will peak around 60 μm: a factor 3 in temperature for a factor 10^3 in energy density or luminosity. At long wavelengths in the submillimeter and millimeter, the intensity should decrease like $I_\nu \simeq \lambda^{-4}$. At wavelengths shorter than this peak the exponential cut off of the Planck function leaves little power emitted in the mid infrared (5 to 25 μm), except in regions very close to stars which should carry only a very small fraction of the integrated luminosity of a galaxy.

The emission spectrum of diffuse cirrus clouds only heated by the diffuse interstellar radiation field is shown in figure 1 (Boulanger et al., 2000). The peak emission and the long wavelength behaviour fit with expectations. On the other hand about one third of the power is concentrated in the region 5 to 50 μm far more than what is expected from the modified black body at the equilibrium temperature; this emission contains a set of prominent features between 5 and 18 μm. *ISO* has shown that many infrared galaxies were showing this behaviour in their integrated spectra.

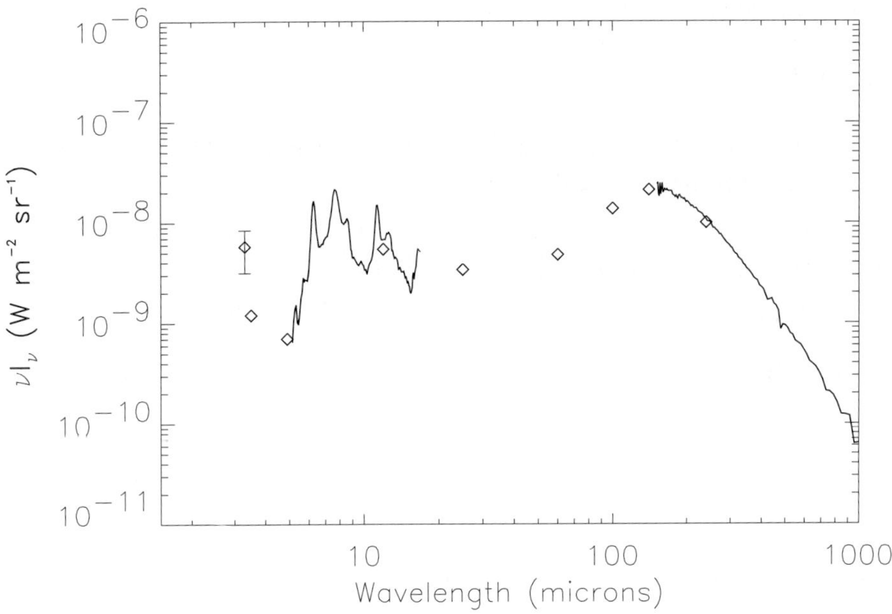

Figure 1. Diffuse interstellar cirrus emission for a column density N(HI)=10^{20} cm^{-2} (from Boulanger et al., 2000)

Mayo Greenberg (Greenberg, 1976) proposed a new mode of emission for interstellar dust grains when they are small enough. The idea of very small interstellar dust grains has been proposed by Platt (Platt, 1956) earlier but the emission was not discussed at that time. Mayo made the point that the heat capacity could be small enough and the probability to absorb a photon small enough that the dust grain could be transiently heated to high temperature and radiate before absorbing the next photon. This mechanism was invoked by K. Sellgren (Sellgren, 1983) to explain reflection nebulae observations and Leger & Puget, 1984 proposed that large Polycyclic Hydrocarbons (PAHs) molecules could be responsible for the so called "unidentified infrared bands". Puget et al., 1985 proposed that this could also explain the mid infrared emission of the Galactic disc and predicted the cirrus emission now observed.

Mayo Greenberg made another related far reaching prediction: the grains in dense molecular clouds opaque to optical radiation are very cold (5 to 10 K). Molecules can condense on these grains in ice mantles in which chemistry induced by cosmic rays and UV radiation when the grains reach the surface of the cloud produces larger molecules and ultimately organic solid material. Experiments in Mayo's laboratory demonstrated this mechanism and showed the famous "yellow stuff" residue. We know today that the PAHs are very underabundant in dense clouds but are overabundant with respect to the diffuse medium of the surfaces of molecular clouds to disappear again in HII regions. This is very likely to be the sign of the physics described by Mayo Greenberg. The organic material seen in absorption at 3.4 μm (aliphatic hydrocarbons, Willner et al., 1979) is seen in emission at 3.28 μm (aromatic hydrocarbons) at the surface of clouds and it has been shown that aliphatic hydrocarbons transform into aromatic ones when heated at high enough temperature. We are thus probably watching directly the mantle evaporation and transformation of the yellow stuff predicted by Mayo into free large PAH molecules (Greenberg et al., 1972).

The SED of infrared galaxies has thus important properties for the observations of redshifted ones:

- The long wavelength spectrum is very steep ($\simeq \lambda^{-4}$)
- A substantial fraction of the energy comes out in a set of features between 5 and 9 μm
- The SED peaks near 100 μm leading to "negative K-corrections" in the submillimeter
- The main PAH feature leads to "negative K-corrections" in the 10 to 30 μm range

The "negative K-corrections" makes redshifted infrared galaxies easier to observe. This effect is particularly strong at long wavelengths: at 1 mm the flux of a ULIRG radiating 10^{12} $L_{\odot}$ is seen as a source with a constant flux of 1 mJy for $1.5 < z < 10$!

3. The Cosmic Infrared Background (CIB)

The CIB is the energy content of the universe today produced by galaxies at all redshifts and seen as an isotropic extragalactic background radiation. It was predicted by Patridge and Peebles (1967) that observations of such a background gives a powerful tool to constrain cosmological evolution especially if you observe at wavelengths larger than a maximum of emission. Partridge & Peebles, 1967 made that point for the near infrared beyond the peak of the stellar emission of galaxies ($\lambda > 1\mu m$).

The detection of the CIB was the major objective of the DIRBE experiment aboard COBE. In fact it was by using the FIRAS spectrometer data that the CIB was first detected at long wavelengths: $\lambda > 200\mu m$ (Puget et al., 1996). The CIB has also been detected by DIRBE at 2.4, 3.5, 100, 140, 240 μm (see Hauser & Dwek, 2001 for a review). The Cosmic background is also measured in the optical (Berstein et al., 2002). In the mid infrared the interplanetary zodiacal dust emission is so strong that only upper limits were obtained by DIRBE. The combination of number counts by ISOCAM at 15 μm (see Cesarsky & Elbaz, 2003) and by *Spitzer* at 24 μm (Papovich et al., 2004) with the observations of TeV gamma ray emission from distant AGNs gives a good measurement of the background at these wavelengths. The full cosmic background spectrum is shown in figure 2 (from Gispert et al., 2000).

The optical and infrared cosmic backgrounds are well separated and the first surprising result is that the power in the infrared part is equal or larger than the power in the optical one although we know that locally, the infrared output of galaxies is only one third of the optical. This implies a much stronger evolution of the infrared luminosity of infrared galaxies than of optical ones.

A second important property to note is that the slope of the long wavelength part of the CIB: $B_\nu \simeq \lambda^{-2}$ is much less steep than the long wavelengths spectrum of galaxies. This implies that the millimeter CIB is not due to the millimeter emission of the galaxies making the bulk of the emission at the peak of the CIB ($\simeq 170\mu m$).The millimeter part of the CIB must thus be dominated by galaxies at rather high redshift for which the SED peak has been shifted to the sub-millimeter. The millimeter CIB contains information on the total energy output by these galaxies and about their spatial distribution which cannot be reached by observation of individual galaxies before very deep and large surveys are available at these wavelength which probably requires ALMA. The implications in terms of energy output have been drawn by Gispert et al., 2000. The infrared production rate per comoving unit volume (1) evolves faster between redshift zero and 1 than the optical one and (2) has to stay constant at higher redshifts up to redshift 3 at least.

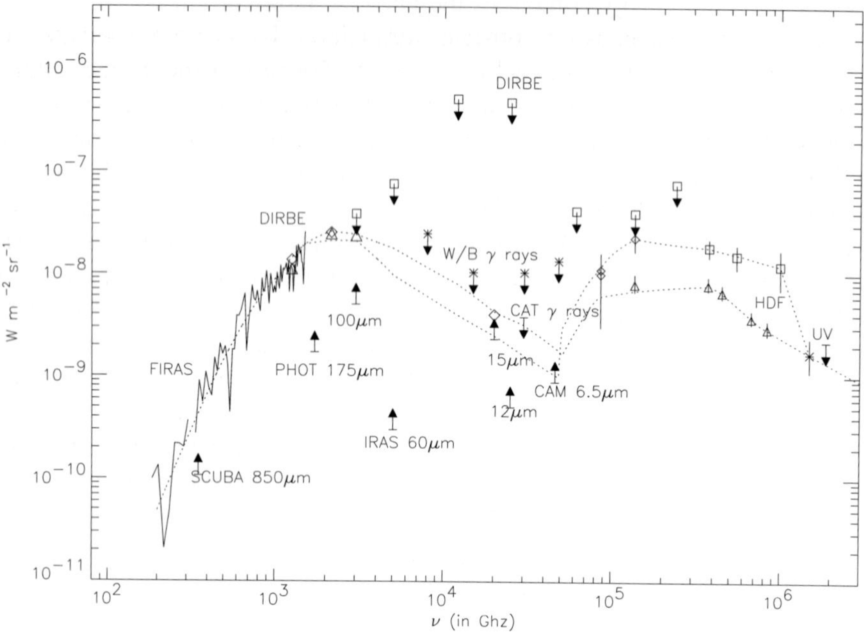

Figure 2. Cosmic Background from the UV to the millimeter wavelength (from Gispert et al., 2000)

4. The luminosity function of infrared galaxies and its evolution

The first deep ISOCAM surveys in the Hubble Deep Field brought spectacular results (see Cesarsky & Elbaz, 2003 for a review):

- A rather small density of sources detected at 15 μm had an average brightness which was about 20% of the optical brightness of the galaxies detected by the Hubble Space Telescope
- These infrared galaxies were filling more than half of the CIB at this wavelength when compared to the upper limits coming from the TeV Gamma rays
- Most of these sources were not associated with nearby galaxies
- The redshift distribution showed that they are concentrated in the redshift range 0.5 to 1 with a maximum at 0.8

- These sources are high luminosity sources with high infrared to optical ratio. The average luminosity was $3 \times 10^{11} L_\odot$.

This showed that the infrared output at redshift 0.8 is not dominated by the infrared part of the SED of the galaxies making the bulk of the optical output energy but by a small subclass in number of high luminosity infrared galaxies. These were similar to the local starburst galaxies detected by IRAS but they dominate the total output. The luminosity function is dominated by galaxies 30 times more luminous than the optical $L_\star$ galaxies. The peak in the redshift distribution results from the combination of two factors: the strong evolution in numbers of these galaxies and the negative K-correction due to the block of PAH features centered around 8 μm which peaks at redshift 0.8 for the 15 μm bandpass. The most luminous infrared galaxies and in some cases the anisotropies of the cosmic background were also detected by ISO at 170 μm, SCUBA at 850 μm and IRAM at 1.3 mm. Lagache et al., 2003 showed that combining (1) the number counts for the brightest sources and (2) the CIB spectrum and its anisotropies, constrains strongly the typical SED of infrared galaxies, the luminosity function and its evolution with redshift. Figure 3 shows the typical SED and figure 4 the luminosity function as a function of redshift. It has a number of implications:

- The infrared output energy is dominated by galaxies of luminosity increasing with redshift from $2 \times 10^{10} L_\odot$ locally to $3 \times 10^{11} L_\odot$ at z=1 and $3 \times 10^{12} L_\odot$ at redshift larger (2 to 3).

- The infrared galaxies are dominated by galaxies with a starburst type SED.

This last conclusion agrees with the spectroscopic studies done with *ISO* by eg. the Garching group who concluded that only 15% of the luminosity is due to AGN activity.

Very recently the first results from the *Spitzer* observatory brought already very interesting new elements on this question although much more is to be expected in the coming years. One observational aspect already mentioned above is the negative K-correction due to the PAH features. It was predicted by Lagache et al., 2003 that, if infrared galaxies keep a similar SED as the one they have up to redshift 1 up to redshift 2.5 the counts in the 24 μm *Spitzer*-MIPS band will be affected in a similar way as the 15 μm ISOCAM counts had been. The critical redshift range in which the main PAH block of features is redshifted in the 24 μm band is $2 < z < 3$. If the galaxies dominating the luminosity function are, as predicted, ultraluminous ones with $L \simeq 3 \times 10^{12} L_\odot$ this should lead to a maximum in the log N-logS plot (normalized to the Euclidean one) below 1 mJy. The observations from Papovich, show clearly a maximum at about 300 μJy. Figure 5 (from Lagache et al., 2004) shows how

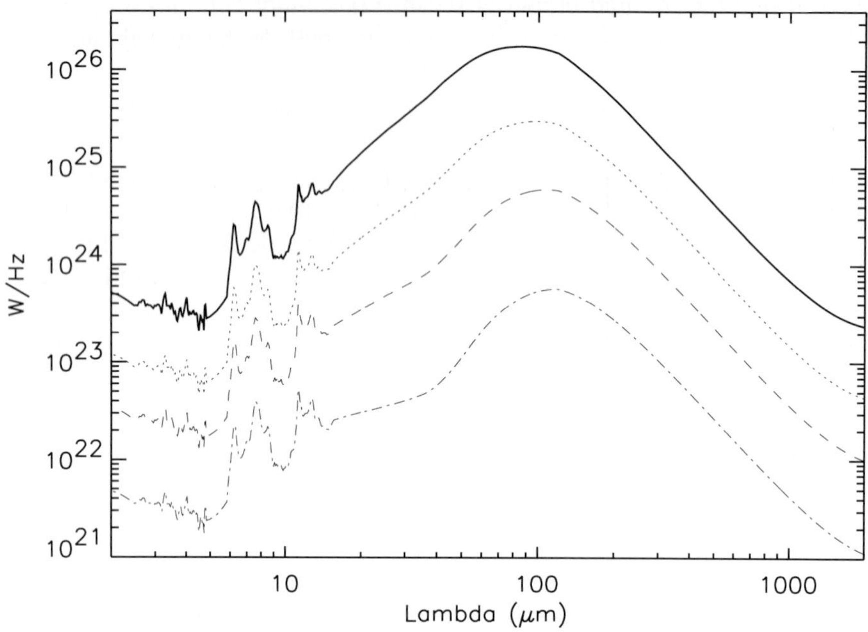

Figure 3. Spectral energy distribution of infrared galaxies. Starburst model spectra for different luminosities: L=3. 10^{12} $L_\odot$ (continuous line), L=5 10^{11} $L_\odot$ (dotted line), L=10^{11} $L_\odot$ (dashed line) and L=10^{10} $L_\odot$ (dotted-dashed line). From Lagache et al. 2003.

galaxies in different redshift bands contribute to the counts. The role of the galaxies with redshift around 2 is clear. This is an indirect indication that the SED has not changed very much between redshift 1 and 2. This needs to be confirmed by statistical spectroscopic studies sampling properly the dominant part of the luminosity function.

5. Open questions

If the presence of strong PAH features is confirmed in the typical spectra of infrared galaxies at redshift beyond 2 which dominate the energy output at this time, it will shed new light on the chemical evolution in galaxies and the early production of large amounts of organic molecules in the universe.

The other open question which is reinforced by the recent *Spitzer* results is related to the luminosity function of infrared galaxies. When going to larger redshifts the evidence is growing that the luminosity function is dominated by more luminous galaxies. As infrared galaxies dominate the energy output at these redshifts, this shows that the bulk of the star formation takes place in

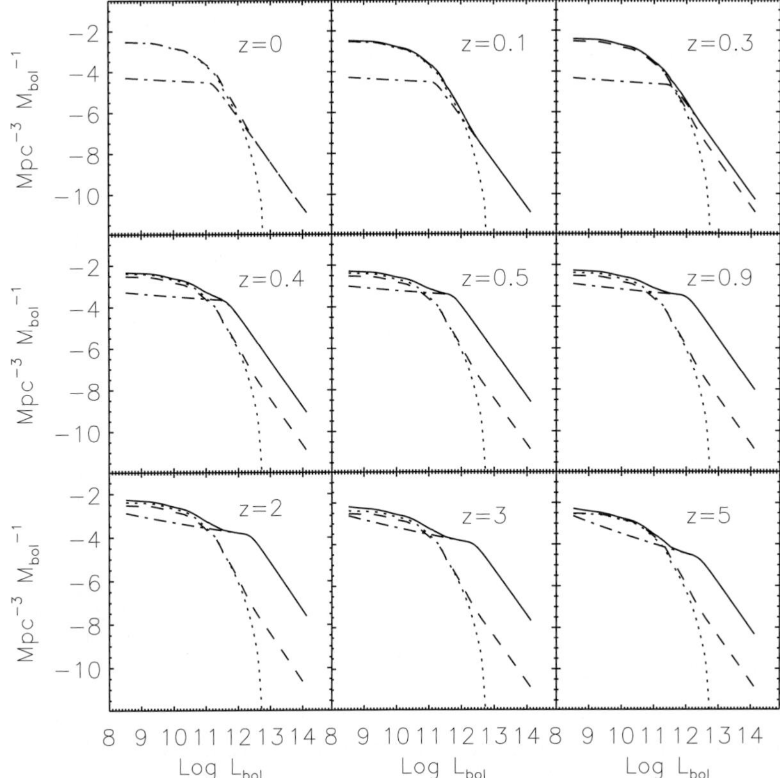

Figure 4. Co-moving evolution of the luminosity function. The dotted line is for the normal galaxies and the dotted-dashed line, for the starburst galaxies. The continuous lines corresponds to both starburst and normal galaxies and the dashed line is the LF at z=0 for comparison (from Lagache et al. 2003).

high luminosity objects. It is unlikely that such large luminosities could be associated with low mass galaxies. In fact the ISOCAM galaxies had been showed to be massive ones (Cesarsky & Elbaz, 2003).

In the standard model of structure and galaxy formation, dwarf galaxies are the first ones to collapse at high z and the mass function of collapsed objects is dominated by low mass galaxies with an evolution where the mass function is dominated more and more by massive galaxies as the universe expands. This opposite evolution of the mass function and of the luminosity function and of the existence of a class of galaxies with rather small number density contributing the largest fraction of the energy output has strong implications.

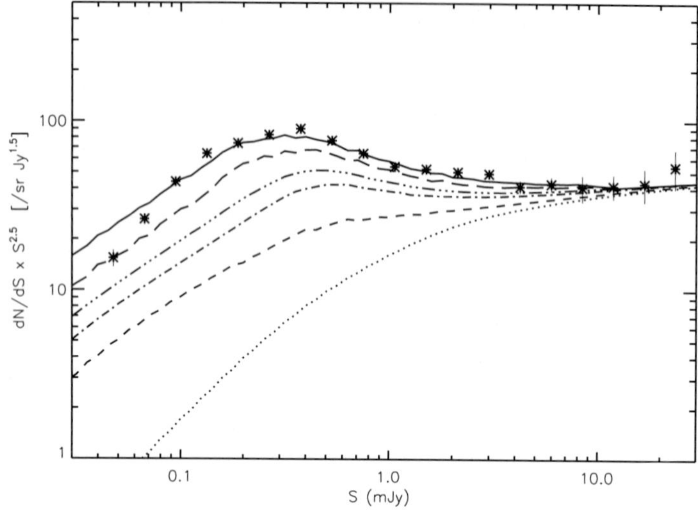

Figure 5. Redshift contribution to the number counts at 24 μm. The dot, dash, dash-dot, dash-3 dot, long-dash correspond to the number counts up to redshifts 0.3, 0.8, 1, 1.3 and 2 respectively. The continuous line is the whole redshift contribution. The model is from Lagache et al. (2004) and the data are from Papovich et al. (2004).

The star formation history in galaxies has probably three parts:

- Typically 40% of the radiation comes from galaxies following a smooth evolution with the energy output mainly in the optical and near IR,
- The second one is the starbust component concentrated in a small fraction of the lifetime during which the luminosity is 10 to 100 times larger and the output is mostly in the infrared.
- The third component is the one associated with the AGN activity which is probably also dominant only during limited periods but carries only 10 to 20% of the total radiated energy.

This picture is mostly constrained by what we now know on the cosmic background from galaxies and the identification of its dominant components.

Infrared galaxies are absent from the largest collapsed structures at present (galaxy clusters) thus are anti-biased with respect to the dark matter. They are often associated with mergers and if merging is a dominant mechanism for galaxy construction, the evolution with time of infrared galaxies is probably the best tracer we have of this activity. It is likely that at redshifts larger than 1, infrared galaxies show a strongly biased distribution with respect to the dark matter when structures and massive galaxies are just forming as merger rate should go as the square of the local density.

All of this needs to be investigated by studying the evolution of the luminosity function and the statistical properties of the spatial distribution of infrared galaxies. *Spitzer* will be great for such studies but will be strongly limited by confusion for redshift beyond 2-2.5. The study of the structure of the CIB is the tool to study the spatial distribution of infrared galaxies at larger redshifts with Planck and Herschel. The ALMA surveys will be needed to get the final answer to these questions.

Acknowledgments

It is a pleasure to thanks the sponsors of this Conference, the Anglo American Chairmans Fund, for their financial support of this J. Mayo Greenberg Lecture.

References

Bernstein R.A. ,Freedman W.L., Madore B.F., 2002, Ap.J. 571, 56

Boulanger F., et al, 2000, ESA-SP 455, 91

Cesarsky C. and Elbaz D., 2003, Science 300, 270

Genzel R. and Cesarsky C., 2000, Ann. Rev. Astron. Astrophys. 38, 761

Gispert R., Lagache G., Puget J. L., 2000, A&A 360, 1

Greenberg J.M., 1976, in Far Inrared astronomy, Oxford, Pergamon PressLtd, p299

Greenberg J.M., et al., 1972, Mem. Soc. Roy. Sci. Liège, 6e série, tomme III, 425.

Hauser M., and Dwek E., 2001 Ann. Rev. Astron. Astrophys. 37, 249

Lagache, G., Dole, H., and Puget, J.-L.2003, MNRAS 338, 551

Lagache G. , Dole H., Puget J.L. et al, 2004, Ap. J. Sup. in press

Léger A. and Puget J.L., 1984, A&A 137, L5

Li, A. and Greenberg, J. M., 2003, Solid state astrochemistry, NATO Science Series II: Mathematics, Physics and Chemistry, Vol. 120

Papovich C., Dole H., et al, 2004, Ap. J. Sup. in press

Partridge R. B. and Peebles P. J. E. , 1967, Ap.J. 148,377

Platt, 1956, ApJ, 123, 486

Puget J.L., Léger A. and Boulanger F., 1985, A&A 142, L19

Puget J.L., Abergel A., Bernard J.P., et al., 1996, A&A 308, L5

Sellgren, K., 1983, Ph. D. thesis

Willner et al., 1979, ApJ, 229, L65

THE PHYSICAL EVOLUTION OF MASS AND DUST IN DISTANT GALAXIES

Brigitte Rocca-Volmerange[1,2]
[1] *Institut d'Astrophysique de Paris, 98bis Bd Arago, F-75014 Paris, France,*
[2] *Universite Paris XI, Bât 333, F-91405 Orsay Cedex, France*

Abstract Main clues on galaxy evolution are derived independently from stellar masses in the K-band and from dust thermal emissions at 12μm, both for high-z populations of galaxies.

As a first result (Rocca-Volmerange et al, 2004), the most massive galaxies known as radio powerful ellipticals, trace in the K-band Hubble diagram the $10^{12}M_{\odot}$ critic mass of the fragmentation predicted by the dissipative self gravitational models (Rees & Ostriker, 1977). Because these massive galaxies were already formed at z =4, they required an extremely short mass accumulation duration, implying strong constraints on the two time scales $\tau_{freefall}$ and $\tau_{cooling}$. Finally the measured size of the cloud, ionized by these massive radio sources, is also compatible with the critic radius predicted by the theory.

On another wavelength domain dominated by dust, we are publishing the results of a deep survey observed at 12 μm with the ISOCAM camera (Rocca-Volmerange et al. and Seymour et al., preprints). The covered field is the ESO-Sculptor redshift survey (de Lapparent et al 1997) area. The limit flux, reaching 0.2mJy, allows to follow the evolution of the brightest galaxies and starbursts at high redshifts. The 12 μm source counts are interpreted by averaging inhomogeneities (the survey area is $\simeq$800 arcmin^2). We take into account cosmological and evolution corrections computed by a new version of the code PÉGASE.3 (Fioc et al, in preparation), extended to the far-IR. As a second result, the evolution scenarios which allowed to significantly interpret the multispectral faint galaxy counts in the optical, also give predictions in the far-infrared compatible with the 12 μm observations, making robust our scenarios of galaxy evolution. They confirm that the main evolving component in this wavelength range is largely dominated by actively star forming spiral galaxies. More investigations at high-z are required to understand if only the most massive galaxies form by dissipative gravitational collapse. In the present status, the current evolution scenarios of spiral galaxies remain compatible with both the gravitational collapse and/or the hierarchical merging. The perspectives of multi spectral surveys are built to solve this debated question.

Keywords: Galaxy: evolution- stellar content- high-redshift Infrared: galaxies

D. Block et al. (eds.), Penetrating Bars through Masks of Cosmic Dust, 523–532.

1. Introduction

The general view that galaxies form as gas collapses into dark matter haloes, shock heats, is characterized by two different time-scales of respectively radiative cooling and self-gravitation. Star formation appears by gas condensation, regulated by supernovae heating and massive black holes. Gas exchanges may also influence by infall, cooling-flows or extra galactic winds, the mass accumulation, gas density and the star formation activity. This classical scenario characterized by the two main physical time-scales ($\tau_{freefall}$ and $\tau_{cooling}$) has been estimated by Rees and Ostriker, 1977 and Silk, 1977. Then other fundamental processes of interactions, mainly guided by the CDM cosmological theory (Blumenthal et al, 1984), proposed the evolution by hierarchical merging of massive haloes collapsed from the amplified primordial fluctuations. Between the two models, time-scales of galaxy mass evolution strikingly differ. In particular, the mass accumulation by merging of normal nearby galaxies is predicted at a peak of star formation around z=1 (i.e. time-scales of $\simeq$ 10Gyrs) while the bulk of stellar population of the old elliptical galaxies would only require a time-scale $\leq$1Gyr to form. Moreover the nearby massive galaxies of $10^{12}M_{\odot}$ formed by recent merging process would appear blue at z=0 due to active star formation, once more incompatible with red elliptical galaxies. The evolved red populations of massive elliptical galaxies are better in agreement with dissipative self-gravitation models suffering an efficient cooling. Because of the legendary degeneracies: age-metallicity, metallicity-velocity dispersion and redshift-brightness, it has been difficult to identify the dominant galaxy processes which form nearby galaxies. The double gain of observing fainter galaxies and improving the spectral resolution of SEDs, give the hope to rapidly solve the debate for all types of galaxies.

The redshift range $0 < z < 4$, in particular for $z > 1.5$, arguably hosts the bulk of star formation and AGN activity in the history of the Universe, and may also be the era during which the Universe's properties change more rapidly than any time since reionization. During this epoch, the comoving space density of bright QSOs and of submm galaxies shows a strong peak. Stellar population of galaxies along the Hubble sequence also develops during this era, in particular the rapid star formation that characterizes the most massive elliptical galaxies at higher redshifts, then begins to decline and quiescent development of spiral disks become possible. Our knowledge of the properties of galaxies during this epoch is in rapid progress both at lower redshifts (and high spectral resolution) and at higher redshifts (and low spectral resolution). Paradoxally it appears easier to spectroscopically analyze the high-z galaxies at $z > 3$ than galaxies in the redshift desert $1.4 < z < 2.6$, due to the lack of strong emission lines in the optical window. A rapidly increasing sample of high-z galaxies has been discovered at $z > 3$ with the help of their huge emission lines. Most of them are embedded in host elliptical galaxies, populated with evolved stellar populations and may also show evidences of recent star formation activities: the most distant radio galaxies as 4C41.17 (z=3.8) show traces of recent star formation (van Breugel, et al, 1998). The most powerful telescopes (VLT, Keck and HST) are now able to discover stellar populations at redshifts $z > 4$. With robust galaxy evolution models, we may measure ages and masses of these populations, and the corresponding baryonic mass of their neutral gas reservoir.

Hereafter, we present two new studies: one is constrained by the measurements of evolved stellar masses, up to z $\simeq$ 4 and predicts the corresponding baryonic masses of galaxies at the highest epochs. The other aims to fit faint source counts of the deep ISOCAM survey (Rocca-Volmerange et al, in preparation) observed at 12μm (LW10 filter) at the depth of the Guaranted-Time Deep Survey (Elbaz et al, 1999).

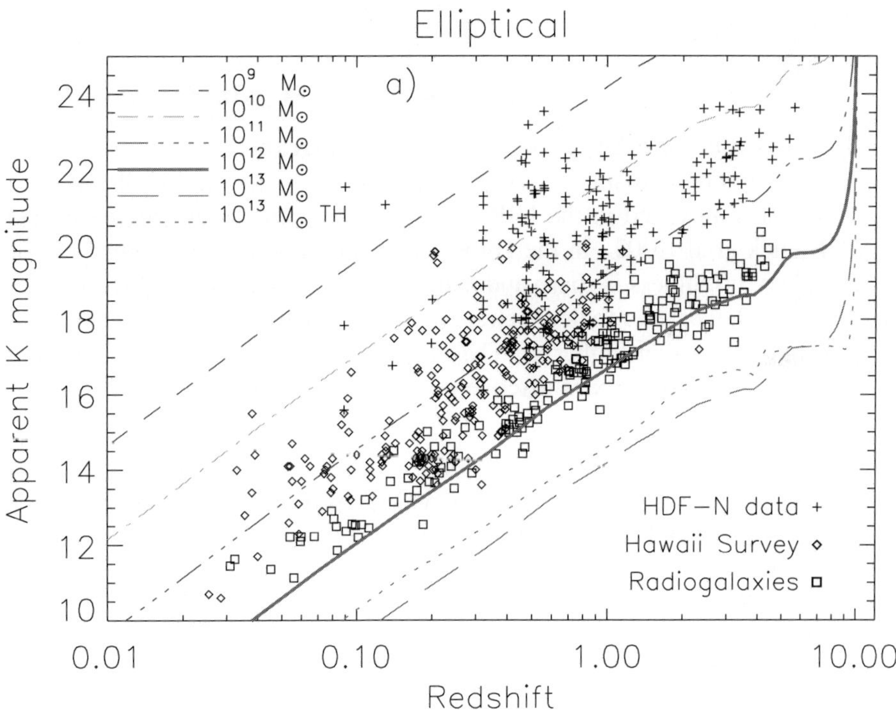

Figure 1. Observations of galaxies and radio galaxies in the K-band and comparison with predictions of PEGASE models of elliptical galaxies for various values of the input parameter M_{bar} (the mass of the initial reservoir): the K-z limit fits the 10^{12}M$_\odot$ model, a theoretical limit of fragmentation (Rocca-Volmerange et al, 2004).

We performed the analysis of the K-band and 12μm photometry by using our spectrophotometric evolutionary model PEGASE, well calibrated on spectral energy distribution (SEDs) of nearby templates (Fioc & Rocca-Volmerange, 1999). From the main SED features (cut-off, slope changes), it becomes possible to predict valid photometric redshifts (Le Borgne & Rocca-Volmerange, 2002) which are in excellent agreement with spectroscopic redshifts, up to z=4. That is the first evidence that we may be confident in our evolution scenarios. Moreover the multi-spectral faint galaxy counts, including the deepest HDFs and Hawaian surveys, were also reproduced from the far-UV to the near-Infrared (Fioc & Rocca-Volmerange, 1999) with these scenarios confirming their robustness. The new version of our code PEGASE , already proposed to the community (http://www.iap.fr/pegase), has been recently extended to the far-IR, allowing to follow the dust-metal component and the nebular emission in coherency with the stellar component on a large scale of wavelengths. Objectives are to understand the main parameters of galaxy formation and the evolution scenarios by types respecting the multi spectral emissions (stars+ ionized gas+ dust). The main ouputs will be the variation of stellar , baryonic, dust and ionized gas masses as a function of time.

2. Observational evidences of 0< z <4 galaxy masses

1 The bright limit from the K-z Hubble diagram

The sample of distant field and radio galaxies, including all the deep samples observed in the NIR K-band photometry is assembled by De Breuck et al, 2002. In the Hubble K-band diagram, the upper limit of galaxy brightness is sharp and mainly traced by nearby massive galaxies at $z \simeq 0$, whatever their spectral types, and by hosts of the most powerful radio galaxies at higher z. The most distant sample , up to z >4, are powerful radio galaxies discovered with the largest telescopes (VLT and Keck).

The main feature of the so-called K-z relation of radio galaxies is its continuity on the $0 < z < 4$ range, which has been puzzling for years. By varying either the cosmological parameters (H_0, q_0 and Λ_0) (Longair & Lilly , 1984, and more recently Inskip et al, 2002) or radio powers from the 3CR to the 6C (Eales & Rawlings, 1996; Eales et al, 1997) and to the 7C (Willott et al, 2003) catalogues, several tentatives of interpretations only justified the dispersion of the sequence but they do not explain the extended coverage on the $0 < z < 4$ range, neither the physical origin of the sequence.

2 Sizes and masses of the most distant radio sources

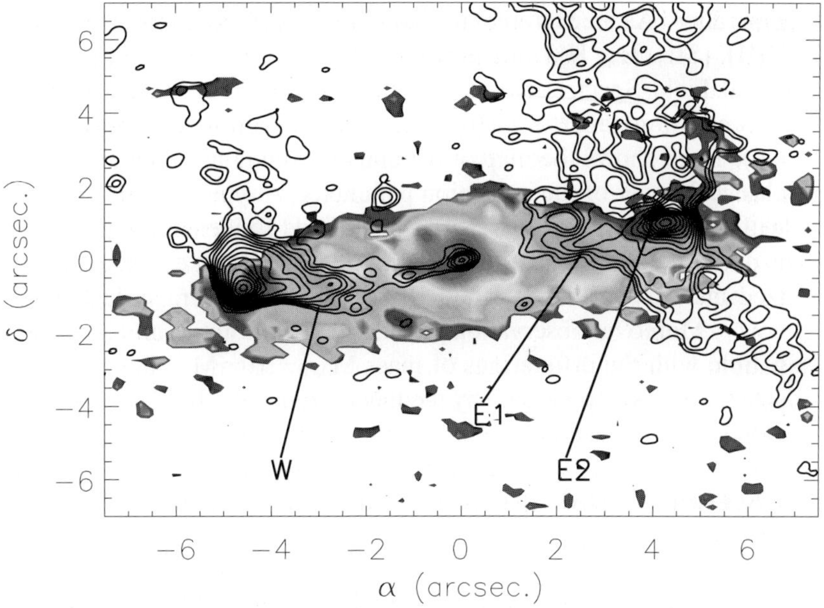

Figure 2. Map of the emission line ratio [OIII]5007A/H_β for the galaxy 3C171 (z=0.238) with the integral field spectrograph OASIS/CFHT (Rocca-Volmerange & Moy, 2003). High resolution VLA isophotes (Hardcastle, 2003) are superimposed. The ionisation parameter regularly decreases from the center to the boundaries, favoring photoionization in the central part and shocks at external boundaries. The limits of the ionized area are characteristics of the overpressured cocoon predicted by Begelman & Cioffi, 1989.

Radio galaxy luminosities are generally high, typically $L \simeq 3$ to $5\,L_*$ for both radio-loud and radio-quiet galaxy hosts (Papovich et al. 2001). Radio galaxy hosts are located in place of elliptical galaxies in the fundamental plane (Kukula et al. 2002). At high z, mass estimates of radio galaxy hosts are rare and generally estimated to a few $10^{11} M_\odot$ without evolution. For extended radio galaxies, the bulk of star formation could happen during the free-fall time while the active nucleus is fueled. The black hole and the stellar mass would then simultaneously grow. Another possibility is based on the alignement of the radio and UV-optical axes, evoking star formation triggered by the impact of the jet on the intervening gas clouds. Before to reach the possibility to detect in-situ star formation sites, we can tentatively explain the origin of the ionised gas in the huge lobes of distant radio galaxies with the help of specific diagnostic diagrams. Maps of the local ionized gas of galaxies are built by using Integral Field Units (IFU) settled on the most powerful telescopes (VLT). Emission lines are extracted from all the individual spectra. The local identification of several lines allow to build local emission line ratios. As an example, we built maps of the main lines ([OIII],[OI], [NII], [SII], H_α, H_β) of the radio galaxy 3C171 (z=0.238) with the Integral Field Spectrograph OASIS when it was still mounted at CFHT (Rocca-Volmerange & Moy, 2003, Rocca-Volmerange et al. submitted). Fig.2 shows the map of the emission line ratio [OIII]5007A/H_β, a classical tracer of the ionisation parameter. A typical symetrical cocoon with sharp boundaries is clearly identified. Evidence of star formation is not quite confirmed along the radio jet. A rough estimate of the ionized mass reachs about a few $10^{12}M_\odot$ with an average density of 100 cm^{-3}. Moreover observations of distant radio sources are not only compatible with the estimations of mass $M_{crit}= 10^{12}M_\odot$, but also with the size r_{crit}=75kpc predicted by the theory, typically the size of the ionized cloud.

3. Scenarios and time-scales of mass evolution

Scenarios of galaxy evolution are built with the help of the code of spectral evolution PEGASE (free access) . Star formation rate is a gas density dependent power law with time-scales of gas consumption increasing from ellipticals (star formation time scale of $\simeq$1Gyr) to spirals (star formation time scale of $\simeq$10 Gyrs). For all types of galaxies, time-scales of star formation are fitted to predict SEDs at z=0 in agreement with statistical observations. The passive evolution of stars is constrained by basic models of internal structures. The evolution of star formation rate is regulated by gas exchanges: infall from the initial gas reservoir and strong galactic winds from supernovae heating. Note that our scenarios do not assume further inflow of cool gas from the halo. Several IMFs were adopted, all of them will confirm our main results. More details of modeling are given in Rocca-Volmerange et al, 2004.Fig.3 and Fig. 4 present, respectively for elliptical (a) and spiral (b) galaxies, the evolution of the stellar mass $M_{\rm stars}$, gas mass $M_{\rm gas}$ and galaxy mass $M_{\rm gal}$ (stars + gas + remnants) as function of age and z with standard cosmology.

4. Interpretation of the K-z relation

We adopt the cosmological model of Friedmann-Lemaître with standard values of parameters H_0, $\Omega_{\rm M}$, Ω_Λ values: 65km.s^{-1}.Mpc^{-1}, 0.3, 0.7. Other cosmologies are analyzed in Rocca-Volmerange et al, 2004. Figure 1 compares the observed K-z distribution of galaxies with the predictions of models for ellipticals. Sequences of constant $M_{\rm bar}$, respectively from 10^9 to 10^{13} $M_\odot$, are computed for the two scenarios. The IMF is from Rana & Basu, (1992) for all sequences, except for the sequence 10^{13} $M_\odot$ also shown for the top-heavy IMF.

The galaxy distribution is entirely covered by the variety of K-z sequences with $10^9 < \mathrm{M_{bar}}/\mathrm{M_{sol}} < 10^{12}$. $\mathrm{M_{bar}}$ is the first parameter which drives the Hubble K-band diagram. Galaxies from the deep optical surveys are fitted either with $10^{10\ to 12}\,\mathrm{M}_\odot$ spirals or less massive $10^{9.5\ to 11}\,\mathrm{M}_\odot$ ellipticals. At $z > 0.3$, the predictions of high z spiral models are incompatible with observations. We show that the observed limit is continuously fitted by the sequence $\mathrm{M}_{bar,max} \simeq 10^{12}\mathrm{M}_\odot$ (Fig. 1), estimated with the galaxy evolution code PEGASE by following the evolving stellar mass in the K-band. $\mathrm{M}_{bar,max}$ is the initial baryonic mass of the reservoir cloud including infall and winds. The result is robust and is related to the physical interpretation of galaxy formation model by self-gravitational collapse regulated by cooling processes (Rees & Ostriker, 1977). $\mathrm{M}_{bar,max}$ is the critical mass M_{crit} of fragmentation, a limit

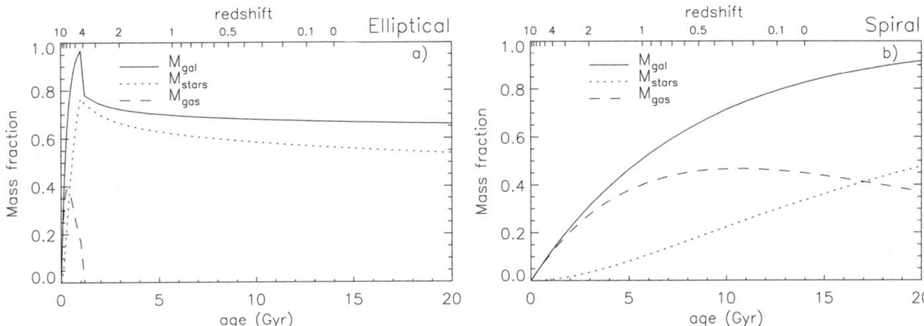

Figure 3. The evolution of stellar, gas and total masses of elliptical galaxies as a function of age and z. Normalisation at $1\mathrm{M}_\odot$ of the initial cloud. The formation redshift z_{for}=10. Star formation starts with infall at age 0 and stops with galactic winds at age 1 Gyr.

Figure 4. The evolution of stellar, gas and total masses of spiral galaxies as a function of age and z. Normalisation at $1\mathrm{M}_\odot$ of the initial cloud. The formation redshift is z_{for}=10. Star formation starts with infall at age 0. No galactic winds.

between the two regimes respectively dominated by gravitation and cooling. Because super massive black holes ($10^{8\ to\ 9}\mathrm{M}_\odot$) are observed in distant radio galaxies (Willott et al, 2003), the cooling process might be specifically efficient in these sources. Moreover the link of massive stellar populations with supermassive black holes is supported by observational relations between black hole masses and absolute blue luminosities or velocity dispersion (Ferrarese & Merritt 2000; Gebhardt et al. 2000; Magorrian et al. 1998). However, it is too early to identify the true nature of the relation between stellar and black hole masses (Silk and Rees, 1998).

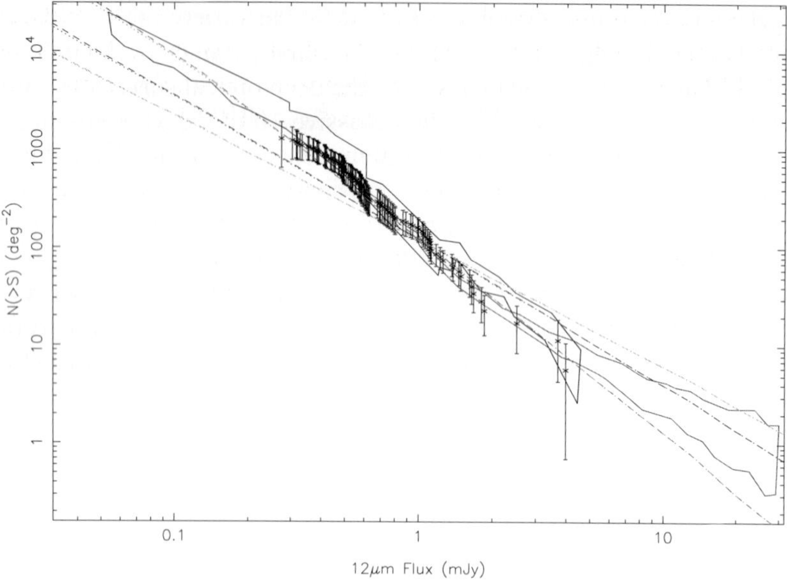

Figure 5. Faint counts in the ESO-Sculptor field at 12μm with ISOCAM (Rocca-Volmerange et al, and Seymour et al, in preparation) are compared to the previous results of deep surveys (black empty frame, Elbaz et al. 1999)- The comparison with the galaxy counts predicted by the evolutionary model PEGASE.3 (Fioc et al, 2004) confirms the rapid evolution of the spiral population. Several luminosity functions, all derived from Fang et al, 2000, are shown (various dashed-points), see text for details

5. Dust evolution from 12μm galaxy counts with ISOCAM

The mid-infrared is known to be an ideal wavelength domain to follow the evolution of dust produced by powerful starbursts, generally driven by galaxy merging and by the active nucleus at high redshifts. The dust evolution M_{dust}(t) in scenarios is fitted on metallicity evolution $Z(t)$.

Deep infrared surveys were performed with the camera ISOCAM (Cesarsky et al, 1996) on the ISO satellite. Elbaz et al, 1999 and references therein, present some preliminary results of galaxy counts at 15μm, down to the cluster Abell 2390 (Altieri et al, 1999). A deep mid-infrared survey at 12μm was observed on the exceptional area of 775 arcmin^2 in the ESO-Sculptor area. This field is also observed in deep CCD photometry BVR_c (Arnouts et al, 1997) and in the NIR (DENIS and 2MASS). Most of the ISO field is common to the redshift survey (de Lapparent et al, 2003). Two joined papers gather the galaxy counts (Rocca-Volmerange et al, in preparation) and the data reduction and analysis (Seymour et al, in preparation). The complete list of 142 sources detected with high significance is presented above an integrated flux of 0.2mJy.

The interpretation is now possible with a new version of the code PEGASE.3, extended to the dust emission in the far-IR domain (Fioc et al. 2004, in preparation). PEGASE.3 computes the cosmological and evolution corrections for all galaxy types in coherence with the optical and NIR emissions.

Fig. 5 presents the comparison of source counts from the ISO-ESS survey, with preliminary results of other surveys (Elbaz et al. 1999). The z=0 luminosity functions are from Rush et al, 1993, Fang et al, 2000 inside the error bars, the IRAS 12μm filter is equivalent to ISOCAM LW10 12μm filter. The predictions of faint galaxy counts at 12μm remarkably fit the ISO-ESS deep counts. The adopted fractions of galaxy types are similar to those used for the optical counts(Fioc & Rocca-Volmerange, 1999). The best fit of evolutionary models confirms the 12μm emission is mainly dominated by the starburst process, active in gas-rich spiral galaxies (star formation rate is proportionnal to the gas density). More details are given in Rocca-Volmerange et al. 2004.

6. Summary

We proposed two analyses of galaxy evolution, respectively based on the K-band and on 12μm wavelength domains. The brightest K band population is dominated by the bulk of low mass red stars of massive ellipticals of $10^{12}M_\odot$. At high redshifts, the evolution and k- corrections are taken into account. Several important results: i) The maximum luminosity sequence traces the fragmentation limit of galaxies ii) at z=4, radio galaxies of $10^{12}M_\odot$ already formed, confirming the theoretical collapse model for this galaxy population. Mass and size of radio galaxies tracing this extreme population, are in agreement with observations. This result favors the idea that active nucleus could play an efficient role in the dissipation process.

The evolution of the high redshift dust emission at 12μm from distant galaxies is interpreted in the ESO-Sculptor Survey, measured through the LW10 filter of the ISOCAM camera. The 12μm population is dominated by massive spiral populations, actively starbursting, according to the interpretation by PEGASE.3. The agreement of the two interpretations make our scenarios robust and constraining for galaxy formation models. Future multi spectral source counts will confirm these results.

References

Altieri B., Metcalfe, L., Kneib, J.P. et al. , 1999, A & A, 343, 65

Arnouts, S.; de Lapparent, V.; Mathez, G.; Mazure, A.; Mellier, Y.; Bertin, E.; Kruszewski, A., 1997, A& A Sup Series, 124, 163

Begelman, M.C. , Cioffi, D.F., 1989, ApJ, 345, 21

Blumenthal, G.R. Faber, S.M., Primack, J. R., Rees, M. J., 1984, Nature, 311, 517

Cesarsky, C. J.; Abergel, A.; Agnese, P.; Altieri, B.; Augueres, J. L.; Aussel, H.; Biviano, A.; Blommaert, J.; Bonnal, J. F.; Bortoletto, F.; and 56 coauthors, 1996, A&A, 315, 32

De Breuck, C., van Breugel, W., Stanford, S.A., et al., 2002, AJ, 123, 637
de Lapparent, V.; Galaz, G.; Bardelli, S.; Arnouts, S., 2003, A&A, 404, 831
de Lapparent, V.; Galaz, G.; Arnouts, S.; Bardelli, S.; Ramella, M. The Messenger (1997), Vol 89, pp 21-28
Dey, A., Spinrad, H., 1996, ApJ, 459, 133
Eales,S., Rawlings,S., 1996, ApJ, 460, 68
Eales, S., Rawlings,S., Law-Green, D., Cotter, G., Lacy, M, 1997, MNRAS, 291, 593
Elbaz, D.; Cesarsky, C. J.; Fadda, D.; Aussel, H.; Desert, F. X.; Franceschini, A.; Flores, H.; Harwit, M.; Puget, J. L.; Starck, J. L.; and 4 coauthors, 1999, A & A, 351, 37
Fang, F., Shupe, D. L.; Xu, C., Hacking, P., 1998, ApJ, 500, 693
Ferrarese, L., Merritt, D., 2000, ApJ, 539, L9
Fioc, M., Rocca-Volmerange, B., 1999, A&A, 351, 869
Gebhardt, K. et al., 2000, ApJ, 539, L13
Genzel, R., Baker, A., Tacconi, L., et al., 2003, ApJ, 584, 633
Hardcastle,M.J., 2003, MNRAS, 339, 360
Inskip, K., Best, P., Longair, M., et al., 2002, MNRAS, 329, 277
Kukula, M. J., Dunlop, J. S., McLure, R. J., 2002, NewAR, 46, 171
Lagache, G., Puget, J. L., Gispert, R., 1999, Ap&SS, 269, 263
Le Borgne, D., Rocca-Volmerange, B., 2002, A&A, 386, 446
Longair, M. S., Lilly, S.J., 1984, JApA, 5, 349
Magorrian, J., et al., 1998, AJ, 115, 2285
Papovich, C., Dickinson, M. Ferguson, H., 2001, ApJ, 559, 620
Rana, N. C., Basu, S., 1992, A&A, 265, 499
Rees, M., Ostriker, J., 1977, MNRAS, 179, 541
Rush, Brian; Malkan, Matthew A.; Spinoglio, Luigi, 1993, ApJS, 89, 1
Rocca-Volmerange, B., Moy, E., Adam, G., 2004, submitted
Rocca-Volmerange, B., Moy, E., 2003, NewAR, 47, 267
Rocca-Volmerange, B., Le Borgne D., De Breuck C., Fioc M., Moy E., 2004, A & A, 415, 931
Silk, J., 1977, ApJ, 211, 638
Silk, J., Rees, M.J., 1998, A & A, 331, L1
van Breugel, W., Stanford, S., Spinrad, H., et al., 1998, ApJ, 502, 614
Willott, C. J.; McLure, R. J.; Jarvis, M. J., 2003, ApJ, 587L, 15
Zirm, A. W., Dickinson, M., Dey, A., 2003, ApJ, 585, 90

THE WARM, COLD AND VERY COLD DUSTY UNIVERSE

Aigen Li
Theoretical Astrophysics Program, University of Arizona, Tucson, AZ 85721

Abstract We are living in a dusty universe: dust is ubiquitously seen in a wide variety of astrophysical environments, ranging from circumstellar envelopes around cool red giants to supernova ejecta, from diffuse and dense interstellar clouds and star-forming regions to debris disks around main-sequence stars, from comets to interplanetary space to distant galaxies and quasars.

These grains, spanning a wide range of sizes from a few angstroms to a few micrometers, play a vital role in the evolution of galaxies as an absorber, scatterer, and emitter of electromagnetic radiation, as a driver for the mass loss of evolved stars, as an essential participant in the star and planet formation process, as an efficient catalyst for the formation of H_2 and other simple molecules as well as complex organic molecules which may lead to the origins of life, as a photoelectric heating agent for the interstellar gas, and as an agent shaping the spectral appearance of dusty systems such as protostars, young stellar objects, evolved stars and galaxies.

In this review I focus on the dust grains in the space between stars (interstellar dust), with particular emphasis on the extinction (absorption plus scattering) and emission properties of *cold* submicron-sized "classical" grains which, in thermal equilibrium with the ambient interstellar radiation field, obtain a steady-state temperature of $\sim$15–25 K, *warm* nano-sized (or smaller) "ultrasmall" grains which are, upon absorption of an energetic photon, transiently heated to temperatures as high as a few hundred to over 1000 K, and the possible existence of a population of *very cold* (<10 K) dust. Whether dust grains can really get down to "temperature" less than the 2.7 K cosmic microwave background radiation temperature will also be discussed. The robustness of the silicate-graphite-PAHs interstellar dust model is demonstrated by showing that the infrared emission predicted from this model closely matches that observed for the Milky Way, the Small Magellanic Cloud, and the ringed Sb galaxy NGC 7331.

Keywords: Interstellar dust – extinction – absorption – infrared emission

1. Introduction

The space between the stars (interstellar space) of the Milky Way Galaxy and other galaxies is filled with gaseous ions, atoms, molecules (interstellar gas) and tiny dust grains (interstellar dust). The first direct evidence which pointed to the existence of interstellar gas came from the ground-based de-

D. Block et al. (eds.), Penetrating Bars through Masks of Cosmic Dust, 535–559.

tection of Na and Ca^+ optical absorption lines (Hartmann 1904).[1] This did not gain wide acceptance until Struve (1929) showed that the strength of the Ca^+ K-line was correlated with the distance of the star. The true interstellar nature of this gas was further supported by the detection of the first interstellar molecules CH, CH^+ and CN (Swings & Rosenfeld 1937, McKellar 1940, Douglas & Herzberg 1941). The presence of dark, obscuring matter in the Milky Way Galaxy was also recognized at the beginning of the 20th century (e.g. see Barnard 1919). Trumpler (1930) first convincingly showed that this "obscuring matter" which dims and reddens starlight consists of small solid dust grains. The dust and gas are generally well mixed in the interstellar medium (ISM), as demonstrated observationally by the reasonably uniform correlation in the diffuse ISM between the hydrogen column density $N_{\rm H}$ and the dust extinction color excess or reddening $E(B-V) \equiv A_B - A_V$: $N_{\rm H}/E(B-V) \approx 5.8 \times 10^{21}\,{\rm mag^{-1}\,cm^{-2}}$ (Bohlin et al. 1978), where A_B and A_V are the extinction at the B (λ=4400 A) and V (λ=5500 A) wavelength bands. >From this correlation one can estimate the gas-to-dust ratio to be $\sim$210 in the diffuse ISM.[2] Despite its tiny contribution to the total mass of a galaxy,[3] interstellar dust has a dramatic effect on the physical conditions and processes taking place within the universe, in particular, the evolution of galaxies and the formation of stars and planetary systems (see the introduction section of Li & Greenberg 2003).

In this review I will concentrate on the radiative properties of interstellar dust. I will distinguish the dust in terms of 3 components: **cold dust** with steady-state temperatures of $15\,{\rm K} \lesssim T \lesssim 25\,{\rm K}$ in thermal equilibrium with the solar neighbourhood interstellar radiation field, **very cold dust** with $T < 10\,{\rm K}$ (typically $T \sim 5\,{\rm K}$), and **warm dust** undergoing "temperature spikes" via single-photon heating. I will first summarize in §2 the observational constraints on the physical and chemical properties of interstellar dust. I will then discuss in §3 the heating and cooling properties of the warm and cold interstellar dust components. In §4 I will demonstrate the robustness of the silicate-graphite-PAHs interstellar grain model by showing that this model closely reproduces the observed infrared (IR) emission of the Milky Way, the Small Magellanic Cloud, and the ringed Sb galaxy NGC 7331. The possible existence of a population of very cold dust (with $T < 10\,{\rm K}$) in interstellar space will be discussed in §5. Whether ultrasmall grains can really cool down to $T < 2.7\,{\rm K}$ and appear in absorption against the cosmic microwave background (CMB) radiation will be discussed in §6.

2. Observational Constraints on Interstellar Dust

Our knowledge of interstellar dust regarding its composition, size and shape is mainly derived from its interaction with electromagnetic radiation: attenuation (absorption and scattering) and polarization of starlight, and emission of IR and far-IR radiation. The principal observational keys, both direct and indirect, used to constrain the properties of dust are the following:

- **(1) Interstellar Extinction.** Extinction is a combined effect of absorption and scattering: a grain in the line of sight between a distant star and the observer reduces the starlight by a combination of scattering and absorption (the absorbed energy is then re-radiated in the IR and far-IR).
  - ñ The wavelength dependence of interstellar extinction – "interstellar extinction curve", most commonly determined from the "pair-method",[4] rise from the near-IR to the near-UV, with a broad absorption feature at about $\lambda^{-1}\approx 4.6\,\mu\mathrm{m}^{-1}$ ($\lambda\approx 2175$ A), followed by a steep rise into the far-UV $\lambda^{-1}\approx 10\,\mu\mathrm{m}^{-1}$. $\longrightarrow$ **The extinction curve tells us the size (and to a less extent, the composition) of interstellar dust.** Since it is generally true that a grain absorbs and scatters light most effectively at wavelengths comparable to its size $\lambda\approx 2\pi a$ (Krugel 2003), there must exist in the ISM a population of large grains with $a\gtrsim\lambda/2\pi\approx 0.1\,\mu\mathrm{m}$ to account for the extinction at visible wavelengths, and a population of ultrasmall grains with $a\lesssim\lambda/2\pi\approx 0.016\,\mu\mathrm{m}$ to account for the far-UV extinction at λ=$0.1\,\mu\mathrm{m}$ (see Li 2004a for details).
  - ñ The optical/UV extinction curves show considerable regional variations. $\longrightarrow$ **Dust on different sightlines have different size distributions (and/or different compositions).**
  - ñ The optical/UV extinction curve in the wavelength range of $0.125 \leq\lambda\leq 3.5\,\mu\mathrm{m}$ can be approximated by an analytical formula involving only one free parameter: $R_V\equiv A_V/E(B-V)$, the total-to-selective extinction ratio (Cardelli, Clayton, & Mathis 1989), whereas the near-IR extinction curve ($0.9\,\mu\mathrm{m}\leq\lambda\leq 3.5\,\mu\mathrm{m}$) can be fitted reasonably well by a power law $A(\lambda)\sim\lambda^{-1.7}$, showing little environmental variations.

    The Galactic mean extinction curve is characterized by $R_V\approx 3.1$. Values of R_V as small as 2.1 (the high latitude translucent molecular cloud HD 210121; Larson, Whittet, & Hough 1996; Li & Greenberg 1998) and as large as 5.6 (the HD 36982 molecular cloud in the Orion nebula) have been observed in the Galactic regions. More extreme extinction curves have been reported for extragalactic objects.[5] $\longrightarrow$ **The optical/UV extinction curve and the value of R_V depend on the environment**: lower-density regions have a smaller R_V, a stronger 2175 A hump and a steeper far-UV rise ($\lambda^{-1} > 4\,\mu\mathrm{m}^{-1}$), implying smaller grains in these regions; denser regions have a larger R_V, a weaker 2175 A hump and a flatter far-UV rise, implying larger grains.
  - ñ In the Small Magellanic Cloud (SMC), the extinction curves of most sightlines display a nearly linear steep rise with λ^{-1} and extremely weak or absent 2175 A hump (Lequeux et al. 1982; Prevot

et al. 1984). $\longrightarrow$ **Grains in the SMC are smaller than those in the Milky Way diffuse ISM as a result of either more efÝcient dust destruction in the SMC due to its harsh environment of the copious star formation associated with the SMC Bar or lack of growth due to the low-metallicity of the SMC, or both.** Regional variations also exist in the SMC extinction curves.[6]

ñ The Large Magellanic Cloud (LMC) extinction curve is characterized by a weaker 2175 A hump and a stronger far-UV rise than the average Galactic extinction curve (Nandy et al. 1981; Koornneef & Code 1981).[7]

- **(2) Interstellar Polarization.** For a non-spherical grain, the light of distant stars is polarized as a result of differential extinction for different alignments of the electric vector of the radiation.

ñ The interstellar polarization curve rises from the IR, has a maximum somewhere in the optical and then decreases toward the UV. $\longrightarrow$ **This tells us that (1) some fraction of interstellar grains are both non-spherical and aligned by some process; (2) the bulk of the aligned grains responsible for the peak polarization (at $\lambda \approx$0.55 μm) have typical sizes of $a \approx \lambda/2\pi \approx$0.1 μm; and (3) the ultrasmall grain component responsible for the far-UV extinction rise is either spherical or not aligned.**

ñ The optical/UV polarization curve $P(\lambda)$ can be closely approximated by the "Serkowski law", an empirical function: $P(\lambda)/P_{\rm max}$ $= \exp[-K \ln^2(\lambda/\lambda_{\rm max})]$, where the only one free parameter, $\lambda_{\rm max}$, is the wavelength where the maximum polarization $P_{\rm max}$ occurs, and K is the width parameter: $K \approx 1.66 \lambda_{\rm max} + 0.01$; $\lambda_{\rm max}$ is indicative of grain size and correlated with R_V: $R_V \approx (5.6 \pm 0.3)\lambda_{\rm max}$ ($\lambda_{\rm max}$ is in micron; see Whittet 2003). $\longrightarrow$ **The sightlines with larger $\lambda_{\rm max}$ are rich in larger grains and have larger R_V for their extinction curves.** A close correlation between $\lambda_{\rm max}$ and the size of the aligned grains [e.g., $\lambda_{\rm max} \approx 2\pi a(n-1)$ for dielectric cylinders of radius a and refractive index n] is predicted by interstellar extinction calculations.

ñ The near-IR (1.64 μm$<\lambda<$5 μm) polarization is better approximated by a power law $P(\lambda) \propto \lambda^{-\beta}$, with $\beta \simeq 1.8 \pm 0.2$, independent of $\lambda_{\rm max}$ (Martin & Whittet 1990, Martin et al. 1992).

- **(3) Interstellar Scattering.** Scattering of starlight by interstellar dust is revealed by reflection nebulae (dense clouds illuminated by embedded or nearby bright stars), dark clouds (illuminated by the general interstellar radiation field [ISRF]), and the diffuse Galactic light ("DGL"; starlight scattered off the general diffuse ISM of the Milky Way Galaxy illuminated by the general ISRF). The scattering properties of dust grains (albedo = ratio of scattering to extinction, and phase function) provide a means of constraining the optical properties of the grains and are therefore indicators of their size and composition and provide diagnostic tests for dust models.

ñ The albedo in the near-IR and optical is quite high ($\sim$0.6), with a clear dip to $\sim$0.4 around the 2175 A hump, a rise to $\sim$0.8 around $\lambda^{-1} \approx$6.6 μm^{-1}, and a drop to $\sim$0.3 by $\lambda^{-1}\approx$10 μm^{-1}; the scattering asymmetry factor almost monotonically rises from $\sim$0.6 to $\sim$0.8 from $\lambda^{-1}\approx$1 μm^{-1} to $\lambda^{-1}\approx$10 μm^{-1} (see Gordon 2004). $\longrightarrow$ **An appreciable fraction of the extinction in the near-IR and optical must arise from scattering; the 2175 A hump is an absorption feature with no scattered component (see §2.4); and ultrasmall grains are predominantly absorptive.**

ñ Surprisingly high near-IR albedo has been reported for several regions: $\sim$0.86 at the K$'$ band ($\lambda\approx$2.1 μm) for the prominent dust lane in the evil eye galaxy NGC 4826 (Witt et al. 1994), $\sim$0.9 at the K$'$ band for the dust in the M 51 arm (Block 1996), and $\sim$0.7 at the J ($\lambda\approx$1.26 μm) and H ($\lambda\approx$1.66 μm) bands and $\sim$0.6 at the K ($\lambda\approx$2.16 μm) band for the Thumbprint Nebula (Lehtinen & Mattila 1996), in comparison with $\sim$0.2 at λ=2.2 μm predicted by the conventional dust models (Draine & Lee 1984; Li & Greenberg 1997). $\longrightarrow$ This implies that for NGC 4826 and M 51 (1) a population of grains at least $\sim$0.5 μm in radii, twice as large as assumed by standard models, may exist in these environments and are responsible for the near-IR scattering; and/or (2) the measured high K$'$ surface-brightness and the deduced high albedo may in part be caused by the thermal continuum emission from stochastically heated ultrasmall grains (Witt et al. 1994; Block 1996). For the Thumbprint Nebula, the high near-IR albedo is readily explained by grain growth (larger-than-average grain sizes) and the accretion of an ice mantle (see Pendleton, Tielens, & Werner 1990; §8 in Li & Greenberg 1997).

ñ Scatterings of X-rays by interstellar dust have also been observed as evidenced by "X-ray halos" formed around an X-ray point source by small-angle scattering. The intensity and radial profile of the halo depends on the composition, size and morphology and the spatial distribution of the scattering dust particles (see Dwek et al. 2004 for a review). The total and differential cross sections for X-ray scattering approximately vary as a^4 and a^6, respectively, **the shape and intensity of X-ray halos surrounding X-ray point sources therefore provide one of the most sensitive constraints on the largest grains along the sightline,** while these grains are gray at optical wavelengths and therefore the near-IR to far-UV extinction modeling is unable to constrain their existence.

A recent study of the X-ray halo around Nova Cygni 1992 by Witt, Smith, & Dwek (2001) pointed to the requirement of large interstellar grains ($a\sim$0.25–2 μm), consistent with the recent Ulysses and Galileo detections of interstellar dust entering our solar system (Grun et al. 1994; Frisch et al. 1999; Landgraf et al. 2000). But Draine & Tan (2003) found that the silicate-graphite-PAH model with the dust size distributions derived from the near-IR to far-UV extinction modeling (Weingartner & Draine 2001a) and IR emission modeling (Li & Draine 2001b) is able to explain the observed X-ray halo.

- **(4) Spectroscopic Extinction and Polarization Features: The 2175 A Extinction Hump ñ the strongest spectroscopic extinction feature.**

ñ Its strength and width vary with environment while its peak position is quite invariant: the central wavelength of this feature varies by only $\pm$0.46% (2σ) around 2175 A (4.6 μm^{-1}), while its FWHM varies by $\pm$12% (2σ) around 469 A ($\approx$1 μm^{-1}).

ñ Its carrier remains unidentified 39 years after its first detection (Stecher 1965). It is generally believed to be caused by aromatic carbonaceous (graphitic) materials, very likely a cosmic mixture of polycyclic aromatic hydrocarbon (PAH) molecules (Joblin, Léger & Martin 1992; Li & Draine 2001b; Draine 2003a).

ñ For most sightlines, this feature is unpolarized. So far only two lines of sight toward HD 147933 and HD 197770 have a weak 2175 A polarization feature detected (Clayton et al. 1992; Anderson et al. 1996; Wolff et al. 1997; Martin, Clayton, & Wolff 1999). Even for these sightlines, the degree of alignment and/or polarizing ability of the carrier should be very small (see Li & Greenberg 2003).

ñ Except for the detection of scattering in the 2175 A hump in two reflection nebulae (Witt, Bohlin, & Stecher 1986), the 2175 A hump is thought to be predominantly due to absorption, suggesting its carrier is sufficiently small to be in the Rayleigh limit.

ñ The detections of this feature in distant objects have been reported by Malhotra (1997) in the composite absorption spectrum of 96 intervening Mg II absorption systems at $0.2<z<2.2$; by Cohen et al. (1999) in a damped Lyα absorber (DLA) at $z=0.94$; by Toft, Hjorth & Burud (2000) in a lensing galaxy at $z=0.44$; by Motta et al. (2002) in a lensing galaxy at $z=0.83$; and very recently by Wang et al. (2004) in 3 intervening quasar absorption systems at $1.4\lesssim z\lesssim 1.5$.

- **(5) Spectroscopic Extinction and Polarization Features: The 9.7 μm and 18 μm (Silicate) Absorption Features ñ the strongest IR Absorption features.**

ñ Ubiquitously seen in a wide range of astrophysical environments, these features are almost certainly due to silicate minerals: they are respectively ascribed to the Si-O stretching and O-Si-O bending modes in some form of silicate material (e.g. olivine $Mg_{2x}Fe_{2-2x}SiO_4$).

ñ The observed interstellar silicate bands are broad and relatively featureless. $\longrightarrow$ **Interstellar silicates are largely amorphous rather than crystalline.** Li & Draine (2001a) estimated that the amount of $a<1\,\mu$m crystalline silicate grains in the diffuse ISM is $<5\%$ of the solar Si abundance.[8]

ñ The strength of the 9.7 μm feature is approximately $\Delta\tau_{9.7\,\mu\rm m}/A_V\approx 1/18.5$ in the local diffuse ISM. $\longrightarrow$ **Almost all Si atoms have been locked up in silicate dust, if assuming solar abundance for the ISM.**[9]

ñ The 9.7 and 18 μm silicate absorption features are polarized in some interstellar regions, most of which are featureless. The only exception is AFGL 2591, a molecular cloud surrounding a young stellar object, which displays a narrow feature at 11.2 μm superimposed on the broad 9.7 μm polarization band, generally attributed to annealed silicates (Aitken et al. 1988).

- **(6) Spectroscopic Extinction and Polarization Features: The 3.4 μm (Aliphatic Hydrocarbon) Absorption Feature.**
  - ñ Widely seen in the diffuse ISM of the Milky Way Galaxy and other galaxies (e.g. Seyfert galaxies and ultraluminous infrared galaxies, see Pendleton 2004 for a recent review), this strong absorption band is attributed to the C-H stretching mode in **aliphatic hydrocarbon dust**. Its exact nature remains uncertain, despite 23 years' extensive investigation with over 20 different candidates proposed (see Pendleton & Allamandola 2002 for a summary). So far, the experimental spectra of hydrogenated amorphous carbon (HAC; Schnaiter, Henning & Mutschke 1999, Mennella et al. 1999) and the organic refractory residue, synthesized from UV photoprocessing of interstellar ice mixtures (Greenberg et al. 1995), provide the best fit to both the overall feature and the positions and relative strengths of the 3.42 μm, 3.48 μm, and 3.51 μm subfeatures corresponding to symmetric and asymmetric stretches of C–H bonds in CH_2 and CH_3 groups. Pendleton & Allamandola (2002) attributed this feature to hydrocarbons with a mixed aromatic and aliphatic character.
  - ñ The 3.4 μm band strength for interstellar aliphatic hydrocarbon dust is unknown. If we adopt a mass absorption coefficient of $\kappa_{\rm abs}(3.4\,\mu{\rm m}) \sim 1500\,{\rm cm}^2\,{\rm g}^{-1}$ (Li & Greenberg 2002), we would require ~ 68 ppm C to be locked up in this dust component to account for the local ISM 3.4 μm feature ($\Delta\tau_{3.4\,\mu{\rm m}}/A_V \approx 1/250$; Pendleton et al. 1994).
  - ñ **This feature is ubiquitously seen in the diffuse ISM while never detected in molecular clouds**. Mennella et al. (2001) and Munoz Caro et al. (2001) argue that this can be explained by the competition between dehydrogenation (destruction of C-H bonds by UV photons) and rehydrogenation (formation of C-H bonds by H atoms interacting with the carbon dust): in diffuse clouds, rehydrogenation prevails over dehydrogenation; in dense molecular clouds, dehydrogenation prevails over rehydrogenation as a result of the reduced amount of H atoms and the presence of ice mantles which inhibits the hydrogenation of the underlying carbon dust by H atoms while dehydrogenation can still proceed since the UV radiation can penetrate the ice layers.
  - ñ Whether the origin of the interstellar aliphatic hydrocarbon dust occurs in the outflow of carbon stars or in the ISM itself is a subject of debate. The former gains strength from the close similarity between the 3.4 μm interstellar feature and that of a carbon-rich protoplanetary nebula CRL 618 (Lequeux & Jourdain de Muizon 1990; Chiar et al. 1998). However, the

survival of the stellar-origin dust in the diffuse ISM is questionable (see Draine 1990).

- ñ So far, **no polarization has been detected for this feature** (Adamson et al. 1999).[10] Spectropolarimetric measurements for both the 9.7 μm silicate and the 3.4 μm hydrocarbon features for the same sightline would allow a direct test of the silicate core-hydrocarbon mantle interstellar dust model (Li & Greenberg 1997), since this model predicts that the 3.4 μm feature would be polarized if the 9.7 μm feature (for the same sightline) is polarized (Li & Greenberg 2002).

- **(7) Spectroscopic Extinction and Polarization Features: <u>The Ice Absorption Features.</u>**
  - ñ Grains in dark molecular clouds (usually with A_V>3 mag) obtain ice mantles consisting of H_2O, NH_3, CO, CH_3OH, CO_2, CH_4, H_2CO and other molecules (with H_2O as the dominant species), as revealed by the detection of various ice absorption features (e.g., H_2O: 3.1, 6.0 μm; CO: 4.67 μm; CO_2: 4.27, 15.2 μm; CH_3OH: 3.54, 9.75 μm; NH_3: 2.97 μm; CH_4: 7.68 μm; H_2CO: 5.81 μm; OCN^-: 4.62 μm; see Boogert & Ehrenfreund 2004 for a review).
  - ñ Polarization has been detected in the 3.1 μm H_2O, the 4.67 μm CO and 4.62 μm OCN^- absorption features (e.g. see Chrysostomou et al. 1996).

- **(8) <u>The Extended Red Emission: Dust Photoluminescence.</u>** The "Extended Red Emission" (ERE), widely seen in reflection nebulae, planetary nebulae, HII regions, the Milky Way diffuse ISM, and other galaxies, is a far-red continuum emission in excess of what is expected from simple scattering of starlight by interstellar dust (see Witt & Vijh 2004). The ERE is characterized by a broad, featureless band between $\sim$ 5400 A and 9500 A, with a width 600 A $\lesssim$ FWHM $\lesssim$ 1000 A and a peak of maximum emission at 6100 A $\lesssim \lambda_p \lesssim$ 8200 A, depending on the physical conditions of the environment where the ERE is produced.
  - ñ The ERE is generally attributed to photoluminescence (PL) by some component of interstellar dust, powered by UV/visible photons with a photon conversion efficiency $\eta_{PL} \gg 10\%$ (Gordon, Witt, & Friedmann 1998). $\longrightarrow$ **The ERE carriers are very likely in the nanometer size range** because nanoparticles are expected to luminesce efficiently through the recombination of the electron-hole pair created upon absorption of an energetic photon, since in such small systems the excited electron is spatially confined and the radiationless transitions that are facilitated by Auger and defect related recombination are reduced (see Li 2004a).
  - ñ The ERE carrier remains unidentified. Various candidate materials have been proposed, but most of them appear unable to match the observed ERE spectra and satisfy the high-η_{PL} requirement (Draine 2003a; Li & Draine 2002a; Li 2004a; Witt & Vijh 2004). Promising candidates include PAHs (d'Hendecourt et al. 1986) and silicon nanoparticles (Ledoux et al. 1998, Witt et al. 1998, Smith & Witt 2002), but both have their own problems (see Li & Draine 2002a).

- **(9) Spectroscopic Emission Features: The 3.3, 6.2, 7.7, 8.6, 11.3** μm **ìUnidentiÝed Infrared (UIR) Emission features.** The distinctive set of "UIR" emission features at 3.3, 6.2, 7.7, 8.6, and 11.3 μm are seen in a wide variety of objects, including planetary nebulae, protoplanetary nebulae, reflection nebulae, HII regions, photodissociation fronts, circumstellar envelopes, and external galaxies.
  - ñ **These ìUIRî emission features are now generally identiÝed as vibrational modes of PAHs** (Léger & Puget 1984; Allamandola, Tielens, & Barker 1985): C–H stretching mode (3.3 μm), C–C stretching modes (6.2 and 7.7 μm), C–H in-plane bending mode (8.6 μm), and C–H out-of-plane bending mode (11.3 μm).[11] **The relative strengths and precise wavelengths of these features are dependent on the PAH size and its ionization state** which is controlled the starlight intensity, electron density, and gas temperature of the environment (Bakes & Tielens 1994, Weingartner & Draine 2001b, Draine & Li 2001).
  - ñ **Stochastically heated by the absorption of a single UV/visible photon (Draine & Li 2001; Li 2004a), PAHs, containing** $\sim$ **45** ppm **C, account for** $\sim$ **20% of the total power emitted by interstellar dust in the Milky Way diffuse ISM (Li & Draine 2001b).**
  - ñ **The excitation of PAHs does not require UV photons;** long wavelength (red and far-red) photons are also able to heat PAHs to high temperatures so that they emit efficiently at the UIR bands. This is because **the PAH electronic absorption edge shifts to longer wavelengths upon ionization and/or as the PAH size increases.**[12]
  - ñ No polarization has been detected for the PAH emission features (Sellgren, Rouan, & Léger 1988).
  - ñ The **PAH absorption features** at 3.3 μm and 6.2 μm have been detected in both local sources and Galactic Center sources (Schutte et al. 1998; Chiar et al. 2000). The strengths of these features are in good agreement with those predicted from the astronomical PAH model (Li & Draine 2001b).
  - ñ **PAHs can be rotationally excited** by a number of physical processes, including collisions with neutral atoms and ions, "plasma drag", and absorption and emission of photons. It is shown that these processes can drive PAHs to rapidly rotate, with rotation rates reaching tens of GHz. The rotational electric dipole emission from these spinning PAH molecules is capable of accounting for the observed "anomalous" microwave emission (Draine & Lazarian 1998a,b; Draine 1999; Draine & Li 2004a).

- **(10) IR Emission from Interstellar Dust.** Interstellar grains absorb starlight in the UV/visible and re-radiate in the IR. The IR emission spectrum of the Milky Way diffuse ISM, estimated using the IRAS 12, 25, 60 and 100 μm broadband photometry, the DIRBE-COBE 2.2, 3.5, 4.9, 12, 25, 60, 100, 140 and 240 μm broadband photometry, and the FIRAS-COBE 110 μm$<\lambda<$3000 μm spectrophotometry, is characterized by a modified black-body of $\lambda^{-1.7}B_{\lambda}$(T=19.5 K) peaking at $\sim$130 μm in the wavelength range of 80 μm$\lesssim\lambda\lesssim$1000 μm, and a substantial amount of emission at $\lambda\lesssim$60 μm which far exceeds what would be expected from dust at $T\approx$20 K (see Draine 2003a). In addition, spectrometers aboard the IRTS (Onaka et al. 1996; Tanaka et al. 1996) and ISO (Mattila et al. 1996) have shown that the diffuse ISM radiates strongly in emission features at 3.3, 6.2, 7.7, 8.6, and 11.3 μm.

ñ The emission at $\lambda\gtrsim$60 μm accounts for $\sim$65% of the total emitted power. $\longrightarrow$ There must exist a population of **ìcold dustî** in the size range of $a>$250 A, heated by starlight to equilibrium temperatures 15 K$\lesssim T\lesssim$25 K and cooled by far-IR emission (see Li & Draine 2001b).

ñ The emission at $\lambda\lesssim$60 μm accounts for $\sim$35% of the total emitted power. $\longrightarrow$ There must exist a population of **ìwarm dustî** in the size range of $a<$250 A, stochastically heated by single starlight photons to temperatures $T\gg$20 K and cooled by near- and mid-IR emission (see Li & Draine 2001b; Li 2004a).

- **(11) Interstellar Depletions.** Atoms locked up in dust grains are "depleted" from the gas phase. The dust depletion can be determined from comparing the gas-phase abundances measured from optical and UV absorption spectroscopic lines with the assumed reference abundances of the ISM (total abundances of atoms both in gas and in dust; also known as "standard abundances", "interstellar abundances", and "cosmic abundances"). The total interstellar abundances are usually taken to be solar, although Snow & Witt (1996) argued that interstellar abundances are appreciably sub-solar ($\sim$70% solar). Interstellar depletions allow us to extract important information about the composition and quantity of interstellar dust:

ñ In low density clouds, Si, Fe, Mg, C, and O are depleted. $\longrightarrow$ **Interstellar dust must contain an appreciable amount of Si, Fe, Mg, C and O.** Indeed, all contemporary interstellar dust models consist of both silicates and carbonaceous dust.

ñ From the depletion of the major elements Si, Fe, Mg, C, and O one can estimate **the gas-to-dust mass ratio to be $\sim$165.**[13]

ñ In addition to the silicate dust component, there must exist another dust population, since **silicates alone are not able to account for the observed amount of extinction relative to H** although Si, Mg, and Fe are highly depleted in the ISM. Even if all Si, Fe, and Mg elements are locked up in submicron-sized silicate grains, they can only account for $\sim$60% of the total observed optical extinction.[14]

3. Warm and Cold Dust in the ISM

As summarized in §2, the interstellar extinction, polarization, scattering, the near, mid, and far-IR emission and the 3.3–11.3 μm PAH emission features point to the existence of two dust populations in interstellar space:

- There exists a population of large grains with $a \gtrsim 250$ Å. Illuminated by the general interstellar radiation field (ISRF), these grains, defined as ì**cold dust**î, obtain equilibrium temperatures of $15\,\text{K} \lesssim T \lesssim 25\,\text{K}$ and emit strongly at wavelengths $\lambda \gtrsim 60\,\mu$m. These grains are responsible for the near-IR/optical extinction, scattering, polarization and the emission at $\lambda \gtrsim 60\,\mu$m.

  The equilibrium temperature T for a large grain of spherical radius a is determined by balancing absorption and emission:

  $$\int_0^\infty C_{\rm abs}(a,\lambda) c u_\lambda d\lambda = \int_0^\infty C_{\rm abs}(a,\lambda) 4\pi B_\lambda(T) d\lambda \ , \qquad (1)$$

  where $C_{\rm abs}(a,\lambda)$ is the absorption cross section for a grain with size a at wavelength λ, c is the speed of light, $B_\lambda(T)$ is the Planck function at temperature T, and u_λ is the energy density of the radiation field. In Figure 1 we display these "equilibrium" temperatures for graphitic and silicate grains as a function of size in environments with various UV intensities.
- There exists a population of ultrasmall grains with $a \lesssim 250$ Å. These grains have energy contents smaller or comparable to the energy of a single starlight photon. As a result, a single photon can heat a very small grain to a peak temperature much higher than its "steady-state" temperature and the grain then rapidly cools down to a temperature below the "steady-state" temperature before the next photon absorption event. Stochastic heating by absorption of starlight therefore results in transient "temperature spikes", during which much of the energy deposited by the starlight photon is reradiated in the IR – because of this, we call these ultrasmall grains ì**warm dust**î.[15] These grains are responsible for the far-UV extinction and the emission at $\lambda \lesssim 60\,\mu$m (including the 3.3–11.3 μm PAH emission features), dominate the photoelectric heating of interstellar gas (see §2.5 in Li 2004a), and provide most of the grain surface area in the diffuse ISM.

  Since ultrasmall grains will not maintain "equilibrium temperatures", we need to calculate their temperature (energy) probability distribution functions in order to calculate their time-averaged IR emission spectrum. There have been a number of studies on this topic since the pioneering work of Greenberg (1968). A recent extensive investigation was carried out by Draine & Li (2001). We will not go into details in this review, but

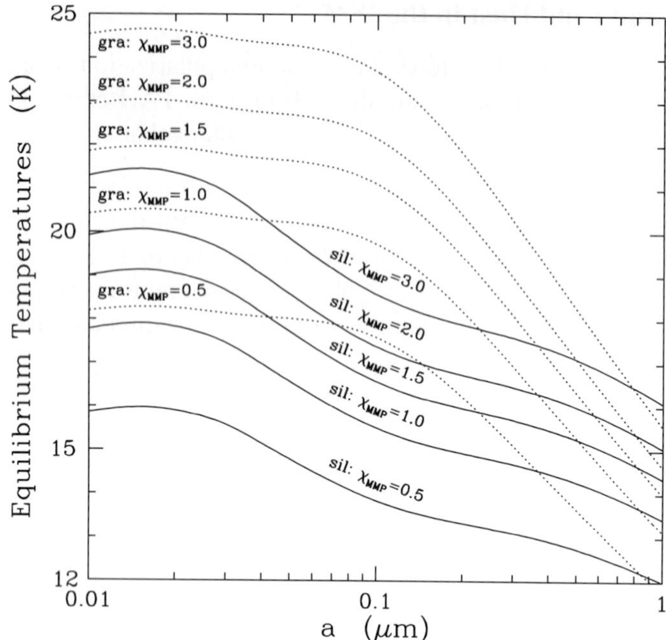

Figure 1. Equilibrium temperatures for graphite (dotted lines) and silicate grains (solid lines) in environments with various starlight intensities (in units of the Mathis, Mezger, & Panagia 1983 [MMP] solar neighbourhood ISRF). Taken from Li & Draine (2001b).

just refer those who are interested to Draine & Li (2001) and a recent review article of Li (2004a).

For illustration, we show in Figure 2 the energy probability distribution functions $dP/d\ln E$ (where dP is the probability that a grain will have vibrational energy in interval $[E, E + dE]$) for PAHs with radii $a = 5, 10, 25, 50, 75, 100, 150, 200, 300$A illuminated by the general ISRF. It is seen that very small grains ($a \lesssim 100$ A) have a very broad $P(E)$, and the smallest grains ($a < 30$ A) have an appreciable probability P_0 of being found in the vibrational ground state E=0. As the grain size increases, $P(E)$ becomes narrower, so that it can be approximated by a delta function for $a > 250$ A.[16] However, for radii as large as a=200 A, grains have energy distribution functions which are broad enough that the emission spectrum deviates noticeably from the emission spectrum for grains at a single "steady-state" temperature T, as shown in Figure 3. For accurate computation of IR emission spectra it is therefore important to properly calculate the energy distribution function $P(E)$, including grain sizes which are large enough that the average thermal energy content exceeds a few eV.

■ Nano-sized *interstellar* diamond and TiC grains have been identified in primitive meteorites based on isotopic anomaly analysis. Exposed to the general ISRF, these grains

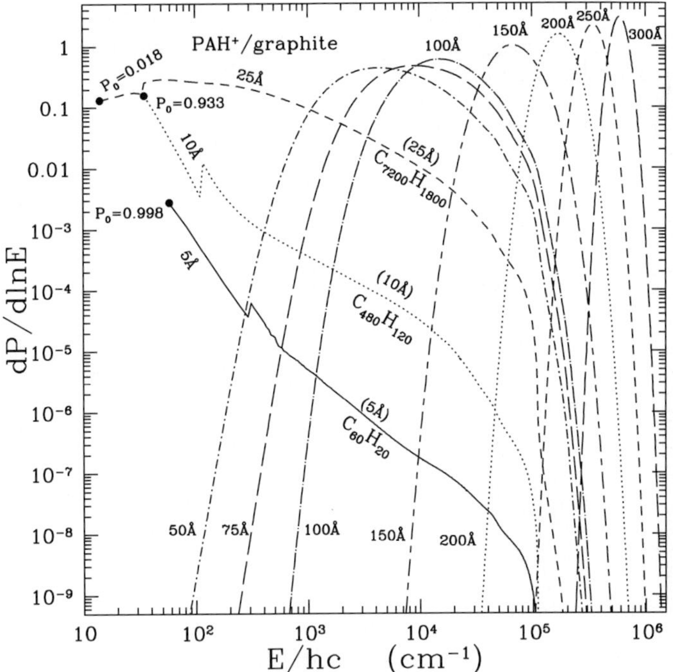

Figure 2. The energy probability distribution functions for charged carbonaceous grains (a = 5 A [$C_{60}H_{20}$], 10 A [$C_{480}H_{120}$], 25 A [$C_{7200}H_{1800}$], 50, 75, 100, 150, 200, 250, 300 A) illuminated by the general ISRF. The discontinuity in the 5, 10, and 25 A curves is due to the change of the estimate for grain vibrational "temperature" at the 20th vibrational mode (see Draine & Li 2001). For 5, 10, and 25A a dot indicates the first excited state, and P_0 is the probability of being in the ground state. Taken from Li & Draine (2001b).

are subject to single-photon heating and should of course be classified as "warm dust". But we should note that they are not representative of the bulk interstellar dust (see §5.4 and §5.5 in Li 2004a). The carriers of the ERE (see §2.8) and the 2175 A extinction hump (see §2.4) are also in the single-photon heating regime. Therefore, they can also be classified as "warm dust". As a matter of fact, the latter is attributed to PAHs (see §2.4). On the other hand, the Ulysses and Galileo spacecrafts have detected a substantial number of large interstellar grains with $a \gtrsim 1\,\mu$m (Grun et al. 1994), much higher than expected for the average interstellar dust distribution. In the ISM, these should of course be considered as "cold dust" or "very cold dust" with $T < 10$ K. But we note that the reported mass of these large grains is difficult to reconcile with the interstellar extinction and interstellar elemental abundances.

4. The Silicate-Graphite-PAHs Interstellar Dust Model

Various models have been proposed for interstellar dust (see Li & Greenberg 2003, Li 2004a, Draine 2004 for recent reviews). In general, these models fall into three broad categories: the silicate core-carbonaceous mantle model (Li

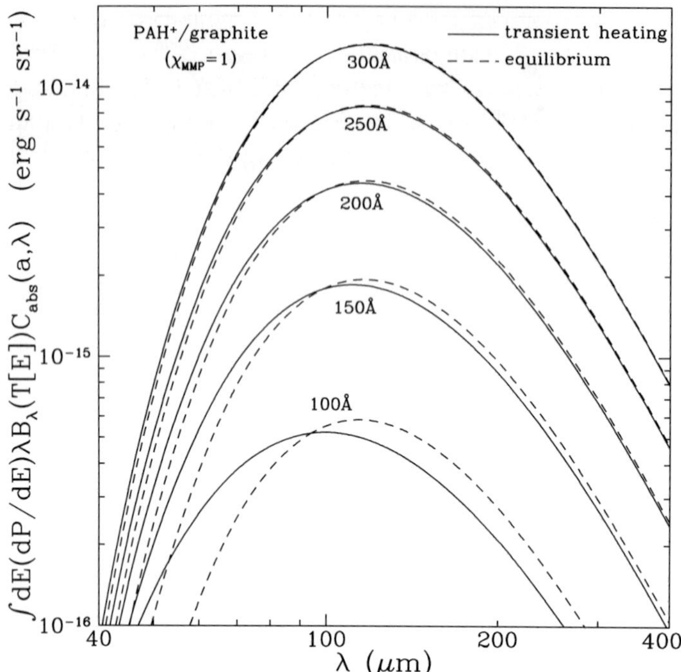

Figure 3. Infrared emission spectra for small carbonaceous grains of various sizes heated by the general ISRF, calculated using the full energy distribution function $P(E)$ (solid lines); also shown (broken lines) are spectra computed for grains at the "equilibrium" temperature T. Transient heating effects lead to significantly more short wavelength emission for $a \lesssim 200$ A. Taken from Li & Draine (2001b).

& Greenberg 1997), the silicate-graphite-PAHs model (Li & Draine 2001b, Weingartner & Draine 2001a) and the composite model (Mathis 1996, Zubko, Dwek & Arendt 2004). In this review I will focus on the IR emission calculated from the silicate-graphite-PAHs model and refer those who are interested in a detailed comparison between different models to my recent review articles (Li 2004a, Li & Greenberg 2003).

The silicate-graphite-PAHs model, consisting of a mixture of amorphous silicate dust and carbonaceous dust – each with an extended size distribution ranging from molecules containing tens of atoms to large grains $\gtrsim 1\ \mu$m in diameter, is a natural extension of the classical silicate-graphite model (Mathis, Rumpl, & Nordsieck 1977; Draine & Lee 1984). We assume that the carbonaceous grain population extends from grains with graphitic properties at radii $a \gtrsim 50$ A, down to particles with PAH-like properties at very small sizes.

With the temperature (energy) probability distribution functions calculated for ultrasmall grains undergoing "temperature spikes" and equilibrium temperatures calculated for large grains illuminated by the local ISRF, the silicate-graphite-PAHs model with grain size distributions consistent with the observed

R_V=3.1 interstellar extinction (Weingartner & Draine 2001a), is able to reproduce the observed near-IR to submillimeter emission spectrum of the diffuse ISM, including the PAH emission features at 3.3, 6.2, 7.7, 8.6, and 11.3 μm. This is demonstrated in Figure 4 and Figure 5 for the high-latitude "cirrus" cloud and 2 regions in the Galactic plane (see Li & Draine 2001b for details).

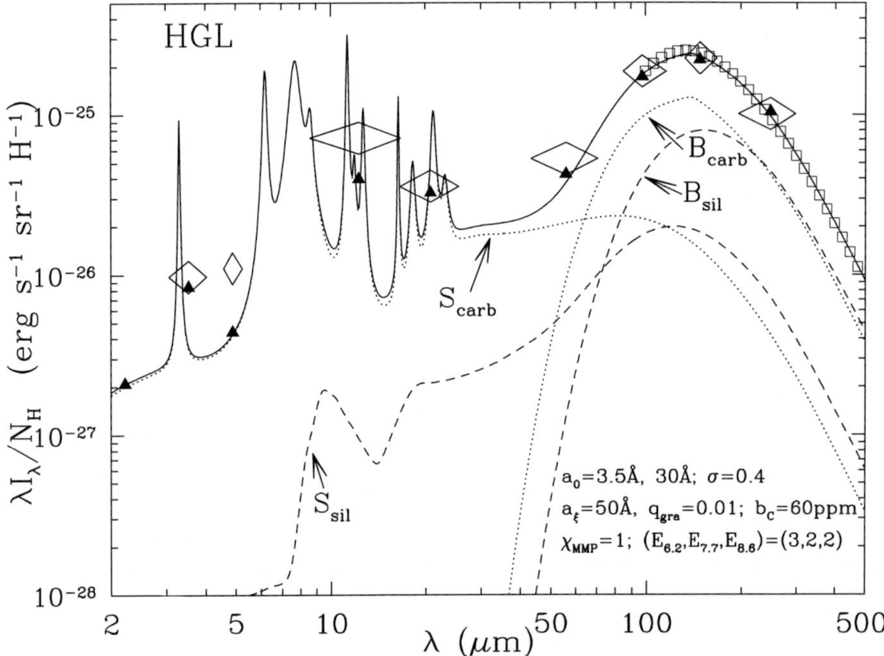

Figure 4. Comparison of the model to the observed emission from the diffuse ISM at high galactic latitudes ($|b| \geq 25^\circ$). Curves labelled B_{sil} and B_{carb} show emission from "big" ($a \geq 250$ A) silicate and carbonaceous grains; curves labelled S_{sil} and S_{carb} show emission from "small" ($a < 250$ A) silicate and carbonaceous grains (including PAHs). Triangles show the model spectrum (solid curve) convolved with the DIRBE filters. Observational data are from DIRBE (diamonds) and FIRAS (squares). Taken from Li & Draine (2001b).

The silicate-graphite-PAHs model, with size distributions consistent with the SMC Bar extinction curve (Weingartner & Draine 2001a), is also successful in reproducing the observed IR emission from the SMC (Li & Draine 2002c), as shown in Figure 6. The dust in the SMC is taken to be illuminated by a distribution of starlight intensities. Following Dale et al. (2001), we adopt a simple power-law function for the starlight intensity distribution. The SMC, with a low metallicity ($\sim 10\%$ of solar) and a low dust-to-gas ratio ($\sim 10\%$ of the Milky Way), has a very weak or no 2175 A extinction hump in its extinction curves for most sightlines (see §2.1) and very weak 12 μm emission (see Fig. 6) which is generally attributed to PAHs, supporting the idea of PAHs as the carrier for the 2175 A extinction hump.[17]

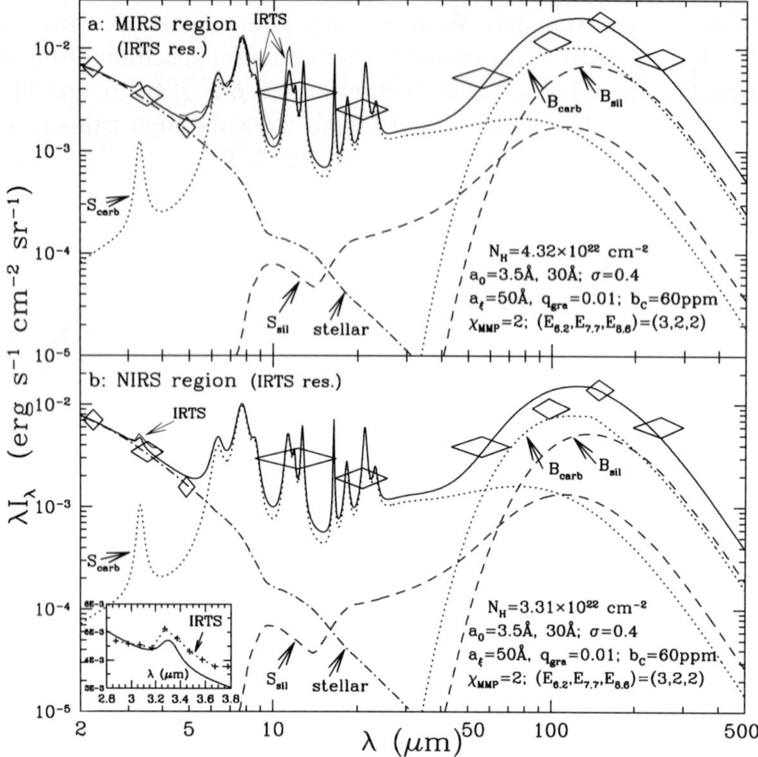

Figure 5. Infrared emission from dust plus starlight for two regions in the Galactic plane: (a) the MIRS region ($44^\circ \le l \le 44^\circ 40'$, $-0^\circ 40' \le b \le 0^\circ$), and (b) the NIRS region ($47^\circ 30' \le l \le 48^\circ$, $|b| \le 15'$). The starlight intensity heating the dust has been taken to be twice the MMP ISRF. The solid curve shows the overall model spectrum; triangles show the model spectrum convolved with the DIRBE filters. DIRBE observations are shown as diamonds. For the MIRS field we show the IRTS MIRS 5–12 μm spectrum (thin solid line). For the NIRS field we show the IRTS NIRS 2.8-3.9 μm spectrum (thin solid line, also shown as cross-dotted curve in inset). Taken from Li & Draine (2001b).

Very recently, the silicate-graphite-PAHs model has also been successfully applied to NGC 7331, a ringed Sb galaxy. Using the same set of dust parameters determined for the Milky Way diffuse ISM (R_V=3.1; Weingartner & Draine 2001a), as shown in Figure 7, this model fits the IR emission observed by the IRAC instrument at 3.6, 4.5, 5.8 and 8 μm and the MIPS instrument at 24, 70 and 160 μm aboard the Spitzer Space Telescope and the 450 and 850 μm SCUBA submillimeter emission observed by JCMT, both for the ring and inside star-forming region and for the galaxy as whole (see Regan et al. 2004 for details). The model also closely reproduces the observed 6.2, 7.7, 8.6, 11.3 and 12.7 μm PAH emission features (see Fig. 2 of Smith et al. 2004).

5. Very Cold Dust or How Cold Could Galaxies Be?

In the 1996 South African "*Cold Dust*" Symposium (Block & Greenberg 1996), the possible existence of a population of **very cold dust** (with equilib-

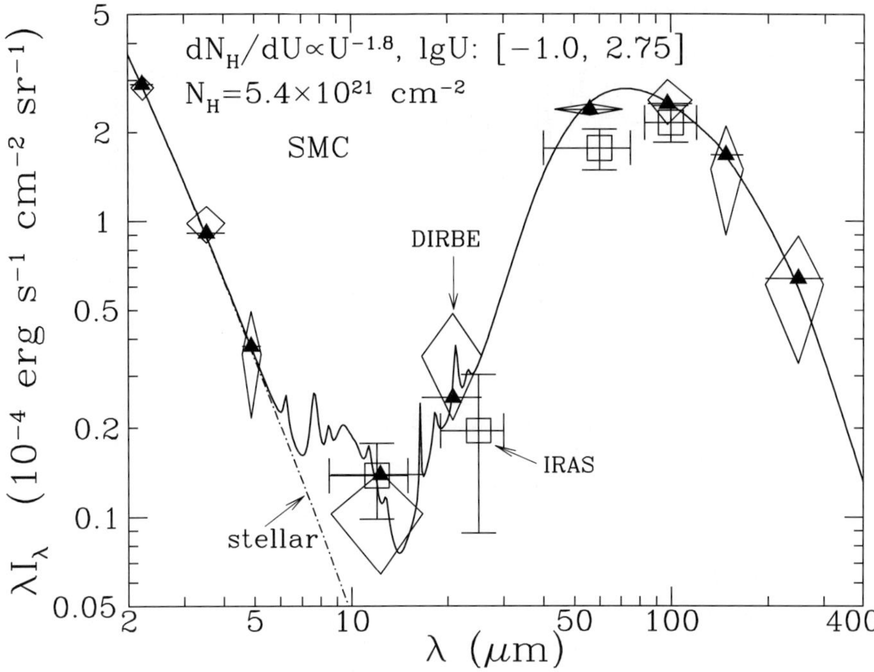

Figure 6. Comparison of the model (solid line) to the observed emission from the SMC obtained by COBE/DIRBE (diamonds) and IRAS (squares) averaged over a 6.25 $\deg^2$ region including the optical bar and the Eastern Wing. Triangles show the model spectrum convolved with the DIRBE filters. Stellar radiation (dot-dashed line) dominates for $\lambda \lesssim 6\,\mu$m. Grains are illuminated by a range of radiation intensities $dN_{\rm H}/dU \propto U^{-1.8}$, $0.1 \leq U \leq 10^{2.75}$, with $N_{\rm H}^{\rm tot} \approx 5.4 \times 10^{21}\,{\rm cm}^{-2}$. Taken from Li & Draine (2002c).

rium temperatures $T < 10\,$K) in interstellar space was a subject received much attention. In his invited paper titled "*How Cold Could Galaxies Be?*" published in the proceedings for that symposium, Mike Disney wrote "... *An eminent cosmologist once advised me to forget all about very cold dust because the T^4 law ensures that it cannot emit, and therefore by implication cannot absorb, much radiation. He sounded plausible, as Cosmologists are apt to sound, but he was in fact totally wrong, as Cosmologists are apt to be.*"

Disney (1996) argued that for grains with higher IR and far-IR emissivities, they can achieve rather low temperatures. At first glance, this appears plausible as can be seen in Eq.(1): for a given interstellar radiation field, grains with fixed UV and optical absorption properties would obtain lower temperatures if their long wavelength emissivities are enhanced. Therefore, Disney (1996) wrote "... *[Since] we are not confident about their [interstellar grains] size distribution and their emissivities, particularly at long wavelength, ...* **we have to keep our minds open to the possible existence of a signiÝcant amounts of very cold (<10 K) material in spiral galaxies.**"

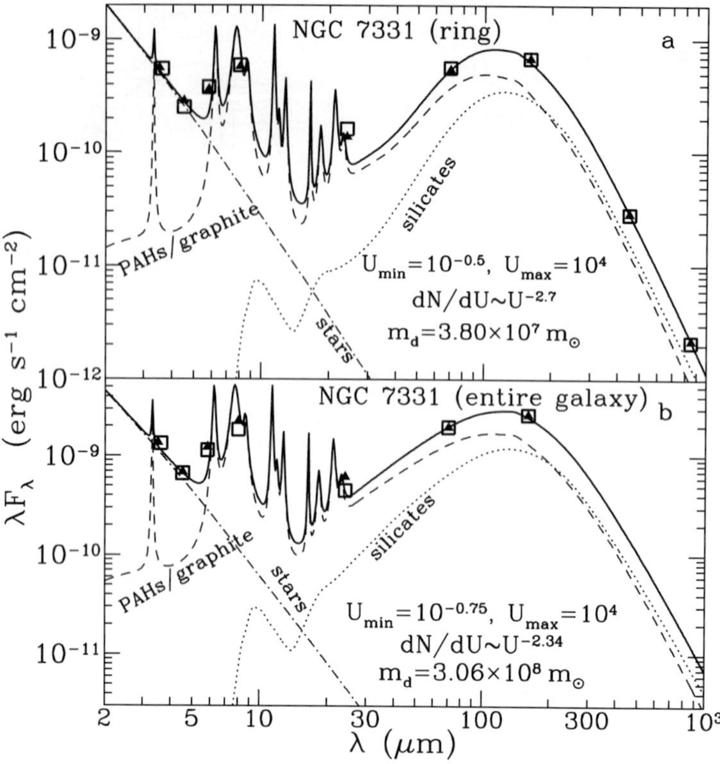

Figure 7. IR emission and model fits to the NGC 7331 ring (a) and the entire galaxy (b). The thick solid lines and triangles are the model-predicted fluxes, and the squares are the observed fluxes. The broken curves indicate the contributions of the different model components. Taken from Regan et al. (2004).

So far, the detection of very cold dust ($T < 10$ K) in interstellar space on galactic scales have been reported for various objects: NGC 4631, a low metallicity ($\sim 1/2$ of solar) interacting galaxy with $T \sim 4$–6 K (Dumke, Krause, & Wielebinski 2004); NGC 1569, a low metallicity ($\sim 1/4$ of solar) starbursting dwarf galaxy with $T \sim 5$–7 K (Galliano et al. 2004); inactive spiral galaxies UGC 3490 with $T \sim 9$ K (Chini et al. 1995), NGC 6156 with $T \sim 8.6$ K, and NGC 6918 with $T \sim 9.4$ K (Krugel et al. 1998); and several irregular and blue compact dwarf galaxies with $T < 10$ K in the Virgo Cluster (Popescu et al. 2002). Very cold dust with $T \sim 4$–7 K has also been detected in the Milky Way Galaxy (Reach et al. 1995; also see Boulanger et al. 2002). This component is widespread and spatially correlated with the warm component (16–21 K). By comparing the dust mass calculated from the IRAS data with the molecular and atomic gas masses of 58 spiral galaxies, Devereux & Young (1990) argued that the bulk of the dust in spiral galaxies is <15 K regardless of the phase of the ISM.

How can dust get so cold? In literature, suggested solutions include (1) the dust is deeply embedded in clumpy clouds and heated by the far-IR emission from "classical grains" (Galliano et al. 2003; Dumke et al. 2004);[18] (2) the dust has unusual optical properties (e.g. fractal or porous grains with enhanced submm and mm emissivity; Reach et al. 1995; Dumke et al. 2004). **However, as discussed in detail by Li (2004b), while the former solution appears to be inconsistent with the fact that the very cold dust is observed on galactic scales, the latter violates the Kramers-Kronig dispersion relation (Purcell 1969; Draine 2003b; Li 2003b),** except for extremely elongated conducting dust (Li 2003a). Perhaps the submm and mm excess emission (usually attributed to very cold dust) is from something else? To avoid the "temperature" problem, one can adopt the **Block direct method** to identify and characterize the presence and distribution of the cold and very cold dust: using near-IR camera arrays and subtracting these from optical CCD images. This method measures the dust extinction cross section and does not require the knowledge of dust temperature (see Block 1996; Block et al. 1994a,b, 1999).

6. Can Dust Get Down to 2.7 K and Appear in Absorption against the CMB?

Ultrasmall grains spend most of their time at their vibrational ground state during the interval of two photon absorption events (§3 and Fig. 2; Draine & Li 2001). In the 1996 South African "*Cold Dust*" Symposium, it was heatedly argued that these grains could obtain a vibrational temperature less than the 2.7 K temperature of the CMB so that they could be detected in absorption against the CMB (Duley & Poole 1998). However, based on detailed modeling of the excitation and de-excitation of these grains, we found that even though these grains do have a large population in the vibrational ground state, nevertheless the vibrational levels are sufficiently excited that the grains would appear in emission against the CMB with brightness temperature $\lesssim 9$ K (see Draine & Li 2004b for details).

Acknowledgments I thank G.J. Bendo, D.L. Block, F. Boulanger, B.T. Draine, R.C. Kennicutt, J.I. Lunine, D. Pfenniger, J.L. Puget, M.W. Regan, and C. Yuan for helpful discussions. I am grateful to D.L. Block and the organizing committee for inviting me to this stimulating symposium. I also thank my advisors B.T. Draine and the late J.M. Greenberg for guiding me to this fascinating field – cosmic dust.

Notes

1. Struve & Elvey (1938) appear to be first to show that the interstellar gas is mostly hydrogen.

2. Let interstellar grains be approximated by a single size of a (spherical radius) with a column density of $N_{\rm d}$. The gas-to-dust mass ratio is

$$\frac{m_{\rm gas}}{m_{\rm dust}} \approx \frac{\left(\mu_{\rm H} + [{\rm He/H}]_{\rm ISM}\,\mu_{\rm He}\right) N_{\rm H}}{(4/3)\,\pi a^3 \rho_{\rm d} N_{\rm d}} \approx \frac{1.4\mu_{\rm H}\,N_{\rm H}}{(4/3)\,\pi a^3 \rho_{\rm d} N_{\rm d}} \tag{2}$$

where $\mu_{\rm H}$ and $\mu_{\rm He}$ are respectively the atomic weight of H and He; $[{\rm He/H}]_{\rm ISM}$ is the interstellar He abundance (relative to H), which is taken to be that of the solar value, ~ 0.1; $\rho_{\rm d}$ is the mass density of interstellar dust. The hydrogen (of all forms)-to-dust column density $N_{\rm H}/N_{\rm d}$ can be derived from

$$\frac{N_{\rm H}}{E(B-V)} = \frac{N_{\rm H}}{A_V/R_V} = \frac{R_V\,N_{\rm H}}{1.086\,\pi a^2 Q_{\rm ext}(V)\,N_{\rm d}} \tag{3}$$

where R_V is the total-to-selective extinction ratio, $Q_{\rm ext}(V)$ is the dust extinction efficiency at V-band ($\lambda = 5500$ A). Therefore, the gas-to-dust ratio can be readily estimated from

$$\frac{m_{\rm gas}}{m_{\rm dust}} \approx 1.14\,\mu_{\rm H}\,\frac{Q_{\rm ext}(V)}{a\rho_{\rm d}}\,\frac{1}{R_V}\,\frac{N_{\rm H}}{E(B-V)} \approx 212\left(\frac{Q_{\rm ext}[V]}{1.5}\right)\left(\frac{0.1\,\mu{\rm m}}{a}\right)\left(\frac{2.5\,{\rm g\,cm^{-3}}}{\rho_{\rm d}}\right). \tag{4}$$

If we take canonical numbers of $R_V{\approx}3.1$, $Q_{\rm ext}(V){\approx}1.5$, $a{=}0.1\,\mu{\rm m}$, $\rho_{\rm d}{\approx}2.5\,{\rm g\,cm^{-3}}$, $m_{\rm gas}/m_{\rm dust}$ would be around 210.

3. In our Milky Way Galaxy, interstellar matter (gas and dust; $7 \pm 3 \times 10^9 {\rm M_\odot}$), contributes roughly $\sim 20\%$ of the total stellar mass ($4 \pm 2 \times 10^{10} {\rm M_\odot}$). Therefore, the mass fraction of interstellar dust is just $\sim 0.1\%$ in our Galaxy ($\sim 1 \pm 0.2 \times 10^{11} {\rm M_\odot}$ within 10 kpc, Kennicutt 2001)!

4. In this method, the wavelength dependence of interstellar extinction is obtained by comparing the spectra of two stars of the same spectral type, one of which is reddened and the other unreddened.

5. Falco et al. (1999) found $R_V{\approx}1.5$ for an elliptical lensing galaxy at $z_l{\approx}0.96$, and $R_V{\approx}$ 7.2 for a spiral lensing galaxy at $z_l{\approx}0.68$. Wang et al. (2004) found the extinction curves for two intervening quasar absorption systems at $z{\approx}1.5$ to have $R_V{\approx}0.7,1.9$.

6. For example, there is at least one line of sight (Sk 143=AvZ 456) with an extinction curve with a strong 2175A hump detected (Lequeux et al. 1982; Prevot et al. 1984; Bouchet et al. 1985; Thompson et al. 1988; Gordon & Clayton 1998). This sightline passes through the SMC wing, a region with much weaker star formation (Gordon & Clayton 1998). The sight lines which show no 2175A hump all pass through the SMC Bar regions of active star formation (Prevot et al. 1984; Gordon & Clayton 1998).

7. Strong regional variations in extinction properties have also been found in the LMC (Clayton & Martin 1985; Fitzpatrick 1985,1986; Misselt, Clayton, & Gordon 1999): the sightlines toward the stars inside or near the supergiant shell, LMC 2, which lies on the southeast side of the 30 Dor star-forming region, have very weak 2175 A hump (Misselt et al. 1999).

8. Kemper, Vriend & Tielens (2004) found that crystalline fraction of the interstellar silicates along the sightline towards the Galactic Center is $\sim 0.2\%$.

9. The amount of Si (relative to H) required to deplete in dust to account for the observed 9.7 μm feature strength is

$$\left[\frac{\rm Si}{\rm H}\right] = \frac{\Delta\tau_{9.7\,\mu{\rm m}}}{N_{\rm H}}\,\frac{1}{\kappa_{\rm sil}^{\rm abs}(9.7\,\mu{\rm m})\,\mu_{\rm sil}} = \frac{\Delta\tau_{9.7\,\mu{\rm m}}}{A_V}\,\frac{A_V}{N_{\rm H}}\,\frac{1}{\kappa_{\rm sil}^{\rm abs}(9.7\,\mu{\rm m})\,\mu_{\rm sil}} \tag{5}$$

where $\kappa_{\rm sil}^{\rm abs}(9.7\,\mu{\rm m})$ is the silicate mass absorption coefficient at λ=9.7 μm; $\mu_{\rm sil}$ is the silicate molecular weight. With $\kappa_{\rm sil}^{\rm abs}(9.7\,\mu{\rm m}){\approx}2850\,{\rm cm^2\,g^{-1}}$ and $\mu_{\rm sil}{\approx}172\mu_{\rm H}$ for amorphous olivine $\rm MgFeSiO_4$, the local diffuse ISM ($\Delta\tau_{9.7\,\mu{\rm m}}/A_V{\approx}1/18.5$, $A_V/N_{\rm H}{\approx}5.3{\times}10^{-22}\,{\rm mag\,cm^2}$) requires $\left[\frac{\rm Si}{\rm H}\right]{\approx}35\,{\rm ppm}$.

10. Hough et al. (1996) reported the detection of a weak 3.47 μm polarization feature in the Becklin-Neugebauer object in the OMC-1 Orion dense molecular cloud, attributed to carbonaceous materials with diamond-like structure. See Li (2004a) and Jones & d'Hendecourt (2004) for a detailed discussion on interstellar diamond.

11. Other C–H out-of-plane bending modes at 11.9, 12.7 and 13.6 μm have also been detected. The wavelengths of the C–H out-of-plane bending modes depend on the number of neighboring H atoms: 11.3 μm for solo-CH (no adjacent H atom), 11.9 μm for duet-CH (2 adjacent H atoms), 12.7 μm for trio-CH (3 adjacent H atoms), and 13.6 μm for quartet-CH (4 adjacent H atoms).

12. Li & Draine (2002b) have modeled the excitation of PAH molecules in UV-poor regions. It was shown that the astronomical PAH model provides a satisfactory fit to the UIR spectrum of vdB 133, a reflection nebulae with the lowest ratio of UV to total radiation among reflection nebulae with detected UIR band emission (Uchida, Sellgren, & Werner 1998).

13. Let $[X/H]_{\odot}$ be the interstellar abundance of X relative to H (we assume interstellar abundances to be those of the solar values: $[C/H]_{\odot}\approx 391$ parts per million [ppm], $[O/H]_{\odot}\approx 501$ ppm, $[Mg/H]_{\odot}\approx 34.5$ ppm, $[Fe/H]_{\odot}\approx 34.4$ ppm, and $[Si/H]_{\odot}\approx 28.1$ ppm [Sofia 2004]); $[X/H]_{gas}$ be the amount of X in gas phase ($[C/H]_{gas}\approx 130$ ppm, $[O/H]_{gas}\approx 375$ ppm; Fe, Mg and Si are highly depleted in dust: $[Fe/H]_{gas}\approx 1$ ppm, $[Mg/H]_{gas}\approx 2$ ppm, and $[Si/H]_{gas}\approx 2$ ppm [Sofia 2004]); $[X/H]_{dust}$ be the amount of X contained in dust ($[C/H]_{dust} = [C/H]_{\odot} - [C/H]_{gas}\approx 261$ ppm, $[O/H]_{dust}\approx 126$ ppm, $[Mg/H]_{dust}\approx 32.5$ ppm, $[Fe/H]_{dust}\approx 27.1$ ppm, $[Si/H]_{dust}\approx 32.4$ ppm). Assuming H/C=0.5 for interstellar carbon dust, the gas-to-dust mass ratio is

$$\frac{m_{\rm gas}}{m_{\rm dust}} \approx \frac{1.4\,\mu_{\rm H}}{\sum_{\rm X}\,[{\rm X/H}]_{\rm dust}\,\mu_{\rm X}} \approx 165\,. \tag{6}$$

where the summation is over Si, Mg, Fe, C and O, and $\mu_{\rm X}$ is the atomic weight of X in unit of $\mu_{\rm H} \approx 1.66 \times 10^{-24}$ g.

14. Assuming all Si, Mg, and Fe elements of solar abundances are condensed in silicate dust of a stoichiometric composition $MgFeSiO_4$ with a characteristic size $a\approx 0.1\,\mu$m, the contribution of the silicate dust to the optical extinction is

$$\begin{aligned}\left(\frac{A_V}{N_{\rm H}}\right)_{\rm sil} &\approx 1.086\,\pi a^2 Q_{\rm ext}(V) N_{\rm sil}/N_{\rm H} \\ &\approx \frac{1.086\,\pi a^2 Q_{\rm ext}(V)\left(\sum_{\rm X=Si,Mg,Fe}[{\rm X/H}]_{\odot}\,\mu_{\rm X} + 4\,[{\rm Si/H}]_{\odot}\,\mu_{\rm O}\right)\mu_{\rm H}}{(4/3)\,\pi a^3 \rho_{\rm sil}} \\ &\approx 3.2\times 10^{-22}\,{\rm mag\,cm^2}\,. \end{aligned} \tag{7}$$

where $N_{\rm sil}$ is the column density of silicate dust, $\rho_{\rm sil}\approx 3.5\,{\rm g\,cm^3}$ is the mass density of silicate material, and $Q_{\rm ext}(V)$ is the visual extinction efficiency of submicron-sized silicate dust which is taken to be $Q_{\rm ext}(V)\approx 1.5$.

15. The idea of transient heating of very small grains was first introduced by Greenberg (1968). This process was not observed until many years later when the detection of the near-IR emission of reflection nebulae (Sellgren, Werner, & Dinerstein 1983) and detection by IRAS of 12 and 25 μm Galactic emission (Boulanger & Perault 1988) were reported.

16. This is because for large grains individual photon absorption events occur relatively frequently and the grain energy content is large enough that the temperature increases induced by individual photon absorptions are relatively small.

17. Li & Draine (2002c) placed an upper limit of $\sim$0.4% of the SMC C abundance on the amount of PAHs in the SMC Bar. But we note that the PAH emission features have been seen in SMC B1#1, a quiescent molecular cloud (Reach et al. 2000). For this region, Li & Draine (2002c) estimated that $\sim$3% of the SMC C abundance to be incorporated into PAHs.

18. Siebenmorgen et al. (1999) argued that the dust embedded in UV-attenuated clouds within the optical disk of typical inactive spiral galaxies cannot become colder than $\sim$6 K.

References

Adamson, A.J., Whittet, D.C.B., Chrysostomou, A., Hough, J.H., Aitken, D.K., Wright, G.S., & Roche, P.F. 1999, ApJ, 512, 224

Aitken, D.K., Roche, P.F., Smith, C.H., James, S.D., & Hough, J.H. 1988, MNRAS, 230, 629

Allamandola, L.J., & Hudgins, D.M. 2003, in Solid State Astrochemistry, ed. V. Pirronello, J. Krelowski, & G. Manico (Dordrecht: Kluwer), 251

Allamandola, L.J., Tielens, A.G.G.M., & Barker, J.R. 1985, ApJ, 290, L25

Allamandola, L.J., Hudgins, D.M., & Sandford, S.A. 1999, ApJ, 511, L115

Anderson, C.M., et al. 1996, AJ, 112, 2726

Bakes, E.L.O., & Tielens, A.G.G.M. 1994, ApJ, 427, 822
Barnard, E.E. 1919, ApJ, 49, 1
Block, D.L. 1996, see **BG96**, 1
Block, D.L., & Greenberg, J.M., ed., 1996, New Extragalactic Perspectives in the New South Africa (Dordrecht: Kluwer) (hereafter **BG96**)
Block, D.L., Puerari, I., Frogel, J.A., Eskridge, P.B., Stockton, A., & Fuchs, B. 1999, Ap&SS, 269, 5
Block, D.L., Bertin, G., Stockton, A., Grosbol, P., Moorwood, A.F.M., & Peletier, R.F. 1994a, A&A, 288, 365
Block, D.L., Witt, A.N., Grosbol, P., Stockton, A., & Moneti, A. 1994b, A&A, 288, 383
Bohlin, R.C., Savage, B.D., & Drake, J.F. 1978, ApJ, 224, 132
Boogert, A.C.A., & Ehrenfreund, P. 2004, see **WCD04**, 547
Bouchet, P., Lequeux, J., Maurice, E., Prevot, L., & Prevot-Burnichon, M. L. 1985, A&A, 149, 330
Boulanger, F., & Perault, M. 1988, ApJ, 330, 964
Boulanger, F., Bourdin, H., Bernard, J.P., & Lagache, G. 2002, in EAS Publ. Ser., Vol. 4, Infrared and Submillimeter Space Astronomy, ed. M. Giard, J.P. Bernard, A. Klotz, & I. Ristorcelli (Paris: EDP Sciences), 151
Cardelli, J.A., Clayton, G.C., & Mathis, J.S. 1989, ApJ, 345, 245
Chiar, J.E., Pendleton, Y.J., Geballe, T.R., & Tielens, A.G.G.M. 1998, ApJ, 507, 281
Chiar, J.E., et al. 2000, ApJ, 537, 749
Chini, R., Krugel, E., Lemke, R., & Ward-Thompson, D. 1995, A&A, 295, 317
Chrysostomou, A., Hough, J.H., Whittet, D.C.B., Aitken, D.K., Roche, P. F., & Lazarian, A. 1996, ApJ, 465, L61
Clayton, G.C., & Martin, P.G. 1985, ApJ, 288, 558
Clayton, G.C., et al. 1992, ApJ, 385, L53
Cohen, R.D., Burbidge, E.M., Junkkarinen, V.T., Lyons, R.W., & Madejski, G. 1999, BAAS, 31, 942
Dale, D.A., Helou, G., Contursi, A., Silbermann, N.A., & Kolhatkar, S. 2001, ApJ, 549, 215
Devereux, N.A., & Young, J.S. 1990, ApJ, 359, 42
Disney, M. 1996, see **BG96**, 21
d'Hendecourt, L.B., Leger, A., Olofson, G., & Schmidt, W. 1986, A&A, 170, 91
Douglas, A.E., & Herzberg, G. 1941, ApJ, 94, 381
Draine, B.T. 1990, in ASP Conf. Ser. 12, The Evolution of the Interstellar Medium, ed. L. Blitz (San Francisco: ASP), 193
Draine, B.T. 1999, in Proc. of the EC-TMR Conf. on 3K Cosmology, ed. L. Maiani, F. Melchiorri, & N. Vittorio (Woodbury: AIP), 283
Draine, B.T. 2003a, ARA&A, 41, 241
Draine, B.T. 2003b, in The Cold Universe, Saas-Fee Advanced Course Vol. 32, ed. D. Pfenniger (Berlin: Springer-Verlag), 213
Draine, B.T., & Lee, H.M. 1984, ApJ, 285, 89
Draine, B.T., & Lazarian, A. 1998a, ApJ, 494, L19
Draine, B.T., & Lazarian, A. 1998b, ApJ, 508, 157
Draine, B.T., & Li, A. 2001, ApJ, 551, 807
Draine, B.T., & Li, A. 2004a, in preparation
Draine, B.T., & Li, A. 2004b, in preparation
Draine, B.T., & Tan, J.C. 2003, ApJ, 594, 347
Duley, W.W., & Poole, G. 1998, ApJ, 504, L113
Dwek, E., Zubko, V., Arendt, R.G., & Smith, R.K. 2004, see **WCD04**, 499
Falco, E.E., et al. 1999, ApJ, 523, 617

Fitzpatrick, E.L. 1985, ApJ, 299, 219
Fitzpatrick, E.L. 1986, AJ, 92, 1068
Frisch, P.C., et al. 1999, ApJ, 525, 492
Galliano, F., Madden, S.C., Jones, A.P., Wilson, C.D., Bernard, J.-P., Le Peintre, F. A&A, 407, 159
Gordon, K.D. 2004, see **WCD04**, 77
Gordon, K.D., & Clayton, G.C. 1998, ApJ, 500, 816
Gordon, K.D., Witt, A.N., & Friedmann, B.C. 1998, ApJ, 498, 522
Greenberg, J.M. 1968, in Stars and Stellar Systems, Vol. VII, ed. B.M. Middlehurst, & L.H. Aller, (Chicago: Univ. of Chicago Press), 221
Greenberg, J.M., Li, A., Mendoza-Gomez, C.X., Schutte, W.A., Gerakines, P.A., & de Groot, M. 1995, ApJ, 455, L177
Grun, E., Gustafson, B.A.S., Mann, I., et al. 1994, A&A, 286, 915
Hartmann, J. 1904, ApJ, 19, 268
Hough, J.H., Chrysostomou, A., Messinger, D.W., Whittet, D.C.B., Aitken, D.K., & Roche, P.F. 1996, ApJ, 461, 902
Joblin, C., Leger, A., & Martin, P. 1992, ApJ, 393, L79
Jones, A.P., & d'Hendecourt, L.B. 2004, see **WCD04**, 589
Kemper, F., Vriend, W.J., & Tielens, A.G.G.M. 2004, ApJ, 609, 826
Kennicutt, R.C. 2001, in Tetons 4: Galactic Structure, Stars and the Interstellar Medium, ed. C.E. Woodward, M.D. Bicay, & J.M. Shull (San Francisco: ASP), 2
Koornneef, J., & Code, A.D. 1981, ApJ, 247, 860
Krugel, E. 2003, Physics of Interstellar Dust (Bristol: IoP)
Krugel, E., Siebenmorgen, R., Zota, V., & Chini, R. 1998, A&A, 331, L9
Landgraf, M., Baggaley, W.J., Grun, E., Kruger, H., & Linkert, G. 2000, J. Geophys. Res., 105, 10343
Larson, K.A., Whittet, D.C.B., & Hough, J.H. 1996, ApJ, 472, 755
Ledoux, G., et al. 1998, A&A, 333, L39
Leger, A., & Puget, J.L. 1984, A&A, 137, L5
Lehtinen, K., & Mattila, K. 1996, A&A, 309, 570
Lequeux, J., & Jourdain de Muizon, M. 1990, A&A, 240, L19
Lequeux, J., Maurice, E., Prevot-Burnichon, M.-L., Prevot, L., & Rocca-Volmerange, B. 1982, A&A, 113, L15
Li, A. 2003a, ApJ, 584, 593
Li, A. 2003b, ApJ, 599, L45
Li, A. 2004a, in ASP Conf. Ser. 309, Astrophysics of Dust, ed. A.N. Witt, G.C. Clayton, & B.T. Draine (San Francisco: ASP), 417
Li, A. 2004b, to be submitted to ApJ
Li, A., & Draine, B.T. 2001a, ApJ, 550, L213
Li, A., & Draine, B.T. 2001b, ApJ, 554, 778
Li, A., & Draine, B.T. 2002a, ApJ, 564, 803
Li, A., & Draine, B.T. 2002b, ApJ, 572, 232
Li, A., & Draine, B.T. 2002c, ApJ, 576, 762
Li, A., & Greenberg, J.M. 1997, A&A, 323, 566
Li, A., & Greenberg, J.M. 1998, A&A, 339, 591
Li, A., & Greenberg, J.M. 2002, ApJ, 577, 789
Li, A., & Greenberg, J.M. 2003, in Solid State Astrochemistry, ed. V. Pirronello, J. Krelowski, & G. Manico (Dordrecht: Kluwer), 37
Malhotra, S. 1997, ApJ, 488, L01
Martin, P.G., & Whittet, D.C.B. 1990, ApJ, 357, 113

Martin, P.G., Clayton, G.C., & Wolff, M.J. 1999, ApJ, 510, 905
Martin, P.G., et al. 1992, ApJ, 392, 691
Mathis, J.S. 1996, ApJ, 472, 643
Mathis, J.S., Mezger, P.G., & Panagia, N. 1983, A&A, 128, 212
Mathis, J.S., Rumpl, W., & Nordsieck, K.H. 1977, ApJ, 217, 425
Mattila, K., et al. 1996, A&A, 315, L353
McKellar, A. 1940, PASP, 52, 187
Mennella, V., Brucato, J.R., Colangeli, L., & Palumbo, P. 1999, ApJ, 524, L71
Mennella, V., et al. 2001, A&A, 367, 355
Misselt, K.A., Clayton, G.C., & Gordon, K.D. 1998, ApJ, 515, 128
Motta, V., et al. 2002, ApJ, 574, 719
Muñoz Caro, G.M., Ruiterkam, R., Schutte, W.A., Greenberg, J.M., & Mennella, V. 2001, A&A, 367, 347
Nandy, K., Morgan, D.H., Willis, A.J., Wilson, R., & Gondhalekar, P. M. 1981, MNRAS, 196, 955
Onaka, T., et al. 1996, PASJ, 48, L59
Pendleton, Y.J. 2004, see **WCD04**, 573
Pendleton, Y.J., & Allamandola, L.J. 2002, ApJS, 138, 75
Pendleton, Y.J., Tielens, A.G.G.M., & Werner, M.W. 1990, ApJ, 349, 107
Pendleton, Y.J., Sandford, S.A., Allamandola, L.J., Tielens, A.G.G.M., & Sellgren, K. 1994, ApJ, 437, 683
Popescu, C.C., Tuffs, R.J., Völk, H.J., Pierini, D., & Madore, B.F. 2002, ApJ, 567, 221
Prévot, M.L., Lequeux, J., Prevot, L., Maurice, E., & Rocca-Volmerange, B. 1984, A&A, 132, 389
Purcell, E.M. 1969, ApJ, 158, 433
Reach, W.T., Boulanger, F., Contursi, A., & Lequeux, J. 2000, A&A, 361, 895
Reach, W.T., et al. 1995, ApJ, 451, 188
Regan, M.W., et al. 2004, ApJS, in press
Schnaiter, M., Henning, Th., & Mutschke, H. 1999, ApJ, 519, 687
Schutte, W.A., et al. 1998, A&A, 337, 261
Sellgren, K., Rouan, D., & Léger, A. 1988, A&A, 196, 252
Sellgren, K., Werner, M.W., & Dinerstein, H.L. 1983, ApJ, 271, L13
Siebenmorgen, R., Krügel, E., & Chini, R. 1999, A&A, 351, 495
Smith, J.D., et al. 2004, ApJS, in press
Smith, T.L., & Witt, A.N. 2002, ApJ, 565, 304
Snow, T.P., & Witt, A.N. 1996, ApJ, 468, L65
Sofia, U.J. 2004, see **WCD04**, 393
Stecher, T.P. 1965, ApJ, 142, 1683
Struve, O. 1929, MNRAS, 89, 567
Struve, O., & Elver, C.T. 1938, ApJ, 88, 364
Swings, P., & Rosenfeld, L. 1937, ApJ, 86, 483
Tanaka, M., et al. 1996, PASJ, 48, L53
Thompson, G.I., Nandy, K., Morgan, D.H., & Houziaux, L. 1988, MNRAS, 230, 429
Toft, S., Hjorth, J., & Burud, I. 2000, A&A, 357, 115
Trumpler, R.J. 1930, PASP, 42, 214
Uchida, K.I., Sellgren, K., & Werner, M.W. 1998, ApJ, 493, L109
Wang, J., Hall, P.B., Ge, J., Li, A., & Schneider, D.P. 2004, ApJ, 609, 589
Weingartner, J.C., & Draine, B.T. 2001a, ApJ, 548, 296
Weingartner, J.C., & Draine, B.T. 2001b, ApJS, 134, 263
Whittet, D.C.B. 2003, Dust in the Galactic Environment (2nd ed; Bristol: IoP)

Witt, A.N., & Vijh, U.P. 2004, see **WCD04**, 115
Witt, A.N., Bohlin, R.C., & Stecher, T.P. 1986, ApJ, 305, L23
Witt, A.N., Clayton, G.C., & Draine, B.T., ed., 2004, ASP Conf. Ser. 309, Astrophysics of Dust (San Francisco: ASP) (hereafter **WCD04**)
Witt, A.N., Gordon, K.D., & Furton, D.G. 1998, ApJ, 501, L111
Witt, A.N., Smith, R.K., & Dwek, E. 2001, ApJ, 550, L201
Witt, A.N., Lindell, R.S., Block, D.L., & Evans, R. 1994, ApJ, 427, 227
Wolff, M.J., Clayton, G.C., Kim, S.H., Martin, P.G., & Anderson, C.M. 1997, ApJ, 478, 395
Zubko, V.G., Dwek, E., & Arendt, R.G. 2004, ApJS, 152, 211

TURBULENCE AND GALACTIC STRUCTURE

Bruce G. Elmegreen
IBM Research Division, T.J. Watson Research Center, 1101 Kitchawan Road, Yorktown Hts., NY 10598, USA

Abstract Interstellar turbulence is driven over a wide range of scales by processes including spiral arm instabilities and supernovae, and it affects the rate and morphology of star formation, energy dissipation, and angular momentum transfer in galaxy disks. Star formation is initiated on large scales by gravitational instabilities which control the overall rate through the long dynamical time corresponding to the average ISM density. Stars form at much higher densities than average, however, and at much faster rates locally, so the slow average rate arises because the fraction of the gas mass that forms stars at any one time is low, $\sim 10^{-4}$. This low fraction is determined by turbulence compression, and is apparently independent of specific cloud formation processes which all operate at lower densities. Turbulence compression also accounts for the formation of most stars in clusters, along with the cluster mass spectrum, and it gives a hierarchical distribution to the positions of these clusters and to star-forming regions in general. Turbulent motions appear to be very fast in irregular galaxies at high redshift, possibly having speeds equal to several tenths of the rotation speed in view of the morphology of chain galaxies and their face-on counterparts. The origin of this turbulence is not evident, but some of it could come from accretion onto the disk. Such high turbulence could help drive an early epoch of gas inflow through viscous torques in galaxies where spiral arms and bars are weak. Such evolution may lead to bulge or bar formation, or to bar re-formation if a previous bar dissolved. We show evidence that the bar fraction is about constant with redshift out to $z = 1$, and model the formation and destruction rates of bars required to achieve this constancy. Bar dissolution has to be accompanied by rapid bar reformation to get the constant bar fraction. This reformation is consistent with numerical simulations by Block et al. (2002), but it may not be possible according to models by Regan & Teuben (2004). The difference between these simulations is partly the result of a difference in the two models for gas viscosity, which depends on the approximations used to represent turbulence. In the Regan & Teuben model, a constant observed bar fraction implies that bars do not dissolve significantly in a Hubble time.

Keywords: Turbulence, Interstellar Matter, Star Formation, Galactic Structure, Galaxy Bars

D. Block et al. (eds.), Penetrating Bars through Masks of Cosmic Dust, 561–579.

1. Introduction

Interstellar turbulence plays a major role in the structure and dynamics of interstellar gas, and through its influence on star formation, energy dissipation, and viscosity, has a strong impact on galactic structure. Conversely, galactic structure has an impact on turbulence through the energy that comes from spiral instabilities. A recent review of ISM turbulence is in Elmegreen & Scalo (2004) and Scalo & Elmegreen (2004), and a review of star formation in a turbulent medium is in Mac Low & Klessen (2004).

Interstellar turbulence is difficult to model numerically because a large number of resolution elements are required and there are many terms in the equations. For example, Wada et al. (2002) simulated the inner regions of a 2D disk with self gravity, star formation and supernovae, but no magnetic fields. Shukurov et al. (2004) did a 3D simulation with magnetic fields but no gravity. de Avillez & Breitschwerdt (2004) modeled the vertical structure in a galaxy disk at high resolution.

Several points about ISM turbulence are important for this discussion.

(1) The cool neutral medium is usually supersonic on scales larger than several tenths of a parsec and therefore easily compressed in the converging parts of turbulent flows.

(2) The magnetic field follows and resists this compression, although shocks still occur parallel and perpendicular to the field if the gas speed exceeds the Alfvén speed. When the gas moves slower than this, compression occurs mostly in the parallel direction (e.g., Ostriker, Stone & Gammie 2001).

(3) The energy density of motion in supersonic turbulence dissipates rapidly, in about the crossing time of any flow region, regardless of the presence of a magnetic field (Mac Low et al. 1998; Stone, Ostriker & Gammie 1998; Mac Low 1999; Padoan & Nordlund 1999). This implies that bulk ISM velocities should dissipate in $\sim$ 20 My, the flow time over the disk thickness, and on much shorter times for smaller scales. Consequently, ISM motions require constant stirring on the scale of the disk thickness.

(4) Clouds, cloud cores, and other ISM structures, as well as whole star-forming regions, should maintain their form and activity for only one or two crossing times if their gas is supersonically turbulent. Clouds or cores where thermal motions dominate might live much longer if they are not unstable to gravitational collapse.

A crossing time corresponds to the shortest possible formation time for stars, clusters, associations, and star complexes. Short is in a relative sense though, not in the sense of an absolute number of years, because the bigger structures take longer absolute times to form. Short formation times are measured directly by the duration of star formation (Elmegreen & Efremov 1996; Ballesteros-Paredes, Hartmann & Vázquez-Semadeni 1999; Elmegreen 2000;

Hartmann et al. 2001). It follows that internal energy sources are not required to generate the turbulence observed in most clouds. It can usually be attributed to residual energy left over from the compressive processes that formed the clouds, in addition to gravitational binding energy for the collapsing clouds.

Turbulence in *incompressible* flows can span a wide range of velocities and spatial scales. Fast flows on large scales contain slower flows on smaller scales in a cascade that extends all the way from the energy source down to the energy dissipation in molecular collisions. The power spectrum of squared-velocity in one dimension is close to a power law with an index of $-5/3$ for incompressible flows. This was predicted by Kolmogorov (1941) using energy conservation for a downward cascade. The power spectrum of any squared quantity in 1D is the energy spectrum, $E(k)dk$. The energy spectrum is related to the power spectrum $P(k)dk$ observed in higher dimensions D by $E(k)dk = P(k)dk^D$. This picture of a uniform cascade of turbulent energy in space and velocity involves interactions between velocity wavenumbers that are comparable in magnitude – sometimes referred to as local in wavenumber space. This is because large flows or eddies carry along the smaller eddies, in a Galilean invariant sense, without much interaction between the two. Possible complications from non-locality in incompressible turbulence were discussed by Zhou, Yeung & Brasseur (1996).

(5) Supersonic ISM turbulence has non-local energy flow in shock fronts when it carries energy from large scales directly to the atomic mean free path without passing through an intermediate cascade. Energy in supersonic turbulence also gets transferred between solenoidal, compressional, and thermal modes. Boldyrev (2002) predicted an energy spectrum $E(k) \sim k^{-1.74}$ by assuming MHD turbulence is mostly solenoidal (incompressible) with Kolmogorov scaling, and that the most dissipative structures are shocks (see also Boldyrev, Nordlund & Padoan 2002). Cho & Lazarian (2002) separated the shear (incompressible) and the fast and slow compressible modes in a compressible MHD simulation for the case where thermal pressure is much less than magnetic pressure and found little coupling between the incompressible and compressible parts. For purely solenoidal driving, the shear modes had $k^{-5/3}$ energy scaling for velocity and field, as did the density in the weakly coupled slow mode; the weakly coupled fast mode had $k^{-3/2}$ scaling for velocity, magnetic field, and density. Cho & Lazarian (2003) got the same result for the high pressure case.

(6) Supersonic turbulence can also carry energy from small to large scales, as when an explosion produces a large moving shell.

There are many observations of power spectra for ISM column densities and emission fluctuations (see review in Elmegreen & Scalo 2004). They are all approximately power laws with a slope of around -3 ± 0.2 in 2D maps (i.e., slightly steeper than the 2D Kolmogorov slope of $-8/3$). These power laws often extend from the smallest observable scale to the largest, including the whole galaxy if data are available, as for the HI maps of the SMC (Stanimirovic 1999) and LMC (Elmegreen, Kim, & Staveley-Smith 2001). Local HI emission (Dickey et al. 2001) shows the same structure.

Another important difference between the Kolmogorov model of incompressible turbulence and ISM turbulence is the spatial scale for energy input. In the Kolmogorov model, kinetic energy is applied on some large scale and it causes motions on smaller and smaller scales down to the dissipation length, where the advection rate equals the dissipation rate. In contrast,

(7) ISM motions are stirred frequently and over a very wide range of scales by various types of sources.

On large scales, turbulent energy comes from gravitational and magnetic instabilities, gravitational scattering of nearby clouds, cloud-disk impacts, and superbubbles. On intermediate scales it comes from supernovae, stellar winds, and expanding HII regions. On small scales it comes from low mass stellar winds and gravitational wakes, Kelvin-Helmholtz and other fluid instabilities, and possibly cosmic-ray streaming, although that is probably more important in the ionized medium. Reviews of these processes are in Norman & Ferrara (1996), Mac Low & Klessen (2004), and Elmegreen & Scalo (2004).

For the ISM, energy sources are so close together in time and space that the gas reaction to one source of energy is usually interrupted by another source before the first fully dissipates. For example, a swing-amplified gravitational instability might make a spiral arm and drive motions in the gas on a kpc scale, but star formation inside this arm and self-gravity inside smaller clouds will drive other motions on smaller scales before the original spiral arm energy is dissipated. This driving energy is not necessarily partitioned into a power law in wavenumber space, or at least not the same power law as the turbulent energy it creates. Thus the resulting power law for ISM turbulent energy will in general be a combination of the distribution of scales for the input energy and the distribution of scales for the non-linear gas reaction to this input. Only on scales much smaller than the smallest input, or for times that are significantly removed from the last input event, can a state of pure gas turbulence be realized. This might apply to scintillation observations, for example (see Scalo & Elmegreen 2004).

One observational implication of this multi-scale agitation is that the ISM often resembles a network of shells or spiral arm fragments with most of the cool neutral matter along the shells or in the arms. Shells dominate the structure of the LMC (Kim et al. 1999) and other small galaxies (e.g., Ho II: Puche et al. 1992) where shear is low, while shells (Brand & Zealey 1975; Heiles 1979) and spiral arms tend to dominate the structure when shear is high (high shear rate means in comparison to the shell growth rate or the instability growth rate). The shells in the LMC are even somewhat self-similar, spanning a range of scales over a factor of at least 10 (Elmegreen, Kim, & Staveley-Smith 2001). Flocculent spiral arms are self-similar too, combining into a power law power spectrum (Elmegreen, Elmegreen, & Leitner 2003). Thus shells and spiral arms are intimately related to turbulence because both are drivers of turbulence and both are structural reactions to turbulence (Wada, Spaans, & Kim 2000).

There is a similar multi-scale aspect to energy *dissipation* in the ISM, making it different again from incompressible turbulence:

(8) Energy dissipation in the neutral ISM covers a wide range of scales, ranging from decompression regions downstream of spiral arms or in disk-halo outflows, to ion-neutral slip viscosity in all structures with field strength gradients, to continuous dissipation in smooth magnetic shocks, to thin hydrodynamic shocks. A similar range of scales is involved with dissipation of turbulence in the ionized component (see review in Elmegreen & Scalo 2004).

The wide range of scales for both energy input and dissipation underscores the difference between ISM turbulence and the standard Kolmogorov picture of energy cascade from large scale input to small scale dissipation.

In the next section, we outline several aspects of interstellar turbulence as they are related to galactic structure. Given the current level of uncertainty about the nature of ISM turbulence, our view on many of these issues is rapidly evolving.

2. Turbulence and Star Formation

Stellar and galactic power sources in the ISM compress interstellar gas and form clouds directly, often in shock fronts, while shock-shock collisions and secondary shocks inside these fronts make further structures down to very small scales (0.1 pc or less). The first generation shocks could be from a spiral density wave or swing amplified instabilities, or it could be from an explosion and subsequent shell. The secondary shocks are inside the compressed clouds and in the hot cavities between them. If the overall medium is significantly self-gravitating or if a cloud is significantly self-gravitating, then some of the secondary shocks inside these regions can produce gas that is also significantly self-gravitating, and this gas can collapse into smaller clouds, cloud cores, or stars before the ambient turbulence shears and distorts the region out of existence. A review of these star-formation processes is in MacLow & Klessen (2004), while simulations are in Gammie et al. (2003), Y. Li et al. (2003), P.S. Li et al. (2004) and elsewhere. A different type of simulation is in papers by Bate and collaborators (e.g., Bate, Bonnell & Bromm 2003), who find collapse in turbulent gas down to very small scales and masses where the optical depth becomes large. Then star formation proceeds upwards in mass from there, as a result of accretion. In either case, cloud turbulence defines the primary cloud structure and self-gravity inside this structure leads to star formation.

An important result of turbulence simulations in isothermal gas, as might apply to molecular cloud cores, is that the probability distribution function (pdf) for density is approximately a log-normal (e.g., Li, et al. 2004), with a power law tail in gravitationally collapsing regions at high density (Klessen 2000). The log-normal indicates that the gas density changes randomly and multiplicatively by successive compressions and rarefactions (Vázquez-Semadeni 1994). When the gas becomes sufficiently self-gravitating, this random, two-directional process stops and the gas density increases monotonically until a star forms. Observations suggest that the point of no return occurs at a density of about 10^5 molecules cm^{-3}. This may be just a coincidence for the local regions that are usually observed, or it may be the result of specific processes that change at this density, promoting collapse. For example, at around 10^5 cm^{-3}, big charged grains start to decouple from the magnetic field, molecules freeze onto grains, and the turbulent speed drops to about the sound speed, making further compressions difficult (see review in Elmegreen 2000).

The density pdf may be integrated over the high density tail to find the volume fraction of gas at high density, f_V. In the log-normal of Wada & Norman (2001), which may be representative of galaxy disks, the volume fraction at a density above 10^5 times the average density is equal to $f_V = 10^{-9}$. Similarly, the mass fraction above 10^5 times the average density is $f_M \sim 10^{-4}$. If turbulence establishes the density sub-structure in clouds, and if gas collapses only at densities greater than 10^5 times the average, then this 10^{-4} should be the fraction of the ISM mass that goes into stars in each turbulent crossing time on the small scale, where the collapse occurs. The star formation rate per unit volume on a galactic scale is then

$$SFR/V = f_V \epsilon \rho_5 \left(G\rho_5\right)^{1/2}, \quad (1)$$

where ϵ is the fraction of the gas mass inside a dense clump that goes into a star or stars in a dynamical time, f_V is again the volume fraction of the whole ISM above this density, and ρ_5 is the high density where collapse becomes inevitable, taken here to be $10^5 \mu$ cm^{-3} for mean molecular weight $\mu \sim 4 \times 10^{-24}$ g. The basic rate of star formation used in this equation is the dynamical rate from gravity, $(G\rho)^{1/2}$. Star formation usually proceeds at about this rate with fairly high efficiency in a cloud core, $\epsilon \sim 0.3 - 0.5$ (e.g., Matzner & McKee 2000). Setting the density at 10^5 cm^{-3} and using $f_V = 10^{-9}$ gives a galaxy wide average star formation rate of 10^{-5} M$_\odot$ pc^{-3} My^{-1}.

This is the same rate that comes from the Kennicutt (1998) observation, which gives a star formation rate per unit area of $SFR/A \sim 0.033\Sigma\Omega$ for mass column density Σ and galaxy rotation rate Ω. This observation applies to all galaxies in Kennicutt's survey, to the inner parts of these galaxies and to starburst galaxies. To convert this to a rate per unit volume, we use the fact that the local density is always about the tidal density, $\rho_{tid} = 3\Omega^2/\left(2\pi G\right)$, assume a flat rotation curve, and assume an exponential disk. Then integrating over the disk from the center to an outer edge at four exponential scale lengths gives an average star formation rate per unit volume of

$$SFR/V \sim 0.012\rho\,(G\rho)^{1/2} \tag{2}$$

The result for average density $\rho \sim 1\mu$ g cm^{-3} is the same as the estimate above, 10^{-5} M$_\odot$ pc^{-3} My^{-1}.

This exercise suggests that the star formation rate in essentially all galaxies, averaged over large enough areas, is determined by the available gas (the ρ term) turning into stars on the local dynamical rate, $(G\rho)^{1/2}$, with an efficiency of ~ 0.01. This low average efficiency is understood for a turbulent medium to be the result of turbulent fragmentation: a small but predictable fraction of the gas is at a density high enough to form stars, which is about 10^5 times the average ρ. The rest of the gas has a density lower than this, making it more easily distorted by random turbulent flows.

Turbulence not only partitions the gas into clumps, but it also locates these clumps in a certain fashion, giving the overall ISM a hierarchical structure with large clumps containing small clumps in many levels of the hierarchy. If stars form in the densest regions of this gas, then they should have the same hierarchical structure. This is widely observed to be the case. Zhang, Fall, & Whitmore (2001) showed that clusters in the Antenna galaxy are correlated in a power-law fashion (i.e., hierarchically clumped together) for scales less than $\sim$ 1 kpc. Elmegreen & Elmegreen (2001) found fractal structure in the star fields of many galaxies and showed that the fractal dimension is about the same as for the gas. Earlier work by Feitzinger & Braunsfurth (1984), Feitzinger & Galinski (1987), Elmegreen & Efremov (1996), and Efremov & Elmegreen (1998) also found fractal patterns for young stars.

Entire galaxies also have correlated optical structure from a combination of star formation and dust extinction. Power spectra of azimuthal profiles of the optical light from galaxies show power-law forms with slopes comparable to $-5/3$ (Elmegreen, Elmegreen & Leitner 2003), which is the slope for the velocity power spectrum in Kolmogorov turbulence. The power spectrum of the optical light in NGC 5055 is essentially the same as the power spectrum of HI in the LMC, as shown in Figure 1. Models including randomly positioned foreground stars, hierarchically distributed clusters in the galaxy, plus hierarchically distributed field-star light in the galaxy reproduce the observations well (Elmegreen et al. 2003).

Different nomenclature is often applied to various parts of this hierarchy, ranging from star complexes such as Gould's Belt on the largest scales (Efremov 1995) to OB associations and subgroups on smaller scales. These regions are all likely to be part of the same physical processes involving gravitational

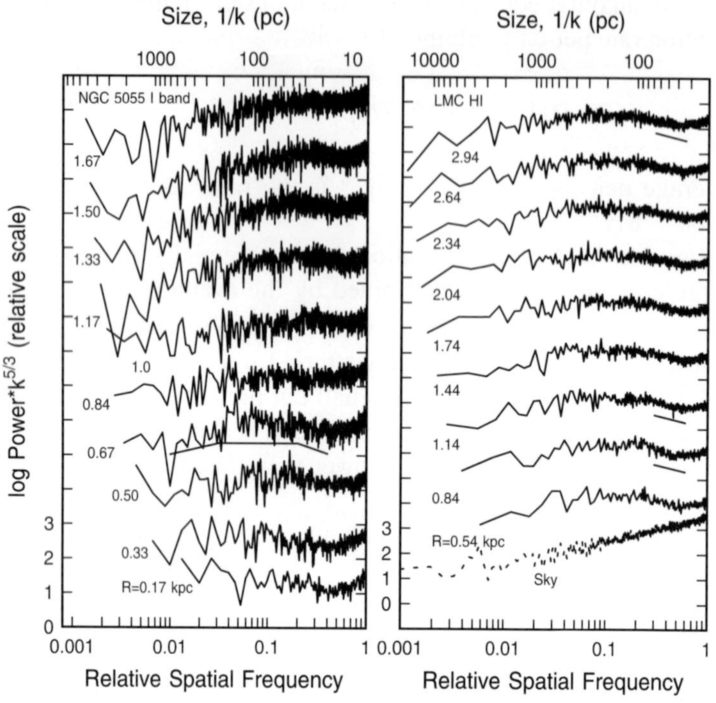

Figure 1. Power spectra of the azimuthal profiles of HI emission from the LMC (right) and optical I-band emission from NGC 5055 (left). The power spectra are multiplied by $k^{5/3}$ to flatten them for easy inspection if they are near the Kolmogorov spectrum of $k^{-5/3}$. The LMC gas and NGC 5055 star formation have remarkably similar power spectra, and power spectra that are also similar to that of incompressible turbulence. This result suggests that star formation distributions are regulated by turbulent processes. (Figure from Elmegreen 2004.)

instabilities, turbulent fragmentation, and direct compression from stellar pressures.

Recall the discussion in the introduction where the driving sources for ISM turbulence were noted to be widespread and overlapping in space and time. This means that shell formation and cloud compression by high pressure events are sources for turbulence as well as sources for triggering star formation. Star formation operates only in the densest parts of the ISM, and all of these stirring processes occur at much lower densities. Thus the organization to star formation on large scales may not be important for the overall star formation rate. Nevertheless this stirring affects the spatial distribution of the clusters that eventually form, often placing them along the rims of shells or in comet-heads.

Elmegreen (2002) summarized this dichotomy by saying that turbulence fragments the gas into dense pieces, but these pieces are constantly buffeted and compressed by the same energy sources that drive this turbulence, triggering star formation in a high fraction of cases. As long at this turbulence makes a regular density pdf, the overall star formation rate is determined in large part by the fraction of the gas at high density. The rate that comes from this fraction is about the same as the large-scale rate that comes from gravitational instabilities at the average density of the ISM. The details of the stirring and local triggering processes are relatively unimportant (see also Wada & Norman 2001).

Another way to visualize this is to consider the bottleneck to star formation in galaxies. The rate-limiting processes are all on the large scale where the density is low and the dynamical time long. On this scale, the onset of star formation is primarily by gravitational processes, such as swing-amplified instabilities. However, stars actually form on a much smaller scale. The link between these two scales is provided by turbulence. The small dense clumps made by this turbulence each turn into stars very quickly, and have virtually no consequence, individually, to the processes on the large scale because they represent only a very small fraction of the total gas mass (10^{-4}). However, these small clumps are continuously regenerated by the turbulence on larger scales, forming more stars, and this regeneration continues for the dynamical time on the large scale. In each dynamical time on the large scale, 100 cycles of core regeneration occur, turning $100 \times 10^{-4} = 1\%$ of the gas into stars. This gives the Kennicutt (1998) star formation law for all galactic regions, including starbursts.

3. Turbulence and Disk Accretion

Turbulence produces viscosity, which leads to disk accretion. If the viscous time is proportional to the star formation time, then a gas disk can evolve toward an exponential profile (Lin & Pringle 1987; Yoshii & Sommer-Larsen 1989; Saio & Yoshii 1990; Gnedin, Goodman & Frei 1995; Ferguson & Clarke 2001). This process does not appear to be self-regulating, however. If the viscous time is much less than the star formation time, then there is the type of feedback that is needed for regulation: the disk accretes, the density increases, the star formation rate per unit area increases, and the two rates come into balance. However, if the viscous time is much greater than the star formation time, then there is no feedback: star formation simply removes the gas by turning it into stars, and the accretion stops. Modern cosmology simulations produce exponential disks from the start (Robertson et al. 2004), without needing an evolutionary process involving accretion and star formation, so viscous production of exponential disks is not necessary.

Disk accretion brings gas to the centers of galaxies, leading to bar destruction if the accreted gas mass is large enough (Hasan & Norman 1990; Pfenniger & Norman 1990; Bournaud & Combes 2002; Debattista et al. 2004). Negative torques from viscous accretion outside the bar region are usually overcome by positive bar and spiral arm torques there, which drive a net outflow. Accretion inside the bar region is driven mostly by bar torques, with positive or negative pressure gradients that contribute to these torques. The net accretion rate depends on the energy loss in addition to the torques. This energy loss is uncertain but likely to be rapid. Energy from star formation can restore some of the internal gas energy downstream, and the pressure from this energy, pushing against the front side of the spiral or bar, can restore some of the lost angular momentum.

The equation of motion for a fluid has a contribution to the time derivative from viscosity that can be written $\nu\nabla^2 v$, from which the accretion time may be estimated to be D^2/ν for inverse gradient distance D. The viscous coefficient ν is from turbulence rather than molecular collisions, and its value is unknown. If it can be represented by the product of a length and a speed, $\nu = \lambda c$, then λ might be the outer scale for the correlated motions and c the rms turbulent speed on that scale. Both of these quantities are highly uncertain, particularly because the ISM may have a 3D type of turbulence on scales smaller than the disk thickness and a 2D turbulence on larger scales (Elmegreen, Kim, & Staveley-Smith 2001). A good guess for λ might be the disk thickness itself, in which case $\lambda \sim 200$ pc. Then the velocity dispersion is the rms speed of most of the gas mass, which is ~ 5 km s^{-1}. These parameters give an accretion time over distance D:

$$t_{acc} \sim \frac{10 \text{ Gy } (D/\text{kpc})^2}{(\lambda/200 \text{ pc}) \left(c/5 \text{ km s}^{-1}\right)}. \tag{3}$$

This is a very long time, even longer if we consider that the real length for the disk gradient is the exponential scale length, which is typically several kpc. Numerical simulations of galaxy disks with gas represented by discrete particles that have longer mean free paths than the disk thickness (averaged over an orbit) can have shorter viscous accretion times than this, possibly leading to spurious effects.

Given the likely small value for the viscous coefficient, disk accretion is dominated by gravitational torques produced in spiral arms and bars. Spirals and bars transfer angular momentum from the inner disk to the outer disk (Lynden Bell & Kalnajs 1972). Bars produce accretion in the inner part, often to a nuclear ring (Regan & Teuben 2004), and outflow in the outer part, often to an outer resonance ring (Schwarz 1981). Bars are much stronger perturbers than spirals because typically only barred galaxies have outer resonance rings (Buta & Combes 1996). The lack of outer resonance rings in non-barred galaxies seems to imply that these galaxies never had a significant bar in their past. This may imply there are relatively few galaxies that have dissolved bars (see Section 4).

4. Turbulence, Viscosity, and Bar Dissolution

The evolution of galaxies over a Hubble time has been studied extensively by simulations but has only recently been observed directly through deep images with the Hubble Space Telescope. Young galaxies often appear physically small and at high restframe surface brightness (Bouwens & Silk 2002), although cosmological dimming allows us to see only the highest surface brightness members of a sample. Galaxies with normal sizes are also present at high z (Simard et al. 1999; Ravindranath et al. 2004). Of interest is the process of galaxy growth and the redistribution of mass inside galaxies during growth. Turbulent viscosity and dissipation play important roles in this redistribution.

Chain galaxies (Cowie, Hu, & Songalia 1995) are interesting because they appear to be unique to high z. They are linear structures with several large bright clumps and no exponential disk or bulge. If they are edge-on disk galaxies (Dalcanton & Schectman 1996; Reshetnikov, Dettmar, & Combes 2003; Elmegreen, Elmegreen, & Sheets 2004a; Elmegreen, Elmegreen, & Hirst 2004b), then their local analogues do not have clumps that extend nearly as far in the vertical direction (e.g., Hoopes, Walterbos, & Rand 1999). The large clump size implies high-speed turbulent motions if the clumps are self-gravitating. The lack of spirals in the face-on counterparts implies that turbulence dominates shear during star formation (Elmegreen, et al. 2004b).

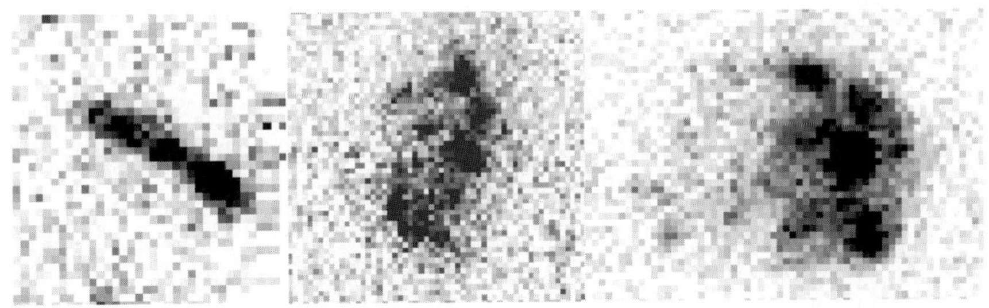

Figure 2. Three high-redshift galaxies from the background field of the Tadpole galaxy are juxtaposed to suggest that chain galaxies (like that on the left) are edge-on versions of clump clusters (like that on the right). This projection is confirmed by the distribution of width-to-length ratios, which is flat as in normal disk galaxies. Neither chains nor clump clusters have exponential disks or prominent red bulges near their centers. Irregular high-z galaxies like this have giant blue clumps that suggest star formation occurred in gas rich disks with large turbulent speeds, comparable to several tenths of the rotation speed. Turbulence in young galaxies could be the result of high galactic accretion rates. (From Elmegreen et al. 2004b.)

Figure 2 shows three galaxies in the deep field of the Tadpole galaxy (Tran et al. 2003) where we found 69 chain galaxies with three or more giant clumps (as shown on the left in the figure), 58 other linear structures with one or two clumps, and 87 tight clusters of clumps that looked like face-on versions of the chain galaxies (as shown in the middle and right frames of the figure). None of these objects have exponential disks or bright red clumps in their centers that could be bulges. The colors and magnitudes of the clumps and of the whole galaxies in these samples are all about the same, and the distribution of the width-to-length ratio is flat down to a lower limit of ~ 0.2. Such a flat distribution is appropriate for circular disks and similar to that for local spiral galaxies in the RC3 (de Vaucouleurs et al. 1991; Elmegreen et al. 2004b). The implication of these results is that clump-clusters are probably face-on versions of chain galaxies.

The giant clumps in these galaxies are blue and likely to be star-forming regions. Their diameters are ~ 500 pc and they are spaced from each other by several kpc, which is several tenths of the disk diameter. The fact that there are just a few giant clumps per galaxy, along with their dominance of the disk light and the lack of obvious spiral arms, suggests they formed by gravitational instabilities in a medium that is mostly gas and has a relatively high turbulent speed (Noguchi 1999; Immeli et al. 2003, 2004). If we set the Jeans length $1/k_J = c^2/(\pi G\Sigma)$ equal to $1/4$ the galaxy radius, R, for gas surface density Σ, then we need a turbulent speed $c \sim 2V(\Sigma/\Sigma_T) \sim 0.3V$ to $0.5V$ for orbit speed V, where Σ_T is the total effective surface density, including dark matter, that contributes to the rotation. At this turbulent speed, the instability time is comparable to the orbit time and collapsing regions should be spun up by Coriolis forces unless there is a magnetic field in the disk. The brightness of the clumps compared to the underlying disk suggests they are among the first generations of star formation. Presumably some will eventually merge to form an exponential disk or a bulge (Noguchi 1999).

The origin of the turbulence in these galaxies is not clear. Shells or other reactions to star formation are not evident, so most of the turbulent energy may come from the galaxy formation process itself.

Barred galaxies at high z are also prominent in this field. We found 22 clearly barred galaxies out to $z = 1$, complete with grand design spiral arms, bulges, and exponential disks. There were also another 21 galaxies that looked barred on contour plots, which showed an inner isophotal twist. The bar fraction was determined as a function of the ratio of axes and compared with the local bar fraction. Bars were less prominent at high inclinations, more so than local bars at equally high inclinations; the difference is probably the result of poor resolution for the distant bars. We estimated that perhaps twice as many bars were lost to inclination effects at high z than locally. The bar fraction was also determined as a function of z using photometric redshifts from Benitez et

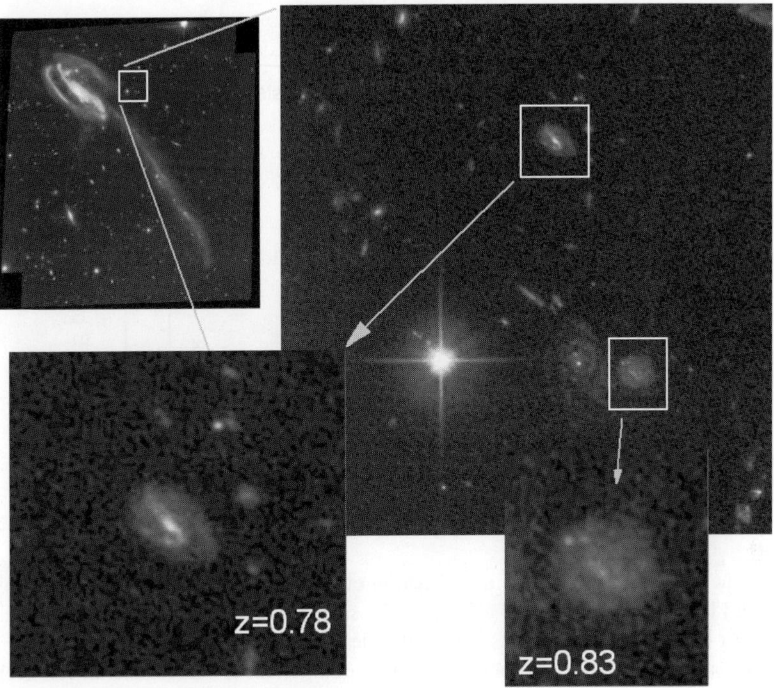

Figure 3. The Hubble Space Telescope Advanced Camera for Surveys field of the Tadpole galaxy (Tran et al. 2003) is shown in two successive stages of blow-ups revealing normal-looking barred galaxies. This field contains 22 clear barred galaxies like this and 21 additional barred galaxies that are so small their bars show up mostly as inner twists in isophotal contours. The abundance of bars at high z is important for studies of bar dissolution following gas accretion.

al. (2004). This fraction is about constant out to $z \sim 1$ and equal to 0.2 to 0.3. Corrected for inclination, the bar fraction is about the same as the local fraction, which is ~ 0.4 in B band depending on Hubble type (Elmegreen et al. 2004c). Figure 3 shows two bars in successive blow-ups of the Tadpole field. Figure 4 shows the bar fraction as a function of redshift.

We suggested that a constant bar fraction with z over the last ~ 8 Gy (out to $z = 1$) offers no evidence for bar dissolution over a Hubble time (Elmegreen et al. 2004c). If bars dissolved, then they had to reform, as suggested by Block et al. (2002). However, if non-merger interactions preferentially formed bars, rather than destroyed them (Noguchi 1987; Gerin, Combes & Athanassoula 1990; Berentzen et al. 2004), and if such interactions were more frequent in the past because of the higher galaxy density, as is likely, then the bar formation

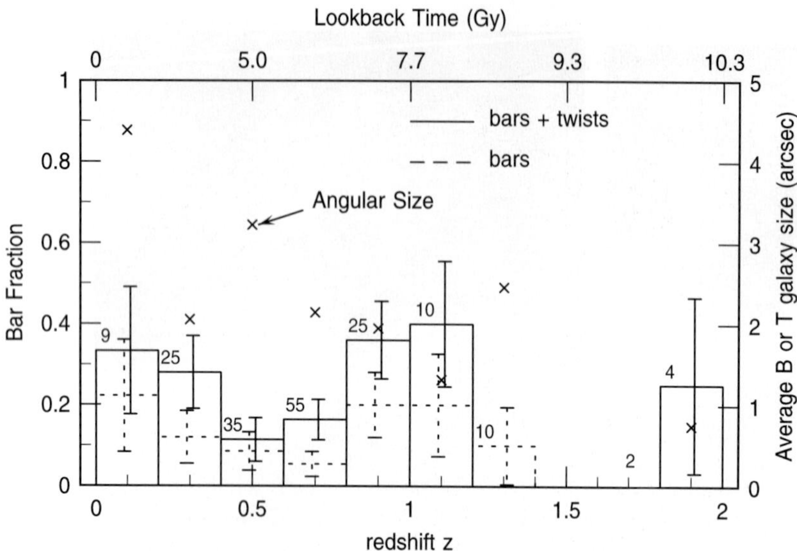

Figure 4. The bar fraction is shown as a function of redshift z for galaxies in the Tadpole field. The solid line histograms are for all bars in the sample, including those which show up only on contour plots. The dashed lines are for the clearest cases of bars. The bar fraction is about constant, suggesting either that bar dissolution is unimportant or bars regenerate relatively quickly once they dissolve. Bar regeneration implies that accreted gas in the outer disk can make its way to the center where it can drive a new bar instability, according to Block et al. (2002). Such accretion without a bar depends on turbulent viscosity and on gravitational torques from spiral arms. (From Elmegreen et al. 2004c.)

rate was higher in the past. To maintain a near-constant bar fraction, this means either that the dissolution rate had to be much higher in the past, or that most bars formed early in the Universe and did not dissolve. Regan & Teuben (2004) suggest, for example, that gas accretion in a bar stops at an ILR ring and does not get to the center, in which case the bar would not dissolve.

A simple model illustrates how bar dissolution is possible if the dissolution rate is proportional to the internal formation rate and an additional formation rate from collisions is proportional to the square of the co-moving density out to $z = 2$. Suppose bars form by collisions at the rate

$$\mathcal{F}_{col} = \mathcal{I}_0 \, (1+z)^6 \, / \, (1+2)^6 \tag{4}$$

for constant $\mathcal{I}_0$. Here we normalize to the rate at $z = 2$. If this were the only bar formation process and there were no bar destruction, then $\mathcal{I}_0 = 0.51$ Gy^{-1} gives a bar fraction of $f = 0.4$ today. Suppose bars also form by internal

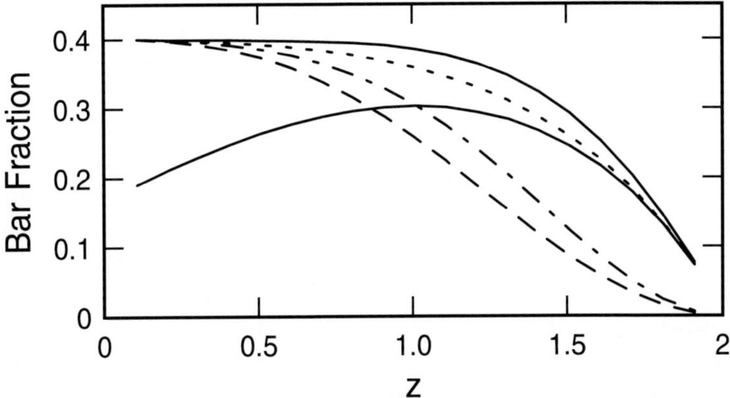

Figure 5. A simple model of bar formation and destruction by various processes is shown to illustrate how prompt bar reformation following internal dissolution can maintain a constant bar fraction out to $z \sim 1$. The dotted line forms bars by galaxy interactions from $z = 2$ to today and has no bar dissolution. The dashed line has bar formation by internal processes with a rate that increases with time at first and then decreases exponentially with an e-folding time of 0.1 Hubble time. The dot-dashed line has the same internal process with a bar dissolution rate proportional to the formation rate. The solid line with a constant bar fraction out to $z = 1$ includes all three processes. All of these examples were tuned to give a bar fraction of 0.4 today. The solid line with a bar fraction that increases out to $z = 1$ has bar formation by interactions and bar dissolution by internal processes with a time scale of 10 Gy. Faster dissolution makes the rise to $z = 1$ larger. This latter case illustrates that even a small amount of bar dissolution cannot operate without continuous re-formation if the bar fraction is to be constant out to $z = 1$.

processes at a rate given by

$$\mathcal{F}_{int}(z) = \frac{\mathcal{F}_0 (t - t_2)}{0.1\tau} e^{(t-t_2)/(0.1\tau)}. \tag{5}$$

This formation rate increases at first and then decreases with an exponential time scale of 0.1 times the age of the Universe, τ; t_2 is the time at $z = 2$. If this were the only bar formation process, then $\mathcal{F}_0 = 0.385$ Gy^{-1} gives a bar fraction of $f = 0.4$ today. Finally, suppose the bar destruction rate is proportional to the internal formation rate:

$$\mathcal{F}_{des} = \mathcal{D}_0 \mathcal{F}_{int}. \tag{6}$$

If bars only formed at the internal process with $\mathcal{F}_{int} = 0.385$ Gy^{-1}, and if $\mathcal{D}_0 = 1$, then the final bar fraction would be 0.32 with 9% of all formed bars having been destroyed.

To solve for the bar fraction f we use the equation

$$\frac{df}{dt} = \mathcal{F}_{col}(1-f) + \mathcal{F}_{int}(1-f) - \mathcal{F}_{des} * f. \quad (7)$$

We solve this numerically using the conversion from t to z in a standard ΛCDM Universe (see formulae in Elmegreen et al. 2004c). The results are shown in Figure 5 for several parameter values. The dotted line is the bar fraction as a function of z with only collisions operating and no destruction using $\mathcal{I}_0 = 0.51$ Gy^{-1}. The dashed line is for internal bar formation only, no destruction, and $\mathcal{F}_0 = 0.385$ Gy^{-1}. The dot-dashed line is for internal bar formation and destruction with $\mathcal{F}_0 = 0.605$ Gy^{-1} and $\mathcal{D}_0 = 1$. All of these were tuned to give a bar fraction of 0.4 today, as was the model for the top solid line, which has all three processes acting together using $\mathcal{I}_0 = 0.51$ Gy^{-1}, $\mathcal{F}_0 = 0.385$ Gy^{-1}, and $\mathcal{D}_0 = 1.85$. Only the solid and dotted lines are constant out to $z \sim 1$. For the case with all three processes, the time integral of the destruction rate, which is the last term in equation 7, equals -0.33, meaning that this fraction of all the galaxies had a bar dissolve. The other solid line in figure 5 is for a case with $\mathcal{I}_0 = 0.51$ Gy^{-1}, no internal formation ($\mathcal{F}_0 = 0$), and a constant destruction rate with $\mathcal{F}_{des} = 0.1$ Gy^{-1}. This last example illustrates how even a small amount of bar destruction (i.e., with a bar dissolution timescale of 10 Gy) and no reformation gives a noticeably rising bar fraction out to $z = 1$.

5. Conclusions

Turbulence is related to galactic structure through the star formation rate and morphology and through viscous forces and ISM energy dissipation. ISM turbulence is a complicated process involving thermal and magnetic pressures plus self-gravity, with frequent and multi-scale energy input and dissipation. Between bursts of energy input and when self-gravity is unimportant, MHD turbulence has several properties that carry over from Kolmogorov turbulence, including the power spectrum of the incompressible part of the flow (the shear or solenoidal part).

Star formation is not just the result of gravitational collapse in a turbulent medium because many sources of pressure, such as HII regions and supernovae, make clouds independently of turbulence and compress the turbulence-made clouds further, triggering additional star formation. A high fraction of all star formation may be triggered in this way. Inside cloud cores, turbulent compression and self-gravity may dominate stellar compression. Also, during the formation of spiral arms by gravitational instabilities and the formation of giant molecular clouds by turbulence and self-gravity, stellar pressures may be unimportant because they are relatively rare on these scales.

Long-term disk evolution also depends on turbulence through its effect on gas viscosity. Turbulent viscosity should be much smaller than simulated viscosity with sticky particles unless the particle collisions are frequent and highly dissipative. Turbulent dissipation in the ISM is so rapid that the entire energy content on the scale of the disk thickness has to be replaced every few disk crossing times. There are apparently enough energy sources with close enough spacings to do this.

When strong bars or spirals are present, global disk accretion is dominated by gravitational torques from these objects and by torques at galactic shock fronts. Accretion by turbulent viscosity is much slower. The appearance of bars at high z with rather normal abundance among disk galaxies suggests either that bars formed early in the Universe and were not easily destroyed, as suggested by Regan & Teuben (2004), or that the bars which were destroyed were promptly replaced by new bars, as suggested by Block et al. (2002). An important difference between these two simulations is the treatment of viscosity. The appearance of irregular galaxies with giant star-forming regions suggests that turbulent velocities were large during the first Gigayear in a galaxy's life.

Acknowledgments

B.G.E. acknowledges support from NSF Grant AST-0205097. Helpful discussions with Debra Elmegreen are appreciated.

References

Athanassoula, F. 1984, Phys.Reps., 114, 319

de Avillez, M.A., Breitschwerdt, D. 2004, Astroph. Sp. Sci. 289, 479

Ballesteros-Paredes, J., Hartmann, L., Vázquez-Semadeni, E. 1999, ApJ, 527, 285

Bate, M.R., Bonnell, I.A., Bromm, V. 2003, MNRAS, 339, 577

Benitez, N. et al. 2004, ApJS, 150, 1

Berentzen, I., Athanassoula, E., Heller, C.H., Fricke, K.J.2004, MNRAS, 347, 220

Bertin G., Lin C.C., Lowe S.A. & Thurstans R.P. 1989, ApJ, 338, 78

Block, D. L., Bournaud, F., Combes, F., Puerari, I., & Buta, R. 2002, A&A, 394, L35

Boldyrev, S. 2002, ApJ, 569, 841

Boldyrev, S., Nordlund, A., & Padoan, P. 2002, Phys. Rev. Lett., 89, 031102

Bournaud, F., & Combes, F. 2002, A&A, 392, 83

Bouwens, R., & Silk, J. 2002, ApJ, 568, 522

Brand, P.W.J.L., & Zealey, W.J. 1975, A&A, 38, 363

Buta, R.C. & Combes, F. 1996, Fundam. Cos. Phys., 17, 95

Cho, J., Lazarian, A. 2002, Phys. Rev. Lett., 88, 245001

Cho, J., Lazarian, A. 2003, MNRAS, 345, 325

Cowie, L., Hu, E., & Songalia, A. 1995, AJ, 110, 1576

Dalcanton, J.J., & Schectman, S.A. 1996, ApJ, 465, L9

Debattista, V.P., Carollo, C.M., Mayer, L., & Moore, B. 2004, ApJ, 604, L93

Dickey, J,M., McClure-Griffiths, N.M., Stanimirovic, S., Gaensler, B.M., Green, A.J. 2001, ApJ, 561, 264

Efremov, Y.N. 1995, AJ, 110, 2757
Efremov, Y. N., & Elmegreen, B. G. 1998, MNRAS, 299, 588
Elmegreen, B.G. 2000, ApJ, 530, 277
Elmegreen, B.G. 2002, ApJ, 577, 206
Elmegreen, B.G. 2004, in Star Formation in the Interstellar Medium,ed. F. Adams, D. Johnstone, D. Lin and E. Ostriker, Astron. Soc. Pacific Conf. Series, in press.
Elmegreen, B.G. & Efremov, Y. 1996, ApJ, 466, 802
Elmegreen, D.M., & Elmegreen, B.G. 2001, AJ, 121, 1507
Elmegreen, B,G., Kim, S., Staveley-Smith, L. 2001, ApJ, 548, 749
Elmegreen, B.G. & Scalo, J. 2004, ARAA, 42, in press
Elmegreen, B.G., Elmegreen, D.M., & Leitner, S.N. 2003, ApJ, 590, 271
Elmegreen, B. G., Leitner, S. N., Elmegreen, D. M., Cuillandre, J.-C. 2003, ApJ, 593, 333
Elmegreen, D.M., Elmegreen, B.G., & Sheets, C.M. 2004a, ApJ, 603, 74
Elmegreen, D.M., Elmegreen, B.G., & Hirst, A.C. 2004b, ApJ, 604, L21
Elmegreen, D.M., Elmegreen, B.G., & Hirst, A.C. 2004c, ApJ, 612, in press
Feitzinger, J.V., & Braunsfurth, E. 1984, A&A, 139, 104
Feitzinger, J. V., & Galinski, T. 1987, A&A, 179, 249
Ferguson, A.M.N., & Clarke, C. J. 2001, MNRAS, 325, 781
Gammie, C.F., Lin, Y.-T., Stone, J.M., & Ostriker, E.C. 2003, ApJ, 592, 203
Gerin, M., Combes, F., & Athanassoula, E. 1990, A&A, 230, 37
Gnedin, O.Y., Goodman, J., & Frei, Z. 1995, AJ, 110, 1105
Immeli, A., Samland, M., & Gerhard, O. 2003, in Galactic and Stellar Dynamics, Proceedings of JENAM 2002, ed. C. M. Boily, P. Pastsis, S. Portegies Zwart, R. Spurzem & C. Theis, EAS Publications Series, Volume 10, p. 199.
Immeli, A., Samland, M., Gerhard, O., & Westera, P. 2004, A&A, 413, 547
Hartmann, L., Ballesteros-Paredes, J., & Bergin, E.A. 2001, ApJ, 562, 852
Hasan, H., & Norman, C. 1990, ApJ, 361, 69
Heiles, C. 1979, ApJ, 229, 533
Hoopes, C.G., Walterbos, R.A.M. & Rand, R.J. 1999, ApJ, 522, 669
Kennicutt, R.C. 1998, ApJ, 498, 541
Kim, S., Dopita, M.A., Staveley-Smith, L., Bessell, M.S. 1999, AJ, 118, 2797
Klessen, R.S. 2000, ApJ, 535, 869
Kolmogorov, A.N. 1941, Proc. R. Soc. London Ser. A, 434, 9
Li, Y., Klessen, R.S., Mac Low, M.-M., 2003, ApJ, 592, 975
Li, P. S., Norman, M.L., Mac Low, M.-M., Heitsch, F. 2004, ApJ, 605, 800
Lin, D.N.C., & Pringle, J.E. 1987, ApJ, 320, L87
Lynden-Bell, D., & Kalnajs, A.J. 1972, MNRAS, 157, 1
Mac Low, M.-M., Klessen, R.S., Burkert, A., Smith, M.D. 1998, Phys. Rev. Lett., 80, 2754
Mac Low, M.-M., 1999 ApJ 524, 169
Mac Low, M.-M., Klessen, R.S. 2004, Rev. Mod. Phys., 76, 125
Matzner, C.D., & McKee, C.F. 2000, ApJ, 545, 364
Noguchi, M. 1987, MNRAS, 228, 635
Noguchi, M. 1996, ApJ, 514, 77
Norman, C.A., & Ferrara, A. 1996, ApJ, 467, 280
Ostriker, E. C., Stone, J. M., Gammie, C.F. 2001, ApJ, 546, 980
Padoan, P., & Nordlund, A. 1999, ApJ, 526, 279

Pfenniger, D., & Norman, C. 1990, ApJ, 363, 391
Puche, D., Westpfahl, Brinks, E., & Roy, J-R. 1992, AJ, 103, 1841
Ravindranath, S., et al. 2004, ApJ, 604, L9
Regan, M.W., & Teuben, P.J. 2004, ApJ, 600, 595
Reshetnikov, V., Dettmar, R.-J., & Combes, F. 2003, A&A, 399, 879
Robertson, B., Yoshida, N., Springel, V., Hernquist, L. 2004, ApJ, 606, 32
Saio, H., & Yoshii, Y. 1990, ApJ, 363, 40
Scalo, J., & Elmegreen, B.G. 2004, ARAA, 42, in press.
Schwarz, M.P. 1981, ApJ, 247, 77
Shukurov, A., Sarson, G.R., Nordlund, A., Gudiksen, B., Brandenburg, A. 2004, Astroph. Space Sci., 289, 319
Simard, L., Koo, D.C., Faber, S. M., Sarajedini, V. L., Vogt, N. P., Phillips, A. C., Gebhardt, K., Illingworth, G. D., & Wu, K. L. 1999, ApJ, 519, 563
Stanimirovic, S., Staveley-Smith, L., Dickey, J.M., Sault, R.J., Snowden, S.L. 1999, MNRAS, 302, 417
Stone, J. M., Ostriker, E. C., Gammie, C. F. 1998, ApJ, 508, L99
Tran, H. et al., 2003, ApJ, 585, 750
de Vaucouleurs, G., de Vaucouleurs, A., Corwin, H., Buta, R., Paturel, G., & Fouque, P. 1991, Third Reference Catalogue of Galaxies, New York: Springer-Verlag
Vázquez-Semadeni, E. 1994, ApJ, 423, 681
Wada, K., Spaans, M., & Kim, S. 2000, ApJ, 540, 797
Wada, K., Norman, C.A. 2001, ApJ, 547, 172
Wada, K., Meurer, G., Norman, C.A. 2002, ApJ, 577, 197
Yoshii, Y., & Sommer-Larsen, J. 1989, MNRAS, 236, 779
Zhang, Q., Fall, S. M., & Whitmore, B. C. 2001, ApJ, 561, 727
Zhou, Y., Yeung, P.K., Brasseur, J.G. 1996, Phys. Rev. E, 53, 1261

Comments from J. Kormendy: Observations of elliptical galaxies also imply that cold gas falls toward the center more quickly than we can easily understand. Many elliptical galaxies have tiny regular dust disks at radii of a few arcsec, commonly aligned with their minor axes (see Kormendy & Djorgovski, 1989, ARAA, 27, 235 for a review). We have good reasons to believe that the gas was accreted – for example, it is frequently in retrograde rotation. In general, the accretion victim is likely to have been a disk galaxy or a Magellanic irregular. The stars phase-mix away in shells and ripples at large radii. With respect to these, the gas clearly finds the center very quickly. I wonder whether the implied dissipation is related to the dissipation that you need to bring accreted gas to the inner parts of disks.

CHAOS IN SPIRAL GALAXIES

Frank H. Shu[1], Sukanya Chakrabarti[2] and Greg Laughlin[3]
[1]*National Tsing Hua University, Hsinchu 30013, Taiwan, ROC,* [2]*Department of Astrophysics, University of California, Berkeley CA, 94720,* [3]*Lick Observatory, University of California, Santa Cruz, CA 95064*

Abstract We review spiral density-wave theory and its realtionship to the disordered appearance of population I objects in some disk galaxies. We discuss mechanisms proposed for the formation of (a) feathers by gravitational instability behind galactic shocks, (b) branches by the action of ultraharmonic resonances, (c) spurs by reflection of leading waves off sharp features induced by nonlinear dredging, and (d) flocculence by the chaos produced from overlapping resonances. We conclude that disorder arises in spiral galaxies not so much from disorderly causes as from too sensitive a response of the interstellar medium to an orderly but nonlinear spiral gravitational field.

Keywords: galactic structure, spiral galaxies, interstellar gas dynamics

1. Introduction

Measurements of the rotation curves of spiral galaxies in the 1950s (e.g., Burbidge, Burbidge, & Prendergast 1959) showed that the angular speed of circular motion Ω decreases with radial distance ϖ from the galactic center in a similar fashion as already deduced for the Milky Way system (Lindblad 1925, 1927; Oort 1927, 1952). This knowledge led to discussions of the winding dilemma of material spiral arms in a field of differential galactic rotation (e.g., Oort, Kerr, & Westerhout 1957; see also the prescient remarks of Alexander 1857), and to the proposal in the 1960s of spiral density waves as a means to resolve this problem (Lindblad 1960, Lin & Shu 1964).

Following Kalnajs (1973), but with a few addendums, Figure 1 gives a pictorial representation of the basic idea of density-wave theory. Because of epicyclic oscillations at a frequency κ (= $\sqrt{2}$ times Ω for flat rotation curves where $\Omega(\varpi) \propto \varpi^{-1}$), the orbit of a star or a gas cloud around the center of a galaxy will generally not form a closed figure. Viewed in a frame that rotates at pattern frequency $\Omega_p > 0$, however, the orbit can be made to close. If the major axes of closed orbits of different sizes line up in a single direction, the collection of orbits will define an oval distortion, or a barred galaxy. If the ma-

D. Block et al. (eds.), Penetrating Bars through Masks of Cosmic Dust, 581–600.

jor axis of closed orbits of increasing sizes are systematically twisted forward or backward in azimuthal angle, the collection of orbits will define a leading or trailing spiral pattern, or an ordinary spiral galaxy. For the barred or spiral pattern not to wind up but merely to undergo a rigid-body rotation at a constant pattern speed Ω_p at all ϖ, self-gravity must come into play so as to make the non-axisymmetric disturbance a normal-mode of the system.

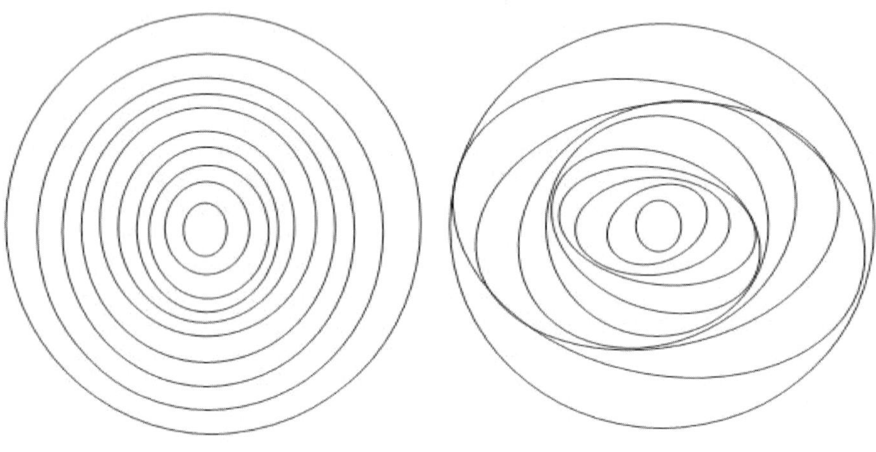

Figure 1. (a) Disk galaxy with oval distortion. (b) Disk galaxy with spiral structure.

Depending on the properties of the axisymmetric state on top of which the oval distortion or spiral structure is a (small) perturbation, non-axisymmetric disturbances can grow spontaneously as self-excited normal modes of the system. In particular, an outward transfer of angular momentum by the gravitational torques of a bar or a spiral is energetically favorable for the system (Lynden-Bell & Kalnajs 1972).

The most powerful global instabilities rely on the swing mechanism for wave amplification (Goldreich & Lynden-Bell 1965, Julian & Toomre 1966; see the review of Bertin & Lin 1996). In Figure 2, we use a WKBJ description to describe the group propagation of a packet of spiral waves and to localize the surface densities of energy E_w and angular momentum J_w of quasi-steady wave trains (e.g., Toomre 1969, Shu 1970, Mark 1971). For small-amplitude waves, E_w and J_w are both proportional to the square of the wave amplitude in such a way that $E_w = \Omega_p J_w$. Outside corotation, where the wave pattern

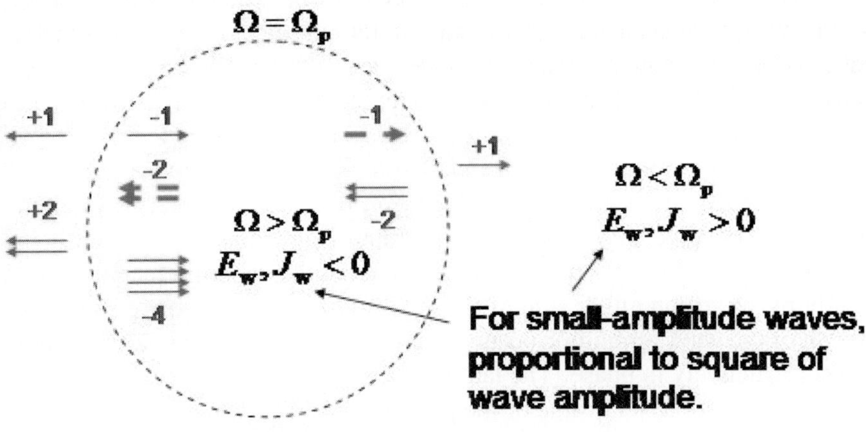

Figure 2. Over-reflection across co-rotation barrier.

rotates faster than the matter, $\Omega_p > \Omega$, E_w and J_w are both positive. Inside corotation, where the pattern rotates slower than the matter, $\Omega_p < \Omega$, E_w and J_w are both negative. The change in sign of E_w and J_w across the corotation circle where $\Omega = \Omega_p$ yields a macroscopic astrophysical example where it is possible to get something for nothing, a state of affairs made popular by the ideas of cosmic inflation.

Imagine starting off in a disk galaxy with (short) trailing spiral waves, and suppose that the two set of waves carry -1 and $+1$ units of energy or angular momentum, respectively, inside and outside of corotation. The total perturbation is created therefore by removing a slight amount of angular momentum from the interior of the galaxy and depositing it in the exterior (through the creation of the trailing-spiral density-waves), with the global change summing to zero. The short trailing spiral waves propagate away from the corotation circle, with the inwardly propagating trailing waves transforming into leading (short) spiral waves after they pass through (or reflect from) the galactic center. The latter also carry -1 units of angular momentum but they propagate outwards toward the corotation circle. As the short leading waves reflect off or cross the corotation circle (usually by tunneling across it), they are transformed by the action of "swing" into trailing spiral waves. The transmitted wave carries $+X$ units of angular momentum (where X is some positive number), and to conserve angular momentum (and energy), the reflected wave must carry $-(1 + X)$ units of angular momentum. In Figure 2, we assume $X = 1$, but it may much exceed 1 if the conditions for "swing" are too favorable (see Toomre

1977 or Binney & Tremaine 1981). The reflected $-(1+X)$ units of trailing spiral waves interior to corotation propagate to the galactic center and upon propagating through it (or equivalently, reflecting off it) again become leading spiral waves, now carrying $-(1+X)$ units of angular momentum. Upon transmission and swing into $+X(1+X)$ units of trailing spiral waves outside corotation, angular momentum conservation in the swing process now over-reflects a packet of trailing spiral waves inside corotation carrying $-(1+X)^2$ units of angular momentum. At any time in the disk, the total wave angular momentum or wave energy is 0, but each round-trip from and to the corotation circle produces a $(1+X)$-folding of the energy and angular momentum carried locally by waves on either side of the corotation circle. A steady exponential growth of the wave amplitude occurs everywhere if the waves inside corotation add constructively in phase. The latter condition yields an eigenvalue determination for the numerical value of the pattern speed Ω_p.

The potentially rapid growth of swing-amplified spiral disturbances in the stellar disk raises the danger of the runaway increase of stellar random velocities that could self-eliminate such unstable normal modes (see the review of Toomre 1977). Zhang (1996) has suggested however that stellar disks may be sufficiently dissipative in their own right to saturate the effect. Independent of such stellar dissipation, Roberts & Shu (1972; see also Kalnajs 1972) pointed out that the inducement of collisional shocks in the interstellar medium could provide a practical limitation to the unrestrained growth of spiral disturbances. Even spiral gravitational fields of relatively small amplitude (say, 5% of the axisymmetric gravitational field) would yield a response in the interstellar gas that is highly nonlinear (Roberts 1969). In the classic depiction, a two-armed spiral shockwave arises with peak gas-density response (corresponding to the dust lanes of a spiral galaxy) that is slightly out of phase with respect to the minimum of the spiral gravitational potential of the disk stars. If the gravity associated with the gas (ignored in Roberts' calculations) were considered, this shift would induce a back-reaction that could offset the continued growth of the spiral wave in the disk stars.

Roberts (1969) also speculated that the compression experienced by the interstellar medium behind galactic shocks could trigger gravitational instability and star formation after a time delay downstream from the dust lanes of $\sim 3\times 10^7$ years. How this triggering works is a matter of some controversy (e.g., Shu et al. 1972; Allen, Atherton, & Tilanus 1986; Elmegreen 1993).

However, even the most beautiful grand-design spirals, such as M51 or M100, exhibit spiral substructures – branches, feathers, and sometimes (leading) spurs – that belie the simplest Roberts (1969) description. The discrepancy has grown with the observational discovery (e.g., Elmegreen 1980, Elmegreen & Elmegreen 1990, Block et al. 1994, Block & Wainscoat 1991) that near-infrared images of spiral galaxies, which emphasizes the distribution of mature disk stars, look considerably smoother and more regular than blue-light images, which emphasize the distribution of young OB stars and their associated giant H II regions, and which are closely correlated with radio maps of the atomic and molecular gas distributions out of which the OB stars are born. In the most puzzling cases, the blue- or visible-light images can appear quite flocculent in spiral galaxies whose infrared images have an almost grand-design quality to them.

2. Possible Explanations of Spiral Substructure

It is generally accepted today that grand-design spirals in the near-infrared represent quasi-stationary normal modes in the disk stars more or less as proposed by C. C. Lin and coworkers from 1964 onwards. The more ragged pattern seen in blue light reflects the response of the interstellar medium (and the young stars newly born from it), which is subject to a more diverse set of forces than are the mature disk stars. Apart from universal gravitation, the gas experiences also magnetic fields, interstellar turbulence, and various radiative and gas-dynamic interactions with forming and dying stars. Given this complexity, it is not surprising that two extreme viewpoints have developed to explain spiral substructure.

The first view holds that irregular structures have irregular causes. The original depiction of the "swing" process as "shearing bits and pieces" of spiral arms by Goldreich & Lynden-Bell (1965) and even Julian & Toomre (1966) basically had a rather chaotic causal picture in mind. The complementary description in terms of a steady amplification of quasi-steady wave-trains depicted in Figure 2 came later (Mark 1976, Goldreich & Tremaine 1979; for full-disk calculations and simulations, see Shu et al. 2000). Consistent with the thought that "irregular structures have irregular causes," Seiden and Gerola (1978) proposed that spiral substructure results from "propagating star formation" triggered by stellar explosions (see also Roberts & Hausman 1984). More generally, Block et al. (1994, 1996) advocated that blue- and visible-light images of spiral galaxies look so different from their near-infrared images because the dynamics of gas and stars are decoupled.

The second view holds that irregular structures might have regular causes. Shu, Milione, & Roberts (1973, hereafter SMR) suggested branch formation as a nonlinear phenomenon associated with ultraharmonic resonances. An intermediate position is represented by Roberts' speculation in 1969 of gravitational instability behind galactic shock to create new-born stars. A more rigorous treatment of the idea in non-magnetic (Balbus & Cowie 1985, Balbus 1988) and magnetic contexts (Lynden-Bell 1965; Elmegreen 1993; Kim & Ostriker 2002, hereafter KO) shows that the mechanism is more suitable for the formaton of feathers as transient rather than permanent condensations.

Chakrabarti, Laughlin, & Shu (2003, hereafter CLS) propose that branches, spurs, and feathers all form as a result of nonlinear dynamics. In this philosophy, the distributions of interstellar gas and the population I stars which form from this gas appear so irregular compared to the distribution of older disk stars that support the underlying spiral gravitational field, not because the gas is too weakly coupled to this field, but because it is *too strongly coupled.*

3. Feathers

By examining the pitch angles and widths of feathers, Elmegreen (1980) argued that they are related to (unstable) density waves. Balbus & Cowie (1985) studied the local linear stability of spiral arms against a background of nonlinear flow defined by the Roberts (1969) two-armed shockwave solutions. They proposed that the axisymmetric criterion (Toomre 1964) needs to be modified by adopting an effective Q in spiral arms defined by

$$Q_{\rm sp} \equiv Q_g \mathcal{C}^{-1/2} \qquad \text{where} \qquad Q_g \equiv \frac{\kappa c_g}{\pi G \Sigma_g}. \tag{1}$$

In equation 1, $\mathcal{C} \equiv \Sigma_{\rm max}/\Sigma_g$ is the compression ratio of peak surface density $\Sigma_{\rm max}$ to the azimuthally-averaged value Σ_g, and Q_g is the usual Toomre Q parameter for a gas disk with $c_g(\varpi)$ being its effective sound speed. (A commonly adopted assumption of many simulations is an isothermal equation of state with c_g = a single constant.) In other words, the relevant local Q in spiral arms does not simply replace Σ_g in the definition for Q_g by $\Sigma_{\rm max}$, which would introduce a factor $\mathcal{C}^{-1}$ for $Q_{\rm sp}$. The net destabilizing effect is smaller by $\mathcal{C}^{1/2}$ because the stabilizing effect of local spin (or vorticity $\propto \kappa$) produces a compensating factor $\propto \mathcal{C}^{1/2}$. Thus, if $Q_g \approx 3$, then a shockwave compression by an order of magnitude or more is sufficient to reduce the effective Q in spiral arms to locally unstable levels, $Q_{\rm sp} < 1$.

Balbus (1988) suggested that quasi-periodic, closely spaced feathers may develop for wavenumbers roughly parallel to the spiral arms. Following an earlier linear stability analysis by Lynden-Bell (1966), Elmegreen (1993) and Kim & Ostriker (2001) pointed out that magnetic braking reduces the stabilizing effects of local spin-up. KO followed up these suggestions with nonlinear MHD simulations. Their calculations are carried out in two dimensions, but within the local approximation of a shearing sheet. They find that magnetic fields primarily in the flow direction of the sheet does indeed help to destabilize it, but the effect is somewhat complex since the compression factor $\mathcal{C}$ behind galactic shocks, for a given spiral gravitational field, is also decreased by the presence of magnetization (e.g., Roberts & Yuan 1970). Nevertheless, the MHD development of quasi-regularly spaced feathers, jutting out roughly perpendicular to the main direction of spiral arms, is extremely impressive (see Figure 3). These simulations strongly support the contention that feathers

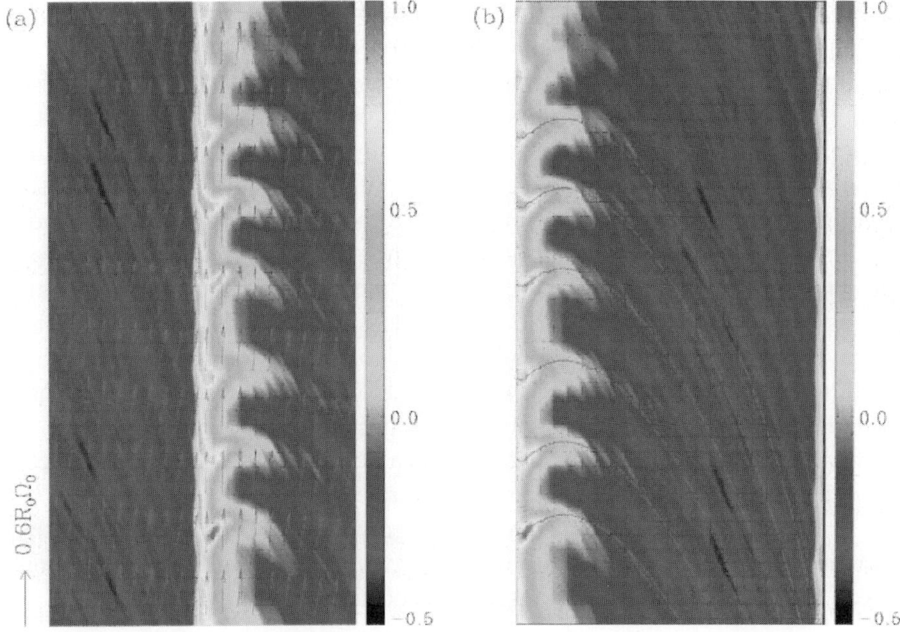

Figure 3. Magnetically mediated feather formation behind galactic shocks, from Kim & Ostriker (2002).

are transiently induced, gravitationally unstable, density waves behind galactic shocks.

4. Branches and Ultraharmonic Resonances

A phenomenon not amenable to a local analysis is ultraharmonic resonance. Resonances (especially weak ones) take many orbit cycles to accumulate their full effects. The shearing sheet (Goldreich & Lynden-Bell 1965, Julian & Toomre 1966) retains the dynamics of shear and rotation (through Coriolis effects), but neglects the spatial curvature. This limit process, $\varpi \rightarrow \infty$, cannot then account for previous passages at any given azimuthal location.

Strictly speaking, ultraharmonic resonances refer to oscillatory forcing at an integer multiple > 1 of some natural oscillation period, or equivalently, at a fraction of some natural oscillation frequency. The effect is a nonlinear phenomenon and does not arise for sinusoidal forcing of a linear oscillator at any frequency other than the natural frequency. However, if the oscillator is nonlinear or the forcing is nonlinear, then resonance can still occur. As an example, one can cause a child on a swing to oscillate with ever larger amplitude by pulling her (to mimic gravity) once per cycle in beat with the natural rhythm of the swing. One can accomplish the same task, not by pulling her toward the back each time she enters the upswing, but, say, every second time (or every n-th time). For the effects to accumulate at small amplitude, one must apply relatively quick jerks and not tug in a slow sinusoidal way; otherwise whatever pulls in the upswing are undone by similar pulls during the downswing. In other words, resonance is possible because nonlinear forcing at, say, twice the natural period will usually contain a first harmonic which is equal to the natural period. Less obviously, forcing by a single sinusoid at a period equal, say, to twice the natural period of oscillation can also create an ultraharmonic resonance if the oscillator is a nonlinear one (such as a galactic orbit displaced appreciably from a circular trajectory).

For forcing of interstellar gas by a spiral gravitational field whose azimuthal variation is a single sinusoid with periodicity m, SMR found in a slightly nonlinear, asymptotic analysis that resonances occur at radii ϖ where

$$\frac{m(\Omega_p - \Omega)}{\kappa} = \pm \left(\frac{1}{n^2} + x \right)^{1/2} \quad n = 1, 2, 3, \ldots, \quad \text{with } x \equiv k^2 c_g^2 / \kappa^2. \tag{2}$$

In the definition for x, k is the radial wavenumber associated with the (tightly wrapped) background spiral wave. The numerator of the left-hand side of equation 2 represents the Doppler-shifted frequency at which matter, rotating at angular speed Ω, meets waves with crests in m spiral arms, rotating at angular speed Ω_p. The denominator is the natural frequency κ of radial oscillation of material perturbed away from a circular orbit. Except for the (usually small) pressure shift embodied in the term x, resonance occurs when $m(\Omega_p - \Omega)$ is $\pm 1/n$ times the natural frequency κ.

The condition with $n = 1$ on the right-hand side of equation 2 corresponds to a Lindblad resonance and is a linear phenomenon. The condition with $n = 2$ corresponds to the first ultraharmonic resonance and arises in quadratic order of the wave amplitude, etc. We label resonances as "inner" if the minus sign is adopted in equation 2 and "outer" if the plus sign is adopted. SMR found a major branching of the basic $m = 2$-armed spiral response to occur at the inner $n = 2$ resonance.

The pattern of ultraharmonic resonances is as follows. The first ultraharmonic resonance $n = 2$ lies farthest from the corotation circle, but not as far as the inner and outer Lindblad resonances. As n increases, the locations of successively higher ultraharmonics march closer (on both sides), but never reach the corotation radius (for $x > 0$). The limit points, $n \to \infty$, correspond to where the Doppler-shifted phase speed of the wave in the radial direction, $m|(\Omega_p - \Omega)/k|$, is equal to the sound speed c_g. Defining i to the inclination angle of the spiral arms relative to the circular direction, and noting for tightly wrapped spiral waves, $\sin i \approx \tan i \equiv m/|k|\varpi$, we can also interpret the $n \to$ limit points as where the unperturbed relative velocity of gas and pattern perpendicular to spiral arms, $\varpi|(\Omega_p - \Omega)\sin i|$, becomes sonic. Thus, even spiral waves of very small amplitude can induce a transonic crossing at radii corresponding to large n ultraharmonics. They therefore easily induce galactic shocks. For a practical model of the spiral structure of our own Galaxy, SMR showed (see their Fig. 9) that the critical forcing amplitude required to produce secondary and primary shockwaves in the interstellar gas takes a dip near the location of the major ultraharmonic resonances $n = 2$, $n = 3$, etc; rising up at radii between the the resonances, but never reaching critical values that are more than about $\mathcal{F} = 1\%$ of the axisymmetric field.

When the background forcing is periodic but not a single sinusoid in its azimuthal variation, then m in equation 2 should be replaced by mj because of the j-th harmonic of the basic angular periodicity m, where $j = 1, 2, 3, \ldots$ (CLS). This generalization allows many additional radii ϖ to satisfy the condition of ultraharmonic resonance. In particular, large j and moderate n resonances can formally lie close to large n and moderate j resonances. These radii are all discrete and countable only at infinitesimal wave amplitude. When the spiral forcing has finite strength, the resonances broaden (in a similar sense that spectral lines are never delta functions in frequency but have finite widths) and have a spread of radii in which the gas response is enhanced (see Fig. 9 of SMR, also Artymowowicz & Lubow 1992). In particular, the largest spiral field that can be imposed that still allows steady flow solutions with the insertion of m or more shockwaves at a radius midway between the $n = 2$ and the $n = 3$ ultraharmonic resonance in the model of SMR (which was constructed with only $j = 1$ forcing) is $\mathcal{F} \approx 3\%$. SMR misleadingly speculate that the problem might lie in the asymptotic (WKBJ-like) approximation used in their mathematical analysis, or in the neglect of the effects of the self-gravitation of the gas. Because chaos theory was not yet well developed, they did not consider a more exciting resolution: namely, that the actual solution might become *time-dependent*, because (ultraharmonic) resonance-overlap leads to *chaos*.

When only a single resonance is present, a nonlinear oscillator like a pendulum will generally find a way to orient the midpoint of its swing to align with the midpoint of the periodic forcing. Pulls during half the cycle are then cancelled by pulls over the other half of the cycle. For galactic orbits forced by a spiral or a bar, the long axis will simply align in steady state with the direction of the potential minimum (or potential maximum) so as to undo over part of the orbit what is done in another part. The dissipation implicit in galactic shock-waves makes gas trajectories irreversible. In these circumstances (see Roberts & Shu 1972), the nonclosure of gas streamlines leads to dredging effects with consequences for inducing spurs that will be described below. When one has two or more overlapping resonances (which is dependent on the forcing amplitudes being large enough to create appreciable width), it becomes generally impossible to solve the oscillator's difficulties with an alignment strategy. The long axis of the orbit cannot point in two or more different directions simultaneously. As soon as the oscillator tries to follow the dictates of one resonance, it begins to accumulate secular influences from the other resonance. The situation is akin to having two bosses – one can go crazy under such circumstances. So also will a nonlinear oscillator generally go chaotic, sooner or later, when it comes under the influence of two or more overlapping resonances.

5. Results of Numerical Simulations

To test these theoretical ideas, CLS performed a series of numerical hydrodynamical simulations. The simulations were carried out with the gaseous surface density Σ_g being a fixed fraction $f = 0.1$ of the hypothetical surface density $\Sigma \propto \varpi^{-1}$ needed by a hypothetical "full disk" to support a given flat rotation curve, $\varpi\Omega =$ constant. Two cases for Q_g are considered, $Q_g = 1.3$ and $Q_g = 2.48$. Spiral forcing of different fractional field strength $\mathcal{F}$ compared to the axisymmetric gravitational field were considered, as imposed by two spiral planforms (see Fig. 3 of CLS). We call these planforms, "log" and "SYL," respectively after logarithmic spirals and Shu, Yuan, & Lissauer (1985). Although specific pattern speeds are chosen to run simulations, the actual values are not important apart from setting a concrete linear scale (e.g., the radius of the corotation circle) in a problem that would otherwise be scale free.

Simulations with relatively high spiral-field strengths (say, $\mathcal{F}$ larger than about 3%) at low Q_g (= 1.3) cannot proceed for very many disk revolutions (near the corotation circle) without developing runaway gravitational instabilities behind galactic shocks. To avoid such problems requires low $\mathcal{F}$, or high Q_g, or both. When disks evolve for many revolutions, nonlinear dredging produces density pileups that can pull spiral arms locally into continued gravitational collapse.

Figure 4 shows a log case ("linear" forcing) when $Q_g = 1.3$ and $\mathcal{F} = 2.5\%$ with parameters chosen so that the rotation period at unit radius is 250 Myr, and the corotation circle has dimensionless radius 1.14. Most of the dynamic range is focused on the region outside of corotation in these simulations. The gas response is smooth and resembles the imposed spiral forcing 95 Myr into the simulation. After 477 Myr, the branching (marked by arrows) at the outer $n = 2$ ultraharmonic is well developed.

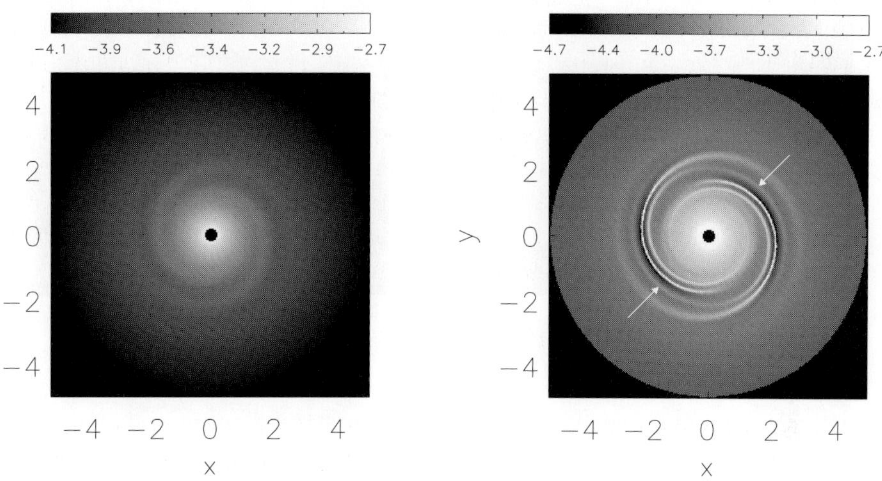

Figure 4. (a) Log spiral planform, $Q = 1.3$, $\mathcal{F} = 2.5\%$, snapshot at 95 Myr. (b) Snapshot at 477 Myr, development of branches marked by arrows.

Figure 5 shows the same case ran at higher spiral forcing $\mathcal{F} = 5\%$. Again, the gas response is smooth and resembles the imposed spiral forcing after 95 Myr, except the response amplitude is expectedly higher than before. After 477 Myr, branching at the outer $n = 3$ ultraharmonic is evident as well as the branching at the outer $n = 2$ ultraharmonic. In addition, regions of especially high postshock compression have developed feathering. Because of our more limited spatial resolution at small scales and perhaps because of our neglect of the effects of the interstellar magnetic field, the feathering is less impressive than the examples illustrated by KO (cf. Fig. 3).

Figure 6 shows a larger $Q_g = 2.48$ case run at the relatively high spiral forcing of $\mathcal{F} = 15\%$. The pattern speed has been raised so that corotation now occurs at a dimensionless radius of 2.13 to resolve better the regions interior to corotation. As before, the response is smooth and resembles the imposed "linear" (log) forcing after 95 Myr. However, so much nonlinear dredging occurs, with gas spiraling in toward the inner $n = 2$ ultraharmonic, that trailing spiral waves propagating toward this region partially reflect into leading spurs

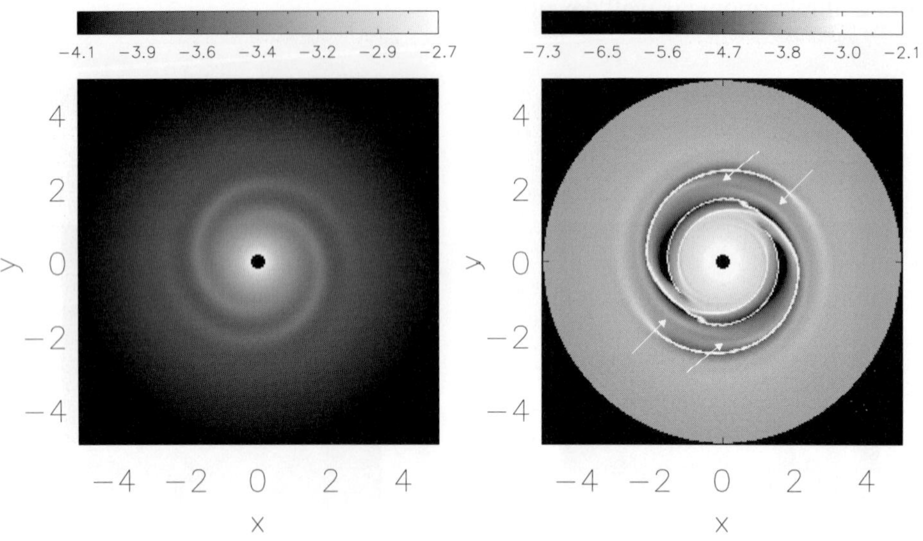

Figure 5. (a) Log spiral planform, $Q = 1.3$, $\mathcal{F} = 5\%$, snapshot at 95 Myr. (b) Snapshot at 477 Myr, development of branches marked by arrows.

(marked by arrows), as well as continuing in transmission at somewhat reduced amplitudes as trailing spiral waves into the central regions.

Figure 7 repeats the high Q_g simulation of Figure 6 but with "nonlinear" (SYL) spiral forcing at a more modest $\mathcal{F} = 7\%$. At both times 95 Myr and 477 Myr, less change in the shape of the response occurs than we might have expected from the more open-spiral forcing of the SYL planform. We attribute the effect to the intrinsic preference of the self-gravity of the gas for shorter arm spacings. The modest forcing amplitude evidently does not produce enough nonlinear dredging at 477 Myr for us to see noticeable internal reflections into leading spurs. We remark that sponge boundary conditions have been deliberately introduced by CLS in all their simulations to prevent wave propagation through the center into leading waves. This is done to suppress any residual swing amplification that might remain in the system despite parameters in the "active" part of the problem (the gas disk) that discourage such distractions.

Figure 8 repeats the low Q_g simulation of Figure 4, but with reduced spiral forcing $\mathcal{F} = 1.3\%$ to allow runs to longer simulation times without runaway gravitational instabilities. Trailing spiral waves now have enough time to propagate well beyond even the outer Lindblad resonance as essentially acoustic disturbances. This accounts for the regular outer pattern of tightly wound spi-

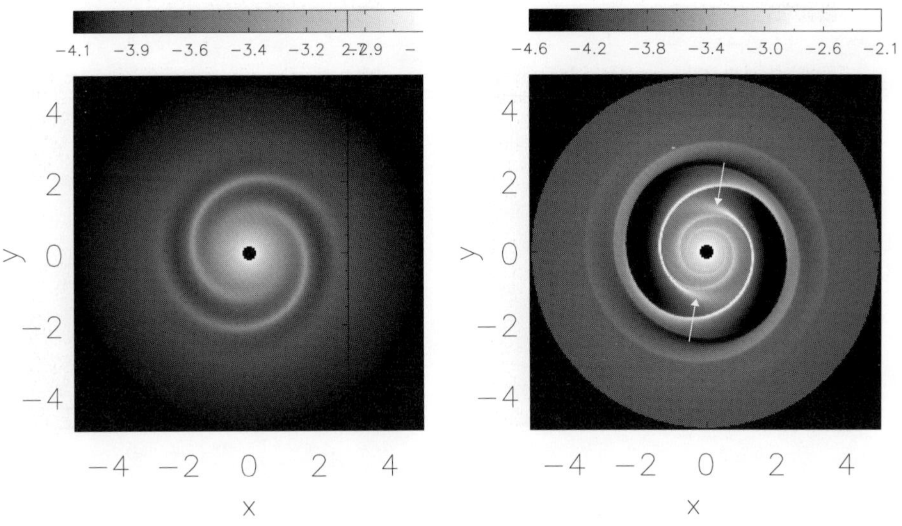

Figure 6. (a) Log spiral planform, $Q = 2.48$, $\mathcal{F} = 15\%$, snapshot at 95 Myr. (b) Snapshot at 477 Myr, development of spurs marked by arrows.

ral arms in both the 576 Myr and 2870 Myr frames. Some irregularities are seen in the 2870 Myr frame that are not so apparent in the 576 Myr frame, but one cannot say that a large amount of true flocculence or chaos is present in either image at low and linear spiral forcing.

Figure 9 shows a high Q_g simulation at larger (but still linear in the logarithmic spiral sense) forcing $\mathcal{F} = 3.5\%$ and with a higher pattern speed to emphasize better the regions interior to corotation. Much of the coupling to trailing acoustic waves beyond the outer Lindblad resonance is damped by the conditions of the simulations, so our attention is not distracted as before by very tightly wrapped spiral arms. This simulation demonstrates that flocculence does not arise at modest spiral forcing if that forcing is restricted to a single sinusoid in azimuthal angle.

Figure 10 is carried out with nonlinear SYL forcing; otherwise the parameters are the same as in Figure 9. Despite the more open spacing of spiral arms underlying the imposed forcing in Figure 10, the response is more *tightly* wound at 576 MYr, and much more flocculent at 2870 Myr than in Figure 9. This example presents the best case for our hypothesis that overlapping ultraharmonic resonances is the cause underlying chaotic gas response and flocculent spiral structure.

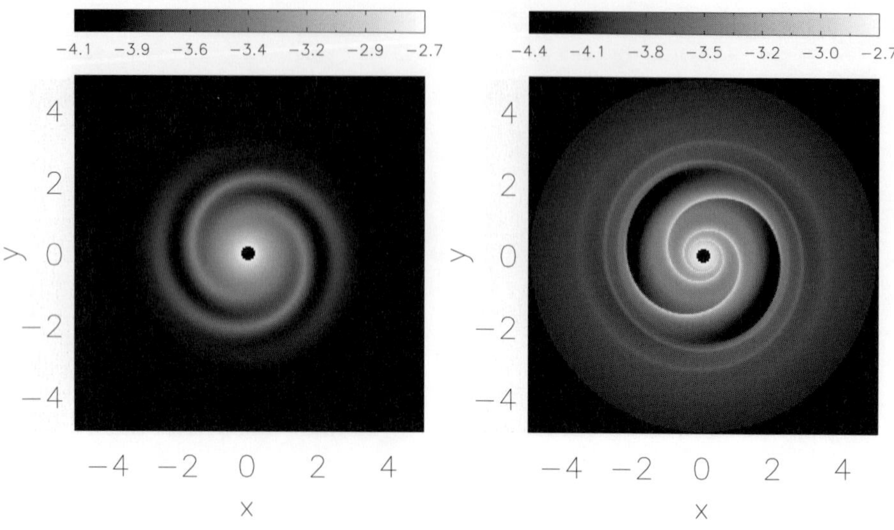

Figure 7. (a) SYL spiral planform, $Q = 2.48$, $\mathcal{F} = 7\%$, snapshot at 95 Myr. (b) Snapshot at 477 Myr.

Figure 11 shows the same case as Figure 10 but carried out with yet higher SYL forcing $\mathcal{F} = 5\%$. This simulation demonstrates that even $Q_g = 2.48$ does not suffice to save the system over the long run from wild time-dependent behavior if the background spiral forcing is reasonably high. Notice that the chaotic component of the response is not present at the relatively short time of 576 Myr despite the presence of very strong density compression behind galactic shocks, perhaps because overlapping ultraharmonic resonances have not yet had time to accumulate their effect. Thus, a paradoxical interpretation may exist for highly chaotic and irregular systems in blue light with underlying grand-design infrared spirals. These systems may be the best indicators that the background spiral structures in the disk stars are not, as previously thought, themselves transient or non-existent, but, in fact, have lifetimes that perhaps extend to billions of years.

6. Summary

The unrestrained gravitational collapse of local spiral arms in some simulations clearly indicates that important physical processes, such as feedback from forming stars or the limiting effects of interstellar magnetic fields, have

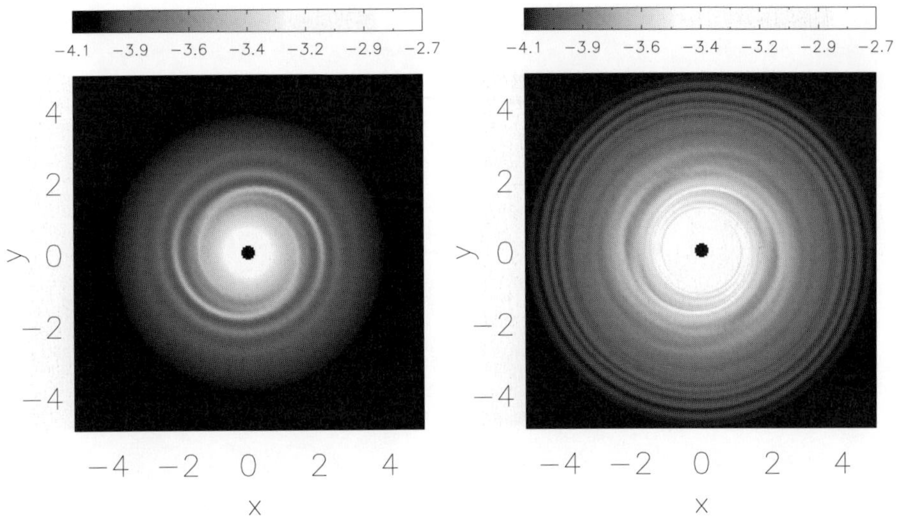

Figure 8. (a) Log spiral planform, $Q = 1.3$, $\mathcal{F} = 1.3\%$, snapshot at 576 Myr. (b) Snapshot at 2870 Myr, mild development of feathering.

been left out of the CLS simulations. Nevertheless, we can still draw plausibly strong conclusions from the results of these calculations.

First, we agree with the assessment made by KO that *transient* local gravitational instabilities behind galactic shocks explain the phenomenon of feathering. Magnetic fields probably both mediate the onset of the instability and limit its ultimate degree of compression. The typical delay of $\sim$ 30 Myr cited by Roberts (1969) between the formation of dust lanes and the ultimate breakout of giant H II regions created by the formation of OB stars is probably related to the galactic-dynamics of this instability process. The delay of 1 Myr or so observed for the formation of molecular cloud cores and the appearance of individual stars in nearby molecular clouds involve small-scale effects (see the reviews of Evans 1999 and Lada & Lada 2003) that are, at best, only indirectly related to the larger events occurring on the scale of spiral arms.

Second, we agree with the original suggestion of SMR, and many but not all subsequent assessments, that the branching of major spiral arms are to be associated with the action of ultraharmonic resonances, with the most important of these being the $n = 2$ ultraharmonic (either inner or outer). We have also vindicated their guess that the self-gravity of the gas is a crucial ingredient at maintaining the compressive action of ultraharmonic resonances. Previous in-

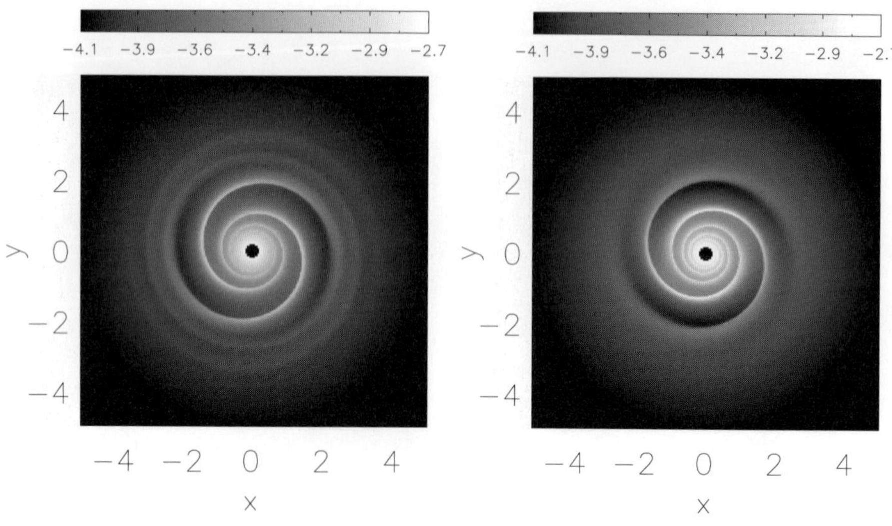

Figure 9. (a) Log spiral planform, $Q = 2.48$, $\mathcal{F} = 3.5\%$, snapshot at 576 Myr. (b) Snapshot at 2870 Myr.

vestigations that come to dissimilar conclusions (e.g., KO) under-estimated the importance of whole-disk simulations to accumulate the azimuthally periodic effects of ultraharmonic resonances.

Third, the simulations of CLS show that leading spurs can arise as a result of internal reflections off sharp features produced in the interstellar gas distribution. Leading spiral waves therefore have a separate possible origin than propagation through galactic centers. Since the effect is distinctly nonlinear, it is hard to assess what impact this finding might have on the operation of the swing mechanism in the amplification of global normal modes. All such effects were deliberately suppressed by the simulation protocol of CLS.

Fourth, although definitive proofs are hard to obtain via numerical simulation, the CLS calculations make a plausible case for the role of overlapping ultraharmonic resonances as a primary cause of chaotic behavior in galactic spiral structure. In this view, flocculence arises in spiral galaxies not because the interstellar gas is decoupled from the same gravitational dynamics as the disk stars, but because it is too strongly coupled to that dynamics.

Finally, we note that acceptance of the conclusions of items two, three, and four above implies an underlying belief in the hypothesis of quasi-stationary spiral structure propounded forty years ago by Lin & Shu (1964). The time

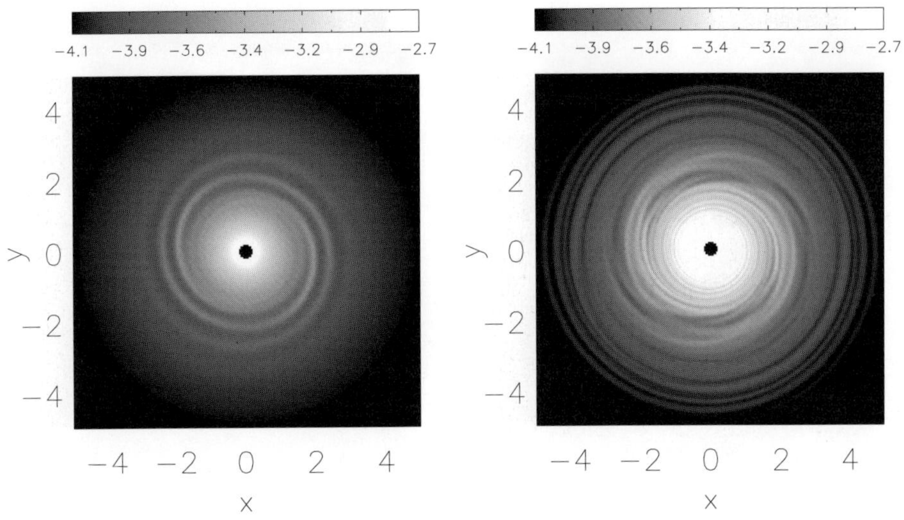

Figure 10. (a) SYL spiral planform, $Q = 2.48$, $\mathcal{F} = 3.5\%$, snapshot at 576 Myr. (b) Snapshot at 2870 Myr, development of flocculent gaseous structure.

has arrived for *complete* dynamical simulations that take into account the active responses of both the star disk and the gas disk (with embedded magnetic fields) if we are to realize the full vision of the hypothesis of quasi-stationary spiral structure.

References

[1] Alexander, S. 1857, AJ, 2, 95

[2] Allen, R. J., Atherton, P. D., & Tilanus, R. P. J. 1986, Nature, 319, 296

[3] Artymowicz, P., & Lubow, S. 1992, ApJ, 389, 129

[4] Balbus, S. A. 1988, ApJ, 324, 60

[5] Balbus, S. A., & Cowie, L. L. 1985, ApJ, 297, 61

[6] Bertin, G., & Lin, C. C. 1996, Spiral Structure in Galaxies: A Density Wave Theory (MIT Press)

[7] Binney, J., & Tremaine, S. 1987, Galactic Dynamics (Princeton: Princeton Univ. Press)

[8] Block, D. L., Bertin, G., Grosbol, P., Moorwood, A. F. M., & Peletier, R. F. 1994, A&A, 288, 365

[9] Block, D. L., Elmegreen, B. G., & Wainscoat, R. J. 1996, Nature, 381, 674

[10] Block, D. L., & Wainscoat, R. J. 1991, Nature, 353, 48

[11] Burbidge, E. M., Burbidge, G., & Prendergast, K. 1959, ApJ, 130, 739

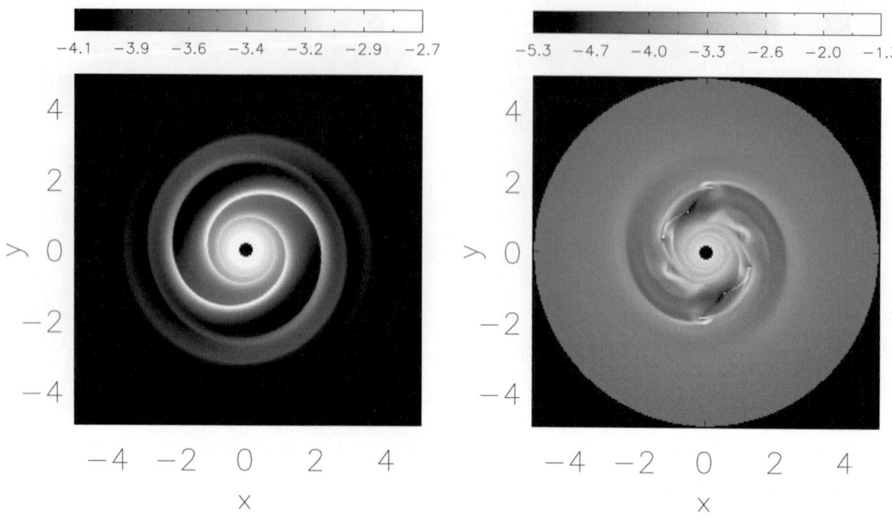

Figure 11. (a) SYL spiral planform, $Q = 2.48$, $\mathcal{F} = 5\%$, snapshot at 576 Myr. (b) Snapshot at 2870 Myr, development of chaotic and flocculent gaseous structures.

[12] Chakrabarti, S., Laughlin, G., Shu, F. H. 2003, ApJ, 596, 220 (CLS)

[13] Elmegreen, B. G. 1993, in Protostars and Planets III, ed. E. Levy & J. Lunine (Tucson: Univ. Arizona Press), 97

[14] Elmegreen, B. G., & Elmegreen D. M. 1990, ApJ, 355, 52

[15] Elmegreen, D. M. 1980, ApJ, 242, 528

[16] Evans, N. J. 1999, ARAA, 37, 311

[17] Goldreich, P., & Lynden-Bell, D. 1965, MNRAS, 130, 125

[18] Goldreich, P., & Tremaine, S. 1979, ApJ, 233, 857

[19] Julian, W. H., & Toomre, A. 1966, ApJ, 146, 810

[20] Kalnajs, A. J. 1972, ApL, 11, 41

[21] Kalnajs, A. J. 1973, PASAu, 2, 174

[22] Kim, W. T., & Ostriker, E. C. 2001, ApJ, 559, 70

[23] Kim, W. T., & Ostriker, E. C. 2002, ApJ, 570, 132 (KO)

[24] Lada, C. J., & Lada, E. A. 2003, ARAA, 41, 57

[25] Lin, C. C., & Shu, F. H. 1964, ApJ, 140, 646

[26] Lindblad, B. 1925, ApJ, 62, 191

[27] Lindblad, B. 1927, MNRAS, 87, 553

[28] Lindblad, P. O. 1960, Stockholm Obs. Ann., bd. 21, no. 4

[29] Lynden-Bell, D. 1966, Observatory, 86, 57

[30] Lynden-Bell, D., & Kalnajs, A. J. 1972, MNRAS, 157, 1
[31] Mark, J. W. K. 1976, ApJ, 205, 363
[32] Oort, J. H. 1927, BAN, 3, 275
[33] Oort, J. H. 1952, ApJ, 116, 233
[34] Oort, J. H., Kerr, F. T., & Westerhout, G. 1958, MNRAS, 118, 379
[35] Roberts, W. W. 1969, ApJ, 158, 123
[36] Roberts, W. W., & Hausman, M. A. 1984, ApJ, 282, 106
[37] Roberts, W. W., & Shu, F. H. 1972, ApL, 12, 49
[38] Roberts, W. W., & Yuan, C. 1970, ApJ, 161, 887
[39] Seiden, P. E., & Gerola, H. 1978, ApJ, 223, 129
[40] Shu, F. H. 1970, ApJ, 160, 99
[41] Shu, F. H., Laughlin, G., Lizano, S., & Galli, D. 2000, ApJ, 535, 190
[42] Shu, F. H., Milione, V., Gebel, W., Yuan, C., Goldsmith, D. W., & Roberts, W. W. 1972, ApJ, 173, 557
[43] Shu, F. H., Milione, V., & Roberts, W. W. 1973, ApJ, 183, 819 (SMR)
[44] Shu, F. H., Yuan, C., & Lissauer, J. J. 1985, ApJ, 291, 356 (SYL)
[45] Toomre, A. 1964, ApJ, 139, 121
[46] Toomre, A. 1969, ApJ, 158, 899
[47] Toomre, A. 1977, ARAA, 15, 437
[48] Zhang, X. 1996, IAU Symp. 169, Unsolved Problems of the Milky Way, eds. L. Blitz & P. Teuben, (Kluwer), p. 517

David Block: May I ask your thoughts on a comment I received, that your paper with Chakrabarti & Laughlin stands on one unjustified assumption, i.e. that the gas you consider evolves under “imposed rigidly rotating spiral potentials". The comment continued: “Is it not true to say that only barred galaxies (i.e. barred modes) can be justified without the help from the gas; instead normal unbarred spirals (i.e. unbarred modes) must require the cooperation (which you may call coupling) between gas and stars.”

Frank Shu: The comment is correct, except for the word “unjustified.” Whether one decides to include the back reaction of the gas response on the background stellar distribution, for either bars or spirals, is a matter of ultimate self-consistency, not of being “justified to make the approximation or not.” It is no more or less justifed in the case of bars than in the case of ordinary spirals. Of course, one would prefer to include the back reaction of the stellar disk, and the dark-matter halo, and any nearby galaxies, ... The question is whether any new qualitative features emerge from such improvements. (I think there is one important feature, but it is not the one the questioner is thinking about.)

David Block: You’ve aroused my deep curiosity with your last sentence in parentheses, Frank!

Frank Shu: OK, here's the thinking. Taking an absolutely spatially ordered, time-invariant, background force field is the *worst* case scenario for inducing a disordered response. Yet, even in this worst case scenario, the gas response often becomes flocculent.

Where taking the back reaction of the gas into account does become important is something that was covered in my talk, and is also alluded to briefly at the end of the paper by Chakrabarti, Laughlin & Shu. Without the back reaction, which gives a weakening of the stellar density wave, swing-amplified spiral modes in the stellar disk, will tend to runaway exponentially. This catastrophic runaway, with its attendant pumping of the stellar (Toomre) Q, is what eventually kills the mode in many N-body simulations. The nonlinear dissipation associated with the gaseous response (not a feature of most computations) can saturate the growth of the stellar mode, as speculated upon originally by Roberts and Shu (1972), and lead to the justification of the original Lin-Shu hypothesis of quasi-stationary spiral structure.

OBTAINING STATISTICS OF TURBULENT VELOCITY FROM ASTROPHYSICAL SPECTRAL LINE DATA

A. Lazarian
University of Wisconsin-Madison, Dept. of Astronomy

Abstract Turbulence is a crucial component of dynamics of astrophysical fluids dynamics, including those of ISM, clusters of galaxies and circumstellar regions. Doppler shifted spectral lines provide a unique source of information on turbulent velocities. We discuss Velocity-Channel Analysis (VCA) and its offspring Velocity Coordinate Spectrum (VCS) that are based on the analytical description of the spectral line statistics. Those techniques are well suited for studies of supersonic turbulence. We stress that a great advantage of VCS is that it does not necessary require good spatial resolution. Addressing the studies of mildly supersonic and subsonic turbulence we discuss the criterion that allows to determine whether a traditional tool for such a research, namely, Velocity Centroids are dominated by density or velocity. We briefly discuss the use of higher order correlations as the means to study intermittency of turbulence. We discuss observational data available and prospects of the field.

Keywords: turbulence, molecular clouds, MHD

1. What can be learned from fluctuations of velocity?

As a rule astrophysical fluids are turbulent and the turbulence is magnetized. This ubiquitous turbulence determines the transport properties of interstellar medium (see Elmegreen & Falgarone 1996, Stutzki 2001, Cho et al. 2003) and intracluster medium (Inogamov & Sunyaev 2003, Sunyaev, Norman & Bryan 2003, see review by Lazarian & Cho 2004a), many properties of Solar and stellar winds, accretion disks etc. One may say that to understand heat conduction, transport and acceleration of cosmic rays, propagation of electromagnetic radiation in different astrophysical environments it is absolutely essential to understand the properties of underlying turbulence. The mysterious processes of star formation (see McKee & Tan 2002, Elmegreen 2002, Pudritz 2001) and interstellar chemistry (see Falgarone 1999 and references therein), shattering and coagulation of dust (see Lazarian & Yan 2003 and references therein)

D. Block et al. (eds.), Penetrating Bars through Masks of Cosmic Dust, 601–611.

are also intimately related to properties of magnetized compressible turbulence (see reviews by Elmegreen & Scalo 2004).

From the point of view of fluid mechanics astrophysical turbulence is characterized by huge Reynolds numbers, Re, which is the inverse ratio of the eddy turnover time of a parcel of gas to the time required for viscous forces to slow it appreciably. For $Re \gg 100$ we expect gas to be turbulent and this is exactly what we observe in HI (for HI $Re \sim 10^8$).

Statistical description is a nearly indispensable strategy when dealing with turbulence. The big advantage of statistical techniques is that they extract underlying regularities of the flow and reject incidental details. Kolmogorov description of unmagnetized incompressible turbulence is a statistical one. For instance it predicts that the difference in velocities at different points in turbulent fluid increases on average with the separation between points as a cubic root of the separation, i.e. $|\delta v| \sim l^{1/3}$. In terms of direction-averaged energy spectrum this gives the famous Kolmogorov scaling $E(k) \sim 4\pi k^2 P(\mathbf{k}) \sim k^{5/3}$, where $P(\mathbf{k})$ is a *3D* energy spectrum defined as the Fourier transform of the correlation function of velocity fluctuations $\xi(\mathbf{r}) = \langle \delta v(\mathbf{x}) \delta v(\mathbf{x} + \mathbf{r}) \rangle$. Note that in this paper we use $\langle ... \rangle$ to denote averaging procedure.

The example above shows the advantages of the statistical approach to turbulence. For instance, the energy spectrum $E(k)dk$ characterizes how much energy resides at the interval of scales $k, k + dk$. At large scales l which correspond to small wavenumbers k (i.e. $l \sim 1/k$) one expects to observe features reflecting energy injection. At small scales one should see the scales corresponding to sinks of energy. In general, the shape of the spectrum is determined by a complex process of non-linear energy transfer and dissipation. For Kolmogorov turbulence the spectrum over the inertial range, i.e. the range where neither energy injection nor energy dissipation are important, is characterized by a single power law and therefore self-similar. Other types of turbulence, i.e. the turbulence of non-linear waves or the turbulence of shocks, are characterized by different power laws and therefore can be distinguished from the Kolmogorov turbulence of incompressible eddies.

In view of the above it is not surprising that attempts to obtain spectra of interstellar turbulence have been numerous. In fact they date as far back as the 1950s (see von Horner 1951, Munch 1958, Wilson et al. 1959). However, various directions of research achieved various degree of success (see Kaplan & Pickelner 1970, a review by Armstrong, Rickett & Spangler 1995). For instance, studies of turbulence statistics of ionized media were more successful (see Spangler & Gwinn 1990) and provided the information of the statistics of plasma density at scales 10^8-10^{15} cm. This research profited a lot from clear understanding of processes of scintillations and scattering achieved by theorists (see Narayan & Goodman 1989). At the same time the intrinsic limitations of the scincillations technique are due to the limited number of sampling directions, and relevance only to ionized gas at extremely small scales. Moreover, these sort of measurements provide only the density statistics, which is an indirect measure of turbulence.

Velocity statistics is much more coveted turbulence measure. Although, it is clear that Doppler broadened lines are affected by turbulence, recovering of velocity statistics was extremely challenging without an adequate theoretical insight. Indeed, both velocity and density contribute to fluctuations of the intensity in the Position-Position-Velocity (PPV) space.

2. What do velocity centroids tell us?

Let us consider "unnormalized" velocity centroids:

$$S(\mathbf{X}) = \int v_z \, \rho_s(\mathbf{X}, v_z) \, dv_z, \tag{1}$$

where ρ_s is the density of emitters in the PPV space[1].

In the case of emissivity proportional to the first power of density and provided that the turbulent region is thin for its radiation, analytical expressions for structure functions[2] of centroids, i.e. $\left\langle [S(\mathbf{X_1}) - S(\mathbf{X_2})]^2 \right\rangle$ were derived in Lazarian & Esquivel (2003, henceforth LE03). In that paper the following criterion for centroids to reflect the statistics of velocity[3] was established:

$$\left\langle [S(\mathbf{X_1}) - S(\mathbf{X_2})]^2 \right\rangle \gg \langle V^2 \rangle \left\langle [I(\mathbf{X_1}) - I(\mathbf{X_2})]^2 \right\rangle, \tag{2}$$

where $I(\mathbf{X})$ is the intensity $I(\mathbf{X}) \equiv \int \rho_s dV$ and the velocity dispersion $\langle V^2 \rangle$ can be obtained using the second moment ot the spectral lines:

$$\langle V^2 \rangle \equiv \langle \int_{\mathbf{X}} V^2 \rho_s dV \rangle / \langle \int_{\mathbf{X}} \rho_s dV \rangle. \tag{3}$$

LE03 proposed to subtract the right hand sight of the expression (2) from the left hand sight of (2) to obtain *modified velocity centroids* that may still reflect velocity statistics even when ordinary centroids are dominated by density contribution. A numerical study in Esquivel & Lazarian (2004) confirmed that the criterion given by eq (2) is correct and revealed that for MHD turbulence simulations it holds for turbulence with Mach number less than 2. This was consistent with an earlier analysis of a different set of MHD turbulence data by Brunt & Mac Low who observed that velocity centroids poorly present velocity statistics for high Mach number turbulence. Esquivel & Lazarian (2004) studied modified velocity centroids and the statistical errors arising from subtracting the corresponding large numbers in order to find the residual fluctuations. They conclude that at the moment velocity centroids can be identified as a technique to study subsonic turbulence as well as very mildly supersonic turbulence.

Table 1. A summary of analytical results for channel map statistics derived in LP00.

Slice thickness	Shallow 3-D density $P_n \propto k^n, n > -3$	Steep 3-D density $P_n \propto k^n, n < -3$
2-D intensity spectrum for thin slice	$\propto K^{n+m/2}$	$\propto K^{-3+m/2}$
2-D intensity spectrum for thick slice	$\propto K^n$	$\propto K^{-3-m/2}$
2-D intensity spectrum for very thick slice	$\propto K^n$	$\propto K^n$

thin means that the channel width $<$ velocity dispersion at the scale under study
thick means that the channel width $>$ velocity dispersion at the scale under study
very thick means that a substantial part of the velocity profile is integrated over
m is the power-law index of velocity structure function, i.e. $\langle (v(x+r) - v(x))^2 \rangle \sim r^m$.
K is a 2D wavevector in the plane of the slice, k a 3D wavevector.

3. How can supersonic turbulence be studied?

What do channel maps tell us?

Power spectra of fluctuations measured within narrow ranges of velocities have been observed by different authors at various times (see Green 1993). Those power spectra were guessed to be related to underlying turbulence, but what exactly those observed spectra mean was completely unclear.

This problem was addressed in Lazarian & Pogosyan (2000, henceforth LP00) who found the relation between the the spectrum of intensity fluctuations in channel maps and underlying spectra of velocity and density. They found that the power index of the intensity fluctuations depends on the thickness of the velocity channel (see Table 1).

It is easy to see that both steep and shallow underlying density the power law index *steepens* with the increase of velocity slice thickness. In the thickest velocity slices the velocity information is averaged out and it is natural that we get the density spectral index n. The velocity fluctuations dominate in thin slices, and the index m that characterizes the velocity fluctuations, i.e. $|\delta v| \sim l^{m/2}$, can be obtained using thin velocity slices (see Table 1). Note, that the notion of thin and thick slices depends on a turbulence scale under study and the same slice can be thick for small scale turbulent fluctuations and thin for large scale ones. The formal criterion for the slice to be thick is that *the dispersion of turbulent velocities on the scale studied should be less than the velocity slice thickness*. Otherwise the slice is *thin*.

One may notice that the spectrum of intensity in a thin slice gets shallower as the underlying velocity get steeper. To understand this effect let usconsider turbulence in incompressible optically thin media. The intensity of emission in a slice of data is proportional to the number of atoms per the velocity interval given by the thickness of the data slice. Thin slice means that the velocity dispersion at the scale of study is larger than the thickness of a slice. The increase of the velocity dispersion at a particular scales means that less and less energy is being emitted within the velocity interval that defines the slice. As the result the image of the eddy gets fainter. In other words, the larger is the dispersion at the scale of the study the less intensity is registered at this scale within the thin slice of spectral data. This means that steep velocity spectra that correspond to the flow with more energy at large scales should produce intensity distribution within thin slice for which the more brightness will be at small scales. This is exactly what our formulae predict for thin slices (see also LP00).

The result above gets obvious when one recalls that the largest intensities within thin slices are expected from the regions that are the least perturbed by velocities. If density variations are also present they modify the result. When the amplitude of density perturbation becomes larger than the mean density, both the statistics of density and velocity are imprinted in thin slices. For small scale asymptotics of thin slices this happens, however, only when the density spectrum is shallow, i.e. dominated by fluctuations at small scales.

Is spatial resolution necessary to study statistics of velocity?

Spatial resolution is essential for centroids and channel maps. However, the velocity fluctuations are also imprinted on the fluctuations of intensity along the velocity coordinate direction. The corresponding 3D PPV spectra were derived in LP00. Table 3 in LP00 states that two terms, one depending only on velocity and the other depending both on velocity and density, contribute to the spectrum measured along velocity coordinate. for steep density the intermediate scaling therefore is $k^{2n/m}$ (n is negative), while the small scale asymptotics scales as $k^{-6/m}$. If the density is shallow the situation is refersed, namely, at larger scales $k^{-6/m}$ assymptotics dominates, while $k^{2n/m}$ assymptotics is present at smaller scales. The transition from one assymptotics to another depends on the amplitude of density fluctuations. Therefore both density and velocity statistics can be restrored from the observations this way. If the measurements are done with an instrument of poor spatial resolution these are the expected scalings. A further study of these interesting regime is done in Chepurnov & Lazarian (2004). The only requirement for the VCS to operate is for the turbulence to be supersonic and for the instrument to have an adequate *spectral* resolution.

4. What is the effect of absorption?

The issues of absorption were worrisome for the researchers from the very start of the research in the field (see Munch 1958). The erroneous statements about the effects of absorption on the observed turbulence statistics are widely spread in the literature. For instance, a fallacy that absorption allows to observe density fluctuations localized in the thin surface layers of clouds, i.e. 2D turbulence, exists (see discussion in LP04).

Using transitions that are less affected by absorption, e.g. HI, allows frequently to avoid the problem. However, it looks foolish to disregards the wealth of spectroscopic data only because absorption is present. A study of absorption effects is given in LP04. There it was found that for sufficiently thin slices the scalings obtained in the absence of absorption still hold provided that the absorption on the scales under study is negligible. A similar criterion is valid for the VCS. From the practical point of view, absorption imposes an upper limit on the scales for which the statistics can be recovered.

If integrated intensity of spectral lines is studied in the presence of absorption non-trivial effects emerge. Indeed, for optically thin medium the spectral line integration results in intensity reflecting the density statistics. LP04 showed that this may not be any more true for lines affected by absorption. Depending on the spectral index of velocity and density fluctuations the contributions from either from density or velocity dominate the integrated intensity fluctuations. When velocity is dominant a very interesting regime for which intensity fluctuations show universal behavior, i.e. the power spectrum $P(K) \sim K^{-3}$ emerges. If density is dominant, the spectral index of intensity fluctuations is the same as in the case an optically thin cloud. Conditions for these regime as well as for some more interesting intermediate asymptotic regimes are outlined in LP04.

5. How can intermittency be studied?

Velocity and density power spectra do not provide a complete description of turbulence. Intermittency of turbulence (its variations in time and space) and its topology in the presence of different phases are not described by the power spectrum. Recent numerical research that employed higher order correlation functions (Muller & Biskamp 2000, Cho, Lazarian & Vishniac 2002b, 2003) showed them to be a promising tool. For instance, the distinction between the old Iroshnikov-Kraichnan and the GS95 model is difficult to catch using power spectra with a limited inertial range, but is quite apparent for fourth order statistics. The difference in physical consequences of whether the turbulence dissipates in shocks or in intermittent vortices may be very substantial. Recent research (see review by Lazarian & Cho 2004) suggests that dissipation of interstellar motions via vortices is very important. Although the scaling of vortex intermittency with the Reynolds number R is not clear, it is suggestive that the intermittency is increasing with R.

Higher order statistics obtained from observational data were reported for observed velocity[4] in Falgarone et al. (1994) and for density in Padoan et al. 2003. According to Falgarone & Puget (1995) and Falgarone et al. (1995), the intermittency in vorticity distribution can result in the outbursts of localized dissipation that make tiny regions within cold diffuse clouds chemically active. This is an extremely important conclusion that stimulates more intensive studies of higher moment statistics from observations.

6. What are the niches for different techniques?

Astrophysical fluids demonstrate turbulent motions at very different Mach numbers. The "old and good" velocity centroids are shown (LE03, Esquivel & Lazarian 2004) to be reliable only at low Mach numbers. This makes some of the earlier results obtained, e.g. for hypersonic motions in HI using velocity centroids (see Meville-Deschenes et al. 2003), as well as some results on molecular clouds somewhat questionable. Observational testing whether the criterion (2) is satisfied is essential for a confident use of centroids to study velocity statistics.

For supersonic turbulence VCA and VCS present the best bet at the moment. The analytical description of intensity fluctuations in PPV space obtained in LP00 and LP04 allows to reliably separate velocity and density contributions to the observed fluctuations. VCS looks to be the most promising tool as it requires only frequency resolution to be adequate. It opens an avenue for studies of turbulence in poorly resolved objects, e.g. for extragalactic research.

Another limitation on the use of the techniques arises from the density-velocity correlations in the data. LP00 analytically studied the effect of velocity-density correlations for the VCA and concluded that even for the maximal possible level of correlation, the scaling of intensity fluctuations in thin velocity channels is not affected. Further numerical studies in Lazarian et al. (2001) and Esquivel ct al. (2003) confirmed that the velocity-density correlations present in compressible MHD flows do not change the statistics obtained with the VCA. Our ongoing research shows that velocity centroids seem to be more affected by the velocity-density correlations, but this is not a dominant effect for at least for the low Mach number flows for which the centroids are applicable. Currently we study the effect of correlations arising from gravity.

For studies of higher order statistics velocity centroids present the best bet for the moment. This places limitations on the data sets to be studied (e.g. the criterion (2) should be satisfied), and stimulates the search for an alternative techniques.

We also would like to stress that the different techniques discussed are complementary. For instance, it was noted in LE03 that the velocity centroids are sensitive only to solenoidal motions, while VCA is affected by both potential and solenoidal motions. Therefore combining the two techniques it should be possible to study the effects of compressibility in the astrophysical flows. In addition, different techniques are affected differently by gas temperature. This allows to get insight into the temperature distribution along the line of sight.

Obtaining the statistics of velocity turbulence may provide sometimes unexpected bonuses. For instance, the damping scale of turbulence can be determined if the velocity dispersion is known at the injection scale. This scale could provide a standard yardstick for finding the distances to clouds, e.g. to high velocity clouds.

7. What do observations tell us?

Application of the VCA to the Galactic data in LP00 and to Small Magellanic Cloud in Stanimirovic & Lazarian (2001) revealed spectra of 3D velocity fluctuations consistent with the Kolmogorov scaling. LP00 argued that the same scaling was expected for the magnetized turbulence appealing to the Goldreich-Shridhar (1994) model. Esquivel et al. (2003) used simulations of MHD turbulent flows to show that in spite of the presence of anisotropy caused by magnetic field the expected scaling of fluctuations is Kolmogorov. Studies by Cho& Lazarian (2002, 2003) revealed that the Kolmogorov-type scaling is also expected in the compressible MHD flows. These studies support MHD turbulence model for SMC.

Studies of turbulence are more complicated for the inner parts of the Galaxy, where (a) two distinct regions at different distances from the observer contribute to the emissivity for a given velocity and (b) effects of the absorption are important. However, the analysis in Dickey et al. (2001) showed that some progress may be made even in those unfavorable circumstances. Dickey et al. (2001) found the steepening of the spectral index with the increase of the velocity slice thickness. They also observed the spectral index for strongly absorbing direction approached -3 in accordance with the conclusions in LP04.

21-cm absorption provides another way of probing turbulence on small scales. The absorption depends on the density to temperature ratio ρ/T, rather than to ρ as in the case of emission. However, in terms of the VCA this change is not important and we still expect to see emissivity index steepening as velocity slice thickness increases, provided that velocity effects are present. In view of this, results of Deshpande et al. (2001), who did not see such steepening, can be interpreted as the evidence of the viscous suppression of turbulence on the scales less than 1 pc. The fluctuations in this case should be due to density and their shallow spectrum $\sim k^{-2.8}$ may be related to the damped magnetic structures below the viscous cutoff (Cho, Lazarian & Vishniac 2002b).

Studies of velocity statistics using velocity centroids are numerous (see O'Dell 1986, Miesch & Bally 1994, Miesch, Scalo & Bally 1999). The analysis of observational data in Miesch & Bally (1994) provides a range of power-law indexes. Their results obtained with structure functions if translated into spectra are consistent with $E(k) = k^{\beta}$, where $\beta = -1.86$ with the standard deviation of 0.3. The Kolmogorov index $-5/3$ falls into the range of the measured values. L1228 exhibits exactly the Kolmogorov index -1.66 as the mean value, while other low mass star forming regions L1551 and HH83 exhibit indexes close to those of shocks, i.e. ~ -2. The giant molecular cloud regions show shallow indexes in the range of $-1.9 < \beta < -1.3$ (see Miesch et al. 1999). It worth noting that Miesch & Bally (1994) obtained somewhat more shallow indexes that are closer to the Kolmogorov value using autocorrelation functions. Those may be closer to the truth as in the presence of absorption in the center of lines, minimizing the regular velocity used for individual centroids might make the results more reliable. Whether the criterion given by (2) is satisfied for the above data is not clear. If it is not satisfied, which is quite possible as the Mach numbers are large for molecular clouds, then the spectra above reflect the density rather than velocity. If the criterion happen to be satisfied a more careful study taking absorption effects into account is advantageous.

Apart from testing of the particular scaling laws, studies of turbulence statistics should identify sources and injection scales of the turbulence. Is turbulence in molecular clouds a part of a large scale ISM cascade (see Armstrong et al. 1997)? How does the share of the energy within compressible versus incompressible motions vary within the Galactic disk? There are examples of questions that can be answered in future.

Acknowledgments A.L. acknowledges the support of Center for Magnetic Self-Organization in Laboratory and Astrophysical Plasmas. Fruitful discussions with Alexey Chepurnov and Dmitry Pogosyan are acknowledged.

Notes

1. Traditionally used centroids include normalization by the integral of ρ_s. This, however does not substantially improve the statistics, but makes the analytical treatment very involved (Lazarian & Esquivel 2003).
2. Expressions for the correlation functions are straightforwardly related to those of structure functions (Monin & Yaglom 1975). The statistics of centroids using correlation functions was used in a later paper by Levier (2004).
3. LE03 showed that the solenoidal component of velocity spectral tensor can be obtained from observations using velocity centroids in this regime.
4. Whether in all cases the used centroids reflected the actual velocity statitics is not sure because the criterion by Lazarian & Esquivel (2003) has not been applied to the data.

References

Armstrong, J. W., Rickett, B. J., & Spangler, S. R. 1995, ApJ, 443, 209
Brunt, C.M., Mac Low, M.M. 2004, ApJ, 604, 196
Chepurnov, A. & Lazarian, A. 2004, ApJ, submitted

Cho, J., Lazarian, A. 2002, Phy. Rev. Lett., 88, 245001 (CL02)

Cho, J., Lazarian, A. 2003a, MNRAS, 345, 325

Cho, J., Lazarian, A., Honein, A., Knaepen, B., Kassinos, S., & Moin, P. 2003, ApJ, 589, L77

Cho, J., Lazarian, A., & Vishniac, E. 2002a, ApJ, 564, 291

Cho, J., Lazarian, A., & Vishniac, E. 2002b, ApJ, 566, L49

Cho, J., Lazarian, A., & Vishniac, E. 2003b, ApJ, 595, 812

Deshpande, A.A., Dwarakanath, K.S., Goss, W.M., 2000, ApJ, 543, 227

Elmegreen, B. 2002, ApJ, 577, 206

Elmegreen, B. & Falgarone, E. 1996, ApJ, 471, 816

Elmegreen, B. & Scalo, J. 2004, ARA&A, in press

Esquivel, A. & Lazarian, A. 2004, ApJ, submitted

Esquivel A., Lazarian A., Pogosyan D., & Cho J. 2003, MNRAS, 342, 325

Falgarone, E. 1999, in *Interstellar Turbulence*, ed. by J. Franco, A. Carraminana, CUP, (henceforth *Interstellar Turbulence*) p.132

Falgarone, E., Lis, D. C., Phillips, T. G., Pouquet, A., Porter, D. H., Woodward, P. R. 1994, ApJ, 436, 728

Falgarone, E., Panis, J.-F., Heithausen, A., Perault, M., Stutzki, J., Puget, J.-L., Bensch, F. 1998, A& A, 331, 669

Falgarone, E, & Puget, J.-L. 1995, A& A, 293, 840

Falgarone, E., Pineau des Forets, G., & Roueff, E. 1995, A&A, 300, 870

Goldreich, P. & Sridhar, S. 1995, ApJ, 438, 763

Green, D.A. 1993 MNRAS, 262, 328

von Horner, S. 1951, Zs.F. Ap., 30, 17

Inogamov, N.A. & Sunyaev, R.A. 2003, Astronomy Letters, 29, 791.

Kolmogorov, A. 1941, Dokl. Akad. Nauk SSSR, 31, 538

Kaplan, S.A. & Pickelner, S.B. 1970

Lazarian, A. & Cho, J. 2004, Ap&SS, in press

Lazarian, A. & Esquivel, E. 2003, ApJ, 592, L37 (LE03)

Lazarian, A. & Pogosyan, D. 2000, ApJ, 537, 720 (LP00)

Lazarian, A. & Pogosyan, D. 2004, ApJ, in press (LP04)

Lazarian, A. & Pogosyan, D., & Esquivel, A. 2002, in *Seeing Through the Dust*, ASP Conf. Proc. Vol 276, eds by A. R. Taylor et al. (San Francisco), p. 182

Lazarian, A. & Pogosyan, D., Vazquez-Semadeni, E., & Pichardo, B. 2001, ApJ, 555, 130

Lazarian, A. & Yan, H. 2003, in "Astrophysical Dust" eds. A. Witt & B. Draine, APS, in press

Levier, F. 2004, ApJ, in press

McKee, Christopher F.; Tan, Jonathan C. 2002, Nature, 416, 59

Miesch, M.S., & Bally, J. 1994, ApJ, 429, 645

Miesch, M. S., Scalo, J., & Bally, J. 1999, ApJ, 429, 645

Miville-Deschenes, M.A., Levrier, F., & Falgarone, E. 2003, ApJ, in press,

Miville-Deschenes, M.A., Joncas, G., Falgarone, E., & F. Boulanger, 2003, A&A, in press, astro-ph/0306570

Monin, A.S., & Yaglom, A.M. 1975, Statistical Fluid Mechanics: Mechanics of Turbulence, vol. 2, The MIT Press

Muller, W.-C. & Biskamp, D. 2000, Phys. Rev. Lett., 84(3), 475

Narayan, R., & Goodman, J. 1989, MNRAS, 238, 963

Munch, G. 1958, Rev. Mod. Phys., 30, 1035

O'Dell, C.R. 1986, ApJ, 304, 767

Ossenkopf, V. & Mac Low, M.-M. 2002, A&A, 390, 307

Padoan, P., Boldyrev, S., Langer, W., & Nordlund, A. 2003, ApJ, 583, 308
Pudritz, R. E. 2001, From Darkness to Light: Origin and Evolution of Young Stellar Clusters, ASP , Vol. 243. Eds T. Montmerle and P. Andre. San Francisco, p.3
Stutzki, J. 2001, Astrophysics and Space Science Supplement, 277, 39
Stanimirovic, S. & Lazarian, A. 2001, ApJ, 551, L53
Spangler, S.R., & Gwinn, C.R. 1990, ApJ, 353, L29
Sunyaev, R.A., Norman, M.L., & Bryan, G.L. 2003, Astronomy Letters, 29, 783.
Wilson, O.C., Munch, G., Flather, E.M., & Coffeen, M.F. 1959, ApJS, 4, 199

SHURE

ESTIMATING POWER SPECTRA OF GALAXY STRUCTURE: CAN STATISTICS HELP?

Philip B. Stark
Department of Statistics, University of California, Berkeley, CA 94720-3860 USA

Abstract Spectrum estimates based on the raw periodogram (squared modulus of the Fourier transform) can be improved upon, at least according to criteria that matter to statisticians. Estimates based on averages of power spectrum estimates derived using orthogonal weight functions (*multitaper* estimates) can have better bias and variance properties. The raw periodogram also can be "post-processed," using wavelet shrinkage or other smoothing techniques, to produce estimates that have more desirable statistical properties than the periodogram, and that are visually more pleasing. Subject to assumptions that can be difficult to test, these statistical techniques might improve inferences about the underlying physics, depending on which features of the spectrum matter physically. Differences between the periodogram and wavelet-shrinkage estimates are quite noticeable for the angular power spectrum of starlight from NGC 4622. However, estimates of the slope of a power law model for the spectrum are nearly identical.

Keywords: Multitaper, wavelet shrinkage

1. Introduction

Angular power spectra of light from galaxies as a function of radius from the galactic center contain information about the kinematics and dynamics of galactic formation and structure. Some recent articles that use power spectra of galactic light or HI emission to answer questions about the dynamics of galaxies estimate the power spectrum of galaxy structure using the periodogram–the squared modulus of the Fourier transform (e.g., Elmegreen et al., 2001; Elmegreen et al., 2002).

The periodogram spectrum estimate has statistically undesirable properties; for example, it is not statistically consistent in many problems (e.g., Percival and Walden, 1993). (An estimator is *consistent* if it converges in probability to the truth as the number of data grows.) Moreover, when stationary time series are observed over a finite window, the periodogram is the squared modulus of the Fourier transform of the data convolved with a sinc function. The large "ringing" sidelobes of the sinc function produce broad-band leakage

D. Block et al. (eds.), Penetrating Bars through Masks of Cosmic Dust, 613–617.

in the periodogram spectrum estimate, especially if the underlying spectrum has peaks. Statisticians often assert that more sophisticated techniques can produce better spectrum estimates than the squared modulus of the Fourier transform—at least when the data are noisy or when the data are a realization of a stochastic process whose power spectral density is sought (e.g., Percival and Walden, 1993). "Better" typically means having smaller mean integrated squared error—a combination of bias and variance—but other criteria are also used.

2. Multitaper power spectrum estimates

In estimating power spectra from partial observations (data with temporal or spatial gaps, time series with limited observing times), *multitaper spectrum estimates* can be helpful. Multitaper techniques (Percival and Walden, 1993) apply several orthogonal apodizing filters to the data, find the periodogram of each filtered data set, then average those quadratic spectrum estimates. The filters are designed to minimize broad-band leakage, a form of bias. Combining estimates derived using orthogonal filters reduces the variance of the estimate. Komm et al., 1999 and Fodor and Stark, 2000 apply multitapering to spectrum estimation problems in helioseismology.

3. Wavelet Shrinkage

Estimating angular power spectra of galaxies is a bit different from the problem described above: the data are observed on circles, essentially without gaps (but for contamination by nearby stars, etc.), so the signals are periodic. *Wavelet shrinkage denoising* can be applied in this situation. David Donoho, Ian Johnstone and others have shown (under stylized assumptions) that a straightforward estimator is asymptotically optimal for estimating a wide class of functions observed with additive Gaussian noise (e.g., Donoho et al., 1995; Donoho and Johnstone, 1998). The estimator consists of transforming the observations into their empirical wavelet decomposition, discarding wavelet coefficients that are small compared with a threshold related to the estimated noise level, possibly decreasing the remaining coefficients, and finally inverting the wavelet transform from those "shrunken" wavelet coefficients. This procedure is a nonlinear smoother that can adapt to spatially varying roughness of the underlying function.

Under some technical assumptions, wavelet shrinkage improves estimates of the logarithm of the spectrum from the log periodogram (e.g., Moulin, 1994; Gao, 1996). (The logarithm equalizes the variances of the estimates at different frequencies.) In some spectrum estimation problems, multitapering and wavelet shrinkage can be combined to good effect (Walden et al., 1995). For an application to helioseismology, see Komm et al., 1999.

4. Results: Wavelet Shrinkage Spectrum Estimation

Profs. D. Block and R. Buta provided me a Hubble Space Telescope optical image of NGC 4622. I estimated the location of the center of the galaxy as the point of maximum intensity, pixel $(507, 594)$, then masked an annulus of data at distances between 100 and 109 pixels from the center, corresponding to distances of about 2.181kpc to 2.377kpc. I divided the annulus into 512 equiangular bins, and averaged the data in each bin (there were 7–20 pixels in each). Figure 1 shows the result. WaveLab (Buckheit et al., 1995) is a library of MATLAB functions for wavelet analysis, including implementations of wavelet shrinkage. I used WaveLab to "denoise" the log periodogram. Figure 2 shows the log periodogram and the wavelet-denoised periodogram, using symmlets of degree 8, leaving the lowest 4 levels alone, and using Stein's unbiased risk estimate for thresholding. Figure 3 shows the resulting spectra after taking the antilog and multiplying by $k^{5/3}$, corresponding to a power-law spectrum with exponent $-5/3$. The wavelet-denoised spectrum is much smoother in most places, but not everywhere, illustrating the adaptive nature of wavelet shrinkage.

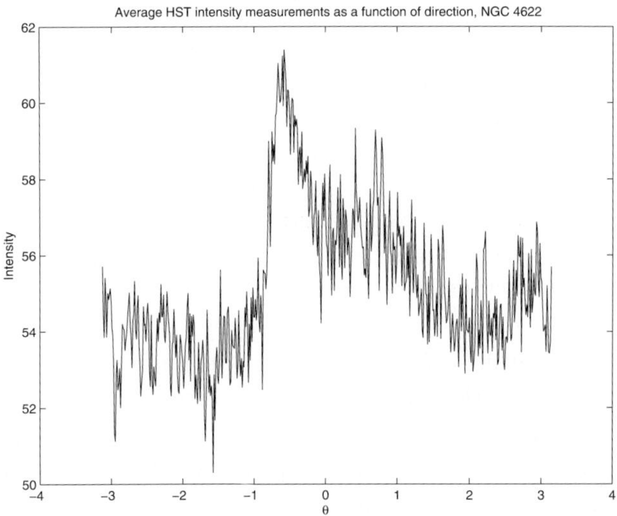

Figure 1. Average HST intensity measurements for NGC 4622 at a distance of about 2.181kpc to 2.377kpc, as a function of direction

Do the differences matter? That depends on many things, including *which* parameters are physically interesting. If the parameters are overall summaries, such as the exponent of a power-law model, the differences probably do not matter much. For example, I fit regression lines to the log periodogram and the denoised log periodogram to estimate the exponent in a power-law model for

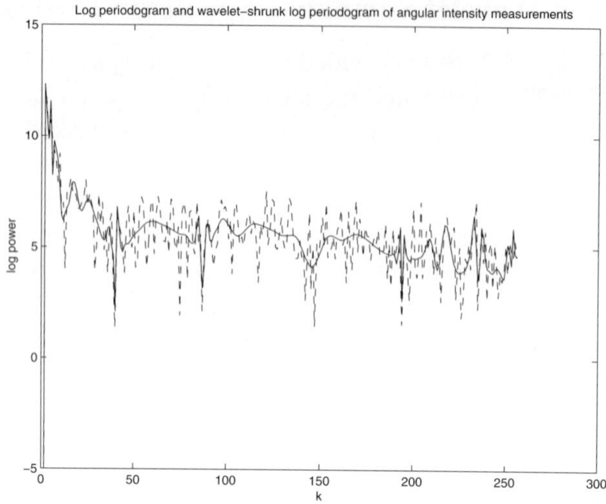

Figure 2. Log periodogram of the data in figure 1, and wavelet-denoised log periodogram. Y-axis is truncated at -5 and 15. The solid curve is the wavelet-denoised estimate. See text for the parameters used in wavelet shrinkage.

the spectrum. The results differ slightly but not significantly. The formal uncertainties of the regression coefficients for the denoised spectrum are smaller, as one would expect from its greater smoothness, but a simulation study would give a better assessment of the bias and uncertainty in the results. This is but

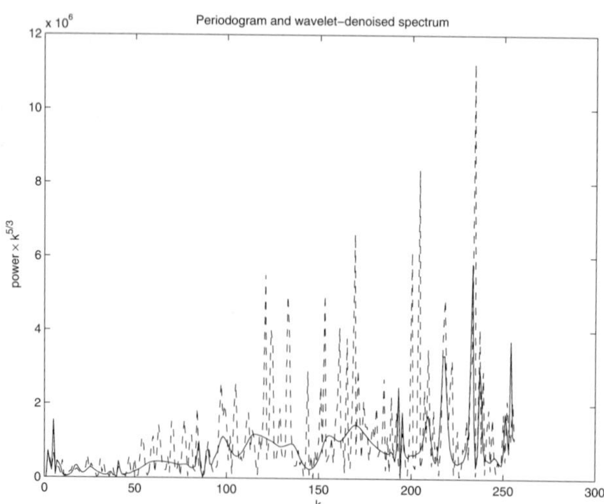

Figure 3. Periodogram and wavelet-denoised spectrum estimate of data shown in Fig. 1, both multiplied by $k^{5/3}$. The solid curve is the wavelet-denoised estimate.

one example, and hardly conclusive. (I did repeat the numerical experiment for an annulus of data bounded by radii of 200 and 209 pixels; the result was similar.) On the other hand, inferences about the existence of or location of a particular peak in the spectrum probably are much more sensitive to the estimation method.

Acknowledgments

I am grateful to Prof. David Block for encouraging me to look at this problem and for helpful conversations; to Prof. Block and Prof. Ron Buta for HST image data of NGC 4622; and to Dr. Bruce Elmegreen for angular swath data for NGC 5055. I am very grateful to the scientific and local organizers of the conference for the chance to participate.

References

Buckheit, J., Chen, S., Donoho, D., Johnstone, I., and Scargle, J. (1995). WaveLab. Technical report, Stanford University.

Donoho, D. and Johnstone, I. (1998). Minimax estimation via wavelet shrinkage. *Ann. Stat.*, 26:879–921.

Donoho, D., Johnstone, I., Kerkyacharian, G., and Picard, D. (1995). Wavelet shrinkage: asymptopia? (with discussion). *J. Roy. Stat. Soc., Ser. B*, 57:301–369.

Elmegreen, B., Kim, S., and Staveley-Smith, L. (2001). A fractal analysis of the H I emission from the Large Magellanic Cloud. *Ap. J.*, 548:749–769.

Elmegreen, D., Elmegreen, B., and Eberwein, K. (2002). Dusty acoustic turbulence in NGC 4450 and NGC 4736. *Ap. J.*, 564:234–243.

Fodor, I. and Stark, P. (2000). Multitaper spectrum estimation for time series with gaps. *IEEE Trans. Signal Processing*, 48:3472–3483.

Gao, H.-Y. (1996). Choice of thresholds for wavelet shrinkage of the log spectrum. Technical report, MathSoft, Seattle, WA.

Komm, R., Gu, Y., Stark, P., and Fodor, I. (1999). Multitaper spectral analysis and wavelet denoising applied to helioseismic data. *Ap. J.*, 519:407–421.

Moulin, P. (1994). Wavelet thresholding techniques for power spectrum estimation. *IEEE Trans. Signal Process.*, 42:3126–3136.

Percival, D. and Walden, A. (1993). *Spectral Analysis for Physical Applications: Multitaper and Conventional Univariate Techniques*. Cambridge, Cambridge.

Walden, A., E.McCoy, and Percival, D. (1995). Spectrum estimation by wavelet thresholding of multitaper estimates. Technical report, Dept. of Mathematics, Imperial College of Science, Technology and Medicine, London.

FROM $Z > 6$ TO $Z \sim 2$: UNEARTHING GALAXIES AT THE EDGE OF THE DARK AGES

Garth Illingworth & Rychard Bouwens
UCO/Lick Observatory, University of California, Santa Cruz, CA 95064

Abstract Galaxies undergoing formation and evolution can now be directly observed over a time baseline of some 12 Gyr. An inherent difficulty with high–redshift observations is that the objects are very faint and the best resolution (HST) is only ~ 0.5 kpc. Such studies thereby combine in a highly synergistic way with the great detail that can be obtained for nearby galaxies through "archaeological" studies. Remarkable advances are being made in many areas, due to the power of our observatories on the ground and in space, particularly the unique capabilities of the HST ACS. Three new developments are highlighted. First, is the derivation of stellar masses for galaxies from spectral energy distributions (SEDs) using HST and now Spitzer data, and dynamical masses from both sub–mm observations of CO lines and near–IR observations of optical nebular lines like Hα. A major step has been taken with evidence that points to the $z \sim 2-3$ LBGs having masses that are a few $\times 10^{10}\ M_{\odot}$. Second is the discovery of a new population of red, evolved galaxies, again at redshifts $z \sim 2-3$ which appear to be the progenitors of the more massive early–type galaxies of today, with dynamical masses around a few $\times 10^{11}\ M_{\odot}$. Third are the remarkable advances that have occurred in characterizing drop–out galaxies (LBGs) to $z \sim 6$ and beyond, less than 1 Gyr from recombination. The HST ACS has played a key role here, with the dropout technique being applied to i and z images in several deep ACS fields, yielding large samples of these objects. This has allowed a detailed determination of their properties (e.g., size, color), and meaningful comparisons against lower–redshift dropout samples. The use of cloning techniques has overcome many of the strong selection biases that affect the study of high redshift populations. A clear trend of size with redshift has been identified, and its impact on the luminosity density and star formation rate estimated. There is a significant, though modest, decrease in the star formation rate from redshifts $z \sim 2.5$ out through $z \sim 6$. The latest data also allow for the first robust determination of the luminosity function at $z \sim 6$. Last, but not least, the latest UDF ACS (optical) and NICMOS (near–IR) data has resulted in the detection of some galaxies at $z \sim 7-8$.

Keywords: Galaxy Formation, Galaxy Evolution, High Redshift Galaxies

D. Block et al. (eds.), Penetrating Bars through Masks of Cosmic Dust, 619–636.

1. Watching Galaxies Form and Grow

Direct observation of galaxies in their formative stages is proving to be an extremely powerful approach for understanding how galaxies form and grow. The key physical processes are, in principle, directly observable. However, the primary challenge for direct observations is that high redshift galaxies are faint and small, and the rest frame observations are often in the UV, a spectral region that has been poorly characterized locally – though current observations by GALEX are beginning to provide the needed $z \sim 0$ benchmark.

A key consideration, of particular relevance for this workshop, is that studies of high redshift galaxies are complementary to what can be learned from nearby galaxies – although strongly synergistic may be a more appropriate characterization. The ability to do both extraordinarily detailed "archaeological" studies of nearby galaxies at one epoch (now) contrasts with the much more superficial characterization that can be carried out for distant galaxies. The resolution at high redshift, $6 - 8$ kpc arcsec^{-1} from $z \sim 0.5$ to $z \sim 6+$, is dramatically worse than can be achieved at $z \sim 0$, and presents a serious challenge, even with HST. The discussion that has taken place at this workshop about bars in $z \sim 1+$ galaxies is an example of how small changes in resolution (from WFPC2 to ACS) can lead to significant differences in conclusions about the properties of high redshift galaxies. Nonetheless the opportunity to measure the properties of galaxies directly, even at the global level, to within < 1 Gyr from the Big Bang, is exciting and valuable, even if the details are mostly lacking.

In the talk by Ken Freeman (this volume), a further example was given of how observations at high and low redshift combine to challenge our view of galaxy evolution. The thin disk of our galaxy appears to have been undisturbed for some 10 Gyr, corresponding to a relatively quiescent life since $z \sim 2$. Yet large disks appear to become rare around $z > 1.2$. Are we missing many such disks in our observations? Or were we just "lucky" in the Milky Way....?

A very interesting constraint on the nature of high redshift galaxies comes from the cosmic baryon budget as discussed by Fukugita, Hogan and Peebles (1998). They compiled a census of where the baryons are at $z \sim 0$. The vast majority are in gas or plasma (83%). Of the 17% locally that are in stars, 73% are in spheroids (bulges/ellipticals), 25% are in disks and 2% are in late-type galaxies. Thus a key issue at high-redshift (at $z > 1 - 2$) is identifying and characterizing spheroid (bulge and elliptical) buildup. Most of the star formation at high redshift must be part of the buildup of spheroids, with different classes of objects (LBGs, evolved $J - K_s$ objects, SCUBA sources etc) exemplifying different phases of this process. This is particularly true if the 25% in disks was assembled relatively late – maybe after $z \sim 1.5$?.

2. Observations at High Redshift

The observational goals of high–redshift galaxy studies have largely been to establish global properties, such as luminosity, sizes, colors, structure (scale lengths, shapes, etc), dust content, kinematics, etc. From these we hope to determine the galaxy luminosity function, stellar (from SEDs) and dynamical mass distributions, luminosity density and its evolution, star formation rate evolution, mass buildup, merging rate, AGN role, etc.

The lack of resolution and the inherent faintness of high–redshift galaxies results in a "morphology challenge". This basic problem is further exacerbated by our difficulty in obtaining high spatial resolution observations in the rest–frame optical. Our high resolution images (HST ACS) are essentially all in the optical, which is in the rest–frame UV at $z \gtrsim 1.5$. At these wavelengths, evolved populations become faint and the effect of dust is magnified. Comparisons (morphology, structure, shape, etc.) with low redshift optical samples are then subject to biases. High spatial resolution IR ($1 - 10+$ μm) imaging such as that from WFC3 on HST, and later from JWST, are key steps that will alleviate this problem. Spitzer of course, will provide remarkably valuable data on high–redshift objects, but its very low spatial resolution precludes any measurement of structure.

From the perspective of high–redshift galaxy studies one of the most significant developments of the last decade has been a much needed refinement of the cosmological parameters. The large uncertainties in timescale that permeated galaxy formation discussions in the 1980s and 1990s have largely evaporated. The WMAP results (Bennett et al. 2003), combined with other constraints, have enabled us to compare timescales from local observations with redshift "ages" with increased confidence.

With $H_0 = 71, \Omega_m = 0.27, \Omega_\lambda = 0.73, t_0 = 13.665$ Gyr, it is of interest to note a few timescales and epochs relevant for galaxy formation and evolution. Some useful numbers: $z \sim 6$ corresponds to an age of ~ 1 Gyr; redshift 5 is when the universe is $\sim$10% of its current age; redshift 2 is at $\sim$25%, or about 3.4 Gyr from recombination; and redshift 0.8 corresponds to an age ~ 7 Gyr, when the universe was just half its present age. At the earliest times for galaxies, the universe appears to be reionized by $z \sim 6$, <1 Gyr from recombination, with reionization likely starting at $z \sim 15 - 20$ (Kogut et al. 2003). This requires that the first UV-bright stars occurred within $0.2 - 0.3$ Gyr of recombination. Since the first QSOs are seen at $z \sim 6 - 7$, deep potential wells must be in place by that time (within ~ 0.5 Gyr of the first stars forming!).

By $z \sim 1+$, all the significant elements of the classic Hubble sequence seem to be in place (though not necessarily in the same proportions as today). This is rather striking. We tend to be rather cavalier about the $z \sim 1$ epoch, treating it as but a "slightly modified version of now". Yet $z \sim 1 - 1.5$ is around $8 - 9$ Gyr ago, more than half the age of the universe. It is fascinating to think that the galaxy population of today was largely in place by $z \sim 1$, and that this buildup happened between $z \sim 6$, or somewhat before, and $z \sim 1.5 - 1$, or over a timespan of only a few Gyr.

Another way of describing this change is to note that around $z \sim 1.3$ is when the morphological characteristics of galaxies appear to undergo a change to much less structured forms (even taking into account the tendency of UV observations to enhance star–forming regions relative to older, smoother populations). Much less regular galaxies, like those common in the deepest HST images (the HDFs and the UDF), appear to become the dominant forms. For galaxies, one could think of three epochs: "reionization" at $z > 6$, the "weird" ages from $\sim 1 < z < 6$, and the "normal" galaxy epoch from $z \sim 1$ onwards. Dramatic examples of galaxies at higher redshifts can be seen in the HST UDF images (see, e.g., the color plate in this volume from the UDF). A fascinating recent result is that the star formation rate (and mass buildup) in galaxies is approximately constant from $z \sim 6+$ to $z \sim 1$ (the first 50% of time) and then decreases significantly ($5 - 10\times$) to $z \sim 0$.

A well-known, but nonetheless extremely critical observational problem for studies of high–redshift galaxies is that surface brightness goes as $(1 + z)^4$. This surface brightness dimming dramatically impacts the detectability of galaxies at high redshift (corresponding to $\sim 600\times$ – or 7 mag – from $z \sim 0$ to $z \sim 5$).

3. HST ACS

We are fortunate to have a number of powerful new tools with which carry out the needed observations. The remarkable new spectrographs on large ground-based telescopes (DEIMOS, VIRMOS, IMACS, for example) have revolutionized our ability to take large numbers of redshifts. HST, Chandra and Spitzer each contribute unique and important information on the nature of high–redshift objects. Arguably though, it is the Advanced Camera on HST (Ford et al. 2003) that is providing the most valuable and unique data at this time (though the impact of Spitzer will undoubtedly grow). High–redshift galaxies are small, and therefore it has only been through the capabilities of HST and the ACS that we have been able see structural details. In short, the advent of the HST ACS has greatly increased our ability to "watch galaxies form and grow". The sensitivity, resolution and excellent filter set have provided us with images from which large samples of high–redshift galaxies can be derived. Of particular interest are those galaxies with red enough $i - z$ colors to qualify as i–dropouts – galaxies at redshifts $z \sim 6$, within 1 Gyr of recombination. Such objects have been the focus of a number of papers over the last year (e.g., Bouwens et al. 2003b, Stanway et al. 2003, Yan et al. 2003, Dickinson et al. 2004). Spectroscopic confirmation is beginning to appear (e.g., Bunker et al. 2003, Dickinson et al. 2004, Stanway et al. 2004), but such observations are challenging, as Weymann et al. (1998) demonstrated with their $z = 5.6$ object, which took over 7 hours of integration on Keck.

Given the crucial and central role that HST plays in studies of high–redshift galaxies, it is worthwhile listing those fields that are playing a starring role in our collective efforts to push the frontiers on the properties of distant galaxies. In addition to the time-honored WFPC2 HDF–N and HDF–S fields, there are now several HST ACS datasets with deep observations in multiple filters (broadly B, V, i, z). For the highest redshift galaxies the key filters are the two that are new on HST, the ACS i and z filters, which are perfect for detecting $z \sim 6$ galaxies, i.e., i–dropouts. The new ACS fields are the Great Observatories Origins Deep Fields (GOODS) CDF–S and HDF–N, the Hubble Ultra-Deep Field (UDF), and the two UDF–Parallel fields UDF–Ps (which are deeper than the original HDFs). Most of the UDF has also been imaged with NICMOS in J_{110} and H_{160}, giving a very deep dataset that complements the ACS optical filter data. The UDF data is deep, with a 5σ limiting magnitude of $\sim 30 - 31$ AB mag in B, V, i, z and ~ 27.5AB mag in J_{110}, H_{160} in the UDF–IR image.

4. High-Redshift Galaxies - Issues/Questions

The breadth of studies on high redshift galaxies can be exemplified by noting some of the questions which are actively being addressed in current observational programs:

- When did the first galaxies begin to grow (at $z > 7$, by $z \sim 10 - 15$?)
- When were the first giant Es assembled?
- What are the sub-mm (SCUBA) galaxies? – and are they an important contributor to the mass density?
- Are we missing a significant fraction of assembling galaxies (dusty reddened objects – a key area for Spitzer)?
- Do we really have a good estimate of the star formation history of the universe SFR(z)? Are our dust estimates right?
- How/when do disks form – and how common are large disks at $z \sim 1.5$ or even 2+? What is the role of bars in evolution at high redshift?
- What is the mass assembly history? We see light, but what really counts is mass (and that is really important for the connection to theory).
- Why are black hole masses and galaxy velocity dispersions so tightly related? When did the first massive black holes buildup in galaxies?
- What was the role of AGNs in early galaxy evolution?

• What are the $z \sim 1$ progenitors of the ~50% (?) of todays E/SO population that have formed since then?

5. High-Redshift Galaxies - Current Frontiers

The capabilities of the current generation of large ground-based telescopes (like Keck, the VLT, Gemini and Magellan), combined with the Great Observatories (HST, Chandra, Spitzer) have led to the initiation of numerous studies of the intermediate–to–high–redshift universe. On–going examples, which were discussed at this workshop, are the VIRMOS, DEEP, GEMS, COSMOS and GOODS surveys. They will provide great insight into the nature of the universe at $z \sim 0.7-1.4$. Complementary to these studies of field galaxies are programs on clusters like that described by at this meeting by Holland Ford. This ACS GTO team program focuses on rich clusters of galaxies out to $z \sim 1.3$, with an extension to protoclusters at high–redshift, as described by George Miley.

A key part of these surveys is the characterization of galaxies at intermediate redshift through kinematical measures (velocity dispersion and rotation curves) of very large samples, as described by David Koo for example, for the DEEP survey. A number of groups are focusing on deriving fundamental plane parameters for both field and cluster galaxies out to $z \sim 1.3$, and beyond.

A subject of great interest is understanding the nature of the powerful sub–mm sources (e.g., SCUBA sources) and establishing their redshifts and properties. Considerable progress is occurring in this area, but truly astonishing gains will come about when ALMA comes on line. LIRGS and ULIRGS at high redshift will continue to be the focus of considerable effort in the future.

The nature of Lyman break galaxies (LBGs) at $2 < z < 4$ is starting to become clearer, as IR and kinematical data begin to establish their stellar and dynamical masses. Many important studies are continuing on these sources. A related activity is assessing the properties of galaxies in the redshift "desert" at $1.5 < z < 2.2$. Recent results, for both star–forming and evolved galaxies in this redshift range, will prove to be very valuable in finally establishing the star formation history of the universe across the full redshift range.

These studies, and others, will benefit from major new datasets, particularly from Spitzer (e.g., the GTO programs, and Legacy surveys like GOODS) and HST deep fields (like the GOODS CDF–S and HDF–N, UDF, UDF–Ps, and UDF–IR from NICMOS), combined with Chandra data, and with ground–based spectra from 8-m class telescopes (and then from ALMA).

Spitzer, in particular, will provide a valuable addition to the photometry (optical and IR) now available for large samples of galaxies, extending the wavelength baseline by a factor of 10 or more, thereby improving our leverage to constrain the stellar mass. The IR data will largely free us from the tyranny of uncertain M/Ls that abound in purely optical (rest–frame UV) studies of star–forming (or recently star–forming) galaxies. Some initial studies demonstrated the value of measuring the stellar mass buildup history of galaxies (e.g., Dickinson et al. 2003). Galaxy stellar masses have also recently been derived from early Spitzer IRAC observations (Barmby et al. 2004).

It is clear that a vast number of surveys and studies are under way, in a broad area, and it is impractical to give more than a superficial overview of them in this review. Instead we will concentrate on developments in three areas at the forefront. These all involve very recent developments on galaxies at $z > 2$. They include, in addition to the stellar masses derived from optical and near-IR rest-frame SEDs mentioned above, the dynamical masses from kinematical observations of $z > 2$ LBG galaxies (e.g., Erb et al. 2003, Erb et al. 2004, Genzel et al. 2003); the nature of the red, evolved population of galaxies at high redshift, the $J - K_s$ galaxies of Franx et al. (2003) and van Dokkum et al. (2003, 2004); and the very high redshift samples of LBGs, the dropout samples at $z \sim 6$, and beyond from HST ACS (and NICMOS for the $z \sim 7$ galaxies), e.g., as discussed by Bouwens et al. (2003b), Stanway et al. (2003), and Yan et al. (2003).

The current frontier for high redshift objects is at $z \sim 6$ (the ACS UDF and NICMOS UDF–IR images together have extended the dropout sources to redshifts 7 and beyond, but the samples are small). Rapid changes in the properties of high redshift galaxies must occur beyond $z \sim 6$ and so careful characterization of objects, even those separated by small intervals of time, is an important aspect of the study of $z \sim 3 - 6+$ galaxies. There is great value in having large samples of $z \sim 3 - 5$ objects to contrast with the $z \sim 6$ galaxies. Though only $0.2 - 1.0$ Gyr later in cosmic history, $z \sim 3 - 5$ galaxies are larger and much better characterized than $z \sim 6$ galaxies, providing key baseline information needed to evaluate evolutionary changes with redshift.

6. Masses of High-Redshift galaxies

Star formation is a continuing challenge for galaxy formation and evolution models. The physics is complex, and not well constrained by the observational evidence. There has been significant progress in the parameterization of the key steps in the models. However, enough discrepancies exist between the model (semi–analytic or hydro code) predictions of light and the observed photometric properties of galaxies that it behooves the observational astronomers to do all they can to measure mass scales for galaxies as a function of redshift. The more direct coupling with theory will be a key step in understanding the buildup of galaxies. Fortunately, a number of observational programs are beginning to give greater insight into the mass scales of galaxies at key epochs.

It is important to distinguish direct dynamical measures of mass from those that rely on SED fits and assumptions about the IMF to derive "stellar" masses.

Both approaches have value, but significant assumptions go into the latter, which may not be valid in all cases or at all stages of the buildup of a galaxy (the IMF may change with time in star–forming regions). The challenges of actually measuring gas or stellar kinematics in high–redshift galaxies means that the derivation of stellar masses from SEDs is more widely used at this time. But both techniques provide valuable insights and are becoming increasingly common (see, e.g., Rudnick et al. 2003, Shapley et al. 2004).

For SED–fit stellar masses of high–redshift galaxies, most results have relied on rest–frame UV–optical observations, and were therefore subject to sizable uncertainties due to the effect on star formation on the measured M/L ratios. This approach was nonetheless a valuable first step, and demonstrated the value of deriving the evolution of the global stellar mass density from $z \sim 3$, i.e., deriving the the mass buildup history of the universe. For example, Dickinson et al. (2003) used the HDF–N data, from the WFPC2 U_{300} to the NICMOS H–band IR data down to 26.5 mag (AB), determined photometric redshifts and rest-frame B mag for a wide range of redshifts, and then derived stellar masses.

This approach to deriving stellar masses will mature as data from Spitzer become available for large samples. The IRAC data from Spitzer, with its simultaneous 5.12 x 5.12 arcmin images at 3.6, 4.5, 5.8, and 8μm, will allow fluxes of high redshift galaxies like LBGs to be determined at rest–frame near–IR ($\sim 2\mu$m). This will greatly reduce concerns about the uncertain M/L – since the recent star formation history will have significantly lower influence on the flux at 2μm. As noted above, the first such results are beginning to appear from the Spitzer IRAC GTO team. Barmby et al. (2004) carried out SED fits for Spitzer IRAC observations of LBGs. These were combined with optical data and Bruzual and Charlot (2003) models to estimate stellar masses. Since the Spitzer IRAC data corresponded to rest–frame $1-2\mu$m there is greater confidence in the derived "stellar masses" than those derived just from the optical (though, reassuringly, the Spitzer IRAC results are consistent with the earlier rest–frame optical results, e.g., Papovich et al. 2001). Interestingly, the typical stellar galaxy masses were found to be about $2-4\times10^{10}$ $M_{\odot}$ – characteristic of current–day massive bulges.

While much will be done over the coming years using SEDs from Spitzer the most valuable observations will be direct dynamical measures of mass from kinematic observations. These have the potential to allow direct comparison of the mass scales of galaxies at different redshifts with the very large body of data that is now available at $z \sim 0$.

A recent paper by Genzel et al. (2003) exemplifies the power of sub–mm observations, a field which will grow dramatically when ALMA comes on line. IRAM interferometer observations were made of the SCUBA source SMM J02399–0136 at $z = 2.8$ in the CO (3–2) line. This object is extended in a

Keck image. The sub–mm velocity–position diagram showed a structure characteristic of a rotating disk. The "two–peaked" line profile looked much like those seen in disk galaxies at low redshift from HI radio observations. What was particularly striking about this object was the magnitude of the rotation: $v_{rot} \sim 420$ km s^{-1} implying a mass of some 3×10^{11} $M_\odot$ within ~ 8 kpc at $z = 2.8$! This is a strikingly massive galaxy. While challenging at this time, such observations should be routine with ALMA.

The recent advent of efficient long–slit IR spectrographs on large telescopes is also providing opportunities for studies of the kinematic motions in high–redshift galaxies. Using emission lines in high–redshift LBGs to derive velocities for the gas has long been a goal, but the large outflows in these strongly star–forming objects (first seen in the spectra obtained of the strongly–lensed $z = 4.92$ galaxy G1 in the $z = 0.33$ cluster Cl1358+62 – see Franx et al. 1997) has limited the value of any measures derived from UV lines like Lyα.

Fortunately the optical nebular lines like Hα and [NII] provide velocities that are likely to be more representative of the dynamically–induced velocity field in the galaxy. But these fall in the near–IR at $1.2 - 2\mu$m for $z \sim 2 - 3$ galaxies and so require near–IR spectroscopy. As noted above this is now possible with ISAAC on the VLT and NIRSPEC on Keck, for example, and allows the measurement of velocities to give dynamical mass estimates.

A key paper in this area was that of Erb et al. (2003). They took Hα observations of 16 $z \sim 2 - 2.6$ LBG galaxies with Keck NIRSPEC and VLT ISAAC. They derived "rotation" curves for a sample of objects that were extended along the spectrograph slit. No particular effort was made in these observations to put the slits along the long axes of the objects (it is not clear, however, in these high–redshift star forming objects that the "long axis" of the light distribution corresponds to the major axis of the mass distribution, since the star forming regions may be scattered about the galaxy in a quite non–uniform way). Nonetheless they found that the emission lines were extended and tilted, characteristic of those seen in rotating galaxies. They derived "rotation velocities", as well as velocity dispersions from the widths of the Hα line. To derive masses they needed a characteristic radius, and used the mean half–light radius that was determined for a subset of the sample for which HST images were available. The half–light radius was about 0.2 arcsec, corresponding to $\sim$1.6 kpc at the redshift of this sample ($z \sim 2.3$). While the effect of outflows on the measured kinematics remains to be determined, it is expected that the optical nebular lines will be much less affected than the UV lines like Lyα. Ultimately one will need high S/N spectra to compare the emission line velocities with the interstellar and photospheric absorption line velocities as a check on the outflow contribution to the emission lines.

The current kinematical data give typical masses of a $2-6 \times 10^{10}$ $M_\odot$. This is strikingly similar to the stellar masses estimated by Barmby et al. (2004)

from Spitzer SED fits! This work is being continued with further results given by Erb et al. (2004).

7. Red, Evolved Galaxies at $z \sim 2 - 3$

One of the most striking developments in the last two years in the field of high–redshift galaxies has been the discovery of an evolved population of galaxies at $z \sim 2 - 3$. These objects were detected from very deep near–IR JHK imaging in the HDF–S and in a well–studied distant cluster field, MS1054–03. Their SEDs suggested that they were relatively unreddened, evolved galaxies at high redshift (Franx et al. 2003). Their SEDs were best fit with early–type populations. Galaxies with evolved stellar SEDs that are red in $J - K_s$ (> 2.3) were considered likely, from models, to have redshifts around $z \sim 2 - 3$. An initial spectroscopic study (van Dokkum et al. 2003) supported these inferences. These first papers suggested that these objects are likely to be progenitors of current–day early–type galaxies.

These objects are now being characterized with increasing detail (see, e.g., Forster Schreiber et al. 2004). A recent development has been to use velocity measurements (line widths) to establish their mass scales. This was done with near–IR spectroscopy again, using NIRSPEC on Keck to measure the width of Hα in four of these objects (van Dokkum et al. 2004). The ISAAC VLT images of the galaxies with line widths showed that they are resolved with characteristic sizes $r_e \sim 0.5$ arcsec. Given measured velocity dispersions and a length scale, a characteristic mass could be determined. It was found to be high, $2 - 5 \times 10^{11}$ $M_{\odot}$, some $10\times$ that of the LBG galaxies. These $J - K_s$ galaxies (or DRGs – distant red galaxies) are larger, redder and more massive than LBGs. Strikingly, they are massive even compared to present day samples of early–type galaxies. A comparison of the 7 DRGs with measured velocity dispersions (and hence masses) with ~20,000 local SDSS ellipticals shows that they fall at the high end of the SDSS distribution (which extends to about 10^{12} $M_{\odot}$). The median SDSS elliptical mass is a little over 10^{11} $M_{\odot}$). As expected, the DRGs have a dynamical M/L_B that is substantially larger than that of the LBGs, ~ 1 $vs.$ <0.2.

An estimate of their mass density at $z \sim 2 - 3$ shows that their contribution is comparable to the LBGs at that redshift. Clearly the DRGs are an important contributor to the overall galaxy population at high redshift.

8. Dropout Galaxies – to $z \sim 6$ and Beyond....

Over the last 8 years the HDFs have played a central role in the study of high–redshift galaxies. Recently the HST ACS has been used to obtain high quality multi–band data on a number of fields, which rival or exceed the HDFs in their value for detecting and characterizing high–redshift galaxies. Several

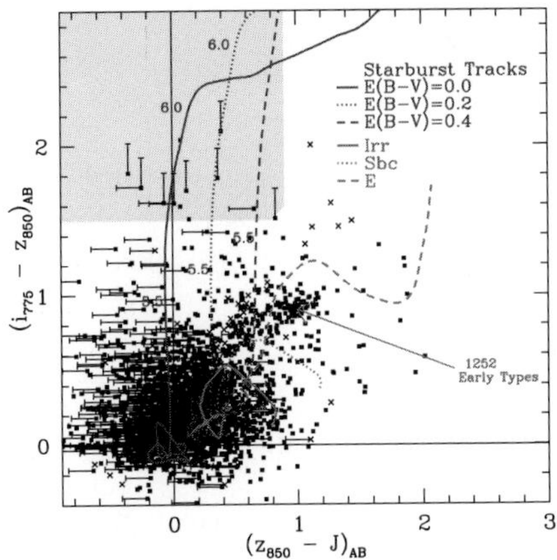

Figure 1. Selection of i–dropouts in the $(i - z)(z - J)$ two–color plane. The optical data is from the HST ACS images of the RCDS 1252–2927 cluster (Rosati et al. 1998). The IR data is from ISAAC on the VLT. The selection limits (particularly the $(i - z) > 1.5$ cut – see Bouwens et al. 2003b) returns $z \sim 6$ galaxies with minimal contamination (an estimated 11% contamination rate).

of these fields have been the used extensively for identifying samples of high–redshift dropout galaxies over the last year. The most important of these are the GOODS fields, the UDF, and the UDF–Parallels (UDF–Ps), though we will also mention a distant cluster field around RCDS 1252–2927 used for some dropout work. All have excellent HST ACS i_{775} and z_{850} data, while the GOODS, the UDF and UDF–Ps fields also have deep B_{435} and V_{606} data. Near–IR data also is of great value in isolating the highest redshift samples, and for minimizing contamination, though this tends to be a minor problem for conservatively–chosen dropout samples. The excellent VLT ISAAC IR data (Lidman et al. 2004) in RCDS 1252–2927 makes a substantial contribution to the selection of i–dropouts. Similar data is available in some of these fields, particularly the UDF, with the NICMOS UDF–IR data (Thompson et al. 2004), and the GOODS CDF–S field which has extensive VLT ISAAC data.

The selection of dropout galaxies is routinely done in the two–color plane. An example for $z \sim 6$ i–dropouts is shown in Fig 1 for the RCDS 1252–2927 field (from Bouwens et al. 2003b). The ACS data reaches typically to $z_{850,AB} \sim 27.3$ mag (6σ), while the ground–based IR data goes impressively deep, down to $J_{AB} = 25.7$ and $K_{s,AB} = 25.0$ mag (5σ). The fraction of $z \sim 6$ objects in the IR footprint of RCDS 1252–2927 is impressively small, only 12 out of ∼3000 galaxies (0.3%). Even so the estimated contamination of the i–dropouts is only about 11%. A number of these candidates have been observed with Keck and the VLT and confirmed to be at $z \sim 6$. A total of 23 $z \sim 6$ galaxies were found in the four ACS pointings of the RCDS 1252–2927 field, giving a surface density of 0.5 ± 0.2 i–dropouts per square arcmin to $z_{AB} = 26.5$ mag (though this surface density appears to be a larger than the cosmic average). The objects are very small, though nevertheless resolved,

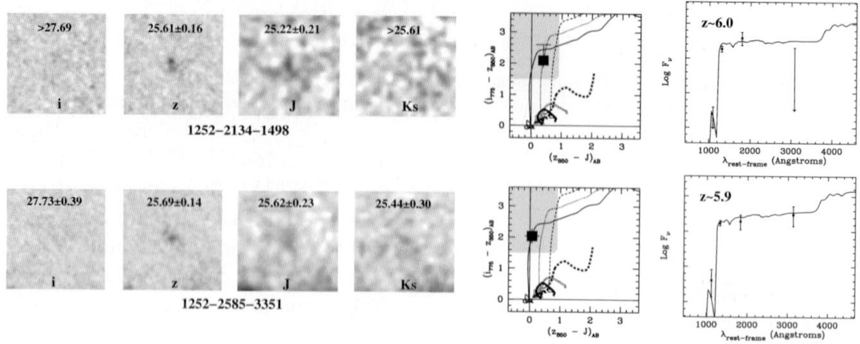

Figure 2. Images ($3'' \times 3''$) in i, z, J, K_s of $z \sim 6$ objects, along with two–color schematics (showing starburst tracks as a function of redshift for different reddenings - see Fig 1), and starburst galaxy SEDs (10^8 Gyr), with the best fit redshift. The sources are all in RCDS 1252–2927. The magnitudes given for the sources are AB mag.

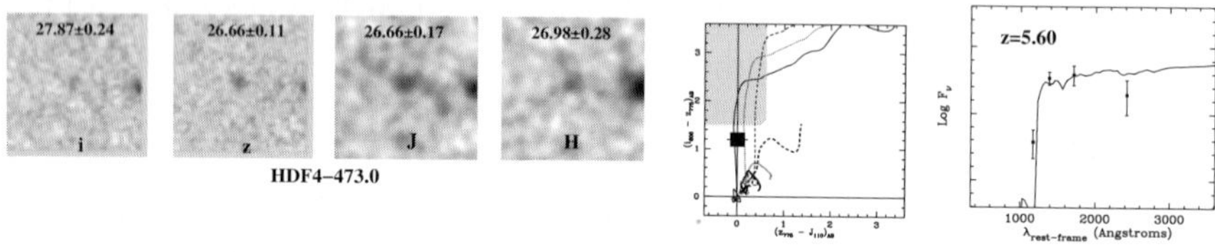

Figure 3. As in Fig 2, but for the Weymann et al. (1998) galaxy in the HDF–N whose redshift was measured to be $z = 5.6$ from over 7 hours of integration with the LRIS spectrograph on Keck. The redshift determined from the photometric data was also $z = 5.6$.

with typical half-light radii of $0.15''$ or ~ 0.9 kpc. In this particular field, the $z \sim 6$ objects reach down to $\sim 0.3L_{*,z=3}$ (Steidel et al. 1999).

Two of the brighter i–dropouts from the RCDS 1252–2927 field are shown in Fig 2, along with their location in the two–color plane, and SED fits used to establish the redshifts. The ACS i and z data from the HDF–N also allowed for a search for i–dropouts. A reassuring result was that the Weymann et al (1998) object in the HDF–N was given a photometric redshift $z \sim 5.6$, quite consistent with its spectroscopic value of $z = 5.60$. It was very close to meeting our i–dropout criterion (its $i - z = 1.2$ color was just a little too blue). While not a true i–dropout, this consistency suggested that our selection was yielding bona–fide high redshift objects. Other spectroscopic results (Bunker et al. 2003 and Dickinson et al. 2004), and our own ongoing Keck programs, have only served to strengthen our confidence in the dropout approach.

While the RCDS 1252–2927 field provided a significant sample of i–dropouts (with a good assessment of the contamination), the best samples of brighter dropouts come from the two ACS GOODS fields, CDF–S and HDF–N (see Giavalisco et al. 2004). From these fields, Bouwens et al. (2004b) derived a large number of B, V and i–dropouts, augmenting them with a smaller but very useful sample of U–dropouts from the HDF–N and HDF–S fields so that a self–consistent differential analysis could be applied across a large redshift range, $z \sim 3$ to $z \sim 6$. Even with relatively conservative selection criteria, Bouwens et al. (2004c) derive 1235 $z \sim 4$ B–dropouts, 407 $z \sim 5$ V–dropouts, and 59 $z \sim 6$ i–dropouts. These samples go as faint as 0.2, 0.3, 0.5 $L_{*,z=3}$ (using the Steidel et al. 1999 value for $L_{*,z=3}$), respectively, with 10σ limiting magnitudes of 27.4 in the $i_{775,AB}$ band and 27.1 in the $z_{850,AB}$ band.

The large samples and wide areal coverage of the GOODS fields are nicely complemented by the two UDF–Ps obtained in parallel with the deep NICMOS images of the UDF. These fields have overlapping ACS images on a 45″ grid with 9 orbits each in B and V, 18 orbits in i and 27 orbits in z (as well as 9 orbits with the grism). They reach impressively faint, to 28.8, 29.0, 28.5 and 27.8 mag (10σ) in $B_{435}, V_{606}, i_{775}$, and z_{850} AB-mag, respectively – or to $0.1 - 0.2L_{*,z=3}$. The UDF itself is an impressive addition to these fields, taking the limits to $< 0.1L_{*,z=3}$.

9. Dropout Galaxies: Evolution

A major issue with deriving the evolution of galaxy properties at high redshift is systematic error – primarily through the many selection effects that can influence the nature of the samples, even when derived from very similar datasets. Of these the $(1+z)^4$ surface brightness dimming is the dominant effect, but many others affect the derived samples (e.g., size evolution, color evolution, definition of selection volumes, data properties as a function of redshift, filter band, and instrument, etc.). To treat these effects, we compare our highest redshift samples with "cloned" projections of our lower redshift samples (e.g., Bouwens et al. 1998; Bouwens et al. 2003a), allowing us to contrast intrinsic evolution from changes brought about by the selection process itself.

One of the key results is that of size evolution. An excellent illustration of this is provided in Fig 4a with data from the UDF showing that there is a clear decrease in size with redshift for objects of fixed luminosity (Bouwens et al. 2004b). While a more rigorous demonstration of this is given in Fig 4b, this trend has now been demonstrated in a variety of ways with both the GOODS data and the UDF–Ps data (Ferguson et al. 2004; Bouwens et al. 2004a; Bouwens et al. 2004c). The preferred size scaling from the best dataset, the UDF, is $(1+z)^{-1}$, but the measured size scalings have ranged from $(1+z)^{-1}$ to $(1+z)^{-1.5}$ in the above studies and may depend upon luminosity.

A major goal of these studies is to extend the constraints on the luminosity density and the star formation rate with redshift to higher redshifts $z \sim 6$

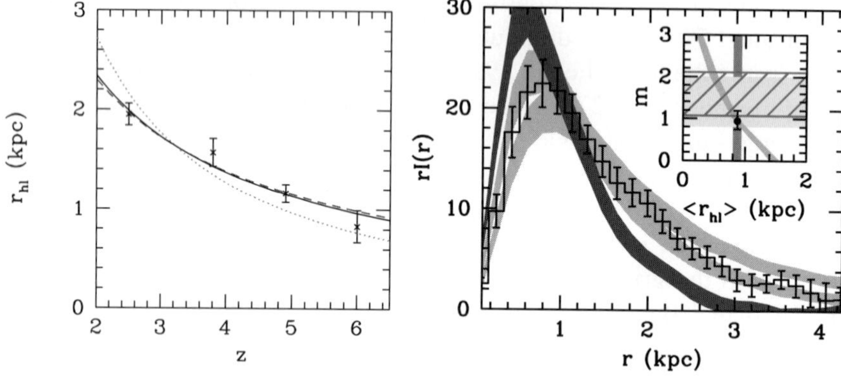

Figure 4. (Left) The mean half–light radius (measured from growth curves and corrected for PSF effects) versus redshift for objects of fixed luminosity $(0.3 - 1.0L_{*,z=3})$. Data $(\pm 1\sigma)$ are from the $z \sim 2.5$ HDF-N + HDF-S U-dropout sample and UDF B, V, and i–dropout samples plotted at their mean redshifts $z \sim 3.8$, ~ 4.9, and ~ 6.0, respectively. The dotted line shows the $(1+z)^{-1.5}$ scaling expected assuming a fixed circular velocity and the dashed line shows the $(1+z)^{-1}$ scaling expected assuming a fixed mass (Mo et al. 1998). A least squares fit favors a $(1+z)^{-1.05\pm0.21}$ scaling (solid black line). This comparison is not unbiased since objects are not selected or measured to the same surface brightness threshold. The UDF is nevertheless deep enough at these magnitudes to minimize these biases. (Right) A more rigorous derivation. The mean radial flux profile determined for the 15 intermediate magnitude $(26.0 < z_{850,AB} < 27.5)$ objects from our UDF i–dropout sample compared against that obtained from similarly–selected U–dropouts cloned to $z \sim 6$ with different size scalings: $(1+z)^0$, $(1+z)^{-1}$, and $(1+z)^{-2}$. The best fit is at $(1+z)^{-1}$. The inset establishes this more accurately, and shows how the mean size of the projected U–dropouts vary as a function of the $(1+z)^{-m}$ size scaling exponent m (a correction is made for PSF effects). Since the mean half–light radius is 0.87 ± 0.07 kpc (shown as a vertical band), this suggests a value of $0.94^{+0.25}_{0.19}$ for the scaling exponent m. Significantly tighter constraints are possible on the size (surface brightness) evolution from the UDF data (Bouwens et al. 2004b) than was possible in our previous study with the UDF–Ps data (Bouwens et al. 2004a: hatched region) and GOODS (Bouwens et al. 2004c: shaded region), though these probe slightly different ranges in luminosity.

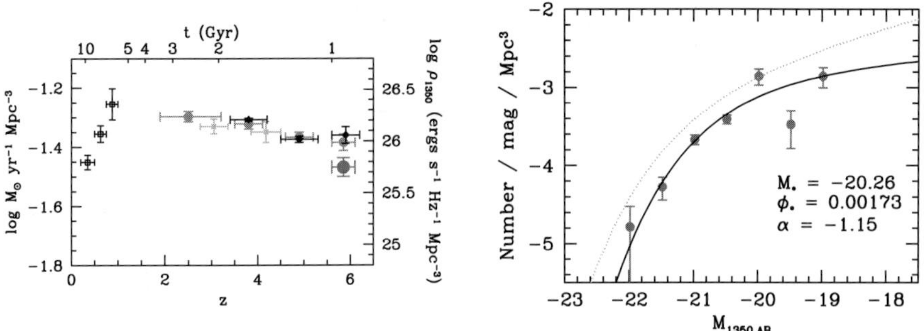

Figure 5. (Left). Star formation rate evolution (dust–free) with redshift and age (top), integrated down to $0.2L_{*,z=3}$. The rest–frame UV continuum luminosity density is given on the right axis for the high redshift values (Bouwens et al. 2004c). Note the small Δt from $z \sim 6$ to $z \sim 3$. A Salpeter IMF is used to convert the luminosity density to SFR (see Madau et al. 1998). The four solid circles from $z = 2.5$ to $z = 6$ are the values from the GOODS data (Bouwens et al. 2004c). Other determinations are Lilly et al. (1996 – open squares), Steidel et al. (1999 – crosses), Bouwens et al. (2004a – solid circles at $z = 6$) and Giavalisco et al. (2004 – solid diamonds). The Thompson et al. (2001) values are similar to those shown here. The low point at $z = 6$ includes the effect of size evolution on the $z \sim 6$ Bouwens et al. (2004a) value (indicating how significant this effect can be). (Right). The rest–frame continuum UV luminosity function (at 1350 A) at $z \sim 6$ from the GOODS field (for $M_{1350,AB} < -19.7$) and the UDF–Ps. The best fit values for a Schechter luminosity function are shown on the figure. The Steidel et al. (1999) $z \sim 3$ luminosity function (dotted line) is also shown. A preliminary analysis from the UDF (Bouwens et al. 2004d) suggests that the $z \sim 6$ faint end slope is at least as steep, if not steeper, than the slope found at $z \sim 3$, e.g., $\alpha = -1.6$.

and beyond. A related goal is to improve the constraints at lower redshifts ($z \sim 2 - 5$). These new datasets are proving to be of great value for these two goals, as Fig 5a demonstrates, showing several of the more recent estimates which have been made on the (dust–free) star formation rate out to $z \sim 6$. These data also permit a derivation of the luminosity function to significantly fainter than $L_{*,z=3}$, as was done by Bouwens et al. (2004a) with the GOODS + UDF–Ps data (Fig 5b). While much work is still in progress, the UDF is allowing us to make significant improvements to these measures, particularly at $z \sim 6$, where it provides a significant check on both the incompleteness and faint end slope (Bunker et al. 2004; Bouwens et al. 2004b; Bouwens et al. 2004d).

10. Dropout Galaxies: $z \sim 7 - 8$ Galaxies

The UDF promises to be a resource comparable to the original HDFs. NICMOS J_{110} and H_{160} IR data were also taken covering a substantial fraction of

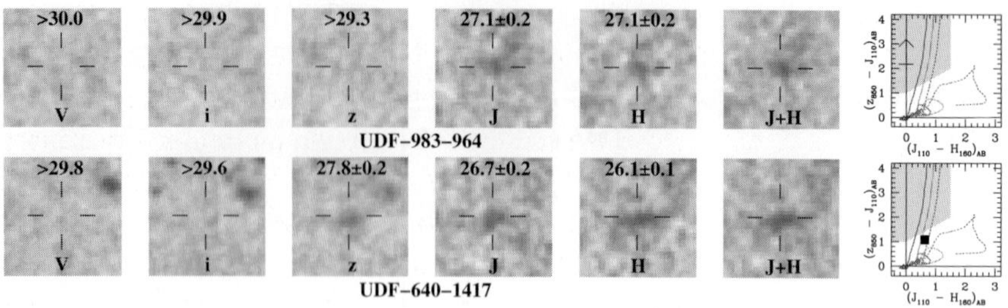

Figure 6. Images ($3'' \times 3''$) in the $V_{606}, i_{775}, z_{850}, J_{110}, H_{160}$ bands (AB mags) of a $z \sim 7\text{–}8$ object (a z_{850}–dropout), plus a very red $(z_{850} - J_{110})_{AB} = 1.1$ object (bottom) which nearly met our selection criteria and could be a reddened starburst at $z \sim 6.5$ (or, improbably, a reddened early–type at $z \sim 1.6$) (see Bouwens et al. 2004e). The $J_{110} + H_{160}$ image for each object is included, along with its position in color–color space, in the two rightmost panels. The sources are in the UDF–IR field (Thompson et al. 2004).

the UDF (Thompson et al. 2004). Together these datasets will provide great insight into the nature of high-redshift galaxies, given the wide wavelength coverage (especially when the Spitzer CDF–S data also become available) and the great depth. In particular, they allow a search for z–dropouts, i.e., galaxies at redshifts $z > 7$. Remarkably, there are indications that we have detected such galaxies (see Bouwens et al. 2004e). An example of such a detection is shown in Fig 6, along with that for a slightly lower redshift galaxy (one at $z \sim 6.5$). The detections in Bouwens et al. (2004e) suggest that luminous galaxies are in place at $z \sim 7 - 8$, and that the UV luminosity density, while down compared to that at $z \sim 6$ is still quite significant.

11. Summary – High Redshift Galaxies at $z \sim 2 - 7$

There has been remarkable progress on characterizing galaxies at high redshift, particularly since the ACS came on line after the HST servicing mission SM3B in 2002. Significant numbers of evolved $z \sim 2 - 3$ galaxies have been detected. These $J - K_s$ galaxies, or DRGs, appear to be the progenitors of ellipticals and massive bulges (at one stage of their development). Substantial progress has been made on determining mass scales from kinematics and from stellar populations (SEDs). This is challenging but the future holds great promise with observations from HST, Spitzer, ALMA, and ground-based 8–10 m IR spectrographs. Large samples of LBG (strongly star-forming) galaxies at $z > 2$ to $z \sim 6$ have been detected. The ACS GOODS + HDF U, B, V, i dropout samples extend to ~27 AB mag ($\sim 0.2 - 0.5 L_{*,z=3}$) and total some 1900 galaxies, of which ~200 are $z \sim 2.5$ U–dropouts, ~1230 are $z \sim 4$ B–dropouts, ~410 are $z \sim 5$ V–dropouts, and ~60 are $z \sim 6$ i–dropouts. The UDF, UDF–Ps and other fields increase the i–dropout sample to over 200. Our cloning analysis of several datasets including the GOODS CDF–S & HDF–N, and the UDF plus the UDF–Ps fields has shown that there is a systematic decrease in the size of dropout galaxies as a function of redshift ($2 < z < 6$). The best fit is $(1+z)^{-1}$, but the precise scaling may depend on luminosity, ranging from $(1+z)^{-1}$ to $(1+z)^{-1.5}$. This implies a $\sim 2 - 3\times$ decrease in galaxy size from $z \sim 2.5$ to $z \sim 6$. Our recent studies on the UDF–Ps + GOODS fields indicated that there is a $\sim 2.5\times$ increase in the rest–frame continuum UV luminosity density from $z \sim 6$ to $z \sim 3.8$ (within a period of $\sim$1 Gyr). The uncorrected star formation rate density at $z \sim 6$ was just $0.38 \pm 0.08\times$ the star formation rate density at $z \sim 3.8$. New data from the GOODS, the UDF–Ps, and UDF fields are permitting an i–dropout luminosity function to be constructed down to $< 0.1L_{*,z=3}$ ($\sim$ 29 AB mag). Studies indicate that these galaxies could provide the needed reionizing flux at $z \sim 6$ and earlier. Remarkably, HST ACS and NICMOS data in the UDF have led to the likely detection of (a few) $z \sim 7 - 8$ z–dropout galaxies.

Acknowledgments

We would like to thank the organizers for an excellent meeting in a wonderful country. We particularly appreciate the Anglo American Chairman's Fund for sponsoring the Conference. We acknowledge the remarkable advances that have come about because of HST and its amazing imagers, particularly the ACS. We regret the decision by NASA to cancel SM4 that would lead to the premature death of HST. We hope that a mission, astronaut or robotic, to add the SM4 instruments and extend Hubble's life, comes about. We owe a lot to our team members on the ACS GTO team* and the UDF–IR team**, and particularly the PIs, Holland Ford and Rodger Thompson (*ACS GTO team: Holland Ford, Txitxo Benitez,Tom Broadhurst, Piero Rosati, Marijn Franx, Marc Postman, Brad Holden, Rick White, John Blakeslee, Dan Magee, Gerhardt Meurer plus many other team members; **UDF-IR team: Rodger Thompson, Mark Dickinson, Marijn Franx, Pieter van Dokkum, Adam Riess, Xiaohui Fan, Dan Eisenstein, Marcia Rieke). Support from NASA grant NAG5–7697 and NASA/STScI grant HST–GO–09803.05–A is gratefully acknowledged. ACS was developed under NASA contract NAS5–32865.

References

Barmby, P., et al. ApJS., in press, astro-ph/0405624 (2004).
Bennett, C. L., et al. ApJS., **148**, 97 (2003).
Bouwens, R., Broadhurst, T. and Silk, J. ApJ., **506**, 557 (1998).
Bouwens, R., Broadhurst, T., & Illingworth, G. ApJ., **593**, 640 (2003a).
Bouwens, R. J., et al. ApJ., **595**, 589 (2003b).
Bouwens, R. J., et al. ApJL., **606**, L25 (2004a).
Bouwens, R. J. et al. ApJL., in press (2004b).

Bouwens, R. J. et al. ApJ., submitted (2004c).
Bouwens, R. J. et al. ApJ., in preparation (2004d).
Bouwens, R. J. et al. ApJ., in preparation (2004e).
Bruzual, G. & Charlot, S. MNRAS, **344**, 1000 (2003).
Bunker, A. J., et al. MNRAS, **342**, L47 (2003).
Dickinson, M., et al. ApJL., **600**, L99 (2004).
Dickinson, M., Papovich, C., Ferguson, H. C., & Budavari, T. ApJ., **587**, 25 (2003).
Erb, D. K., et al. ApJ., **591**, 101 (2003).
Erb, D.K., et al. ApJ., in press, astro-ph/0404235 (2004).
Ferguson, H.C., et al. ApJL., **600**, L107 (2004).
Ford, H. C. et al. Proc. SPIE, **4854**, 81 (2003).
Forster Schreiber, N.M., ApJ., in press (2004).
Franx, M., et al. ApJL., **486**, L75 (1997).
Franx, M., et al. ApJL., **587**, L79 (2003).
Fukugita, M., Hogan, C. J., & Peebles, P. J. E. ApJ., **503**, 518 (1998).
Giavalisco, M., et al. ApJL., **600**, L93 (2004).
Genzel, R., et al. ApJ., **584**, 633 (2003).
Kogut, A., et al. ApJS., **148**, 161 (2003).
Lidman, C., et al. A.& A., in press (2004).
Lilly, S.J., Le Fevre, O., Hammer, F., & Crampton, D. ApJ., **460**, L1 (1996).
Madau, P., Pozzetti, L. & Dickinson, M. ApJ., **498**, 106 (1998).
Mo, H. J., Mao, S., & White, S. D. M. MNRAS, **295**, 319 (1998).
Papovich, C., Dickinson, M., & Ferguson, H. C. ApJ., **559**, 620 (2001).
Rosati, P., et al. ApJ., **492**, L21 (1998).
Rudnick, G., et al. ApJ., **599**, 847 (2003).
Shapley, A.S., et al. ApJ., in press, astro-ph/0405187 (2004).
Stanway, E. R., Bunker, A. J., & McMahon, R. G. MNRAS, **342**, 439 (2003).
Stanway, E. R., et al. ApJL., **604**, L13 (2004).
Steidel, C.C., et al. ApJ., **519**, 1 (1999).
Thompson, R. I., Weymann, R. J., & Storrie-Lombardi, L. J. ApJ., **546**, 694 (2001).
Thompson, R. I., et al., in preparation, (2004).
van Dokkum, P.G., et al. ApJL., **587**, L83 (2003).
van Dokkum, P.G., et al. ApJ., in press, astro-ph/0405482 (2004).
Weymann, R. J., et al. ApJL., **505**, L95 (1998).
Yan, H., Windhorst, R. A., & Cohen, S. H. ApJL., **585**, L93 (2003).

BIRDS
of
SOUTH
AFRIC

MASKS IN THE MILKY WAY

K.C. Freeman
Research School of Astronomy & Astrophysics, Mt Stromlo Observatory, The Australian National University, Canberra

Abstract The dust mask is the most obvious of the masks that obscure our view of the Milky Way. But there are others, including the mass mask, kinematical masks, and structural, chemical and dynamical masks. In each case, there are ways to penetrate the mask.

Keywords: Milky Way, galaxies, dust, galaxy dynamics, chemical abundances

1. Introduction

Various kinds of masks affect our perception and understanding of the Milky Way. Interstellar dust is certainly one such mask. On the large scale, it obscures the central bulge of our galaxy: the boxy structure of the bulge was not properly revealed until the first near-infrared images became available. But dust is not the only mask that obscures our view of the Galaxy. For example, our inability to image mass rather than light clearly gives us a very incomplete perception of the size and structure of the Milky Way. In this talk, I will discuss some of these masks that contribute to concealing the Galaxy from our view.

2. Overview of our Galaxy

Here is a brief overview of our Galaxy as we know it, shown as a cartoon in Figure 1. The thin disk is the defining component, as in other spirals. Around the thin disk is the thick disk, a disklike structure of larger scaleheight and lower mass than the thin disk, which is also very common in other disk galaxies. A small bulge is at the center of the Milky Way, and there is a very extended roughly spherical low-density metal-poor halo which includes most of the globular clusters and other classic population II objects. The whole system is enveloped in a very massive dark halo which contains more than 90% of the mass of the galaxy and extends far beyond the visible components.

The mass of the Galaxy (dark + luminous) out to a radius of 50 kpc is about $5 \times 10^{11} M_\odot$; its total mass is about $2 \times 10^{12} M_\odot$ (Wilkinson & Evans 1999), Sakamoto *et al.* 2003).

D. Block et al. (eds.), Penetrating Bars through Masks of Cosmic Dust, 639–653.

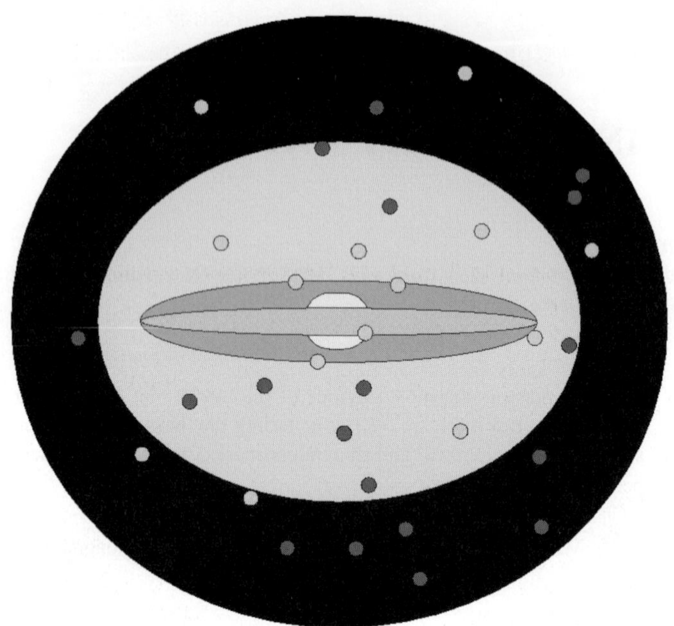

Figure 1. Cartoon of the Milky Way, showing the stellar thin and thick disks, the central bulge, the metal-poor stellar halo and the dark halo (From Freeman & Bland-Hawthorn 2002).

The masses of the stellar components are about $6 \times 10^{10} M_\odot$ for the disk + thick disk, $2 \times 10^{10} M_\odot$ for the bulge and only $1 \times 10^9 M_\odot$ for the metal-poor halo. For more details, see Freeman & Bland-Hawthorn 2002).

The bulge, thick disk and metal-poor of our Galaxy are primarily old components, with ages in the range 10-13 Gyr. The thin disk has had a roughly uniform star formation history continuing to the present time (*e.g.* Rocha-Pinto *et al.* 2000); the oldest thin disk stars near the sun have ages of about 10 Gyr from white dwarfs (Oswalt *et al.* 1996, Legget et al 1998) and old subgiants (Edvardsson *et al.* 1993, Sandage et al 2003).

3. The Dust Mask and the Bulge

The dust mask is the most apparent of the masks that we encounter in the Milky Way. It obscures the bulge, particularly the northern half, and the real nature of the bulge is not at all evident in optical light. However, when imaged in the NIR, as in the COBE/DIRBE image shown in Figure 2, the small boxy bulge and inner disk can be clearly seen.

de Vaucouleurs (1983) classified our Galaxy as SAB(rs)bc. He penetrated the mask of dust by using a variety of morphological and kinematical properties based on his knowledge of the properties of other galaxies. This classification was not taken very seriously at the time. However, some years later, the bar indeed emerged from the observed near-IR asymmetry of the inner light distribution (which is apparent in Figure 2), stellar photometry of bulge clump stars, the kinematics of gas and planetary nebulae etc.

Figure 2. Infrared image of the MIlky Way taken by the DIRBE instrument on board the COBE satellite. Note the small boxy bulge.

Launhardt et al (2002) give a description of the properties of the galactic bulge, and we can compare it with the rather different bulge of M31. In Figure 3, we see

(i) contours of the near-IR light distribution, in which the asymmetry of the bulge is obvious. The bulge appears thicker at positive longitudes, and this is attributed to the barlike nature of the bulge; the end at positive longitudes is closer to the sun and hence appears thicker.

(ii) the minor axis surface brightness distribution of the galactic bulge. Our bulge is a small bulge with an exponential light distribution, typical of the bulges of later-type spirals. Later-type galaxies mostly have near-exponential bulges, rather than $r^{1/4}$ bulges, which is usually interpreted as indicating that their bulges are not merger products but are more likely generated by disk instability (*e.g.* Courteau et al 1996). A nuclear disk is seen as the bright feature in the innermost region of the bulge profile.

We can contrast the exponential bulge of the Milky Way with the bulge of M31. The surface brightness distribution of the bulge of M31, as derived from surface photometry and star counts by Pritchet & van den Bergh (1994), shows an $r^{1/4}$ structure over many magnitudes in surface brightness. This is typical of the larger bulges of early type galaxies, which are believed to form through merger events and subsequent violent relaxation.

Boxy bulges, as in our Galaxy, are associated with bars, which in turn are believed to come from bar buckling instability of disks. See Combes & Sanders (1981) for the theory, and Bureau & Freeman (1999) for the observational evidence. The boxy bar/bulge of our Galaxy is about 3.5 kpc long, with an axial ratio of 1:0.3:0.3, pointing about 20° from the sun-center line into first quadrant (*e.g.* Bissantz & Gerhard 2002).

The galactic bulge rotates, like most other bulges. Figure 4 shows the rotational velocity against longitude for K giants and planetary nebulae from Beaulieu et al (2000), and also the velocity dispersion against longitude. The velocity dispersion of the inner disk, from Lewis & Freeman (1989), is shown as a horizontal line. We see that the velocity dispersions of the inner disk and bulge are fairly similar. It is not easy to separate the inner disk and bulge kinematically.

The age and metallicity of the galactic bulge have been studied by many authors. A recent photometric study of bulge stars at $(l, b) = (0.3, -6.2)^{\circ}$ (Zoccali *et al.* 2003) shows clearly that the bulge is an old population with a metallicity distribution extending from about [Fe/H] = $+0.2$ to -1.8. There is no evidence for a younger population. The minor-axis metallicity gradient of the bulge, compiled from several sources, is shown in Figure 5. The Zoccali *et al.* value fits well into this gradient.

Near the center of the bar/bulge is a younger population, on a scale of about 100 pc. This is the nuclear stellar disk, with a mass of about $1.5 \times 10^9 M_{\odot}$ (Launhardt *et al.* 2002), which can be seen as a small central feature above the exponential surface brightness distribution shown in Figure 3. There is also a young nuclear stellar cluster with a mass of about $2 \times 10^7 M_{\odot}$ in the central 30 pc. About 70% of the luminosity of these central components comes from young main sequence stars.

In summary, the galactic bulge, as we now see it through the dust mask, is not a dominant feature of the Milky Way; it contributes only about 25% of the stellar light. The bulge is often described as metal-rich, but the metallicity distribution is in fact quite broad, with only a weak extension to [Fe/H] > 0. The bulge is probably an evolutionary structure of the disk, rather than a feature of galaxy formation in the early universe (*e.g.* Kormendy & Kennicutt 2004).

4. The Mass Mask

Like most galaxies, the Milky Way would look very different if we were able to image it in mass rather than light. The stellar mass of our Galaxy is about $8 \times 10^{10} M_{\odot}$. The total mass (dark + luminous) is about $10^{10} \times R$(kpc):

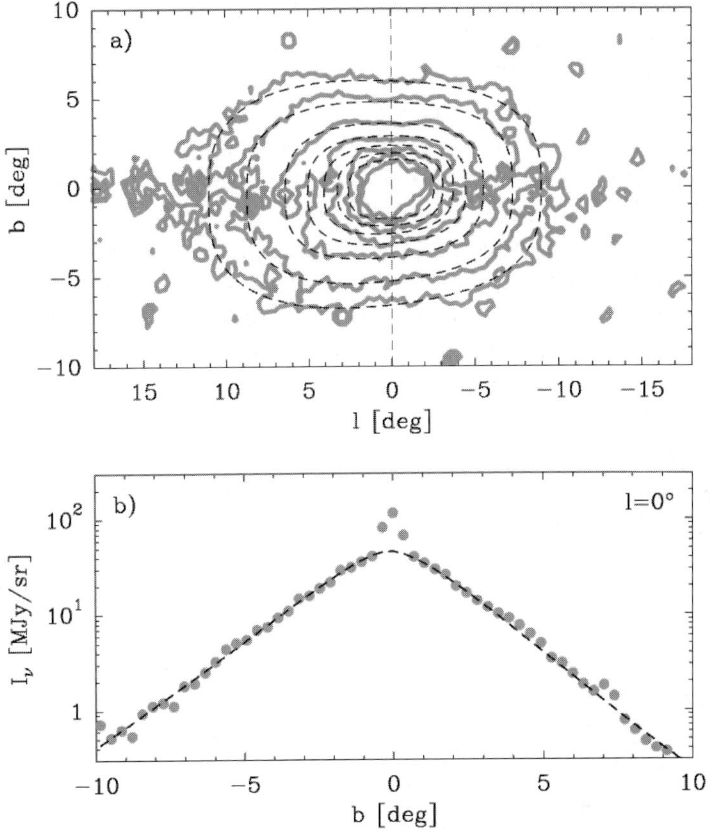

Figure 3. *Upper:* Contours of the near-IR light distribution of the galactic bulge. Note the boxy asymmetric shape of the bulge. *Lower:* the minor axis surface brightness distribution of the bulge. Note its exponential form. The central feature is a bright nuclear disk. From Launhardt *et al.* (2002).

e.g. out to 50 kpc, the enclosed mass is about $5 \times 10^{11} M_\odot$ (Wilkinson & Evans 1999).

The total galactic mass is probably about $2 \times 10^{12} M_\odot$, derived from the M31 and Leo I timing arguments and least action studies of the Local Group and is consistent with the dynamics of satellite systems around other spirals (*e.g.* Prada *et al.* 2003). Indications from simulations and observations of satellite systems are that the density distribution of dark halos falls as r^{-3} at large radii. With this density distribution, the galactic halo probably extends

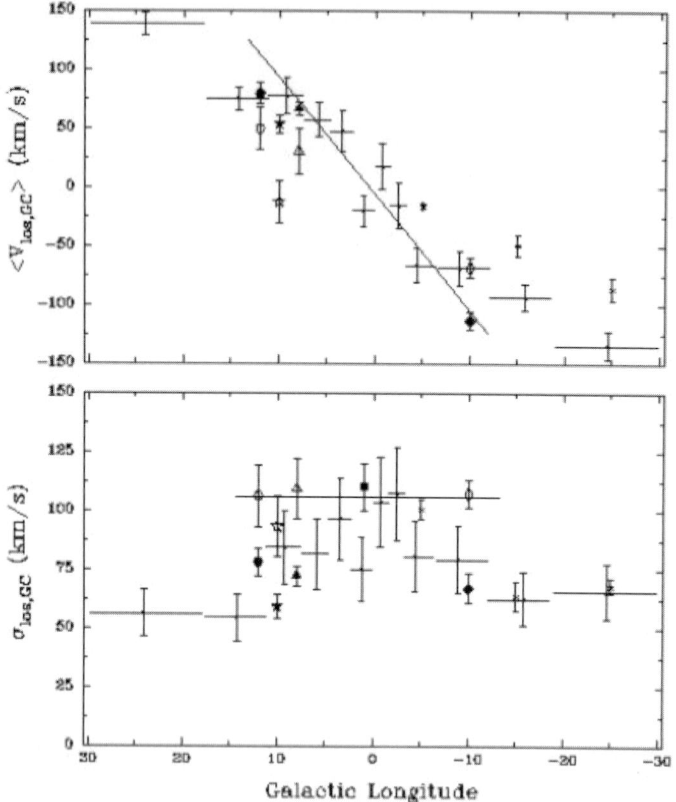

Figure 4. *Upper:* rotation of the galactic bulge, and *Lower:* the velocity dispersion of the galactic bulge, from kinematics of red giants and planetary nebulae. (From Beaulieu *et al.* 2000). The horizontal line shows the velocity dispersion of the inner disk.

out to a radius of about 300 kpc, almost half way to M31. This radius is much larger than the extent of any directly measured rotation curves, so these "timing arguments" give a realistic lower limit on the total mass of the dark halo.

The significance of the mass mask is seen when we look at the mass to luminosity ratio M/L and the ratio of dark to luminous mass. The luminosity of the Galaxy is about $2 \times 10^{10} L_{\odot}$. With the masses given above, the total M/L ratio of the Galaxy is about 100, and the ratio of total mass to stellar mass is about 25.

Satellite systems of disk galaxies give a useful estimate of the total mass and extent of the dark halos. Individual galaxies have only a few observable satellites each, but combining observations of many satellite systems gives a

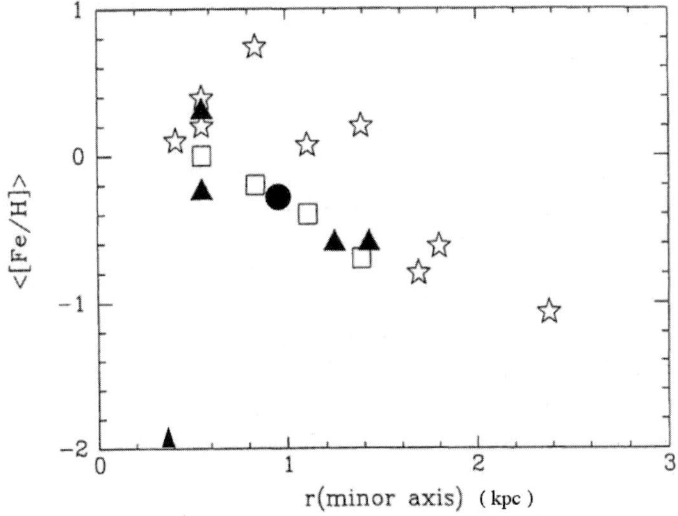

Figure 5. Abundance gradient, towards the minor axis of the galactic bulge, for a range of spectroscopic and photometric indicators. The filled circle shows the Zoccali *et al.* (2003) data. (From Minniti *et al.* 1995).

measure of the mass of a typical dark halo. For example, Prada et al (2003) looked at the kinematics of about 3000 satellites around about 1000 galaxies. They find that the halos typically extend out to about 300 kpc and the density distribution at large radius is steeper than expected for isothermal systems: $\rho(r) \sim r^{-3}$, like most cosmological models including the NFW halos. The total M/L ratios are typically 100-150, comparable with the estimate of 100 for our Galaxy. (The Prada galaxies are bright systems, similar to the Galaxy).

How different would the Galaxy look if we were able to image it in mass rather than light ? We expect it would look lumpy and much larger. CDM simulations (*e.g.* Moore et al 1999) predict scale-free lumpy dark halos. Much substructure would be seen in the dark matter of the Local Group, although this has not yet been associated with similarly large numbers of dwarf galaxies in light.

It is not clear yet whether such dark mattter substructure is really present. There are far less small luminous objects seen nearby than expected from simulations (*e.g.* Jerjen 2004), which suggests that the dark lumps should be mostly baryon-poor. If the lumpy halos turn out to be incorrect, then the concept of non-Newtonian gravity (*e.g.* MOND) lurks in the background and may be the best mask of all.

5. The Kinematics of the Stellar Halo: a Dynamical Mask

The stellar halo of the Galaxy is a slowly rotating pressure-supported subsystem. For halo stars with [Fe/H]< -1.7, the systemic rotation is 30 to 50 km s^{-1}, and the velocity dispersion is (140,105,95) km s^{-1} in the (U,V,W) directions. For the more metal-rich halo stars with [Fe/H] > -1.7, the rotation increases with metallicity, probably due to the increasing contribution of the thick disk (*e.g.* Chiba & Beers 2000).

The halo is not dynamically well-mixed. It shows substructure; the long dynamical timescales in the outer halo allow the survival of identifiable debris of past accretion events. The tidally disrupting Sagittarius dwarf is a striking example. It can be seen in projection in configuration space (Majewski *et al.* 2003), and even more clearly in velocity space (Ibata *et al.* 1995). A large fraction of the halo stars in the galactic meridional plane may be associated with the Sgr debris (Morrison *et al.* 2001).

The tidal streams from the currently disrupting Sgr dwarf are interesting, but the ancient streams from small objects accreted long ago into the halo could be even more interesting, because they would allow us to probe the formation process of the galactic halo. These old streams are too faint to see in configuration space, but they can be unmasked in phase space, for example in (R_G, V_G) space (*e.g.* Harding *et al.* 2001: here R_G, V_G are galactocentric radius and radial velocity) or in integral space: *i.e.* the space of integrals of the motion for stellar orbits, like the stellar energy and angular momentum (E, L_z). Helmi *et al.* (1999) showed how the debris of halo objects accreted 12 Gyr ago can be readily recovered within 6 kpc of the sun from their distribution in integral space as determined from GAIA kinematics.

Accretion is believed to be important for building the stellar halo, but it is not clear yet how much of the halo comes from discrete accreted objects (debris of objects formed at high z) relative to stars formed during the baryonic collapse of the Galaxy. Recent simulations of pure dissipative collapse (*e.g.* Samland et al 2003) suggest that the halo may have formed mainly through a lumpy collapse, with only about 10% of its stars coming from accreted satellites. In any case, we may be able to trace the debris of these lumps and accreted satellites from their phase space structure, and so unmask the details of how the stellar halo was built up.

6. Structural Masks: the Thin and Thick Disks

Discovered only around 1980, the thick disks of our Galaxy and other spirals were masked from our recognition by the brighter thin disk. Like most of the masks, we can see around them with effort and measurement. The thick disk emerged from stellar photometry and star counts for the Milky Way and from accurate deep surface photometry (for other galaxies).

Most spirals (including our Galaxy) have a second thicker disk component. In some galaxies, it is easily seen. NGC 4762 is an unusual S0 galaxy which was one of the first in which the thick disk was identified (Tsikoudi 1980). In this system, the thick disk is bright and can be seen directly in photographic images Figure 6 shows two different stretches of the DSS image of NGC 4762 that illustrate the thin and thick disks of this galaxy.

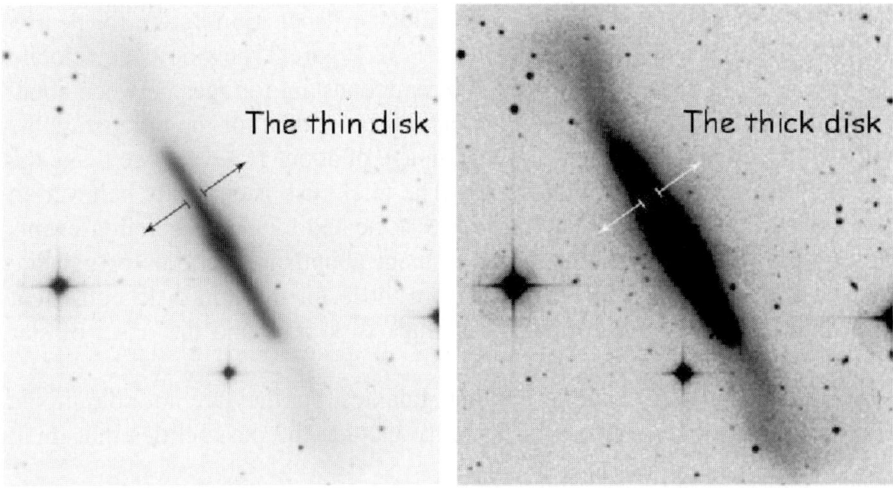

Figure 6. Images of the SO galaxy NGC 4762. The left image shows the thin disk and bulge. The deeper image on the right shows its more extended thick disk. The base of the arrows in each image show the height above the plane at which the thick disk becomes brighter than the thin disk (Tsikoudi 1980).

The galactic thick disk has a scale height from star counts of about 800 to 1200 pc, compared to the $\sim$ 300 pc scale height of the old thin disk. Its density is between 4 and 10% of the nearby thin disk, and its mean abundance [Fe/H] is about -0.7, with a tail of metal-poor stars extending down to [Fe/H] $\simeq -2.2$, and its stars appear to be older than 10 Gyr. The thick disk rotates quite rapidly, lagging the local standard of rest by about 30 km s^{-1} (Chiba & Beers 2000). Its velocity dispersion in the solar neighborhood in the (U,V,W) directions is (46,50,35) km s^{-1}, and its scale length is believed to be between 3.5 and 4.5 kpc, similar to that of the thin disk. Little is known about the radial extent of the galactic thick disk. This is an important question: if the thick disk really is the heated early thin disk, then the radial extent of the thick disk would indicate the size of the thin disk at the time of the heating event that led to the formation of the thick disk.

In most other galaxies, deep photometry is needed to see the thick disk. Dalcanton & Bernstein (2002) made BRK surface photometry of 47 late-type edge-on galaxies. They find that all are embedded in a flattened low surface brightness red envelope or thick disk. The age of these thick disks, estimated from the integrated BRK colors, is $>$ 6 Gyr and the population appears to be not very metal-poor, like the thick disk of the Milky Way.

Formation of the thick disk is a nearly universal feature of the formation of disk galaxies. The thick disk appears to be a discrete component of the galaxies (rather than just the hot tail of the thin disk). The discrete nature of the galactic thick disk is clearly seen from the age-velocity dispersion relation for nearby stars as measured by Edvardsson *et al.* (1993). Figure 7 show how the velocity dispersion of old disk stars appears to remain constant for ages between about 2 and 10 Gyr, as expected from dynamical theory; then for the older stars the velocity dispersion doubles abruptly at an age of about 10 Gyr as we make the transition from thin to thick disk stars. The thick disk is currently believed to have formed from the early stellar thin disk heated by early accretion events rather than by secular heating. The element abundance patterns of galactic thick disk stars also appear different from those for the thin disk, consistent with time delay between formation of thick disk stars and the onset of star formation in the current thin disk

Thick disks are very common in disk galaxies, but they are not ubiquitous. The formation pictures discussed recently include the possibilities that thick disks are:

- a normal part of disk settling (*e.g.* Samland et al 2003)
- accretion debris (*e.g.* Abadi et al 2003, Walker et al 1996)
- the early stellar thin disk, heated by accretion events such as the ω Cen accretion event (*e.g.* Bekki & Freeman 2003).

To summarize the current understanding of the thick disk, it appears that thick disks formed long ago, more than 10 Gyr ago in the Milky Way. They are chemically and dynamically distinct from the thin disk, and may have formed by the heating of the early stellar thin disk in an epoch of merging which ended more than 6 Gyr ago.

If the heating by accretion picture is correct, the thick disk may be one of the most significant components for studying galaxy formation, because it presents a kinematically recognizable 'snap-frozen' relic of the (heated) early disk. I use the description 'snap-frozen' because secular heating thereafter is unlikely to affect its dynamics significantly. Its stars are energetic enough to spend most of their orbital period away from the galactic plan and its heating processes, so we see the thick disk much as it was immediately after its formation.

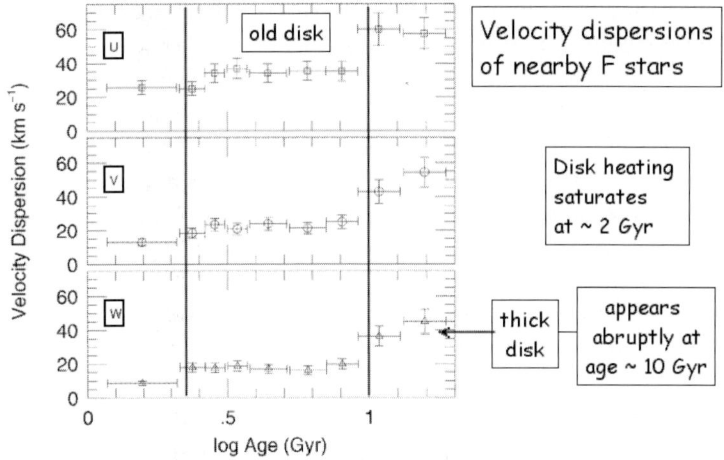

Figure 7. Velocity dispersion componenents (U,V,W) as function of stellar age for F stars in the solar neighborhood. Note how the velocity dispersion remains approximately constant for stellar ages between about 2 and 10 Gyr (the old disk) and then rises abruptly for stars older than 10 Gyr (the thick disk). From Edvardsson *et al.* (1993).

7. Chemical and Dynamical Masks : Retrieving Hidden Substructure

The galactic halo shows dynamical substructure, which is believed to be the debris of the small accreted objects that built up the halo. The Sgr dwarf is just the most recent example of such events. The galactic disk also shows kinematical substructure. These are called moving stellar groups. They are usually observed as stars in the solar neighborhood with common motions and sometimes common chemical properties. Some groups are apparently quite old.

Why would we expect to see substructure in the old disk ? Star formation is believed to occur in clumps of 10^4 to $10^6 M_\odot$. Some moving groups are debris of such star-forming aggregates in the disk. These dissolving aggregates will appear as moving groups. They are fossils of star forming events in the disk. Other groups may be debris of infalling objects, whose orbits have been circularised and dragged down into the galactic plane by dynamical friction (*e.g.* Walker *et al.* 1996). Simulations of the formation of an early-type spiral, by Abadi *et al.* 2003, show that many of the oldest stars in galactic disks may be the debris of such accreted objects. The disk heating processes and the shuffling of stellar angular momenta by spiral structure (*e.g.* Binney & Sellwood 2002) mean that the dynamical identity of some such aggregates will be lost. However we may be able to recognise them chemically if they were initially fairly homogeneous in their chemical abundances.

The nearby Hipparcos stars are dominated by three major moving groups, as shown in Figure 8. The Sirius and Hyades groups or streams comprise mainly early-type stars and are probably the remains of star-forming aggregates in the disk. The Hercules group is defined by mainly later-type stars. Dehnen (2000) and Fux (2001) argue that this group is related to a local kinematic disturbance associated with the galactic bar.

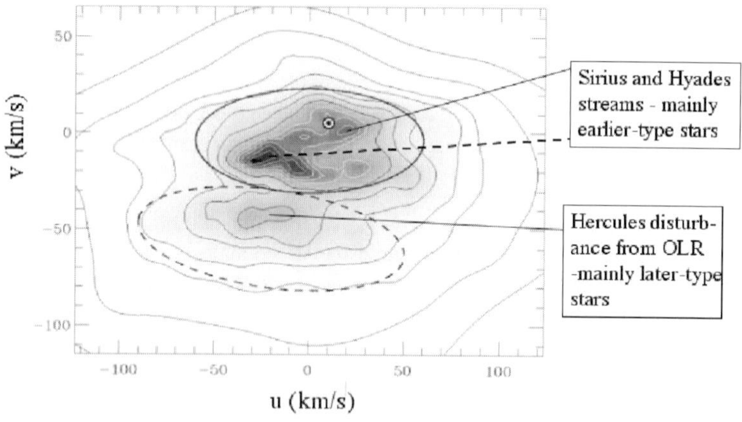

Figure 8. Contours of stellar density in the kinematic (U,V) plane for nearby Hipparcos stars (from Dehnen 2000). The three major moving groups are shown.

The solar neighborhood also includes sparser substructures that are much older. Examples of such groups include the thin disk HR 1614 group (Feltzing 2000), with an age of about 2 Gyr and an abundance [Fe/H] $\sim$ 0.2, and the Arcturus group (Eggen 1971, Navarro *et al.* 2004) whose nature is not yet clear. With a velocity V = -116 km s^{-1} relative to LSR and an abundance [Fe/H] ~ -0.6 for Arcturus itself, the Arcturus group could be a thick disk group, or it may be the debris of an accreted object as argued by Navarro *et al.* (2004). These two moving groups are nice examples of substructures surviving in the galactic disk. The chemical homogeneity of the moving groups is an important diagnostic of their origin; Gayandhi de Silva and Mary Wilson are working on this question at RSAA. The properties of the moving groups in the disk will become very interesting with RAVE and GAIA, as much larger kinematical and chemical samples of disk stars become available.

8. Recovering Ancient Star-forming Aggregates in the Thick Disk: Chemical Tagging

Star formation in the disk is believed to occur in clumps of 10^4 to $10^6 M_\odot$. It is likely that the disk also include the debris of satellites which have been accreted and tidally disrupted. We have the possibility to recover the remains of these various kinds of star forming systems from long ago. Probably some of the dynamical information about these ancient systems has been lost, but some is preserved because secular heating of the thick disk is not very important.

We would like to reconstruct the ancient star-forming aggregates of the thick disk, which have largely dispersed by phase-mixing (not heating). They will probably be structurally invisible (*i.e.* invisible in configuration space) but we may be able to identify them in integral space (as for halo) and by their chemical properties. This leads us to the concept of chemical tagging.

Chemical tagging relies on similarities in the detailed chemical abundances of thick disk stars ([Fe/H], [α/Fe], r- and s-process elements) to tag them to common ancient star-forming aggregates; see Freeman & Bland-Hawthorn (2002). The detailed abundance pattern reflects the chemical evolution of the gas from which the aggregates formed. Different supernovae provide different yields, depending on their mass, metallicity, detonation details, ejected mass etc. This is seen very clearly in the large scatter in the element ratios for metal-poor stars (*e.g.* Wallerstein *et al.* 1997). The large observed scattered in [X/Fe] for the more metal-poor stars suggests that the neutron capture elements in these stars are the products of only a few nucleosynthesis events; this is confirmed by simulations.

A primary requirement for chemical tagging is that the star-forming aggregates have unique chemical signatures or abundance profiles, *i.e.* that the star-forming gas clouds are chemically well mixed. High resolution studies of the Hyades cluster are encouraging. Paulson *et al.* (2003) and de Silva *et al.* (2004) find a very small dispersion in the element abundances of Hyades stars: the rms spread in [Fe/H] is about 0.04 and the intrinsic spread in the heavier elements is similar, below the level of the measurement errors.

How many distinct enrichment sites might there have been for the thick disk (about 10% of thin disk by mass) ? Say there was one SN event per 100 years (as in thin disk now). The duration of star formation in thick disk was about 1 Gyr, so there were probably about 10^7 enrichment sites during the star formation phase of the thick disk. Establishing millions of independent chemical signatures may appear too difficult. However, more than 60 of the elements arise from neutron capture processes. Say that only 30 of these elements were detectable and we could measure only two distinct abundance levels for each. That would already give more than 10^9 independent cells in the chemical abundance space. A potential problem is that the heavy element production varies in lockstep from site to site; some elements are seen to correlate, but not all (*e.g.* Burris *et al.* 2000).

This is the kind of methodology we can look forward to using in the near future to probe disk and thick disk substructure, as very large samples of stellar kinematics and abundances become available. What observational developments are on the horizon ? The RAVE program, to acquire stellar radial velocities and abundances at a spectral resolution of about 8000, is already well under way. It is scheduled to produce data for 100,000 stars in its first phase, and about 50 million stars in its second phase. The Gemini Wide Field Spectrograph program has the potential to generate high resolution abundance data for about one million stars. The GAIA mission is expected to generate space motions and intermediate resolution abundance data for up to 300 million stars.

References

Abadi, M. *et al.* 2003. ApJ, 597, 21.
Beaulieu, S. *et al.* 2000. AJ, 120, 855.
Bekki, K, Freeman, K. 2003. MNRAS, 346, L11
Binney, J., Sellwood, J. 2002. MNRAS, 336, 785.
Bissantz, N., Gerhard, O. 2002. MNRAS, 330, 591.
Bureau, M., Freeman, K. 1999. AJ, 118, 126.
Burris, D. *et al.* 2000. ApJ, 544, 302
Chiba, M., Beers, T. 2000. AJ, 119, 2843.
Combes, F., Sanders, R. 1981. A&A, 96, 164.
Courteau, S. *et al.* 1996. ApJ, 457, L73.
Dalcanton, J., Bernstein, R. 2002. AJ, 124, 1328.
Dehnen, W. 2000. AJ, 119, 800.
De Silva, G. *et al.* 2004. In preparation.
de Vaucouleurs, G. 1983. ApJ, 268, 451.
Edvardsson, B. *et al.* 1993. A&A, 275, 101.
Eggen, O. 1971. PASP, 83, 271.
Feltzing, S., Holmerg, J. 2000. A&A, 357, 153.
Freeman, K., 1991. In "Dynamics of Disk Galaxies", ed B. Sundelius, Gøteborg, p 15,
Freeman, K., Bland-Hawthorn, J. 2002. ARA&A, 40, 487.
Fux, R. 2001. A&A, 373, 511.
Harding, P. *et al.* 2001. AJ, 122, 1397.
Helmi, A. *et al.* 1999. In "The Third Stromlo Symposium: The Galactic Halo", ed Gibson, B., Axelrod, T. Putman, M., ASP Conference Volume 165, p 125.
Ibata, R. *et al.* 1995. MNRAS, 277, 781.
Kormendy, J., Kennicutt, R, 2004. ARA&A, 42, in press.
Jerjen, H. 2004. Seminar presentation.
Launhardt, R. *et al.* 2002. A&A, 384, 112.
Leggett, S. *et al.* 1998. ApJ, 497, 294.
Lewis, J., Freeman, K. 1989. AJ, 97, 139.
Majewski, S. *et al.* 2003. ApJ, 599, 1082.
Moore, B. *et al.* 1999. ApJ, 524, L19.
Morrison, H. *et al.* 2001. ApJ, 555, L37.
Navarro, J. *et al.* 2004. ApJ, 601, L43.

Oswalt, T. *et al.* 1996. Nature, 382, 692.
Paulson, D. *et al.* 2003. AJ, 125, 3185
Prada, F. *et al.* 2003. ApJ, 598, 260.
Pritchet, C., van den Bergh, S. 1994. AJ, 107, 1730.
Quillen, A., Garnett, D. 2000. astro-ph/0004210.
Rocha-Pinto, H. *et al.* 2000. A&A, 358, 869.
Sakamoto, T. *et al.* 2003. A&A, 397, 899.
Samland, M. *et al.* 2003. A&A, 399, 961.
Sandage, A. *et al.* 2003. PASP, 115, 1187.
Tsikoudi, V. 1980. ApJS, 43, 365.
Walker, I. *et al.* 1996. ApJ, 460, 121.
Wallerstein, G. *et al.* 1997. Rev. Mod. Phys., 69, 995.
Wilkinson, M., Evans, N.W. 1999. MNRAS, 310, 645.
Zoccali, M. *et al.* 2003. A&A, 399, 931

Comments from J. Kormendy: People often hope that our Galaxy is a typical example of galaxy evolution, because it can be observed in more detail than any other. But in fact, your review emphasizes that our Galaxy is very atypical of the galaxies that form by hierarchical clustering. Almost none of the stars or stellar components that you discussed are consistent with galactic assembly via major mergers. So our Galaxy is an example of the well-known problem of how, in a hierarchical clustering Universe, pure disk galaxies can form.

Comment from Ken Freeman: Absolutely!

THE HIERACHICAL FORMATION OF THE GALACTIC DISK

Julio F. Navarro
CIAR and Guggenheim Fellow, Department of Physics and Astronomy, University of Victoria, Victoria, BC, Canada.

Abstract I review the results of recent cosmological simulations of galaxy formation that highlight the importance of satellite accretion in the formation of galactic disks. Tidal debris of disrupted satellites may contribute to the disk component if they are compact enough to survive the decay and circularization of the orbit as dynamical friction brings the satellite into the disk plane. This process may add a small but non-negligible fraction of stars to the thin and thick disks, and reconcile the presence of very old stars with the protracted merging history expected in a hierarchically clustering universe. I discuss various lines of evidence which suggest that this process may have been important during the formation of the Galactic disk.

Keywords: Milky Way, galaxy formation, hierarchical clustering, cosmology, dark matter.

1. Introduction

Stellar disks are traditionally viewed as ensembles of stars on nearly circular orbits, dynamically fragile to fluctuations in the gravitational potential brought about by accretion events and mergers. Indeed, it is common to use the oldest stars in the Milky Way disk to date the time of its last major merger. Indications that star formation in the solar neighbourhood has been ongoing for most of the age of the Universe, coupled with the presence of metal-poor, old stars on disk-like orbits in the vicinity of the Sun, suggest that the Galaxy has suffered few, if any, important accretion events in the past $\sim$ 10 Gyr. This relatively peaceful evolution is at odds with the rather hectic merging activity expected in hierarchically clustering structure formation models (see, e.g., Steinmetz & Navarro 2000).

Recent numerical simulations (Abadi et al 2003a,b) suggest one way of reconciling the flurry of early mergers expected in a ΛCDM universe—the current paradigm of structure formation models—with the presence of an old disk component in the Milky Way. These simulations show that satellite accretion events may contribute stars not only to the spheroid, but also to the disk com-

D. Block et al. (eds.), Penetrating Bars through Masks of Cosmic Dust, 655–659.

ponent of a galaxy. These stars make up the core of satellites whose orbits are eroded, circularized, and brought into the plane of the galaxy before being disrupted. Such events contribute upwards of $\sim 10\%$ of stars in the disk, and the majority of old disk stars.

In this contribution I describe several lines of evidence that suggest that, indeed, the Milky Way disk may contain a number of stars which originate in the tidal debris of disrupted satellites. If confirmed by further scrutiny, this evidence would demonstrate that mergers and accretion events have been responsible for shaping the disk of the Milky Way, as well as its spheroid, and would provide strong support to the hierarchical mode of galaxy assembly envisioned in the ΛCDM scenario.

2. Tidal debris in the disk of the Milky Way

Since the bulk of the thin disk of the Galaxy is made up of stars that formed *in situ* following the collapse of a gaseous, dissipative component, left-over debris from past accretion events is most easily identified in samples of stars that minimize contamination by the thin disk. Thus, tidal debris should show more prominently in metal-deficient star samples, since these are likely to contain stars that formed before the merging activity abated, as well as in samples collected above, below, or in the outskirts of, the Galactic plane.

One prime example of this is the discovery by the SDSS team of a "ring-like" structure in the direction of the anti-galactic center (Yanny et al 2003). Further studies have confirmed that this is a dynamically coherent structure that spans a large arc on the plane of the Galaxy located at roughly 18 kpc from the Galactic center. Numerical simulations show that these "tidal arcs" occur naturally during the disruption of satellites on orbits coplanar with the disk (Helmi et al 2003), and have identified features that distinguish this accretion interpretation from other competing scenarios, such as spiral arms or the resonant response of the disk to the influence of the Magellanic Clouds or of the bar in the Galactic bulge.

One example of "tidal arcs" found in the simulation of Abadi et al (2003a,b) is shown in Figure 1. (Details may be found in Helmi et al 2003.) The arc circumscribed azimuthally between $\phi = 100^o$ and $\phi = 210^o$ is caused by the apocentric "wrapping" of the inner tidal arm stripped from the satellite after a recent pericentric passage. The arc is fed by a continuous stream of particles escaping the satellite. There is a clear gradient in the energies of particles in the arc; the most bound are most advanced along the arc (i.e. larger ϕ), which is reflected in a clear velocity gradient across the arc, shown in the middle panel of Figure 1.

This arc is a short-lived transient feature that weakens as particles of different energy phase mix throughout the disk. Thus, if this interpretation is correct,

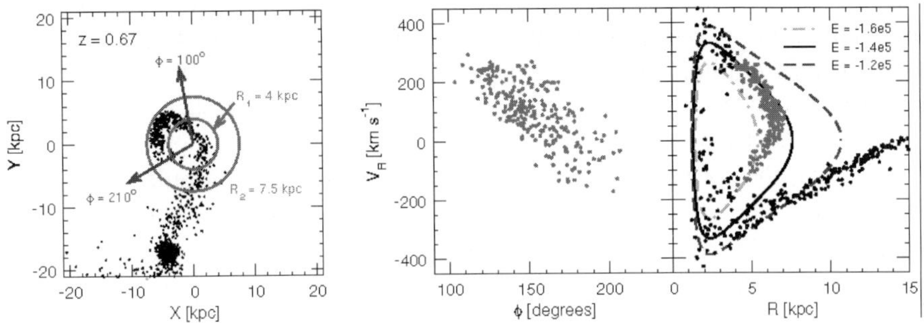

Figure 1. Snapshot of the disruption of a satellite on an orbit roughly coplanar to the disk, taken from the simulation of Abadi et al. (2003a,b). Left panel shows a "face on" projection of the debris shortly after the first pericentric passage. The main galaxy is at the center of coordinates, but only the stars in the satellite are shown for clarity. Middle panel shows, in a cylindrical coordinate system, the (galactocentric) radial velocity versus azimuthal angle ϕ for particles in the arc. Right hand panel shows radial velocity and distance for all satellite particles. Curves in this panel indicate the loci of particles with three different values of the binding energy, E, and angular momentum equal to the average of particles in the arc. See Helmi et al (2003) for details.

one the SDSS "ring" to be dynamically "young", and that the parent (disrupting) satellite may be still lurking somewhere in the Galactic disk. Recently, there have been intriguing suggestions that oddities in the distribution of disk stars in the direction of Canis Major may be due to the core of a disrupting dwarf rather than to the presence of a warp in the outer Galactic disk (Martin et al 2004). I conclude that the evidence in favour of interpreting the SDSS "ring" as debris from a (probably ongoing) accretion event is, if not conclusive, at least very compelling.

Are there other examples of this process? We have begun to scrutinize dynamically coherent structures in samples of stars that favour metal-deficient stars, in particular the compilations of Beers et al (2000, B00) and Gratton et al (2003, GCCLB). Figure 2a shows the distribution of specific angular momentum of stars in the B00 and GCCLB samples. These compilations favour metal-poor stars, and therefore contain mainly stars in the solar neighbourhood that belong to the traditional spheroid and thick disk components. As one can see in Figure 2a, the J_z-distribution is not smooth, and is punctuated by a number of local "peaks", even when only stars with orbits confined to the plane are considered.

These ensembles of stars with common rotation speeds are reminiscent of the dynamical substructures that Eggen (1986 and references therein) identified as "moving groups" in the solar neighbourhood. Eggen proposed that these groups were the late stages in the disruption of open clusters, recognizable locally as ensembles of stars with negligible dispersion in their rotation speed. Unfortunately, the lack of accurate distances for most candidates complicated the analysis and prevented proper assessment of his hypothesis. We (Navarro, Helmi & Freeman 2003) have reanalyzed Eggen's hypothesis for the case of the "Arcturus group".

Arcturus, the fourth brightest star in the night sky, is a Population II giant whose odd kinematics has been documented for centuries. It is very close to the Sun (hence its apparent brightness), and moves on an orbit confined to the Galactic plane, yet its rotation speed is only *half* that of the Sun. The vertical line labelled "Arcturus" in Figure 2a shows that there is a significant excess (over a smooth distribution) of stars of similar angular momentum as Arcturus in the B00 and GCCLB compilations.

Although there is a peak in the J_z distribution of stars in the B00 compilation that coincides with Arcturus, further inspection shows that these stars span a wide range of metallicities, casting doubt on Eggen's interpretation of this group as a disrupting star cluster. Furthermore, the dispersion in rotation speed of the group far exceeds the ~ 0.5 km s^{-1} that would be characteristic of a disrupting cluster. The metal abundances of stars in the group, however, show a distinct pattern, as shown in Figure 2b. This figure shows the enrichment in α elements ([α/Fe]) as a function of iron abundance, in solar units ([Fe/H]), for all stars in the GCCLB compilation (open circles). (α element abundances are not available for stars in the B00 catalog.)

The solid circles in Figure 2b correspond to those stars in the bin labelled "Arcturus" in Figure 2a. The trend defined by these stars is distinct from that traced by all GCCLB stars. The Arcturus group candidates define a narrow path in the [α/Fe] vs [Fe/H] plane well approximated by one of the simple closed-box self-enrichment models of Matteucci & Francois (1989, see solid line in Figure 2b). Thus, GCCLB stars with J_z comparable to Arcturus form a dynamically and chemically coherent group of stars which might be naturally identified with debris from a disrupted satellite galaxy. Further corroborating evidence is provided by the discovery by Gilmore, Wyse & Norris (2002) of a population of stars above and below the plane of the Galaxy with angular momentum similar to Arcturus, which they interpret as the debris of a tidally disrupted satellite.

The discovery of evidence for past (and perhaps ongoing) accretion onto the disk of the Milky Way suggests a reassessment of the traditional scenarios for the assembly and enrichment of stars in the solar neighbourhood. Extrapolating boldly, one may even argue that *most* metal-deficient stars (disk and spheroid) in the solar neighbourhood have been contributed by various accretion events throughout the life of the Galaxy. Confirming such assertion would clarify the role of mergers and accretion events in the formation of the Galactic

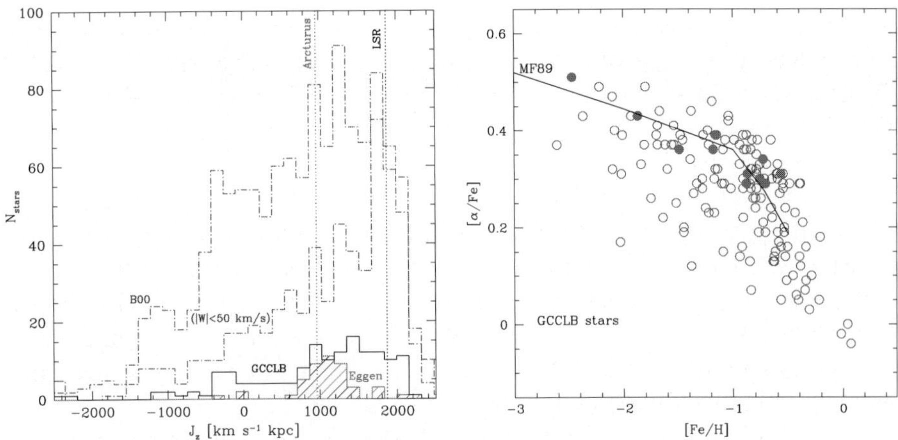

Figure 2. (a) Angular momentum distribution of stars in the Beers et al (2000, B00) catalog (top two histograms); in the Gratton et al (2003, GCCLB) compilation (solid histogram), as well as that of stars identified as "Arcturus group" candidates by Eggen (1986, shaded histogram). (b) α-enhancement ([α/Fe]) vs iron abundance ([Fe/H]) for all stars in the GCCLB compilation. Stars with angular momentum similar to Arcturus are shown as solid circles, and compared to a simple closed-box self-enrichment model (Matteucci & Francois 1989).

disk and would lend strong support to hierarchical theories of galaxy formation.

I wish to thank my collaborators, in particular Amina Helmi, Mario Abadi, Andres Meza, and Matthias Steinmetz, for allowing me to reproduce results of our collaborative program here.

References

Abadi, M. G., Navarro, J. F., Steinmetz, M., & Eke, V. R. 2003a, ApJ, 591, 499

Abadi, M. G., Navarro, J. F., Steinmetz, M., & Eke, V. R. 2003b, ApJ, 597, 21

Beers, T. C., et al 2000, AJ, 119, 2866 (B00)

Eggen, O. J. 1998, AJ, 115, 2397

Gilmore, G., Wyse, R. F. G., & Norris, J. E. 2002, ApJL, 574, L39

Gratton, R. G., et al 2003, A&A, 404, 187 (GCCLB)

Helmi, A., Navarro, J. F., Meza, A., Steinmetz, M., & Eke, V. R. 2003, ApJL, 592, L25

Martin, N.F., et al 2004, MNRAS, 348, 12

Matteucci, F. & Francois, P. 1989, MNRAS, 239, 885

Steinmetz, M., & Navarro, J. F. 2002, NewA, 7, 155

Yanny, B. et al. 2003, ApJ, 588, 824

THE SINGS VIEW OF BARRED GALAXIES

Michael W. Regan[1], Michele D. Thornley[2,1], George J. Bendo[3], Bruce T. Draine[4], Aigen Li[3], Daniel A. Dale[5], Charles W. Engelbracht[3], Robert C. Kennicutt[3], Lee Armus[6], Daniela Calzetti[1], Karl D. Gordon[3], George Helou[7], David J. Hollenbach[8], Thomas H. Jarrett[9], Lisa J. Kewley[10], Claus Leitherer[1], Sangeeta Malhotra[1], Martin Meyer[1], Karl A. Misselt[3], Jane E. Morrison[3], Eric J. Murphy[11] James Muzerolle[3], George H. Rieke[3], Marcia J. Rieke[3], Helene Roussel[7], John-David T. Smith[3], and Fabian Walter[12]

[1]*Space Telescope Science Institute, 3700 San Martin Drive, Baltimore, MD 21218*

[2]*Department of Physics, Bucknell University, Lewisburg, PA 17837*

[3]*Steward Observatory, University of Arizona, Tucson, AZ 85721*

[4]*Princeton University Observatory, Princeton, NJ 08544-1001*

[5]*Department of Physics & Astronomy, University of Wyoming, Laramie, WY 82071*

[6]*Spitzer Science Center, Caltech, MS 220-6, Pasadena, CA 91125*

[7]*Caltech, MC 314-6, Pasadena, CA 91101*

[8]*NASA Ames Research Center, MS 245-3, Moffett Field, CA 94035-1000*

[9]*Caltech, MC 320-47, Pasadena, CA 91101*

[10]*Harvard-Smithsonian Center for Astrophysics, 60 Garden Street, Cambridge, MA 02138*

[11]*Dept. of Astronomy, Yale University, New Haven, CT 06520*

[12]*NRAO, PO Box O, Socorro, NM 87801*

Abstract Well-resolved infrared observations of nearby galaxies are of fundamental importance to the study of the processes that affect galactic evolution. Here, we report on the first imaging results from the *Spitzer* Infrared Nearby Galaxies Survey (SINGS) using observations of the Sb galaxy NGC 7331. We present images of NGC 7331 over a large range of wavelengths that allow us to compare the distributions of gas, stars, and dust in unprecedented detail.

1. Introduction

Galaxy evolution depends significantly on the amount and distribution of gas and dust, the locations and amount of star formation, and how the energy

D. Block et al. (eds.), Penetrating Bars through Masks of Cosmic Dust, 661–665.

from star formation is reprocessed by the local environment. The mid- to far-infrared regime is ideal for studying these issues, providing access to a large fraction of the bolometric luminosity (Soifer, Neugebauer, & Houck 1987, as well as tracers of stars, dust and gas. To conduct a comprehensive study of the local distribution and effects of star formation in nearby galaxies, we have undertaken the Spitzer Infrared Nearby Galaxies Survey (SINGS) (Kennicutt et al. 2003), studying a diverse sample of 75 nearby galaxies with the Spitzer Space Telescope.

The evolution of a spiral galaxy is strongly influenced by non-axisymmetric structures within the galaxy. The most influential of these structures are stellar bars, which are commonly believed to be able to drive gas to smaller radii. By driving the gas to smaller radii, a stellar bar can smooth out chemical gradients, create star forming rings (Regan & Teuben 2003), and fuel central starbursts (Heller & Shlosman 1994). These effects can in turn change the Hubble type of the galaxy (Friedli & Benz 1993; Norman et al. 1996) and even destroy the bar itself(Hasan & Norman 1990). However, as we can only observe the current state of a galaxy and must infer what occurred in the past and will occur in the future, a combination of observations and models is essential for *quantifying* the effectiveness of bars in galaxy evolution.

The 3.6-8 micron filters available on the Spitzer InfraRed Array Camera (IRAC, (Fazio et al. 2004)) provide a unique and efficient tool for measuring the resolved distribution of mass and the morphology of the interstellar medium in a barred galaxy. As part of the validation of the instrumental performance and the SINGS observing strategy, we obtained observations of the spiral galaxy NGC 7331 with Spitzer in December 2003. NGC 7331 is a relatively highly inclined (i=77deg, (Garcia-Gomez et al. 2002)) Sb galaxy at a distance of 15.1 Mpc (Hughes et al. 1998).

2. Observations of NGC 7331

We show an example of the capabilities of Spitzer imaging in Figure 1. Here we show three images of the NGC 7331 at optical (V-band) and the two short wavelength channels of IRAC (3.6 microns and 4.5 microns) (Regan et al. 2004). The optical image shows a prominent dust lane across the near side of the bulge. Meanwhile, the IRAC images have the same stellar morphology but the dust lane disappears. Regan et al. (2004) showed that the hot dust contributions at these wavelengths is less than 10% making them excellent tracers of the stellar light and thus the mass in the galaxy. The IRAC images are better than K-band observations because the dust extinction at 3.6 and 4.5 microns is 1/2 and 1/5 of K-band, respectively (Rieke & Lebofsky 1985). Determining an accurate mass distribution requires both a high signal to noise image of the galaxy, unaffected by dust, and a good measurement of the sky level. Our SINGS observing strategy provides both.

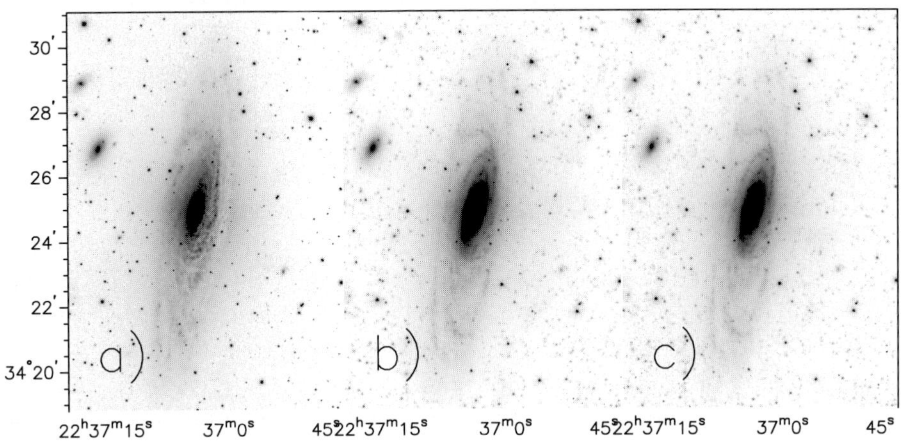

Figure 1. SINGS image of the galaxy NGC 7331. a) A V-band image showing the dust lane the crosses the near side of the bulge. b) An IRAC 3.6 micron image showing the same stellar distribution but without the dust lane. c) An IRAC 4.5 micron image that also shows that this wavelength is also a good tracer of the stellar light in a galaxy.

In Figure 2 we show how the long wavelength channel (8 microns) of IRAC is a good match to the CO map (Regan et al. 2001). In this central region there is almost no atomic gas. Therefore, the 8 micron image appears to be an excellent tracer of the total ISM column density. Thus, we will be able to use both the IRAC 8 micron and the MIPS images of hot dust to map the ISM distribution in barred galaxies. By comparing the distribution of the ISM in barred and unbarred galaxies we will be able to better quantify the effect that the bars have on the radial distribution of the ISM.

3. Summary

Imaging of barred galaxies as part of the SINGS project will provide a new and powerful method of studying barred galaxies. Short wavelength IRAC imaging will provide high fidelity images of the stellar content of the galaxies that can be used to generate mass models of the galaxies. Longer wavelength IRAC images of PAH emission and MIPS images of dust will provide high resolution maps of the ISM in the galaxies that can be used to quantify the effect of the bars on the evolution of the host galaxies.

References

Fazio, G. G., et al. 2004, ApJS, in press

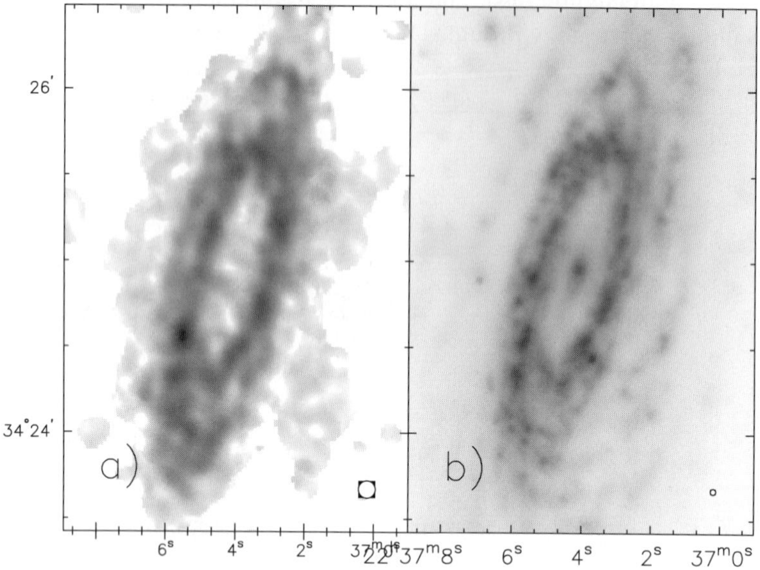

Figure 2. A comparison of two tracers of the ISM in NGC 7331. a) A CO 1-0 image from BIMA SONG (Regan et al. 2001) showing the ring of molecular gas at a radius of 6 kpc. b) An 8 micron image from IRAC showing the same ring at higher resolution (∼2 arcseconds). Here we are primarily seeing emission from Polycyclic Aromatic Hydrocarbons (PAHs). This is evidence that the 8 micron emission may be tracing the total ISM content.

Friedli, D. & Benz, W. 1993, A&A, 268, 65
Garcia-Gomez, C., et al. 2002, A&A, 389, 68
Hasan, H. & Norman, C. 1990, ApJ, 361, 69
Heller, C. H. & Shlosman, I. 1994, ApJ, 424, 84
Hughes, S. M. G., et al. 1998, ApJ, 501, 32
Kennicutt, R. C., et al. 2003, PASP, 115, 928
Norman, C. A., Sellwood, J. A., & Hasan, H. 1996, ApJ, 462, 114
Regan, M. W., et al. 2004, ApJS, in press
Regan, M. W. & Teuben, P. J. 2003, ApJ, 582, 723
Regan, M. W., Thornley, M. D., Helfer, T. T., Sheth, K., Wong, T., Vogel, S. N., Blitz, L., & Bock, D. C.-J. 2001, ApJ, 561, 218
Rieke, G. H. & Lebofsky, M. J. 1985, ApJ, 288, 618

Comments from J. Kormendy: I don't want to disagree with you about secular evolution in oval disks, but the galaxies appear to do so. Observational evidence for pseudobulges is very strong in oval galaxies. One of the best examples is NGC 4736, which has a bright, rotation-dominated pseudobulge and a well defined "nuclear" star-formation ring. Of course, NGC 4736 may have been barred in the past – its disk brightness profile is like that of a lens – so much of the pseudobulge may have been built by a now-defunct bar. But the star formation ring is happening now. And the molecular gas concentration observed by the BIMA SONG has a relatively short star-formation timescale. So these features cannot be caused by a defunct bar but rather are consistent with secular evolution driven by the oval disks.

QUANTIFYING BAR STRENGTH: MORPHOLOGY MEETS METHODOLOGY

Paul B. Eskridge
Department of Physics & Astronomy, Minnesota State University, Mankato MN 56001, USA

Abstract A set of objective bar-classification methods have been applied to the Ohio State Bright Spiral Galaxy Survey (Eskridge et al. 2002). Bivariate comparisons between methods show that all methods agree in a statistical sense. Thus the distribution of bar strengths in a sample of galaxies can be robustly determined. There are very substantial outliers in all bivariate comparisons. Examination of the outliers reveals that the scatter in the bivariate comparisons correlates with galaxy morphology. Thus multiple measures of bar strength provide a means of studying the range of physical properties of galaxy bars in an objective statistical sense.

Keywords: galaxies: fundamental parameters — galaxies: spiral — galaxies: statistics — galaxies: structure — infrared: galaxies

1. Introduction

The existence and importance of bars in spiral galaxies has been recognized for the entire history of extragalactic astronomy (Curtis 1912). Hubble (1936) included the barred (SB) and unbarred (S) classes as the tines of the tuning fork. de Vaucouleurs (1959) first noted that "barredness" is not a binary state, and invented the class SAB for galaxies that are detectably, but not strongly barred. The relevance of observed wavelength to the detection of bars was first pointed out by Hackwell & Schweizer (1983) who discovered a strong bar in their H-band image of the optical SAB NGC 1566. The most extensive demonstration of this issue to date is that of Eskridge et al. (2000). One of the shortcomings of the above-cited work is that it uses subjective bar classification. Subjective classification is a useful starting point, but to move toward a physical understanding of morphology requires the general adoption of an appropriate, quantifiable and objective metric. The problem is that bars do not lend themselves naturally to scalar quantification. There have been a large number of "bar-strength quantifiers" proposed and implemented by different groups, and all of them measure somewhat different things (e.g., Elmegreen &

D. Block et al. (eds.), Penetrating Bars through Masks of Cosmic Dust, 667–671.

Elmegreen 1985; Ohta et al. 1990; Martin 1995; Wozniak et al. 1995; Abraham & Merrifield 2000; Buta & Block 2001).

This presentation reports on the progress of a study to intercompare a set of different bar-strength quantifiers using a single large, and statistically complete sample of nearby spiral galaxies (the OSU sample – see Eskridge et al. 2002). The main purposes of the study are two-fold:
1) To determine how well different bar-strength classifiers compare to one another and study the sorts of situations in which they fail to agree.
2) To address how reliably we can study the bar fraction of high-z galaxies. This follows from a comparison of the results from techniques that are most demanding of the observational data with those that are the least demanding.

Section 2 presents the methods included in this study. Section 3 presents some of the bivariate relations found. Section 4 includes a discussion of the findings to date and a summary of issues yet to be fully investigated.

2. Summary of Methods

The most data intensive method included in this study is the Q_g "disk-torque" measure of Buta et al. (2004). It is a refinement of the Q_b "bar-torque" measure of by Buta & Block (2001). Both of these methods are pure NIR methods by definition. The goal of these methods is to express the strength of the bar in terms of its gravitational effect on the surrounding disk. In principle, this is an excellent physical means of measuring bar strength. In practice, the methods requires assumptions about disk structure that are difficult to verify. They also require high spatial resolution, high S/N images, and are very laborious to implement.

Both of these methods are based on Fourier-mode decomposition. In its simplest guise, the Fourier-mode method amounts to measuring the strength of the constant-phase $m = 2$ mode in an image. This study uses analysis by D. Elmegreen and her students, following Elmegreen & Elmegreen (1985).

A method proposed by Abraham & Merrifield (2000) measures bar strength by the maximum ellipticity of the inclination-corrected disk. It is thus most sensitive to long, thin bars. The method was refined by Whyte et al. (2002), who also applied it to the OSU survey sample. This study uses the results of Whyte et al. (2002).

The complete study will consider the neural-net method of Odewahn et al. (2002), developed for automated classification of high-redshift galaxies. It is the most forgiving of data quality. These results are not yet available.

3. A Selection of Bivariate Results

Figures 1–3 show a selection of bivariate comparisons amongst Q_g (Buta et al. 2004), the H-band m=2 Fourier mode amplitudes (from D. Elmegreen), and the B- and H-band Abraham bar parameter F_{bar} (Whyte et al. 2002). In all cases, statistically significant correlations exists (at better than the 10^{-4} level from a variety of tests), but with substantial scatter and large outliers. The dashed lines are the ordinary least-squares bisectors.

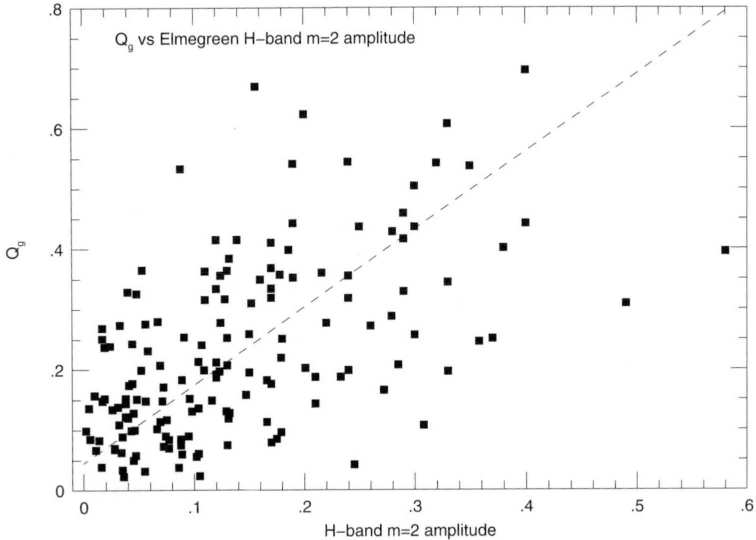

Figure 1. H-band $m = 2$ Fourier mode against Q_g.

4. Summary and Future Work

It is a great reassurance that all the methods examined here provide statistically self-consistent results, even when comparing different methods in different wave-bands (NIR torque methods compared with B-band bar ellipticity, as an example). Thus any of the methods discussed here can provide a good statistical means of measuring bar strength. This argues that bar-strength analysis of high-redshift galaxy samples can, at least in principle, be compared with analysis of nearby systems even when the observations are different rest-frame wavelengths, and at very different spatial sampling. The inclusion of neural-net techniques to the current study promises to be a very interesting advance.

A more careful examination of the bivariate results reveals that even very similar methods tend to have significant scatter between them, and the occasional very substantial outlier. The existence of these outliers actually gives us the opportunity to study the failure modes of the various techniques under consideration. As an example, the two systems with the largest H-band F_{bar} for their Q_g values (the points on the bottom right of Fig. 2b) are NGC 3166 and NGC 4772. Both these galaxies have H-band morphologies of SB0 (Eskridge

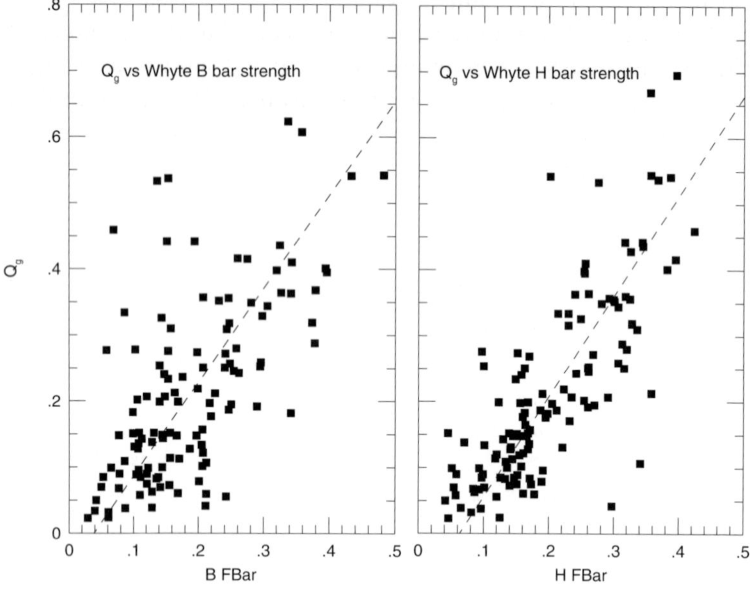

Figure 2. Q_g vs a) H-band F_{bar} and b) B-band F_{bar}.

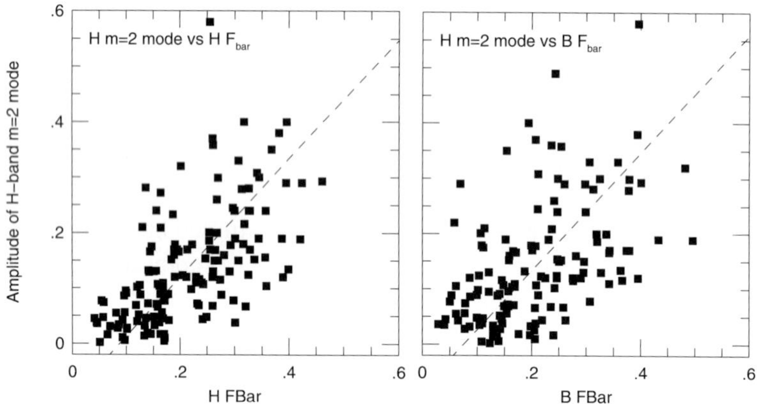

Figure 3. H-band Fourier $m = 2$ mode vs a) H-band F_{bar} and b) B-band F_{bar}.

et al. 2002), with relatively large bulges and weak disks. The F_{bar} statistic is very good at describing such bars, but Q_g is not. This is due to the relatively strong bulges causing the effective torque from the bars to be small. In the same plot, the two galaxies with the largest Q_g for their H-band F_{bar} values are NGC 1042 and NGC 3319. Both these systems have grand design spiral patterns and short bars. The shortness of the bars appears to drive the low F_{bar} values. But the combination of the bars and the strong spiral patterns results in a large disk torque. A true Q_b analysis, as discussed in Buta et al. (2003), will improve the sensitivity of torque-based analysis to pure bar components. The scatter in the bivariate plots is generally correlated with morphological patterns in the sample. The use of multiple measures of bar strength thus offers us the opportunity to study full range of the physical properties of bars in an objective, statistical sense.

Acknowledgments

This study is made possible by the work of the authors of Block et al. (2002), Buta et al. (2004) and Whyte et al. (2002), and D. Elmegreen and her students. I thank them for their efforts. Thanks are also due to David Block and Ken Freeman for inviting me, and for organizing the conference. The OSU survey was made possible by grants AST-9217716 and AST-9617006 from the US National Science Foundation. This work is supported by Minnesota State University and by NASA grant HST-GO-09892.02.

References

Abraham, R.G. & Merrifield, M.R. (2000) AJ, 120, 2835

Block, D.L., Bournaud, F., Combes, F., Puerari, I. & Buta, R. (2002) AAp, 394, L35

Buta, R. & Block, D.L. (2001) ApJ, 550, 243

Buta, R., Block, D.L. & Knapen, J. (2003) AJ, 126, 1148

Buta, R., Laurikainen, E. & Salo, H. (2004) AJ, 127, 279

Curtis, H.D. (1912) PASP, 24, 227

de Vaucouleurs, G. (1959) Handb. Phys., 53, 275

Elmegreen, B.G. & Elmegreen, D.M. (1985) ApJ, 288, 438

Eskridge, P.B., Frogel, J.A., Pogge, R.W., Quillen, A.C., Davies, R.L., DePoy, D.L., Houdashelt, M.L., Kuchinski, L.E., Ramirez, S.V., Sellgren, K., Terndrup, D.M., & Tiede, G.P. (2000) AJ, 119, 536

Eskridge, P.B., Frogel, J.A., Pogge, R.W., Quillen, A.C., Berlind, A.A., Davies, R.L., DePoy, D.L., Gilbert, K.M., Houdashelt, M.L., Kuchinski, L.E., Ramirez, S.V., Sellgren, K., Stutz, A., Terndrup, D.M., & Tiede, G.P. (2002) ApJS, 143, 73

Hackwell, J.A. & Schweizer, F. (1983) ApJ, 265, 643

Hubble, E.P. (1936) *The Realm of the Nebulae* (New Haven: Yale University Press)

Martin, P. (1995) AJ, 109, 2428

Odewahn, S.C., Cohen, S.H., Windhorst, R.A. & Phillip, S.(2002) ApJ, 568, 539

Ohta, K., Hamabe, M. & Wakamatsu, K.-I. (1990) ApJ, 357, 71

Wozniak, H., Friedli, D., Martinet, L., Martin, P. & Bratschi, P. (1995) AApS, 111, 115

Whyte, L.F., Abraham, R.G., Merrifield, M.R., Eskridge, P.B., Frogel, J.A., & Pogge, R.W. (2002) MNRAS, 336, 1281

(constant M/L)

ESTIMATION OF BAR STRENGTHS FROM NEAR-IR IMAGES

Heikki Salo[1], Eija Laurikainen[1] and Ronald Buta[2]
[1]*Division of Astronomy, Dept.of Phys. Sciences, Univ. of Oulu, FIN-90014, Finland,*
[2]*Univ. of Alabama, Tuscaloosa, USA*

Abstract The relative tangential force amplitude provides an useful measure for the bar-associated gravitational perturbation in galaxies. However, the estimation of gravitational field from near-IR images involves several uncertainties. Some of these uncertainties, as well as our current efforts in analyzing the effects of image noise and the artificial bulge stretch are summarized, illustrated by examples with both observational and synthetic images.

Keywords: Galaxies: barred

1. Introduction

The ratio of the tangential force amplitude relative to the mean axisymmetric radial force, F_T/F_R, estimated from near-IR images, is a useful tool for analyzing bar-associated gravitational perturbations in galaxies. Especially, the maximum value of this force ratio, Q_g, gives a convenient single number characterizing the strength of the bar. This gravitational torque method (GTM) was first applied to a large sample of galaxies by Buta and Block (2001), using Cartesian force evaluation by Quillen et al. (1994). A similar approach, but with polar force evaluation method by Salo et al. (1999), has recently been applied to samples drawn from the 2 Micron All Sky Survey (2MASS) and the Ohio State University Bright Galaxy Survey (OSUBGS, Eskridge et al. 2002), with the main interest on the relation between bars and nuclear activity (Laurikainen et al. 2002, 2004a) and on the distribution of bar strengths in spiral galaxies (Buta et al. 2004).

Application of GTM involves several uncertainties which need to be carefully analyzed. The main assumptions are: (i) the near-IR light distribution traces the mass distribution, i.e. a constant M/L is assumed, (ii) the vertical density distribution can be approximated by some simple functional form, like an exponential profile, with some proper scale parameter h_z. The assumption of constant M/L certainly fails in the outer parts of the disks due to dark halos.

D. Block et al. (eds.), Penetrating Bars through Masks of Cosmic Dust, 673–677.

Nevertheless, in the bar region where the maximum Q_g typically occurs, the effect of halos is generally insignificant for bright galaxies, as shown in Buta et al. (2004) by applying the correlation between galaxy luminosity and dark halo contribution found by Persic et al. (1996). The effect of unknown vertical structure is more problematic: although the exact functional form is not crucial (see e.g. Laurikainen and Salo 2002), the derived Q_g depends significantly on the assumed vertical scale height. We have tried to reduce this uncertainty by connecting h_z to the radial scale length h_r, using Hubble-type dependent empirical h_z/h_r ratios derived by de Grijs (1998). Typically, the uncertainty due h_z is $\pm 5\%$ for late type spirals, increasing to perhaps $\pm 25\%$ for early type spirals (Laurikainen et al. 2004b).

The accurate orientation parameters are also crucial for the derived perturbation strengths. In our analysis of OSUBGS galaxies the inclinations and position angles were estimated from deep optical B-band images. The resulting uncertainties in Q_g are typically only a few percents (Laurikainen et al. 2004b). However, for some of the galaxies our estimates of orientation parameters showed large differences in comparison to RC3. Other concern in interpreting Q_g is the possible contribution from spiral arms besides the bar. Using a bar-spiral decomposition method, Buta, Block and Knapen (2003) estimated this to result in about 4% uncertainty in Q_g.

In the following two additional aspects of force evaluation are briefly examined: the sensitivity of GTM to a reduced image quality (Section 2), and the effect of artificial bulge stretch during image deprojection (Section 3).

2. Image quality: smoothing by azimuthal Fourier decomposition

In Cartesian integration the deprojected galaxy image is directly converted to gravitational potential via 2-D FFT, with the vertical density profile incorporated to the convolution kernel. Similar procedure can be applied in polar coordinates, after decomposing the surface density into azimuthal Fourier components. For good quality data exactly same force profiles are obtained (see Laurikainen and Salo 2002), but differences emerge for poorer quality images (Fig. 1). In particular, the Cartesian method can yield spurious large F_T/F_R in the noisy outer parts of the disk; these are absent in the polar method due to the smoothing involved in using only low order m-components. The polar method is also fairly robust against reduced image resolution, checked by re-sampling the original NOT image shown in Fig. 1 into successively poorer resolutions. Remarkably, essentially the same Q_g is obtained, as long as the distance of the peak F_T/F_R corresponds to at least about 10 image pixels. However, this result is likely to depend on the smoothness of the density distribution and the S0 galaxy used as an example may be somewhat atypical.

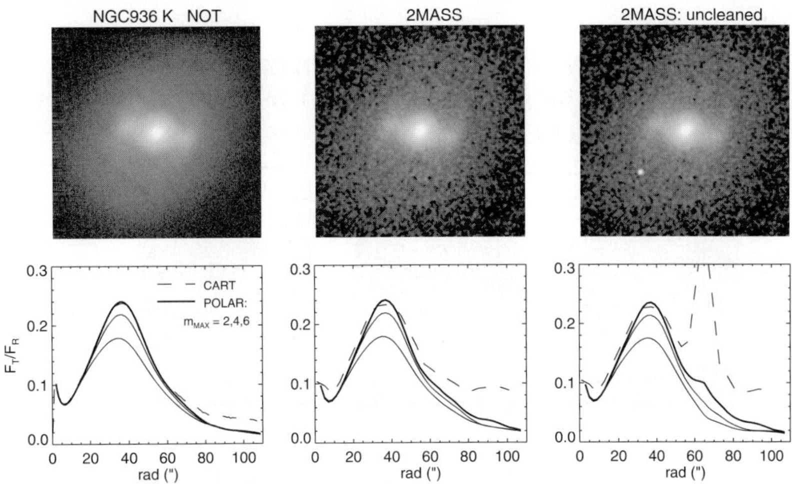

Figure 1. Comparison of polar and Cartesian force integration. The *upper left frame* shows the Ks band image of NGC 936 observed at the Nordic Optical Telescope, while the *middle frame* is from 2MASS survey (mosaic from the Large Galaxy Atlas). Both frames cover $220'' \times 220''$ and have been cleaned from foreground stars. The *lower row* shows the F_T/F_R profiles, evaluated both with Cartesian (dashed lines) and with polar integration (solid lines indicate profiles when various numbers of even Fourier components are included, with $m_{max} = 2, 4, 6$). The main difference is the larger F_T/F_R in the outer disk for the Cartesian method, arising due to increased noise level; this artifact is practically absent in the polar method. The difference becomes even more pronounced if there are defects in the image: an exaggerated example is given in the *rightmost frame*, containing one uncleaned foreground star.

3. Bulge separation

An important concern in the application of GTM is how to prevent the artificial bulge stretch during deprojection from affecting the derived Q_g. As a first approach, we have devised a method where the bulge is approximated with a seeing-convolved spherical density distribution, separated from the galaxy image before the disk deprojection and the calculation of disk forces. For this purpose a three-component 2D decomposition is used, including a Ferrers bar besides an exponential disk and a Sersic bulge. Although the bar model is rather crude, its inclusion to the fit is often crucial to prevent bar light from being assigned into unrealistically large bulges (see Peng et al. 2002, who also stress the insufficiency of 2-component models). Comparison to force evaluations without any special treatment of bulges shows that the overall effect of bulge stretch is small (Laurikainen et al. 2004b, Buta et al. 2004). However, in individual cases correcting for this effect is important, especially for systems with a large inclination.

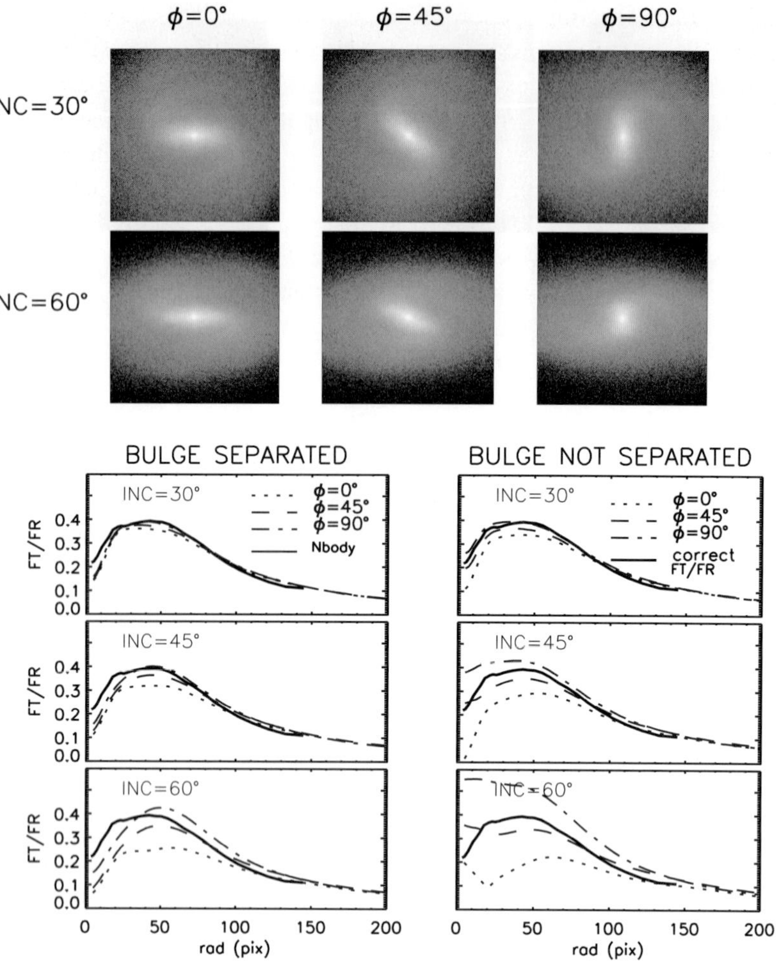

Figure 2. Tests with synthetic data. *Upper frames* display an N-body simulation, viewed from different inclinations and directions with respect to the bar (ϕ denotes the angle between the bar major axis and the line of nodes). The 3-D simulation model consisted of an analytical Plummer-bulge and a self-consistent exponential disk with $N = 2 \cdot 10^6$ particles, performed with the method of Salo and Laurikainen (2000), with mass ratio $M_{\rm bulge}/M_{\rm disk} = 0.25$ and $r_{\rm eff}/h_r \approx 0.5$, where $r_{\rm eff}$ is the effective radius of the bulge and h_r is the final radial scale length of the disk. The length of the bar is about $2.5h_r$ (about 65 pixels in the snapshots). In *lower frames* the solid curve indicates the actual FT/FR profile in the dynamical simulation, while the dashed lines display the force ratios derived from synthetic images, assuming an exponential vertical profile with a constant $h_z = h_r/4$ (estimated from dynamical simulation for the bar region). *In the left* the spherical bulge component has been subtracted from the simulated image before the deprojection and force calculations: the deviations from the correct force profile are small even for INC$= 45^\circ$. *In the right* no bulge separation was made, and the artificial bulge stretch leads to much more significant deviations: in the case of viewing along the bar ($\phi = 90^\circ$) forces are overestimated (bulge deprojects parallel to the bar), while the opposite takes place when viewing perpendicular to the bar ($\phi = 0^\circ$).

Fig. 2 illustrates the importance of bulge separation, using synthetic N-body simulation data viewed from various inclinations. We have deliberately chosen an example where the extent of the formed bar is large in comparison to bulge. Even in this case, treating the bulge light in a similar manner than the disk leads to large deviations from the true force profile. Depending on the viewing orientation with respect to bar, the bulge stretch may either increase or decrease the apparent force maximum. When the spherical shape of the bulge is taken into account, the deviations are significantly smaller, and follow from the vertical structure of the disk and bar. The system shown in Fig. 2 is rather thick, with $h_r/h_z \approx 4$, mimicking an early type spiral; for later spirals the deviations after bulge subtraction would be smaller.

However, note that in the example shown the disk orientation, the h_z/h_r and the spherical bulge profile are based on the values known from the dynamical model, whereas in a realistic case observational uncertainties in these parameters will degrade the results. Also, in the case of non-spherical bulges the applied bulge correction would be only approximative.

4. Summary

The calculation of F_T/F_R force ratios provides a straightforward method for characterization of perturbations associated with bars and spirals, provided that the vertical extent of the disk can be estimated. In particular, azimuthal smoothing implicit in polar method makes the results fairly robust against image resolution and noise. This makes the method feasible also for larger z's, provided that disk orientation and scale length can be reliably estimated. Treatment of spherical bulge helps to remove effects of artificial bulge stretch, but further improvements might still be needed in order to apply the method for thick systems, or to systems with triaxial bulges.

References

Buta, R., and D. L. Block (2001). ApJ, 550, 243
Buta, R., Block, D., and J. Knapen (2004). AJ 126, 1148.
Buta, R., Laurikainen, E., and H. Salo (2004). AJ 127, 279.
de Grijs, R. (1998). MNRAS, 299, 59
Eskridge, P. et al. (2002). ApJS, 143, 73
Laurikainen, E., Salo, H., and P. Rautiainen (2002). MNRAS, 331, 880
Laurikainen, E., and H. Salo (2002). MNRAS, 337, 118
Laurikainen, E., Salo, H., and R. Buta (2004a). ApJ, 607, 103
Laurikainen, E., Salo, H., Buta, R. and S. Vasylyev (2004b). in preparation
Peng, C., Ho, L., Impey, C., and H.-W. Rix (2002). AJ 124, 266
Persic, M., Salucci, P., and F. Stel (1996). MNRAS 281, 27
Salo, H. et al. (1999). AJ 117, 792
Salo, H. and E. Laurikainen. (2000). MNRAS 319, 393
Quillen, A. C., Frogel, J. A. and R. Gonzales (1994). ApJ, 437, 162

BAR STRENGTHS MEASURED FOR THE OSUBGS SAMPLE: ACTIVE VS. NON-ACTIVE GALAXIES

Eija Laurikainen[1], Heikki Salo[1] and Ronald Buta[2]
[1] *Division of Astronomy, Dept. of Phys. Sciences, Univ. of Oulu, FIN-90014, Finland,*
[2] *Univ. of Alabama, Tuscaloosa, USA*

Abstract Bar fractions and bar strengths are compared for active and non-active galaxies using a sample of 180 galaxies from the Ohio State University Bright Galaxy Survey (OSUBGS). The identification of bars and calculation of the perturbation strengths are made using Fourier techniques. We found that bars in AGNs are massive, but have weak perturbation strengths due to dilution of forces by massive bulges. Also, active galaxies were found to avoid the strongest bars.

Keywords: Galaxies: barred; galaxies: active

1. Introduction

Large scale bars are known to be efficient in triggering gas inflows, manifesting as nuclear/circumnuclear star formation, indicated for example by a large number of circumnuclear rings with significant star formation in the inner rings (Knapen 2004). However, the connection is less clear for Seyferts and LINERs (AGNs): although Simkin, Su and Schwartz (1980) payed attention to the prominence of bars in Seyfert galaxies, most studies have shown that the connection between bars and Seyfert type nuclear activity is statistically insignificant.

Some of the latest studies in the near-IR indicate that Seyferts might have more bars than the non-active systems (Knapen et al. 2000), and a large number of outer rings (Hunt and Malkan 1999), typically related to bars. But it has also been discussed by Kormendy and his coworkers that AB-type bars, generally assumed to be weak bars, might actually be ovals or lenses. This highlights the importance of using some well defined criteria for bars capable of detecting also the weakest bars. We have re-investigated the role of bars on nuclear activity using a large well defined sample of spiral galaxies, applying Fourier methods for calculating bar strengths (Laurikainen, Salo and Buta, 2004; Laurikainen, Salo, Buta Vasilyev, 2004).

D. Block et al. (eds.), Penetrating Bars through Masks of Cosmic Dust, 679–683.

2. Quantification of Bar Strengths by Fourier Methods

We use a sample of 180 galaxies based on the Ohio State University Bright Galaxy Survey (OSUBGS, Eskridge et al. 2002, EFP), limiting to disk inclinations less than 65^0. In order to estimate the perturbation strengths (Q_g) a gravitational bar torque method (GTM) is used where the ratio of the maximum tangential force to the mean axisymmetric radial force is calculated as a function of radius ($Q_T(r)$). The method is described by Salo in this conference (see also Buta). The structural decomposition is part of the refined GTM and Fig. 1 shows the importance of including a bar model to the fit. Fig. 2 shows how GTM offers a reasonable tool to identify bars: in a barred galaxy strong m=2 and m=4 amplitudes of density (A_2 and A_4) appear in the bar region inducing a peak in the $Q_T(r)$-profile. Bars were distinguished from spiral arms mainly by assuming that the m=2 phase is maintained nearly constant in the bar region, and from ovals due to the ellipticity of the bar/oval.

The Fourier method picks up mainly SB-type bars as defined visually by EFP in the near-IR. However, AB-type structures in galaxies seem to be either ovals or spiral arms inducing non-axisymmetirc forces similar to those found in SA-type galaxies.

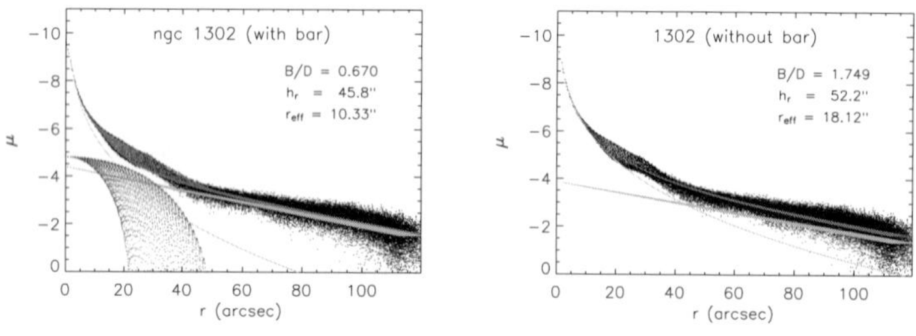

Figure 1. Bulge/disk (B/D) ratios are estimated applying 2-dimensional decomposition to H-band images: the disks are assumed to be exponential while the bulges are described by a Sersic's function. A crude bar model in terms of Ferrer's function can also be included to the fit, in order to prevent bar light from being attributed to the fitted bulge (see also Peng et al. 2002)

3. Bars in active and non-active Galaxies

Active galaxies (Seyferts, LINERs and HII/starburst galaxies) were found to have bars more frequently than the non-active systems (72 % vs. 55 %) confirming some earlier results for Seyfert galaxies in the near-IR (Knapen et al. 2000; Laine et al. 2002). In our study active and non-active galaxies were not matched in morphological type, but the fraction of Fourier bars appeared to be similar in early and late-type galaxies. Quite unexpectedly, all activity types

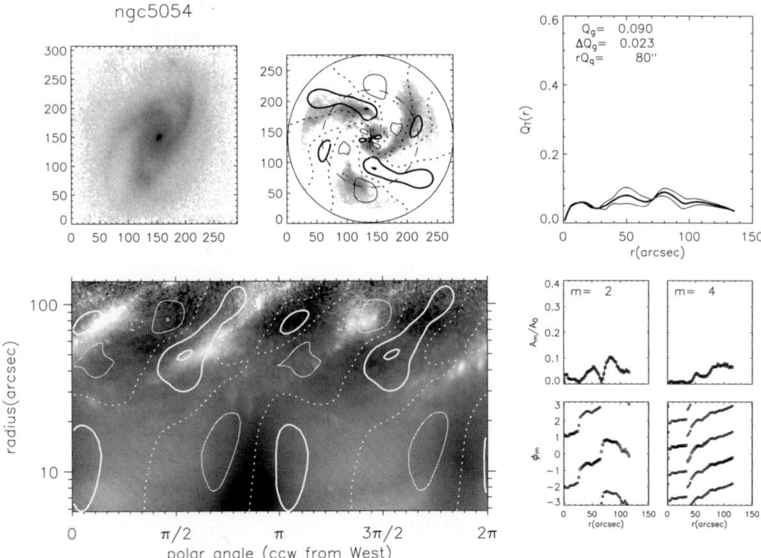

Figure 2. The *upper left corner* shows the original H-band image, and *the middle upper panel* the deprojected image, where the m=0 component is subtracted. Overlayed on the image is the "butterfly pattern", which shows the regions of the maximum tangential forces in contours divided to intervals of 0.1 bar strength units. The dotted lines indicate the regions where the tangential forces change the sign. The *upper right corner* shows the radial Q_T-profile. The *lower left panel* shows the butterfly plot in log-polar coordinates. The contours and dotted lines are the same as in the upper middle panel. The right lower corner shows the m=2 and m=4 Fourier density amplitudes and phases.

were found to have similar bar fractions. AGNs also have more outer rings than HII/starburst galaxies or the non-active systems, confirming the earlier result by Hunt and Malkan (1999).

The large fraction of bars and rings in active galaxies thus seems to indicate that bars might play an important role for feeding the nuclei. However, we also found that bars in Seyferts and LINERs have *weaker* perturbation strengths than bars in HII/starburst or in the non-active systems, confirming the earlier result by Laurikainen et al. (2002). This is clearly related to the Hubble type: it is well known that Seyferts and LINERs appear mainly in early Hubble types, whereas HII/starburst galaxies and the non-active systems are more concentrated to later types. We found that Q_g *increases* towards later type galaxies (Buta et al. 2004), but also that the A_2 Fourier amplitude *decreases* towards later type galaxies (see Figs. 3 and 4). The explanation for this kind of behavior is the mass of the bulge, which dilutes the non-axisymmetric forces corresponding even 5 bar torque classes, as defined by Buta and Block (2001). The number of outer rings is also known to be correlated with the Hubble type, the rings appearing mainly in early-type galaxies. Quite interestingly, we found that the number of outer rings increases almost linearly with the B/D-ratio. A natural explanation for this correlation would be that due to the small bulges in late-type galaxies, the bar pattern speeds are sufficiently small to place the OLR outside the galaxy disk.

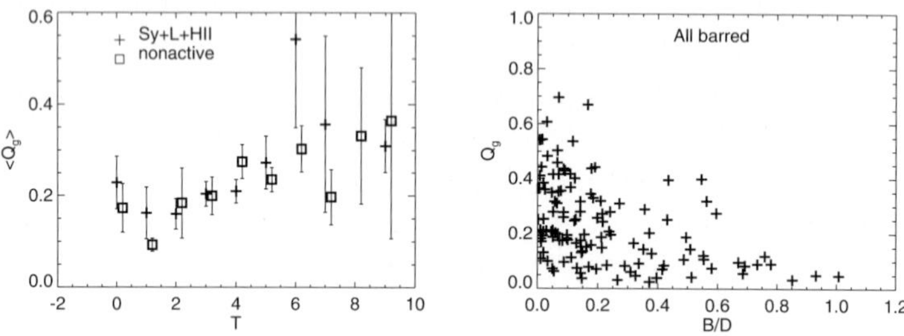

Figure 3. *Left panel*: the mean maximum relative perturbation strength, Q_g, versus RC3 type index. The vertical bars are the mean errors. *Right panel*: Q_g versus B/D-ratio for all barred galaxies in the sample identified by the Fourier method.

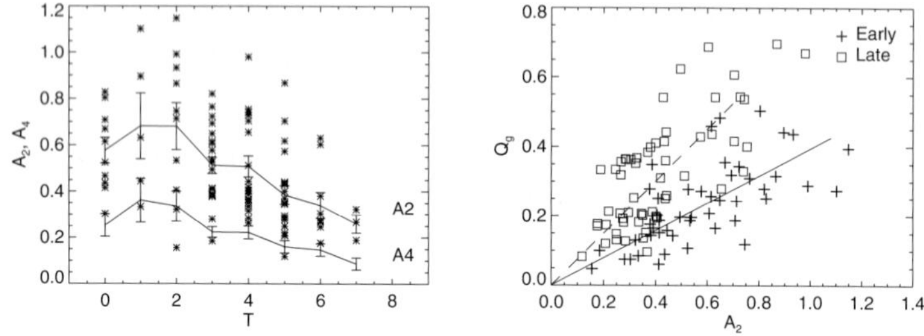

Figure 4. *Left panel*: the m=2 and m=4 amplitude of density (A_2 and A_4 versus RC3 type index. The lines indicate the mean values. *Right panel*: Q_g versus A_2: also shown are linear fits for early (T=0-3) and late-type (T=4-9) galaxies, as defined in RC3.

4. Are Bars in active and non-active Galaxies similar?

We can at least tell in which kind of bars *AGNs do not reside*: they avoid bars with the strongest perturbation strengths or bars in which the perturbation strength is strong near to the nucleus (moderate or large Q_g, small r_{Qg}/h_r). Bar induced perturbation strengths in these galaxies are strong partly due to the small bulges, but the bars are also massive in terms of having strong A_2 amplitudes. Based on the analysis in our study alone it is hard to say whether these galaxies have no AGN because they lack massive bulges, or because the perturbation strengths are strong near to the nucleus. In any case it is surprising that these galaxies largely lack also nuclear/circumnuclear star formation, although strong bars are expected to be efficient triggers of gas inflow.

There are also some characteristics of bars that seem to *increase the probability of having an AGN*: we found Seyfert or LINER type nuclear activity in 9 out of 10 galaxies having bars with thick inner sectionswith thin outer ends. Characteristic for these bars is that they are extremely massive compared with bars in other galaxies of similar morphological types. These bars reside mainly in early Hubble types having strengths that are somewhat stronger than in early-type galaxies in general. There also seems to be a general trend showing that bars in the *active early-type galaxies* are less centrally concentrated than bars in the *non-active early-type galaxies*.

5. Summary

We found that dilution by massive bulges in early-type galaxies largely explains the *weak perturbation strengths* in Seyferts and LINERs, in comparison to those found in HII/starburst galaxies or in the non-active systems. Massive bulges are probably also an explanation to the large number of outer rings in AGNs. Bars in early-type galaxies (and in AGNs) are also found to be at the same time long and massive (large A_2) and have weak perturbation strengths (small Q_g). Although the Hubble type seems to explain most of the characteristics of bars in active galaxies, we also found that AGNs appear preferably in galaxies having thick bars with thinner outer ends . There is also a weak evidence showing that bars in AGNs are flatter than bars in their non-active counterparts. Also, all active galaxies avoid massive bars in late-type galaxies having strong tangential forces near to the nucleus.

References

Buta, R., Block, D. (2001). ApJ, 550, 243
Buta, R., Laurikainen, E., Salo, H. (2004). AJ, 127, 279
Eskridge, P. et al. (2002). ApJS, 143, 73
Hunt, L. K., Malkan, M. A. (1999). ApJ, 516, 660
Knapen, J. (2004). A &A, in press
Knapen, J., Shlosman, I., Peletier, R. F. (2002). ApJ, 529, 93
Laine, S. et al. (2002). ApJ, 567, 97
Laurikainen, E., Salo, H., Rautiainen, P. (2002). MNRAS, 337, 1118
Laurikainen, E., Salo, H., Buta, R. (2004). ApJ, 607, 103
Laurikainen, E., Salo, H., Buta, R., Vasilyev, S. (2004). submitted to MNRAS
Peng, C. Y., Ho, L., Impey, C. D., Rix, H.. (2002). AJ, 124, 266
Simkin, S. M., Su, H. J., Schwartz, M. P. (1980). ApJ, 237, 404

GLOBULAR CLUSTERS: GALACTIC AND INTERNAL MOTIONS*

Ivan R. King
Astronomy Department, University of Washington
Box 351580, Seattle, WA 98195-1580
king@astro.washington.edu

Abstract Motions of, and in, globular clusters can be observed from the ground and with the Hubble Space Telescope. This review covers types of observations and some of the results that they produce. Interactions of the Galactic bar with cluster orbits can in principle produce useful information, but the practical outlook is pessimistic.

Keywords: Globular clusters: general, astrometry, Galaxy: kinematics and dynamics

1. Introduction

Ever since the days of Shapley and of Baade, globular clusters have been a touchstone for the understanding of the structure of our Galaxy, as well as for the study of stellar populations and evolution. In our present discussions of bars in galaxies, the globulars can, through their orbital motions, serve as a kinematical probe of the gravitational field of the bar that crosses the center of our own Milky Way. Before treating the motions, however, I will describe the ways in which they are measured. This will naturally introduce some discussion of astrometry with the Hubble Space Telescope, and I will devote the remaining parts of my paper to describing some of the work that my colleagues and I have been doing on globular clusters with HST.

I will begin with ground-based observations of globular-cluster motions. After a description of the techniques that make that work possible, I will move to absolute proper motions of clusters as measured from HST images. Then will come a discussion of globular-cluster orbits, first in general and then in relation particularly to the Galactic bar. Next I will move to internal motions in, and dynamics of, globular clusters, along with their relationship to the Galactic

*Based on observations with the NASA/ESA Hubble Space Telescope, obtained at the Space Telescope Science Institute, which is operated by AURA, Inc., under NASA contract NAS 5-26555.

D. Block et al. (eds.), Penetrating Bars through Masks of Cosmic Dust, 685–701.

distance scale. Finally I will discuss the HR diagram, with particular emphasis on observational study of the hydrogen-burning limit at the bottom of the main sequence.

I should emphasize that this is collaborative work with a group of colleagues. The one who deserves the most credit is Jay Anderson, of Rice University. He has developed most of the methods on which our results depend. Other important contributors are Giampaolo Piotto and Luigi (Rolly) Bedin of the University of Padua, and Adrienne Cool of San Francisco State University.

2. Ground-Based Studies of Motions

Radial velocities

Radial velocities of globular clusters have been measured with slit spectrographs, with cross-correlators, and with Fabry-Perot imaging. Early measurements were of the integrated cluster light, but most modern measures are made on individual stars; the average of a large number of stars can give an accuracy considerably better than 1 km s^{-1}.

The February 2003 version of the Harris (1996) on-line catalog of globular-cluster data gives radial velocities of 139 of the 150 clusters that it contains, with stated errors that range from 0.1 to 57 km s^{-1}; most of them are better than 2 or 3 km s^{-1}.

One caution about the use of radial velocities to study internal motions in globular clusters: to get a sample of substantial size one has to go to faint stars, whose radial velocities have one serious drawback. The image of each star is overlaid also by diffuse cluster light, which has the mean velocity of the cluster. The fainter the star, the more this cluster light tends to pull the star's velocity toward the mean velocity of the cluster. Corrections for this effect are problematical.

Proper motions

Bulk motions. An increasing number of absolute proper motions of clusters have been measured, the increase being largely due to the efforts of Dinescu et al. (1997, 1999a, 1999b, 2003), who are using the Yale proper-motion survey of the southern sky to measure motions with respect to galaxies, or, in heavily obscured fields, to Hipparcos-based reference stars. The last of their four papers gives a compilation of proper motions for 38 clusters, including their own measurements and others taken from the literature.

Internal motions. There have been only a few ground-based studies of internal proper motions in globular clusters. The most notable is by van Leeuwen et al. (2000), who measured about 50 plate pairs of ω Centauri, taken by the Yale–Columbia southern refractor with a time separation of about 50 years.

They were able to measure good proper motions in the outer parts of the cluster, but our more accurate and reliable HST measurements show that the van Leeuwen et al. proper motions near the cluster center have much too high a velocity dispersion.

Similar measurements were made on several northern clusters, using plate pairs taken by the Yerkes refractor with a time interval of nearly a century (Cudworth & Rees 1990, Rees 1992, Rees 1993, Peterson, Rees, & Cudworth 1995). The measurements sufficed to distinguish cluster stars from field stars, and yielded absolute proper motions of three clusters (which have since been superseded by better measures). For M4 and M92 internal dispersions of proper motion were compared with the dispersions of radial velocities to yield distance estimates, but these were not accurate enough to have heavy weight.

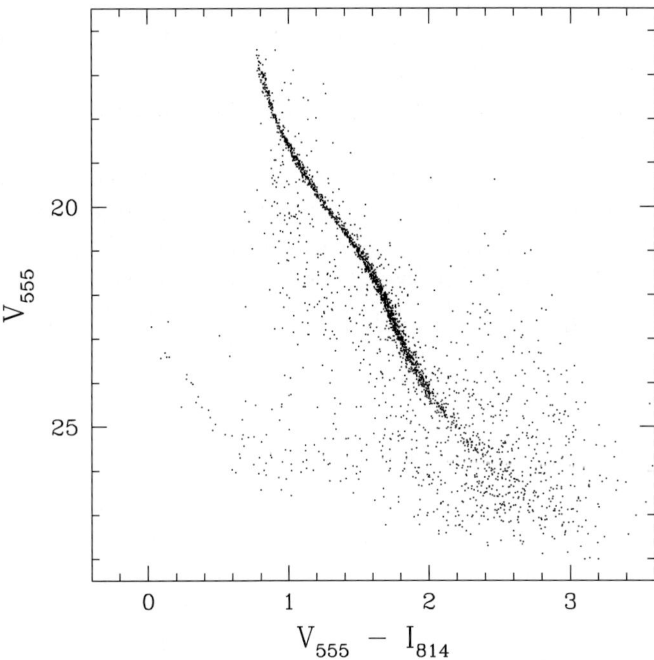

Figure 1. HST color–magnitude diagram of NGC 6397. Note how the bottom end of the main sequence is overwhelmed by field stars.

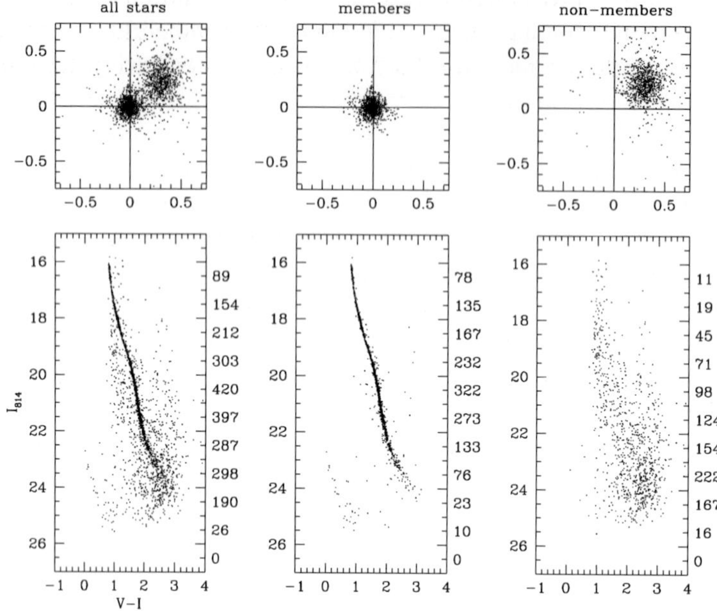

Figure 2. Proper motions (above) and color–magnitude diagrams (below) in NGC 6397. Left, all stars; center, cluster stars; right, field stars.

3. The Role of the Hubble Space Telescope

HST and its capabilities

Although the Hubble Space Telescope was launched in 1990, it began functioning at its full capability only in 1994, after the first repair mission had corrected the spherical aberration of the primary mirror. At that time its workhorse camera was Wide Field and Planetary Camera 2, until a later repair mission in 2002 installed, additionally, the Advanced Camera for Surveys, which has a more favorable throughput that takes it more than a magnitude fainter than WFPC2. I will be discussing results here from both cameras.

As for motions, HST has done rather few radial velocities of stars; except for cases where stars have to be separated from near neighbors, radial velocities are better done from the ground. So where HST is concerned I will be talking about proper motions and what can be done with them.

Our astrometric techniques

We got into HST astrometry as a sort of accident. Our real aim was to study faint color–magnitude diagrams of globular clusters. But the first good one that we got (Fig. 1) was utterly frustrating. We had the faint stars, but for the last 1.5–2 magnitudes we couldn't tell which stars were cluster members and which belonged to the field. But I did a little arithmetic and realized that if we re-imaged the cluster after two or three years, we would be able to use proper motions to do the separation.

The result was even better than we had hoped (Fig. 2). We got a clear separation, as a result of which we could see that there were plenty of field stars at a magnitude where the cluster seemed to be petering out. This is how we got into the hydrogen-burning-limit business that I will be discussing later.

The next step on our road to astrometry was even more accidental. My old friend Georges Meylan had a project to find the fastest-moving stars in 47 Tucanae by taking images 2 years apart, but he had lost the members of his team who were going to make the measurements. So he asked if we could do it. When Jay Anderson began measuring the images, he found that he could not only detect the fast-moving stars; he could measure proper motions for *all* of the stars.

This put a premium on accuracy, and we set out to improve ours. Or rather, Jay set out to improve it, with occasional suggestions from me. To jump to the bottom line, Figure 3 shows how well we are able to do.

How were we able to reach this level of accuracy? There were a number of steps. First was to learn how to measure star positions reliably. In doing this the first major problem is that the position that you measure for a star can very easily have errors that depend on where the center of the star falls with respect to pixel boundaries. The designers of cameras, out of the desire to cover a large field, often undersample the point spread function (PSF) of the star images, and this exacerbates the problem, which we call *pixel-phase error*. Figure 4 illustrates the errors that we had at the beginning, when using position-measuring methods that we had thought were sound.

We have found that the key to avoiding pixel-phase error is to use an extremely accurate PSF. To do this, we had to re-examine what a PSF really is. The general view is that the telescope produces an intrinsic PSF and that the instrument then convolves its own signature with that, to produce the observed PSF. This is in fact true, but it is quite misleading. The only thing that really matters is the set of PSFs that we see at all possible pixel phases. The concept that we introduced (Anderson & King 2000) is the *effective PSF*, which is a continuous function of (x, y), the position of the pixel center relative to the center of the star. Instead of a PSF made up of pixels, we work with a continuous ePSF, whose values express the fraction of the light of a normalized PSF that will fall in a pixel centered at (x, y).

The process of deriving an accurate ePSF is far from trivial. First we need a set of well-dithered images—that is, images displaced by fractions of a pixel in each coordinate. From these, we derive a mean position for each star, so that

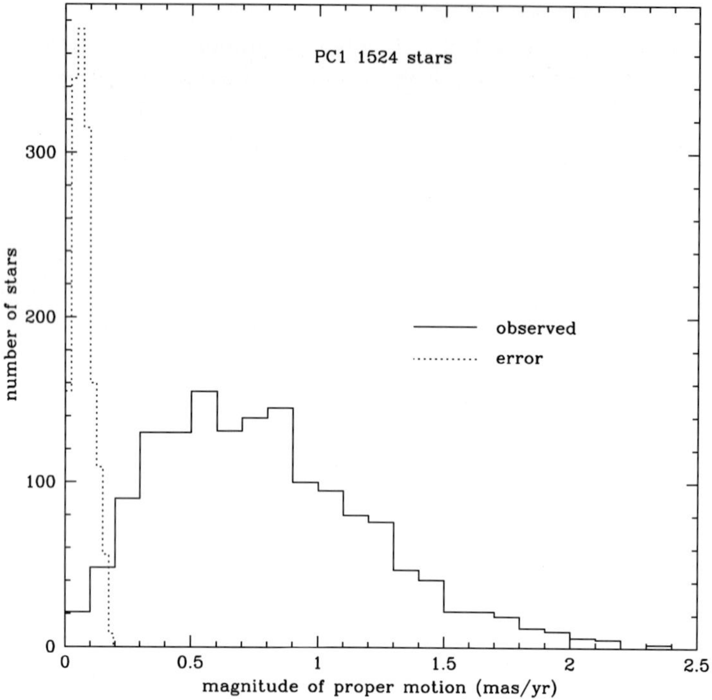

Figure 3. Magnitudes of our measured proper motions in 47 Tuc, and of their errors.

each individual measurement leads to a residual. We then test for pixel-phase error by plotting the residuals against the pixel phase of the star image from which it came.

As for derivation of the ePSF, it is an iterative process in which we alternately combine all well-exposed star images to improve the PSF, and then use the new PSF to improve the positions, so that in the next round of combining images the individual image shifts will be more accurate. The process converges in a dozen iterations or fewer.

The other major problem of precise astrometry is correcting for geometrical distortions. After allowing for a "stitching error," in which the details of the manufacturing process caused every 34th row of each of the WFPC2 chips to be 3% narrower than the others (Anderson & King 1999), we faced the main problem of distortion, which is purely geometrical. Designers of optical instruments normally devote all their effort to creating sharp, symmetrical images—without caring much where they land; as a result, distortions tend to be quite large. At the corners of a WFPC2 chip the distortion amounts to as

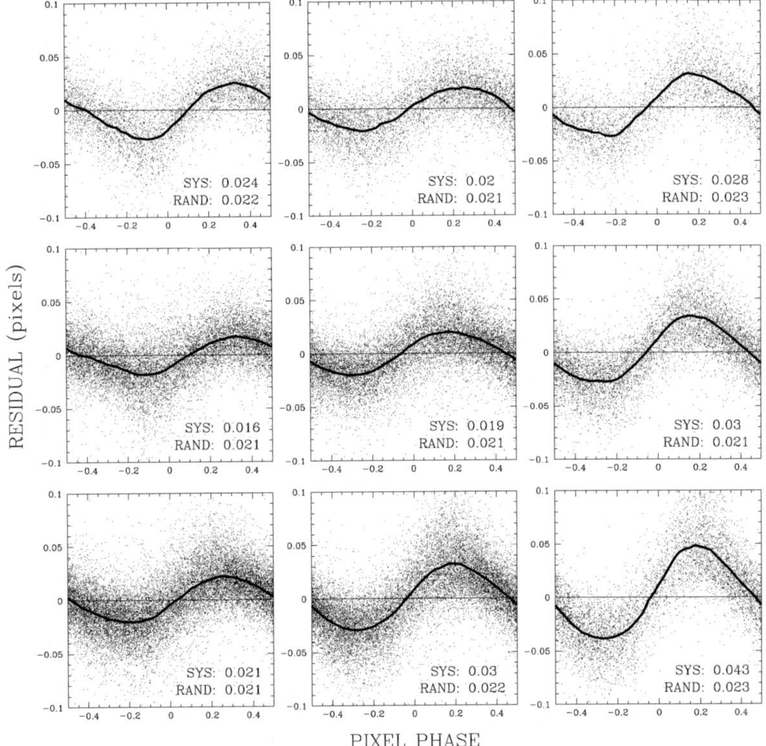

Figure 4. Residuals in x plotted against pixel phase, for 9 different regions of the WF2 chip. Our later methods flatten out these curves completely.

much as 5 whole pixels; for ACS the corresponding number is 50. The distortion in ACS also skews the field; the rectangular chips project on the sky into rhomboids with corner angles of 82° and 98°.

Deriving the corrections for geometric distortion is laborious; it requires a set of images of the same field taken at positional offsets and with rotations. (Without rotated images, skewing cannot be measured.) Again, we derive a mean position for each star, and examine all the residuals as a function of position on the chip. From such data sets we have derived distortion corrections for WFPC2 with a precision of 0.02 pixel everywhere (Anderson & King 2003a), and for ACS within 0.01 pixel everywhere (Anderson & King, STScI Instrument Science Reports, in preparation). Unfortunately, however, the presence of refractive elements makes the corrections wavelength-dependent, and they have not yet been derived with this accuracy for all filters.

Absolute proper motions of clusters measured with HST

Armed with techniques of accurate measurement and precise reduction, we set out to measure proper motions of globular-cluster stars. HST is able to do this more accurately than can ground-based observations, because of its superior resolving power and scale.

A first problem to consider is the absolute proper motion of a cluster as a whole. Here we have the advantage of being able to use the average motion of all the stars that we measure in a cluster; the problem, however, is in having an absolute reference standard.

In two cases we have been able to find special circumstances that provide an extragalactic reference standard. In one of the fields observed in M4 we used multicolor photometry to identify a quasi-stellar object, and using this single QSO as a reference object we were able to measure the absolute proper motion of M4 (Bedin et al. 2003).

The second special case is 47 Tucanae, which lies in front of the outer reaches of the Small Magellanic Cloud. Measurement of the average proper motion of 47 Tuc stars with respect to several hundred background stars of the SMC has yielded a relative proper motion of high precision (Anderson & King 2003b). Unfortunately, however, the absolute proper motion of the SMC is not well enough known, so that that of 47 Tuc suffers from the same uncertainty of 0.2 mas yr^{-1}. Interestingly, NGC 362 is capable, in principle, of a similar measurement, but adequate HST material does not exist.

The obvious reference standard for HST is, as on the ground, small images of galaxies. Our own research group has not yet attempted to measure galaxies. Kalirai et al. (2004), however, have done this in M4. Interestingly, the errors quoted for the absolute proper motion of M4 by Kalirai et al. from HST galaxy measurements, by us from a single QSO, and by Dinescu et al. (1999a) from ground-based galaxy measurements, are all comparable. But it is my strong belief that galaxy motions can be measured with HST with considerably more accuracy than they have been so far, so that HST will yield the best absolute proper motions—where the necessary images exist.

4. The Orbits of Globular Clusters in the Galaxy

Given the present-day position and velocity of a globular cluster, and the potential field of the Milky Way, one can calculate the orbit of the cluster, in past and future time. One purpose of such orbits is to see where the cluster goes, in order to estimate the effect of tidal shocks on the depletion of its population. Another purpose is to use the orbits as possible tracers of the Galactic potential.

(Note that I will deal here only with real orbits. A number of studies have been based on radial velocities alone, allowing in some statistical way for the fact that two of the three components of the velocity are being ignored. With a single exception, I will not discuss results that come from such a partially blind approach.)

The existing orbits, however, do not seem reliable enough for intensive study. There are difficulties both in the cluster data and in the Galactic model. First, there is considerable uncertainty in the present-day position of any cluster with respect to the Galactic center. Even though the Sun–center distance is settling down nicely to a value just under 8 kpc (Eisenhauer et al. 2003), the distances of individual clusters from the Sun are uncertain at at least a 10% level. Second, we observe from a moving platform, whose velocity around the Galactic center is still somewhat uncertain—although improved measurements of the Oort-constant difference $A - B$ suggest an optimism similar to that about the Sun–center distance. Third, the Galactic potential is probably the worst bugaboo. Different orbit calculations have been made with different Galactic models, and are thus difficult to compare. Much worse is the fact that nearly all calculations have been done in a smooth, axisymmetric Galactic potential—hardly a reasonable assumption to make in a meeting that is studying bars!

Orbits in an axisymmetric Galaxy

For completeness I will mention the published orbits that exist. A set of 25 orbits is given by Dauphole et al. (1996), who show meridional sections that are quite interesting (subject to the reservations that I have expressed above). Odenkirchen et al. (1997) recalculated 15 of these orbits after reducing the proper motions to the Hipparcos system; they use a different potential, however. In that second paper they do not plot the orbits.

In all cases it is the proper motions that limit the size of the sample. Dinescu et al. (1999b) list proper motions for 38 clusters, so that more orbits could be calculated if it were desired.

Matching orbits to the spatial distribution

An insightful new approach to globular-cluster orbits was introduced by Dauphole & Colin (1995). Noting that a cluster spends most of its orbital time near its apocentric distance, they argued that the striking lack of globular clusters beyond 40 kpc from the Galactic center implies that few or no cluster orbits should have apocenters in that region. They used this condition as a constraint on the Galactic potential.

Their idea can be generalized, by calculating orbits for as many clusters as possible, and deriving a density distribution from the entirety of each orbit, with each point weighted by the time that the cluster spends there. The sum of these should resemble the observed Galactic density distribution of those clusters. A disagreement would throw into question the potential in which the orbits had been calculated, and the need for agreement could then constrain the Galactic potential.

The effect of the bar on globular clusters

The dynamics of bars are discussed elsewhere in this volume, far more competently than I could do it. In any case, what matters here is the potential, not the dynamics.

In a paper that is still in press at this writing, Pichardo, Martos, & Moreno (2004) discuss the effect of a Galactic bar on the orbits of globular clusters that happen to venture into the region of the bar. Since the exact form of the bar across the center of our Milky Way is uncertain, they use three different versions of the Galactic potential, each including a bar. The results are most interesting. With the kind permission of the authors, Figure 5 reproduces one of their results. Successive rows are meridional sections of orbits calculated in an axisymmetric Galactic potential and in their three different versions of the potential of a barred Milky Way.

It is clear that the presence of a bar greatly increases the tendency of the orbit of a cluster to go chaotic. Moreover, the two lowest orbits in the middle column show that we will not know the behavior of orbits in the vicinity of a bar until we know the form of the bar much better than we do now.

In an earlier paper (Pichardo et al. 2003) the same research group showed that even the gravitational influence of spiral arms can seriously affect the nature of orbits in the Galaxy. These two results cast severe doubt on the likelihood that the motions of globular clusters are going to shed a great deal of light on the potential of the Milky Way.

A related question is, what does the bar do to the clusters themselves? In a paper that is perhaps an exception to the rule against working from radial velocities alone, Long, Ostriker, & Aguilar (1992) conclude that the addition of a bar is not likely to have a serious effect on the rate of destruction of globular clusters by Galactic forces.

Possible relationship of clusters to the bar

A number of authors have speculated about whether some globular clusters could actually belong to the bar structure. (Again I will leave out papers that ignore the proper motions and rely on radial velocities alone.) The one paper that is outstanding in this respect is by Dinescu et al. (2003), who examine the positions and motions of six clusters, all of which appear to be within 3 kpc of the Galactic center. These clusters are NGC 6266, 6304, 6316, 6723, 6528, and 6553, and they all have small space velocities that are unlikely ever to carry them much farther from the center. The authors do not actually compute

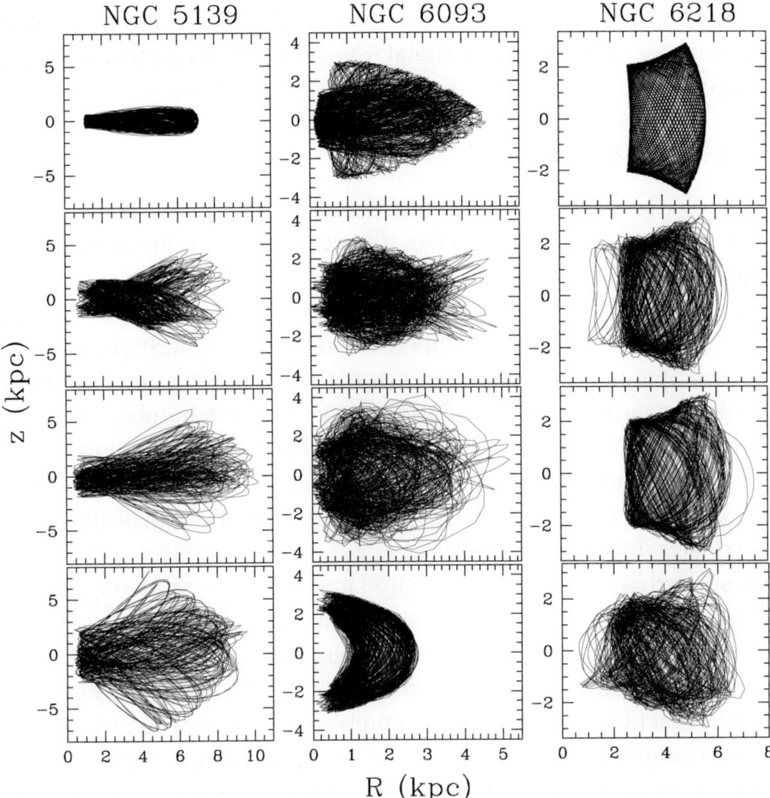

Figure 5. Meridional sections of orbits of three globular clusters in four different Galactic potentials. The top row uses an axisymmetric potential; the others use three different models that contain a bar. (Courtesy of Pichardo et al. [2004].)

orbits, however, promising them for a later paper instead. But they argue that NGC 6528, at least, is a member of the bar, and that several of the others could be so also.

The weakness in their study, however, is that not only are the velocities subject to the uncertainties of true velocity that were mentioned in an earlier subsection; worse, the uncertainties of cluster distances—exacerbated here by large and uncertain values of interstellar absorption—make their true positions relative to the Galactic center even more uncertain. Whether we shall ever know if any globular clusters are really a part of the bar seems a dubious prospect to me.

5. Measuring Internal Motions in Globular Clusters

Given our ability to measure proper motions of individual stars in globular clusters, and thus the internal dispersion of proper motions, two applications immediately suggest themselves. One is the measurement of cluster distances by comparing the dispersions of radial velocities and of proper motions, and the other is study of the internal dynamics of the clusters.

Application to the distance scale

In principle the measurement of the distance of a globular cluster by comparing the dispersions of radial velocities and proper motions is simple and straightforward. The two dispersions are measures of the same quantity in linear and angular units, respectively, and comparison of the two immediately gives the distance. Furthermore, this method is capable of high accuracy, because in principle the only source of error is the sampling error of the dispersions, which in each case is $1/\sqrt{2N}$, where N is the number of stars measured. If we had 10,000 radial velocities and 10,000 proper motions (an ideal but by no means unrealistic example), the resulting distance could be accurate to 1%—in an astronomical community where the present uncertainty in globular-cluster distances is 10–15%.

As mentioned in Section 2, this method has already been attempted from the ground, but the proper-motion errors of measurement far outweigh the sampling error, and the results are not good enough to be significant.

Although calculating cluster distances in this way seems straightforward in principle, in practice there are serious complications, the worst of them caused by the rotations of globular clusters, which are typically large enough to perturb the velocity dispersions quite seriously. In 47 Tuc, for example—the cluster on which our group is working most intensively, the velocity dispersion is about 10 km s^{-1} in each coordinate, while the line-of-sight rotation is ± 5 km s^{-1} (Meylan & Mayor 1986) and the plane-of-the-sky rotation is comparable in size (Anderson & King 2003b).

The rotation affects the dispersion both of radial velocities and of proper motions. The effect on the radial velocities is obvious; for the proper motions the effect is more subtle. It is caused by a mean motion across the line of sight that is in opposite directions in front of and behind the cluster. In NGC 2808, for example, we find that this effect makes the equatorial dispersion of proper motions 15% larger than the polar one.

In addition, the velocity dispersion in a cluster is not radially constant. To a first approximation it is so indeed, but when carefully measured it shows variations that will vitiate our calculations if the radial velocities and proper motions are not measured in the same place. Moreover, anisotropy of velocity dispersions affects radial and transverse motions differently.

For all these reasons, the velocity dispersions cannot simply be compared; they must instead be fitted to a dynamical model of a cluster. In this case multi-mass King models will not suffice, as they did for Pryor et al. (1989, 1991) in using dispersions of radial velocities to determine masses of clusters. No King model has enough adjustable parameters to fit the velocity data that are now available for many globular clusters.

What is needed instead is Schwarzschild's orbit-library method. In this method the cluster is represented by choosing appropriate numbers of stars to populate each orbit of a library that covers the parameter space of possible orbits. The orbits are calculated in a potential chosen to represent that of the cluster. The numbers are adjusted until a satisfactory fit is achieved with the observed distributions of density and velocity. Our colleague Francesca De Angeli, at Padua, is at present developing such a model for 47 Tuc, to fit the extensive data that we have on that cluster. At the same time a group under Tim de Zeeuw is modeling similarly extensive data on ω Centauri.

Jay Anderson and I are following a different approach to a cluster distance; we have images from which we hope to measure the trigonometric parallax of 47 Tuc, with respect to its background of SMC stars. Even though we are unlikely to achieve better than 10% accuracy, the measurement of a parallax of 1/4 milliarcsec will be a new *tour de force* in astrometry. What has been delaying us is that since we got our WFPC2 material, the ACS group at STScI has chosen our field as the standard calibration field for ACS, so that new material of higher quality is accumulating, and we have to decide when we have enough to begin our measurements.

Anisotropy of motions in globular clusters

Still another application of proper motions is that they give us a chance to check anisotropy—the disparity between radial and tangential motions. In this respect proper motions are much better than radial velocities. With radial velocities what one has to do is to compare the radial drop-off of velocity dispersion with what various models say it ought to be, and choose the amount of anisotropy that fits best. Proper motions give it directly; one simply looks at off-center fields and compares the dispersions in the radial and tangential directions.

When we first did this, for an outer field in 47 Tuc, we got a surprise. In all dynamical models of clusters that have anisotropic motions, it is assumed that the velocity ellipsoid consists of similar surfaces. What we found is that anisotropy is stronger at higher than at lower velocities; thus the models are wrong and need to be redone.

Another surprise came in ω Centauri, which is so little relaxed that it should have retained its original anisotropy. Everyone "knows" that the formation of a globular cluster, whether by expansion or by collapse, should leave it with a strong anisotropy. But in ω Cen we found almost none. Again the conventional picture needs to be rethought.

Rotation of globular clusters

As previously mentioned, globular clusters rotate significantly. Of observed flattenings (White & Shawl 1987), typical values of the ellipticity ($\epsilon = [a - b]/a$, where a and b are the semi-major and semi-minor axes, respectively) are 0.1. Allowance for random orientations suggests that a typical value of the 3-D ellipticity of a globular cluster is 0.2. According to simple manipulation of the virial tensor (King 1961), the rotational kinetic energy of a typical globular cluster is thus nearly half of the kinetic energy of its random internal motions.

In two clusters the third component of rotation, in the plane of the sky, has been measured. For 47 Tuc Anderson & King (2003b) measured the mean motion of 47 Tuc stars with respect to Small Magellanic Cloud stars in the background, in two fields on opposite sides of the cluster. For only one other cluster has the plane-of-the-sky rotation been measured. The ground-based study of ω Cen described in Section 2 produced a curve of angular velocity as a function of radius, but with an unknown zero point, since the orientation of each plate was unknown. The authors set the zero point by requiring that the shape of the resulting rotation curve resemble as closely as possible that of the curve derived from radial velocities by Meylan & Mayor (1986). The resulting rotation vector suggests that the axis of ω Cen is inclined by about 45° and that the true ellipticity of the cluster is about 0.4.

It seems unlikely that the plane-of-the-sky rotation will be measured for any other clusters. No other cluster has a plate series like that of ω Cen, and the only other cluster with SMC stars in its background is NGC 362, which is so much farther than 47 Tuc that its internal motions are expected to be too small to be measured with sufficient accuracy.

6. Problems of Color–Magnitude Diagrams

Peculiarities in Omega Centauri

The color–magnitude diagram of ω Centauri is the most peculiar among all globular clusters. It had long been known that there is a spread of color on the red giant branch, but accurate modern photometry has resolved the RGB into three distinct components (Ferraro et al. 2004). More recently we have shown (Bedin et al. 2004) that the middle main sequence is split into two distinct parallel branches, and Norris (2004) has suggested that the entire set of sequences can be explained as a superposition of three populations with suitably chosen values of metallicity and helium content.

It has been suggested that the stars of the most metal-rich RGB have a bulk motion significantly different from that of the other stars (Pancino et al. 2003), but this result has been challenged by Platais et al. (2003). We now have much more accurate proper motions than those used by Ferraro et al., and we see no difference in motion at all.

Why should ω Cen be so different from other clusters? It has often been suggested that ω Cen is a merger product, or the remaining nucleus of a dwarf spheroidal captured and mostly destroyed by the Milky Way, or both (see various papers in van Leeuwen, Hughes, & Piotto 2002).

The hydrogen-burning limit

Kumar (1963) pointed out long ago that there is a lower limit to the mass that a star can have. This limit is set by the ability of the pre-stellar configuration to create a high enough central temperature to ignite the hydrogen-burning nuclear reactions that will make it a star, rather than a brown dwarf. (Kumar had used the term "black dwarf.") Even though Kumar used what would now be considered to be an unrealistic helium abundance, he predicted, as do more

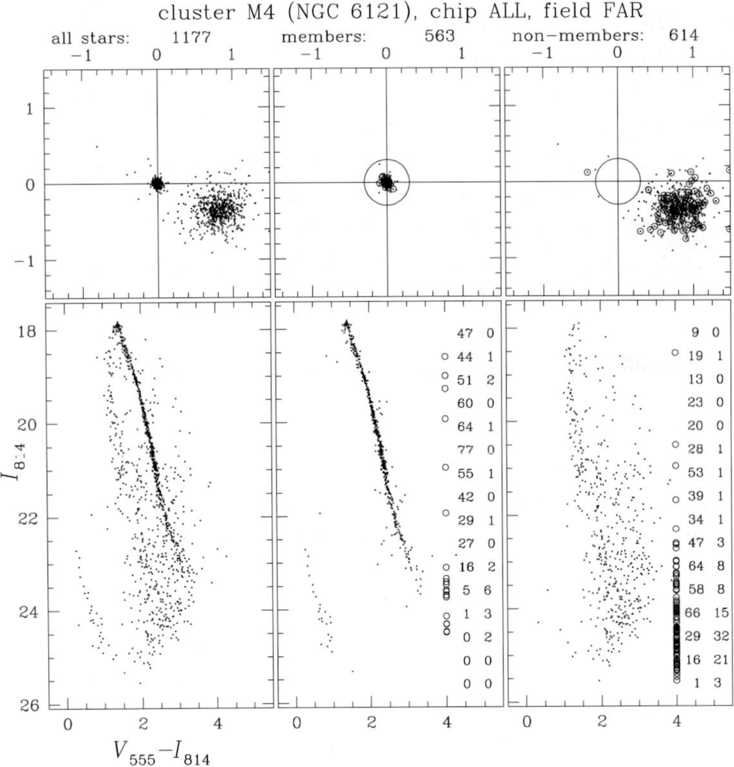

Figure 6. Separation diagram for M4, similar to Fig. 2 for NGC 6397. Open circles are stars detected in I but not in V. Those brighter than $I_{814} = 23$ are due to artifacts on diffraction spikes of bright stars.

modern theories, that the hydrogen-burning limit would lie near $0.07m_\odot$ for stars of solar metallicity and near $0.09m_\odot$ for metal-poor stars. These masses correspond to such faint absolute magnitudes that until recently observational study of the hydrogen-burning limit was beyond reach.

Observationally, the hydrogen-burning limit is not a sharp limit at all, since we do not observe mass but rather luminosity. A brown dwarf that is significantly below the mass limit cools rapidly, and after the $>$10 Gyr age of a globular cluster its luminosity is pitifully low. One that is close to the limit, however, cools much more slowly and will, even at the present time, have a luminosity not far below those of the faintest stars of the main sequence. The practical result is that a tiny range of mass maps into a large range of luminosity. Since the number of stars per unit interval of mass does not change rapidly, the number per unit interval of absolute magnitude becomes very small. In other words, in the observed quantity, magnitude, the signature of the hydrogen-burning limit is a strikingly steep drop in the luminosity function. The main sequence does not end; instead it peters out into near-nothingness.

Figure 2 showed this phenomenon in NGC 6397. For M4, whose distance modulus is half a magnitude larger, we extended our reach by including the stars that are detected in the I band but not in V (Fig. 6). Those of them whose proper motions label them as cluster members must be members of the cluster main sequence; if they were white dwarfs they would be seen in V. Again we see the signature of the hydrogen-burning limit.

With the deeper penetration of ACS, we can extend our work to more clusters. We have programs in progress for the enigmatic ω Centauri and for 47 Tucanae, which will extend our study into a range of higher metallicity ([Fe/H] = -0.7), and also for the rich open cluster NGC 6791, whose [Fe/H] is +0.4.

7. Summary and Conclusions

In this brief romp through motions of clusters, and motions in clusters, I hope I have shown how much our knowledge has progressed in recent years. But I conclude with regret that cluster motions are not likely to contribute to an understanding of the central bar of the Milky Way.

Acknowledgments

I thank Dr. Barbara Pichardo for directing me to the literature of the orbits of globular clusters in the vicinity of the Galactic bar, and for supplying me with Figure 5. My work is supported by grants from the Space Telescope Science Institute, particularly GO 9444 and GO 9815. I am especially grateful to the Anglo-American Chairman's Fund for support at the conference.

References

Anderson, J., & King, I. R. (1999) P.A.S.P., 111, 1095.

Anderson, J., & King, I. R. (2000) P.A.S.P., 112, 1360.

Anderson, J., & King, I. R. (2003a) P.A.S.P., 115, 113.

Anderson, J., & King, I. R. (2003b) A.J., 126, 772.

Bedin, L. R., Anderson, J., King, I. R., & Piotto, G. (2001) Ap.J., 560, L75.

Bedin, L. R., Piotto, G., King, I. R., & Anderson, J. (2003) A.J., 126, 247.

Bedin, L. R., Piotto, G., Anderson, J., Cassisi, S., King, I. R., Momany, Y., & Carraro, G. (2004) Ap.J., 605, L125.

Cudworth, K. M., & Rees, R. (1990) A.J., 99, 1491.

Dauphole, B., & Colin, J. (1995) A&A, 300, 117.

Dauphole, B., Geffert, M., Colin, J., Ducourant, C., Odenkirchen, M., & Tucholke, H.-J. (1996) A&A, 313, 119.

Dinescu, D. I., Girard, T. M., van Altena, W. F., Mendez, R. A., & Lopez, C. E. (1997) A.J., 114, 1014.

Dinescu, D. I., Girard, T. M., van Altena, W. F., & Lopez, C. (1999a) A.J., 117, 277.

Dinescu, D. I., Girard, T. M., & van Altena, W. F. (1999b) A.J., 117, 1792.

Dinescu, D. I., Girard, T. M., van Altena, W. F., & Lopez, C. E. (2003) A.J., 125, 1373.

Eisenhauer, F., Schodel, R., Genzel, R., Ott, T., Tecza, M., Abuter, R., Eckart, A., & Alexander, T. (2003) Ap.J. 597, L121.

Ferraro, F. R., Sollima, A., Pancino, E., Bellazzini, M., Straniero, O., Origlia, L., & Cool, A. M. (2004) Ap.J., 603, L81.

Harris, W. E. (1996) A.J., 112, 1487.

Kalirai, J. S., et al. (2004) Ap.J., 601, 277.

King, I. (1961) A.J., 66, 68.

King, I. R., Anderson, J., Cool, A. M., & Piotto, G. (1998) Ap.J., 492, L37.

Long, K., Ostriker, J. P., & Aguilar, L. (1992) Ap.J., 388, 362.

Meylan, G., & Mayor, M. (1986) A&A, 166, 122.

Norris, J. E. (2004) Ap.J. Letters, submitted

Odenkirchen, M., Brosche, P., Geffert, M., & Tucholke, H.-J. (1997) New Astr., 2, 477O.

Pancino, E., Seleznev, A., Ferraro, F. R., Bellazzini, M., & Piotto, G. (2003) M.N.R.A.S., 345, 683.

Peterson, R. C., Rees, R. F., & Cudworth, K. M. (1995) Ap.J., 443, 124.

Pichardo, B., Martos, M., Moreno, E., & Espresate, J. (2003) Ap. J., 582, 230.

Pichardo, B., Martos, M., & Moreno, E. (2004) Ap. J., 609, July 1 issue (in press); astro-ph/0402340.

Platais, I., Wyse, R. F. G., Hebb, L., Lee, Y.-W., & Rey, S.-C. (2003) Ap.J., 591, L127.

Pryor, C., McClure, R. D., Fletcher, J. M., & Hesser, J. E. (1989) A.J., 98, 596.

Pryor, C., McClure, R. D., Fletcher, J. M., & Hesser, J. E. (1991) A.J., 102, 1026.

Rees, R. F. (1992) A.J., 103, 1573.

Rees, R. F. (1993) A.J., 106, 1524.

van Leeuwen, F., Le Poole, R. S., Reijns, R. A., Freeman, K. C., & de Zeeuw, P. T. (2000) A&A, 360, 472.

van Leeuwen, F., Hughes, J. D., & Piotto, G. (eds.) (2002) *ω Centauri a Unique Window into Astrophysics* (ASPCS 265). San Francisco: Astron. Soc. Pacific.

White, R. E., & Shawl, S. J. (1987) Ap.J., 317, 246.

EVOLUTION OF SELF-GRAVITATING GAS DISKS DRIVEN BY A ROTATING BAR POTENTIAL

Chi Yuan[1,2] and David C.C. Yen[1,3]
[1]*Institute of Astronomy & Astrophysics, Academia Sinica, Taipei, Taiwan,*[2]*Physics Department, National Taiwan University, Taipei, Taiwan,* [3]*Mathematics Department, Fu Jen Catholic University, Hsinchung, Taipei, Taiwan*

Abstract It is well known that a rotating bar potential can transport angular momentum to the disk and hence cause the gas disk to re-distribute its mass. The process leads to disk evolution. And, in particular, it results in fuelling AGNs and starburst ring activities. In this paper, we will present the numerical simulations to show how this mechanism works. The problem, however, is quite complicated. We classify our simulations according to the type of Lindbald resonances and try to single out their individual roles in affecting the dynamics and evolution of the disk. We will show several movies to demonstrate various features of the calculations. Among many interesting results, we identify origin of the starburst rings and the dense circumnuclear molecular disks in the galactic center to the instability of the disk. Unlike most of the other simulations, the self-gravitation of the disk is emphasized in this study.

Keywords: Bar-driven density waves, resonance excitation, disk instability, numerical simulations

1. Introduction

It has been almost 30 years since the first work on numerical simulation of bar-driven spirals in a galactic disk was published (Sanders and Huntley 1976, Huntley, Roberts and Sanders 1978). It looks strange after all these years that we are still working on the same problem. I think there are at least four reasons to justify our persistence of researching on the subject.

The first one has to do with the nature of the problem. The problem is exceedingly complicated. It depends on the galactic rotation, the bar force, the rotation speed of the bar, the initial gas distribution, the $p-\rho$ (pressure-density) relation, the sound speed and the viscosity of the disk. Furthermore, if we want to do it right, we must add the z-dimension. To cover the entire parameter space and bring out the essential physics is a formidable job. It certainly has not yet been satisfactorily accomplished.

D. Block et al. (eds.), Penetrating Bars through Masks of Cosmic Dust, 703–712.

The second is the numerical methods, which have undergone revolutionary changes during the past 30 years. People have repeatedly put new techniques and novel ideas into the code development and discovered new results and hence faced new challeges. None of the popular codes available today in the astrophysics community, as far as we know, has proven to be problem free. The fact that there is still room for improvement has kept people trying it again with new codes.

The third is the new observations. The discovery of AGNs and starburst rings, and the possible identification of supermassive blackholes in the galactic nuclei, and the revelation the fine details of the galactic central regions by high-resolution optical and radio observing facilities have stimulated new interests in this old field.

The fourth is the theory of resonance excitation is much better understood today. The pioneer work by Goldreich and Tremaine (1979) has laid the solid foundation for this field. Non-linear asymptotic theory for galactic problems has developed by Yuan and Cheng (1991), and Yuan Kuo (1997). Some of the results can be verified by the numerical simulations. This parallel approach in theoretical study is a healthy development and encourages further research efforts on both fronts.

In this paper, we take up the numerical simulation work again with our new codes, the Antares codes, which, we believe, is capable of handling this problem more accurate and more reliable than any of the codes used for this problem. To simplify the physics, we choose to treat only the gas-dust disks in the central part of the galaxies. It is a much clear-cut problem, since the gas disk there is almost entirely decoupled from the stellar component. This can be understood from the dispersion velocity of stars and gas. The effective sound speed of the gas, due to its dissipative nature, is about 10 km/s, while the random dispersion speed of the stars is an order of magnitude higher in the central regions. Therefore, the stellar population has little disk component and is mainly in the bulge. So, unlike the outer parts of the galactic disk, we are in a rare situation to deal a pure gasdynamic problem alone. The central gas-dust disks are not only ideal for the theory, but also the unique test ground, to test the theory with recent high-resolution observations of galactic centers.

In what follows, we first set the stage for the calculations, by specifying the bar potential, the rotation curves, and initial distribution of the gas density (section 2). The numerical methods, the Antares code, are described in section 3. Then, we present results according to the Lindblad resonances (sections 4). In section 5, we discuss three issues related to the calculations: the Toomre instability of the disk, the starburst rings, and the dense circumnuclear molecular disks. Other remarks on the calculations are also to be found there.

2. The Basic Galactic Model and the Bar Potential

In the subsequent discussions, we adopt two model rotation curves. For the first one, we choose the one used by Elmegreen & Elmegreen (1990), which can be expressed as

$$V = V_0 \frac{r}{r^B + r^{1-A}}, \tag{1}$$

where r is a dimensionless radius, normalized by 1 kpc, $V_0 = 300\ km/s$, and $B = 0$ and $A = -0.1$. This rotation curve rises slowly from the center and reaches a nearly constant value about 220 km/s outside. Since $V \sim r$ and the epicyclic frequency $\kappa \simeq 2\Omega$ near the galactic center, the $\Omega - \kappa/2$ curve drops to zero at the center and has a maximum in the central region. When the pattern speed of the bar, Ω_p, is lower than the maximum, there will be three Lindblad resonance. Moving out from the center, there are inner inner, outer inner and outer Lindblad resonance (IILR, OILR and OLR). This is shown in the right panel of Figure 1 (OLR out of the range, not shown here). If Ω_p is greater than the maximum of $\Omega - \kappa/2$, it only intersects with the $\Omega + \kappa/2$ curve, hence only one resonance, the OLR, also shown in the right panel of Figure 1. The rotation curve is depicted in the left panel of Figure 1.

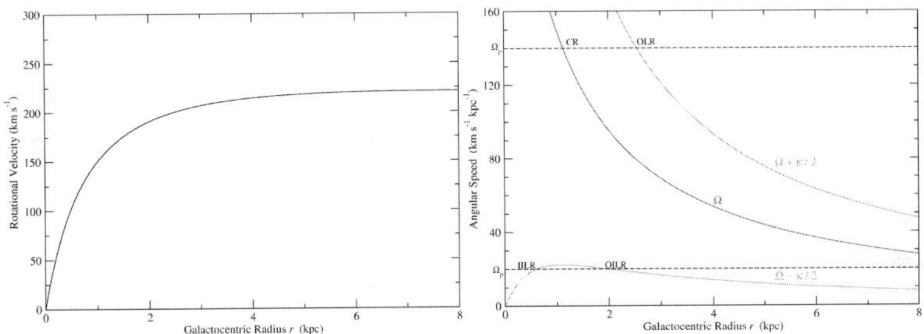

Figure 1. Elmegreen rotation curve (slowly rising). The left panel is the rotation speed in km/s vs. radius in kpc. In the right panel, the middle curve is angular velocity Ω, and the top curve and bottom curve are respectively the $\Omega + \kappa/2$ and $\Omega - \kappa/2$, all in km/s-kpc. The two horizontal lines represent speeds of the bar rotation. The radii of the intersections with the $\Omega + \kappa/2$ curve and the $\Omega - \kappa/2$ curve are the locations of Lindblad resonances.

The second rotation curve is the nearly flat one, which is

$$V = V_0 (\frac{r}{r + \epsilon})^{1/2}, \tag{2}$$

with $\epsilon = 0.01$. It rises rapidly from the center, representing high concentration of mass in the center. In this case, the $\Omega - \kappa/2$ curve does not have a local maximum, hence no IILR. The pattern speed of the bar, Ω_p, would intersect with $\Omega \pm \kappa/2$ curves, resulting two Lindblad resonances, the OILR and OLR.

If Ω_p is sufficiently low, only the OILR will be in the galactic domain. Thus, we have another case of a single Lindblad resonance, as shown in Figure 2.

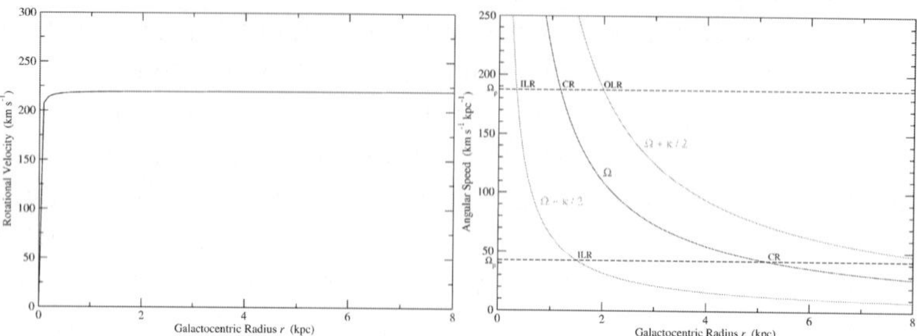

Figure 2. Nearly flat rotation curve (fast rising). The left panel is the rotation speed in km/s vs. radius in kpc. In the right panel, the middle curve is angular velocity Ω, and the top curve and bottom curve are respectively the $\Omega + \kappa/2$ and $\Omega - \kappa/2$, all in km/s/kpc. The two horizontal lines represent speeds of bar rotation. The intersects on the top and bottom curves are the locations of Lindblad resonances.

The bar potential is taken to be

$$\mathcal{V} = \Psi \cos\theta, \tag{3}$$

with

$$\Psi = \Psi_0 \frac{r^2}{(a^2 + r^2)^2}, \tag{4}$$

where a is at the potential minimum. This potential has the property that it goes to zero as r^2 and approaches r^{-2} for large r. Thus, the bar force at $r = 0$ is zero and behaves as r^{-3} (not r^{-2}) when r is large. In other words, the axisymmetric component of the bar potential has been removed.

The initial gas density is set to be constant for the cases of non-self-gravitating disks, in which we simply take $\sigma/\sigma_0 = 1$, where σ is the surface density. For the self-gravitating disks, we use

$$\sigma = \sigma_0 e^{-(\frac{r}{r_0})^2}, \tag{5}$$

where σ_0 is the initial surface density at the center and usually we take a value of 50 $M_\odot - pc^{-2}$. The value of r_0 is specified by $\sigma = (1/5)\sigma_0$ at $r = 3$ kpc.

3. The Antares Codes

The Antares Codes have two main systems: the relaxation codes and the high-order Godunov codes. Both are available in Cartesian and polar coordinates. At present, the codes are written for 2-dimensional Euler equations coupled with the Poisson equation for an infinitesimally thin disk and the gas is assumed isothermal. The relaxation codes were written earlier and are slightly more dissipative than the high-order Godunov codes, but less time consuming. The high-order Godunov codes are based on the idea to calculate the flux on the interfaces according to the values obtained from the exact Riemann solution, while the relaxation codes use interfacial values obtained from an approximate Riemann solution. Both codes adopt the van Leer type limiters and dependent variables are piecewise linear from cell to cell . For both codes, discretization only applies to the space variables, so they are semi-discretization codes. We advance discretized system of equations in time by 2nd order Runge-Kutta method (RK2). The contributions of the source terms enter through RK2.

Cartesian coordinates avoid the inner boundary conditions for the problem, which is most convenient for the galaxy problems. But they are less natural coordinates for rotational problems, so a force balance problem emerges. We have to make special effort to treat the balance law problem to ensure their performance. Polar coordinates do not have the force balance problem, but they have singularity problem at the center. For the time being we simply impose an inner boundary and cut the center off to circumvent the problem. Radiative inner boundary condition, i.e., non-reflecting, is hard to maintain when the inner boundary is close to the center and when the conditions on it are time-dependent, a case we must confront with in bar-driven galactic disk problems. This is why our Cartesian coordinate codes are better performing ones for the galaxy problems.

The Poisson equation for disk problems is intrinsically 3-dimensional. We are thus not to solve the complete potential function problem, rather simply to calculate the force in the plane by direct integration. The force at each cell can be written as a double summation of the product of surface density and a convolution kernel. Using fast Fourier transform, we can reduce the entire computation for force from $N^2 \times N^2$ to $N^2(\ln N)^2$. We are making it of second-order accuracy and parallelized now.

All our codes used in the calculations must pass three tests: (1) Exact smooth solutions remain 2nd order accuracy as long as our simulation lasts. (2) Basic states remain unchanged for as long as our simulation runs (the balance law problem) (3) Non-reflection boundary condition holds. Although codes are still two-dimensional and without many advanced features such as AMR, we think they are among the most accurate and reliable codes available today.

4. Numerical Results

In this section, we present our numerical simulations according to the resonance, first the single resonance and then the combination of them. At the same time, we point out the relevant observations that they may be responsible for. When we refer to the theory in the following context, we mean the nonlinear asymptotic theory of resonantly excited spiral density waves in galaxies developed by us (Yuan & Cheng 1991; Yuan & Kuo 1997).

The OLR and Starburst Rings We use the Elmegreen rotation curve and a fast rotating bar such that only OLR exists (See Figure 1). The result is a pair of tightly wound spirals being generated at the OLR with no other features present, exactly as our theory predicts. The ring-like spirals are in close resemblance to the starburst ring seen in NGC4313. However, when self-gravity of the disk is included, the spirals become unstable and develop into chaos. This is because the rapid increase of surface density, σ, in the narrow spiral-ring region forces the Toomre's $Q = a\kappa/\pi G\sigma$ there to go under 1 and thus turn that region into instability. These are shown in Figure 3.

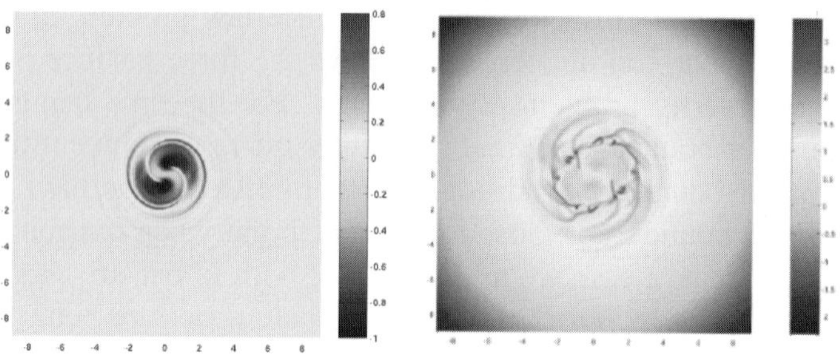

Figure 3. Surface density for single OLR. Spiral waves are excited at 2 kpc by a fast bar potential, lying horizontally. The high density spiral regions would lead to starbust activities as shown in the right panel is when the self-gravity of the disk is introduced.

The OILR and the Oval Inner Rings Resonance excitation at OILR generates a pair of open spirals spanning over almost the entire galaxy initially. The spirals quickly change into a pair of tightly wound oval-shaped spirals near the resonance. If OILR is located at a radius, say, of 2 kpc, the spirals look like the open spirals of NGC5248 (Jogee et al 2002) after 1 rotation of the bar, and like the starburst ring in NGC1512 after 4 rotation of the bar. The evolutionary sequence is shown in Figure 4.

The OLR-OILR Combination If the OILR is close to the center, an oval disk with spirals imbedded in it is formed near the center. The disk is gravitationally stable even with extremely high surface density, upto $10^3\ M_\odot/pc^2$. This is because the high values of epicyclic frequency κ near the center cancel out the high surface density so Toomre's Q remains greater than 1. The result explains the origin of the circumnuclear molecular disk seen in many nearby galaxies. The OLR does not affect the result in the center. However, as just explained, the spiral arms may become unstable due to the sharp rise of their

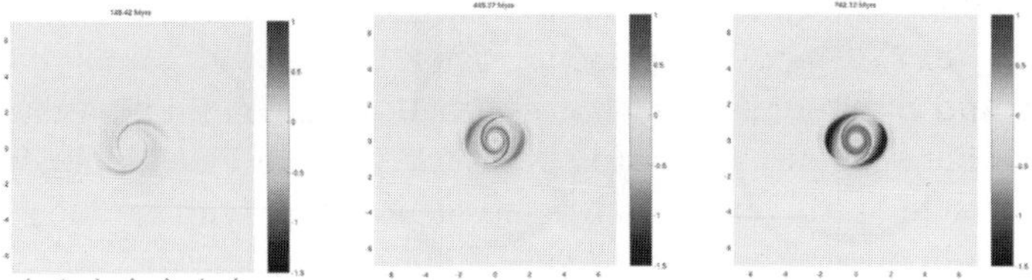

Figure 4. Surface density for single OILR. Spiral waves are excited at 2 kpc. The three panels are sequence of evolution at 1, 3, and 5 turns of the bar. Steady state is reached at after 4 turns.

surface density while the epicyclic frequency remains low. The instability leads to starburst ring activities. The waves excited at both resonances usually do not interfere, since the waves at OLR propagate outward while those at OILR propagate inward. Thus the combination is simply the superposition of the last two cases. This is the situation for NGC1068 (Bruhweiler et al 2001) and for the Milky Way with the 3-kpc arm outside (Yuan & Cheng 1991) and a dense circumnuclear molecular disk at the center (Jackson et al 1996). Shown in Figure 5 is a pair of tightly wound spirals generated at 1.5 kpc and a dense circumnuclear molecular disk of spiral-oval shape forms within 300 pc.

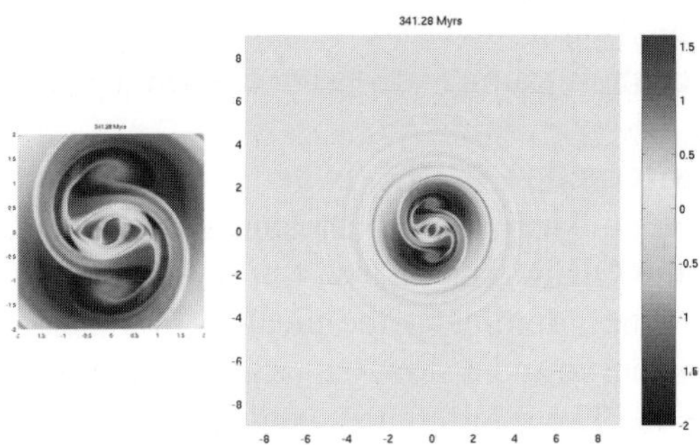

Figure 5. OLR-OILR combination. Spiral patterns excited by a bar at OLR (at 1.5 kpc) and at OILR (at 0.3 kpc) are shown here after 10 rotation of the bar. On the left is the magnified central region in which the extremely dense oval disk with spirals imbedded in it is shown in the center. We believe it can be identified as the observed high-density circumnuclear molecular disk. The diamond shape ring may be interpreted as the one seen in NGC6782.

The IILR and the Leading Spirals As expected from the non-linear asymptotic theory, the resonance at the IILR will produce a pair of leading spirals. Since the IILR is necessarily close to the center, the thickness and the curva-

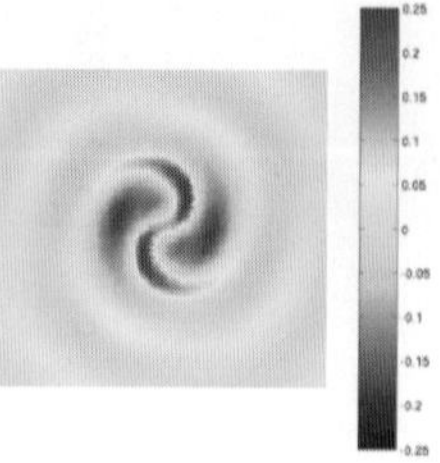

Figure 6. Surface density for single IILR. Leading spiral waves are excited at 0.15 kpc. The steady state is reached after 1 bar rotation. Some modulation on the waves pattern is seen. The size of the frame is 4x4 in kpc. Leading spiral results are expected by the theory, but single IILR is probably a rare case in nature.

ture effects come in and the true meaning of the resonance, which is based on the linear analysis in a plane geometry, may have to be modified. This case may not be seen in reality, unless IILR can be moved to a distance beyond, say, 0.4 kpc from the center (See the discussion below). In figure 6, we show a theoretical result of leading spirals excited for single IILR at 0.15 kpc after 10 bar revolutions. Steady state is reached after 1 bar rotation. Unlike the others cases, leading spirals extend all the way to the outer boundary which is set at 9 kpc in the calculation. Spiral waves are not smooth. Some modulation is seen, indicating that minor short trailing waves are excited in the calculation.

The IILR-OILR Interaction If IILR is not so close to the center, then the true resonance may exit. There are two possible cases. In both cases, the IILR is located close to the OILR. This is true for any realistic rotation curve and therefore their interaction dominates the picture. In the first case, we use the Elmegreen's rotation curve and a slowly rotating bar (See the lower part of the left panel in Figure 1). The interference, between the outgoing leading spiral waves generated at the IILR and the inward going trailing waves excited at the OILR, produces some surprising results. A bar of gas is formed. Spirals which are connected to the tips of the bar change from leading to trailing periodically. And this goes on for at least 20 rotations of the bar (See Figure 7). This case is important since the gas bar is in phase with the imposed bar potential of stellar origin. It reinforces the bar potential, in stead of weakening it. (2) In the second case, we assume the disk rotates as a rigid-body rotation near the center, such that the central disk becomes a wave-forbidden region. The location of the IILR then is pushed out to the outer boundary of the rigid-body rotation. For the present case, we let IILR be located at 1 kpc and OILR at 1.8 kpc. Unlike the previous case, a ring will be formed. See Figure 8.

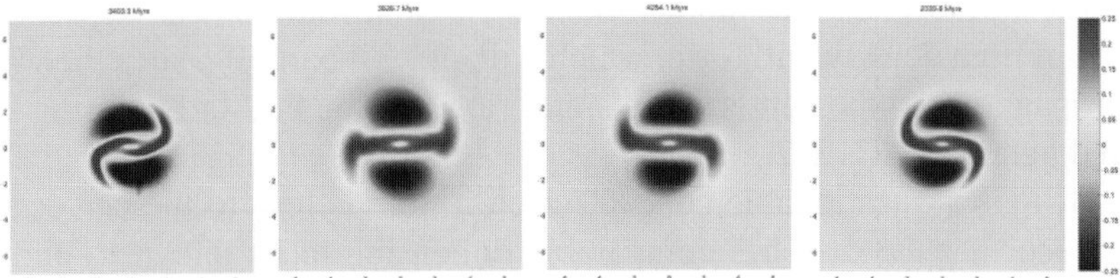

Figure 7. IILR-OILR interaction. Strong interaction between leading spirals excited at IILR of 0.6 kpc and trailing spirals at OILR of 2.0 kpc. A prominent gas bar is formed, with spirals at its tips changing periodically from leading to trailing for at least 20 turns of the bar, an unexpected result. Note the gas bar reinforces the imposed stellar bar, lying horizontally.

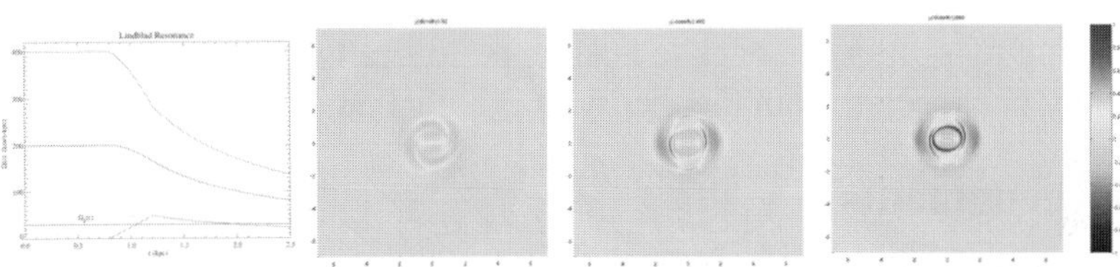

Figure 8. IILR-OILR interaction. In this case, within 0.8 kpc the disk rotates like a rigid body. Thus, the IILR is at 1 kpc and OILR at 1 kpc, shown on the left-most panel. Formation of a ring is seen left to right in the surface density diagrams at 1, 5 and 10 turns of the bar.

5. Summary

The inclusion of the self-gravitation of the disk has opened a new horizon to us. We not only see the structure and evolution of the disk, but also witness the instability of the disk and the development into chaos. We identify the starburst ring phenomenon to the **instability** of the spiral arms about 1-2 kpc from the center and the super dense circumnuclear molecular disk to the **stability** of the oval-spiral disk formed at the center. The former is a result of resonantly excited spiral waves at the OLR, while the latter, at the OILR, when it is sufficiently close to the center (a few hundred parsecs). Often both features are present in the same galaxy, such as in NGC1068 and the Milky Way. In such cases, the rotation curve must rise rapidly from the center. This implies a high concentration of matter at the center and is capable of hosting a fast nuclear bar (Yuan & Kuo 1997).

If the OILR is sufficiently away from the center (3 kpc), it will produce an open spiral pattern in close resemblance to that in NGC5248, in the first couple turns of the bar. In order to have OILR sufficiently out, the bar must be large and its rotational speed must be small. This can only happen in galaxies with a major bar or oval distortion. However, as the disk evolves, the matter in the disk will lose angular momentum and move inward. Then, a spiral-ring structure of high density will form, which will, at least partially, keep the waves from propagating to the center. This spiral ring may be the starburst ring seen in barred galaxy NGC1512.

The IILR is a questionable concept when it is too close to the center, say, 100 pc such that the disk has comparable dimension to its thickness and the linear theory no longer holds. But if it so happens that they can be placed to a distance of 400 pc or greater, it then would make sense to talk about the IILR. In such cases, IILR and OILR are close by and interact with each other. It may form a ring, a superposition of trailing and leading spirals in the case the rotation curve behaves like a rigid-body rotation near the center. Or, it may form a bar-spiral feature in gas, similar to the major bar structure in the nearby barred galaxies. The spirals which connects to the ends of the bar flip from leading to trailing and back to leading periodically. This is a phenomenon completely unexpected from the theory. It is a challenge to the theorists.

Acknowledgments

The authors wish to thank I-Liang Chern, Wei-Cheng Wang, Paul Woodward, and Chao-Chin Yang for their contributions to the code development, and Lien-Hsuan Lin and Yang-Shyang Li for their support on graphical presentation. The work is in parts supported by a Key Project of Academia Sinica and by NSC grant NSC92-2112-M-001-068.

References

Bruhweiler, F. C., Miskey, C. L.; Smith, A. M., Landsman, W., Malumuth, E. (2001) "Ionization, Extinction, and Spiral Structure in the Inner Disk of NGC 1068", Astrophys. J., 546, 866.

Elmegreen, B.G., & Elmegreen, D.M. (1990), "Optical tracers of spiral wave resonances in galaxies - Applications to NGC 1566",Astrophys. J.,355,52.

Goldreich, P. & Tremaine, S. (1979), "The Excitation of Density Waves at the Lindblad And Corotation Resonances by an External Potential", *Astrophys. J.*, 233, 857.

Huntley, J.M., Sanders, R.H., & Roberts, W.W. (1978), "Bar-driven Spiral Waves in Disk Galaxies, Astrophys. J.,221,521.

Jackson, J. M., Heyer, M. H., Paglione, T. A. D., Bolatto, A, D. (1996), "HCN and CO in the Central 630 Parsecs of the Galaxy", Astrophys. J., 456, L91.

Jogee, S., Shlosman, I., Laine, S., Englmaier, P., Knapen, J. H.; Scoville, N., & Wilson, C. D. (2002), "Gasdynamics in NGC 5248: Fuelling a Circumnuclear Starburst Ring of Super-Star Clusters", Astrophys. J., 575, 156.

Sanders, R.H., & Huntley, J.M. (1976), "Gas Response to Oval Distortion in Disk Galaxies", Astrophys. J.,209,53

Yuan, C. & Cheng, Y. (1991), "Resonance excitation of spiral density waves in a gaseous disk. II - A nonlinear theory and application to the 3 kiloparsec arm", Astrophys. J., 376, 104.

Yuan, C. & Kuo, C.L. (1997), "Bar-driven Spiral Density Waves and Accretion in Gaseous Disks", Astrophys. J., 486, 750.

STELLAR DISK TRUNCATIONS: WHERE DO WE STAND?

M. Pohlen[1], J. E. Beckman[1], S. Hüttemeister[2], J. H. Knapen[3], P. Erwin[1], and R.-J. Dettmar[2]
[1] *Instituto de Astrofísica de Canarias*
[2] *Astronomisches Institut, Ruhr-Universität Bochum*
[3] *Centre for Astrophysics Research, University of Hertfordshire*

Abstract In the light of several recent developments we revisit the phenomenon of galactic stellar disk truncations. Even 25 years since the first paper on outer breaks in the radial light profiles of spiral galaxies, their origin is still unclear. The two most promising explanations are that these 'outer edges' either trace the maximum angular momentum during the galaxy formation epoch, or are associated with global star formation thresholds. Depending on their true physical nature, these outer edges may represent an improved size characteristic (e.g., as compared to D_{25}) and might contain fossil evidence imprinted by the galaxy formation and evolutionary history. We will address several observational aspects of disk truncations: their existence, not only in normal HSB galaxies, but also in LSB and even dwarf galaxies; their detailed shape, not sharp cut-offs as thought before, but in fact demarcating the start of a region with a steeper exponential distribution of starlight; their possible association with bars; as well as problems related to the line-of-sight integration for edge-on galaxies (the main targets for truncation searches so far). Taken together, these observations currently favour the star-formation threshold model, but more work is necessary to implement the truncations as adequate parameters characterising galactic disks.

1. Introduction

The structure of galactic disks is of fundamental importance for observationally addressing the formation and evolution of spiral galaxies. By measuring scalelengths for galaxy samples at different redshifts, it is for example possible to directly address the evolution of galactic disks using the scalelength as a tracer of their sizes (e.g. Labbé et al. 2003). To a first approximation, the radial light distribution of disks is well described by an exponential decline (Freeman 1970). However, for 25 years (since van der Kruit 1979), we have know that this exponential light distribution does not extend to arbitrarily large radii. Early, deep imaging using photographic plates showed that the disks are

D. Block et al. (eds.), Penetrating Bars through Masks of Cosmic Dust, 713–722.

radially truncated at the outer parts (Fig.1). While truncations were difficult to measure at that time – still true today for deriving their exact shapes – it has become progressively easier to detect these *outer edges* with modern CCD equipment. In the optical they occur at galactocentric distances of roughly 2-6 exponential scalelengths (cf. next Sect.). Truncations could be used as an additional, intrinsic parameter measuring the total size of galactic disks (maybe replacing the widely used, but purely arbitrary D_{25} definition), and truncations may be linked to the formation and evolution of disk galaxies in general. Despite the fact that they have been known for so long, we do not have a secure understanding of the origin for these outer edges, so their application as a fundamental disk characteristic is not yet possible. In this overview we will summarise what is currently known about truncations and describe the quest to unveil their true origin.

2. Where are normal galaxies truncated?

To understand their origin and use them as characteristic galaxy parameters (e.g. in comparative studies of different galaxy samples), one has to empirically establish *where* 'normal disks' are typically truncated – conventionally expressed in terms of radial scalelength – and *what* this truncation looks like.

The edge-on view

In their seminal series of papers on the 3-D light distribution in edge-on galaxies, van der Kruit & Searle (e.g. 1981a, 1981b) measured the ratio of truncation radius (R_{tr}) to radial scalelength (h) as 4.2 ± 0.6 for a sample of seven nearby, large galaxies using photographic plates. They found that stellar disks have "rather sharp edges", justifying the terms *cut-off* and *truncation* they used to describe them. Barteldrees & Dettmar (1994) collected more sensitive CCD images for a much larger sample of 27 edge-on galaxies – later reanalysed and slightly enlarged by Pohlen et al. (2000a) – finding significantly lower ratios of $R_{\rm tr}/h$, down to 2.9 ± 0.7 in the latter study. The Groningen group followed with their CCD survey providing a mean ratio of 3.8 ± 1.0 for a pilot sample of four galaxies (de Grijs et al. 2001). This lies in between the former two extremes, but clearly confirms the explicitly larger scatter observed by Pohlen et al. (2000a) compared to van der Kruit & Searle (1981b). Since line-of-sight (LOS) integration makes edge-on galaxies ideal candidates to search for truncations, Pohlen (2001) obtained new data, trying to finally settle the true nature of the truncations by using significantly deeper exposures and selecting objects as undisturbed as possible. This study showed without doubt that the previously used concept of a sharp truncation does not really hold, as already suggested in de Grijs et al. (2001). It is just possible to fit a sharply truncated model to this new data set (cf. Fig.1a): Pohlen (2001) derived a value of

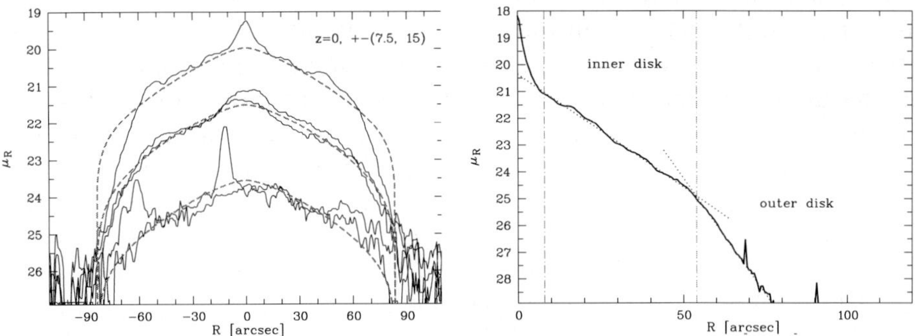

Figure 1. *(a) Left panel:* Edge-on view: Radial surface brightness profiles *(solid lines)* of the edge-on galaxy NGC 522 together with the best fitting sharply-truncated model *(dashed line)* *(b) Right panel:* Face-on view: The azimuthally averaged radial surface brightness profile of the face-on galaxy NGC 5923 *(solid line)*. In addition to the inner part ($<$ 8 arcsec) dominated by the bulge component an inner exponential slope *(inner dotted line)* is clearly visible out to the break radius at $R_{\rm br} = 54''$followed by another outer exponential part *(outer dotted line)*.

$R_{\rm tr}/h = 3.5 \pm 0.8$ using 56 galaxies, which is similar to the values of 3.6 ± 0.6 from the full analysis of 20 galaxies from the Groningen CCD survey (Kregel et al. 2002). However, it has become clear that the radial profiles are far better described by a *two-slope*, or *broken exponential* structure, characterised by inner and outer scalelengths and a well-defined break radius. The deepest data show that LOS integration – a fairy godmother for detecting truncations – is at the same time the major drawback in analysing them in detail (Pohlen 2001). Since disks are only empirically described by an exponential – typically over at most a few scalelengths – the edge-on geometry of such a two-slope disk, with additional dust obscuring the major axis, complicates an already delicate fitting process. Edge-on, there are always uncertainties in where to mark the break radius (cf. Kregel et al. 2002), and in particular the calculation of the true, unprojected scalelength is heavily model dependent. Within a sharply truncated model used by e.g. Pohlen et al. (2000b) there is a strong coupling between the two parameters $R_{\rm tr}$ and h. However, fitting an infinite exponential model – excluding the truncation region – does not really solve the problem, especially for galaxies which exhibit an early break (in terms of scalelength) and a steep outer decline. The LOS integration will distribute the measured light along the assumed uniformly exponential decline so the best determined fitting scalelength will be systematically too small (e.g. Pohlen et al. 2004). A simpler approach is that of Pohlen (2001), who avoided 3D, LOS issues by fitting the observed profile with a 1-D, two-slope model, finding $R_{\rm br}/h_{\rm in} = 2.5 \pm 0.8$ from 37 galaxies. However, the fitted scalelengths in this case are not necessarily the intrinsic scalelengths, both because dust extinction increases the measured scalelength and because the scalelength of a 1D fit is larger than the intrinsic scalelength by as much as 20% (e.g., de Grijs 1997).

The face-on view

The only way to settle the question of where the breaks in the radial profiles really occur is to go back to face-on galaxies as done earlier by van der Kruit (1988). Therefore, Pohlen et al. (2002) took deep, carefully flatfielded exposures of three face-on galaxies. The face-on geometry is practically independent of possible LOS effects caused by an integration across the multicomponent disk and much less affected by dust. These data quantitatively establish the smooth truncation behaviour of the radial surface brightness profiles, which is best described by a two-slope or broken exponential model, characterised by inner $h_{\rm in}$ and outer $h_{\rm out}$ exponential scalelengths separated at a relatively well defined break radius $R_{\rm br}$ (cf. Fig.1b). The mean value for the distance independent ratio of break radius to inner scalelength – which marks the start of the truncation region – is $R_{\rm br}/h_{\rm in} = 3.9 \pm 0.7$ for the three galaxies. This value is significantly larger than the value for the edge-on sample, but this is expected given the systematic errors in the scalelength measurement in the edge-on case described above. It is slightly lower compared to $R_{\rm tr}/h = 4.5 \pm 1.0$ for the 16 face-on galaxies obtained by van der Kruit (1988). This small shift could be due to the different methods used to derive the truncation – only estimated from isophote maps by van der Kruit (1988) – or the sample sizes and selections.

The combined view

In summary, the radial profiles show a two-slope or broken exponential behaviour with an inner and outer disk separated by a rather sharp break radius at $2.5 \lesssim R_{\rm br}/h_{\rm in} \lesssim 5.8$, taking the mean value from the three face-on galaxies of Pohlen et al. (2002) and the scatter from the larger edge-on sample of Pohlen (2001). However, it is clear that truncations – seen in different data sets, using different methods, and analysed by different groups – are *real*, in contrast to the recent claim of Narayan & Jog (2003). It is not a matter of "questioning their very existence", but rather of determining the shape of the profiles beyond the break radius, which is admittedly still difficult to measure and a "puzzle" to explain.

3. Are all disk galaxies truncated?

The data of Pohlen (2001) suggest that a very high fraction –more than 79%– of mostly late-type edge-on disks are truncated. The remaining galaxies are either still disturbed (despite the careful selection), show an additional major structure not accounted for in a single component disk – such as outer envelopes or a strong bar –, or (in one case) have S/N that is too low. Kregel et al. (2002) (see also Kregel 2003) support a rather high frequency of radial truncation features for a similar late-type sample of 34 edge-on galaxies. They find successful truncation fits for 20 galaxies (59%), whereas most of the other galaxies are rejected for technical reasons such as superimposed foreground stars or low S/N. The latter is the main problem of this sample in respect to truncations yielding only lower limits of typically $R_{\rm tr}/h \gtrsim 4.5$ for the other galaxies and just two cases of $R_{\rm tr}/h \gtrsim 6.0$. However, just recently Erwin et al. (2004), analysing a large sample of intermediately inclined, early-type galaxies, found seven barred galaxies without measured truncations beyond six scalelengths. Although the edge-on samples certainly also contain barred galaxies, we may ask if only barred galaxies (sometimes) lack truncations. It is true that the detailed studies by Barton & Thompson (1997) and Weiner et al. (2001), which showed disks without truncation extending to at least seven and ten scalelengths, respectively, dealt with two strongly barred face-on galaxies. But the large collection of surface brightness profiles for unbarred or only weakly barred Sb–Sc galaxies by Courteau (1996) shows several cases of profiles which extend to at least six scalelengths without truncation, and the sample of Erwin et al. (2004) also includes one unbarred Sa galaxy (UGC 3580) with $R_{\rm tr}/h \gtrsim 5.8$. It is still puzzling why the large edge-on samples – which have the advantage of LOS integration – are missing many of the "untruncated" galaxies (or galaxies with truncations beyond six scalelengths) seen in face-on samples. From the few late-type, face-ons in Pohlen et al. (2002) and the large collection of early-type, face-ons in Erwin et al. (2004) we know that azimuthal averaging does not play a significant role in smoothing out truncations. If untruncated disks really do exist, then models which attempt to explain truncations also need to explain why some disks are *not* truncated.

4. What is the origin of truncations?

Despite the recent success in disentangling where the disks are truly truncated, the nature of galactic disk truncations is far from understood and their place in theories of disk galaxy formation is still under debate. The proposed hypotheses span a rather wide range of possibilities, though there are clearly two favoured explanation: the *collapse model* and the *threshold model.* Van der Kruit (1987) deduced a direct connection to the galaxy formation process describing the truncations as remnants from the early collapse. The truncation reflects here the maximum angular momentum of the protogalaxy, which is completely independent of the hierarchical merging history. On the other hand, Kennicutt (1989) following Toomre (1964) has suggested a dynamical critical star-formation threshold ($\Sigma_{\rm c}$) – recently addressed in a different version by Schaye (2004) – for thin, rotating, isothermal gas disks. If this persists over sufficient time it should produce a visible turnover or truncation in the observed stellar luminosity profile at that radius. Here the truncation radius corresponds to the radial position where $\Sigma_{\rm gas}(R) = \Sigma_{\rm c}$. More recently, van den Bosch (2001) combined the collapse and threshold models. While there is a truncation of the cold gas which reflects the maximum specific angular momentum of the baryonic mass that has cooled, the truncation radius of the *stars* reflects the presence of a star-formation threshold. Alternatively, Zhang & Wyse (2000) and Ferguson & Clark (2001) have presented evolutionary scenarios – typified as viscous evolution models – which show features surprisingly comparable to the observed two-slope structure. However, according to Ferguson & Clark (2001) the breaks just reflect the initial conditions in the gas and tend to be smeared out while the galaxy evolves with time. Battaner et al. (2002) have proposed a connection with large scale galactic magnetic fields. This approach seems controversial, since it also explains flat rotation curves without dark matter; nevertheless, it is able to explain e.g. possible $R_{\rm tr}/h$ variations with morphological type or rotational velocity. The formation of a truncated stellar disk by tidal interaction, although appealing, is in contradiction to the radial spreading shown by Quinn et al. (1993) in their numerical N-body simulations of satellite mergers. However, we are still missing an up-to-date N-body+SPH *(stars and gas)* simulation addressing a possible connection between truncations and interaction.

5. Can observations decide yet?

The star-formation threshold models of Schaye (2004) and van den Bosch (2001) both predict a correlation of the ratio $R_{\rm tr}/h$ with the central surface brightness. Low-surface brightness galaxies should have lower values, as indeed indicated by current observations (cf. Fig.2a or Kregel 2003). The collapse model does not account for such a correlation since it predicts instead a fixed value for all galaxies, allowing for only very small scatter. According to the threshold models a similar trend should be observed in relation to the mass of the galaxy. Low-mass systems (characterised e.g. by smaller rotational velocities) should have smaller values for the ratio of truncation radius to radial scalelength, as indicated by Pohlen (2001) (cf. Fig.2b). Additional support for a correlation with mass comes from recent work on nearby dwarf galaxies. Hunter (2002), Simon et al. (2003), and Hidalgo et al. (2003) all find similar two-slope profiles with rather low values for $R_{\rm br}/h_{\rm in}$. The currently available large edge-on samples do not sufficiently cover the Hubble sequence to address the prediction by van den Bosch (2001) that early-type, bulge dominated galaxies (Sa to S0) should have $R_{\rm tr}/h \lesssim 2$. However, in the large sample of moderately inclined, early-type, barred galaxies of Erwin et al. (2004), most galaxies do not follow this relation. Only a small subset does, but these truncations are far better described as bar-related OLR breaks (cf. Sect.6). This

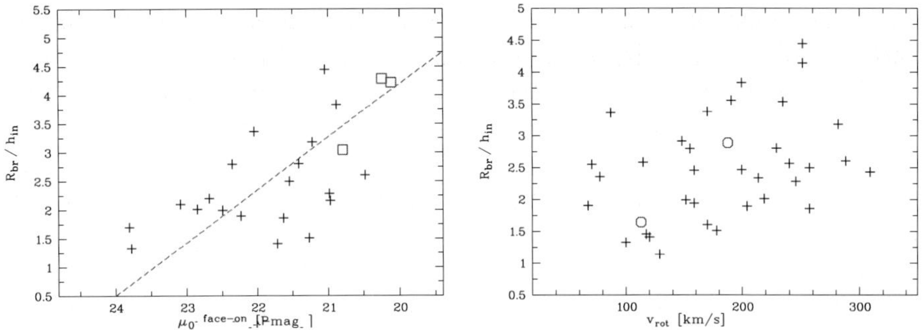

Figure 2. *(a) Left panel:* Ratio of break radius to inner scalelength ($R_{\rm br}/h_{\rm in}$) versus face-on central surface brightness. *Crosses:* edge-ons by Pohlen (2001). *Squares:* face-ons by Pohlen et al. (2002). *Dashed line:* threshold prediction by Schaye (2004) according to Kregel (2003) *(b) Right panel:* $R_{\rm br}/h_{\rm in}$ versus maximum rotational velocity $v_{\rm rot}$. *Crosses:* edge-ons by Pohlen (2001). *Circles:* two face-on galaxies with known $v_{\rm rot}$ from the sample of Pohlen et al. (2002) shifted to account for the mean systematic error of the two samples (cf. Sect.2).

failure of the promising threshold model of van den Bosch (2001) could be due to its rather general implementation of the bulge formation process. Kennicutt (1989) (see also Martin & Kennicutt 2001), measuring radial Hα distributions, observed that the star-formation rate drops abruptly at a finite radius. Unfortunately only a few galaxies with both Hα and optical breaks are currently available for testing the one-to-one correlation expected in a simple threshold scenario. Current data does indicate such a correlation (cf. Fig.3a or Hunter 2002), while beyond the break the decline in the Hα luminosity seems to be more rapid than the stellar one. We know that disk galaxies have radial colour gradients: the disk gets bluer towards the edge (de Jong 1996). This implies that disks get radially younger – arguing for an inside-out formation scenario – in the sense that their inner regions are older and more metal rich than their outer regions (Bell & de Jong 2000). However, this does not explain the truncation radius as the radius to which the disk has now grown (e.g. Larson 1976), since the stars beyond the break are indeed younger but still many Gyrs old. This is confirmed by Davidge (2003), who studied the resolved stellar populations of NGC 2403 and found that while young stars are restricted to the inner parts, AGB stars observed in the outer parts suggested that star-formation has occurred there during intermediate epochs of 1 Gyr or more. The measured B- and R-band profiles of the face-on galaxy NGC 5923 – which trace the combined star-formation history – indicate that the radial blueing extends only out to the optical break, which is at the same radius in both bands (cf. Fig.3b). Outside the break the outer disk has a constant blue colour $(B-R) = 0.9$, which argues for either a single population or a coherent star-formation history.

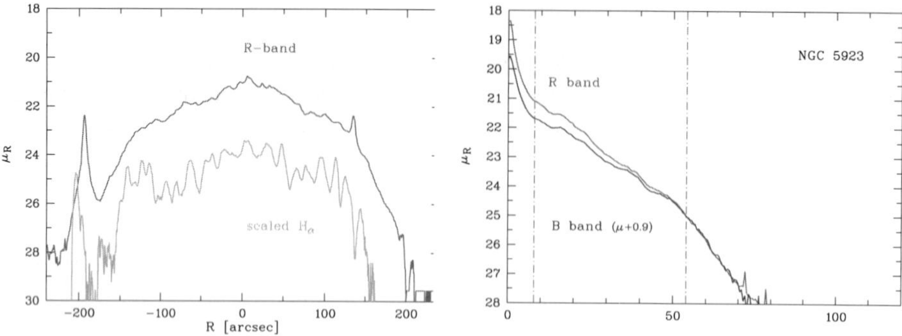

Figure 3. *(a) Left panel:* Comparison of R-band major axis profile (6.2′ vertically averaged) of UGC 7321 and a scaled version of the continuum subtracted Hα . *(b) Right panel:* The B- and R-band azimuthally averaged radial surface brightness profiles of the face-on galaxy NGC 5923.

NIR data may help to determine the question if the old stellar population – tracing the mass of the galaxy – changes across the break. Data by Florido et al. (2001) show for the first time that truncations are also present in NIR images and that they are rather sharp. It turned out that these truncations are significantly smaller compared to the optical ones and it would be interesting to confirm these observations independently (cf. Pohlen 2001).

6. Are bars an alternative solution?

Many barred galaxies possess rings in their outer disk produced by the bar's outer Lindblad resonance (OLR). While some of these rings are seen in radial profiles as mere bumps on an underlying exponential disk, Erwin et al. (2004) found that a large fraction of their barred, early-type galaxies exhibit breaks associated with the OLR. Many of these *OLR breaks* resemble the classical truncations described above, but located further inside. This implies that the bar is able to re-shape the initial gas disk (maybe replenished by later infall) out to substantial radii (≈ 1–$2 \times h_{\rm in} \approx 2 \times$radius of bar). Without additional data it is not yet clear if these OLR breaks also occur in more late-type galaxies – known to have smaller and weaker bars – thus providing a true alternative to the now favoured threshold theory (Erwin et al. 2004). However, the three clearly unbarred face-on Sbc–Sc galaxies of Pohlen et al. (2002) argue against OLR breaks as the sole explanation for truncations, if one does not assume that these galaxies once had bars which are now dissolved. The intriguing existence of barred but untruncated galaxies (cf. Sect.3) implies that bars may also *prevent* truncations in some cases.

7. Any open questions?

The smooth nature of truncations clearly shows that there are stars outside the break radius. While the previously considered sharp truncations would agree nicely with the threshold model, a simple critical density does *not* account for the observed two-slope structure. There are two possible explanations for these stars beyond the break radius. They could be either born in situ – out of an initial, maybe viscously redistributed, and already replenished gas disk – or they are stars from the inner disk which have migrated outwards in a kind of diffusion process. Ferguson et al. (1998) already showed for a couple of galaxies that there is local star-formation at large galactocentric distances, probably beyond a break radius, supporting the former possibility. However, both approaches still have to explain *why* the inner and outer disk regions are produced with different, yet still exponential, slopes, *why* the transition zone appears to be rather sharp, and – even more challenging – *why* the outer disk is so large: Pohlen et al. (2002) showed for the face-on galaxies that the break radius is at $R_{\rm br} = \frac{2}{3} R_{\rm lmp}$ using the last measured point ($R_{\rm lmp}$) as a reference. In the case of the Milky-Way ($R_{\rm lmp} \geq 21$ kpc) this implies an inner disk of 14 kpc and an outer disk of at least 7 kpc extra radius! This is significantly larger than the usually assumed values of ≈ 1 kpc previously deduced for a sharp truncation. According to Pérez-Martin (this volume) there seems to be a trend of galaxies being truncated earlier (in terms of h) at higher redshift (consistent with Tamm & Tenjes 2003). So far, none of the described models has addressed such an interesting evolution with redshift.

8. Summary and Outlook

Finally, we want to emphasise that the observed radial profiles show rather sharp break radii. However, they are not sharply truncated – in the sense of an outer radius beyond which no further stars can be found – as is often assumed in the literature. Despite the success of recent studies of the structural parameters, a medium-deep survey of a well selected sample of local, intermediate inclination to face-on disk galaxies along the Hubble sequence is still necessary, to provide ultimately *the reference value* of $R_{\rm tr}\,/h$. The star-formation threshold model seems to be the most promising hypothesis to explain the presence of truncations so far. Although less direct than the collapse model – which directly relates observed truncations to initial conditions – van den Bosch's (2001) model showed how to implement such a star-formation threshold in a collapse model which is able to trace the evolutionary picture. We are still working on additional observational support, such as more Hα-to-optical break comparisons, detailed measurements of the atomic and molecular gas densities out to the break radius, and measurements of the stellar kinematics and population differences inside and outside the break.

References

Barteldrees, A., & Dettmar, R.-J. 1994, A&AS, 103, 475
Barton, I. J., & Thompson, L. A. 1997, AJ, 114, 655
Battaner, E., Florido, E., & Jiménez-Vicente, J. 2002, A&A, 388, 213
Bell, E. F., & de Jong, R. S., 2000, MNRAS, 312, 497
Courteau, S. 1996, ApJS, 103, 363
Davidge, T. J., 2003, AJ, 125, 3046
de Grijs, R. 1997, PhD Thesis, Rijksuniversiteit Groningen, Netherlands
de Grijs, R., Kregel, M., & Wesson, K. H. 2001, MNRAS, 324, 1074
de Jong, R. S. 1996, A&A 313, 45
Erwin, P, Pohlen, M., & Beckman, J., 2004, in prep.
Ferguson, A. M. N., Wyse, R. F. G., Gallagher, J. S., & Hunter, D. A. 1998, ApJ, 506, L19
Ferguson, A. M. N., & Clark, C. J. 2001, MNRAS, 325, 781
Florido, E., Battaner, E., Guijarro, A., Garzón, F., & Jiménez-Vicente, J. 2001, A&A, 378, 82
Freeman K. C. 1970, ApJ, 160, 811
Hidalgo, S. L., Marín-Franch, A., & Aparicio, A. 2003, AJ, 125, 1247
Hunter, D. A., 2002, in: Outer Edges of Dwarf Irr Galaxies, Lowell Workshop, Online-Proceedings
Kennicutt, R. C., 1989, ApJ, 344, 685
Kregel, M., van der Kruit, P. C., & de Grijs, R. 2002, MNRAS, 334, 646
Kregel, M., 2003, PhD Thesis, Rijksuniversiteit Groningen, Netherlands
Labbé, I., et al. 2003, ApJL, 591, L95
Larson, R.B. 1976, MNRAS, 176, 31
Martin, C. L., & Kennicutt, R. C., Jr. 2001, ApJ, 555, 301
Narayan, C. A. & Jog, C. J. 2003, A&A, 407, L59
Pohlen, M., Dettmar, R.-J., & Lütticke, R. 2000a, A&A, 357, L1
Pohlen, M., Dettmar, R.-J., Lütticke, R., & Schwarzkopf, U. 2000b, A&AS, 144, 405
Pohlen, M., 2001, PhD Thesis, Ruhr-Universität Bochum, Germany
Pohlen, M., Dettmar, R.-J., Lütticke, R., & Aronica, G. 2002, A&A, 392, 807
Pohlen, M., Balcells, Lütticke, R., & Dettmar, R.-J. 2004, A&A, in press
Quinn, P. J., Hernquist, L., & Fullagar, D. P. 1993, ApJ, 403, 74
Schaye, J. 2004, ApJ, in press, astro-ph/0205125
Simon, J. D., Bolatto, A. D., Leroy, A., & Blitz, L. 2003, ApJ, 596, 957
Tamm, A. & Tenjes, P. 2003, A&A, 403, 529
Toomre, A. 1964, ApJ, 139, 1217
van den Bosch, F. C. 2001, MNRAS, 327, 1334
van der Kruit, P. C. 1979, A&AS, 38, 15
van der Kruit, P. C. 1987, A&A, 173, 59
van der Kruit, P. C. 1988, A&A, 192, 117
van der Kruit, P. C., Searle, L., 1981a, A&A 95, 105
van der Kruit, P. C., Searle, L., 1981b, A&A 95, 116
Weiner, B. J., Williams, T. B., van Gorkom, J. H., & Sellwood, J. A. 2001, 546, 916
Zhang, B. & Wyse, R. F. G. 2000, MNRAS, 313, 310

ON THE UNIFICATION OF DWARF AND GIANT ELLIPTICAL GALAXIES

Alister W. Graham and Rafael Guzman
Department of Astronomy, University of Florida, Gainesville, FL 32611, USA.

Abstract The near orthogonal distributions of dwarf elliptical (dE) and giant elliptical (E) galaxies in the μ_e–M and μ_e–$\log R_e$ diagrams have been interpreted as evidence for two distinct galaxy formation processes. However, continuous, linear relationships across the alleged dE/E boundary at $M_B = -18$ mag — such as the relationships between central surface brightness (μ_0) and: a) galaxy magnitude (M); and b) light-profile shape (n) — suggest a similar initial formation mechanism. Here we explain how these latter two trends in fact necessitate a different behavior for dE and E galaxies, exactly as observed, in diagrams involving μ_e (and also $<\mu>_e$). Together with other linear trends across the alleged dE/E boundary, such as those between luminosity and color, metallicity, and velocity dispersion, it appears that the dEs form a continuous extension to the E galaxies. The presence of partially depleted cores in luminous ($M_B < -20.5$ mag) Es does however signify the action of a different physical process at the centers ($<\sim$300 pc) of these galaxies.

The common distinction between a dwarf elliptical (dE) galaxy and an (ordinary) elliptical (E) galaxy is whether the absolute magnitude is fainter or brighter than $M_B = -18$ mag respectively ($H_0 = 50$ km s^{-1} Mpc^{-1}, Sandage & Binggeli 1984). By the term dE, we additionally mean objects brighter than -13 B-mag; that is, we are not talking about (Local Group) dwarf spheroidal galaxies, whose range of colors suggest a range of formation processes (e.g., Conselice 2002). The realization that dE light-profiles could be reasonably well described with an exponential function (Faber & Lin 1983; Binggeli, Sandage & Tarenghi 1984) and that bright ellipticals are better fit with de Vaucouleurs' $r^{1/4}$-law helped lead to the notion that they are two distinct families of galaxies (e.g., Wirth & Gallagher 1984, but see Graham 2002). One of the seminal papers supporting this view is Kormendy (1985). By plotting central surface brightness against luminosity, Kormendy showed two relations, almost at right angles to each other: one for the dE galaxies and the other for the luminous elliptical galaxies. Similar diagrams using μ_e, the surface brightness at the effective half-light radius r_e, or $<\mu>_e$, the average surface brightness within r_e, also show two somewhat perpendicular relations (e.g., Capaccioli, Caon,

D. Block et al. (eds.), Penetrating Bars through Masks of Cosmic Dust, 723–728.

& D'Onofrio 1992). These differences are commonly interpreted as evidence for different formation mechanisms, resulting in the belief that a dichotomy exists between the dE and E galaxies. To understand, and in fact resolve, this apparent dichotomy, we must turn to the issue of galaxy structure.

In the past, some authors have restricted the radial extent of galaxy light-profiles (excluding inner and outer parts; e.g, Burkert 1993) or adjusted the sky-background levels (e.g., Tonry et al. 1997) in order to make the $r^{1/4}$ model fit — such was the ingrained belief in this classic model. However, luminosity-dependent deviations from $r^{1/4}$ profiles had been known for some time (e.g., Capaccioli 1984, 1987; Michard 1985; Schombert 1986; Caldwell & Bothun 1987; Kormendy & Djorgovski 1989; Binggeli & Cameron 1991; James 1991). Schombert (1986) recognized the inadequacy of the $r^{1/4}$ model for describing Es, since it only fits the middle $21 < \mu_B < 25$ part of bright galaxy profiles. Kormendy & Djorgovski (1989) noted that the best $r^{1/4}$ fits were for Es with $M_B \sim -21$ mag; brighter and fainter galaxies having a different logarithmic profile curvature than that of the $r^{1/4}$ model. It is these variations in the stellar distribution which have recently provided the key to understanding the true nature of the connection between the dE and E galaxies.

Sersic's (1968) $r^{1/n}$ model can encompass both de Vaucouleurs' $r^{1/4}$ model and the exponential (n=1) model — and a variety of other profile shapes — by varying its 'shape parameter' n. Analyzing a sample of 80 early-type galaxies in the Virgo and Fornax Clusters, Caon, Capaccioli, & D'Onofrio (1993) and D'Onofrio, Capaccioli, & Caon (1994) showed how the elliptical galaxy light-profile shapes vary systematically with measurements of the half-light galactic radii and luminosity obtained independently of the fitted Sersic model. Additionally, numerous studies have demonstrated that dE galaxy profiles are *not* universally exponential, but rather are best fit with a range of Sersic profiles (i.e., n is not always $= 1$; Davies et al. 1988; Cellone et al. 1994; Young & Currie 1994; Durrell 1997; Jerjen & Binggeli 1997; Binggeli & Jerjen 1998; Graham & Guzman 2003). The resulting trend between luminosity and light-profile shape has begun to erase the dichotomy between dwarf and luminous ellipticals. To show how the remaining dichotomies can be eliminated, by explaining the apparently divergent behavior of dE and E galaxies in certain structural parameter diagrams, we will use the compilation of 249 dE and E galaxies presented in Graham & Guzman (2003) and shown here in Fig. 1.

Kormendy's (1985) plot of central surface brightness (μ_0) versus magnitude showed a large discontinuity and gap between the dE and E galaxies. However, there was an absence of galaxies with magnitudes around $M_B = -18 \pm 1$ in that sample, exactly where one might expect to see the two groups connect. If we remove galaxies with $M_B = -18 \pm 1$ from our plot of μ_0 vs. M_B (Fig. 1c), we obtain a figure very much like Kormendy's (1985) Figure 3. Thus we can see that part of the "discontinuity" had arisen from an incomplete sam-

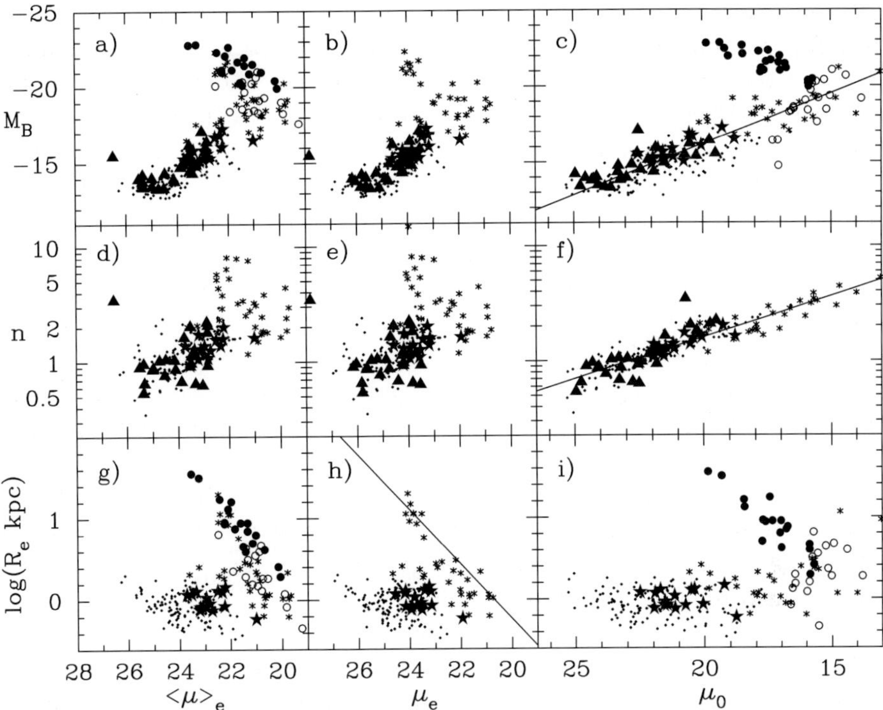

Figure 1. Mean surface brightness within r_e ($<\mu>_e$), surface brightness at r_e (μ_e), and central host galaxy surface brightness (μ_0) versus host galaxy magnitude (M_B), global profile shape (n), and half-light radius (r_e). Due to biasing from the magnitude cutoff at $M_B \sim -13$, the line $M_B = (2/3)\mu_0 - 29.5$ in panel c) has been estimated by eye rather than by a linear regression routine. The line $\mu_0 = 22.8 - 14\log(n)$ in panel f) has also been estimated by eye. The line in panel h) has a slope of 3 and represents the Kormendy (1977) relation known to fit the luminous elliptical galaxies which define the panhandle of this complex distribution (Capaccioli & Caon 1991; La Barbera et al. 2002). Dots represent dE galaxies from Binggeli & Jerjen (1998), triangles are dE galaxies from Stiavelli et al. (2001), large stars are dE galaxies from Graham & Guzman (2003), asterix are intermediate to bright E galaxies from Caon et al. (1993) and D'Onofrio et al. (1994), open circles represent the so-called "power-law" E galaxies from Faber et al. (1997), and the filled circles represent the "core" E galaxies from these same authors. Figure taken from Graham & Guzman (2003). $H_0 = 70$ km s^{-1} Mpc^{-1}.

pling of the intermediate-luminosity galaxies which fill in the apparent gap. Nevertheless, there is still an obvious change in slope in the overall μ_0–M_B relation. This can be explained with the observation that the most luminous elliptical galaxies possess partially evacuated "cores". Their observed central surface brightnesses are thought to be fainter than the original value due to the damage from coalescing supermassive black holes following a galaxy merger. As stressed in Graham & Guzman (2003), the brighter galaxies lying perpendicular to the μ_0–M_B relation defined by the less luminous ($M_B > -20.5$)

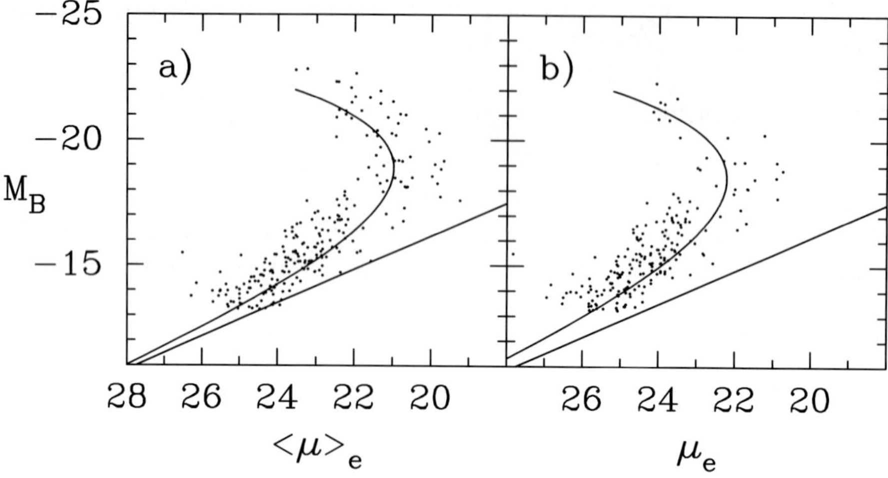

Figure 2. The curved lines show the predicted behavior of galaxies in the a) magnitude–mean surface brightness and b) magnitude–effective surface brightness diagrams. The straight line is from Fig. 1c; the data points are from Fig. 1a and 1b.

elliptical galaxies are all "core" galaxies. The simplest explanation is that all elliptical galaxies follow a linear μ_0–M_B trend, *except* when core formation modifies the central surface brightness of the most luminous ellipticals. Since core formation is thought to be a later event, the initial mechanism(s) of dE and E galaxy formation are likely to be the same.

Another reason why E galaxies are thought to be different from dE galaxies is because they don't follow the same M_B–μ_e and M_B–$<\mu>_e$ relations. Why are these relations apparently different for the dE and E galaxies? This, it turns out, has nothing to do with core formation but is due to the systematic changes in profile shape with galaxy magnitude. Even though many of the galaxies in Fig. 1a–c lack Sersic n measurements, the relationship between magnitude and n (e.g., Fig.10 in Graham & Guzman 2003) can be used to determine a representative value of n for a given M_B. From the Sersic model, we know that $\mu_e = \mu_0 + 1.086b$ and $< \mu >_e = \mu_e - 2.5\log[e^b n\Gamma(2n)/b^{2n}]$, where $b \sim 2n - 1/3$ (e.g., Graham & Colless 1997). From this, the straight line in Fig. 1c transforms into the *curved* relationships between M_B and μ_e, and M_B and $<\mu>_e$ (Fig. 2). Thus, we can reproduce the observed correlations in Fig. 1a and 1b. The different slopes for the dE and E galaxy distributions in these diagrams are merely a consequence of a continuously varying profile shape with galaxy luminosity — they do not imply distinctly different galaxy formation processes for dEs and Es. Using $L = 2\pi r_e^2 <I>_e$, it can be shown that the same mechanism is also behind the different slopes in the $<\mu>_e$–$\log r_e$

and $M - \log r_e$ diagrams. A natural consequence is that the location of dEs and Es in Fundamental Plane type analyses that use I_e, or $<I>_e$, will also be different (Guzman et al. 2004, in prep).

Acknowledgments

We are grateful for funding provided by NASA through grants HST-AR-08750.02-A and HST-AR-09927.01-A administered by the Space Telescope Science Institute. A.G. is also thankful for NSF funding administered by the American Astronomical Society's International Travel Grant Program.

References

Binggeli, B., & Cameron, L.M. 1991, A&A, 252, 27
Binggeli, B., & Jerjen, H. 1998, A&A, 333, 17
Binggeli, B., Sandage, A., & Tarenghi, M. 1984, AJ, 89, 64
Burkert, A. 1993, A&A, 278, 23
Caldwell, N., & Bothun, G.D. 1987, AJ, 94, 1126
Caon, N., Capaccioli, M., & D'Onofrio, M. 1993, MNRAS, 265, 1013
Capaccioli, M. 1984, in New Aspects of Galaxy Photometry, ed. J. Nieto, Springer-Verlag, p.53
Capaccioli, M. 1987, in Structure and Dynamics of Elliptical Galaxies, IAU Symp. 127, Reidel, Dordrecht, p.47
Capaccioli, M., & Caon, N. 1991, MNRAS, 248, 523
Capaccioli, M., Caon, N., & D'Onofrio, M. 1992, MNRAS, 259, 323
Cellone, S.A., Forte, J.C., & Geisler, D. 1994, ApJS, 93, 397
Conselice, C.J. 2002, ApJ, 573, L5
Davies, J.I., et al. 1988, MNRAS, 232, 239
D'Onofrio, M., Capaccioli, M., & Caon, N. 1994, MNRAS, 271, 523
Durrell, P. 1997, AJ, 113, 531
Faber, S.M., & Lin, D.M.C. 1983, ApJ, 266, L17
Faber, S.M., et al. 1997, AJ, 114, 1771
Graham, A.W. 2002, ApJ, 568, L13
Graham, A.W., & Colless, M. 1997, MNRAS, 287, 221
Graham, A.W., & Guzman, R. 2003, AJ, 125, 2936
James, P. 1991, MNRAS, 250, 544
Jerjen, H., & Binggeli, B. 1997, in The Nature of Elliptical Galaxies; The Second Stromlo Symposium, ASP Conf. Ser., 116, 239
Kormendy, J. 1977, ApJ, 218, 333
Kormendy, J. 1985, ApJ, 295, 73
Kormendy, J., & Djorgovski, S. 1989, ARA&A, 27, 235
La Barbera, F., et al. 2003, ApJ, 595, 127
Michard, R. 1985, A&AS, 59, 205
Sandage, A., & Binggeli, B. 1984, AJ, 89, 919
Schombert, J.M. 1986, ApJS, 60, 603
Sersic, J.L. 1968, Atlas de galaxias australes
Stiavelli, M., et al. 2001, AJ, 121, 1385
Tonry J., Blakeslee J.P., Ajhar E.A., Dressler A. 1997, ApJ, 475, 399
Wirth, A., & Gallagher, J.S. 1984, ApJ, 282, 85
Young, C.K., & Currie, M.J. 1994, MNRAS, 268, L11

Comments from J. Kormendy: I continue strongly to believe that spheroidal (Sph) galaxies like Draco and NGC 205 are distinct from the well-defined sequence of elliptical galaxies that extends from M32 roughly to M87. Physical properties and formation physics both are different. I believe that Sphs are physically not small ellipticals and so should not be called dwarf ellipticals. Instead, their properties and formation are related to those of late-type spiral and Irr galaxies. Here are three reasons among many:
(1) In the (almost-)central parameter correlation diagrams, it requires very high spatial resolutioni to study objects in the E-Sph luminosity overlap region. Only a few M32 analogues in Virgo have been measured; they show that M32 is not peculiar but rather connected with giant ellipticals in a continuous fundamental plane sequence. Sph galaxies are not easy to measure; nuclei and faint bulges should be photometrically decomposed from the Sph main body; available resolution may be too low to allow this, and then the Sph parameters will inevitably be incorrect and similar to those of ellipticals. When we plot only those galaxies for which parameters can securely be measured, the Sph and E sequences look quite distinct.
(2) Nevertheless, a few galaxies will partly fill the gap between the E and Sph sequences. However, they are not numerous. The luminosity function of ellipticals peaks in the middle of the sequence; fainter Es are increasingly rare, and M32s are very rare indeed. Similarly, Sph galaxies have luminosity functions that decline steeply toward higher luminosities; faint ones are very common, but ones brighter than M32 are rare. This difference in luminosity functions was used even by Sandage and collaborators to argue* that E and Sph galaxies (which they call dEs) are distinct.
(3) Formation mechanisms of ellipticals and Sphs are clearly different. Ellipticals formed by major merges. The dSph companions of our Galaxy mostly have heterogeneous stellar populations including intermediate-age stars. Five to seven Gyrs ago, about half of these dSphs had not yet had their last burst of star formation and presumably were still dI galaxies. And indeed, Sph and S+Irr galaxies have very similar fundamental plane correlations. My conclusion – and that of many other authors – is that Sph galaxies are late-type galaxies transformed by processes such as starbursts, gas ejection driven by supernova energy feedback, ram pressures tripping, and harassment. So spheroidal and elliptical galaxies look very different, both in their parameter correlations and their formation physics.
* Their work was published in the ESO/OHP workshopon Dwarf Galaxies and in their AJ paper on luminosity functions that was part of their Virgo cluster survey series.

PHOTODISSOCIATION AND THE MORPHOLOGY OF H I IN GALAXIES

Ronald J. Allen
Space Telescope Science Institute
3700 San Martin Drive, Baltimore MD 21218, USA

Abstract Young massive stars produce Far-UV photons which dissociate the molecular gas on the surfaces of their parent molecular clouds. Of the many dissociation products which result from this "back-reaction", atomic hydrogen H I is one of the easiest to observe through its radio 21-cm hyperfine line emission. In this paper I first review the physics of this process and describe a simplified model which has been developed to permit an approximate computation of the column density of photodissociated H I which appears on the surfaces of molecular clouds. I then review several features of the H I morphology of galaxies on a variety of length scales and describe how photodissociation might account for some of these observations. Finally, I discuss several consequences which follow if this view of the origin of HI in galaxies continues to be successful.

Keywords: galaxies: ISM – ISM: clouds – ISM: molecules – ISM: atomic hydrogen – ISM: photodissociation

1. Introduction

I want to begin this talk with a brief discussion of some aspects of the astrophysics of photodissociation regions (PDRs) in order to make it clear that the production of H I from H_2 in the ISM is quite inevitable, and that the cycling of H I $\leftrightarrow$ H_2 is likely to be both continuous and ubiquitous in galaxies. I will then move on to a brief description of some of the features of the H I morphology of galaxies which appear to be amenable to an explanation in terms of PDR astrophysics.

The material presented here is a review of results and views published by myself and by others in previous journal and conference papers. New in this paper are some preliminary results on the effects of radial gradients in the metallicity of the ISM on the overall H I distribution in galaxies in the context of the photodissociation picture described here.

D. Block et al. (eds.), Penetrating Bars through Masks of Cosmic Dust, 731–748.

2. H I from photodissociation of H_2 in the ISM

The physics and astronomy of the photodissociation $\leftrightarrow$ reformation process for molecules in the ISM is a subject of active research, and an excellent review of the field was published a few years ago by Hollenbach & Tielens(1999). The results have been successfully applied to star-forming regions in the Galaxy on length scales of typically 0.1 - 1 pc. The first indication that photodissociation may be operating to affect the *large-scale* morphology (100-1000 pc) of the H I in galaxies was found in M83 by Allen, Atherton & Tilanus(1986), who noticed a clear spatial separation between a particularly well-defined dust lane several kiloparsec long and the associated ridge of H I and H II in the southern spiral M 83. Appropriately for this conference, M 83 is a barred spiral galaxy, and the strong stellar density wave which is apparently driven by the bar has produced large streaming velocities in the gas across the arms, leading to a measureable separation of various phases of the ISM.

Dissociation of H_2 by far-UV photons

It requires a photon of energy ≥ 14.67 eV to lift an H_2 molecule directly from its ground state to the electronic continuum. Such photons will be rare in the ISM since they are strongly absorbed in ionizing H I to H II. For this reason it was initially thought (Spitzer(1948)) that H_2 would be very long-lived in the ISM. However, closer examination of the electronic and vibrational energy-level diagram for H_2 revealed another dissociation channel. In this "fluorescence" process, less energetic photons can raise the H_2 molecule to an excited electronic state. When the molecule decays to the ground electronic state, it can end up in a variety of excited vibrational states. However, vibrational levels above 14 are so energetic that they break the chemical bond holding the H_2 molecule together, and two H I atoms are produced. A quantitative description of this process was first given by Stecher & Williams(1967). It starts when H_2 molecules absorb photons primarily at wavelengths of $\lambda \approx 110.8$ and ≈ 100.8 nm through transitions to the to the electronically-excited Lyman ($X^1\Sigma_g^+ \rightarrow B^1\Sigma_u^+$) and Werner ($X^1\Sigma_g^+ \rightarrow C^1\Pi_u^+$) bands. In the subsequent decay to various vibrationally-excited levels of the ground electronic state, $\sim 10-15\%$ of the H_2 molecules will dissociate into two H I atoms. Considering that even higher electronic states exist in H_2, it is clear that the FUV spectrum over the whole range from 91.2–110.8 nm (13.6–11.2 eV) contributes to the dissociation. In high-UV-flux environments ($G_0 \gtrsim 10^4$, Shull(1978)), photons with wavelengths as long as ≈ 185 nm (≈ 6.6 eV) can continue to create H I by dissociating "pumped" H_2 ($X^1\Sigma_g^+, 2 < v \leq 14$) via additional Lyman- and Werner-band transitions. Verification that this process actually occurs in the ISM was obtained when the predicted UV fluorescence spectrum was first observed by Witt et al.(1989) in the Galactic nebula IC 63.

Using the notation of Sternberg(1988), the rate at which H_2 is dissociated by this process in the ISM per unit volume can be written as:

$$R_{diss} = D\chi \times e^{-\tau_{gr,1000}} \times f_s(N_2) \times n_2, \tag{1}$$

in units of H_2 molecules dissociated cm^{-3} sec^{-1}, where $n_2 = n(H_2)$, and

D	=	the unattenuated H_2 photodissociation rate in the average ISRF,
χ	=	the incident UV intensity scaling factor,
σ	=	the effective grain absorption cross section per H nucleus in the FUV continuum,
$\tau_{gr,1000}$	=	$\sigma(N_1 + 2N_2)$ is the dust grain opacity at $\lambda \sim 100$ nm, and
n	=	$n_1 + 2n_2$ the volume density of H nuclei.

Since $D = 5.43 \times 10^{-11}$ s^{-1} (according to the simplified 3-level model, Sternberg(1988)) the time scale for this process on the surface of a typical GMC ($n_2 \sim 50$ molecules cm^{-3}) illuminated by the ISRF is $\tau_{diss} = (Dn_2)^{-1} \approx 10$ yr! Note that each dissociation produces two H I atoms on the cloud surface, and the appearance of a layer of H I when a FUV radiation field is "switched on" (e.g. from the birth of a new O–B star) is instantaneous compared to most other time scales in the ISM.

Formation of H_2 on dust grains

Formation of H_2 occurs in the ISM most efficiently on dust grains (cf. e.g. Hollenbach & Tielens(1999) and references cited there). The model for the formation rate depends on several parameters (some of which are not accurately known), as well as on the nature of the dust grains (which may vary from place to place in a galaxy). The usual parametrization is:

$$R_{form} = \gamma_2 \times n \times n_1 \tag{2}$$

in units of H_2 molecules cm^{-3} sec^{-1}, and $n = n_1 + 2n_2$. The rate coefficient for unit density, solar metallicity, and 100K kinetic temperature (roughly the ISM in the solar neighborhood) is $\sim 1 - 3 \times 10^{-17}$ cm^3 sec^{-1}. This equation is strongly dependent upon the dust–to–gas ratio ($\delta = A_V/N_H$) and weakly dependent upon the gas temperature (T), since

$$\gamma_2 = 3.0 \times 10^{-18}(\delta/\delta_0)T^{1/2}y_F(T) \text{ cm}^3 \text{ s}^{-1}, \tag{3}$$

where δ_0 refers to the value of δ in the solar neighborhood, and $y_F(T)$ represents the efficiency of H_2 formation. The product $T^{1/2}y_F(T)$ is thought to be constant to within a factor of 2 (Hollenbach et al.(1971)).

The time scale for this process is $\tau_{form} = (2n\gamma_2)^{-1} \approx 5 \times 10^8/n$ yr, and this will be the rate-determining time scale in an equilibrium situation where photodissociation is balanced by reformation on dust grains.

Equilibrium

In recent years, much effort has gone into calculating the level populations of the ro-vibrational lines of the ground state of H_2, and the intensities of the associated quadrupole line emission. These lines can be observed with space missions (e.g. SWAS, ISO) in PDRs in the Galaxy and in the nuclear regions of other galaxies. The H I column density in a PDR is calculated with the same physics used to determine the excitation of the H_2 near-infrared fluorescence lines. While the computations for those lines are rather complicated, the determination of H I column can be obtained from a simplified version of the model. Furthermore the 21-cm line emission from H I is almost always optically thin, so the observations usually yield the H I column directly for comparison with the model. The formation and destruction rates described above are set equal, and the equation solved, in the present case for the H I column density. A logarithmic form for the analytic solution to this equation was first given by Sternberg(1988); see also Appendix A of this paper for a brief derivation. Other relevant references are given in Allen et al.(2004). The model is a simple semi-infinite slab geometry in statistical equilibrium with FUV radiation incident on one side. The solution gives the steady state H I column density along a line of sight perpendicular to the face of the slab as a function of χ, the incident UV intensity scaling factor, and the total volume density n of H nuclei. Sternberg's result is:

$$N(\mathrm{H\,I}) = \frac{1}{\sigma} \times \ln\left[\frac{D\mathcal{G}}{\gamma_2 n}\chi + 1\right], \qquad (4)$$

D	=	the unattenuated H_2 photodissociation rate in the average ISRF,
γ_2	=	the H_2 formation rate coefficient on grain surfaces,
σ	=	the effective grain absorption cross section per H nucleus in the FUV continuum,
χ	=	the incident UV intensity scaling factor,
$N(\mathrm{H\,I})$	=	the H I column density,
n	=	the volume density of H nuclei.

Equation 4 has been developed using a simplified three-level model for the excitation of the H_2 molecule and is applicable for low-density ($n \lesssim 10^4$ cm^{-3}), cold (T $\lesssim$ 500 K), isothermal, and static conditions, and neglects contributions to $N(\mathrm{H\,I})$ from ion chemistry and direct dissociation by cosmic rays. The quantity $\mathcal{G}$ here (not to be confused with G_0 to be defined momentarily) is a dimensionless function of the effective grain absorption cross section σ, the absorption self–shielding function $f_s(N_2)$, and the column density of molecular hydrogen N_2:

$$\mathcal{G} = \int_0^{N_2} \sigma f_s e^{-2\sigma N_2'} \mathrm{d}N_2'.$$

The function $\mathcal{G}$ becomes a constant for large values of N_2 due to self–shielding (Sternberg(1988)). Using the parameter values in this equation adopted by Madden et al.(1993), we have:

$$N(\mathrm{H\,I}) = 5 \times 10^{20} \times \ln[1 + (90\chi/n)]$$

where n is in cm^{-3}. This is a steady state model, with H_2 continually forming from H I on dust grain surfaces, and H I continually forming from H_2 by photodissociation.

Equation 4 is strongly dependent upon the dust–to–gas ratio ($\delta = A_V/N_H$) and weakly dependent upon the gas temperature (T), since

$$\begin{aligned}
\sigma &= 1.883 \times 10^{-21}(\delta/\delta_0)\ \mathrm{cm}^{-2}, \\
\gamma_2 &= 3.0 \times 10^{-18}(\delta/\delta_0)T^{1/2}y_F(T)\ \mathrm{cm}^3\ \mathrm{s}^{-1}, \\
\mathcal{G} &= (\sigma/\sigma_0)^{1/2}\mathcal{G}_O,
\end{aligned}$$

where δ_0, σ_0, and $\mathcal{G}_O$ refer to values in the solar neighborhood, and $y_F(T)$ represents the efficiency of H_2 formation. The product $T^{1/2}y_F(T)$ is thought to be constant to within a factor of 2 (Hollenbach et al.(1971)).

Equation 4 also contains a dependence on the level of obscuration in the immediate vicinity of the FUV source. While the variable χ represents the intrinsic FUV flux associated with the star-forming region, we generally observe an attenuated FUV flux. Assuming any extinction associated with the star-forming region is in the form of an overlying screen of optical depth $\tau(FUV)$, $\chi(\mathrm{observed}) = \chi e^{-\tau(FUV)}$. However, an accurate correction for this effect may be difficult, Since the ISM in the immediate vicinity of the FUV source will have been disturbed by stellar winds and any prior supernovae.

Assuming solar neighborhood values of $\sigma_0 = 1.883 \times 10^{-21}$ cm^2, $D = 5.43 \times 10^{-11}$ s^{-1}, $\gamma_{20} = 3 \times 10^{-17}$ cm^3 s^{-1}, and $\mathcal{G}_O \approx 5 \times 10^{-5}$ (Sternberg(1988)), and neglecting the weak temperature dependence of γ_2, equation 4 becomes:

$$N(\mathrm{H\,I}) = \frac{5 \times 10^{20}}{(\delta/\delta_0)} \ln \left[\frac{90\chi}{n} \left(\frac{\delta}{\delta_0} \right)^{-1/2} + 1 \right], \tag{5}$$

where $\chi = \chi(\mathrm{observed})e^{\tau(FUV)}$. The behavior of $N(\mathrm{H\,I})$ as a function of χ is displayed for $\delta/\delta_0 = 0.2$, $\tau(FUV) = 0$ and $n = 30$, $n = 300$, and $n = 3000$ in Figure 1. These values of δ/δ_0 and $\tau(FUV)$ are appropriate for the outer regions of M101, as discussed in Smith et al.(2000). Values of $N(\mathrm{H\,I}) \gtrsim 10^{22}$ cm^{-2} are not likely to be observed as the atomic gas probably becomes optically thick at this point:

$$N(\mathrm{H\,I}) \quad = \quad 1.82 \times 10^{18} \int_{-\infty}^{\infty} T_s \tau(v) dv$$

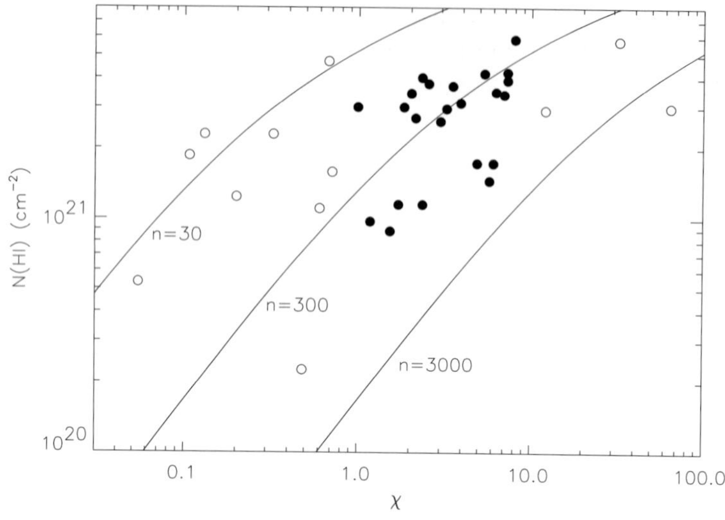

Figure 1. The Relationship between $N(\mathrm{H\,I})$ and χ. The observed and predicted values of $N(\mathrm{H\,I})$ are shown as a function of χ. The modeled behavior of $N(\mathrm{H\,I})$ assumes values of $\delta/\delta_0 = 0.2$ and $\tau(FUV) = 0$, appropriate for the outer regions of M101. The observations are clearly consistent with the physics underlying the photodissociation picture.

$$\begin{aligned} &\rightarrow \quad 1.82 \times 10^{18} T_s \tau \Delta v \\ &\approx \quad 10^{22} \mathrm{cm}^{-2}, \end{aligned}$$

for spin temperatures of $T_s \approx 100$K and profile FWHMs of $\Delta v \approx 20$ km s^{-1} typical of M101 (Braun(1997)), and for optical depths of $\tau \approx 2.5$, corresponding to a ratio between the brightness and kinetic temperatures of $T_B/T_K \approx 0.9$. This value is appropriate for the highest-brightness regions of M101, as indicated in Figure 8a of Braun(1997). Figure 1 also shows the measurements for each of the 35 candidate PDRs in M101 as analysed by Smith et al.(2000). The data indicate that the properties of observed regions in M101 are consistent with photodissociation of an underlying molecular gas of moderate volume density.

Allen et al.(2004) have recently re-examined equation 5 and compared it with the full numerical treatment used in the "standard" Ames model summarized in Kaufman et al.(1999). A conversion of the FUV flux χ used by Sternberg(1988) to the quantity G_0 used by Kaufman et al.(1999) is first required; this is because Sternberg(1988) and Kaufman et al.(1999) use different normalisations for the FUV flux (see Appendix B in Allen et al.(2004)). When distributed sources illuminate an FUV-opaque PDR over 2π sr, the conversion is $\chi = G_0/0.85$ (see Footnote 7 in Hollenbach & Tielens(1999)), resulting in

$90\chi/n = 106G_0/n$. With this change, Allen et al.(2004) fitted the analytic expression for $N(\mathrm{H\,I})$ above to the model computations in the range in which cosmic ray dissociation is not a major contributor, roughly for $G_0 \gtrsim 1$, $n \gtrsim 10$ cm^{-3}. The result is that no consistent improvement is obtained by using any value for the coefficient of G_0 other than the value 106 deduced above, although a modest improvement is obtained by using a slightly larger value for the leading coefficient in the equation, 7.8×10^{20}, corresponding to a value of 1.3×10^{-21} cm^2 for the effective grain absorption cross section. With these small adjustments, the final best-fit equation is:

$$N(\mathrm{H\,I}) = \frac{7.8 \times 10^{20}}{(\delta/\delta_0)} \ln\left[\frac{106G_0}{n}\left(\frac{\delta}{\delta_0}\right)^{-1/2} + 1\right] \mathrm{cm}^{-2}, \qquad (6)$$

where $n = n(HI) + 2n(H_2)$ and G_0 and δ_0 are the (normalized) FUV flux and dust/gas ratio in the ISM of the solar neighborhood. In Figure 2 we show values

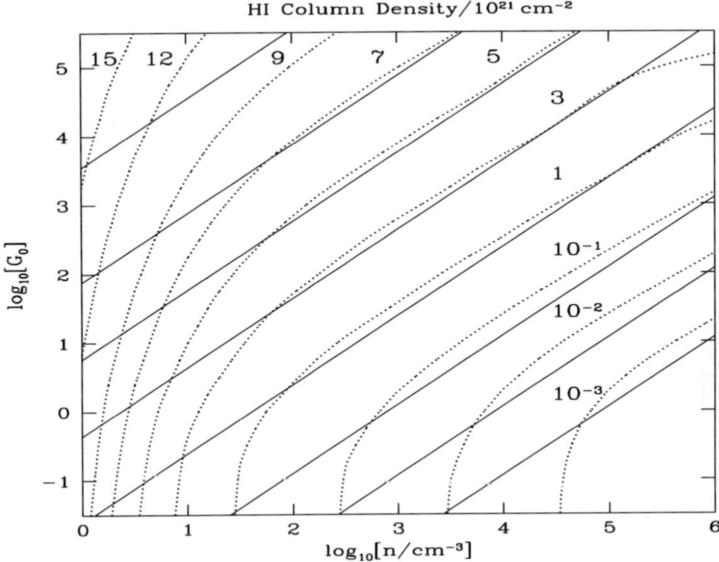

Figure 2. Contours of constant H I column density $N(\mathrm{H\,I})$ in units of 10^{21} cm^{-2} in the PDR as a function of the density n and incident FUV flux G_0 for the standard model parameters in Kaufman et al.(1999) (*dotted lines*), and for the analytic approximation of eq. 6 (*solid lines*). The labeled contour values are for the numerical model; the contours for the analytic model start from 10^{-3} (lower right corner) and increase to 10^{-2}, 10^{-1}, 1, 3, 5, 7, and end at 10 (upper left corner).

from equation 6 plotted as solid lines together with dotted contour lines from the "standard" numerical model for $\delta = \delta_0$. The agreement is generally good over much of the n–G_0 parameter space of interest here; differences occur mainly in the top left corner of the diagram, and at low values of FUV flux. In the top left corner the analytic formula under-predicts the amount of H I column density computed from the standard model by about 30% owing to H I production by ion chemistry reactions such as $H_2^+ + H_2 \rightarrow H_3^+ + H$, $HCO^+ + e^- \rightarrow CO + H$, and $PAH^- + H^+ \rightarrow PAH + H$ (where PAH is a polycyclic aromatic hydrocarbon), which are important at high G_0 and low n. At values of $G_0 \lesssim 1$ the contours of $N(\mathrm{H\,I})$ for the numerical model become vertical; this is because the standard model includes a low level of cosmic ray ionization which contributes a small amount of H I by dissociation of H_2 even for $G_0 = 0$.

It should be noted that both the analytic and the numerical forms depend on several rather crude parameters used to describe the properties of the gas – dust mixture in the ISM (dust cross section, H I sticking coefficients, etc.), and that not all proponents of PDR models use the same values for these parameters. A coordinated effort is presently taking place among the modellers, first to see if different PDR codes can produce the same results when using the same numerical values for parameters (no, they can differ, and sometimes by a lot!), and second, to try to reach some agreement on an acceptable set of values for these parameters.

There are several noteworthy aspects of this equation:

- $N(\mathrm{H\,I})$ depends only on the *ratio* of G_0/n. Low FUV flux, low density environments in galaxies can produce the same column of H I found in high flux, high density environments. The difference will be in the thickness of the H I layer; much thicker layers of H I are associated with the low flux, low density environments;
- at a given n, $N(\mathrm{H\,I})$ increases first linearly with G_0, but then "saturates" and increases only *logarithmically* after that;
- at a given G_0, $N(\mathrm{H\,I})$ *decreases* logarithmically with increasing n, and;
- the H I column *decreases* as the dust/gas ratio increases.

3. Time scales

As discussed in the previous section, the relevant time scale for the production of H I on the surfaces of molecular clouds is the formation time for H_2 on dust grains. The full expression including the dust/gas dependence is:

$$\begin{aligned} \tau_2 &= (2n\gamma_2)^{-1} \\ &= \frac{5 \times 10^8}{(n_1 + 2n_2) \times (\delta/\delta_0)} \text{ yrs.} \end{aligned} \qquad (7)$$

Table 1 gives typical values for τ_2 in several different environments in the ISM, and Table 2 lists a number of time scales set by other processes in galaxy disks.

Table 1. H I production time scales in typical ISM environments.

Environment	n	δ/δ_0	τ (years)
Solar metallicity GMC	100	1	$\sim 3 \times 10^6$
Intercloud gas	10	1	$\sim 3 \times 10^7$
Outer galaxy GMC	100	0.1	$\sim 3 \times 10^7$

Table 2. Other time scales in galaxies.

Situation	Time scale (years)
GMC crossing time $R/\Delta V$	$\sim 7 \times 10^6$
Spiral arm crossing time	$\sim 5 \times 10^7$
B3 star lifetime for FUV production	$\sim 5 \times 10^7$
Galaxy rotation time	$\sim 5 \times 10^8$
Hubble time	$\sim 10^{10}$

It is clear that the time scale for H I production by photodissociation of H_2 is short enough to be relevant. For instance, H I is produced in the same time scales as that of molecular cloud formation (Pringle, Allen & Lubow(2001)) and of the star formation in those molecular clouds (Elmegreen(2000)), and in a tenth of the lifetime of a typical B star for FUV photon producation. This must be the reason why we see H I in regions where young stars form, and why H I appears in "rims" around clusters of B stars in spiral arms. H I also appears in a time short compared to the time it takes for a GMC to cross a spiral arm, and the H I will correspondingly disappear as the FUV production ceases further down stream and the gas reverts to a predominantly molecular form. Even in the far outer parts of galaxies we see that the H I time scale is still only about 10% of the rotation time of the galaxy, so for undisturbed galaxies we can expect an equilibrium to obtain between H I and H_2 even in these sparse environments.

4. Major features in the H I morphology of galaxies

Our picture of the major features of the H I morphology in disk galaxies has closely followed the steady improvements in angular resolution of centimeter-wave radio telescopes, first with filled apertures (Dwingeloo, NRAO 300', Parkes, Arecibo, GBT...), and later with synthesis imaging instruments (Westerbork, VLA, ATNF...).

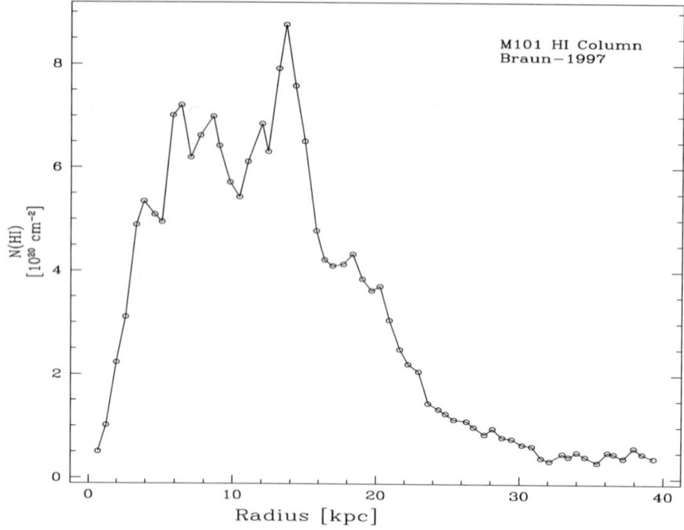

Figure 3. Radial distribution of H I surface brightness for the nearby giant Sc galaxy NGC5457 = M 101 obtained by averaging the H I data in annular elliptical rings. From Braun(1997), adjusted to an assumed distance of 5.4 Mpc. R_{25} for this galaxy is $13.5' \approx 21$ kpc.

Radial distribution of H I surface brightness

A typical averaged H I radial surface brightness profile of a nearby giant Sc galaxy is shown in Figure 3. The main features to notice are:

- the central depression;
- the "flat top", typically at a level of $5 - 10 \times 10^{20}$ cm^{-2}, and;
- the long "tail" to faint levels in the distant outer parts.

What parts of this plot can be explained by the photodissociation picture descibed in the previous section? An explanation for the "flat top" was actually suggested in terms of the photodissociation picture nearly 20 years ago by Shaya & Federman(1987), but I fear that paper has been widely ignored by most workers in the field of extragalactic H I and H_2. A good look at equation 6 makes it clear: owing to the *logarithmic* dependence of N(H I) on the *ratio* G_0/n and the fact that the largest concentrations of young stars also go along with regions of highest gas density, we ought not to expect values of H I column density much in excess of a few times the coefficient in front of the log. So even in the most active starbursting regions of galaxies, we are not likely to observe values of the H I column density much in excess of $\sim few \times 10^{21}$

cm^{-2}. As to the decline in the outer regions, this is likely to be a combined effect of a declining ambient FUV flux and a declining area filling factor, since Smith et al.(2000) have shown that the total gas density n on the surfaces of GMCs located in the neighborhood of massive young stars does not appear to change much with radius.

Finally, the depression in the inner parts can be explained as a consequence of the general increase in the metallicity of the ISM in the inner parts of galaxy disks. In Figure 4 my student colleague Ben Waghorn has fitted equation 6 to the combined radial data on the FUV distribution and the metallicity gradient in M 101. The free parameters are the H I area filling factor (taken here to be 0.3 everywhere), and the GMC gas volume density n (fits shown for 100, 200, and 300 cm^{-3}). We see that, in spite of the strong increase in FUV flux G_0 in the inner parts of the galaxy, the rapid rise of the dust/gas ratio (assumed proportional to the O/H ratio) actually results in a *decrease* in N(H I), as observed, and the quantitative fit is also reasonable. I note here that this result has already been described by Smith et al.(2000) for a small subset of young star clusters in M 101 accounting for only a few percent of the total H I content of the galaxy; what we are now seeing is that the same explanation appears viable for *all* the H I in the galaxy. We have examined nearly a dozen nearby spirals in this way, and find reasonable agreement for about half of them. The other half show an indication that there is more H I present than FUV-related photodissociation can explain. Interestingly, these galaxies nearly all have bright nonthermal radio continuum disks, suggesting that there is a component of the H I being maintained from dissociation by cosmic rays penetrating throughout the GMCs. This work is ongoing.

Spiral structure

In the Introduction to this paper I pointed out that it was thanks to a strong density wave in the southern barred spiral M 83 that the importance of photodissociation in affecting the morphology of galaxies on the large scale was first unmasked. Other studies have followed on M83 and on other galaxies (M 51, M 100; see Smith et al.(2000) for references) and have generally agreed that the initial interpretation in terms of photodissociation remains the most viable option. The separation in the case of M 83 is about 250 pc, and arises because of the difference between the spiral pattern speed and the rotation speed of the gas, coupled with the time for collapse of GMCs and the time that a massive young star lives on the main sequence.

The clear existence of spiral features in the H I distribution of a galaxy external to our own was perhaps first convincingly demonstrated for M 101 by Allen, Goss, & Van Woerden(1973). Although this galaxy apparently does not have a very strong density wave, the H I is arranged in thin spiral segments

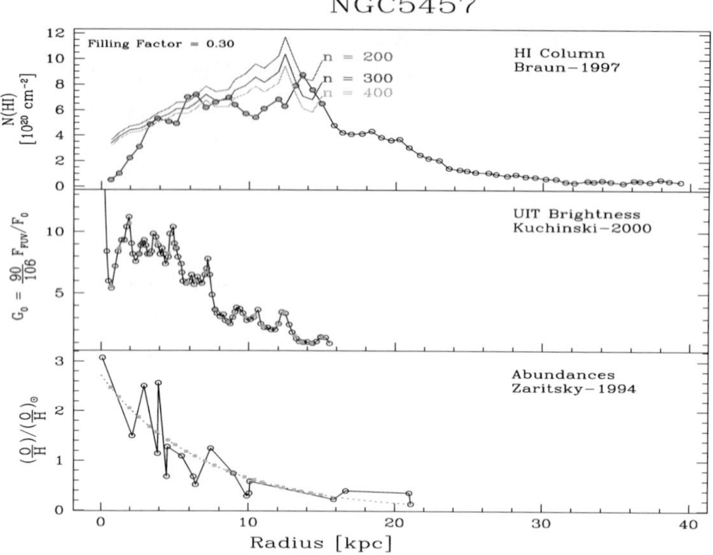

Figure 4. H I column density (top), FUV brightness (middle), and O/H ratio (lower) as a function of radius for M 101. Equation 6 has been used to produce the fits shown in the top panel for an area filling factor of 0.3 and $n = 100, 200, \& \ 300 \text{ cm}^{-3}$. From work in progress by Allen, Waghorn, & Heiner.

which appear in the inner disk and can be traced over a large part of the main body of the galaxy right out to beyond R_{25}. Figure 5 shows the H I image (grey, kindly provided in digital form by R. Braun) with the FUV contours superposed (Smith et al.(2000)). The correspondence is excellent, at least as far out in the disk as the UIT data extend. We expect to see the FUV image and spiral features grow further when the GALEX data become available later this year, and we can confidently predict that the close correspondence with the H I will continue.

H I arcs and blisters

The first study to successfully identify the characteristic PDR "arc" or "blanket" morphology of H I in close association with far-UV sources in a nearby galaxy was carried out on M81 by Allen et al.(1997). The problem is, of course, to obtain sufficient linear resolution ($\sim$ 100 pc) in the H I observations to permit one to identify the morphology of the PDR structures. An important point to note is that the best "correlation" is between the H I and the far-UV, and *not* between the H I and the Hα.

A study similar to that done on M81 but with more quantitative results has been carried out on M101 by Smith et al.(2000), who used VLA-H I and UIT

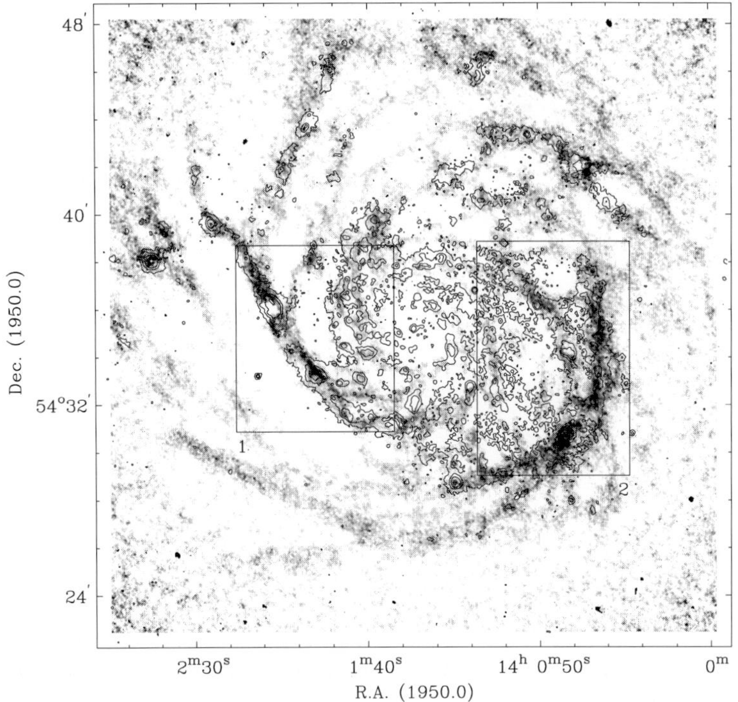

Figure 5. The distribution of atomic hydrogen as seen at a resolution of $6'' \approx 200$ pc in M 101 (Braun(1995)). Thin, prominent spiral arms wind outwards to beyond the optical disk at $R_{25} \approx 13.5'$, with the outer H I arms generally traceable back to arms which start in the main body of the disk. Contours of the Far-UV emission recorded by UIT are superposed. From the PDR analysis by Smith et al.(2000).

far-UV data to identify and measure PDRs over the whole extent of the M101 disk. From these observations they derived the volume density of the H_2 in the adjacent GMCs in the context of the PDR model. Figure 6 shows the best estimate of the H_2 volume densities of GMCs near a sample of 35 young star clusters. The range in density (30 - 1000 cm^{-3}) is typical for GMCs in our Galaxy, lending support to the use of the PDR picture, and also shows little trend with galactocentric distance.

It must be mentioned here that Braun(1997) has offered a different interpretation of the discrete H I-bright features, which he called the “High-Brightness Network” and identified with the “Cold Neutral Medium” phase of the two-phase model for the ISM (Field, Goldsmith, & Habing(1969), see also Wolfire et al.(1995) for a more recent discussion). However, in a recent paper, Wolfire et al.(2003) favor the interpretation of Smith et al.(2000) in terms of PDR-generated H I .

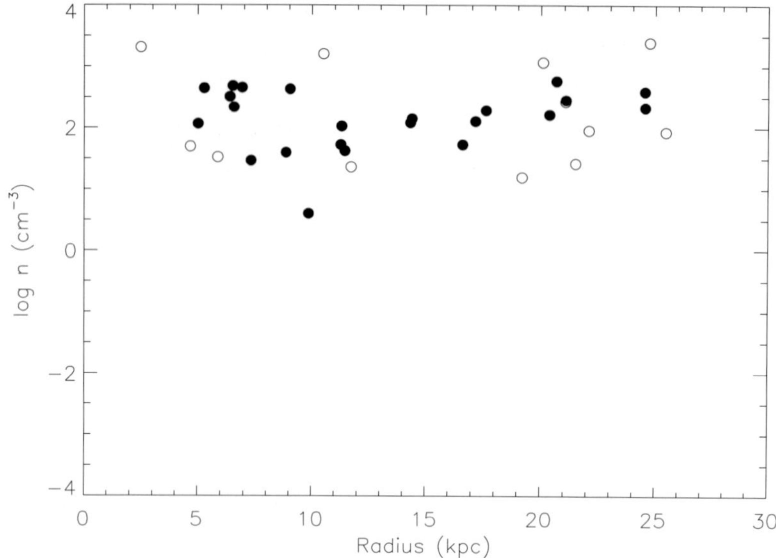

Figure 6. Total gas volume density in GMCs near a sample of 35 young star clusters in M101. This is all H_2 deep within the cloud. See Figure 19 in Smith et al.(2000) for further details.

There is other, IR spectral evidence that PDRs are important for understanding the physics of the ISM in galaxy disks. KAO observations of the 158μm C II line suggest that as much as 70%-80% of the H I in NGC 6946 could be produced by photodissociation (Madden et al.(1993)), and ISO spectra in the mid-IR indicate that the bulk of the mid-IR emission from galaxy disks arises in PDRs (Laurent et al.(1999); Vigroux et al.(1999)).

5. Some implications

So, you say, let's suppose I am right in my view that ***H* I *is not the fuel for the star formation process in galaxies, but merely the smoke from it!*** Why does this matter? Well, the idea that H I directly and quantitatively tracks the main component of the ISM in galaxy disks is a basic tenet of our current view of star formation on the large scale in galaxies, and it may take a few moments to consider the alternatives and the consequences of "shifting the paradigm". I can think of at least 3 consequences at the moment:

- There must be significantly more gas present in galaxies in the form of "cold" H_2. The H I is showing us mainly only the surfaces of molecular clouds, and the PDR model by itself does not provide a prescription for how to go from the H I as a surface phenomena to the H_2 in the volume of the clouds. But we can confidently predict that *more* H_2 will be present than we currently think. H_2 in amounts from 2 - 5 times that of the known H I could probably be "hiding" in galaxy disks without clearly violating any known constraints.

- The far outer parts of galaxy disks, where small amounts of H I appear with only sparse star formation, are prime sites for "hiding" H_2. A focussed effort to find such gas ought to be made. Possibilities for detection may include measurements of dust opacity (it is likely to be too cold to emit any appreciable amounts of Far-IR continuum or line emission, but this needs to be considered carefully[1]), and molecular absorption lines. The anomalous absorption of the Cosmic Microwave Background by the 6 and 2-cm lines of formaldehyde H_2CO are intriguing possibilities which ought to be explored further.

- We are interpreting the "Schmidt Law" for global star formation *backwards*. This is a particularly far-reaching consequence of the photodissociation picture favored here. The situation has been described by Allen(2002). Basically, we have inverted "cause"and "effect" for many years in this discussion, viewing the H I column as the cause, and the star formation rate (e.g. quantified by the FUV flux) as the effect. The observed relationship between these two quantities (roughly a power law on a log-log plot) is called the "Schmidt Law for Global Star Formation", and has been the basis for many attempts to develop a physical theory for large-scale star formation in galaxies involving gravitational instablility in the disk. Such a theory is still incomplete. On the other hand, the PDR picture favored here views the *FUV flux as the cause* and the *H I column as the effect*, and provides a simple explanation for the observed correlation in terms of physics we already know (see Figures 7a and 7b, from Allen(2002)).

6. Summary Remarks

To summarize the latest views on this topic, first a conclusion which has been corroborated by several authors and which by now seems quite solid:

- The H I spiral arms in the inner parts of "grand design" galaxies consist mostly (and perhaps even entirely) of photodissociated H_2.

To this I would add the following points established in papers by myself and my co-workers:

- As well as O stars, B stars born in spiral arms play a major role in this process;

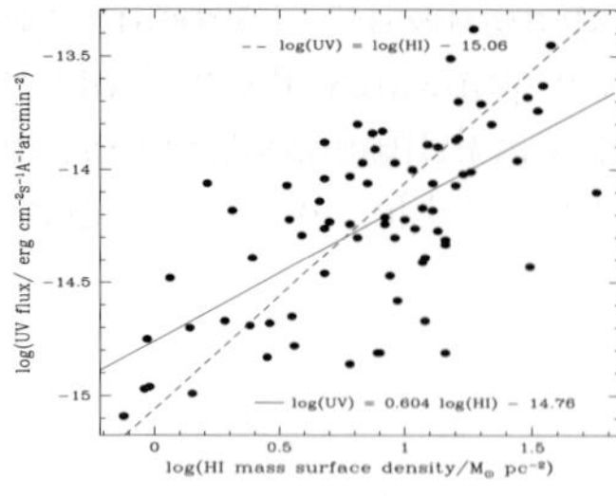

Figure 7a. Data showing a correlation between average observed 21-cm line surface brightness (converted to "H I mass surface density") on the X-axis and the Far-UV surface brightness (taken as a measure of the formation rate of massive stars) on the Y-axis, from Deharveng et al.(1994). "Schmidt Law" fits are shown, with indexes of 1 (dashed line) and 0.6 (solid line).

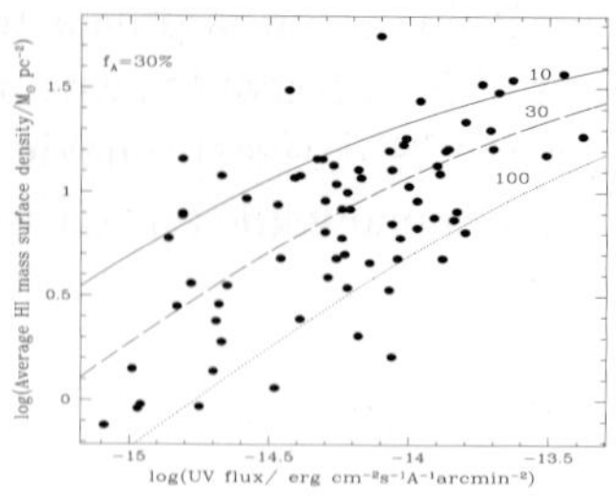

Figure 7b. The data plotted with axes inverted so as to emphasize the explanation in terms of photodissociation. The solid curves are the models of H I production in PDRs described briefly in the text, and are labelled with the proton volume densities of the parent GMCs. The H I area filling factor is assumed to be 0.30 over the disk of the galaxy.

- The PDR morphological signature is widespread in HI when enough linear resolution ($\lesssim$ 100 pc) is available, and;
- since there seems to be no reason to have more than one H I formation mechanism, I conclude that both the inner and the (far) outer H I arms in spirals are photodissociated H_2.

New results on the radial distributions of H I described in this review are tantalizing, but need further work:

- The general shape of the H I radial distributions in many spirals appears to be amenable to explanation in the context of a simple photodissociation model:
  - variations in H_2 density, FUV intensity, and dust/gas ratio can control the appearance of H I, and;
  - Additional H I may be produced by an elevated flux of cosmic rays in some galaxies.

Photodissociation of H_2 can explain a number of features in the morphology of H I in galaxies on scales from 100 pc to 10 kpc. Note that this process is bound to produce H I as long as H_2 and FUV photons are present; the real question we need to answer is "how much?", i.e. what fraction of the total H I content in a galaxy is cycling repeatedly through a molecular phase on time scales which are short compared to the dynamical evolution time of the galaxy? If this fraction proves to be large, then there must be an even larger reservoir of H_2 present to sustain it. Estimating the quantities of cool/cold H_2 hiding in the ISM of disk galaxies is pure guesswork at the moment, but factors of 2 - 5 times that in the form of H I may not be unreasonable, and with a spatial distribution as extended as the H I itself.

Acknowledgments

I am grateful to my colleagues at STScI for their contributions to the stimulating scientific environment we enjoy there, to Hal Heaton and Michael Kaufman for their collaboration on the models described here, and to David Block and others on the SOC for the opportunity to attend this meeting and to present my views on the subject of H I in galaxies.

Appendix: Derivation of the H I column density in equation 4

Equation 4 is obtained by setting the rate of formation of H_2 on grains equal to the rate of destruction through photodissociation by far-UV photons. Using the notation of §2:

$$\begin{aligned} R_{form} \times n \times n_1 &= R_{diss} \times n_2 \\ &= DG_0 \times e^{-\tau_{gr,1000}} \times f_s(N_2) \times n_2. \end{aligned} \tag{A.1}$$

For a simple 1D slab geometry at constant density $n = n_1 + 2n_2$, we can write $n_1 = dN_1/dx$, etc., so that the equilibrium equation becomes:

$$R_{form} \times n \times (dN_1/dx) = DG_0 \times e^{-\sigma(N_1+2N_2)} \times f_s(N_2) \times (dN_2/dx). \tag{A.2}$$

This can be written as a simple first-order separable differential equation in the column densities:

$$e^{+\sigma N_1} \times dN_1 = \frac{DG_0}{Rn} f_s(N_2) e^{-2\sigma N_2} \times dN_2. \tag{A.3}$$

This can be integrated through the entire slab on the half-plane $x \geq 0$ to give:

$$e^{+\sigma N_1}|_{x=\infty} - e^{+\sigma N_1}|_{x=0} = \frac{DG_0}{Rn} \int_0^\infty f_s(N_2) e^{-2\sigma N_2} \sigma dN_2. \tag{A.4}$$

The integral over N_2 on the RHS is just some number, call it $\mathcal{G}$, so that:

$$N_1 = \frac{1}{\sigma} \ln[1 + \frac{DG_0 \mathcal{G}}{Rn}]. \tag{A.5}$$

Notes

1. The paper by Boulanger at this meeting has made a start in this regard.

References

Allen, R.J. 2002, in *Seeing Through the Dust*, eds. A.R. Taylor, T.L. Landecker, & A.G. Willis (ASP Conference Series, Vol. 276), 288

Allen, R.J., Goss, W.M, & Van Woerden, H. 1973, A&A, 29, 447

Allen, R.J., Atherton, P. D., & Tilanus, R. P. J. 1986, Nature, 319, 296

Allen, R.J., Knapen, J. H., Bohlin, R., & Stecher, T. P. 1997, ApJ, 487, 171

Allen, R.J., Heaton, H.I., & Kaufman, M.J. 2004, ApJ, 608, 314.

Braun, R. 1995, A&AS, 114, 409

Braun, R. 1997, ApJ, 484, 637

Deharveng, J.-M., Sasseen, T.P., Buat, V., Bowyer, S., Lampton, M., & Wu, X. 1994, A&A, 289, 715

Elmegreen, B.G. 2000, ApJ, 530, 277

Field, G.B., Goldsmith, D.W., & Habing, H.J. 1969, ApJ, 155, L149

Hollenbach, D.J., Werner, M.W., & Salpeter, E.E. 1971, ApJ, 163, 165

Hollenbach, D.J., & Tielens, A.G.G.M. 1999, Revs. Mod. Phys. 71, 173

Kaufman, M.J., Wolfire, M.G., Hollenbach, D.J., & Luhman, M.L. 1999, ApJ, 527, 795

Laurent, O., Mirabel, I. F., Charmandaris, V., Gallais, P., Vigroux, L., & Cesarsky, C. J. 1999, in *The Universe as seen by ISO*, ed.P. Cox & M.F. Kessler (Noordwijk; ESA), 913

Madden, S.C., Geis, N., Genzel, R., Herrmann, F., Jackson, J., Poglitsch, A., Stacey, G.J., & Townes, C.H. 1993, ApJ, 407, 579

Pringle, J.E., Allen, R.J., & Lubow, S.H. 2001, MNRAS, 327, 663

Shaya, E.J., & Federman, S.R. 1987, ApJ, 319, 76

Shull, J.M. 1978, ApJ, 219, 877

Smith, D.A., Allen, R.J., Bohlin, R.C., Nicholson, N., & Stecher, T.P. 2000, ApJ, 538, 608

Spitzer, L., Jr. 1948, ApJ, 107, 6

Stecher, T.P., & Williams, D.A. 1967, ApJ, 149, L29

Sternberg, A. 1988, ApJ, 332, 400

Vigroux, L., Charmandaris, P., Gallais, P., Laurent, O., Madden, S., Mirabel, F., Roussel, H., Sauvage, M., & Tran, D. 1999, in *The Universe as seen by ISO*, ed.P. Cox & M.F. Kessler (Noordwijk; ESA), 805

Witt, A.N., Stecher, T.P., Boroson, T.A., & Bohlin, R.C. 1989, ApJ, 336, L21

Wolfire, M.G., Hollenbach, D., McKee, C.F., Tielens, A.G.G.M., & Bakes, E.L.O. 1995, ApJ, 443, 152

Wolfire, M.G., McKee, C.F., Hollenbach, D., & Tielens, A.G.G.M. 2003, ApJ, 587, 278

THE TRUE H_2 CONTENT OF SPIRAL GALAXIES

Francois Boulanger
Institut d'Astrophysique Spatiale, Bâtiment 121, Universite Paris XI, 91405 Orsay, France

Abstract Most of the gas in spirals could be lying beyond the optical disk and escape detection through CO observations for being metal poor and cold. We present three perspectives on this hypothesis: (1) Observations of molecular clouds in the SMC which illustrate the limits of CO at tracing molecular gas with low metallicity. (2) Dust observations suggesting the presence of a molecular gas mass beyond the solar circle at least as large as that seen in H I. (3) Evidence for the ubiquitous presence of warm H_2 gas, possibly heated by the localized dissipation of turbulent ISM energy, which can be observed with the Spitzer satellite and could, when better understood, become a tracer of low metallicity molecular gas away from star forming regions.

Keywords: Interstellar gas, Spirals, Dust, H2, metallicity

1. Introduction

The molecular gas content of spirals is poorly determined by existing observations. While atomic gas is directly measured from the 21 cm line, H_2 is inferred from CO emission through a poorly understood conversion factor depending on metallicity, the gas clumpiness and temperature. The H_2 mass to CO luminosity ratio has been empirically determined from observations of the inner, active star forming, parts of the Milky Way and some nearby galaxies. This ratio is not expected to be universal and the amount of H_2 gas in galaxies could be much larger than what is presently inferred from CO observations. In particular, gas traced by the 21 cm line in the outer part of spirals could be tracing photo-dissociated surfaces of clouds which are mostly molecular, although they are not detected in CO (Allen in this volume).

The possible existence of large amounts of hidden interstellar matter in the outer parts of galaxies first emerged from studies of the H I gas dynamics by Bosma (1981) who noted that the surface density of matter needed to explain the flat rotation curves outside the optical disk was roughly proportional to the surface density of atomic gas. This early result has been since confirmed over a larger sample of galaxies. For a set of 24 spirals, Hoekstra et al. 2001 obtained good fits of the H I rotation curves with a mass model in which the

D. Block et al. (eds.), Penetrating Bars through Masks of Cosmic Dust, 749–758.

dark matter surface density is scaled from the observed H I surface density. For most galaxies, the scaling factors are found to cluster in a remarkably narrow range around 10. The gas surface density in H I extended disks could thus be one order magnitude larger than that inferred from the 21 cm line. Hereinafter we refer to this possibility as the H_2 hypothesis.

Pfenniger, Combes and Martinet (1994) were the first to investigate this hypothesis within the topic of this conference: the understanding of galaxy evolution and the Hubble sequence. Obviously, the existence of a large reservoir of gas in the outer parts of spirals would have a major impact on our understanding of star formation and the dynamical and chemical evolution of galaxies. Taking the Milky Way as example, the overall gas mass will be comparable to that of the stars in the disk, but most of the gas will be in the outer parts, not directly associated with on-going star formation. This reservoir could be the main source of the gas flow required to account for the distribution of stellar metallicities with age (Beckman, this volume), to fuel star formation and to reform bars (Combes, this volume).

From the interstellar medium perspective, the obvious questions are how is this gas hidden and how could it be seen? Pfenniger and Combes (1994) proposed that interstellar matter in regions of low UV radiation field, i.e. low star formation activity, could have a fractal distribution with a dimension small enough for most of the mass to be hidden in dense (n $\sim 10^9 \mathrm{cm}^{-3}$) gravitationally bound clumps with high column densities ($N_H \sim 10^{24} \mathrm{cm}^{-2}$). Numerous studies have since investigated the possible ways of detecting these very dense clumps. However, this picture which assumes that interstellar matter is structured by gravity down to the smallest scales does not fit with our present understanding of Galactic molecular clouds. Giant Molecular Clouds (GMCs) and the ISM in general is indeed observed to be structured on all scales but only clouds with masses of $10^4 M_\odot$ and higher are observed to be gravitationally bound (e.g. Heyer et al. 2001). Smaller clouds, including clumps of all sizes within GMCs and small molecular clumps found in cirrus clouds appear not to be massive enough to be bound by self-gravity and are understood as transient structures related to ISM turbulence (Falgarone et al. 1998, Elmegreen and Lazarian in this volume). In this paper we discuss the H_2 hypothesis within this *conventional* view of molecular gas in the Milky Way. We suggest that matter could be hidden in clouds with moderate column densities (Section 2), that it can be traced by dust emission (Section 3) and by the emission from warm H_2 in intermittent structures where the ISM turbulent energy is dissipated (Section 4).

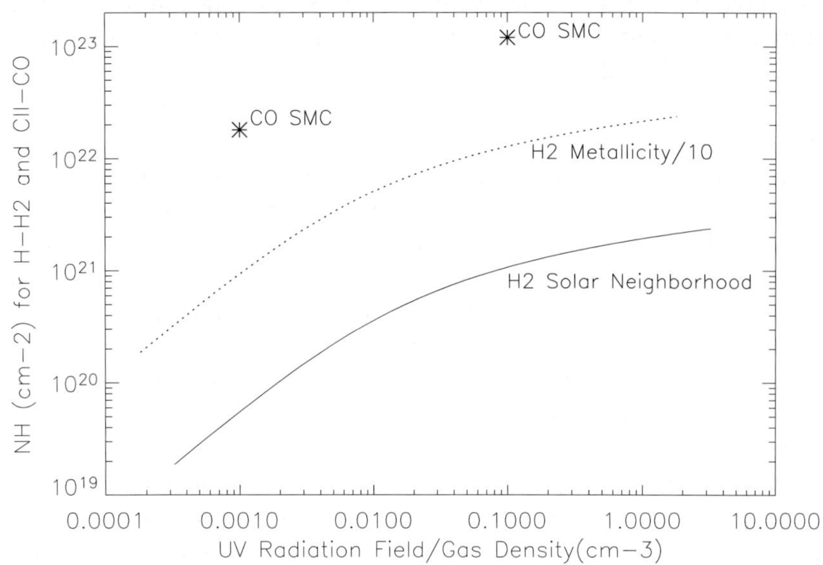

Figure 1. Depth of the H I/H_2 transition layer (hydrogen column density from the cloud edge up to the point at which the gas is half molecular and atomic) as a function of the ratio between the UV radiation field (in Solar Neighborhood units) and the gas density (in cm^{-3}). The solid line is based on Fig. 12 in Draine and Bertoldi (1996) for an H_2 formation rate of $3 \times 10^{-17} cm^3 s^{-1}$ and the standard solar neighborhood extinction curve (R_V = 3.1); for the 1/10 solar metallicity (dotted line) we have scaled down the far-UV extinction and H_2 formation rate by 10. The two stars give the column density at which CO becomes the dominant form of carbon for the two SMC (metallicity 1/10 Solar) models presented by Lequeux et al. (1994).

2. Low Metallicity Molecular Clouds

In low metallicity environments, due to reduced shielding, CO emission is expected to only trace the dense and high column density clumps within molecular clouds (Lequeux et al. 1994). This statement is quantitatively illustrated in Fig. 1 where we have represented the gas column densities at which gas becomes molecular (hydrogen is in H_2 and gas carbon in CO) versus the ratio between the far-UV radiation field and the gas density. This parameter expresses the ratio between the destruction and formation rates, by photodissociation and two-body reactions, respectively. In the Solar Neighborhood, the HI-H_2 transition is observed through UV observations of H_2 electronic transitions in absorption, from the Copernicus and FUSE satellites, to occur for $N_H \sim 5 \times 10^{20}$ cm^{-2} (Rachford et al. 2002). This value corresponds to the Solar Neighborhood curve in Fig. 1 for a gas density $n_H \sim 100\, cm^{-3}$. In lower metallicity gas the H I-H_2 transition occurs at higher column densities because the dust formation rate on grain surfaces is reduced. In Fig. 1, assuming that the H_2 formation rate is proportional to the metallicity, we show that the hydrogen column density for the H I-H_2 transition moves up by roughly a factor 10 for a metallicity 1/10 Solar.

The conversion of gas carbon from C^+ in the diffuse ISM to CO inside molecular clouds occurs deeper into clouds, at column densities from the edge about one order of magnitude larger than H_2. The column density of this critical transition for CO emission also scales inversely to the metallicity for a given ratio between UV field and gas density. Based on the values shown in Fig. 1, it occurs at $N_H > 10^{22}cm^{-2}$, for UV $field/n_H > 10^{-3}cm^3$, while observations of GMCs in the Galaxy show that most of their mass come from area where N_H is in the range 10^{21} to $10^{22}cm^{-2}$ (e.g. Simon et al. 2001). If molecular clouds in the SMC have comparable structure than Galactic GMCs, most of the SMC molecular gas is expected to be CO-poor and thus poorly traced by CO emission.

This expectation is confirmed by a systematic metallicity effect on the CO luminosity (L_{CO}) to virial mass (M_{Vir}) relation from the Milky Way to the LMC and SMC (Rubio et al. 1993, Mizuno et al. 2001). For a given virial mass, the observed CO luminosity is a factor of few smaller in the LMC and a factor 10 for the lower metallicity SMC. Comparing CO and mm dust observations for one SMC molecular cloud away form massive stars, Rubio et al. 2004 have recently found that M_{Vir} computed from the CO size and line width is one order of magnitude smaller than the cloud mass determined from the dust emission. This difference would imply that the CO emission is not tracing the full volume and dynamics of the cloud and thus that M_{Vir} is underestimated. Obviously this early work needs to be extended to other clouds before being generalized but the present result already suggests that the difference in $L_{CO}/M(H_2)$ ratio between the Galaxy and the SMC may be larger than the order of magnitude reduction inferred from the comparison of the L_{CO} and M_{Vir} relations. More precisely the $L_{CO}/M(H_2)$ may vary with the fraction of the cloud mass present in CO emitting clumps. The present conclusion is that we are far from having a reliable estimate of the total H_2 gas in the SMC: not only the masses of the SMC GMCs identified by Mizuno et al. (2001) could be grossly underestimated but more importantly their CO survey could be missing low L_{CO} molecular clouds which as an ensemble could account for most of the gas mass. This possibility is supported by the star formation activity, the numerous young stars seen throughout the SMC bar must form from gravitationally bound molecular clouds, but not by FUSE observations (Tumlinson et al. 2002). This apparent diagreement is not unexpected since the FUSE observations towards hot stars in massive star forming regions sample primarily regions where the UV radiation field is high and where gas is expected to be photo-dissociated and photo-ionized.

The SMC conclusion applies to the outer parts of spirals to the extent that there the ISM is expected, and observed in a few cases (e.g. outer giant H II regions in M101), to be metal-poor. However, the conditions are not exactly similar: in the SMC, active star formation creates a UV field higher than in the Solar Neighborhood while the radiation field in the outer parts of spirals is lower. This will compensate for part of the effect of metallicity on CO emission. But the gas temperature will also be lower which will contribute to reduce the CO emission (Allen et al. 1995).

3. Dust Emission from outer H I Disks

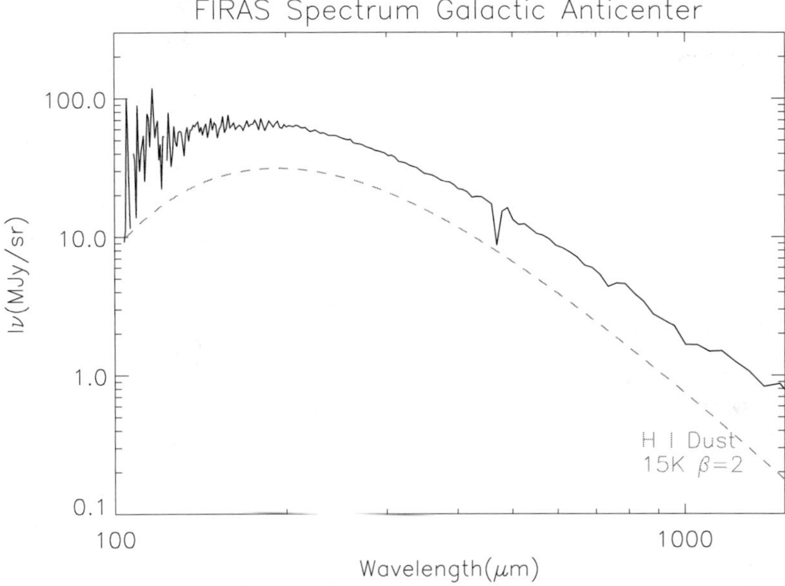

Figure 2. Dust Emission spectrum towards the Galactic anticenter measured by the COBE far-IR absolute spectro-photometer FIRAS. We averaged all FIRAS pixels within $160^\circ < l < 200^\circ$ and $|b| < 1.5^\circ$. The dotted line represents the expected contribution from dust in the atomic gas based on the H I emission, assuming a dust temperature of 15 K and a dust/gas ratio equal to that in the Solar Neighborhood. It represents an upper limit to the contribution of dust in atomic gas that has to be scaled down by the unknown gas metallicity.

Observations of dust emission in galaxies are a rapidly growing field with, after IRAS and ISO, the presence of the Spitzer infrared observatory in space and also the advent of sensitive bolometer instruments on sub-mm/mm ground telescopes. Dust emission is metallicity dependent but unlike CO it traces interstellar matter independently of its density and shielding to UV radiation. It offers thus a mean to look for hidden molecular gas not traced by CO. From the mid to the far-IR the dust emission per dust unit mass scales proportionally to the dust heating rate, i.e. the local intensity of the UV radiation field. At these wavelengths dust emission measure the stellar radiation field which can be related to the galaxy star formation rate rather than its dust content. The sub-mm/mm observations in the Rayleigh Jeans part of the dust spectrum are best suited to measure dust masses. Far-IR observations with enough angular resolution to separate star forming regions from the diffuse ISM emission might also be used to trace dust provided that the dust temperature can be measured by combining brightnesses at two or more wavelengths. The presence of dust in extended H I disks has already been demonstrated through extinction studies (e.g. M 31 by Cuillandre et al. 2001) and the detection of far-IR emission in the outer parts of the edge-on galaxy NGC 891 with ISOPHOT (Popescu et al. 2003). In a near future much more should be learned about dust in galaxies from Spitzer observations.

I chose to illustrate this section of the paper with the far-IR spectrum of the outer Milky-Way disk as measured by the absolute spectro-photometer, FIRAS, on board of the COBE satellite (Fig. 2). Based on H I and CO surveys, the gas seen towards the anticenter is entirely atomic with $\mathrm{N(H\ I)} = 4.5 \times 10^{21}\mathrm{cm}^{-2}$ and $\mathrm{N(H_2)} = 2.5 \times 10^{20}\mathrm{cm}^{-2}$ for the standard Galactic CO emission to H_2 column density ratio. In Figure 2, we show the expected far-IR emission for the H I column density assuming a dust far-IR emissivity per Hydrogen equal to the observed value in the Solar Neighborhood (Boulanger et al. 1996). I assumed the dust temperature to be 15 K, that given by the FIRAS spectrum. This temperature is 17.5 K in the Solar Neighborhood. The change in temperature represents an effective decrease of the stellar radiation field intensity by a factor 3. I consider the dotted curve in Fig. 2 as an upper limit to the contribution of the H I gas to the far-IR dust emission which has to be scaled down by the dust-to-gas ratio, often assumed to scale with the metallicity: a reasonable assumption for the cloud phase where metals are observed to be mostly on grains. Gas in the outer part of the Galaxy is expected to have a lower metallicity than that in the Solar Neighborhood due to a reduced enrichment by stars and accretion of low metallicity gas. For an effective metallicity 1/4 Solar the H_2 column density required to account for the FIRAS emission would be 7 times larger than that of H I. One thus sees that the dust data support the H_2 hypothesis but can not be quantitatively precise about the amount of hidden H_2 in the absence of independent estimates of the gas metallicity.

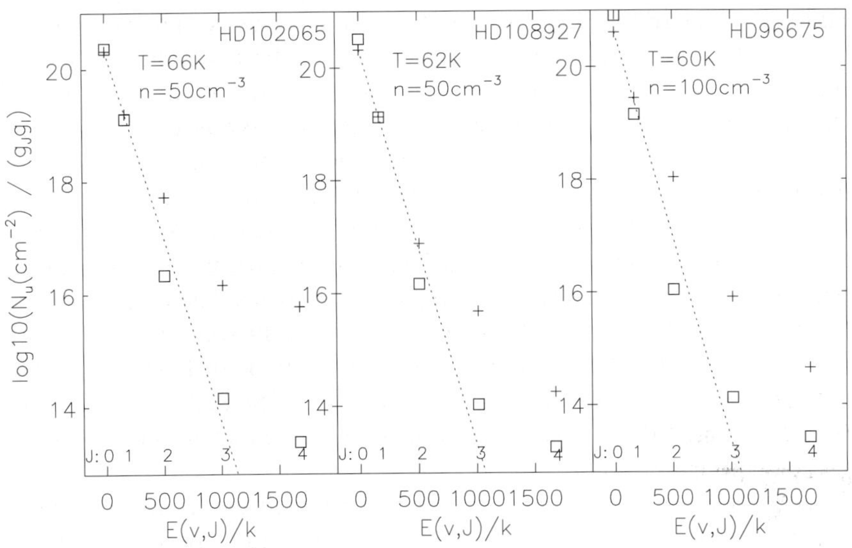

Figure 3. H_2 excitation diagram towards stars behind translucent ($A_V \sim 1$) clouds in Chamaeleon from Gry et al. 2002. The population in the various J levels are derived from absorption spectra covering electronic transitions from H_2. The data is represented by the pluses while the empty squares represent the expected population. The gas densities derived from the modeling are given in the plot. The gas temperature result from the calculation of the balance between heating and cooling processes. Unlike most stars observed by FUSE, these stars are known, from the IRAS sky images, not to intearct with the absorbing cloud which permits to exclude excitation of H_2 to $J \geq 2$ by UV heating and pumping.

4. A new Tracer of H_2

Since the first rotational lines of para and ortho H_2 arise from levels with large energies ($E_{J=2} = 510$ K, $E_{J=3} = 1020$ K), collisional excitation of H_2 is effective only for warm (T> 100 K) molecular gas while the bulk of the H_2 gas in galaxies is known to be cold ($T \sim 10 - 20K$). Molecular gas far away from star forming regions is thus not expected to be seen in emission. This expectation is contradicted by H_2 observations carried out with FUSE and the ISO spectrometer SWS. H_2 absorption lines observed by FUSE, in the direction of late B stars show a fraction of warm H_2 in J $\geq$ 2 rotational states which can not be accounted by UV heating nor pumping (see H_2 excitation diagram and data/model comparison in Fig. 3 taken from Gry et al. 2002). A similar conclusion is reached for a line of sight sampling 18 mag of diffuse gas towards the inner Galaxy and avoiding star forming regions observed with ISO.

The intensities of the S(1)(J=3-1), S(2)(J=4-2) and S(3)(J=5-3) lines are up to one order of magnitude above those predicted by UV heating and pumping of H_2 (Falgarone et al. 2004). In the edge-on galaxy NGC 891, the S(0)(J=2-0) and S(1) H_2 rotational lines observed far from the center of the galaxy are only marginally explained in terms of UV heating (Valentijn & van der Werf 1999). Falgarone et al. (2004) proposed that the warm H_2 gas seen in these observations trace bursts of dissipation of supersonic turbulence, omnipresent in the interstellar medium. It is because the dissipation of turbulence is not evenly distributed in space and time, but concentrated in tiny regions of space, over a short duration ($\approx 10^3$ yr), i.e. its property of intermittency, that the *local* gas heating is large enough to reach the gas temperatures required to excite by collisions the H_2 pure rotational lines. These space-time bursts are modeled as MHD shocks and thin coherent vortices and they temporarily heat a small fraction of the gas ($\approx 1\%$) to temperatures up to 10^3 K (Joulain et al. 1998; Flower & Pineau des Forêts 1998).

H_2 observations are not the only evidence for the existence of warm molecular gas mixed with the cold diffuse ISM. A body of various observations and analyses all converge towards the existence of this elusive molecular gas component. Indirect evidence includes the large abundances observed in the cold diffuse medium of molecules such as CH^+, HCO^+ and OH, for which the formation route requires to overcome endothermicities and barriers of several thousands Kelvin while the thermal energy available in the medium is of the order of 100 K (Lucas & Liszt 1996, Gredel et al. 2002). Last, supersonic turbulence is ubiquitous and its dissipation which has to take part in the progressive infall of gas towards the center of galaxies and star formation at all galactocentric distances, deposits on average a significant amount of energy in the ISM. It is noteworthy that the power per unit surface in the Galaxy provided by the dissipation over a few Gyr of the kinetic energy available in the galactic rotation corresponds to H_2 rotational line intensities of the order of what has been detected at large scale in the H_2 lines in the Milky Way and in NGC 891.

Because it is intermittent, the dissipation of turbulence is an ubiquitous source of excitation of H_2 : unlike UV excitation, it is not confined to the vicinity of young stars. Emission in the pure rotational lines of H_2 would thus be an ubiquitous spectroscopic signature of interstellar matter that could be used to trace matter far from star forming regions, in low metallicity environments where other tracers (CO, dust) fail and where it may become a dominant radiative cooling process.

5. Conclusions

Modeling of H I rotation curves show that it is plausible that outer parts of spirals contain 10 times more gas than that traced by H I emission. If this is true most of the gas in spirals would lye hidden beyond the optical disk. We presented several perspectives on this hypothesis based on recent advances in ISM observations.

- CO and dust observations of SMC molecular clouds suggest that large amounts of H_2 gas can be hidden in low metallicity molecular clouds with a density structure similar to that of Galactic GMCs but CO photo-dissociated outside dense clumps accounting for a small fraction of the cloud masses. This possibility needs to be further tested. Constraints on the surface filling factors of the hidden gas set by extinction studies and absorption spectroscopy need to be considered.

- The Spitzer satellite and ground based sub-mm telescopes will provide a wealth of data on dust in galaxies and be sensitive enough to detect emission from their outer disks. These observations should confirm early evidence, in particular in the Milky Way, for a larger and more diverse set of galaxies. Unlike CO, dust emission probes dust independently of gas density,temperature and shielding to photo-dissociating UV radiation but the quantitative interpretation in terms of gas mass is dependent on metallicity.

- FUSE and ISO H_2 observations of molecular matter far from UV sources, known to be on average cold, reveal amounts of warm H_2 (i.e. in rotational levels $J \geq 2$) than cannot be accounted by heating and pumping by UV nor excitation associated with H_2 formation. The warm H_2 could be tracing localized regions where the turbulent ISM energy is being dissipated. Emission in the mid-IR rotational lines of H_2 would be an ubiquitous and specific, far from UV sources, spectroscopic signature of turbulence dissipation. Spitzer observations are programmed to confirm the early ISO results. The Spitzer spectrometer is expected to be sensitive enough to detect this emission where $\mathrm{N(H_2)} > 10^{22}\mathrm{cm}^{-2}$. These observations may open a new perspective on H_2 in galaxies by permitting to trace low metallicity molecular gas not associated with star formation.

The H_2 hypothesis raises many questions beyond what is discussed in this paper. In relation to the central topic of these proceedings, it is necessary to investigate the impact that the existence of a large reservoir of interstellar matter would have on galaxies evolution and our understanding of the Hubble sequence. How this gas will be able to flow inward? What dynamical processes will trigger/regulate this inward flow which will fuel star formation and impact the dynamical and chemical evolution of galaxies? Is the gas inward flow related to the transfer of Galactic rotational energy into turbulent motions and the dissipation of turbulent energy?

Acknowledgments

This paper borrows from work written in collaboration with Monica Rubio for the CO and Dust emission from SMC molecular clouds (section 2) and Edith Falgarone and Francoise Combes for the proposal of observing hidden H_2 gas through emission in its rotational lines powered by the dissipation of the turbulent ISM energy (section 4).

References

Allen, R. J., Le Bourlot, J.,Lequeux, J.,Pineau des Forêts, G. and Roueff, E.,Ap. J. 1995, Ap. J. 444, 157

Bosma, A. 1981, AJ 86, 1825

Boulanger, F., Abergel, A., Bernard, J. P. 1996, A& A 312, 256

Cuillandre, J.C., Lequeux, J., Allen, R. J., Mellier, Y. and Bertin, E. 2001 ApJ 554, 190

Draine, B. T., and Bertoldi, F. 1996, Ap. J. 468, 269

Falgarone, E., Panis, J.-F., Heithausen, A., Perault, M., Stutzki, J., Puget, J.-L. and Bensch, F. 1998, A& A 331, 669

Falgarone E., Verstraete L., Pineau des Forêts G. and Hily-Blant P. 2004 A&A in press

Flower D. and Pineau des Forêts G. 1998, MNRAS 297, 1182

Gredel R., Pineau des Forêts G. and Federman S.R. 2002 A&A 389 993

Gry, C., Boulanger, F., Nehme, C., Pineau des Forêts, G., Habart, E. and Falgarone, E. 2002, Ap. J. 391, 675

Heyer, M. H., Carpenter J. M. and Snell, R. L. 2001 Ap. J. 551, 852

Hoekstra H., van Albada T.S., Sancisi R. 2001, MNRAS 297, 1182

Joulain K., Falgarone E., Pineau des Forêts G. and Flower D. 1998 A&A 340 241

Lequeux, J., Le Bourlot, J., Pineau des Forêts, G., Roueff, E., Boulanger, F. and Rubio, M. 1994, A& A 292, 371

Lucas R. and Liszt H.S. 1996 A&A 307 237

Mizuno, N., Rubio, M., Mizuno, A., Yamaguchi, R., Onishi, T. and Fukui, Y. 2001, PASJ 53, L45-L49

Pfenniger, D., Combes, F. and Martinet, L. 1994, A& A 285, 79

Pfenniger, D. and Combes, F. 1994, A& A 285, 94

Popescu, C. C., Tuffs, R. J.2003, A& A 410, L21

Rachford, S., Snow, T. P., Tumlinson, J. et al. 2002, Ap. J. 577, 221

Rubio, M., Lequeux, J. and Boulanger, F. 1993, A& A 271, 9

Rubio, M., Boulanger, F., Rantakyro, F. and Contursi, A. 2004, A& A in press

Simon, R. Jackson, J. M., Clemens, D. P., Bania, T. M. and Heyer, M. H. 2001, Ap. J. 551, 747

Tumlinson, J., Shull, J. M., Rachford, B. L. et al. 2002, Ap. J. 566, 857

Valentijn, E. A. and van der Werf, P. P. 1999, 522, L29-L33

X-RAY PERSPECTIVES OF EARLY TYPE GALAXIES

Dong-Woo Kim
Smithsonian Astrophysical Observatory 60 Garden Street, Cambridge, MA 02138, USA

Abstract We present recent results on the X-ray properties of early type galaxies. We construct a bias-corrected X-ray luminosity function of low-mass X-ray binaries (LMXB) and discuss its implication on the nature of LMXBs. We compare LMXBs associated with blue (metal poor), red (metal rich) globular clusters and field LMXBs to understand the X-ray signature of the formation and evolution of galaxies (as well as GCs and LMXBs). With high S/N XMM-Newton spectra, we present well-constrained measurements of 'super-solar' metal abundances in the hot ISM and discuss the relative importance of Type II and Type Ia SNe in the metal enrichment. Finally, we present the evidence of the interaction of the hot ISM with the radio jets/lobes and the ambient ICM and discuss the feedback between AGN and ISM, ram pressure stripping and ICM metal enrichment.

Keywords: Early type galaxies, X-ray, LMXB, ISM, Metal Abundances

1. Introduction

Unprecedented sub-arcsec resolution images and high quality spectra obtained with Chandra and XMM-Newton are providing exciting results on early type galaxies. We are able to directly detect X-ray point sources and now have uncontroversial proof of the existence of populations of X-ray binaries in all E and S0 galaxies and of their large contribution particularly to the emission of X-ray-faint galaxies (see Sarazin et al. 2000; Angellini et al. 2001). In addition to being identified as a point source, their spectral properties (kT $\sim$ 5-10 keV) and radial distribution (following the radial distribution of the stellar, optical light) clearly support that they are mainly LMXBs with an accreting neutron star.

Carefully counting the LMXB contribution (either detected or hidden), and using high S/N XMM-Newton spectra allow us to unambiguously determine spectra parameters (in particular metal abundances) of the hot ISM. The high spatial resolution is also critical in studying detailed sub-structures of the extended hot ISM in early type galaxies. As opposed to the previously observed (and viewed) smooth X-ray emitting hot ISM, the Chandra image is now re-

D. Block et al. (eds.), Penetrating Bars through Masks of Cosmic Dust, 759–768.

vealing fine sub-structures of the hot ISM, such as X-ray valleys in NGC 1316 where the radio jets are in projection propagating (Kim and Fabbiano 2003 - hereafter KF03) and X-ray tails in NGC 7619 where on-going ram-pressure stripping occurs (Kim et al. 2004, in preparation).

In the first part of this paper, we present X-ray properties of LMXBs, in terms of X-ray luminosity function and their connection to the globular clusters. In the second part, we present the results of the hot ISM, its metal abundances and X-ray evidence of its interaction with radio jets/lobes and ICM via ram-pressure stripping.

2. Low-Mass X-ray Binaries (LMXBs)

2.1 X-ray Luminosity Function of LMXBs

With a large number of detected LMXBs, the X-ray luminosity function (XLF) has been established in several early type galaxies. The accurate measurement of XLFs can provide an important clue for our understanding of the accretion physics and the star formation history via the formation and evolution of binary systems in early type galaxies (eg., Fabbiano and White 2003).

It was previously reported that the XLF of LMXBs in X-ray faint early type galaxies (where the hot ISM component is minimal, hence the XLF of LMXBs can be best determined) exhibits a break at the Eddington luminosity ($L_{X,Eddington}$ = 2×10^{38} erg sec^{-1}) of 1.4 $M_\odot$ neutron stars, possibly indicating different populations of LMXBs consisting of neutron stars and black holes (e.g., Sarazin et al. 2000; Irwin et al. 2002). Although this result is very intriguing and could bear an important key to the accretion physics (e.g., Belczynski et al. 2004), KF03 pointed out that the apparent break disappears when incompleteness corrections are applied to LMXBs of NGC 1316, one of the X-ray faint elliptical galaxies. Figure 1 illustrates incompleteness and its impact on the XLF determination, particularly at $L_X \lesssim 2 \times 10^{38}$ erg sec^{-1}.

Correcting for incompleteness with a sample of 14 early type galaxies, Kim and Fabbiano (2004a - hereafter KF04a) confirmed that the individual XLFs are statistically consistent with a single power-law of a (differential) slope β = 1.8 - 2.2 (see Figure 2a). The previously reported break at or near $L_{X,Eddington}$ (the vertical line in Figure 2a) is not required. However, the combined XLF with a reduced statistical error suggests a possible (at a 3 σ significance) presence of an XLF break at L_X = 5 x 10^{38} ergs s^{-1} (the vertical bar in Figure 2b). By independent approaches, Gilfanov (2004) reached similar conclusion on the XLF shape and the break luminosity. It is interesting to note that this break luminosity is consistent with the Eddington limit of the theoretical maximum mass of a neutron star (3 $M_\odot$; Kalogera & Baym 1996). A recent measurement indicates that a neutron star (X7) in 47 Tuc may indeed be more massive than the canonical 1.4 $M_\odot$ (Heinke et al. 2003). Alternatively, a

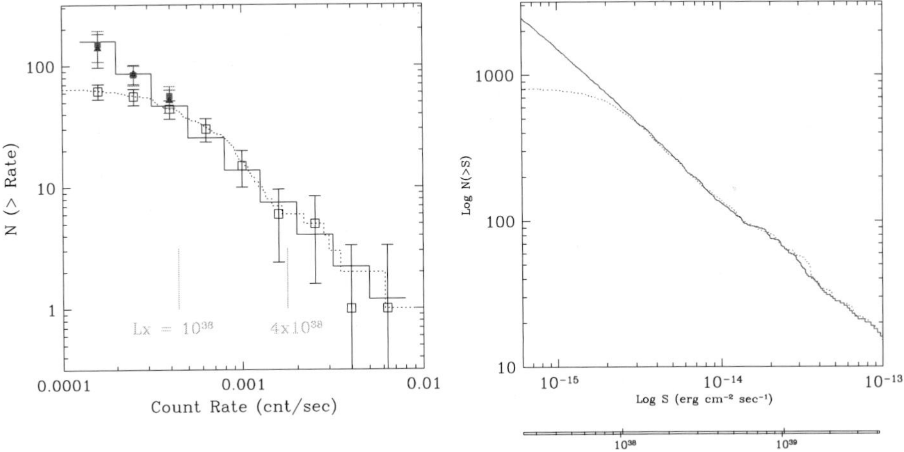

Figure 1. (a) X-ray Luminosity Function of NGC 1316. The binned and unbinned data are denoted by open squares and a dotted line. Three filled squares and triangles at the lowest L_X represent the corrected XLF. (b) Incompleteness of LMXBs of NGC 1316. The input (solid line) and measured (dotted line) XLFs (made with 20,000 simulated sources) are compared. The incompleteness becomes significant at Lx $\lesssim 2 \times 10^{38}$.

He-enriched LMXB could also be consistent with a larger Eddington luminosity (e.g., Grimm et al. 2003). Bildsten and Deloye (2004) suggested that ultracompact binaries with He or C/O white dwarf donors could reproduce the XLF shape observationally determined by KF04a, although some bright LMXBs are certainly not ultracompact binaries. Either way, the Chandra data suggest a more complex picture than previously considered.

The high luminosity portion of the XLF could bring an important clue on the presence/absence of beaming. If the change in slope is real, KF04a result would imply a different population of high luminosity sources: neutron star and black-hole binaries, with different XLFs (as originally suggested by Sarazin 2000, although at a different break luminosity). A step in the XLF would be more indicative of a beaming effect, that may enhance the luminosity of some high accretion rate binaries (King 2002). We also note that the high luminosity population does not resemble that of the ULXs detected in star-forming galaxies, where the XLF slope (with no break) is much flatter (see Fabbiano & White 2003 and refs. therein). The XLF at the higher L_X may critically depend on background/foreground sources. Irwin, Bregman & Athey (2003) found that very bright X-ray sources ($L_X > 2 \times 10^{39}$ ergs s^{-1}) show no concentration toward the galaxies and their number is similar to that of the expected background/foreground sources, suggesting a significant

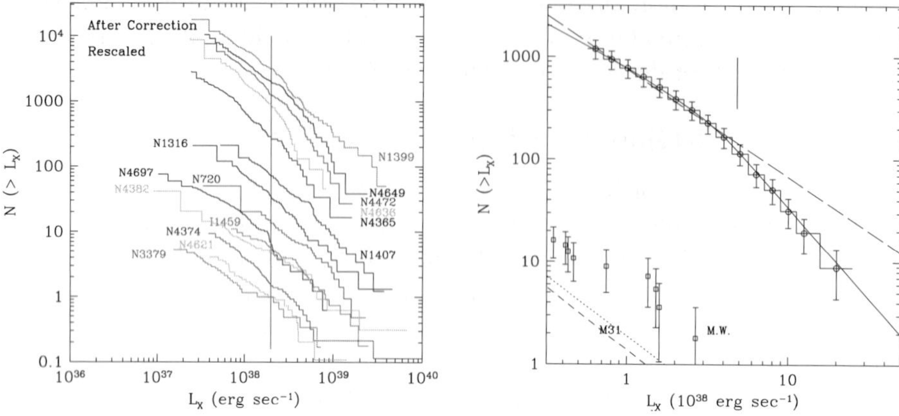

Figure 2. (a) Completeness-corrected XLFs of 14 early type galaxies. They are vertically shifted for visibility. (b) Combined XLF. The best fit single power law and broken power law models are denoted by a dashed line and a solid line. Also plotted are XLFs of bulge sources in the Milky Way and M31.

amount of contamination. This may be consistent with the fact that ULXs are mainly found in the starburst galaxies (eg., in the Antennae, Zezas et al. 2002) and possibly represent the high M (and/or M-dot) tail of the HMXB population which would be absent in early type galaxies (e.g., Gilfanov et al. 2003; however, see also an opposite view by Colbert et al. 2003 and Humphrey et al. 2003).

2.2 Total Integrated L_X(LMXB)

Using the XLFs determined above, we can measure the integrated X-ray luminosity of both detected and undetected sources. Because the XLF is steep, the total X-ray luminosity of LMXBs depends on the undetected lower L_X break of the XLF. Following LMXBs detected in the Milky Way (Grimm et al. 2002) and in the bulge of M31 (Kong et al. 2002), we assume the lower limit to the XLF at 10^{37} erg sec^{-1}. Adopting $M_\odot$(K)=3.33 mag, we obtain the X-ray to optical luminosity ratio:

$$L_X(LMXB)/L_K = 0.20 \pm\ 0.08 \times\ 10^{30} ergs^{-1}/L_{\odot K}$$

The correlation with L_K is slightly tighter than L_B, because the near-IR luminosity may be more appropriate for the old stellar population of early type galaxies and less affected by extinction than the B-band luminosity. In contrast to a large scatter between the total X-ray (the hot ISM + LMXBs) and optical luminosity seen in early type galaxies (e.g., Fabbiano 1989), the correlation between L_X(LMXB) and L_K is statistically significant (with a chance probability of 0.3%). However, there is still non-negligible scatter, $\sim$40% at 1σ

rms (see below for more discussions). The bulges of the Milky Way and M31 follow the general trend of early type galaxies and roughly indicate the upper and lower bounds of the L_X/L_K scatter (see Figure 3a).

2.3 LMXB-Globular Cluster Connection

With the detection of X-ray source populations in early-type galaxies there has been renewed interest in probing the formation of these systems, and in particular in exploring a possible evolutionary link to Globular Clusters (GCs), originally suggested by Grindlay (1984) as the main formation mechanism for Galactic LMXBs. A significant fraction of the sources detected with Chandra in early-type galaxies are associated with GCs (e.g. Angelini et al 2001; Kundu et al. 2002; see compilation in Fabbiano & White 2003). This association has led to the suggestion that GCs may be the birthplace of the entire LMXB population, from where they may be expelled if they receive strong enough formation 'kicks', or may be left behind upon tidal disruption of the parent cluster (see Sarazin et al. 2000; White, Sarazin & Kulkarni 2002).

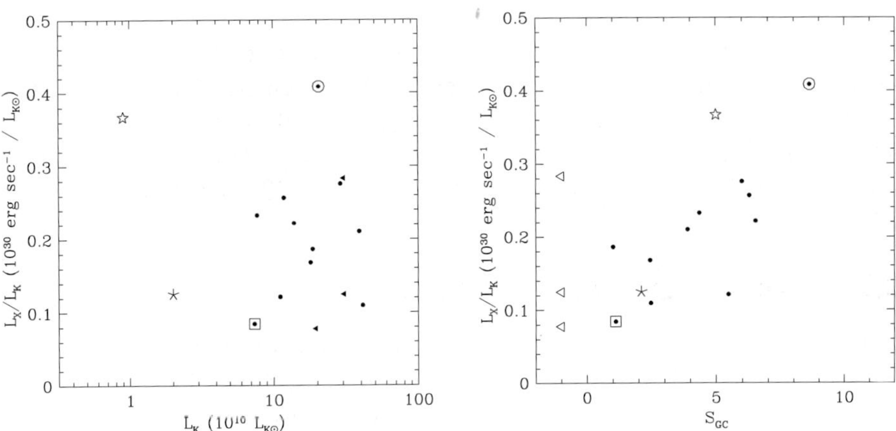

Figure 3. L_X(LMXB) / L_K of 14 early type galaxies is plotted against (a) L_K and (b) the globular cluster specific frequency (S_{GC}). Also plotted are the bulges of the Milky Way (star) and M31 (asterisk).

While there is no obvious dependency of L_X(LMXB)/L_K on the integrated stellar luminosity (Figure 3a), there is a possible correlation with S_{GC} (Figure 3b), which would follow the suggestion of White et al (2002). The Spearman Rank correlation test gives a chance probability of 2%. However, if we exclude NGC 1399, which is the galaxy with the largest SGC (a large circle in Figure 8b), the chance probability increases to 5%.

Using ground and HST observations of 6 giant elliptical Galaxies, Kim, E. et al. (2004; in preparation) have identified ~300 LMXBs associated with

GCs. They confirmed a significantly higher probability (by a factor of 4) to find LMXBs in red (metal-rich) GCs than in blue (metal-poor) GCs, as reported by Kundu et al. (2002) and Sarazin et al. (2003). This indicates that the metal abundance plays a key role in forming LMXBs, e.g., by flatter IMF (Bellazzini et al. 1995) or irradiation induced stellar winds (Maccarone, et al. 2004). The latter model predicts harder X-ray spectra in metal poor clusters. However, in the sample of 6 elliptical galaxies, no significant difference in X-ray spectra from metal rich and poor cluster samples is seen.

Based on the GC-LMXB connection, one might expect the radial distribution of LMXBs is similar to that of GCs, hence flatter than that of the optical halo light. However, Kim et al. (2004) found that LMXBs closely follow the optical halo light than more extended GCs, suggesting a rather complex connection, if any, depending on various factors operating in the LMXB formation in GCs and its subsequent evolution. The close agreement between the radial distributions of the optical light and LMXBs is also seen in NGC 1316 (KF03) and NGC 1332 (Humphrey and Buote 2004). The observed radial behavior of LMXBs may be partly understood, because red GCs are often more centrally concentrated than blue GCs (e.g., Lee et al. 1998). However, there may be other factors, such as variable capture rates as a function of cluster density.

3. The Hot ISM

3.1 Metal Abundances in the Hot ISM

Heavy elements in the hot halos of early-type galaxies are the relic of stellar evolution. Determining their abundance is key to our understanding of these galaxies. In particular, since abundances are related to the supernova yield, they can constrain both the supernova rate (Type Ia and II) and the initial mass function (IMF) of the stellar population (e.g., Renzini et al. 1993). Moreover, these measurements are also important for constraining the evolution of the hot ISM in terms of the energy input from supernovae, which may result in the onset of galactic winds. Yet these measurements are difficult and the results have been controversial (see Fabbiano 1995). While stellar evolution models predicts the super-solar iron abundance, Z_{Fe} = 2-5 times solar (e.g., Arimoto et al. 1997), the ROSAT and ASCA data with a simple thermal model often suggested a hot ISM almost totally devoid of metals in early-type galaxies.

The higher quality Chandra and XMM-Newton data are now showing that the extremely low, sub-solar, Iron abundances suggested by the ROSAT and ASCA analyses can be statistically rejected. In particular, by subtracting a population of 80 discrete sources from the image, KF03 excluded sub-solar metal abundances in the hot ISM of NGC 1316. Applying a two-temperature model of the hot gas to XMM-Newton spectra, Buote (2002) reported the first convincing measurement of super-solar metal abundances (Z_{Fe} = 1.5-2 solar) in the central region of NGC 1399.

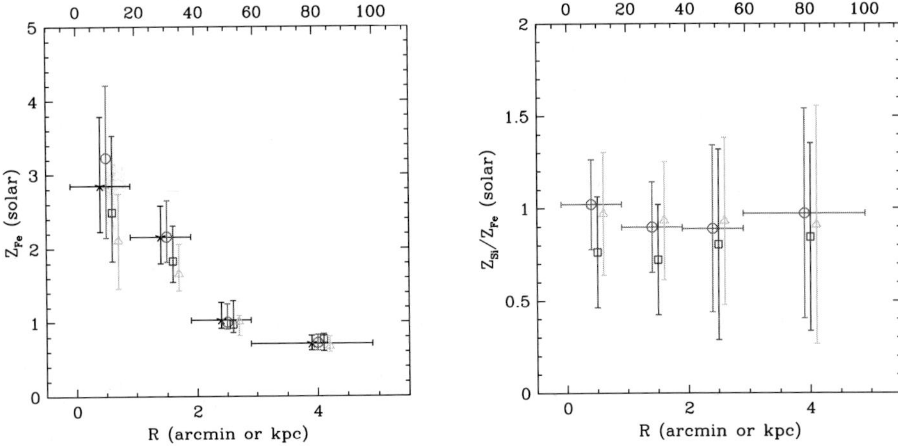

Figure 4. (a) Radial distribution of Fe abundances. (b) Radial distribution of Si to Fe abundance ratio. Different symbols indicate different methods of grouping heavy elements.

Kim and Fabbiano (2004b - hereafter KF04b) considered various systematic effects in analyzing to the X-ray XMM-Newton observations of NGC 507, a dominant elliptical galaxy in a small group of galaxies, and reported 'super-solar' metal abundances (2-3 times solar) inside D_{25} (40 kpc or $2'$ in radius) of both Fe and α-elements in the hot ISM of this galaxy. This is the highest Z_{Fe} reported so far for the hot halo of an elliptical galaxy; this high Iron abundance is fully consistent with the predictions of stellar evolution models, which include the yield of both type II and Ia supernovae. They show that abundance measurements are critically dependent on the selection of the proper emission model. The spatially resolved, high quality XMM spectra provide enough statistics to formally require at least three emission components in each of 4 circum-nuclear concentric shells (within r < $5'$ or 100 kpc): two soft thermal components indicating a range of temperatures in the hot ISM, plus a harder component, consistent with the integrated output of low mass X-ray binaries (LMXBs) in NGC 507. A two-component (thermal + LMXB) model (or even a single temperature model) customarily used in past studies yields a much lower Z_{Fe}, consistent with previous reports of sub-solar metal abundances. This model, however, gives a significantly worse fit to the data (F-test probability < 0.0001). The abundance of α-elements (most accurately determined by Si) is also found to be super-solar. The α-elements to Fe abundance ratio is close to the solar ratio, suggesting that $\sim$70% of the Iron mass in the hot ISM was originated from SNe Type Ia. The α-element to Fe abundance ratio remains constant out to at least 100 kpc (see Figure 4b), indicating that SNe Type II and Ia ejecta are well mixed in a scale much larger than the extent of the stellar body. This result is inconsistent with the scenario of SNe type II driven early galactic winds dominating the large-scale metal enrichment.

3.2 Interaction with Radio Jets and Lobes

NGC 1316 is the first elliptical galaxy for which this type of jet/hot ISM interaction has been reported (Kim, Fabbiano and Mackie 1998). Chandra observations are now revealing many other examples in elliptical galaxies and clusters. Figure 5 shows the direction of radio jets (from the 1.5GHz VLA map in Geldzahler and Fomalont 1984) superposed on the Chandra X-ray image. The radio jets propagate along PA $\sim 120^\circ$ and 320° and slightly bend at their ends toward PA $\sim 90^\circ$ and 270° respectively. It is clearly seen that the radio jets in projection (1) run perpendicular to the direction of the sub-arcmin scale NE-SW elongation of the central X-ray distribution, which is also the direction of the optical figure, and (2) propagate through the valleys of the X-ray emission, confirming the results of the ROSAT HRI observations (Kim et al. 1998). This alignment suggests that the X-ray valleys are cavities in the hot ISM caused by exclusion of hot gas from the volumes occupied by the radio jets. Similar relations between the radio jets/lobes and X-ray features have been reported in the Chandra observations of NGC 4374 (Finoguenov and Jones 2001), Hydra A (McNamara et al. 2000), Perseus cluster, (Fabian et al. 2000) and Abell 4095 (Heinz et al., 2002).

In particular, the hourglass-like feature seen in the core of Abell 4095, is similar to the X-ray feature in NGC 1316, although only one of (possibly) two X-ray cavities in Abell 4095 coincides with the radio emission. Based on the hydrodynamic simulations of Reynolds et al. (2001), Heinz et al. (2002) have interpreted the hourglass feature in Abell 4095 as a sideways expansion of the jet's surface caused by a compression wave (or sonic boom) which would follow after driving a strong shock. However, in NGC 1316 the jets appear to escape the denser halo. It is, therefore, not clear whether a sonic boom could explain the NE-SW elongation. This elongation could simply reflect the hot ISM sitting on the local, spherically asymmetric potential because it also follows the major axis of the galaxy.

Recent Chandra and XMM results on clusters failed to show the long-awaited evidence of cooling. Chandra's discoveries of X-ray cavities indicating an interaction between the radio jets and the hot ISM/ICM supports the idea that AGN/radio-jets may provide the energy feedback necessary to keep the gas from cooling (eg., Fabian 2002).

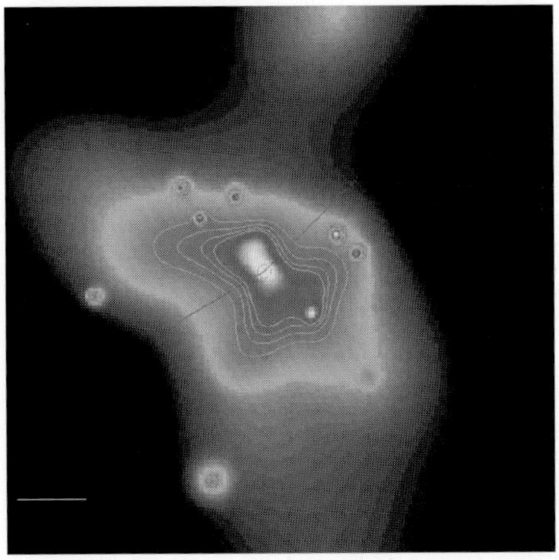

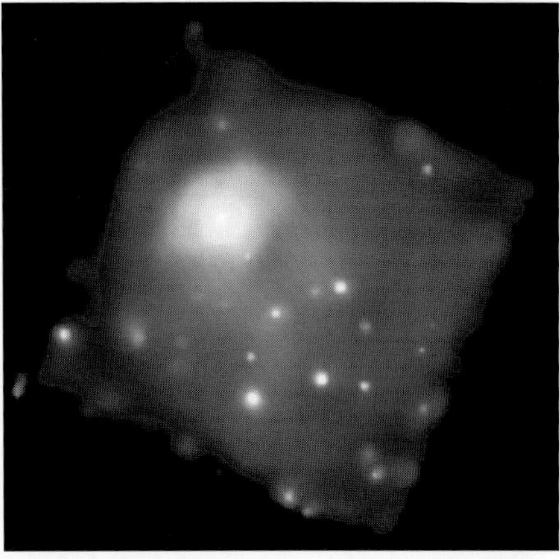

Figure 5. (a) Chandra ACIS image (80"x80") of the central part of NGC 1316. (b) XMM-Newton MOS1+MOS2 image (15'x15') of NGC 7619.

3.3 Interaction with ICM - Ram Pressure Stripping

Ram-pressure stripping has been recognized as one of the important processes in the evolution of galaxies, clusters, and their ISM/ICM (eg., Gunn and Gott 1972). Stripping is also critical for the metal enrichment of the ICM and may be (at least partially) responsible for the large L_X/L_B scatter in early type galaxies (eg., White and Sarazin 1991).

So far, only a few elliptical galaxies (see Fabbiano, Kim and Trinchieri 1992) are known to contain asymmetric, disturbed hot halos: NGC 4406 (Rangarajan et al. 1995; Forman et al. 2001), NGC 4472 (Irwin and Sarazin 1996; Biller et al. 2004 in prep.) and NGC 7619 (Trinchieri, Fabbiano & Kim 1997; Kim et al. 2004, in prep.). Their core-tail structures are stretched ~5-10 times longer than to the opposite direction.

Figure 6 shows the XMM image of NGC 7619 (Kim et al. 2004; in preparation). The X-ray tail is running toward SW. Also noticeable is a sharp discontinuity toward NE. Both the core structure and the direction of the tail consistently indicate that NGC 7619 is moving toward NE. Additionally, a large number of unresolved point-like sources co-exist with the X-ray tail. Some of them contain soft X-ray spectra, reminiscent with those possibly cooler blobs seen at the periphery of the X-ray emitting halo of NGC 507 (Kim and Fabbiano 1995). Most interestingly, we find that Z_{Fe} near the center and at the tail extending from the core are much higher than that of the surrounding regions, indicating we are witnessing the first direct evidence of the metal enrichment process of the ICM by stripping the metal-rich ISM.

References

Angelini, L., Loewenstein, M., & Mushotzky, R. F. 2001, ApJ, 557, L35
Arimoto, N., Matsushita, K., Ishimaru, Y., Ohashi, T., & Renzini, A. 1997, ApJ, 477, 128
Belczynski, K., Kalogera, V., Zezas, A., & Fabbiano, G. 2004, ApJ, submitted.
Bellazzini, M., et al. 1995, ApJ, 439, 687
Bildsten, L., & Deloye, C. J. 2004, ApJ in press. astro-ph/0404234
Buote, D. 2002, ApJ, 574, L135.
Colbert, E. J. M., Heckman, T. M., Ptak, A. F., & Strickland D. K. 2003 astro-ph/0305476
Fabbiano, G. 1989, ARAA, 27, 87
Fabbiano, G. 1995 in Fresh Views of Elliptical Galaxies, A. Buzzoni, eds., ASP Conf. 86, 103
Fabbiano, G., & White, N. E. 2003, astro-ph/0307077
Fabbiano, G., Kim, D.-W., & Trinchieri, G. 1992, ApJS, 80, 531
Fabian, A. C., et al. 2000, MNRAS, 318, L65
Fabian, A. C. 2002, astro-ph/0201386
Finoguenov, A., & Jones, C. 2001, ApJ, 547, L107
Forman et al. 2001 astro-ph/0110087
Geldzahler, B. J., & Fomalont, E. B. 1984, AJ, 89, 1650
Gilfanov, M. 2004, MNRAS in press, astro-ph/0309454
Gilfanov, M., Grimm, H.-J., & Sunyaev, R. 2003, astro-ph/0309725
Grimm, H.-J., Gilfanov, M., & Sunyaev, R. 2002, AA 391, 923.
Grindlay, J. E. 1984, AdSpR, 3, 19
Gunn, J. E. & Gott, J. R. 1972, ApJ, 176, 1
Heinke, C. O. et al. 2003, ApJ, 588, 452
Heinz, S., Choi, Y.-Y., Reynolds, C. S., & Begelman, M. C. 2002, astro-ph/0201107
Humphrey, P. J., et al. 2003, MNRAS, 344, 134
Humphrey, P. J., & Buote, D. 2004, ApJ in press
Irwin, J. A., Athey, A. E., & Bregman, J. N. 2003, ApJ, 587, 356
Irwin, J. A., Sarazin, C. L., & Bregman, J. N. 2002, ApJ, 570, 152
Irwin, J., & Sarazin, C. L. 1996, ApJ, 471, 683
Kalogera, V., & Baym, G. 1996, ApJ, 470, L61
Kim, D.-W., & Fabbiano, G. 1995, ApJ, 441, 182.
Kim, D.-W., Fabbiano, G., & Mackie, G., 1998, ApJ, 497, 699
Kim, D.-W., & Fabbiano, G. 2003, ApJ, 586, 826 (KF03)
Kim, D.-W., & Fabbiano, G. 2004a, ApJ, in press. also in astro-ph/0312104 (KF04a)
Kim, D.-W., & Fabbiano, G. 2004b, ApJ, submitted. also in astro-ph/0403105 (KF04b)
King, A. R. 2002, MN, 335, L13
Kong, A. K. H., et al. 2002, ApJ, 577, 738
Kundu, A., Maccarone, T. J., & Zepf, S. E. 2002, ApJ, 574, L5.
Lee, M. G., Kim, E., & Geisler, D. 1998, AJ, 115, 947.
Maccarone, T. J., Kundu, A., & Zepf, S. E. 2004, astro-ph/0401333
Rangarajan et al. 1995, MN, 277, 1047
Renzini, A., Ciotti, L., D'Ercole, A., & Pellegrini, S. 1993, ApJ, 419, 52
Reynolds, C. S., Heinz, S., & Begelman, M. 2001, ApJ, 549, L179
Sarazin, C. L., Irwin, J. A., & Bregman J. N. 2000, ApJ, L101.
Sarazin, C. L., et al. 2003, astro-ph/0307125
Trinchieri, G., Fabbiano, G., & Kim, D.-W. 1997, AA, 318, 361
White III., R. E., & Sarazin, C. L. 1991, ApJ, 367, 476
White III., R. E., Sarazin, C. L., & Kulkarni, S. R. 2002, ApJ, 571, L23.
Zezas, A., Fabbiano, G., Rots, A, & Murray, S. S. 2002, ApJ, 577, 710

SPIRAL ARM STAR FORMATION IN BARRED GALAXIES

Silvia Baes-Fischlmair[1], Werner W. Zeilinger[1], Juan-Carlos Vega-Beltran[1] and John E. Beckman[2]
[1]*Institut fur Astronomie der Universitat Wien* [2]*Instituto de Astrofisica de Canarias, La Laguna, Spain*

Abstract The two-fold disc-wide symmetry of spiral arms is generally attributed to density waves. Densisty wave theory can also explain a modulation of the star formation pattern along the arms according to the radial relation of the site with respect to the density-wave resonances. The degree of two-fold symmetry is reduced by the presence of a bar component. This is attributed to the increased radial mixing as a result of the non-axisymmetric bar structure. In order to quantify the influence of the bar component we selected a sample of spirals with various bar strengths.

Keywords: Galaxies: Barred, Stars: Formation, Galaxies: Spiral, Galaxies: Structure

1. Sample and Analysis

A sample of spirals with various bar strengths was selected from the Revised Shapley-Ames Catalog of Bright Galaxies (Sandage & Tamman, 1987) and the Third Reference Catalog of Bright Galaxies (de Vaucouleures, et al. 1991). R-band and redshifted Hα +[NII] images of 10 barred galaxies were observed at the Cassegrain focus of the 0.8m IAC80 telescope on Teneriffa of the Teide Obervatory (Spain) using a 1024 $\times$ 1024 pixel CCD. The scale of 0.4325 arcsec/pixel yielded a field of 7.3 $\times$ 7.3 arcmin^2. The typical seeing conditions were about 2.6 arcsec FWHM as measured from stellar images.

The continuum substraction from the Hα+[NII] image was realised by an appropriate scaling of the R-band image. The line-emission images were deprojected via scaling along the semi-minor axis. The forms of the spiral arms were determined through the analysis of the R-band image. Slits of the Hα+[NII] images were extracted along the spiral arms. The single slits were merged to a radial band and a rectified distribution was obtained for each arm. The Hα+[NII] intensity was then integrated perpendicular to the longitudinal axis of an arm yielding a one-dimensional radial intensity profile.

D. Block et al. (eds.), Penetrating Bars through Masks of Cosmic Dust, 769–770.

2. Results

Following Rozas et al. (1997), for each galaxy a cross-correlation parameter for the pixel intensity of the two arms was computed. The vicinity of the bar was excluded for that calculation. A perfect correlation of the intensity distribution would yield a value of 1. The typical cross-correlation parameter for the sample galaxie was found to be < 0.3, indicating a general absence of symmetry in the sample galaxies.

Examples of an "early-type" (NGC 2713) and a "late-type" system (NGC 2525) are presented in Figure 1 showing the asymmetric distribution of the emission-line regions.

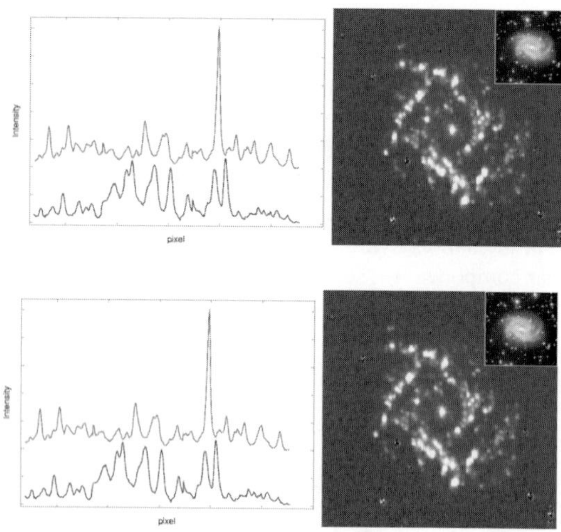

Figure 1. Upper Images: NGC 2525, Lower Images: NGC 2713; Left: radial profile along the spiral arms, Right: deprojected Hα+N[II] image, Upper Right: R band image

References

Rozas, M., Knapen, J.H., Beckman, J.E., 1998, MNRAS, *301*, 631.

Sandage, A. & Tammann, G., 1987, The Revised Shapley-Ames catalog of Bright Galaxies (Univ. Washington)

de Vaucouleures, G., et al., 1991, Third Reference Catalog of Bright Galaxies (Springer Verlag, NY)

Acknowledgments

SBF and WWZ acknowledge the support by the Austrian Science Foundation (project P14783)

GALAXY TYPES AND LUMINOSITY FUNCTIONS IN THE SLOAN DIGITAL SKY SURVEY USING ARTIFICIAL NEURAL NETWORKS

Nicholas M. Ball and Jon Loveday
Astronomy Centre, University of Sussex, Falmer, Brighton, BN1 9QJ, UK

Abstract The galaxy luminosity function bivariate with intrinsic galaxy properties is investigated using the second data release of the Sloan Digital Sky Survey (York et al. 2000, Abazajian et al. 2004). The properties investigated include the morphological T type for 18000 galaxies assigned by an artificial neural network.

Keywords: Artificial neural networks, bivariate galaxy luminosity function

1. Introduction

The galaxy luminosity function (LF) is defined as the comoving number density of galaxies from luminosity L to $L + dL$. The overall LF is well fit by a Schechter function, which can be written as $\phi(L)dL = \phi^* e^{-X} X^{\alpha} \, dX$ where ϕ^* = normalisation, α = faint end slope, $X = L/L^*$, L = luminosity and L^* characteristic L. The dependence of luminosity on other intrinsic galaxy properties is best investigated by calculating the bivariate distribution of luminosity and the other parameter. We use the stepwise maximum likelihood method to calculate the bivariate LF.

2. Results for Various Galaxy Properties

The bivariate LF has been computed for various galaxy properties in the SDSS Data Release 2 to $r \leq 17.6$, including eClass spectral type, inverse concentration index (Petrosian 50% / 90% light radius), restframe colours K corrected to $z = 0.1$, surface brightness and morphological T type with 0 = E, 1 = S0, 2 = Sa, 3 = Sb, 4 = Sc, 5 = Sd, 6 = Irr. The morphological type is assigned by a supervised artificial neural network for 18000 galaxies to $r \leq 15.9$ and an RMS accuracy of ± 0.5 T types, comparable to human experts. One can see a clear correlation between luminosity and the LF shape on morphological type.

D. Block et al. (eds.), Penetrating Bars through Masks of Cosmic Dust, 771–772.

The ellipticals may show a Gaussian LF. Further details of the neural network types are in Ball et al. (2004) and on the LFs in Ball et al. (in preparation).

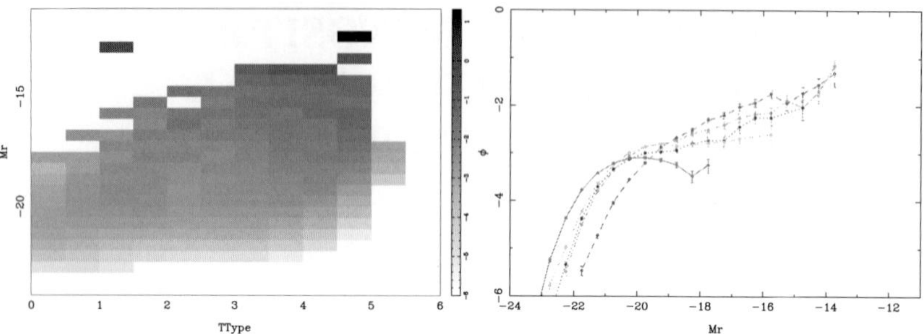

Figure 1. Luminosity function bivariate with morphological T type assigned by a supervised artificial neural network. The left hand panel shows the bivariate LF (increasing from grey to black) as a function of absolute magnitude and type. The right hand panel shows the LF in slices corresponding to T types 0–1 ... 4–5 at bright M_r. There are insufficient galaxies in the 5–6 bin to show up. The error bars are from jackknife resampling.

3. Summary

The SDSS dataset enables the galaxy luminosity function to be measured bivariate with numerous galaxy properties. These include the morphological T type for 18000 galaxies assigned by an artificial neural network. Many further properties can be investigated, and fitting of analytical functions such as a Schechter-Gaussian to the resulting distributions is in progress.

Acknowledgments

The authors wish to thank Drs. M. Fukugita and O. Nakamura for the provision of the eyeball-assigned morphological types used to train the neural network. Funding for the creation and distribution of the SDSS Archive has been provided by the Alfred P. Sloan Foundation, the Participating Institutions, the National Aeronautics and Space Administration, the National Science Foundation, the U.S. Department of Energy, the Japanese Monbukagakusho, and the Max Planck Society. The SDSS Web site is http://www.sdss.org/.

References

Abazajian, K., et al. (2004). "The Second Data Release of the Sloan Digital Sky Survey", ApJ, in press, astro-ph/0403325

Ball, N. M., et al. (2004). "Galaxy Types in the Sloan Digital Sky Survey Using Supervised Artificial Neural Networks", MNRAS, 348, 1038

York, D. G., et al. (2000). "The Sloan Digital Sky Survey: Technical Summary", AJ, 120, 1579

GLOBULAR CLUSTER SYSTEMS AND SUPERMASSIVE BLACK HOLES

Roberto Capuzzo-Dolcetta
Dep. of Physics, Univ. of Roma La Sapienza, P.le A.Moro 2, Roma, Italy

Abstract The evolution of the globular cluster system in a galaxy has relevant implications on the overall galactic structure, both on a large and on a small scale. In particular, the effect of dynamical friction on massive globulars in elliptical galaxies leads to an accumulation of stellar matter in the inner galactic zone that may feed a stellar black hole to make it grow up to supermassive sizes.

Keywords: Clusters: globular, galaxies: elliptical; black holes.

1. Galactic black holes and Globular Cluster Systems

In this report we give a *flash* summary of results on the role of the Globular Cluster System (GCS) of a galaxy on the growth and activity of a central galactic black hole. These results are partly published (Capuzzo-Dolcetta 2004a) and partly going to be published in a forthcoming paper (Capuzzo-Dolcetta 2004b). We refer to the model developed by Capuzzo-Dolcetta (2002).
The GCS is composed by 1000 single mass ($M_0 = 2 \times 10^6$ $M_\odot$) GCs whose orbital velocity dispersion (σ) defines a 'normal' system, as that whose $\sigma = 330$ km s^{-1} is equal to the average σ of stars generating the galactic potential, while the choice of $\sigma = 165$ km s^{-1} corresponds to a 'cold' system and $\sigma = 660$ km s^{-1} to a 'hot ' system. The time behaviour of ρ_* (Fig 1 d) has a correspondence in the time behaviour of the nuclear luminosity, L_n, that shows a fast growth followed by gentler rise up to a short super-Eddington phase followed by a sudden fall to an almost flat level. In the interval of time from 1.3 Gyr (*cold* GCS) to 6 Gyr (*hot* GCS) the b.h. mass grows up to the values of very massive black holes: 1.8×10^9 $M_\odot$, 7.5×10^8 $M_\odot$ and 7.4×10^8 $M_\odot$, in all the three cases studied. A relevant result is that the final black hole mass is quite similar in the various cases studied here: the range of values of the present day central black hole is $0.7 \div 2 \times 10^9$ $M_\odot$.

D. Block et al. (eds.), Penetrating Bars through Masks of Cosmic Dust, 773–774.

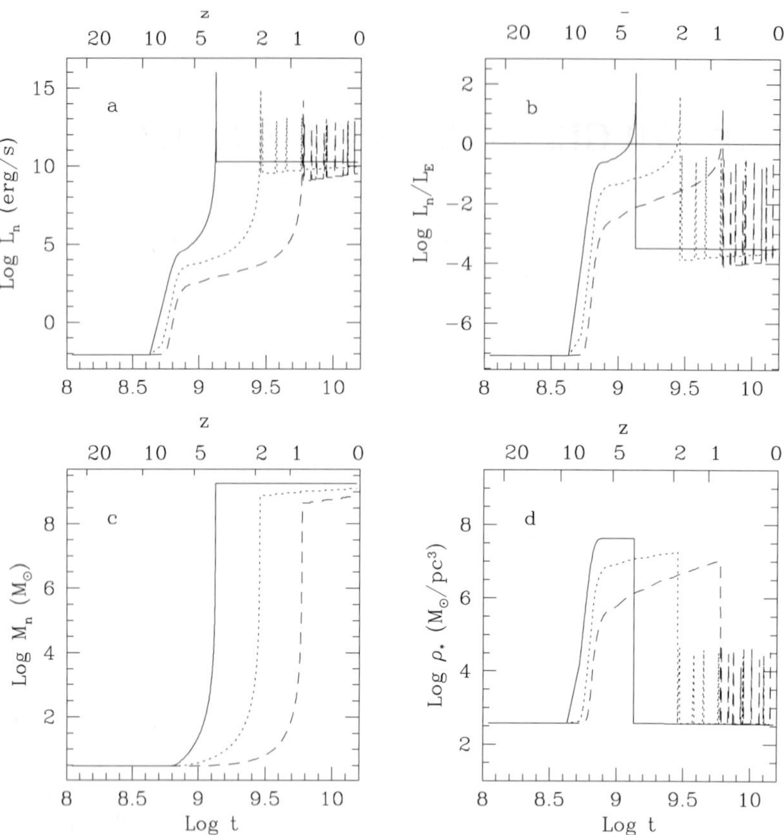

Figure 1. The model refers to a galactic b.h. with initial mass M_{n0} =3 $M_\odot$. Panel a: time evolution of the galactic nuclear luminosity induced by GC mass accretion; solid line refers to the *cold* GCS; dotted to the *normal* GCS; dashed to the *hot* GCS (time is in yr in Log scale; the upper abscissa is the corresponding red-shift in the Einstein-De Sitter model). Panel b: nuclear luminosity in units of Eddington's luminosity. Panel c: time evolution of the galactic nuclear mass. Panel d: time evolution of the supercluster star mass density.

References

Capuzzo-Dolcetta, R. 2002, in E. K. Grebel and W. Brandner , editors, *Proceedings of the International Congress, Modes of Star Formation and the Origin of Field Populations*, ASP Conference Proceedings, Vol. 285, 389

Capuzzo-Dolcetta, R. 2004a, Elliptical galaxy nuclei activity powered by infalling globular clusters, invited talk at the Workshop *Baryons in cosmic structures* held in M.Porzio, october 20-21 2003, to be published in the Proceedings, Giallongo et al. eds.

Galactic Nuclei Activity Powered by Globular Cluster Mass Accretion, 2004b, in preparation

PSEUDOBULGES IN BARRED S0 GALAXIES

Peter Erwin, John E. Beckman, and Juan Carlos Vega Beltran
Instituto de Astrofisica de Canarias, La Laguna, Tenerife, Spain

Abstract We present preliminary results from an ongoing study of the bulges of S0 galaxies. We show that in a subsample of 14 *barred* S0 galaxies, fully half the photometrically defined bulges show kinematic signatures of *pseudobulges* – that is, their kinematics are dominated by rotation. In four of these galaxies, we identify at least two subcomponents in the photometric bulge region: flatter, disk or bar components, assocated with disklike kinematics, and rounder "inner bulges," which appear to be hotter systems more like classical bulges.

Keywords:

1. Introduction

The centers of disk galaxies are conventionally thought to be dominated by a "classical" bulge: a spheroidal, kinematically hot structure similar to a small elliptcal galaxy, which produces the central excess of light above that of the exponential disk. However, Kormendy (1982, 1993) argued that in at least some galaxies the central light came from flattened, disklike components he termed "pseudobulges," possibly the result of bar-driven gas inflow and star formation. At present, very little is known about the demographics of such structures. To address this, we are undertaking a systematic study of bulge morphology and kinematics in early-type disk galaxies.

2. A High Fraction of S0 Pseudobulges

We have performed a preliminary analysis for about two-thirds of the barred S0 galaxies in our sample, using long-slit spectroscopy from the ISIS spectrograph of the 4.2m WHT and kinematic data from the literature (Kormendy 1982; Simien & Prugniel 2000; Caon et al. 2000). The results are shown in Figure 1, which plots $V_{\rm max}/\sigma$ for the photometric bulge regions of fourteen barred S0 galaxies. Half of the galaxies lie on or slightly below the isotropic oblate rotator line, indicating kinematics like those of classical bulges or less luminous ellipticals (e.g., Davies et al. 1983). But the other half lie above, making them kinematic pseudobulges. The seven pseudobulges see here im-

D. Block et al. (eds.), Penetrating Bars through Masks of Cosmic Dust, 775–776.

ply a lower limit of 30% for the entire sample of barred S0s. For four of these seven – NGC 2787, NGC 2950, NGC 3945, and NGC 4371 – we have morphological and/or kinematic evidence that the pseudobulge consists of *two* components (e.g., Erwin et al. 2003): a disklike region (distinct from the galaxy's main, outer disk) and a central region more like a classical bulge.

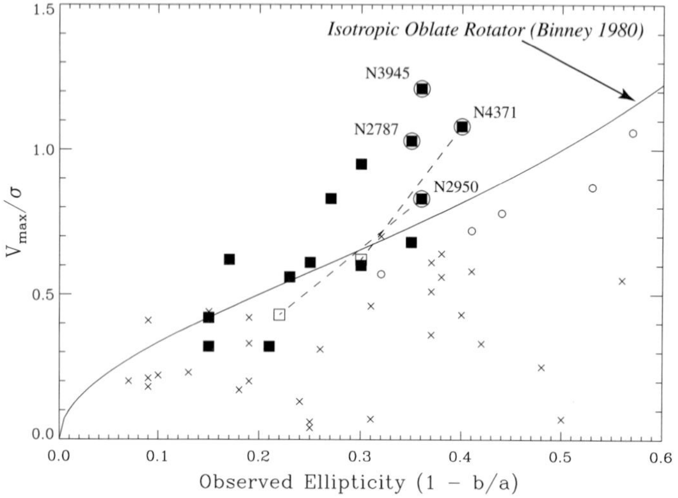

Figure 1. Stellar kinematics of barred S0 bulges: the ratio of maximum velocity $V_{\max}$ to mean velocity dispersion σ in the photometric bulge region versus the ellipticity of that region. The solid line is the relation for isotropic oblate rotators (i.e., "classical" bulges and lower-luminosity ellipticals; Binney 1980); galaxies below it are probably dominated by velocity anisotropy, while galaxies above it are dominated by rotation. The crosses are elliptical galaxies from Davies et al. (1983); the open circles are edge-on bulges in early-type disk galaxies (Kormendy & Illingworth 1982; Jarvis & Freeman (1985). The barred S0 bulges from our sample are filled boxes. Circled bulges have morphological evidence for multiple components in the photometric bulge region; open boxes indicate kinematics for the innermost regions of NGC 2950 and NGC 4371.

References

Binney, J. 1980, MNRAS, 190, 421.

Caon, N., Macchetto, D., & Pastoriza, M. 2000, ApJS, 127, 39.

Erwin, P., Vega Beltran, J. C., Graham, A. W., & Beckman, J. E. 2003, ApJ, 597, 929.

Davies, R. L. et al. 1983, ApJ, 266, 41.

Jarvis, B. J., & Freeman, K. C., 1985, 295, 324.

Kormendy, J. 1982, ApJ, 257, 75.

Kormendy, J., 1993, in *Galactic Bulges, IAU Symposium 153*, ed. H. Dejonghe & H. J. Habing (Dordrecht: Kluwer), 209

Kormendy, J., & Illingworth, G. 1982, ApJ, 256, 460.

Simien, F., & Prugniel, Ph. 2000, A&AS, 145, 263.

BAR AND SPIRAL TORQUES IN THE TRIANGULUM GALAXY M33

Robert Groess[1], David L. Block[1] and Ivanio Puerari[2]
[1]*School of Computational and Applied Mathematics, University of the Witwatersrand, Johannesburg, South Africa,* [2]*Instituto Nacional de Astrofisica, Optica y Electronica, Puebla, Mexico*

Abstract The Triangulum Spiral (M33=NGC 598) presents a daunting task to near-infrared observers, due to its huge spatial extent. From the deepest near-infrared (JHK$_s$) images of M33 yet secured, we identify a bar and calculate the gravitational torque, the bar torque as well as the spiral arm torque. Our methodology reveals the presence of a "class 2" bar, meaning that the tangential force field reaches a maximum of $\sim 20\%$ relative to the axisymmetric radial force field. Furthermore, we are able for the first time to quantify the Reynolds (1927) spiral arm index for M33 (described as 'broad': low arm-interarm contrast), and estimate that the spiral arm torque reaches a maximum of only 15%.

Keywords: Gravitational torque, bar torque, spiral arms, Triangulum Spiral, M33

1. Introduction

The presence of bars and spiral arms in a disk galaxy implies a non – axisymmetric gravitational field (Buta and Block 2001). At least 75% of all spiral galaxies at low z (and perhaps extending to high z) show the presence of an oval/bar (de Vaucouluers 1963, Eskridge et al. 2001) at rest frame J ($1.2\mu m$), H ($1.6\mu m$) and K$_s$ ($2.2\mu m$). The maximum value of the ratio of the tangential force to the mean axisymmetric background radial force is a very useful measure quantifying this field, and defines the gravitational torque parameter Q_g (Sanders and Tubbs 1980, Combes and Sanders 1981). The contribution of the gravitational torque from various components of disk galaxies can be obtained by separating bars and spiral arms using relative Fourier intensity amplitudes. The relative bar and spiral arm torques of M33 are calculated from a special set of deep, 2MASS images, kindly provided to us by Tom Jarrett.

The Triangulum galaxy, M33, extends over one angular degree along its major axis and is particularly plagued by the ubiquity of foreground (Milky Way) stars, about 2500 per square degree down to 15th magnitude at K$_s$ (Block et al. 2004a). A special set of Two Micron All Sky Survey (2MASS) images at J, H and K$_s$ of the M33 region were combined to form a large-area (0.864 deg^2)

D. Block et al. (eds.), Penetrating Bars through Masks of Cosmic Dust, 777–780.

mosaic. The integration time of each of these images was six times that of the nominal 2MASS survey (Block et al. 2004a) and they correspondingly extend approximately 1 magnitude deeper. The foreground Milky Way objects separate from M33 stars both in colour and magnitude at similar galactic latitudes. For a detailed description see Block et al. (2004b).

2. Gravitational Torques

The procedure used to calculate Q_g is called the Gravitational Torque Method (GTM) by Block and Buta (2001) and rests on some basic assumptions:

- Near-IR light traces the mass distribution in a galaxy.
- Galaxies can be deprojected as thin, circular disks.
- The vertical density distribution can be represented as exponential with a scaleheight h_z.

By assuming a constant mass to light ratio across much of the disk, the 2D potential can be derived using Fast Fourier transform techniques (Binney and Tremaine 1987, Quillen, Frogel and Gonzalez 1994). The gravitational torque Q_g measures the strength of the non–axisymmetric potential and can be calculated as the tangential force in terms of the mean radial force as follows:

$$Q_g(R) = F_T(R)/ < F_R(R) > \tag{1}$$

where $F_T(R)$ is the tangential force at each R and $< F_R(R) >$ is the mean radial force at the same R (Combes and Sanders 1981).

Bars are features that would be dominated by even relative Fourier intensity amplitude (I_m/I_0) terms, where m is an integer index (Buta, Block and Knapen 2003). Synthetic reconstruction of the M33 bar using only the even term signatures and applying the GTM, shows the presence of a “class 2” bar ($Q_b^{max} = 0.21$) (see Figure 1). Synthetic reconstruction of the entire Fourier-smoothed image (i.e.: $m = 0, 1, ..., 20$) and subtracting the even terms contribution representing the bar, gives the spiral arms torque on top of the azimuthally symmetric background disk ($m = 0$ component) as $Q_s^{max} = 0.15$. The total gravitational torque is calculated by finding the maximum of the sum of the bar and spiral arms contribution, and is found to be $Q_g^{max} = 0.22$. The values we have calculated assume a vertical scaleheight of $h_z = 325$ pc (see Figure 2).

References

Binney, J., Tremaine, S. 1987, Galactic Dynamics, Princeton: Princeton Univ. Press
Block, D.L et al. 2004a, this volume
Block, D.L. et al. 2004b, A&A Letters (to appear), *astro-ph/0406485*

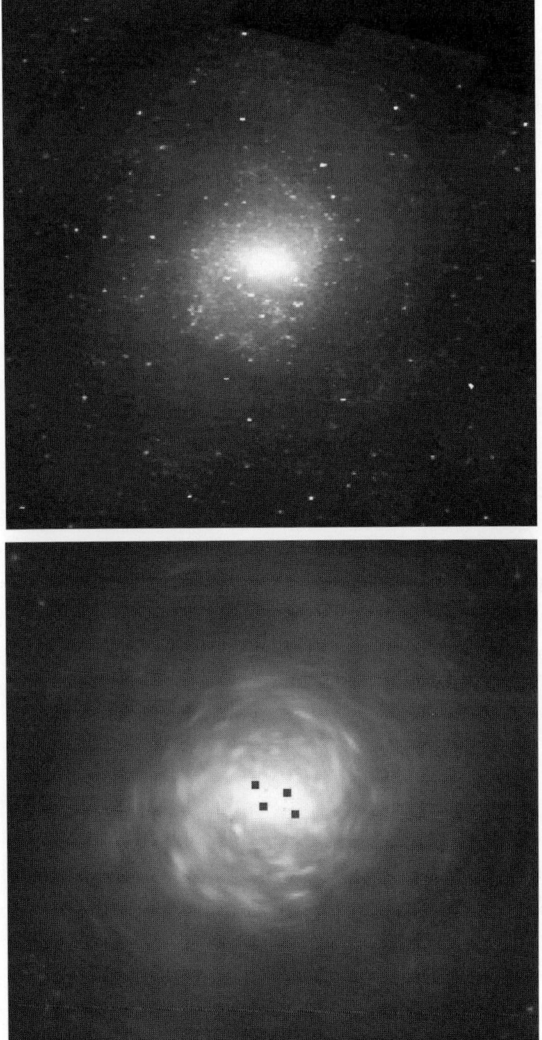

Figure 1. Top: Deprojected, deep 2MASS K_s–band image. Bottom: Synthetic Fourier-smoothed image ($m = 0, 1, ..., 20$). The dots represent the positions of the maximum gravitational bar torque.

Buta, R., Block, D.L. 2001, ApJ, 550, 243.
Buta, R., Block, D.L., Knapen, J.H. AJ 2003, 126, 1148
Combes, F., Sanders, R.H. 1981, A&A, 96, 164.
de Vaucouleurs, G. 1963, ApJS, 8, 31.
Eskridge, P.B., et al. 2000, AJ, 119, 536.
Quillen, A.C., Frogel, J.A., Gonzalez, R. 1994, ApJ, 437, 162

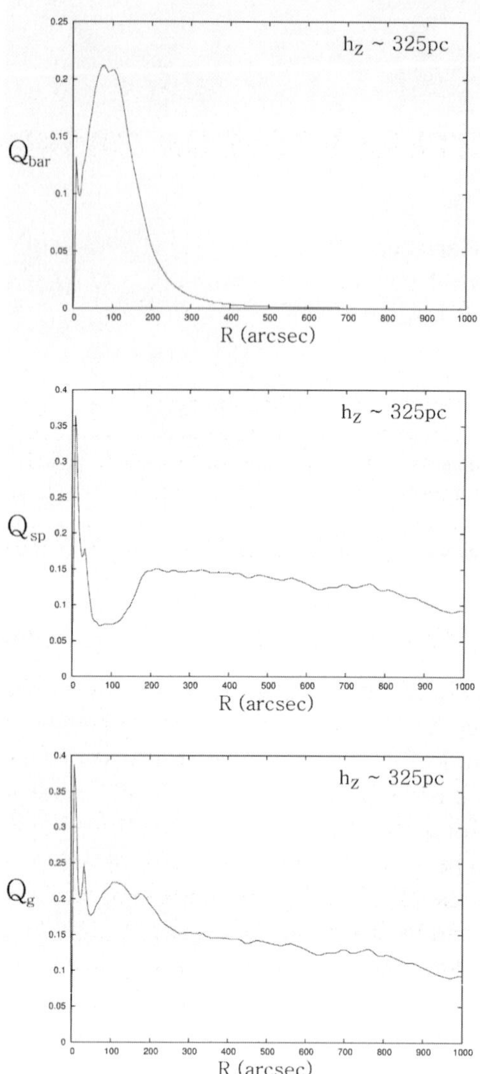

Figure 2. From top to bottom: $Q_b(R)$ (bar+disk), $Q_s(R)$ (spiral+disk), and $Q_g(R)$ (total image).

Reynolds, J.H. 1927, Observatory, 50, 185

Sanders, R.H., Tubbs, A.D. 1980, ApJ, 235, 803.

BAR PARAMETERS FROM Hα OBSERVATIONS

O. Hernandez[1,2], C. Carignan[1], P. Amram[2] and O. Daigle[1,2]
[1]*LAE, Universite de Montreal, C.P. 6128, Succ. centre ville, Montreal, Que. H3C 3J7, Canada*
[2]*OAMP, 2 Place Le Verrier, F–13248 Marseille Cedex 04, France*

Abstract The very first results of the application of the Tremaine-Weinberg (hereafter TW) method on Hα velocity fields are presented to find pattern speeds of galaxies. The technique is used for a sample of four barred galaxies and M51. For two of these, our results are very similar to those obtained using the same technique but based on CO observations. For the three other galaxies, a comparison is provided between this observational method and numerical ones.

1. Tremaine-Weinberg method with Hα velocity fields

The pattern speed Ω_p of bars or spiral pattern is one of the most important parameter to understand the kinematics and the dynamics of spiral galaxies. On one hand, a dynamical method has been proposed by Garcia-Burillo et al. (1993) to model the gas behaviour in the potential derived from an infrared image of the galaxy for a varying range of Ω_p. On the other hand, three observational methods can be used: the inner resonance 4:1 (Elmegreen et al 1992), the sign inversion of the radial streaming motions across corotation (e.g. the Canzian (1993) test) and the TW (1984) method. The latter is derived from the continuity equation. The tracer (stars, HI, CO, etc...) must follow this equation. For Hα, we supposed that, to first order, this equation is satisfied for very short times ($\ll$ dynamical time) and we neglect the internal kinematics of HII regions. In that case, in the galactic frame with respect to the galactic center, the luminosity-weighted mean velocity $<\mathcal{V}_{los}>$ in the strip, divided by the component of the luminosity-weighted mean position $<\mathcal{X}>$ parallel to the line of node, is equal to $\Omega_p \sin i$ ($= <\mathcal{V}_{los}>/<\mathcal{X}>$) .

2. Results: pattern speed of five galaxies

Velocity fields and monochromatic images are obtained using 3D adaptive smoothing (Daigle et al, 2004). The position angle (PA), inclination and systemic velocity have been calculated using GIPSY and KARMA (for the detailed description of the method see Hernandez et al. 2004). The figure shows the

D. Block et al. (eds.), Penetrating Bars through Masks of Cosmic Dust, 781–782.

results obtained for NGC 2903. The fit was performed using a robust Chi-squared model. Ω_p is found to be 66±7 km/s/kpc. The distance has been chosen such that we can compare with Ω_p from other sources. The error calculation takes into account the error on the inclination, on the mean line of sight velocity, on the mean position along the line of node and on the PA. The results for the whole sample are shown in the table. The TW method appears to be efficient to derive Ω_p, even if Hα does not satisfy the continuity equation for a very long time.

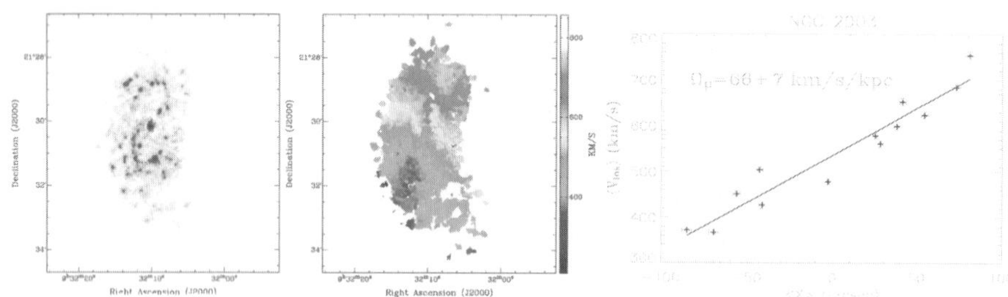

Galaxy name	*Inc* (°)	*PA* (°)	Ω_p *(Hα)* *(km/s/arcsec)*	Ω_p *(Hα)* *(km/s/kpc)*	Ω_p *(CO)* *(km/s/kpc)*	Ω_p *(mv)*[1] *(km/s/kpc)*	*D* *(Mpc)*
NCG 2903	61.4	20	2.3± 0.2	66± 7	n/a	68[a]	7.3
NCG 3359	54	-10	2.0± 0.3	30± 4	n/a	27[b]	13.4
NGC 4321	31.7	153	2.6± 0.3	33± 5	n/a	20[c]	16.1
NGC 6946	38	240	1.1± 0.2	42± 6	39± 9[d]	n/a	5.5
NGC 5194	20	-10	1.8± 0.3	40± 8	38± 7[d]	27[e]	9.5

[a]Helfer et al. 2003 [b]Sempere 1999 [c]Sempere et al. 1995 [d]Zimmer et al 2004 [e]Salo & Laurikainen 2003
[1]mv: modeled values

References

Canzian B., 1993, PASP, 105, 661
Daigle, O., Carignan, C., & Hernandez, O. 2004, in preparation
Elmegreen B. G., Elmegreen D. M., Montenegro L., 1992, ApJS, 79, 37
Garcia-Burillo S., Combes F., Gerin M., 1993, A&A, 274, 148
Helfer T. T., et al., 2003, ApJS, 145, 259
Hernandez, O., Carignan, C., Amram P., & Wozniak H. 2004, in preparation
Tremaine S., Weinberg M. D., 1984, ApJ, 282, L5
Salo H., Laurikainen E., 2000, MNRAS, 319, 393
Sempere M., 1999, Ap&SS, 269, 665
Sempere M. J., Garcia-Burillo S., Combes F., Knapen J. H., 1995, A&A, 296, 45
Zimmer P., Rand R. J., McGraw J. T., 2004, astro-ph/0404365

MOLECULAR GAS IN CLASSICAL ELLIPTICAL RADIO GALAXIES

Jeremy Lim[1], Stephane Leon[2], Francoise Combes[3], and Dinh-V-Trung[1]
[1]*Institute of Astronomy & Astrophysics, Academia Sinica, PO Box 23-141, Taiepei 106, Taiwan*
[2]*Instituto de Astrof«sica de Andaluc«a, CSIC, Apdo. Correos 3004, 18080 Granada, Spain*
[3]*Observatoire de Paris, LERMA, 61 Av. de l'Observatoire, F-75014, Paris, France*

Abstract We report the results of two comprehensive surveys for CO molecular gas in classical (i.e., extended double-lobed) radio galaxies, all of which are classified as elliptical (or perhaps S0) galaxies. Our first survey comprises all twenty-four elliptical galaxies in the revised 3C catalog at redshifts $z \leq 0.031$; only one, 3C 84 (Perseus A), had previously been detected in molecular gas, with a mass of $\sim 10^{10}$ $M_{\odot}$. We detected four galaxies in this sample for the first time. Our second survey comprises all twenty-one classical elliptical (or SO) radio galaxies in the northern hemisphere at $z \leq 0.023$ exceeding a selected radio flux density, including seven of the 3C galaxies (four detected) in our first survey. We deteced another four galaxies in this sample for the first time. Six of the eight galaxies detected have molecular gas masses in the narrow range $2\text{–}4 \times 10^8$ $M_{\odot}$, which is comparable to that in Centaurus A, and in all span the range $1 \times 10^7\text{–}1 \times 10^9$ $M_{\odot}$. All have very broad double-horned line profiles characteristic of a central rotating disk or torus. The remaining galaxies have upper limits in molecular gas mass of a few 10^8 $M_{\odot}$, and in one case a few 10^7 $M_{\odot}$. By contrast, previous surveys of radio galaxies had found molecular gas masses of $10^9\text{–}10^{10}$ $M_{\odot}$, but these surveys preferentially selected IR-bright (i.e., dusty) galaxies, and many of the detections are of systems with compact (nuclear) radio emission hosted by disk or merging galaxies. Our results suggest that the molecular gas in most classical elliptical radio galaxies probably originate from the cannibalism of a relatively low-mass but gas-rich galaxy by the dominant host elliptical galaxy, as proposed by Lim et al. (2000) based on early results from the first survey.

Keywords: Galaxies: active — galaxies: elliptical — galaxies: ISM

1. The Poster

We presented the results from our two surveys, both performed with the IRAM 30-m telescope. The results for just the detected galaxies are summarized in Table 1. We showed their measured CO line profiles, and the dust features imaged in the central regions of these galaxies with the HST (Martel et al. 1999; Verdoes-Kleijn et al. 1999). An example is shown in Figure 1 (two

D. Block et al. (eds.), Penetrating Bars through Masks of Cosmic Dust, 783–784.

others can be found in Lim et al. 2000). For 3C 31, follow-up imaging with the IRAM PdBI reveals that the molecular gas lies in a central rotating disk or torus with a size similar to the dust disk.

Table 1. Radio Galaxies detected in CO Molecular Gas

Name	*Redshift (z)*	*Molecular Gas Mass ($M_\odot$)*
NGC 315	0.016	$(3.03 \pm 0.34) \times 10^8$
NGC 383 (3C 31)	0.017	$(1.06 \pm 0.04) \times 10^9$
NGC 541	0.018	$(2.47 \pm 0.38) \times 10^8$
NGC 3801	0.011	$(3.77 \pm 0.22) \times 10^8$
NGC 3862 (3C 264)	0.022	$(3.09 \pm 0.26) \times 10^8$
NGC 4374 (3C 272.1)	0.003	$(1.40 \pm 0.38) \times 10^7$
NGC 7052	0.016	$(3.45 \pm 0.23) \times 10^8$
UGC 12064 (3C 449)	0.018	$(2.85 \pm 0.23) \times 10^8$

Galactic CO intensity to molecular gas mass conversion, $H_o = 65$ km s^{-1} Mpc^{-1}, and $\Omega_o = 1.0$.

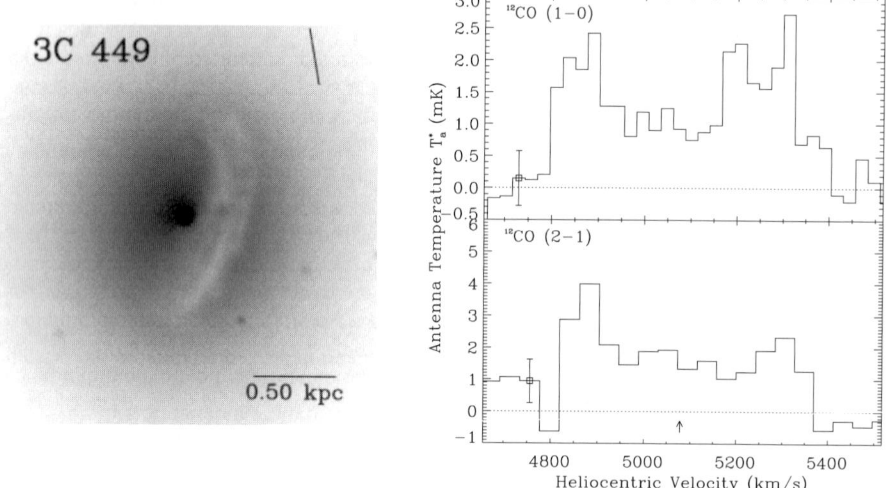

Figure 1. Left panel: Negative grayscale image of the central region of 3C 449, showing what appears to be an inclined disk or torus (from Martel et al. 1999). The dash line indicates the direction of the radio jet. Right panels: Line profiles in both ^{12}CO(1-0) (upper panel) and ^{12}CO(2-1) (lower panel) measured towards the central region of 3C 449. The error bar indicates the $\pm 1\sigma$ uncertainty, and the arrow indicates the optical heliocentric velocity.

References

Lim, J., Leon, S., Combes, F., & Dinh-V-Trung 2000, ApJ, 545, L93
Martel, A. R., & 9 other authors 1999, ApJS, 122, 81
Verdoes Kleijn, G. A., Baum, S. A., de Zeeuw, P. T., & O'Dea, C. P. 1999, ApJ, 118, 2592

BARS AND DUST IN MULTIWAVELENGTH MAPS OF SIMULATIONS OF GALAXY FORMATION

Paola Mazzei[1] and Anna Curir[2]
[1]*Osservatorio Astronomico, Vicolo dell'Osservatorio 3, 35128 Padova, Italy,* [2]*Osservatorio Astronomico, Torino, Italy,*

Abstract Bars are a common feature in simulations of galaxy formation of isolated collapsing systems initially composed of dark matter and gas. They last until 16 Gyr, the final time of our runs. Dust seriously affects bar luminosity, especially in the shorter wavelength spectral range. In more massive systems, showing early-type morphologies and dynamical properties, recent star formation occurs in bars which entail $\sim 42\%$ of their total B luminosity. However such 'intrinsic' bars are strongly affected by dust which reduces their observed B luminosity to $\sim 20\%$ of the total observed , by changing their morphology. In the less massive systems this fraction amounts to only $\sim 6\%$ so that the bar survives to dust effects.

Keywords: Galaxy formation, galaxy evolution, chemo-photometric properties, DM halos

1. Introduction

We perform a large set of Smoothed Particle Hydrodynamic (SPH) simulations including chemo-photometric evolutionary population synthesis models of isolated collapsing systems endowed with triaxial dark matter (DM) halos, gas and the star formation switched on (Mazzei & Curir, 2003; Mazzei, 2003). Cooling, shocks, star formation and in particular feedback are self-consistently accounted for. Our results shed light into the dependence of the system evolution on some parameters so far unexplored such as its total mass, initial geometry and dynamical state. Moreover, the effects of different values of the baryon's to total mass ratio have been investigated. We find intriguing connections between dark and luminous matter. The initial properties of the halo drive the galaxy formation and evolution. We point out important issues concerning the galaxy morphological types. In our framework spiral galaxies arise only in systems with both a total mass lower than $10^{12}\ m\odot$ and a critical value of the baryon's to total mass ratio, around its cosmological value, 0.1. For the same value of such a ratio, the more massive systems host more luminous and redder galaxies, as a consequence of the stronger bursts of star formation arising into

D. Block et al. (eds.), Penetrating Bars through Masks of Cosmic Dust, 785–786.

their halos earlier on. Moreover the star formation does not switch on in the less massive DM halos so that the critical mass to entail a luminous system is $\simeq 10^{11}\, m\odot$. For a given total mass, higher values of such a ratio provide baryons with lower rotational support so that no constraints to the total mass of the system hosting an elliptical galaxy have been derived.
Dust absorption is modeled according to Mazzei et al. (1992), $E(B-V) = N(H)/(4.8 \times 10^{21}) \times (Z_{gas}/0.02)$ where $N(H)$ and Z_{gas} are column density and metallicity of gas.

2. Results

We find that bars are a common feature in simulations of galaxy formation until 16 Gyr, the final time of our runs. Dust seriously affects B maps as shown in Fig. 1 for two models with different total masses. In the more massive case, corresponding to a total B un-reddened magnitude, $M_B = -20$, the bar is about 7 kpc lengths and accounts for 42% of blue intrinsic luminosity, L_B. The dust, very concentrated and distributed in a thin disk, reduces L_B of the bar by 52%, so changing the B morphology of the galaxy. In the less massive case, $M_B = -17.7$, the bar is about 6 kpc lengths and accounts for 27% of L_B. This becomes only 6% including the dust even if dust is devoid in the very inner regions of such a system. So in the more massive case the residual star formation occurs in a 'intrinsic' bar which, for dust effects, changes morphology entailing about 20% of the observed B luminosity of the galaxy ($M_B = -19.7$). In the less massive case, the bar entails only 6% of the observed B luminosity ($M_B = -17.6$) and survives to dust effects.

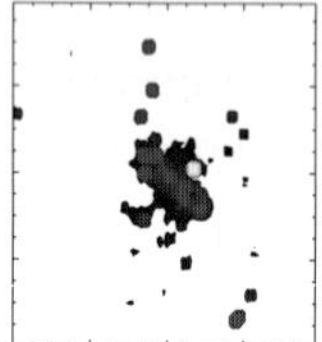
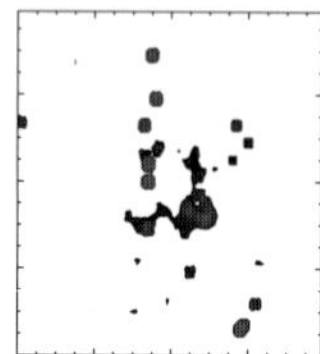
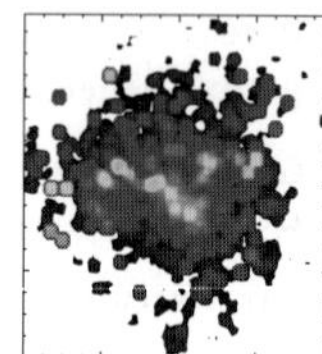
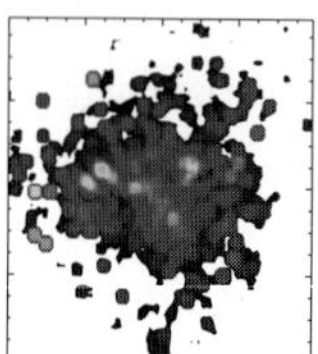

Figure 1. B isophotes (60 contours) at 15.4 Gyr in the rest-frame of systems with total mass $2 \times 10^{12}\, m\odot$, (first and second panel at the left), and $2 \times 10^{11}\, m\odot$, (two right panels). Left panels show xy projection of un-reddened (first) and reddened (second) maps with a luminosity contrast of 100 and a spatial resolution of 0.5 kpc; the same as for right panels.

References

Mazzei, P. (2003). Recent Res. Devel. in Astr. and Astroph., Research Signpost, 1, 457
Mazzei, P, Curir, A. (2003). ApJ 591, 784
Mazzei, P, Xu, C, De Zotti, G. (1992). A&A 256, 45

HOW BARRED IS THE NIR NEARBY UNIVERSE? AN ANALYSIS USING 2MASS

Karin Menendez-Delmestre, Kartik Sheth, Nick Scoville, Tom Jarrett[1], Eva Schinnerer[2], Michael W. Regan[3], and David L. Block[4]
[1]*California Institute of Technology, Pasadena, CA,* [2]*National Radio Astronomy Observatory,* [3]*Space Telescope Science Institute,* [4]*University of the Witwatersrand*

Abstract We determine a firm lower limit to the bar fraction of 0.58 in the nearby universe using J+H+K-band images for 134 spirals from 2MASS. With a mean deprojected semi-major axis of 5.1 kpc, and a mean deprojected ellipticity of 0.45 this local bar sample lays the ground work for studies on bar formation and evolution at high redshift.

The Two Micron All Sky Survey (2MASS) offers a rich source of images for nearby galaxies in K-band, the optimal band for detecting bars. We selected all spiral galaxies (S0/a–Sd) with $i < 65°$, which resulted in a final sample size of 134 spirals. We analysed each galaxy using ellipse-fitting to detect the presence of bars (Fig. 1) and characterise their properties. Our method consists of two signature criteria ($\triangle\epsilon \geq 0.1$, $\triangle PA \geq 10°$) that can be applied to any sample. We find a firm lower limit to the bar fraction of 0.58, which is not significantly different from the fraction inferred from B-band observations (de Vaucouleurs, RC3). This indicates that bar morphology is still recognizable in B-band. However, bar properties are better traced in K-band, where star formation and dust obscuration are less severe. We found that bars occupy in average ~35% of the galaxy disk, with a mean bar axial ratio very close to 2:1 (Fig. 2). A marked dearth of strong bars is in agreement with previous studies (Buta & Block, 2001) and has been interpreted as an indication that bars at the current epoch are a second, third, or later generation (Bournaud & Combes 2002). A slight trend between the bar size and the galaxy size/brightness is also found: larger bars appear to live in larger and more massive galaxies. However, no similar trend with the bar strength was found. For more details on the method and analysis, refer to Sheth et al in this volume.

Our study of bar length and ellipticity sets the stage for the evolution of these properties as a function of redshift. Results from high redshift studies (e.g., Sheth et al. 2003) may then be compared to this nearby sample in order to put more stringent constraints on bar evolution and, ultimately, galaxy evolution.

D. Block et al. (eds.), Penetrating Bars through Masks of Cosmic Dust, 787–788.

References

Buta, R., & Block, D.L. 2001, ApJ, 550, 243

Bournaud, F., & Combes, F. 2002, A&A, 392, 83

Sheth, K., Regan, M.W., Scoville, N.Z., & Strubbe, L.E. 2003, ApJL, 592, 13

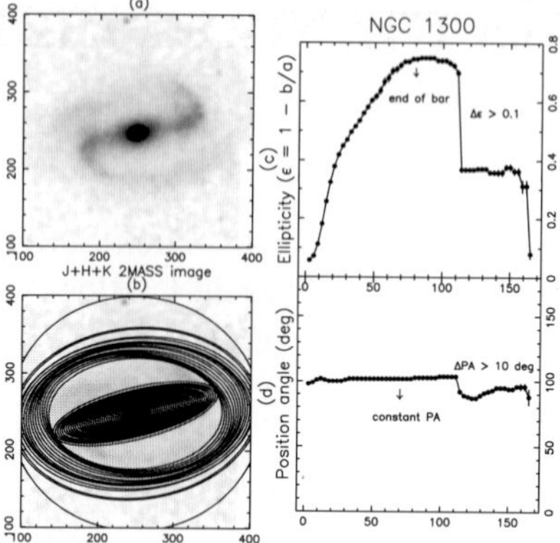

Figure 1. (a) ngc1300 (b) with fitted ellipses. (c) Bar signature: monotonic increase in ellipticity and (d) constant PA followed by sharp change in both parameters.

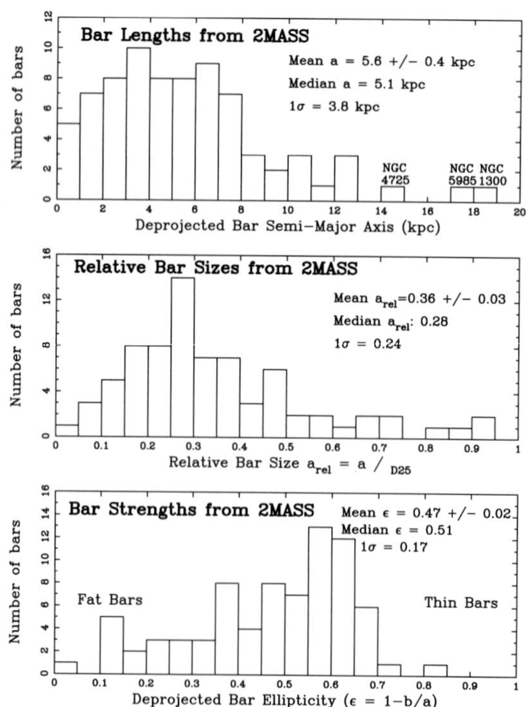

Figure 2. Histograms of bar lengths, relative size and bar strenghts

TRUNCATION OF STELLAR DISKS AT HIGH REDSHIFT

Isabel Perez
Kapteyn Institute, University of Groningen, Postbus 800, Groningen 9700 AV, Netherlands

Abstract We report here the finding that some stellar disks at high redshift are truncated at a shorter radius than disks in the local universe. Surface photometry of a sample of high redshift (0.6<z<1.0) disk galaxies has been carried out to study the shape of the outer disk and the results compared to nearby objects. The stellar disks of 6 out of 16 galaxies show radial truncation. The truncation occurs at $R_{br}/h_R \approx 1.8$ which compared to the value $R_{br}/h_R \approx 4$ for nearby objects reflects systems undergoing rapid evolution.

Keywords: Disk evolution and dynamics, galaxy structure parameters, high-z galaxies

1. Summary and main results from the project

There is evidence that stellar disks are radially truncated beyond a few disk scale lengths (van der Kruit 1979; van der Kruit & Searle 1981). This truncation can be related to a surface density threshold for star formation (Kennicutt 1989). On the other hand, the truncation could also be related to the maximum specific angular momentum of the sphere from which the disk collapsed (van der Kruit 1987).

A study of the truncation radius at redshift ≈1 can enable us to discern between a truncation radius as a relic from galaxy formation or as due to evolution. Such a project has been carried out using HST archived data. A sample of 38 galaxies has been selected with redshifts between 0.6 and 1 from the HDF-S and the HDF-N from the V1.0 GOODS HST/ACS data release.

The galaxies were selected visually to not be strongly interacting and relatively symmetric. Out of the originally selected 38 galaxies, 22 were eliminated from the final list for being too early type, showing bars or clearly disturbed. The galaxies were cleaned from strong foreground and background sources. Ellipse fitting was performed to the light distribution. Averaged profiles of different image segments were also computed. The k-corrections were estimated using the results computed by Fukugita et al. (1995). The obtained azimuthally averaged profiles of three of the truncated galaxies are shown in

D. Block et al. (eds.), Penetrating Bars through Masks of Cosmic Dust, 789–790.

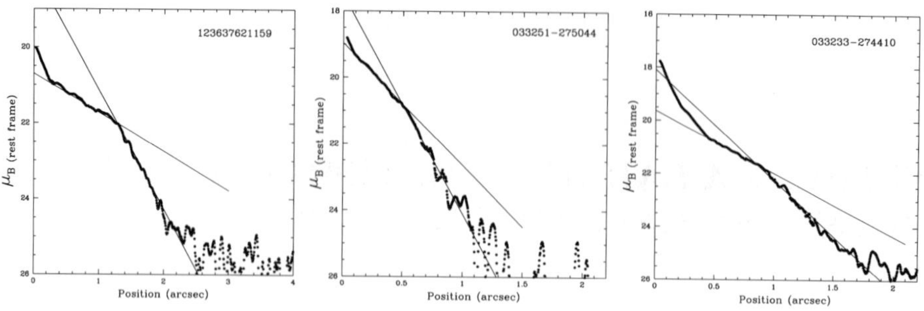

Figure 1. Azimuthally averaged surface brightness profiles for 3 of the 6 galaxies which show truncation. Compare the value of R_{br}/h_R for these high-z galaxies (1.33, 1.82 and 2.16, respectively) with $R_{br}/h_R \approx 4.0$ for nearby objects.

Table 1. 1-D exponential fitting parameters for the sample galaxies.

ID (1)	z (2)	h_R (2)	h_{out} (3)	R_{br} (4)	$\mu_{0,B(rf)}$ (5)	$\mu_{br,B(rf)}$(6)	R_{br}/h_R (7)
123610+621334	0.69	0.59	0.33	0.99	19.7	21.7	**1.68**
123637+621159	0.78	0.99	0.36	1.32	20.6	21.9	**1.33**
033251-275044	0.98	0.28	0.15	0.51	18.9	20.7	**1.82**
033233-274410	0.67	0.45	0.24	0.97	19.6	21.8	**2.16**
123709+622006	1.01	0.36	0.18	0.48	19.2	20.5	**1.35**
123708+621252	0.84	0.51	0.25	1.10	19.6	21.7	**2.15**

(1)Galaxy identification.(2)Inner disk scale length in arcsec.(3)Outer disk scale length in arcsec.(4) Break radius in arcsec.(5)Central surface brightness in the rest frame B-band.(6)Surface brightness at the break radius (rest frame B-band).(7)Ration between (4) and (2).

Fig. 1. In some cases the truncation radius seems to be related to the end of the spiral arm structure. The results from the 1-D exponential fitting of the radial surface brightness profiles are presented in Table 1. The surface photometry shows that the disk truncation does not occur sharply but instead a two-slope model fits better the profiles. This result agrees with the results found for nearby objects (Pohlen 2001 and references within). However, the R_{br}/h_R occurs at $\approx 1.8\pm 0.5$. The transition between the two exponential profiles occurs at a brighter surface density and the central disk surface brightness is a few magnitudes brighter than the Freeman value.

References

Fukugita, M.; Shimasaku, K. & Ichikawa, T. 1995, PASP, 107,945
Kennicutt, R. C. 1989, ApJ, 344, 685
Pohlen, M. 2001, PhD Thesis, Ruhr-University Bochum, Germany
van der Kruit, P.C. 1979, A&A, 38,15
van der Kruit, P.C.& Searle, L. 1981, A&A, 95,105

IS THERE A LARGE STELLAR BAR IN THE LSB GALAXY UGC 7321 ?

Michael Pohlen[1], Marc Balcells[1] and Lynn D. Matthews[2]
[1] *Instituto de Astrofísica de Canarias,* [2] *Harvard-Smithsonian Center for Astrophysics*

Abstract Observations and dynamical modelling suggest that boxy/peanut-shaped (b/p) bulges form through the buckling instability in bars of the parent galaxy disks. Pohlen et al. (2003) present for the first time photometric evidence for peanut-shaped outer isophotes in the edge-on, late-type, low surface-brightness galaxy UGC 7321, a prototype of a purely disk dominated bulgeless galaxy. Here we present first results from our follow-up spectroscopy to search for the presence of a bar through the expected 'figure-8' pattern in the rotation curve.

1. Background

Nearly all disk galaxies have both a stellar disk and a bulge. Nevertheless, there are galaxies which do not have a bulge component. These pure disk galaxies show clearly that the bulge formation is not a necessary outcome of the early galaxy formation process. The low surface brightness galaxy UGC 7321 – a nearby, isolated, "superthin" edge-on galaxy – is an ideal object for studying these disk-dominated, bulgeless galaxies. Although these late-type spirals are believed to exhibit the simplest possible structure, the detailcd studies on UGC 7321 by Matthews et al. (1999), Matthews (2000), and Matthews & Wood (2001) already showed noticeably deviations from a single component, purely exponential disk. Just recently, Pohlen et al. (2003) observed peanut-shaped outer contours in a deep optical image ($\mu \approx 27.2$ R-mag/$\square''$) of UGC 7321. The connection between bars and b/p bulges has been observationally verified by gas kinematics (Kuijken & Merrifield 1995; Bureau & Freeman 1999), and by identifying features in the surface brightness profiles of edge-on galaxies (Lütticke et al. 2000). The peanut in UGC 7321 is obvious in cuts parallel to the major axis (cf. Fig.1a) and if associated with a bar provides first evidence for the thickening process caused by the buckling instability. This would providing valuable clues on the very early phases of the formation of a central dense component by bar-driven mass inflow.

D. Block et al. (eds.), Penetrating Bars through Masks of Cosmic Dust, 791–792.

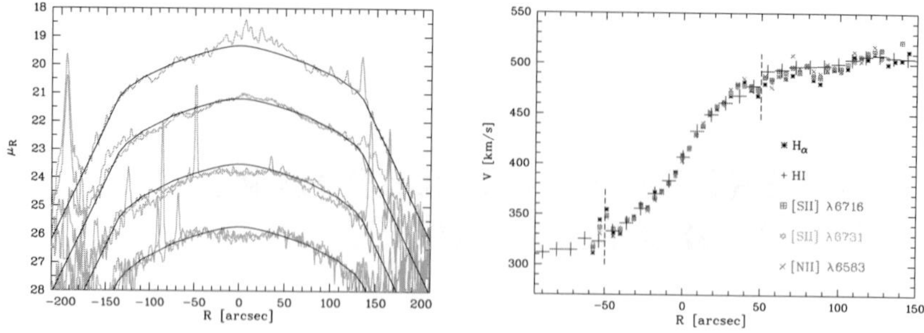

Figure 1. *(a) Left panel:* Radial surface brightness profiles at 0″, ±6″, ±12″and ±18″(shifted by $-2, -1, 0, +1$ mag) overplotted by a single two-slope model (Pohlen et al. 2002). The peanut-shape signature – the central *dip* in the profile – is obvious in the two outermost cuts. *(b) Right panel:* Rotation curve obtained from the emission lines overplotted by the HI data from Uson & Matthews (2003). The *kinks* in the rotation curve appear near the edges of the peanut measured in the optical image marked with the dashed lines.

2. Results

Uson & Matthews (2003) find that the major axis position-velocity profile shows a clearly visible 'figure-8' structure in their recent analysis of HI-data (cf. their Fig. 10). Although thus supporting the photometric evidence by Pohlen et al. (2003) our new ionized gas kinematical measurements *do not* confirm the presence of a bar. Within our resolution of 21 km/s per pixel we do not detect a splitting – typical for the 'figure-8' pattern – of the Hα, sulphur or nitrogen emission lines in the center. However, we find unexpected dips in the rotation curve – symmetric around the center – close to the position of the maximum peanut distortion measured in higher z profiles (cf. Fig.1b). Our conclusion is therefore that either the bar – once producing this peanut structure – is now dissolved or we are missing another mechanism to to create these peanut-shaped contours. An appealing merger scenario is, however, also unlikely since UGC 7321 seems to be quite isolated and the remaining stellar disk is fairly undisturbed for a recent merger.

References

Bureau, M. & Freeman, K. C. 1999, AJ 118, 126

Kuijken K., & Merrifield, M.R. 1995, ApJ, 433, L13

Lütticke, R., Dettmar, R.-J., & Pohlen, M. 2000, A&A, 362, 435

Matthews, L. D., Gallagher, J. S., & van Driel, W. 1999, AJ, 118, 2751

Matthews, L. D. 2000, AJ, 120, 1764

Matthews, L. D. & Wood, K. 2001, ApJ, 548, 150

Pohlen, M., Dettmar, R.-J., Lütticke, R., & Aronica, G. 2002, A&A, 392, 807

Pohlen, M., Balcells, M., Lütticke, R., & Dettmar, R.-J., 2003, A&A, 409, 485

Uson, J. M. & Matthews, L. D. 2003, AJ, 125, 2455

SPIRAL STRUCTURE OF THE MILKY WAY: THE STATE OF AFFAIRS

Thomas Steiman-Cameron[1,3], Mark Wolfire[2] and David Hollenbach[3]
[1]*Astronomy Dept., Indiana Univ., Bloomington, IN 47405, USA,*
[2]*Dept. of Astronomy, Univ. of Maryland, College Park, MD 20742, USA,*
[3]*MS245-3, NASA-Ames Research Center, Moffett Field, CA 94035, USA*

Abstract We have examined all models proposed since the late 1970s for the Galaxy's spiral structure in the context of the all-sky [CII] 158μm and [NII] 205μm cooling line data obtained by the FIRAS instrument of COBE. We present a model which maximizes the agreement between model predictions and FIRAS data.

Keywords: Milky Way, spiral structure

1. Introduction

Over the past twenty-five years more than four dozen models have been proposed for the Milky Way's spiral structure. While these agree that a "global" pattern exists, there is considerable disagreement in the details; even the number of arms is still debated. Underlying much of the uncertainty are difficulties in determining precise spatial locations of spiral tracers. In Steiman-Cameron, Wolfire, & Hollenbach (2004) we propose a new model for the spiral arms and examine existing models in the context of two important cooling lines of the ISM: [CII] 158μm and [NII] 205μm. All-sky intensity maps for these lines were obtained by the Far Infrared Absolute Spectrophotometer (FIRAS) of the Cosmic Background Explorer (COBE) satellite. Both [CII] and [NII] are very sensitive to density, temperature, and incident FUV/UV fluxes, leading emissions from these species to be highly concentrated in the arms. Therefore they provide a powerful tracer of the spiral arms.

Intensity maps of both lines display similar characteristics (*cf.*, Bennett *et al.* 1994). For the sake of brevity, we restrict our discussion here to the [N II] 205 μm transition, a tracer of extended low-density ionized gas. The [N II] 205 μm line intensity in the plane of the Galaxy has maxima on either side of the Galactic center and falls off in a relatively smooth fashion inwards and outwards of these positions (Figure 1). Local maxima in the secular trend are seen at locations thought to represent approximate tangents to spiral arms.

D. Block et al. (eds.), Penetrating Bars through Masks of Cosmic Dust, 793–794.

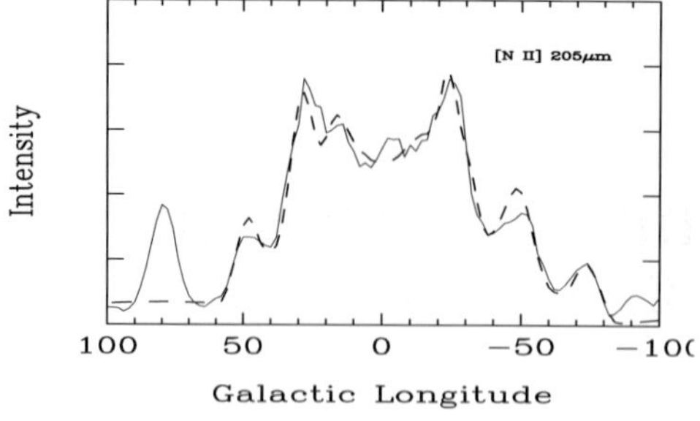

Figure 1. *Left*: Predicted (solid) and observed (dashed) [N II] 205 μm line intensity along the Galactic plane. The feature at $\ell \approx +80^{\circ}$ arises from the Local arm and is not included in the global fit. *Right*: Geometry of the spiral arms producing the fit seen at left. The sun's location is shown.

We constructed 3-dimensional volume emissivity models for $\sim 3 \times 10^{6}$ individual arms. These parametric models specify the arm geometry, orientation, and the manner in which emissivity changes along and away from the arm. Several spiral forms were examined. Line intensities along the Galactic plane were calculated for each model by convolving the model with the FIRAS beam. Least-squares techniques were then used to find the linear combination (lines are optically thin) of two, three, or four-arm intensity profiles from individual models which maximized agreement with the FIRAS data. Only a four-arm model with logarithmic spiral forms was found to be in agreement with the data. The arms are neither symmetric nor identical. Figure 1 compares the observed and model-predicted intensity along the plane and provides a grey-scale depiction of the mid-plane volume emissivity function producing this fit.

To examine the efficacy of existing models, intensities predicted in the Galactic plane by these models were calculated using forms given in the literature. None of the two or more-than-four arm models (excluding the Local arm) are consistent with the FIRAS data. Models with two short and two long arms are unable to fit all inner features. A subset of four-arm models, including Georgelin & Georgelin (1976), later models which build upon this model, and a few additional models, are in reasonable agreement with the FIRAS data. These results lend considerable support to the premise that the Milky Way has four spiral arms that are well represented by the logarithmic spiral form.

References

Bennett, C. L., et al. 1994, ApJ, 434, 587.

Georgelin, Y. M., & Georgelin, Y. P. 1976, A&A, 49, 57.

Steiman-Cameron, T. Y., Wolfire, M., and Hollenbach, D. (2004), submitted to ApJ.

UNCOVERING MORPHOLOGY FROM DUST: A NIR VIEW OF THE INTERACTING GALAXY PAIR NGC 5394/95

Margarita Valdez–Gutierrez[1], Ivanio Puerari[2], Izbeth Hernandez–Lopez[2]
[1] *Departamento de Astronomia, Universidad de Guanajuato, Apartado Postal 144, 36000 Guanajuato, Gto., Mexico.* [2] *Instituto Nacional de Astrofisica, Optica y Electronica, Calle Luis Enrique Erro No. 1, Santa Maria Tonantzintla, 72840, Puebla, Mexico.*

Abstract We present a near infrared 2D Fourier analysis of the interacting pair of galaxies NGC 5394/95 to get insight on its morphology. Our analysis shows that NGC 5394 is a H2β galaxy in the DP classification (Block and Puerari 1999). NGC 5395, in contrast, displays a very complex structure which needs a number of Fourier coefficients to be explained. A tightly wound $m = 1$ (DP class H1α) is the main structure, but other $m = 1$ and $m = 2$ coefficients (suggesting modulation) are also present in the Fourier spectra. The complex structure of NGC 5395 also suggest a strong interaction in the pair. The $m = 1$ coefficients can represent a pseudo ring–type structure, resulting of a collision rather than a passage.

Keywords: Galaxies: Individual (NGC 5394/95) — Infrared: Galaxies — Galaxies: Interactions — Galaxies: fundamental parameters — Galaxies: structure — Galaxies: evolution

Observations in the J, H and K′ passbands were carried out using the near–infrared camera CAMILA (Cruz-Gonzalez et al. 1994) at the 2.1 m telescope of the Observatorio Astronomico Nacional at San Pedro Martir, Baja California, Mexico. CAMILA is based on a NICMOS3 256×256 pixels detector which was used in imaging mode with the focal reducer configuration at f/4.5. The field of view is $3.6' \times 3.6'$ and the plate scale of $0.85''$ pixel^{-1}. Reductions were performed in a standard scheme. The 2D Fourier analysis was performed using the program 2dfft (see Saraiva Schroeder et al. 1994).

In Figure 1 (upper panels) we show the Fourier spectra for K′ for both galaxies in the pair. NGC 5394 shows a classical $m = 2$ spiral structure, while NGC 5395 displays a complex spectra: the main peak is for $m = 1$, showing also a peak at opposite p suggesting spiral arm modulation (Puerari et al. 2000, Elmegreen et al. 1989). The spectrum for $m = 2$ shows a similar behaviour.

D. Block et al. (eds.), Penetrating Bars through Masks of Cosmic Dust, 795–797.

Table 1. Relative amplitude to the main $m = 1$ peak for NGC 5395

	J main peak	J sec. peak	H main peak	H sec. peak	K′ main peak	K′ sec. peak
$m = 1$	1.000	0.540	1.000	0.527	1.000	0.512
$m = 2$	0.514	0.405	0.606	0.495	0.552	0.485

The spectra for the other filters are very similar, and so, they are not shown here. Some relevant values from all spectra for NGC 5395 are displayed in Table 1.

From the Fourier spectra of NGC 5394 and NGC 5395, we can classify the galaxies as H2β (NGC 5394) and H1α (NGC 5395) in the dust penetrated–DP classification of Block and Puerari (1999).

We show the inverse Fourier transform contours for K′ in Figure 1 (bottom panels). For NGC 5394 we show only $m = 2$ (the only relevant Fourier coefficient), while for NGC 5395 we display $m = 1$, $m = 2$ and $m = 1 + 2$. One can see the classical $m = 2$ spiral structure in NGC 5394, while for NGC 5395, a broken ring–type structure is displayed. This complex structure is possibly a result of a strong interaction between the galaxies, more probably a collision rather than a grazing encounter (Kaufman et al. 2002), with NGC 5394 crossing the disk of NGC 5395. In this sense, the encounter between NGC 5394 and NGC 5395 should be classified as "Cartwheel"–type (Puerari 1995) rather than "M51"–type. Further numerical simulations could support the scenario here proposed.

References

Block, D. L., Puerari, I. 1999, A&A, 342, 627

Cruz–Gonzalez, I., Carrasco, L., Ruiz, E., et al. 1994, In: Crawford D. L., Craine E.R. (eds.), Instrumentation in Astronomy VIII, Proc. SPIE 2198, 774

Elmegreen, B. G., Seiden, P.E., Elmegreen, D. M. 1989, ApJ, 343, 602

Kaufman, M., Sheth, K., Struck, C., et al. 2002, AJ, 123, 702

Puerari, I. 1995, PhD Thesis. Observatoire de Marseille (France)

Puerari, I., Block, D. L., Elmegreen, B. G., Frogel, J. A., Eskridge, P. B. 2000, A&A, 359, 932

Saraiva Schroeder, M. F., Pastoriza, M. G., Kepler, S. O., Puerari, I. 1994, A&ASS, 108, 41

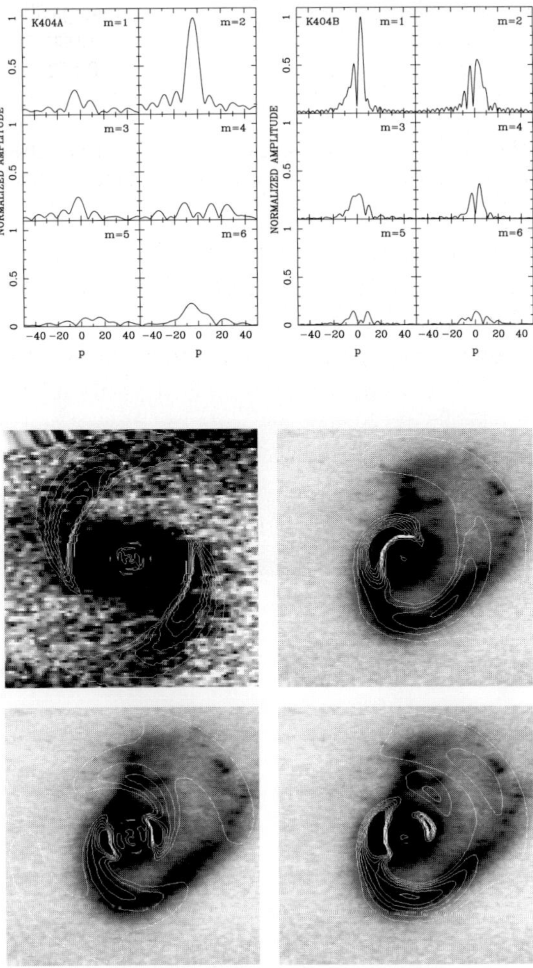

Figure 1. *Upper panels:* 2D Fourier spectra for the K′ image (left panel, NGC 5394; right panel, NGC 5395). The spectra for NGC 5394 show the classical $m = 2$ spiral structure. On the other hand, NGC 5395 displays a more complex spectra (see text for details). *Bottom panels:* Contours of the inverse Fourier transform are overlaid on the deprojected K′ images. Top left: NGC 5394 plus $m = 2$ contours. Top right: NGC 5395 plus $m = 1$ contours. Bottom left: NGC 5395 plus $m = 2$ contours. Bottom right: NGC 5395 plus $m = 1 + 2$ contours.

SCUBA LOCAL UNIVERSE GALAXY SURVEY

Catherine Vlahakis[1], Stephen Eales[1] and Loretta Dunne[2]
[1]*School of Physics and Astronomy, Cardiff University, PO Box 913, Cardiff, CF24 3YB, UK*
[2]*School of Physics and Astronomy, University of Nottingham, Nottingham, NG7 2RD, UK*

Abstract We discuss the progress of the SCUBA Local Universe Galaxy Survey (SLUGS), the first large, statistical sub-mm survey of the local universe. Since our original survey of a sample of 104 *IRAS*-selected galaxies we have recently completed a sample of 78 Optically-Selected galaxies. Since SCUBA is sensitive to the large proportion of dust too cold to be detected by *IRAS* the addition of this optically-selected sample allows us for the first time to determine the amount of cold dust in galaxies of different Hubble types. We detect 6 ellipticals in the sample and find them to have dust masses in excess of $10^7\ M_\odot$. We derive local sub-mm luminosity functions, both directly for the two samples, and by extrapolation from the *IRAS* PSCz, and find excellent agreement.

1. Introduction

Relatively little is known about the sub-mm properties of "normal" galaxies in the local universe – prior to SCUBA there existed only a handful of sub-mm flux measurements or maps. SLUGS is the first, large, systematic survey of the local sub-mm universe. It consists of a sample selected from the *IRAS* Bright Galaxy Sample (Dunne et al., 2000), and a sample selected from the CfA optical redshift survey (Vlahakis et al., in prep., Fig. 2). With the optically-selected sample we seek to understand for the first time key questions such as how the amount of cold dust varies with Hubble type.

2. The 850μm Luminosity Function

Using the *IRAS*-selected sample Dunne et al. produced the first direct estimate of the sub-mm luminosity function (LF). However, since the *IRAS* sample is biased toward galaxies with larger amounts of warmer dust its LF may also be subject to bias. Conversely, the optically-selected sample should, by definition, be free from temperature selection effects.

Here, we derive a direct 850μm LF for the optically-selected sample. However, in order to better constrain the LF at the lower luminosity end we need

D. Block et al. (eds.), Penetrating Bars through Masks of Cosmic Dust, 799–800.

more data points, spanning a greater range of luminosities. We do this by determining an 850μm LF using $\sim$ 10000 galaxies from the *IRAS* PSCz survey. We predict their 850μm luminosities by extrapolating their *IRAS* fluxes using the colour-colour relation from SLUGS (Serjeant and Harrison, 2003). In Fig. 1 we compare the extrapolated (PSCz) and direct LFs and find excellent agreement between all 3 LFs.

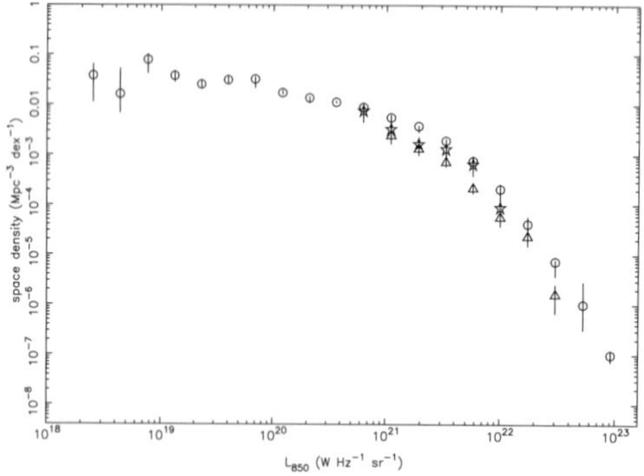

Figure 1. 850μm LF projected from the PSCz, compared with directly measured LF from optically- and *IRAS*-selected samples (circles, stars, triangles respectively).

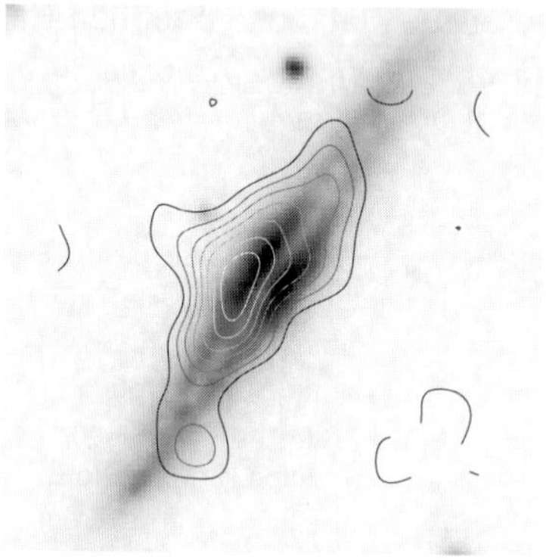

Figure 2. Example from the optically-selected SLUGS: NGC 3987. 850 μm S/N map (1σ contours) overlaid onto *Digitised Sky Survey* optical image.

3. Ellipticals in the Optically-Selected SLUGS

It was once thought that ellipticals were entirely devoid of dust and gas, but optical absorption studies now show that dust is usually present. Dust masses for the $\sim$ 15% of ellipticals detected by *IRAS* have been found to be as much as a factor of 10–100 higher when estimated from their FIR emission compared to estimates from optical absorption, suggesting a diffuse cold dust component (Goudfrooij and de Jong, 1995 and refs. therein, Bregman et al., 1998). At 850μm we detect 6 ellipticals, from a total of 11 ellipticals in the optically-selected sample, and find them to have dust masses in excess of $10^7\ M_{\odot}$. We will investigate this further with SCUBA observations of a larger sample of ellipticals.

References

Bregman Joel N. et al., 1998, ApJ, 499, 670
Dunne Loretta et al., 2000, MNRAS, 315, 115
Goudfrooij P. and de Jong T., 1995, A&A, 298, 784
Serjeant Stephen and Harrison Diana, 2003, astro-ph/0309629
Vlahakis C.E. et al., in prep.

HOW JWST CAN MEASURE FIRST LIGHT, REIONIZATION AND GALAXY ASSEMBLY

Rogier A. Windhorst[1], Haojing Yan[2]
[1] *Department of Physics and Astronomy, Arizona State University, Box 871504, Tempe, AZ 85287, USA* [2] *Spitzer Science Center, California Institute of Technology, MS 100-22, Pasadena, CA 91125, USA*

Abstract We summarize the design and performance of the James Webb Space Telescope that is to be launched to an L2 orbit in 2011, and how it is designed to study the epochs of First Light, Reionization and Galaxy Assembly.

Keywords: James Webb Space Telescope — population III stars — reionization — galaxy formation — galaxy evolution

1. The James Webb Space Telescope and its Instruments

The James Webb Space Telescope (JWST) is a fully deployable 6.5 meter segmented IR telescope (25 m^2 collecting area) optimized for imaging and spectroscopy from 0.6μm to 28μm, to be launched by NASA in 2011 (Mather & Stockman 2000). After its launch, JWST will make a several month journey to the Earth–Sun Lagrange point L2, during which it will be automatically deployed in phases, its instruments will be tested, and it will then be inserted into an L2 halo orbit. JWST has a nested array of sun-shields to keep its ambient temperature at 35-45 K, allowing faint imaging (AB$\lesssim$31.5-32 mag) and spectroscopy (AB$\lesssim$29 mag). From L2, JWST can cover the whole sky in segments that move along in RA with the Earth. It will have an observing efficiency $\gtrsim$70%, and send data back to Earth every day. The JWST science requirements are described by Gardner et al. (2004) and its instruments in the websites below. In summary, JWST has the following instruments:

• **NIRCam:** Near-Infrared Camera made by an UofA + Lockheed + CSA consortium will do imaging from 0.6–5.3μm using a suite of broad-, medium-, and narrow-band filters. NIRCam uses two identical and independently operated imaging modules, with two wavelengths observable simultaneously via a dichroic that splits the beam around 2.35μm. Each of these two channels has an independently operated $2\overset{\prime}{.}2\times4\overset{\prime}{.}6$ FOV. Both channels are Nyquist-sampled: the short wavelength channel at 2μm with $0\overset{\prime\prime}{.}0317$/pixel, and the and long wave-

D. Block et al. (eds.), Penetrating Bars through Masks of Cosmic Dust, 801–804.

length at 4μm with 0$''$.0648 /pixel. NIRCam's ten 2k$\times$2k HgCdTe arrays will be passively cooled.

• **NIRSpec:** The Near-Infrared Spectrograph made by an ESA + GSFC consortium will do spectroscopy with resolving powers of R$\sim$100 in prism mode, of R$\sim$1000 in multi-object mode using a micro-electromechanical array system (MEMS) of micro-shutters that can open slitlets on previously imaged known objects, and of R$\sim$3000 using long-slit spectroscopy. All NIRSpec spectroscopic modes have a $\sim$3$'$.4$\times$3$'$.4 FOV.

• **MIRI:** The Mid-InfraRed Instrument made by an UofA + JPL + ESA consortium will do imaging and spectroscopy from 5–28μm. MIRI is actively cooled by a cryostat and its expected lifetime is at least 5 years. The NIRCam and MIRI sensitivity complement each other straddling 5μm in wavelength, and together allow the first starforming objects to be found to redshifts z=15–20 in $\gtrsim 10^5$ sec (28 hrs) integration times. To see First Light, JWST must observe in near–mid IR, and so need NIRCam at 0.8–5μm and MIRI at 5–28 μm.

• **FGS:** The Fine Guidance Sensor is made by CSA and provide stable pointing at the milli-arcsecond level. It will have sufficient sensitivity and a large enough FOV to find guide stars with $\gtrsim$95% probability at any point in the sky. The FGS will have three simultaneously imaged fields of view of 2$'$.3$\times$2$'$.3, one of which feeds a pure guider channel, one feeds a guider channel plus a long wavelength R$\sim$100 tunable filter channel with light split by a dichroic, and another feeds the short wavelength tunable filter R$\sim$100 channel.

JWST has fully redundant imaging and spectroscopic modes. It will not be serviced at L2, and therefore will undergo an extensive series of ground-testing and thermal vacuum testing in 2008–2009, after its main construction phase in 2004–2008. The main NASA contractor is Northrop Grumman Space Technology ("NGST") in Redondo Beach (CA).

2. Measuring first light, reionization & galaxy assembly

• **First Light:** WMAP (Spergel et al. 2003) has shown that First Light may have happened in two epochs (Cen 2003): (1) Population III stars with 200-300 $M_\odot$ at z$\simeq$15–25 (First Light). These Pop III star clusters and their extremely luminous supernovae should be visible to JWST at z$\simeq$25 to z$\simeq$15; This epoch was likely followed by a second Dark Ages, since Pop III supernovae heated the IGM, which could not cool and form normal Pop II halo stars until z$\simeq$15 to z$\simeq$10; (2) Population II stars (halo stars) form in dwarf galaxies of mass=10^6 to 10^9 $M_\odot$ at z$\simeq$6–10. This will be visible to JWST in the luminosity function (LF) of the first star-forming galaxies from z$\simeq$10 to z$\simeq$6.

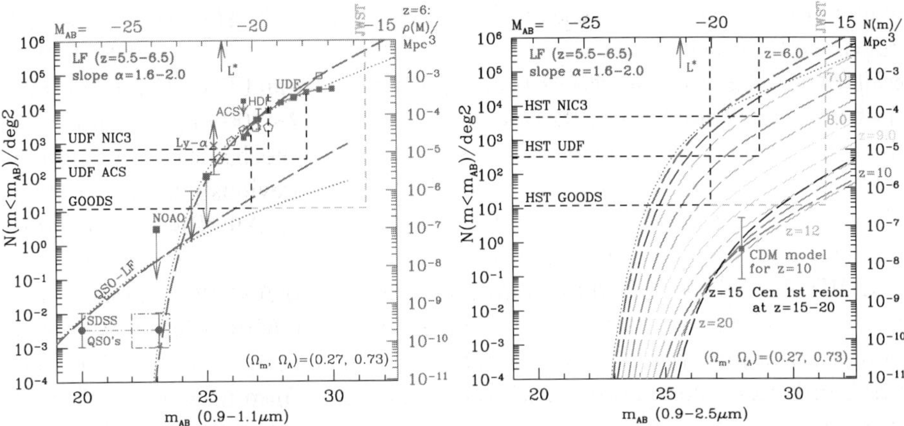

Fig. 1.a (LEFT) The UDF has shown that luminosity function of z≃6 objects is very steep, with faint-end Schechter slope $|\alpha|\simeq1.8$–1.9 (Yan & Windhorst 2004b). Dwarf galaxies and not quasars therefore likely completed the reionization epoch at z≃6 (Yan et al. 2004a). This is what JWST likely will observe in detail. **Fig. 1.b (RIGHT)** HST/ACS can detect objects at z$\lesssim$6.5, but its discovery space A.Ω.Δlog(λ) cannot map the entire reionization epoch. NICMOS similarly is limited to z$\lesssim$8–10. JWST will be able to trace the entire reionization epoch from First Light at z≃20 to the end of the reionization epoch at z≃6.

• **Reionization:** The UDF has shown that luminosity function of z≃6 objects (Yan et al. 2004b) is very steep, with a faint-end Schechter slope $|\alpha|\simeq1.8$–1.9 after correcting for incompleteness of the ACS i-band dropout samples (Fig. 1a). This steep LF may have provided enough UV-photons to complete the reionization epoch at z≃6 (Yan & Windhorst 2004a). Hence, dwarf galaxies and not quasars likely completed the reionization epoch at z≃6. The Pop II stars in dwarf galaxies cannot have started shining pervasively much before z≃6–8, or no neutral H-I would be seen in the foreground of z$\gtrsim$6 quasars (Fan et al. 2003), and so dwarf galaxies likely ramped up their formation gradually from z≃10 to z≃6. This is what JWST will observe in detail.

HST/ACS can detect objects at z$\lesssim$6.5, but its discovery space A.Ω.Δlog(λ) cannot map the entire reionization epoch. NICMOS similarly is limited to z$\lesssim$8–10. JWST will be able to trace the entire reionization epoch from First Light at z≃20 to the end of the reionization epoch at z≃6. With proper survey strategy (area *and* depth), JWST can trace the entire reionization epoch, i.e. detect the first star-forming objects. For this to be successful in realistic or conservative model scenarios, JWST needs to have the quoted sensitivity/aperture (A), field-of-view (FOV=Ω), and wavelength range (0.7-28μm).

• **Galaxy Assembly:** Galaxies of Hubble types formed over a wide range of cosmic time, but with a notable phase transition around z≃1.0: (1) Subgalactic units rapidly merge from z≃7 to z≃1 to grow bigger units; (2) Merger products start to settle as galaxies with giant bulges or large disks around z≃1. These evolved mostly passively since then (as tempered by the cosmological constant, see Windhorst et al. this Vol.), resulting in the giant galaxies that we see today. JWST can measure how galaxies of all types formed over a wide range of cosmic time, by accurately measuring their distribution over rest-frame type as a function of redshift or cosmic epoch. The uncertain rest-frame UV-morphology of galaxies is dominated by young and hot stars, with often copious amounts of dust superimposed. This complicates the comparison with very high redshift galaxies as seen by JWST, although with good images a quantitative analysis of the restframe-wavelength dependent morphology and structure can be made (e.g., Odewahn et al. 2002; Windhorst et al. 2002). JWST can measure how galaxies of all Hubble types formed over a wide range of cosmic time, by measuring their redshift distribution as a function of rest-frame type (Driver et al. 1998). For this, the types must be well imaged for large samples from deep, uniform and high quality multi-wavelength images.

With proper restframe-UV training, JWST can quantitatively measure the evolution of galaxy morphology and structure over a wide range of cosmic time, as following: (1) Most disks will SB-dim away at very high redshifts ($z\simeq$15–20), but they likely formed at $z \lesssim z_{form} \simeq$ 1–2 anyway; (2) High SB structures are visible to $z\simeq$10–15; (3) Point sources (Pop III star clusters and AGN) are visible to $z\simeq$15–20; (4) High SB-parts of mergers/train-wrecks are visible to $z\simeq$10–15. This is what JWST will observe in detail.

The complete Windhorst poster and talk at this conference can be found at www.asu.edu/clas/hst/www/wfpc2/midUV/cy09index.html (click on data).

Acknowledgments

The HST/ACS work was supported from NASA grant GO-9780.* awarded by STScI, which is operated by AURA for NASA under contract NAS 5-26555. RAW acknowledges support from NASA JWST grant NAG 5-12460. RAW thanks the other members of the JWST Flight Science Working Group, the JWST Instrument Teams, and the JWST hardware teams for their hard work on the JWST project.

References

Cen, R., 2003, ApJ, 591, 12

Driver, S. P., et al. 1998, ApJ, 496, L93

Fan, X., et al. 2003, AJ, 125

Gardner, J., Mather, J., Clampin, M., Greenhouse, M., Hammel, H., Hutchings, J., Jakobsen, P., Lilly, S., Lunine, J., McCaughrean, M., Mountain, M., Rieke, G., Rieke, M., Smith, E., Stiavelli, M., Stockman, H., Windhorst, R., & Wright, G. ("the JWST Flight Science Working Group") 2004, Proc. SPIE, Vol. 4014, p. 001–012, in press

http://www.jwst.nasa.gov/

http://ircamera.as.arizona.edu/nircam/

http://ircamera.as.arizona.edu/MIRI/

Mather, J., Stockman, H. 2000, Proc. SPIE Vol. 4013, p. 2-16, in "UV, Optical, and IR Space Telescopes and Instruments", Eds. J. B. Breckinridge & P. Jakobsen (Berlin: Springer)

Odewahn, S. C., et al. 2002, ApJ, 568, 539

Spergel, D. N., et al. 2003, ApJS, 148, 175

http://www.stsci.edu/jwst/

http://www.stsci.edu/jwst/instruments/nirspec/

Windhorst, R. A., et al. 2002, ApJS, 143, 113

Yan, H., & Windhorst, R. 2004a, ApJL, 600, L001

Yan, H., & Windhorst, R. 2004a, ApJL, in press

HIGH RESOLUTION VELOCITY FIELDS IN THE STRONGLY BARRED GALAXY NGC 1530

Almudena Zurita[1,2], Mónica Relaño[3], John E. Beckman[3,4] and Johan H. Knapen[5]
[1]*Universidad de Granada, Dept. de Física Teórica y del Cosmos, 18071–Granada, Spain*
[2]*Isaac Newton Group of Telescopes, 38700–La Palma, Spain*
[3]*Instituto de Astrofísica de Canarias, 38200–La Laguna, Spain*
[4]*Consejo Superior de Investigaciones Científicas, Spain*
[5]*Centre for Astrophysics Research, University of Hertfordshire, Hatfield, AL10-9AB Herts, UK*

Abstract We present Hα emission line mapping of the strongly barred galaxy NGC 1530 at unprecedented angular resolution. Of special interest are our maps of gradients in the non–circular velocity field measured along and across the bar. These trace shocks in the gas, trace the dust lanes and bring out well the zones of maximum shear and compression. We show how these zones are related to the local star formation rate.

Keywords: galaxies: kinematics and dynamics; galaxies: individual (NGC 1530)

1. Introduction

NGC 1530 is a dramatically barred spiral galaxy, well studied in optical and near-infrared for its mass distribution, and in line emission from H I, CO and Hα for its kinematics (e.g. Regan et al. 1996; Downes et al. 1996). Our Fabry–Pérot Hα emission spectroscopy at high angular resolution ($\sim 1''$) over the whole galaxy disc has allowed us to probe its kinematical properties more deeply. The data was obtained with the TAURUS interferometer on the 4.2m WHT in La Palma. After data reduction, the resulting *data cube* of 55 "planes" separated by 18.62 km s^{-1} was used to derive the rotation curve. This was then projected into a two–dimensional model, which after subtraction from the observed velocity field yields a map of the non–circular motions of the ionized gas for the full galaxy. Further details of the data and its analysis are described in Zurita et al. (2004).

2. Velocity gradients and star formation in the bar

The most striking feature of the non–circular velocity map of NGC 1530 is the complex field associated with the bar (Fig. 1e), produced by gas moving in quasi–elliptical orbits aligned with the bar, with projected velocities of ampli-

D. Block et al. (eds.), Penetrating Bars through Masks of Cosmic Dust, 805–806.

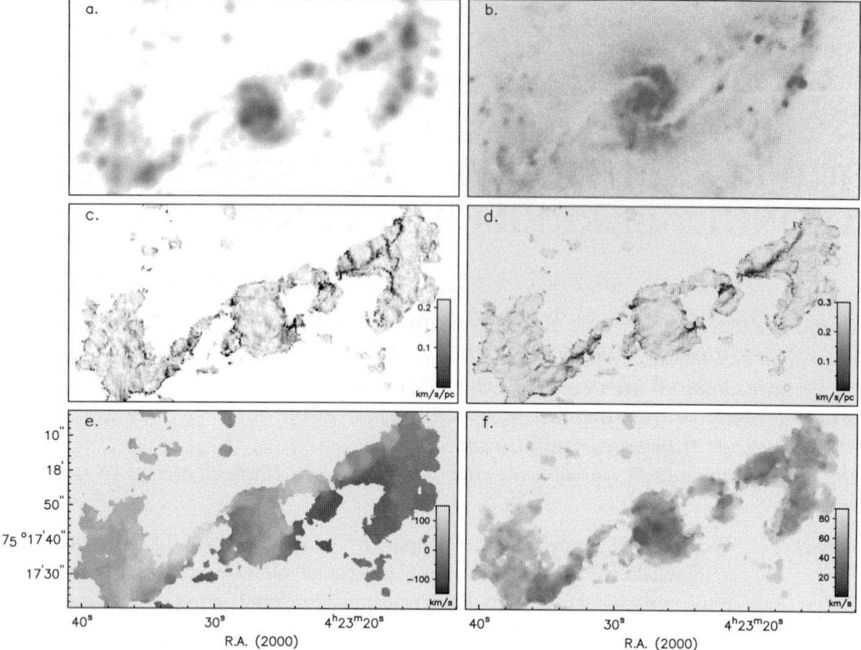

Figure 1. **a)** Intensity map of the Hα emission in zone of the bar. **b)** V band image. Velocity gradients parallel (**c**) and perpendicular (**d**) to the bar. **e)** Non–circular residual velocity map. **f)** Non–thermal velocity dispersion map. All images (a to f) show the same zone of the galaxy.

tude ~120 km s^{-1}, and steep velocity gradients in the bar. From the residual velocity field we generated maps of velocity gradients, parallel and perpendicular to the bar (Figs. 1c and d respectively). There is a striking positional and morphological agreement between the dust lanes (Fig. 1b) and the loci of maximum velocity gradient perpendicular to the bar (Fig. 1d), showing how dust picks out shock–induced flow along the bar. Comparing Fig. 1d with the Hα intensity map (Fig. 1a) we see a complete anti–correlation between shear and local star formation rate. Star forming regions avoid strong shear, but not necessarily high non–circular velocity (cf. Figs. 1a, 1d, 1e), while shocks (seen as strong gradients along the flow in Fig. 1c) act to enhance the local star formation rate (Zurita et al. 2004).

This work was partially supported by the Consejería de Educación y Ciencia de la Junta de Andalucía, Spain. The WHT is operated on the island of La Palma by the Isaac Newton Group in the Spanish Observatorio del Roque de los Muchachos of the IAC.

References

Downes, D., Reynaud, D. Solomon, P. M., & Radford, S. J. E. 1996, ApJ 461, 186
Regan, M., Teuben, P., Vogel, S., & van der Hulst, T. 1996, AJ, 112, 2549
Zurita, A., Relaño, M., Beckman, J. E., & Knapen, J. H. 2004, A&A, 413, 73

EYES TO THE FUTURE: PENETRATING THE DUST MASK – PANEL DISCUSSION

A group of eleven panellists (Block, Buta, Combes, Conselice, Elmegreen, Freeman, Grosbol, Kormendy, Sheth, Shu and Yuan) were allocated approximately 5-7 minutes each, to share their thoughts on galaxy morphology. Our ninety-minute panel discussion was taped, transcribed by Janice Scott, and edited for these Proceedings. Interaction from the audience then followed the Panellists; their questions and comments are also presented here. The Panel session was chaired by Professor R.J. Allen.

1. The Discussion Begins...

Chairman - Ron Allen: I think we all agree that at some point we are trying to understand something about the astrophysics of galaxies. We have to figure out what the correspondences are between the set of morphological parameters "m" on the left side of figure 1, and the set of physical parameters "p" on the right, as indicated by the red arrow and the question mark between them. I also think we want to avoid a philosophical discussion about which of these two approaches is preferable, and concentrate instead on figuring out how we can get the best out of both of them. So without further delay, I'd like to start with our first co-chair, David, and give you the floor for a period of three to five minutes. (off discussion-laughing)

David Block: I'll give a five minute run down here. May I commence by presenting a diagram sent to me by Dr. Allan Sandage [overhead shown]. He presented this diagram in Paris around twenty years ago: in it, he insists that there is a one way path through 'the wall' dividing morphology and theory: specifically, no information must pass from the theorist to the morphologist... yet the theorist must know all that the morphologist has to provide. No doubt this will provide a lot of discussion; it must be a one-way street, Dr. Sandage argues (see also the masterful essay by Sandage, reproduced in this volume). In studying near-infrared morphology, my team of collaborators and I have, over the years, been deeply struck by the emergence of two dominant factors. The first, is the rich duality in spiral structure. Frank Shu also showed magnificent examples in his talk. Great surprises often lie in store when you penetrate masks of cosmic dust, and look 'through the mask'. The second factor which demands my attention, is the ubiquity of only low order Fourier

D. Block et al. (eds.), Penetrating Bars through Masks of Cosmic Dust, 807–834.

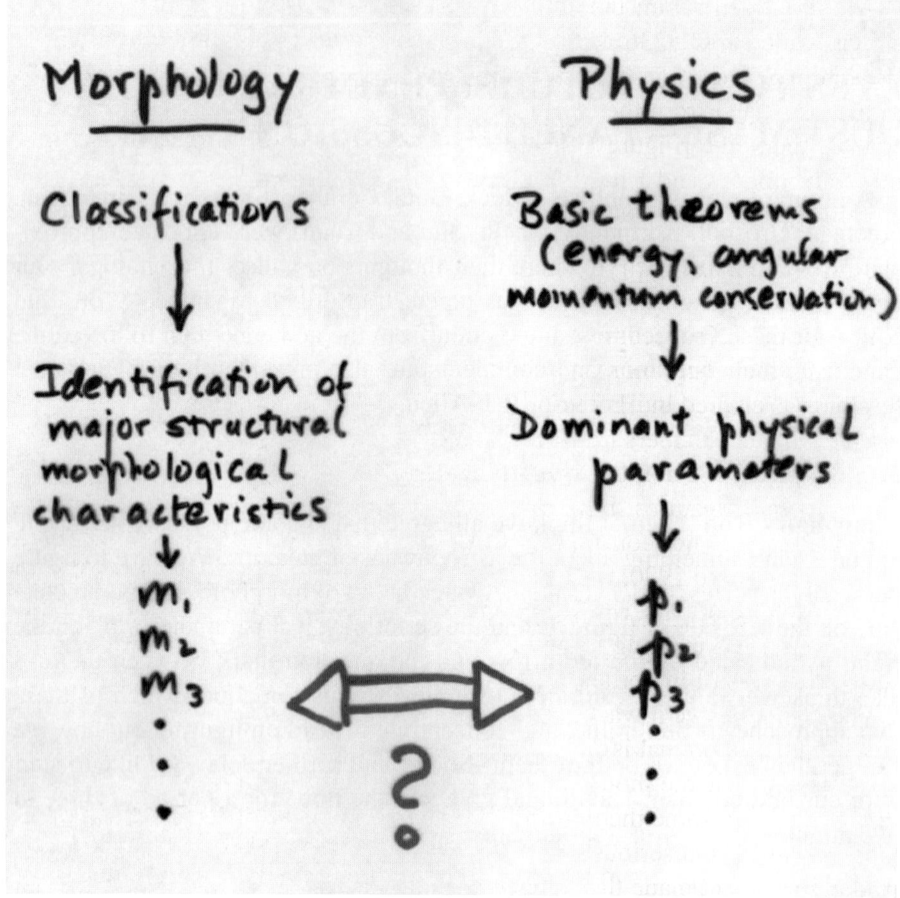

Figure 1. The reader is referred to the introductory remarks by Ron Allen.

modes ($m = 1$, $m = 2$). Modes $m = 3$ or higher are exceedingly rare in the near-infrared regime, as G. Bertin and my team argued our 1994 paper. The scheme presented here [overhead shown, identical to Figure 1 in our paper], summarises, I believe, the simplest near-IR galaxy classification scheme. We characterize spiral galaxies according to the dominant Fourier mode in the dust penetrated regime followed by three dust penetrated arm classes: α, β and γ. At this Conference, Marc Seigar has strikingly demonstrated that these arm classes are related to rotation curve shape. Finally, we include the gravitational torque class. In conclusion, I'd like to draw your attention to the fact that late-type galaxies (classified as such optically) can fall into our α arm class, while galaxies which are classified optically as early-type can fall into the β or even γ bins. The famous barred galaxy, NGC1365, which is Hubble type b, belongs to the very open γ bin; the early-type type a galaxy NGC718 belongs to the β bin, while NGC 5236=M83 (type c) belongs to the tightly wound α class. We find no correlation between dust penetrated arm class and Hubble types, as also expected from the rotation curve studies of Burstein and Rubin. The scheme seen here [Figure 1 of our paper] is perhaps the simplest way which my collaborators and I have found, to take full cognisance of the duality of spiral structure in the dust penetrated near-infrared regime.

Chairman: Thank you David. Now to Ron Buta.

Ron Buta: Yes, in discussing this I wanted to say a little bit about morphology, pre 1990. Back then, morphology was, especially in 1980 and earlier, a purely descriptive subject and I felt that caused a little bit of disdain especially for people who are sticklers for details - like myself. I happen to enjoy the details of morphology - I don't know if I find featureless galaxies as interesting as ones that have features, but I was living in a time when the features, that is the details, were not considered important. The things that were considered important, prior to 1990, were basic classes, like ellipticals, spirals, and maybe S0's. The issue of bars, rings, etc. was not considered important and that came home to me at a meeting fourteen years ago where someone told me that my life's work wasn't important because I was focusing on the details. But I love the details, I believe it was worthwhile, and it never discouraged me.

I want to say on that issue of theory versus observations, that is, that a good classification scheme should be independent of any theory, I know I've violated that rule. When some theoretical models were made at Mount Stromlo observatory, I took them seriously and I went out and looked for the predictions of those models and made the Catalogue of Southern Ringed galaxies. My view is that the purpose of a theory is not simply to explain existing observations but to make predictions that can be tested by further observations. The fact is that in the late 70's, theory became good enough to make predictions about morphology, something I found very intriguing.

I believe now that as we go into the future, we need to continue to try and quantify classification. De Vaucouleurs was always for quantitative classification. This idea about descriptive classification, you try to get away from that, you try to quantify, which is why I've spent a lot of time trying to quantify bar strength.

I also wanted to say, on the issue of infrared classification: is it possible for any of us to take a set of near infrared images, and look at them without any bias towards the Hubble system? It's impossible for us. I can't look at a near infrared image without thinking about Hubble. I know so many galaxies, I think I have their classifications memorized. I can't look at any galaxy without thinking about the Hubble system. So, how do I get away from that bias? How can you start the project fresh with a new view? That's hard.

I think that if you want to compare high and low redshift galaxies, you've got to have some way of matching the resolutions, the number of pixels in the images, and the filters. I really liked IRAC as a way of moving into the future (of morphology) that Giovanni told us about - that's what I want to deal with. We have to be consistent between near and far samples and it's probably going to be very hard.

But just to finish up, on the question of details: when you go to high redshift, you lose the details and you get stuck into that pre-1990 mode. So I worry about that. Thanks.

Chairman: Okay, thanks, Ron. Now to Bruce Elmegreen.

Bruce Elmegreen: Thanks. I think morphology without theory is boring. I think this is why artificial intelligence morphology has not caught on. Hubble had theory in mind: he didn't have a different classification for every different inclination of a disc, he assumed there was just a disc and he could imagine how all the others would look, so he had a theory in mind. I think if we have 100 different morphological systems, that's just fine. This is how we think. If we predict a feature by theory, then there's a natural tendency to bin galaxies into those which have that feature and those which don't and there's your new morphological system. When in the 70's, Debbie and I were interested in density waves, we binned galaxies into those which had them and those which didn't. So you have grand design and flocculent morphologies. Then we could take from there: the star formation rates per unit area were the same in each, so there's some conclusions to make from that. Following Kormendy and Norman, the occurrence of one or the other correlated with bars and companions, so there are conclusions one can make from that too. What I am interested in now is this concept of bar dissolution and reformation, and I see Kartik will tell us more about that in a minute. What I would expect the theoreticians and the simulators to do is to make predictions with their own new morphological classes - what should a new bar look like, what should a dissolved bar look like. It's alright to have the 101st morphological system if we can get an answer to that question. So morphological systems can change with time for me: I can throw away all these old systems, flocculent and grand design for example if I'm not interested in density waves anymore, and get ready for a new morphological system. The fact that Hubble did or didn't do something is of limited interest to me, but only of interest in connection to certain theories. That's my bias, I guess.

Chairman: Okay, thanks Bruce. Kartik, you're up next. Hang on, he seems to have a long cable... (setting up computer connection)...

Kartik Sheth: Well, Ron and Bruce already mentioned a few of the things that I was going to talk about, so let me just add my perspective. A question that I'm asked often is what is a bar? I think that we have so many different ways of defining a bar and it would be helpful if we could come up with a robust, quantitative definition. Since bars come in all shapes and sizes the definition should also try and account for the bar properties. We need to move away from discussing bars with qualitative words like flat, exponential, strong, weak, fat or thin.

One of the things that strikes me as critical as we start to examine the evolution of galaxies and bars at higher redshifts is the need for very good local samples of galaxies. Eskridge and the Ohio State survey have started a nice compilation of galaxies in the near infrared and at this conference Menendez-Delmestre also presented a nice analysis of near-infrared data from the 2MASS Large Galaxy atlas. These samples have on the order of a hundred galaxies, but I think what we need are samples of thousands of galaxies in the infrared for a good, local sample for comparison to high redshifts.

There is a disconnect in the methods applied in galaxy morphology studies by observers at high and low redshifts. Even with the best instruments our images of high redshift galaxies will not have the detail of local galaxies. So it seems futile to try and fit the high redshift Universe into a local classification system. Perhaps we need to work backwards and quantitative parameters (M_{20}, CAS system, Gini coefficients) used at high redshifts should be applied to local samples like the ones I just mentioned.

Finally I would like to stress that one of the most intriguing, outstanding questions is how does the bar fraction and the bars themselves evolve over time? Are bars destroyed? Do they reform? I don't recall who asked the question today, but I thought it was very interesting to also ask does anybody really care? Many of these questions can, in fact, be answered from the ground. An idea that Louis Ho has mentioned to me is using a near-infrared camera, like PANIC to study bars to $z \approx 0.3$. This is feasible on telescopes like Magellan which offer sub-arcsecond seeing regularly. $z \approx 0.3$ is a quarter of the way back to the edge of the Universe and is an adequate time period for investigating the destruction and reformation of bars. I would like to pin down our numerical modeling friends to predictables in their models which can be tested with the plethora of data that is incoming now so that we can begin to address the issue of bar evolution by the time of the next Bars conference in Mexico.

Lastly I would like to note that ultimately we should move towards studying bars using kinematic information like the kind that has been used to study gas kinematics in NGC 1530, NGC 5383 and other nearby galaxies. With the next generation of telescopes and instruments like ALMA and OSIRIS, gas kinematics offers one of the most powerful tools for the study of bars.

Chairman: Thank you Kartik, that's a very nice summary. Chris, are you ready?

Chris Conselice: Can I have your microphone?

Chairman: You sure could, hang on a sec...

Chris Conselice: Okay. So I originally just planned to talk about infrared classifications, and what we can do in the infrared, or what we should do in infrared in the future, but I will briefly address our understanding galaxy morphology in general. I think a major point that a lot of people are realizing is that if you have a classification system for galaxies it has to be applicable to most galaxies. It doesn't have to be applicable to all galaxies, but it should be applicable to most galaxies, and that includes galaxies that we see at high redshift. When you view high redshift galaxies, they are not classifiable on the Hubble sequence. They are not Hubble types, so if you want to know how galaxies are forming, you have to have a classification system, that can accomplish this. Concerning physical morphology - the idea that a morphology should not relate to the physics, I think, is kind of well, kind of silly in a way. If we look at stellar classifications, using spectra of stars, there are very well defined correlations between different classifications and stellar temperatures, luminosities and gravities and that kind of thing. There's a real connection between the morphology of the stellar spectra and the physical properties of stars, and there's no reason why we can't have a similar system for galaxies, and that's sort of what I've been trying to do by developing something quite similar to that, but by measuring the major properties of galaxies, not just of stars, and that's this CAS system that I've talked about already, so I won't do that again. The thing about the infrared, I just wanted to say, is that galaxies, as we've heard, look very different at different wavelengths and this just shows pictures of a spiral galaxy in a nearby universe, from UIT imaging and from Hα and R band imaging. You can see that this galaxy looks quite different, at these different wavelengths. There's an evolved part of this galaxy here, which you don't see in the UV, but you see in the optical. It would be nice to classify all galaxies in the restframe infrared, however I don't think any time soon, at least probably not in our lifetime, will there be a telescope in space that will be able to do high-resolution imaging - rest frame K-band - for galaxies at the highest redshifts, which is a very hard problem to do and there's nothing even coming up that will really do that very well. So I think that the way we have to classify galaxies is based on the optical and the way to do that is through deep infrared imaging. Right now we have only a few NICMOS pointings with HST, and this figure shows, based not on morphology but based on the colours of galaxies, as you go to higher redshift galaxies get very red, and this is just the (R-K) versus redshift distribution in the GOODS field, and you can see that the galaxies are very red and if we're going to study those galaxies we really need to look at them in the rest frame optical, where we can resolve their structures. This figure shows a simulation of bright galaxies in the nearby universe simulated to redshift $z = 1$ on the left and simulated to $z = 2.5$ on the right in the rest frame B-band. If you go to the rest frame UV, these galaxies would essentially disappear and that's the wavelength that most of this deep HST imaging we now have is probing - the frame ultra-violet for galaxies at those redshifts, so we really need wide field deep near infrared imaging. WFC3 would do this - and I certainly hope that flies but if we don't have that we're going to be very limited in our ability to not just classify these galaxies at high redshift but also just identifying them. This figure is a comparison between that simulation at $z = 2.5$ and actual galaxies between $z = 2$ and 3, and you can definitely see that there are indeed differences. First of all you can detect most of these nearby normal Hubble types in the near infrared and this is without including any evolution, most galaxies would certainly be brighter at higher redshift because of evolution and the Hubble sequence certainly is not there at higher redshift. That's it.

Chairman: Thanks, Chris. Now to Frank Shu...

Frank Shu: In his talk David Block mentioned that I wrote first paper on the subject forty years ago. I should therefore mention that I wrote my next-to-last paper on the subject, thirty-one years ago. So it's been a long time since I've been at a conference like this. I must say I've been very impressed by one fact: that in the intervening three decades, the field has moved away from structure and dynamics to evolution. This is a very ambitious, challenging, and worthwhile goal - to really understand the evolution of galaxies. It's something that my generation didn't think about. We didn't think it'd be possible to understand it so soon. For future work, there seems to be two major ideas – one is secular evolution and the other is evolution through mergers. In some sense they represent opposite kinds of science. I'd like to explain what I mean by that.

Secular evolution must take place, especially in bars, and on interesting time scales. We made estimates for normal spiral galaxy torques thirty years ago and found that they wouldn't be important on Hubble times. But if bars form and dissolve, then evolution becomes interesting on timescales of billions of years, and that's very important.

Secular evolution is a kind of science that we are familiar with, a kind that we know and love. We know how to do it; we have equations for it. It's just a matter of solving those equations accurately, right? That's Ron Allen's physics side of things. There are uncertainties, but they are very interesting too. What I find most interesting is the question: what is the dark matter halo made of? What is the equation of motion for the material of the dark halo? Can we use the dynamics that we know about, baryons, to investigate this question – to tell us something about the dynamics and structure of the dark matter halo? That would be very exciting. Just how dissipative a stellar system is, is another very interesting physics question.

Then there's this issue of gas accretion – gas continually falling into disk galaxies. I've been working on star formation using accepted physics equations. One thing that's always bothered me on the galactic scale is that the rate of star formation is such that galaxies should use up all their gas on timescales of billions of years, so we should not have any gas left. Continuing gas infall provides a very natural solution. But then we should see it. It really bothers me that we don't. At least, it doesn't seem to be coming in as atomic hydrogen. Alternatively, if you have 10,000-degree gas settling like a mist, we should also see it. So why don't we see it? I'd like to understand that.

Galaxy evolution through mergers is a different kind of science, mainly driven today, I believe, by cosmological simulations. In the last ten years, cosmology has been tremendously successful, especially with regard to anticipating CMB fluctuations. It's amazing, how accurate the predictions in cosmology have been when we don't know the equations of motion nor the equation of state for 96% of the universe. Just the 4% of the total mass that interacts with light gives us all this information! Most of the success is associated with simple linear theory, which we know how to do very well. When it comes to the non-linear dynamics, as Andy Burkert and George Lake reminded us, the success is less impressive. Why is that? There's this deep suspicion that something fundamental is missing. The standard answer that you will get from the cosmologists is feedback. Feedback from star formation is the magic ingredient. Right. Unfortunately, it's not the kind of star formation I know, where you solve equations from physics. It's a kind of star formation where people invent the equations and the rates and the dependences. Sorry, that's economics, that's not physics!

We should aim higher than that, so that we can make real progress. One of the things I always worry about in star formation is magnetic fields. We haven't yet heard a single talk that had Maxwell's equations in it. It is the only other long range force that we know about, right? When did magnetic fields become important in galaxy formation and evolution? The fact that we see synchroton radiation in radio sources certainly says that in some parts of galaxies, magnetic fields were generated at early times. Magnetic fields are important because they are the dominant way, other than gravitational torques, by which you can transport angular momentum. So if angular momentum is a big problem in galaxy formation, maybe we should look at magnetic fields before we think of exotic forms of dark matter!

But why else do I say evolution through mergers may represent a different kind of science? If you talk about mergers, a small shift in the impact parameter will cause a companion to pass the target in a retrograde or prograde direction. That difference is a classic bifurcation problem. Whether the galaxy comes in direct or retrograde has tremendous influence on the consequent history of the merged object. The result can become completely unpredictable because you are exponentially sensitive to different kinds of initial conditions. This doesn't mean that science is not possible because you can't make predictions – it just means it's a different kind of science. You now have to have very large samples – you're not trying to calculate the behaviour of any individual system; you talk instead of probability distributions and classes of objects.

Let me remind you of the most successful science in this category which resembles the morphological classification of galaxies. Consider the problem faced by evolutionary biologists. They classify different kinds of organisms, for example, animals. They see evolution in the same way we see it, in the historical record – instead of at different redshifts, in their case, in different rock strata. This layer is so many million years old, another layer is so many billions of years old. But how do you connect the fossils found in different rock strata? There's a beautiful theory for this connection; it's called Darwin's theory of evolution. In modern times, Darwin's theory has been extended into the molecular regime, the analog of the relevant physics for biologists. But despite this extension into the deterministic regime, Darwin's theory is still not a predictive theory. It remains an explanatory theory. It can explain, for example, common ancestors. Some became rhinos, some became elephants – they're not really that different. The DNA is almost the same. When it really becomes different is when you consider the dinosaurs, the fishes, or, even more, single cell organisms. Darwin's theory is not predictive, yet no one in biology worries about this, because they know at the mechanistic level there are just too many chance occurrences. How do diploid cells divide into haploid cells (sexcells), etc. There are many, many ways for variations and mutations to take place. There's no way you can see in advance all these chance occurences even though the basic equations at a mechanistic level are, in principle, deterministic. This non-predictiveness does not mean that biology is not scientific. It just means that biology is a different kind of science.

I think we have to be prepared for this eventuality, that there may also be chaotic elements in our subject. Concerning galaxy morphology. I will argue tomorrow that the reason that the optical blue images look so different from the infrared images, is not because the two associated stellar populations are decoupled, but because the gas from which new stars are born is too strongly coupled! It's so strongly coupled to the collective gravitational field that the system becomes chaotic in many circumstances. If that's true, then we really have to take a very different attitude. As we study more and more detail and we see more and more features, we may not be able to explain every feature. It's like the weather, which is also a chaotic system. Make all the measurements you can, still no one can predict the weather seven days from now. You can guess the weather tomorrow with some precision, but you really cannot guess well for a week later, no matter how fine are your observations because of the chaos in the system. We need to be prepared for this in our subject. We may have to philosophically accept a different kind of science. That comment ends my gazing into the crystal ball.

Chairman: Well thanks, Frank, Thank you very much. (applause)...

John Kormendy: I have great respect for the classical morphology that is advocated by Allan Sandage (2004) in his perceptive contribution to this volume. I completely agree that morphology without interpretation is essential early in the history of a subject. This kind of galaxy morphology has worked exceedingly well for many years. But I also believe that we cannot effectively progress beyond a certain point without combining morphology and physics. I think that we have now passed this point.

Classical morphology has been very successful, but this has not happened without blood on the floor. Here is a story that highlights how quickly interpretation becomes relevant. In the 1950s, Morgan invented a classification scheme based on the central concentration of light in galaxies and on their central spectra. He did it entirely as Sandage advocates – with no interpretation. But Morgan's classification has not survived, except for one detail. What survives is the cD class. It refers to first-ranked cluster galaxies that have extraordinarily luminous but low-surface-brightness halos. We believe that these halos are a cluster phenomenon. They consist largely of stars that were stripped from individual galaxies by tidal encounters, and their dynamics are controlled by the cluster's gravitational potential, not by that of the central galaxy. This cluster physics is not included in the Hubble tuning fork diagram, it is important, and it is convenient to have a classification bin for it. But the rest of Morgan's classification scheme has not survived because it correlates with physics less well than the Hubble sequence does. This shows how important it is that a classification lead directly to physical understanding. Hubble implicitly knew this. He chose to classify galaxies using parameters that have become central to our understanding. Sandage recognizes this genius in his description of Hubble classification in the Carnegie Atlas of Galaxies: "Hubble correctly guessed that the presence or absence of a disk, the openness of the spiral-arm pattern, and the degree of resolution of the arms into stars, would be highly relevant. It was an indefinable genius of Hubble that enabled him to understand in an unknown way ... that this start to galaxy classification had relevance to nature itself." In setting up his classification, Hubble made choices with interpretation in mind.

Having said this, I believe that we are now approaching the end of what we can accomplish by separating morphology and physics. I would like to argue that we must break down the wall between morphology and interpretation. Doing this successfully has always been a sign of the maturity of a subject. For example, it has happened in stellar astronomy. I cannot imagine that people who observe and classify phenomena without interpretation would ever have discovered solar oscillations. Without guidance from a theory, how would one ever conceive of the complicated measurements required to see solar oscillations or to use them to study the interior structure of the Sun? In the same way, we need the guidance of a theory to make sense of the bewildering variety of phenomena associated with galaxies and to recognize what is fundamental and what is not. We emphasized this point in Kormendy & Kennicutt (2004, ARA&A, 42, in press), especially in the words kindly quoted by Sandage (2004). In the same spirit, Sandage also opened the door to such a phase when he wrote, in his letter presented at the start of this meeting, that morphology and interpretation must be kept separate "at least until the tension between induction and deduction" gets mature enough. So my aim has been the physical morphology that I have advocated for many years, not (I emphasize) as a replacement for classical morphology – which remains vitally important – but as a step beyond it. Physical morphology is an iteration in detail that is analogous to de Vaucouleurs's iteration beyond the Hubble tuning fork diagram.

Sandage (2004) cautions correctly that this is a step that should be undertaken only with due caution and when the time is ripe. The danger is over-interpretation, or what, with only slightly different emphasis, Sandage calls "hermeneutical circularity". For example, it seems dangerous to me to interpret every wiggle in the light or velocity distribution of a galaxy in terms of one's favorite family of orbits.

I have a second caution that is tangential to the above remarks. It has to do with Chris Conselice's concern that the Hubble sequence applies poorly at large redshifts. The Hubble sequence was invented to describe galaxies as they are now and as they appear at blue wavelengths. There is no guarantee, if morphology successfully distills physics, that it applies to galaxies as they were long ago. Nor must it do so. A biologist who invents a successful classification scheme for butterflies will not be able to classify larvae. This is not a fault of his classification scheme, it just means that no butterfly looks like any larva. The same may be true for galaxies. Kinds of galaxies may have existed in the past that have no counterparts today. No attempt to shoe-horn them into today's Hubble sequence and no attempt to define a single classification scheme that works equally well at all redshifts may work. Similarly, there may be galaxies today that have no counterparts in the past. For example, there must have been a time when no cluster of galaxies was mature enough to have produced a cD. As Frank Shu suggests, galaxies are almost biological in their complexity. This is not the Universe's problem, it is ours. We have to learn to live with it. Thank you.

Chairman: Thank you John, thank you very much. Chi Yuan is up next.

Chi Yuan: Okay, I have a few things to say. I think during this conference, I heard lot about the bar classification. People talked about the major bars and did analysis on them, but not for the central bars. One example is NGC6951, for which both major bar and the nuclear bar are present. The next one is the famous NGC1068 which has no major bar, but if we use wavelet analysis we see clearly that they have a nuclear bar. So I like to know how the nuclear bar and the major bar are really formed. I want to call your attention to the fact that most galaxies have a central gas-dust disc. Okay, this give us a something to work on, because major bars span the entire galaxy and the dispersion velocity of the stars is not much different from the gas out there, so stars and gas disks are very much coupled. For such a system, we need to use stellar dynamics for the stars. This is something nobody has done. For the central bar problem, however, we need to consider gas dynamics only because the gas has low dispersion speed and will form a disc there which is decoupled from the stellar population which has high dispersion speed and forms a spherical bulge. The major bar is necessary to be a slow bar and the central bar can be fast, or can be slow. The next thing I want to say is there are two types of rotation curves: one is a fast rising rotation curve, like this one on top, similar to the Milky Way. The $\Omega-\kappa/2$ curve goes up. The other one is a slowly rising rotation curve, such as that of NGC5248: $\Omega - \kappa/2$ curve drops back to zero at the center. The former can host a fast bar and the latter prefers slow bars. Since major bars are necessarily slowly rotating, so I thought this must indicate something. We may argue the fast rising one implies there is a concentration of matter in the center and the slowly rising one does not. Does this mean that galaxies with a major arm all have slowly rising rotation curves? SoI checked on that. I find that the result is negative. There is no correlation. The next thing I want to mention about is the resonance excitation. This is the very mechanism that is responsible for all spirals excited by bars. But nobody mentioned it in this conference. Different resonance produces different spiral morphology. Moving out from the center, we can have open leading spirals for the inner inner Lindblad resonance, to open trailing spirals for the outer inner Lindblad resonance, and to tightly wound trailing spirals for outer Lindblad resonance. Three types. The last two types of trailing spirals are observed. I would like to point out no matter what orientation the galaxies have, the iso-velocity curves for the tightly wound trailing spirals are always bent inwards along spiral arms, while for open trailing spirals are bent outward around spiral arms, predicted by the theory. This kinematics certainly can be checked by observations. Okay, I stop here, thank you.

Chairman: Preben Grosbol is next.

Preben Grosbol: I basically agree to a large extent with what other people have said in the sense that you can not fully separate morphology from some kind of understanding of the physics. There have to be an interaction. I think an important issue is that one have to be able to test and possibly falsify ideas or theories. If you are not able to do that, it doesn't really matter what you do. If you set up a morphological system which doesn't describe features, which can be used for testing, supported by theory it has little purpose. It is very important to understand the relation between what theories predict and what one can measure quantitatively in real numbers. Artificial binning of measured quantities makes tests less stringent. I think it is dangerous to have very loose schemes which because they are loose are more difficult to use for testing and falsification. The other thing is that one in principle will always be driven by the wish to test ideas. I have long time tried to see if one could falsify or verify the density wave theory and its application to spiral galaxies. This is complicated because there are many functions involved. Many of them (e.g. velocity dispersions) are extremely difficult to measure and therefore introduce significant uncertainties in tests. One then have to go step by step. This is partly done by the morphological classification in the infrared, namely looking on very basic models of disk galaxies, describing their spiral pattern with pitch angle and so on. Next one have to check if the system is fully described with such a simple model and look on the residuals or deviations which do not agree with a simple morphological description of the galaxy. Whether or not such differences can be associated to a physical explanation later is for theorists to find out of. This is similar to the discussions some years ago of the $r^1/4$ law. One first assumed that ellipticals and bulges could be described by this law and first afterwards saw small systematic variations which then lead to the more general 1/n powerlaw. It is important to understand what the basic simple model is and check if it describs the phenomena good enough. If it does not then one have to look on the residuals and try to describe them.

Chairman: Thank you, Preben. Great, thanks. Now to Ken Freeman.

Ken Freeman: I want to make a boring and practical point but it's one that's been bothering me throughout much of our discussion here. It concerns the use

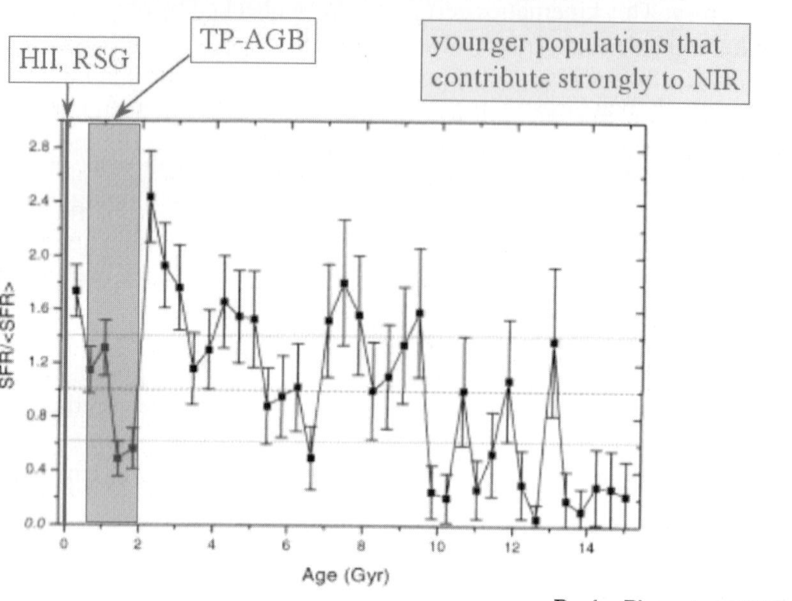

Figure 2. For a discussion, see the contribution by K.C. Freeman.

that we're making of K-band images and the inference that they are showing us the distribution of the stellar mass in spirals.

We use the near-infrared images because of the way they lift the mask of dust. They let us look at the underlying structure of spirals because they best show the distribution of the older disk and old stars. The K-band images are indeed relatively unaffected by dust but there is still an obvious and significant contribution from the young populations. The HII regions are seen in Br-γ, and also the red supergiants and the thermally pulsing AGB stars; Preben referred to this problem yesterday in his talk.

Let's have a quick look at the star formation history in the solar neighbourhood. This is the work of Rocha Pinto et al. (Figure 2). Age is along the horizontal axis and the vertical axis is the normalized star formation rate. The first thing you can see is that the mean star formation rate has been roughly constant so there are a lot of old stars in the disk. Two younger age intervals that contribute strongly the near infrared are marked in red. The HII regions, and the red supergiants are at the very young end. Then there's a bit of a gap and then an intermediate age range (0.6 to 2.0 Gyr) where we get the bright

AGB stars. Those two red regions will contribute pretty strongly to the K-band images, if there are populations of those ages present. In a typical galaxy like ours, you can see from the star formation history that stars of those ages are indeed present.

Okay, so if we want to make a quantitative classification based on the distribution of old stellar populations, or if we want to calculate potential fields from the K-Band images, or if we want to estimate the contribution of old stars to the spiral structure, then it would be much better if we could remove the contribution from these younger populations. The young and old populations have very different mass to light ratios, so contamination from the young population can be very misleading. Let me just show you a couple of examples. Here's M83 - a B-band and a K-band image. And you can see, in the K-band image, the very clear presence of the young HII regions. The sharp structures that we see in this image are probably due as much to HII regions and young stars as they are to the old population. Here's a galaxy that David showed in his talk: NGC2997. The blue and K-band images are on the same scale. This galaxy is a honest, classic, strong two-armed spiral. Now have a look at the K-band image; what are we actually seeing in the K-band? A whole bunch of HII regions are clearly visible, with the young stars that go with them. So the K-band image is really quite strongly affected by a very young population, yet we are interpreting this as old populations.

I think we need to do something about this and I think there are ways to remove the K-band contribution from the young population in our images. The way I'd go about this is to do two-dimensional optical spectrophotometry. Use the Balmer and Paschen emission lines to measure the effective interstellar absorption, point by point, and estimate the contribution to the K-band image from Br-γ. Then I'd use the observed de-reddened SED's, again point by point, to estimate the contribution to the K-band image from young stars (mainly red super giants) by using empirical SED's for LMC clusters and/or synthetic SED's. I believe we must do something about this contamination from young populations before we proceed too far down the line of quantitative classification using K-band images.

Chairman: Thank you, Ken, and it's on to Francoise, Francoise Combes.

Francoise Combes: Okay. It was striking in this series of conferences to see how the classification was becoming more and more quantitative; I don't know why people have remained qualitative before, keeping every sequence of galaxies in bins, Sa, Sb or Sc, according to morphology but always indiscrete and discontinuous boxes. I think it's much better now because of the quantification that has been done in the recent years. So I just put here some ideas how we can select characters to quantify better. Apart from the bulge-to-disk ratio, or gas content, a fine classification can be established on bar strength and asymmetry. As we have discussed a lot in this meeting, I think it's a very nice point to have established the stability status of the galaxies, so we can reconstruct their dynamical history. I will stress also on the kinematical classification because what we want is really the physical state of the galaxy, the mass of the galaxy not only from the luminosity, but also the rotation curve could give us a better mass distribution, in addition to the more obvious gas content, gas-to-star ratio, stellar populations and so on. People have in this panel talked about biology and classification analogy. I was very happy that D. Fraix-Burnet have developped further this analogy with his proposed astrocladistics, this is a new idea, and may be some of the many characters that have to be selected could be these asymmetry and stability parameters to be taken into account as continuous parameters, and we could maybe learn something in regard to the sequence of events in the history of galaxies in analogy with biological evolution.

I'll just pass very quickly on the caveats of such quantitative classification because we have not really discussed these problems: how to deproject galaxies? already difficult at $z = 0$, almost impossible at high redshift, it could create more bars; the problem of dust, of course, the mass to light ratio, that's why the kinematics is very important; and also the asymmetry analysis could be done in density and potential, we have not discussed a lot of transforming densities, we have talked about gravity torques which is the point of view of the potential, complementary.

Well, I wanted also to give some "eyes for the future" because it was the title of the panel and I tried to think of what we are going to do next in future projects. One strong bias in the classification is the filter in which we see galaxies, so of course, people have already noticed in this meeting that we should observe the galaxies in UV now at $z = 0$ to compare to what we see at high redshift (UV in their rest frame), so maybe the GALEX mission will give us a lot more information about that. About kinematics - I wanted to mention here that we have a Large Programme at ESO VLT with the Giraffe spectrometer, to observe a sample in the GOODS fields, so we'll have to select 300 galaxies in which IFU spectroscopy will be completed with a resolution of R=10,000, and in 2D: I insist on this 2D because it is necessary to have the percentage of bars, it's very interesting to have the S distortion of iso-velocities not only velocities along a slit. This Giraffe instrument can study 15 galaxies at the same time with about 20 pixels in one galaxy, at a resolution of 10 km/s, and about $0.4''$ spatially. We have done a lot of models about that and indeed it is possible, even with low resolution at $z = 1$, to know the percentage of bars. Also the rotation curves will give the dark matter content and the evolution of the dark to visible matter with redshift. So in the near future we will have kinematics information with this Large Programme, made by a consortium of French, Italian and German people, it will be available in a few years.

In the future, ALMA will come up in 2007 - so it's not so far from now - we can even go further with the CO line, obtaining 2D velocity fields of galaxies - not so many because the field of view is very small, but this will be also an element to take into account in classifications at high z, since the CO line can be observed in remote starburst galaxies. Further in time, for "eyes inthe very future", the SKA instrument (Square Kilometer Array) will be essential to know the dark matter content of galaxies. Indeed, all what we know today about the dark matter distribution in galaxies is from the HI component, its flat rotation curves well beyond the optical disks. Even with ALMA, we will not have these extended rotation curves, since the CO line stops at the optical disk boundary. With SKA, we will have 2D maps of HI at $z = 2$ and then we could know how the galaxy has been assembled, dark matter or baryons and maybe also the measurement of torques.

Now I want to answer Bruce's point: he said that we must have a grid of models from the theoreticians to really calibrate the evolution at high redshift. We have just now in France (a consortium of four teams) embarked on a project – at least a four years project – to carry on very large simulations – about 100 times more particle than now — about 10^{10} particle simulation or 2048 cube, simulating a Hubble volume - that's why it's called the project HORIZON (we are going to simulate the universe up to the visible horizon), and then with refined mesh codes (like AMR, or multi-zooms code), from large-scale simulation, it is possible to simulate down the scales, until several orders of magnitude, galaxies are reached. The goal is to simulate galaxy formation in a cosmological context. We will study both the cosmic web with all these filaments drizzling into the galaxies and with several zooms like here, the clusters of galaxies and we see now another zoom, with the detailed physics of one spiral galaxies, with several components, dark matter, stars, cold and warm gas, etc. Many physical parameters will be varied, to have a grid of models. This project is run as an instrumental project, like a satellite or ground instrument, and at the end a large database will be available on the Web with thousands and thousands of galaxies with varying parameters. So coming up is really this kind of models and grid of models that we want to compare with observations, here is an example, like a prototype image. To be more realistic, we have put the dust mask in the simulation (by simple radiative transfer) so that we can afterwards penetrate the dust mask again like observers do. So what you see here is a simulation of merger galaxies where we have taken into account the stellar populations (the blue stars are the young stars formed in the merger); we have produced the images in three filters and combined them in a true-color image. In these kinds of simulation we take into account the gas physics, star formation, the yield of heavy elements and dust and a balance of energy, so all the light that it is absorbed by dust is re-radiated in the infrared. We can then produce FIR or submm images to compare with Herschel and ALMA future observations, or JWST, SKA, etc. I just want to leave you with these images where we tried to bring simulations very close to observations.

Chairman: Alright, let's start over here, with Ivan. This is Ivan King.

Ivan King: I'd like to add just a few words about this question of theory and observation. For me it's just as inconceivable for an observer to approach the universe without knowing anything about theory as it is for a theoretician to try to theorize about it without knowing any facts. I think the two have to work together; it's a kind of iterative situation in which we converge on the answer, and if the interaction between theory and observation is good, the convergence can be rapid.

I'd like to give two examples. When Hoyle and Schwarzschild calculated the first evolutionary track, they found that with solar metal abundance they got M67 instead of M3. They knew that there had been some suggestions that high-velocity stars had low abundance, so they lowered the metal abundance in their evolutionary track, and they indeed got something that looked like a globular cluster. That's a good example of theoreticians being guided by knowing something about the observational facts.

On the other side, I think it would be just ridiculous for observers to look at everything in the sky just because it's there. I'd like to give as an example some work that I'm doing at the present time. Theoreticians have predicted that there is a hydrogen-burning limit at the bottom of the main sequence, and I would never be observing the faint ends of globular-cluster main sequences if it hadn't been for that prediction. I regard the purpose of this work as to give the theoreticians the facts that they need in order to proceed. It's very much harder to solve an equation when you don't know the answer, but if you can look in the back of the book and see the answer then you can do it much better. I feel that the observers are giving the theoreticians the answers in the back of the book. Just one more remark: I think Frank Shu gave us an excellent example of the synergy between theory and observation. We have this problem of using up the star forming material; the answer is possibly infall, but here both the theoreticians and the observers are confronted with a disagreement. Either the theoreticians are going to figure out a better scenario or the observers are going to find out how to observe that damned infall!

Chairman: Thank you, Ivan. I think George Lake's comment might follow onto this, so George I'll give you the microphone. George Lake.

George Lake: I agree with Ivan, I think that all the fun is in the red arrows that Ron drew with the question mark. There's a depressing aspect to the philosophy of science as you can either be kind of right so far, or wrong, right but absolutely irrelevant is a third possibility. That third possibility is one the worst. You get there if you are on the "left side" (of Ron Allen's diagram) trying to figure out how many more bins you can put galaxies into or on the "right side" integrating irrelevant equations. In this third possibility, you're not even wrong and that's the thing which I think needs to be strongly avoided.

When I think about some of the questions that people talked about here a lot of them have to do with secular evolution and I'm puzzled by some of them. That namely, galaxies have bars in at least 70% and probably 80% and the big question being asked is how did they dissolve and re-form? They're not taking much time off. So I think the challenge is why do they seem to go away in the simulations, but they don't in the galaxies. That's the problem that needs to be addressed. And also I think that secular evolution is being overestimated. We live in an era when it's just starting to happen and we only see a little bit of it now and we see very, very strong boundaries that constrain it. I talked about one in terms of how velocity correlates very well with Hubble type, another one is that I often hear people say that you know bulges are just ellipticals that just happen to find themselves inside a disc but it's actually not the case. I can ask you one question about a bulge, or spheroid and I can tell you whether it's a bulge or elliptical. Take a look at Kormendy's diagram for V/σ. If the number is above 0.6, it's a bulge, if it's below 0.6, it's an elliptical with 95% probability. Fortunately, you can look at ellipticals inclined but the bulges have to be almost edge on for these observations, so this diagram doesn't confuse things by projection. So there are these very strong boundaries that have to be respected if we're going to try to promote secular evolution.

Chairman: Thanks George. Paul Eskridge.

Paul Eskridge: I have three comments ? not to ? hopefully none of this will sound terribly trillish(ph). The first one is in reaction to something David Block said and that is that optical and near infrared morphology of galaxies correlate very well on average. Near infrared images look a little bit earlier on average and every once in a while they look a lot different but they correlate very well on average. Something I wanted to react ? this is, I'm now going to steal a little bit of my own thunder to respond to something that Kartik said that had to do with the need to find out what a good choice is for a bar strength classifier and the answer to that question is, it doesn't matter, anything will do and I'll talk more about that on Saturday and one last thing, that I wanted to say in response to the point that Chris brought up, having to do with stellar classification, in fact I think the stellar classification scheme is a perfect example of exactly what Sandage had in mind that is you know when Draper and his computers came up with that classification scheme, there was no physics in it at all, they were just saying well the strongest hydrogen lines - we'll call those A and then B and then C and so forth and that's why we have this squirly letter sequence now that we've figured out that it's a temperature function, it's not monotonic with hydrogen line strength, right, that's why we have to have this pneumonic that we all have to (inaudible statement from panel) excuse me. Yeah, that's true also. That's it .

Chairman: Is that it? Thank you Paul. Alright, let's see else... has a hand up... - I see Rogier Windhorst, let's start with Rogier.

Roger Windhorst: Alright. I have only question, I think. On Ron's red arrows, the observables map into theoretical parameters – parameters of the galaxy or the galaxy to be that is. I'm not sure that that's entirely the correct picture we want to work with. If it is true that external factors influence what a galaxy is going to look like – and I think the fundamental plane and κ-space and so on are telling us that the eventual appearance of a galaxy depends on its environment and what surroundings it is living in then and now – perhaps even more importantly then: to what extent has the exponential expansion that Lambda has pumped into the universe at recent redshifts been changing the universe from a merger driven environment to a more quiet, perhaps secular evolving environment? Shouldn't we also be looking at large scales? Large scale environments, cosmology itself, how did Lambda impact galaxy evolution? I haven't heard an answer to that. I think that's the critical issue – Lambda tinkered with the epoch dependent merger rate, and gradually turned it off since redshifts one.

Chairman: This is Kartik Sheth.

Kartik Sheth: That's the reason for doing the COSMOS survey, I mean, the reason for having a two square degree contiguous area is so that we have a good chance of exploring the evolution of galaxies with large scale structure, we have a 50% probability of exploring the largest scale structure.

Rogier Windhorst: Right. It's a very good step in the right direction, but I'd like to hear from the respected theoreticians in the audience – I am completely ignorant about this – I'd like to hear from the theorists, how Lambda itself has been tinkering with galaxy evolution through the epoch dependent merger rate.

Chairman: Anybody like to take that on? George Lake.

George Lake: Well clearly we live in a special epoch where the cosmological Lambda has just started to dominate and so I think that's why it's no surprise that we live in the epoch where things are going from merger dominated to secular evolution because structure formation and collapses are freezing out.

Roger Windhorst: Well isn't that a critical issue that we are dealing with here? If we talk about bar evolution and bar frequency with look back time and all that, or anything evolving with look back time, isn't that what we should be looking at – how Lambda truned off the epoch dependent merger rate – certainly since redshifts of one?

Chairman: Bruce wants to pick this up.

Bruce Elmegreen: How is Lambda related to this? Just the expansion of the universe will take you to a low enough density that most of the galaxies evolved by themselves.

Roger Windhorst: At high redshifts, minor and major mergers go on untempered by Lambda, and are only controlled by gravity and the more modest Einstein-de Sitter like expansion. But at redshifts less than 0.4–1, pieces are still falling into galaxies, but by-and-large the extraexponential expansion induced by Lambda at redshifts below 0.4 drives things that aren't yet close to turn-around forever apart — hence Lambda has been strongly winding down the epoch dependent merger rate since redshifts 1. That must affect everything we see in galaxy evolution since redshifts 1.

Bruce Elmegreen: On larger and larger scales but do galaxies (unclear)

Chairman: George, want to make that comment again?

George Lake: Sure, Lambda accelerates the change from going from a universe that looks like an $\Omega = 1$ universe to a universe where all structures are freezing out. In any low Ω universe, you have the same effect but this is accelerated with the cosmological constant. If Ω were equal to 1 right, then we'd fall into the Virgo cluster and merge with several other things other than Andromeda on the way. But given the fact that structure is frozen out, that's never going to happen, after we merge with Andromeda -that's it - nobody else is coming in.

Chairman: End of story! Anybody else want to take this on or start a new subject? (jokes with delegate off mike). You want to add something? Okay, Chris Conselice.

Chris Conselice: Yes, the? this Roger's point about Lamda is very relevant and the cosmology does affect the way that galaxies will be forming and if you have a (unclear) dominated universe you will have much more mergers happening and the galaxy morphology would be very different and distribution of redshift would be very different and I think simulations to that extent would be very useful to have.

Chairman: Thanks. Somebody's holding a mike up over there, good. Andreas Burkert.

Andreas Burkert: I would like, from a theoretical point of view, to know a little more from the observers. You know we have now moved from the dissipation-less dark matter evolution to dissipative gas dynamics; if we want to understand galaxy formation, we need to understand how gas falls into dark matter halos. Even if we don't understand dark matter, we have an idea of what dark halos look like and we need to understand how stars form. We have no theory of star formation, so we are completely lost – and I think that's the main problem with that. (Comment from panelist inaudible) Yeah, we begin to understand how stars form and we work on that - but we don't have the theory. It's kind of (in some sense) premature to understand galaxies if we don't understand how a star forms inside a molecular cloud. Now I think it's still worth doing, but we need a clear problem. You know, it's too much confusion, if you tell me, if you show me this image and that image! I can do a simulation which sometimes looks like the image [you show me] but I haven't proven that it's the same physics; you know what I mean. Instead of seeing aninfrared picture here and there, I need to see systematics! I don't want a peculiar case: I want something which happens again and again and again and which is a puzzle which is not obvious. If it's obvious again, it's boring - because then I can obviously solve it. So I need something which is a puzzle and which occurs over and over and over and then you ask me the question which physics leads to? can you solve this puzzle? I think these questions haven't yet been found or been identified or clarified. The more [the observers] fire off these questions, the better for the theorists.

Chairman: So hang on Daniel, I was going to give Frank a chance to throw a rock over in Andreas' camp there. Did you want to react any further? Okay, good, Daniel Pfenniger.

Daniel Pfenniger: Concerning this problem of modelling star formation, it seems to me that the major problem before that is to treat the gas in the right way. SPH is much too collisional with respect to the way galaxy behave. The galactic dynamic people often prefer to use sticky particles for a good reason, it's a much less collisional medium but somewhat ad-hoc. For me there is much progress to do in this direction, to understand how to describe interstellar gas on the large scale.

Chairman: Thank you. Jan Palous.

Jan Palous: This is less general but concerns star formation. I had little doubt that bars trigger star formation. During this week I heard that between now and z equal to 1, maybe bars do not change substantially, they do not fade out, they have maybe the same strengths. However I also know the Madau diagram showing that the star formation rate is dramatically decreasing between $z \approx 1$ and $z \approx 0$. This implies that there have to be something else on the top of the bars regulating the star formation rate. May be bars are relevant to star formationg tiggering, but the star formation rate is also sensitive to the total amount of gas available. Bars just provide the hammer, however you have to have something to hammer, which is the gas in the the bar region.

Chairman: We're looking for a bag of nails. I guess. Thank you, Jan. It was back in the back of the room there, yes. I'm afraid I don't remember your name. If you could speak it out.

Tom Steiman-Cameron: I want to go back to the title of the meeting, dealing in particular with bars in the context of the Tuning Fork. By classical definition, the two tongs of the Fork distinguish between barred and unbarred systems. However, it now appears that the classical distinctions are not so clearcut. In particular, many "unbarred" galaxies reveal inner bars and more extensive bars are often "uncovered" when galaxies are observed in the infrared. Thus we now frequently prepend the adjectives "weak", "strong", "nuclear", "inner", and "outer" when referring to bars. At some level, the two tongs of the fork may more appropriately distinguish between strong and weak bars. If this is the case, do weak and strong bars have different origins? I have been led indirectly to this question by trying to understand an apparent dicotomy in how spiral structure is influenced by strong and weak bars. My perception of traditional (strongly) barred spirals is that they are all $m = 2$ (or $m = 1$) systems. Higher order global modes are rarely or never described among these systems. On the other hand, "normal" spirals include numerous well-defined three-arm, four-arm (e.g., M101 and the Milky Way) and even five-arm systems. Many of these, including the Milky Way, are now described as (weakly?) barred. Clearly, strong bars are intimately connected to global spiral structure, but what, if any, is the role of weak bars on spiral structure?

Chairman: Anyone would like to offer a response to that particular question? Oh, sorry, Ron Buta, yeah.

Ron Buta: Yes, I just participated in a project to measure the strengths of bars and spirals, to examine the relationship between the spiral strength and the bar strength and this is using the OSUBGS parameter and we had a wide range of spirals and bars in this sample and the weak bars come in the form of $m = 2$ ovals or just low contrast bars and they don't seem to be related to their spirals. There's no correlation between the spiral torque strength and the bar strength on the weak bars; it's only when the bar is strong with like ? with a QB of

40% or more that there's some relationship. I don't know if the strong bars are driving their spirals, they might be, but for the weak ones they seem not to be related.

Chairman: Okay, John Beckman.

John Beckman: I sympathize strongly with the statement "how can we possibly claim to understand galaxy formation if we don't even understand star formation?" This is a relevant comment in a number of ways; I don't think that the problem is a theorists' problem but an observationalists' problem. I don't think we lack a theory of star formation, I think that we have too many theories of star formation, in other ways there are many postulated ways of forming stars, some correct, some surely wrong, but in order to verify these we have to confront them with observations, which are not easy to obtain as stars tend to form rather quickly and stars with different masses form under different conditions. These are well known problems, but they are difficult observationally for various reasons, one of these being dust penetration. One of the key points about trying to make observations of star formation is that stars form in the dark and in the cold: in the middle of a dust cloud. This has been tackled, and there are researchers working on it, but the fraction of the astronomical community attacking this unsolved problem is going down, and the reason for that is that everybody is rushing out to high z because low z is "geophysics" as we all know, and therefore not glamorous and not worth funding. Astronomers suffer the well-known tendency of rushing to solve a new problem because the old problem is too difficult to solve. In the context of star formation we must begin to apply fine scale gas dynamics, and there has been in general a lack of attention in combining kinematics and morphology to the question. The different weapons to hand, combining observations over a full range of wavelengths have all been mentioned this week, but they need to be applied thoroughly and systematically to star formation. I don't think that anyone would wish to disagree with this point, but it tends to be a truism, which people mouth but don't take seriously. I also have a comment about the relation of theory to observation. I have a revealing anecdote about this. On a specific occasion I was the author of a paper which received strong criticism from a referee, saying that the observations could not be correct because there was not a theory which predicted them. (LAUGHTER) I see that we all understand the absurd nature of this remark. I agree strongly with Frank Shu's statement that astrophysics is a post-dictive science and not, in general, a predictive science. My idea of a good theorist in astrophysics is one who really knows what observations are being made.

Chairman: Okay, I did want to give John Kormendy a chance to respond, I think to the previous question, did I get that right? And then we'll go out further in the audience. So prepare your questions. John.

John Kormendy: I want to comment on weak bars in the context of bar suicide. All week, we have discussed bar demise as though the bar always disappears completely. But this is not what the papers on the subject say. They say that the strength of the bar depends on how far the evolution has proceeded. Bars tolerate central mass concentrations of at least $\sim$ 10 percent of the disk mass. Bigger masses are tolerated if they are fluffy, as in a pseudobulge, and not point-like, as in a supermassive black hole. The bar gets weaker as the central mass grows, but it does not disappear completely unless the central mass gets very large. But bar-driven secular evolution presumably slows down as the bar decays.

Weak bars are seen in many galaxies. We also see some galaxies that contain lens-like components – in other words, features that look like bar relics – in galaxies that have no primary bar. Examples discussed in my talk include NGC 1553 and NGC 4736. But there are more weak bars than barless lenses. So the indications are that bar demise can be a slow process. In many galaxies, it may still be going on.

To me, a weak bar is suggestive evidence for bar suicide. Do we really believe that global disk instabilities or tickling by encounters can make a bar that is a few percent of the disk mass? This is not what happens in simulations. The bars that get made by these processes contain most of the disk mass that is interior to the end of the bar. Would it not be easier to understand the formation of a bar that is a few percent of the disk mass if it formed as a much stronger perturbation and then decayed?

Chairman: Okay thanks John. I'd like to see if there are any people we haven't heard from yet. Yes, please, David Koo.

David Koo: I'd like to make a couple of comments. One – that the idea that Francoise mentioned about HI for high redshift I think also extends to the concept that we need more detailed information, particular much higher spectral resolutions, so that you can actually get to the dispersion of the disc (or at least the relative motions of the components that we see at the high redshifts). I think that would be very relevant. For example, we haven't even discussed thick disks, but I think they obviously are directly related to whatever bars do or do not appear. I could have gone on forten minutes, showing you examples of objects which look like proto-disks – it would be very fascinating to find out what the relative velocities are, using optical, IFU, Giraffe... we might get the first hint that the velocity dispersion say is tens of kilometers, which we could measure today. Whether we can get down to a few – five to ten – that might be more of a challenge. The second comment is actually related also to high z – we are pushing to regimes which I think we shouldn't - that is the identification of what is an object is, when things are assembling themselves in complex environments. This has not been really been settled by photometrists, at least in the high z world, yet. We do not know how to identify single objects. Maybe with plenty of IFU 2D, we might be able to begin to say these pieces belong to this object, this one's part of another object; imagine the difficulty we already have locally to locate Antennae, etc. At high z, I think that gets exacerbateda lot: we may have whole new statistics that are multi-dimensional correlated information that can relate the theory to the observation, so it's not only new classification schemes, but new analysis tools required, to really make these comparisons.

Chairman: It's a good comment, David. Other comments or questions from elsewhere in the room, people we haven't heard from yet? Oh, Shardha, Shardha Jogee.

Shardha Jogee: First I want to respond to what Jan Palous said regarding the Madau plot. The reason we embarked on a study of bar properties at intermediate redshift were two-fold. The first reason was to set constraints on the controversial question of whether bars are long lived or whether they dissolve and reform recurrently over a Hubble time. But the second driver was indeed, as Jan Palous said, to understand what produces the dramatic decline in SFR density from redshfit 1 to 0 in the Madau plot. There are four possibilities for this decline: a decline in the cold gas content, a decline in external triggers (minor mergers, major mergers) or a change in the internal structure of galaxies, where internal structure would involve features such as bars. The current sense we are getting based on analysis of the GEMS data from $z = 0$ to 1, is that we can rule out luminous major mergers as being responsible for the decline in UV-based SFR from $z = 1$ to 0 in the Madau plot. A paper on this is in the works by my co-team member Chis Wolf. Minor mergers and bars are still in the picture. However it is important to note that bars induce circumnuclear SF, but they are not responsible for wide spread 'quiescent' SF over the entire galaxy. This latter mode of SF is seen in quite a few systems (massive spirals) at these redshifts...

The second point I would like to raise pertains to a general wish-list of mine for better dialogue amongst theorists, better dialogue between observers and theorists, and a request to conference organizers to provide panel discussion where observers have the opportunity to talk to a panel of theorists to understand why their predictions do not converge, what they need from observers, and vice-versa. As an observer, one of the prime reasons I come to conferences is to get clear directives from my fellow colleagues who are theorists, as to what observational constraints are needed to nail the big picture. But clear directives require some degree of convergence on part of the theorists... which is lacking right now. At this very conference, I found from talking to many theorists that not only is there is disagreement or general skepticism amoungst theorists on their respective differring results (e.g., on the issue of bar destruction/reformation, bar patterns in isolated and induced bars), but quite often it is not clear to them (or to us observers) why the simulation results are different. Thus better dialogues are needed.

Chairman: I think it's a very good point too.

Didier Fraix-Burnet: There are two goals for classification. The first one is memory recall. It is practical to define some classes and their definitions, rather than millions of galaxies: or billions of living organisms. The second goal is to help find relationships between all the different kinds of objects. One can then understand that it is nonsensical to try to explain, using physics, a classification scheme; that is the origin of the Sandage's wall between morphologists and physicists. One can explain morphological features invoking physics, but one cannot explain the classification. It is really nonsense. So if you want to explain, for example, bars, one can invoke some theory, and then make a classification of different kinds of bars; one can make realistic predictions. Observers will, I think, then be in a good shape to try to test the theory. The second point I would like to make is about the history of classification. I would say astronomers have a huge history, in fact I think astronomy is probably the first job in the world ... or the second one – I don't know. Extragalactic astronomy is very recent in this respect. When I began to work on astrocladistics, I really had to change my mind and do some psychological work also, because the cultural tradition of astronomy is mainly stellar. I think astronomers still have a dream, and Chris Conselice worded that very clearly... there's a dream to classify galaxies as we do for stars. I recall having seen, in a conference of a few years ago, someone trying to make a Hertzsprung-Russel diagram for galaxies! I would like to really agree with Frank Shu and John Kormendy: one knows galaxies are not stars; they are really complex objects, complex in the mathematical sense of the term.

Chairman: That was Didier Fraix-Burnet. Okay we're sort of running out of gas here. Oh, here's another.

Aigen Li: Let me say something different - since we are dreaming of penetrating the dust mask, we really need to understand the MASK ITSELF (namely the properties of dust). However, how much do we know (for sure) about the physics, chemistry, origin and evolution of interstellar dust? Furthermore, dust properties (e.g. dust size and composition) vary from one galaxy (type) to another! Note that we even don't know much about the dust properties in our Milky Way galaxy.

In order to understand the morphology of a galaxy, we of course need to understand the way how dust scatters starlight and re-emit in the infrared. However, we do not completely understand the scattering properties of dust! For example, David Block, our SOC co-chair, found that near infrared scattering properties, the albedo, is as high as 0.9; this cannot be explained by any existing dust models; this would of course affect our view of galaxy morphology. On the other hand, are we sure that we understand the absorption and emission properties of dust? the balance between absorption and emission determines the dust temperature. Recently, many groups of astronomers report the detection of a population of "very cold" interstellar dust in some galaxies with a significant quantity. This "very cold" dust component, with an equilibrium temperature around 5–7 K, locks up a substantial amount of dust mass. If this is true, it would of course change our view of galaxy morphology. So all I want to say is that we should pay much attention to the mask (dust) itself.

Chairman: I would second that, I like cold dust!

Wind down and thanks by Chairman and comment regarding housekeeping.

David Block: Thank you so much everyone on the panel and a deep vote of thanks to everyone in the audience.

CONCLUDING CONFERENCE THOUGHTS ...

Vera C. Rubin
Carnegie Institution of Washington, 5241 Broad Branch Road, NW, Washington DC 20015 USA

EDITORIAL NOTE: Vera Rubin regrettably could not attend our Conference, but she very kindly sent the SOC co-chairs her concluding thoughts, which are printed below. May her husband Bob have many more years of good health.

Dear David and Ken and Ivanio,

Thank you for your generosity in inviting me to add a few words to what I am certain has been a remarkable meeting. The conjunction of Africa, astronomers, wild animals, and a transit of Venus surely has resulted in a wonderful week for all of the participants. And thank you for all your efforts over the past many months which have insured a meeting full of science and adventure.

Those of us who study galaxies should periodically remind ourselves that our field of study is a little less than 100 years old. I hope that important discoveries in the next 100 years will teach us much that we now do not know. I expect that a new understanding of gravity will impact many of our current concepts. We who are consumers of gravitation theory should occasionally worry that our understanding of complex galaxies rests on an incomplete foundation.

If our planet had formed around an M-type star, our eyes would have their peak sensitivity near 1 micron. And on that planet, astronomers would already have developed their schemes for classifying galaxies at 1 micron: by coupling appearance and kinematics. But now that the advancing technology on their planet has made it possible for them to obtain images and spectra of galaxies in shorter wavelengths, they have noticed many new features, and they recognize the need for a broader scheme. This is known as progress.

So as we aim for "dust penetrated near-IR classification schemes", we should recognize that there are many facts that we have not yet uncovered. Our classification scheme should remain general, rather than specific, as our observations catch up with the advancing technology. I like David Block's and Ivanio Puerari's near-IR classification that links pitch angle of the arms with form of the rotation curve. Is this a useful concept? Of course! Given the rapid pace of

D. Block et al. (eds.), Penetrating Bars through Masks of Cosmic Dust, 835–838.

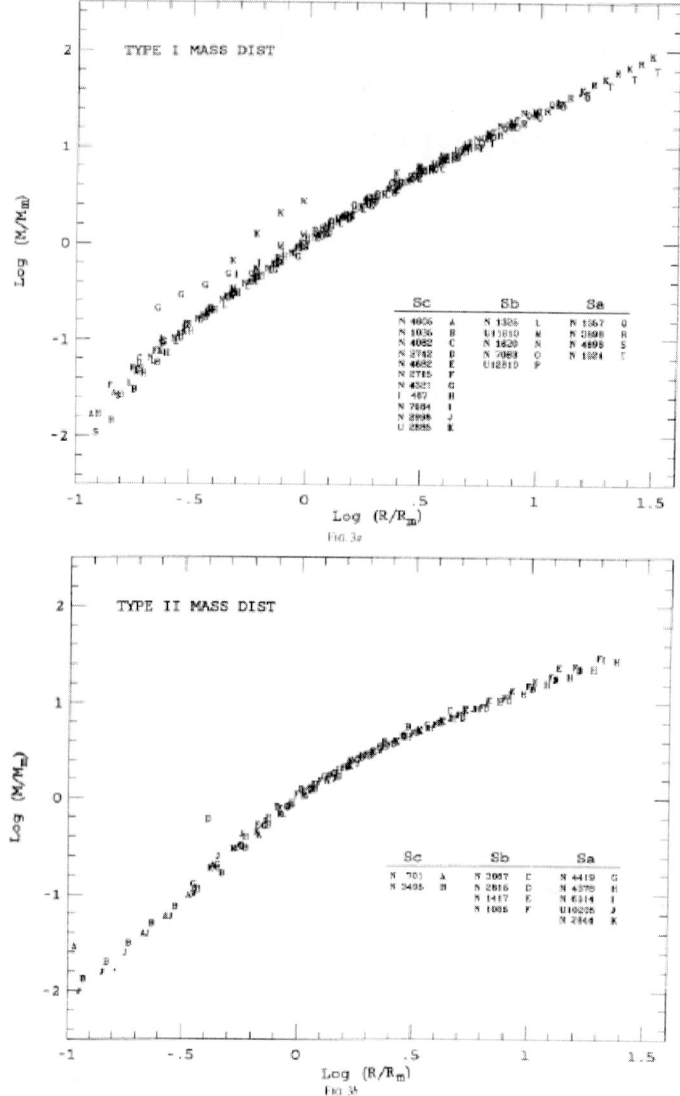

Figure 1. Mass distributions (as inferred from rotation curves) may be delineated into three form families. These form families are not correlated to optical Hubble types; different Hubble types may reside in the same form family. An early hint to the duality of morphology in spiral galaxies. Adapted from Burstein, D. and Rubin, V. 1985, ApJ, 297, 423.

discovery, will it survive as long as Hubble's classification? Perhaps, perhaps not. But for awhile, it will be a valuable tool for learning more about galaxies and hence our universe. That's probably all any of us can ask.

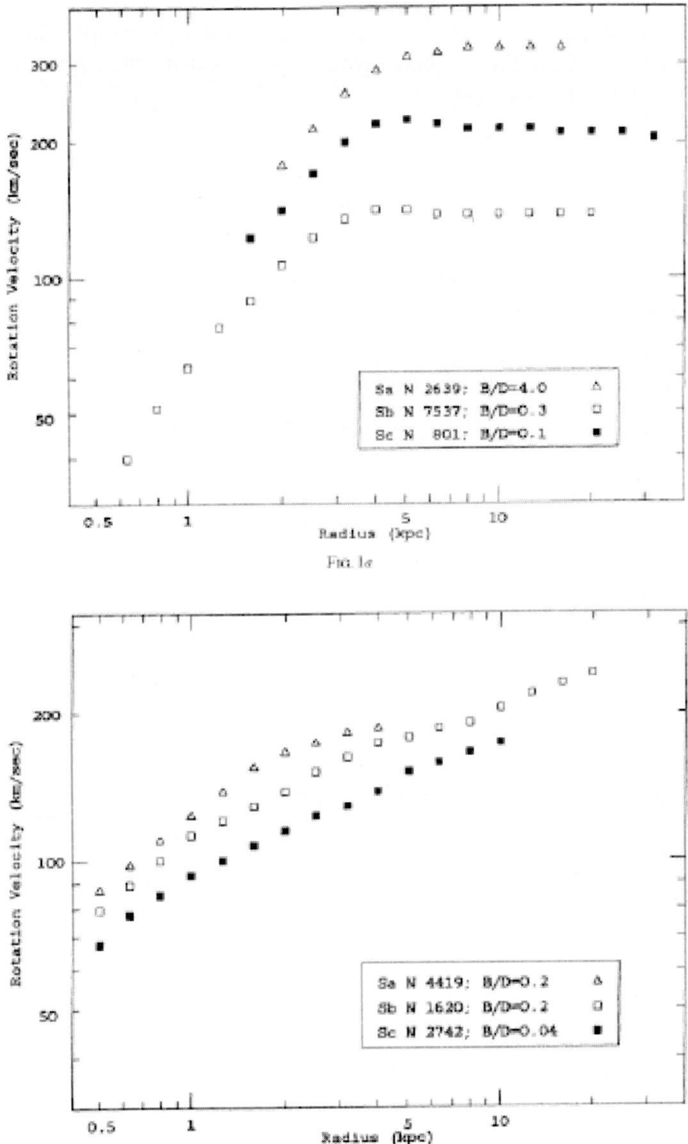

Figure 2. Rotation curve form families: flat (top) and rising (bottom). (Top): The three galaxies at top have a range in bulge-to-disk ratio of ∼40, but their mass distributions (as inferred from their rotation curves) are very similar. (Bottom): As for the top, with a range of 5 in bulge-to-disk ratio. Form families and derived mass distributions are not correlated to optical Hubble bins. Adapted from Burstein, D. and Rubin, V. 1985, ApJ, 297, 423.

My IR credentials are few. When I arrived at the Department of Terrestrial Magnetism in 1965, Kent Ford had constructed a spectrograph, which incorporated an RCA image tube. These tubes enhanced the incoming light by a factor of about 10, for equivalent resolution. He also had a few experimental IR image tubes, which could be used for imaging and spectroscopy. I have lost track of the series of IR images of spirals I took with an image tube camera on the Lowell 42-inch telescope. I do recall that all the galaxies looked generally the same: a pair of diffuse weak arms, and barred centers. This added too many new puzzles, so I put them aside, waiting to be inspired. My fondest hope is that new puzzles continue to inspire us all to learn more about our universe.

THE NOBLEST SCIENTIFIC PROBLEM OF THE AGE: PERSPECTIVES ON THE TRANSITS OF VENUS, 1882 AND 2004

William Sheehan

Editorial Note: Dr W Sheehan, a psychiatrist by profession, is the author of 'The Immortal Fire Within: The life and work of Edward Emerson Barnard' and (with John Westfall) 'The Transits of Venus'. amongst others. He also is a regular consultant to 'Sky and Telescope'.

1. Introduction

On June 8, 2004, a transit of Venus – providentially timed to coincide with the conference "Penetrating Bars through Masks of Cosmic Dust" – occurred. It was the first one witnessed during the lifetimes of any of the participants. The last had taken place on December 6, 1882, when Queen Victoria still sat on the throne of Great Britain, the battle of Majuba had just been fought, and the gold reefs of Witwatersrand still lay undiscovered.

Before June 8, 2004, only five transits had been observed since the invention of the telescope, beginning with Jeremiah Horrocks's and William Crabtree's observation of the transit of 1639. The two 18th century transits saw a host of astronomers dispatched around the world in a bold attempt to implement Edmond Halley's method of using Venus's transits to measure the solar parallax – hence to find the Earth-to-Sun distance and to work out the scale of the Solar System that had been modeled by Johannes Kepler on the basis of his third law of planetary motion (the so-called harmonic law). In 1761, Charles Mason and Jeremiah Dixon – later to achieve fame for surveying the Mason-Dixon line across North America – traveled to Cape Town, where they enjoyed excellent conditions. Captain James Cook inaugurated his explorations of the South Seas by traveling to Tahiti for the transit of 1769. Legentil, Chappe, and many others made names for themselves in pursuing this grand astronomical adventure.

The 1874 transit was the first to be studied photographically. American, British, German, and French teams headed to remote parts of the globe to observe it. The most innovative approach was devised by the French astronomer

D. Block et al. (eds.), Penetrating Bars through Masks of Cosmic Dust, 839–843.

Jules Janssen, who introduced a mechanism built around the gear mechanism of the Colt Revolver to expose a series of images of Venus in the proximity of the Sun's limb onto a curved glass plate - a device that anticipated (and influenced the development of) cinematography.

In the 18th century, the infamous black drop – a ligament joining Venus to the dark sky in the neighborhood of the Sun's limb – frustrated accurate timings of astronomers and explorers. In 1874, the black drop was not generally a serious problem, but there remained serious difficulties in measuring the contacts. The images of Venus and the Sun were softly blurred and the silhouette of the planet seemed to bleed or diffuse into the edge of the Sun rather like a drop of ink into a sheet of blotting paper. Attempts to measure the astronomical unit fell short of expectation. As of 1882, that problem remained, according to the New York Times, "the most important scientific problem of the age."

2. The Transit of 1882

Also according to the New York Times, 1882 was "the first time within the memory of man that the unlearned common people had been permitted to observe a transit, and it is the first revelation of the fact that a transit can be seen through smoked glass."

In City Hall Park in New York, at telescope was erected and the planet exhibited at the price of ten cents a look:

"So great was the rush of people to take a look through it that the services of Park policemen were required to keep them in line awaiting their turn. Once at the telescope a view of a few seconds only was allowed, and by actual count 20 men peered through the glass in five minutes."

Children, like the group here from Harper's Weekly (figure 1), were encouraged to look on that which they would never see again.

At Meriden, Connecticut, the fire-alarm bell rang when the transit began, and schools were closed for the day. A five-year old named Henry Norris Russell viewed the event from Oyster Bay, Long Island, and the sight of Venus against the Sun inspired a life long interest in astronomy – he went on to become the leading American astrophysicist of his day, the director of the Princeton University Observatory and a world-renowned expert on stars.

Simon Newcomb of the U.S. Naval Observatory led a transit expedition to the Huguenot Seminary at Wellington, South Africa (near Cape Town), where an American expatriate Mary Cummings as well as a number of other young ladies arranged too bserve the transit, "in an amateur way". They commanded more accurate results than the professionals. As Newcomb reported: "It was partly the result of good fortune, and partly due to the quickening of the faculties which comes with intense interest. The young ladies of the Seminary took it as a tribute to the greater powers of their own sex". Mary Cummings would

Figure 1. Children, like the group seen here from Harper's Weekly 1882, were encouraged to look on that which they would never see again.

have a medal struck in her honor 122 years later, commemorating her careful observations of the transit of Venus.

3. The transit of 2004

William Harkness, one of Newcomb's colleagues at the U.S. Naval Observatory, reflected on the next transit: "There will be no other till the twenty-first century of our era has dawned upon the earth, and the June flowers are blooming in 2004. What will be the state of science when the next transit season arrives God only knows. Not even our children's children will live to take part in the astronomy of that day."

Now June 8, 2004, the definition of remote futurity for Harkness, has come – and gone.

Not the children of Harkness's generation, but their grandchildren and great-grand children and even great-great grand children, saw Venus plunge into the Sun with the remarkable punctuality of astronomical events. Along with millions of others around the globe, school children and their principals joined amateur and professional astronomers at Bakubung Lodge in Pilanesburg in witnessing the Great Event. When the Sun rose out of the clouds above the burnished yellows and browns of a lovely late-autumn scene on the African bush, Venus was spotted, already present on the Sun, glimpsed as a tiny spot with protective glasses like those used by the children shown in Harper's Weekly or seen as a large black silhouette projected on cardboardor canvas screens with telescopes. Over a period of six hours, it proceeded to track across the Sun for the first time in 122 years.

As they watched Venus's stately advance across the Sun, they no doubt pondered the progress of science since Mary Cummings, Simon Newcomb, William Harkness and all the rest watched the transit of 1882!

The astronomical unit is now known to within a matter meters – not because of refinements of the transit of Venus technique, but by using radio waves whose existence had not yet been demonstrated in 1882, involving measurements of their journeys to and from spacecraft sitting on the surface of Mars. The distances of stars have been accurately determined with the Hipparcos satellite out to hundreds of light years. They in turn have served as stepping-stones in working out the distances in turn to star clusters – to giant variable stars known as Cepheids – and to the still-mysterious objects once simply known as "nebulae" clouds.

In 1882, the Rev. T. W. Webb, the British clergyman-astronomer, referred to the Andromeda nebula as "a mystery in all probability never to be penetrated by man". Its dusky mask has now been penetrated with infrared. The starry nature of the Andromeda nebula – now known as the Andromeda spiral – is known.

What would astronomers in 1882 have made of the fact that by 2004, astronomers would be studying remote systems of stars – Andromedas and Milky Ways – gathering in the far-flung reaches of the universe, their light having traversed some ten billion light years of space and ten billion years of time? Across a distance of some two hundred trillion astronomical units from the Earth!

The next transit of Venus in our time will occur in 2012. It will not be visible from most of South Africa. Then there will be no more until 2117. Will there be a Conference on Barred Galaxies held to coincide with that transit, or will astronomers by then have solved those problems and have moved on to other topics? As Harkness said in 1882: God only knows.

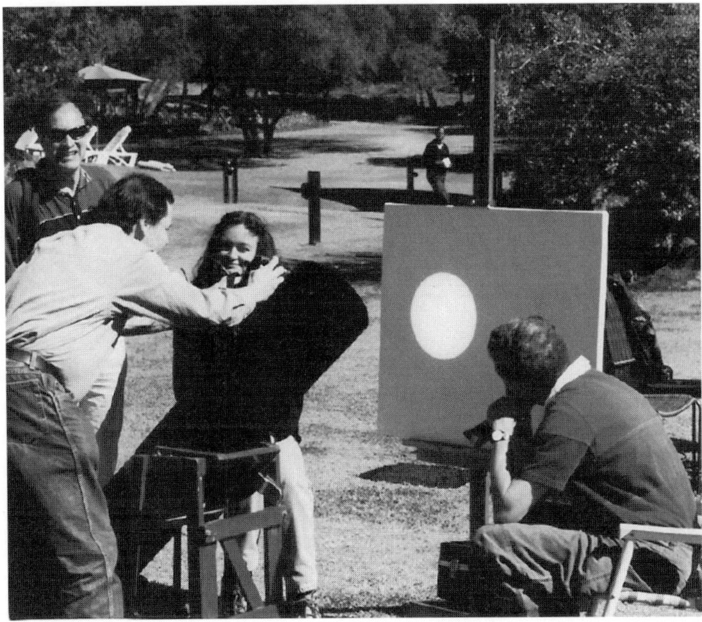

Figure 2. Our telescope operators, showing delegates and students the Venus transit, included Valerie Fraser (facing camera) and Bert van Winsen. Facing the camera at left, with sunglasses, is the author.

Figure 3. Francoise Combes enjoys a detailed view of the 2004 transit from the Pilanesberg, as Trevor Gould (chief telescope co-ordinator) looks on.

DESCRIPTION OF THE COMMEMORATIVE MEDALLION, STRUCK TO CELEBRATE THE TRANSIT OF VENUS, OBSERVED FROM SOUTH AFRICA ON 8 JUNE 2004

Bruce G. Elmegreen
IBM Research Division, T.J. Watson Research Center, 1101 Kitchawan Road, Yorktown Hts., NY 10598, USA

EDITORIAL NOTE: The winning design for the medallion to commemorate the transit of Venus was proposed by the author (BGE). His design was modified by Tommy Sasseen (TS), the die sinker who designed the Kruger Rand. Descriptions of the medallion obverse and reverse are given here, based on the author's speech at the Sun City Gala Dinner on June 9th.

1. Medallion Obverse

When David Block challenged us a year ago to propose a design for a commemorative medallion on the occasion of the transit of Venus, I recalled my image of South Africa from previous visits and realized the theme of the coin should be the richness of life. I viewed this to be the strongest connection between South Africa – the land of abundant exotic life and the origin of humans – and Venus, the Western goddess of Love and Fertility. I wanted a representation of the flora and fauna of South Africa on the medallion in order to compare the dynamics and majesty of life as viewed everyday here with the dynamics and majesty of Venus during her transit across the Sun.

The obverse of the medallion shows two symbolic images of the Sun and two of Venus. The first is the medallion itself, representing the disc of the Sun, along with a thick line for the transit path and a dot for Venus on the path. The second is the image of the Sun with bright rays standing above a line representing the horizon. This shows how the Sun looked to us yesterday morning, at the beginning of the transit. In this second representation, Venus is a Sable antelope, charging across the disc. The Sable antelope was first discovered not far from the Pilanesberg in South Africa. It has long curved horns, which are symbolic of the crescent of Venus as it approaches the Sun. It charges at high speed in this depiction, because Venus moves at the enormous speed of

D. Block et al. (eds.), Penetrating Bars through Masks of Cosmic Dust, 845–847.

126,000 kilometers per hour in its orbit. The horizon shows the flat landscape of the Savannah, which is perhaps the most famous view in South Africa, and a typical tree, like the Acacia Thorn tree, found all along the countryside in Pilanesberg. The main text is "Venus Transits the Sun," while in the disc of the Sun is the Latin inscription, "Soli Deo Gloria" which means "To God alone be the Glory." Beneath the horizon is an African proverb, "Walala Wasala," which means in the ethnic tongue "You snooze, you lose!" – an appropriate phrase for astronomers who make most of their discoveries at night when everyone else sleeps. The date "8 June 2004" and "South Africa' also appear.

2. Medallion Reverse

On the reverse, another theme commemorates the impact that the transit of Venus has on people. We saw yesterday the excitement in the school children and in all of the distinguished visitors who came to our lodge. The image on the medallion is of Mary Elizabeth Cummings, along with historical tracks of Venus across the Sun. Mary Cummings was a student at Mount Holyoke Seminary in Massachusetts (USA). She graduated in 1876 and moved to Wellington South Africa to teach at the newly founded Huguenot Seminary for Girls. Two of her former teachers were there already. In 1881, the 6 inch Fitz refractor telescope that Mary used at Mount Holyoke was moved to Wellington and installed in the Seminary Garden by the Astronomer Royal at the Cape of Good Hope, Sir David Gill.

Because of this American connection, the Director of the US Naval Observatory, Simon Newcomb, came to Wellington to set up his equipment for the 1882 transit of Venus. He instructed Mary Cummings and several other teachers at the Seminary on the reading of a chronometer and showed them how do the scientific experiment of recording the transit with their 6 inch refractor. It seems he even built a transit model in which a metallic disc one foot in diameter moved in and out of a triangular aperture, and placed it a kilometer away from the telescope to give the teachers some practice with realistic atmospheric distortions. When the time came, Mary made better observations than the professionals, to which Newcomb replied that this was "partly the result of good fortune, and partly due to the quickening of the faculties which comes with intense interest, a tribute to the greater powers of their own sex." (Ferguson G. P. - The Builders of Huguenot - 1927; see W.T. Koorts:

http://www.saao.ac.za/ wpk/tov1882/tovwell.html).

Mary Cummings symbolizes the contributions made to astronomy by young people and amateurs. She is particularly famous for her observations of the last transit of Venus in South Africa on December 6, 1882.

3. On African soil

The "Penetrating Bars through Masks of Cosmic Dust" conference was generously sponsored by the Anglo American Chairman's Fund. "We proposed to Anglo American that this conference be held on the African subcontinent to mark the pioneering research done on cosmic dust at the University of the Witwatersrand, South Africa, and to celebrate the occasion of David Block's 50th birthday." (Professor Kenneth C Freeman, FRS.)

4. Acknowledgements

I would like to thank the Anglo American Chairman's Fund for their generous support of the commemorative medallion. I am also grateful to David Block and Trevor Gould for their contributions to the design, and to Tommy Sasseen, for his exquisite artistic rendering of the richness of life on the South African Savannah. Liz Block is thanked for her contribution to the preparation of the coin certificates.

Captions for Colour Plates

Plate 1 (Fig. 1, Block et al.)

Spiral galaxies in the dust penetrated regime are binned according to three quantitative criteria: firstly, Hm, where m is the dominant Fourier harmonic (illustrated here are the two-armed H2 family). Next, follows the dust penetrated pitch angle families α, β or γ (for class α, the pitch angles range from $\sim$ 4-15°; for class β, the deprojected pitch angles range from $\sim$ 15-30°, while open-armed class γ spirals have pitch angles ranging between $\sim$ 35- 75°). Finally, we compute the gravitational torque, which is identical to the bar torque in galaxies presenting a bar. Bar torques are not derived from bar ellipticities but exploit the full gravitational potential of the disk within which it is embedded. Note that early type b spirals (NGC 3992, NGC 2543, NGC 7083, NGC 5371 and NGC 1365) are distributed within all three families (α, β and γ). Hubble type and dust penetrated class are uncorrelated.

Plate 2 (Buta)

An issue of much debate at the Conference, was observational evidence for bar dissolution and bar reformation. Seen here is an HST image of NGC 3081, which *might* be considered in the category of the bar dissolution.

John Kormendy writes: "To me, a weak bar is suggestive evidence for bar suicide. Do we really believe that global disk instabilities or tickling by encounters can make a bar that is a few percent of the disk mass? This is not what happens in simulations. The bars that get made by these processes contain most of the disk mass that is interior to the end of the bar. Would it not be easier to understand the formation of a bar that is a few percent of the disk mass if it formed as a much stronger perturbation and then decayed?"

Plate courtesy Ron Buta.

Plate 3 (Fig. 7, Beckman et al.)

(a) Surface brightness map in Hα of NGC 1530, from a TAURUS data cube from the 4.2m WHT La Palma. (b) Radial velocity map of ionized gas from the peaks of the Hα emission lines across the face of the galaxy, using the same data cube. (c) Contours of Hα surface brightness superposed on a two–dimensional projection of the rotation curve derived from the velocity field shown in (b). (d) Map of the residual, non–circular velocity field, obtained by subtracting off the projected rotation curve in (c) from the complete velocity field in (b). The strong non–circular velocity field aligned with the bar is clearly seen here, as the galaxy inclination causes the flow along one side to be directed towards us, and away from us along the other side of the bar (see text for more details, also see Zurita et al. 2004).

Plate 4 (Fig. 3, Kormendy and Cornell)

Nuclear star formation rings in barred and oval galaxies. Sources: NGC 4314 – Benedict et al. (2002); NGC 4736 – NOAO; NGC 1326 – Buta et al. (2000) and Zolt Levay (STScI); NGC 1512 – Maoz et al. (2001); NGC 6782 – Windhorst et al. (2002) and the Hubble Heritage Program. This figure is from Kormendy & Kennicutt (2004).

Plate 5 (Fig. 4, Kormendy and Cornell)

Nuclear star formation in the unbarred galaxies M51 and NGC 4321 (M100). Dust lanes on the trailing side of the global spiral arms reach in to small radii. As in barred spirals, they are are indicative of gas inflow. Both galaxies have concentrations of star formation near their centers that resemble those in Figure 3. These images are from the *Hubble Space Telescope* and are reproduced here courtesy of STScI.

Plate 6 (Section 5, Fazio et al.)

Mosaic of galaxies at infrared wavelengths taken with the IRAC instrument on the Spitzer Space Telescope. Colors are coded as 3.6 (blue), 4.5 (green), and 8.0μm (red). The galaxies are arranged according to traditional, optical morphological classifications. Starlight appears blue at these mid-infrared wavelengths, while warm dust (emitting in the PAH emission lines) appears red. AGN emission is also red and point-like (e.g., NGC 5548). There is a clear transition from blue, stellar-dominated emission for the early-type galaxies to red, ISM-dominated emission for the late-type galaxies.

Plate 7 (Section 5, Fazio et al.)

Four galaxies observed with the Spitzer Space Telescope. (Top left): NGC 4203 is a lenticular galaxy, presenting striking rings of dust (coded red) in the central domains. (Top right): Masks of dust are well seen in the edge-on galaxy NGC 5746. (Bottom left): NGC 1961 is a late-type spiral with an unusual morphology. (Bottom right): We finally show the nearby spiral NGC 300. The reader is referred to Fazio et al. for further details.

Plate 8 (Fig. 6, Ford et al.)

A composite ACS i,z image of the brighter of the two most conspicuous sub-clusters in CL0152. The field size is 90$''$ ($\sim$ 690 kpc in the restframe) Spectroscopically confirmed members are circled (6″ diameter; $\sim$ 50 kpc in the restframe). There are several thin arcs from lensed background galaxies that are much bluer than the early-type galaxies in the cluster. The lensed galaxy at the center has two components that are mirror images, indicating that the galaxy is very close to a caustic.

Plate 9 (Fig. 7, Ford et al.)

Star-forming galaxies in CL0152 and morphology of the X-ray emitting gas. Spectroscopically confirmed passive galaxies are circled and galaxies with star formation, as indicated by [OII]λ3737 emission, are shown with a "star". The latter are typically spirals and late-type galaxies, while the former are mostly E/S0s. The insets show typical morphologies and spectra. The spatial segregation of the star forming later type galaxies is very striking. The Chandra X-ray isophotes (Demarco et al. 2004; Maughan et al. 2003) are 3, 5, 7, 10, 20 and 30 sigma above the background.

Plate 10 (Top: Fig. 2, Vlahakis et at.; Bottom: Fig. 1, Zurita et al.)

Top: Example from the optically-selected SLUGS: NGC 3987. 850 μm S/N map (1σ contours) overlaid onto *Digitised Sky Survey* optical image.

Bottom: (a) Intensity map of the Hα emission in zone of the bar. (b) V band image. Velocity gradients parallel (c) and perpendicular (d) to the bar. (e) Non–circular residual velocity map. (f) Non–thermal velocity dispersion map. All images (a to f) show the same zone of the galaxy.

Plate 11

Top: SOC member Bruce Elmegreen photographed at our Gala Dinner, with Mrs M. Keeton. Mrs Keeton is CEO of the Anglo American Chairman's Fund, the principal sponsor of our Conference.

Bottom: Venus Transit medallions were handed out at the Gala Dinner by Governor Tito Mboweni and SOC co-chair David Block. Governor Mboweni is the Governor of the Reserve Bank of South Africa.

Plate 12

Top: Trevor Gould photographed at our Gala Dinner. Trevor showed delegates the glories of the African winter skies, each evening after dinner. He also served on the LOC and on the Medallion Committee.

Bottom: SOC co-chair Ken Freeman, delivering his address at Sun City, South Africa. Ken Freeman warmly paid tribute to the sponsors, without whom this Conference would never have seen the light of day.

Plate 13

Top: The 2004 Venus Transit occurred during the week of our conference. This stunning image was secured by Sylvie Beland on June 8, 2004 at 5^h53^m (EDT). The site was the Observatory of "Centre de la Nature" in Laval, Quebec, Canada. She used a Nikon Coolpix 4300 ISO400 at f/2.8, exposure time 1/2 second. For this image, Beland employed a 25mm Lanthanum eyepiece, a Baader filter and her 8-inch f/5 Newton Dobsonian Telescope.

Bottom: The Transit of Venus as photographed by David Finlay. People are seen viewing the transit from the famous Sydney bridge in Australia. Photograph reproduced by permission of *Access All Areas Photography*, at email aaap@internode.on.net.

Plate 14

Top: On the Friday before our Conference, Francoise Combes delivered a popular lecture on extrasolar planets to school children in Kagiso, South Africa. Seen here with Francoise in the school library is student Aifheli Rabambi.

Bottom: A group of students from S G Mafaesa high-school (Kagiso) travelled from Kagiso to view the transit of Venus together with conference delegates. Venus could be identified as a minute black dot against the orange disk of the Sun, through these eclipse glasses. Front left: Tsolofelo Suping; Front right: Ntombi Peters.

Plate 15

Top: David Block photographed with students from the Nirvana Secondary School in Lenasia, South Africa, during the Venus transit. The students seen here also formed part of a special video conference link-up to Cambridge University (UK). Venus took $\sim$ 6 hours to cross the disk of the Sun. At left front is Mrs. Mboweni (snr), family of Reserve Bank Governor Tito Mboweni.

Bottom: Students from S G Mafaesa Secondary school in Kagiso and the Nirvana Secondary School in Lenasia attended a special pre-transit dinner at *The Palace*, Sun City. Photographed at the Palace are Venus Transit lecturer William Sheehan, and students Ntshieni Menaneshe (back) and Mpho Rabambi (front).

Plate 16

Top: The obverse shows the symbolic transit of Venus, using a sable antelope (first discovered not far from the Pilanesberg in South Africa) charging across the Sun, which is shown just above the horizon, during the early morning time of the transit. The horizon line represents the flat horizon of the Savannah, perhaps the most famous landscape of South Africa and has been positioned to represent the path of Venus across the face of the Sun, relative to the edge of the medallion. A typical African tree emphasizes the horizon line, which depicts (not to scale) a silhouette of Venus on the line of transit. The main text is "Venus Transits the Sun," while in the disc of the Sun is the Latin inscription, "Soli Deo Gloria" which means "To God alone be the Glory." Beneath the horizon is an African proverb, "Walala Wasala." This phrase means "You snooze, you lose!" The date '8 June 2004' and 'South Africa' also appear.

Bottom: Mary Cummings and her role in the previous transit of Venus, is the theme of the reverse of the medallion. Her image is displayed against a background of the paths of previous and future transits. Mary was a teacher at the Huguenot Seminary for girls in Wellington South Africa. She was a member of the Class of 1876 at Mt. Holyoke Seminary in Massachusetts (USA) and moved to Wellington to teach at the Huguenot Seminary in 1877. In 1882, just in time for the transit of Venus, the $6''$ Fitz refractor telescope that Mary had used as a student at Mt. Holyoke was moved to Wellington and set up by the Astronomer Royal at the Cape, Sir David Gill. This American connection in South Africa led Prof Simon Newcomb of the US Naval Observatory to set up his own telescope for the Venus transit observations in the Seminary Garden. Mary observed the transit and documented its timing with precise chronometer readings, right alongside her more famous mentor, Newcomb, and is still credited today for her part in this important observation. Mary Cummings symbolizes the essential role played by women in astronomy, and is particularly famous for her observations of the last transit of Venus in South Africa on December 6, 1882.

H2
Gravitational Torque
NGC 2857
NGC 309
NGC 7083
NGC 5236
NGC 718
NGC 5371
NGC 3992
NGC 2543
NGC 1365
α
β
γ

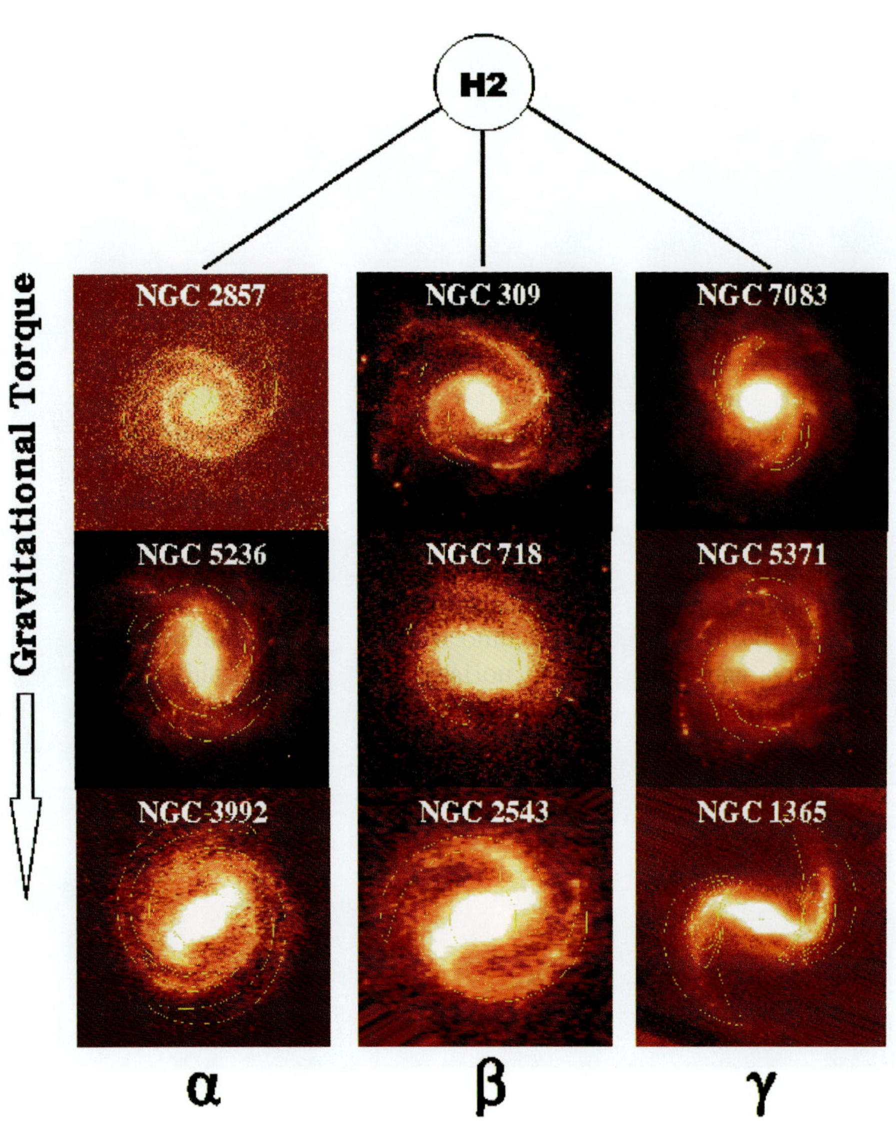

Plate 1

Plate 2

a.
b.
c.
d.
19'10"
18'20"
75°17'30"
16'40"
Dec. (2000)
200
100
0
−100
−200
kr
km/s
40s
30s
4h23m20s
10s
R.A. (2000)

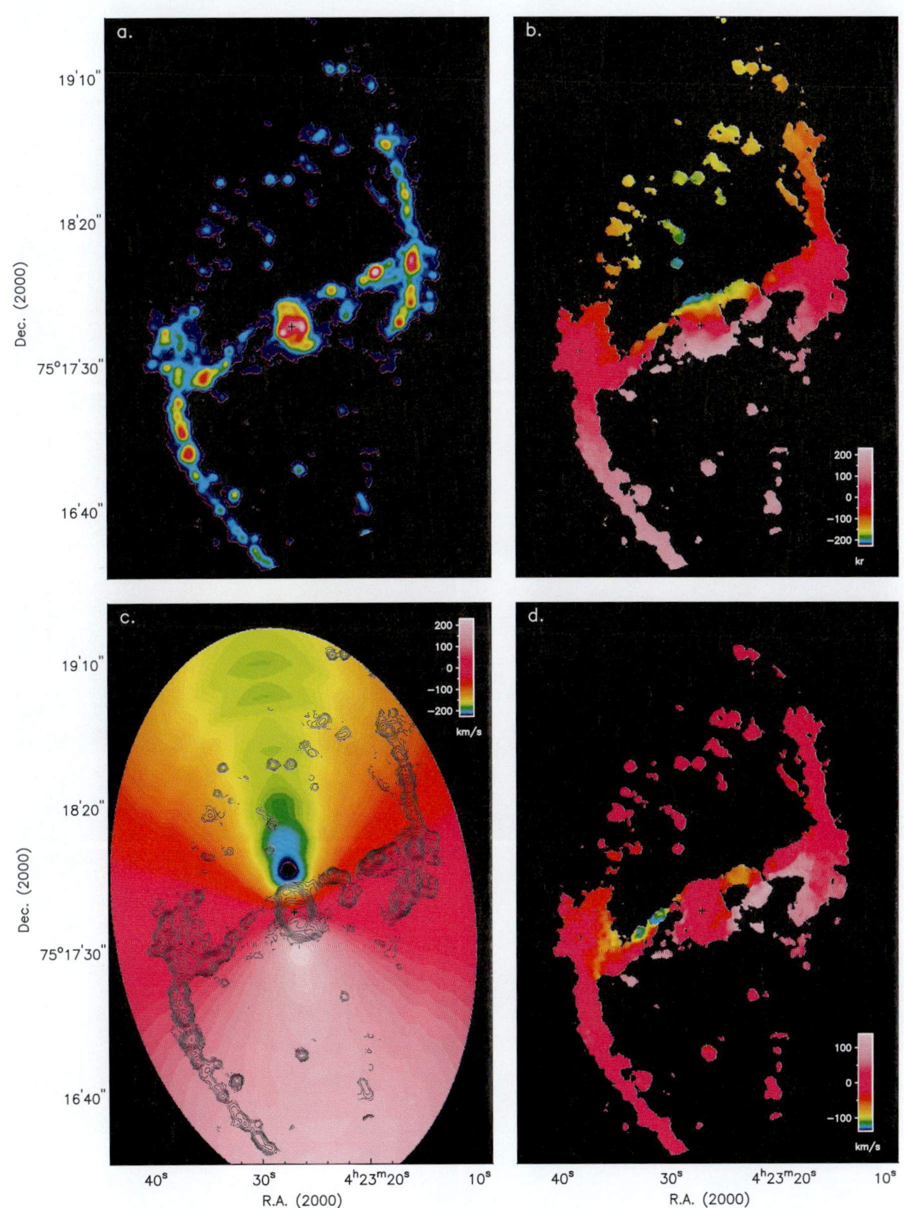

Plate 3

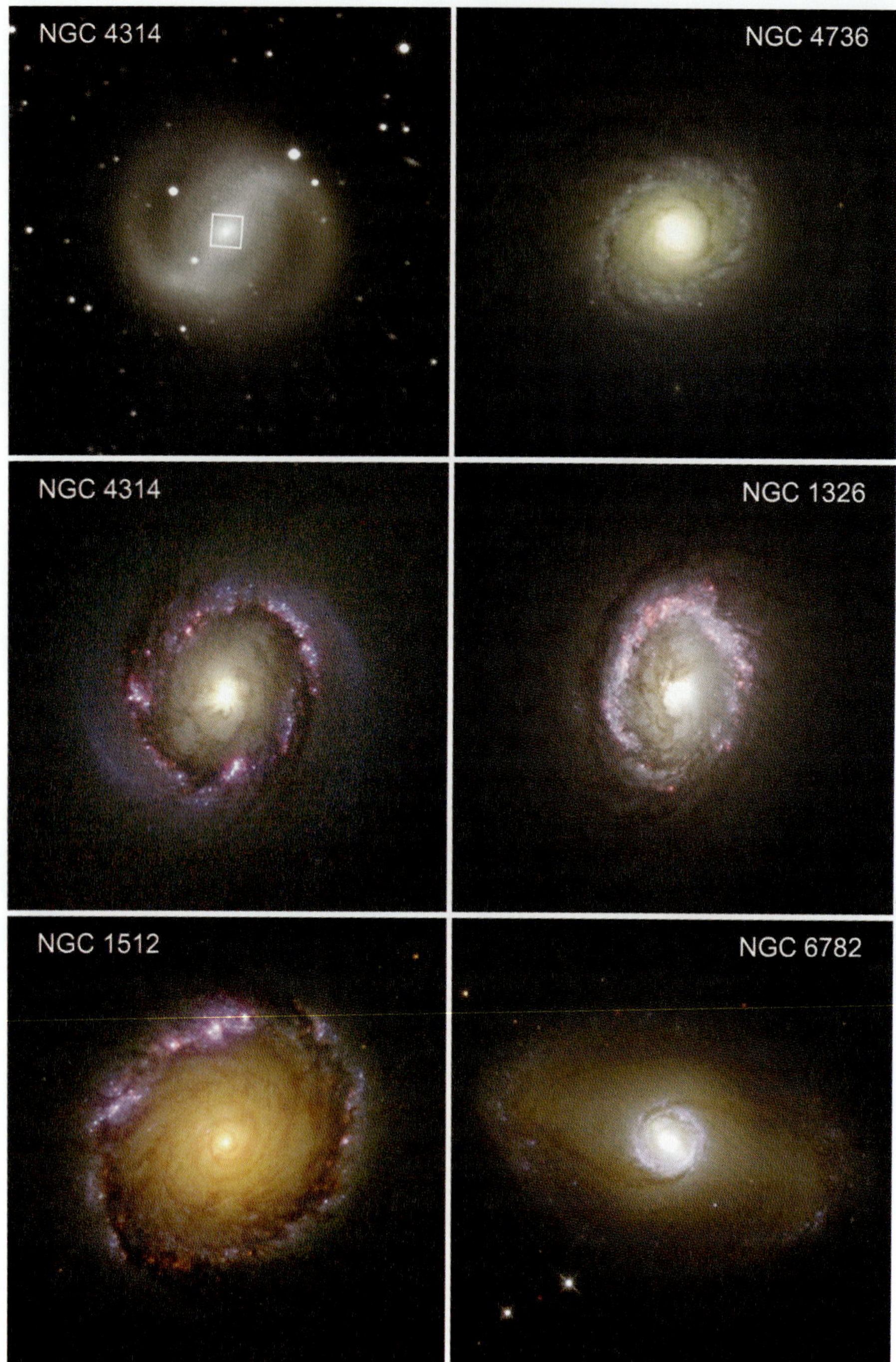

Plate 4

Plate 5

E
S0
Sbc
Sm
S0
Sab
d
Im
NGC 777
NGC 5846
NGC 5813
NGC 5322
NGC 5077
NGC 5079
NGC 1023
NGC 5363
NGC 4203
NGC 6340
NGC 5548
NGC 3031 (M 81)
NGC 5746
NGC 3147
NGC 1961
NGC 5669
NGC 6689
NGC 5668
NGC 300
NGC 5585
NGC 5204
NGC 2366
NGC 1156
NGC 1569
NGC 4038/9 "Antennae"

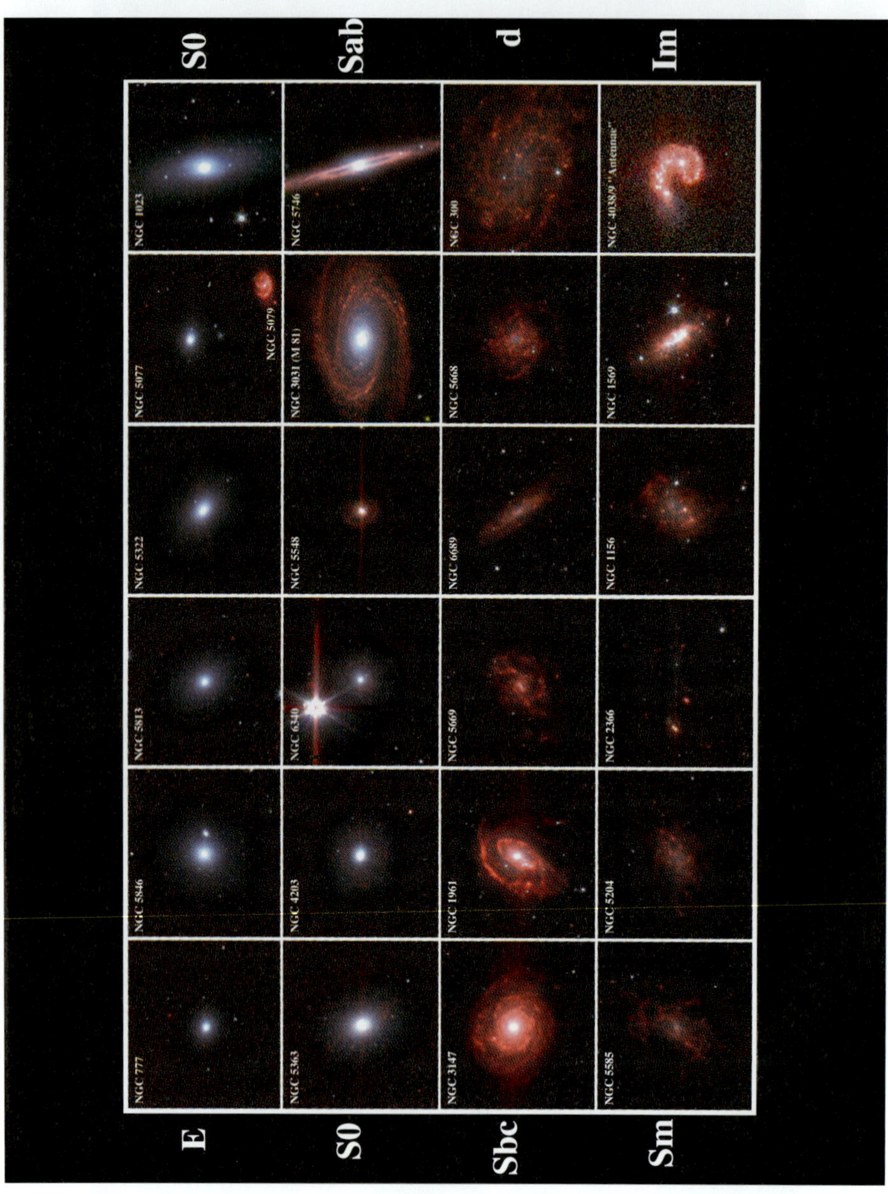

Plate 6

Plate 7

Plate 8

56
57
58
−13° 59'
−14° 00'
01h 52m 50s
45s
40s
35s
Flux(λ)
6500
7000
7500
8000
8500
Wavelength (Å)

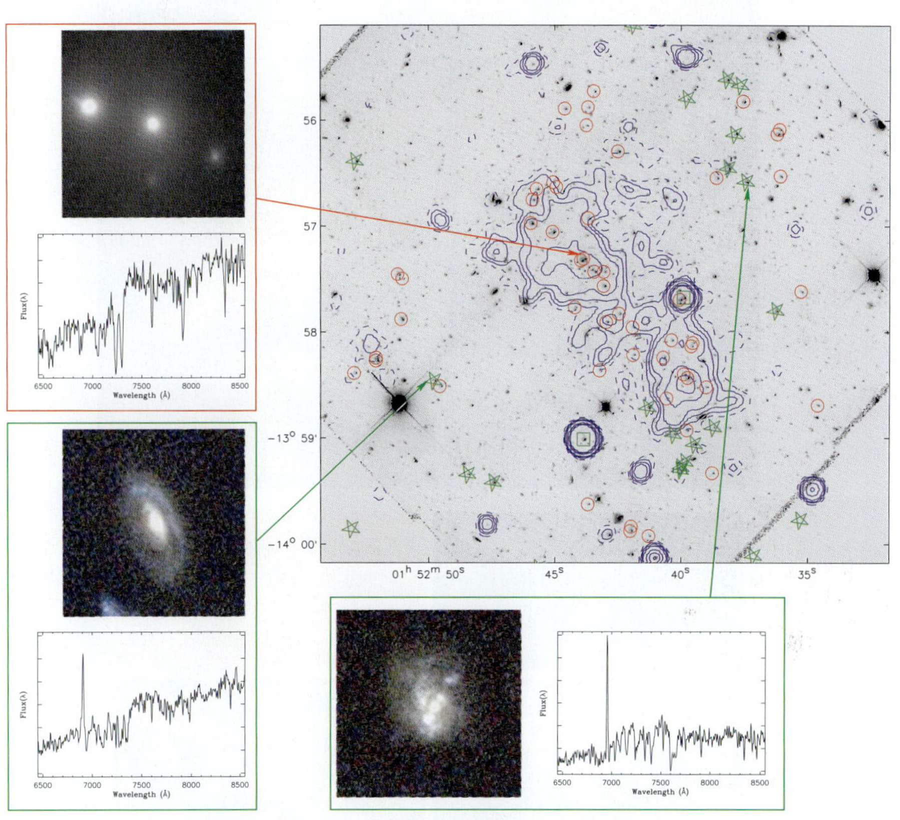

Plate 9

a.
b.
c.
d.
e.
f.
0.15
0.1
0.05
km/s/pc
0.2
0.1
km/s/pc
10"
18'
50"
75°17'40"
17'30"
Dec. (2000)
100
0
-100
km/s
70
60
50
40
30
20
10
km/s
40s
30s
4h23m20s
R.A. (2000)
40s
30s
4h23m20s
R.A. (2000)

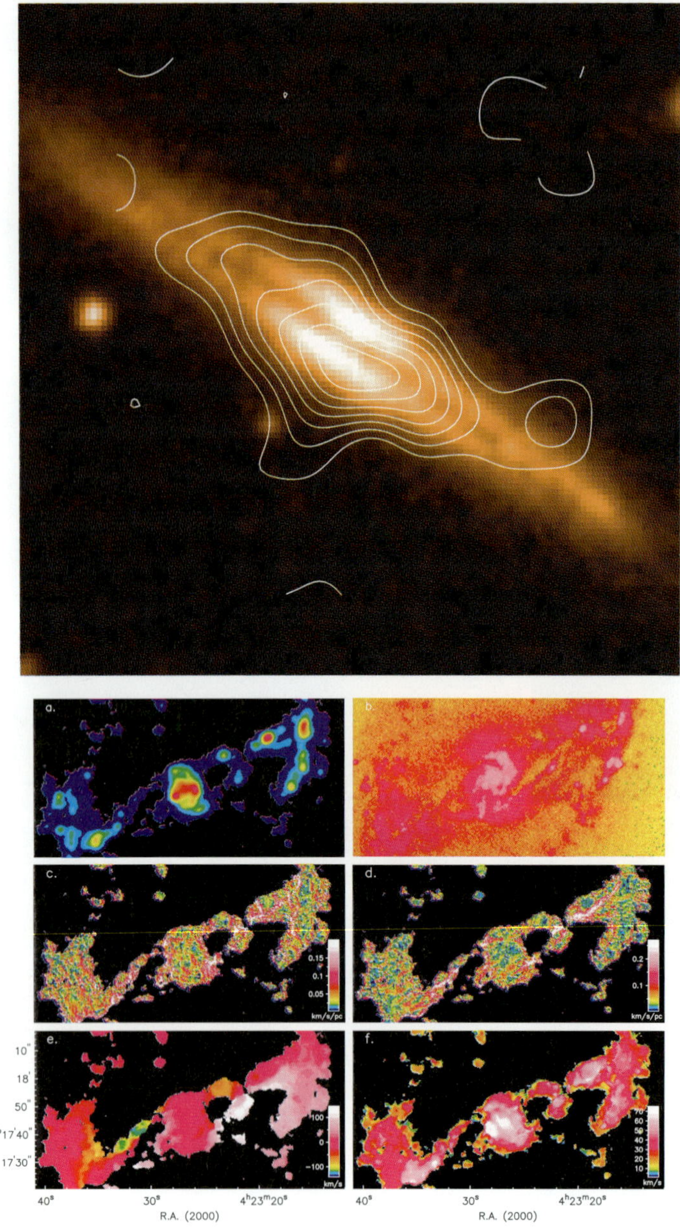

Plate 10

Plate 11

Plate 12

Plate 13

Plate 14

Plate 15

VENUS TRANSITS THE SUN
SOLI DEO GLORIA
BOE
WALALA WASALA
8 JUNE 2004
TS
SOUTH AFRICA

1874
1769
2012
Au 999.9
1882
2004
1761 MARY CUMMINGS
2004
1882

Plate 16

LIST OF PARTICIPANTS

Ronald J. Allen	Space Telescope Science Institute, **USA**
Silvia Baes-Fischlmair	Institut fur Astronomie der Universitat Wien, **Austria**
Nicholas M. Ball	Astronomy Centre, University of Sussex, **UK**
John E. Beckman	Instituto de Astrofisica de Canarias, **Spain**
David L. Block	School of Computational and Applied Mathematics, University of Witwatersrand, **South Africa**
Harry Blom	Kluwer Publishers, **Holland**
Francois Boulanger	Institut d'Astrophysique Spatiale, Universite Paris XI, **France**
Frederic Bournaud	LERMA, Observatoire de Paris, **France**
Sarah Bryan	University of KwaZulu-Natal, Pietermaritzburg, **South Africa**
Martin Bureau	Columbia Astrophysics Laboratory, **USA**
Andreas M. Burkert	University Observatory Munich, **Germany**
Ronald J. Buta	Department of Physics & Astronomy, University of Alabama, **USA**
Gabriela Canalizo	Department of Earth Sciences and IGPP, University of California, **USA**
Roberto Capuzzo-Dolcetta	Department of Physics, Univ. of Roma La Sapienza, **Italy**
Claude Carignan	LAE, Universite de Montreal, **Canada**
Francoise Combes	LERMA, Observatoire de Paris, **France**
Christopher J. Conselice	California Institute of Technology, **USA**
Enrico M. Corsini	Dipartimento di Astronomia, Universita di Padova, **Italy**
Catherine Cress	University of KwaZulu-Natal, Pietermaritzburg, **South Africa**
Barbara Cunow	Department of Mathematics, Applied Mathematics and Astronomy, University of South Africa, **South Africa**
Anna Curir	INAF - Astronomical Observatory of Torino, **Italy**
Elena D'Onghia	Max-Planck-Institut fur extraterrestrische Physik, **Germany**
Bruce G. Elmegreen	IBM Research Division, T.J. Watson Research Center, **USA**
Eric Emsellem	Centre de Recherche Astronomique de Lyon, **France**
Paul B. Eskridge	Department of Physics & Astronomy, Minnesota State University, **USA**
Giovanni G. Fazio	Harvard-Smithsonian Center for Astrophysics, **USA**
Holland Ford	Johns Hopkins University, **USA**

Didier Fraix-Burnet Laboratoire d'Astrophysique de Grenoble, **France**
Ken C. Freeman Mt Stromlo Observatory, RSAA, Australian National University, **Australia**
Uta Fritze – v. Alvensleben Universitätssternwarte Göttingen, **Germany**
Alister W. Graham Department of Astronomy, University of Florida, **USA**
Robert Groess School of Computational and Applied Mathematics, University of Witwatersrand, **South Africa**
Preben Grosbol European Southern Observatory, **Germany**
Luis C. Ho The Observatories of the Carnegie Institution of Washington, **USA**
Kelly Holley-Bockelmann University of Massachusetts, **USA**
Susanne Huttemeister Astronomisches Institut, Ruhr-Universitat Bochum, **Germany**
Leslie K. Hunt Istituto di Radioastronomia/Sezione Firenze, **Italy**
Garth Illingworth UC Observatories/Lick Observatory & Department of Astronomy and Astrophysics, University of California, **USA**
Rolf Jansen Dept. of Physics and Astronomy, Arizona State University, **USA**
Shardha Jogee Space Telescope Science Institute, **USA**
Dong-Woo Kim Smithsonian Astrophysical Observatory, **USA**
Ivan King Astronomy Department, University of Washington, **USA**
Johan H. Knapen Centre for Astrophysics Research, University of Hertfordshire, **UK**
David C. Koo UC Observatories/Lick Observatory & Department of Astronomy and Astrophysics, University of California, **USA**
John Kormendy Department of Astronomy, University of Texas, **USA**
George Lake Washington State University, **USA**
Eija Laurikainen Division of Astronomy, Dept. of Phys. Sciences, Univ. of Oulu, **Finland**
Alex Lazarian University of Wisconsin-Madison, Dept. of Astronomy, **USA**
Olivier Le Fevre Laboratoire d'Astrophysique de Marseille, **France**
Aigen Li Theoretical Astrophysics Program, University of Arizona, **USA**
Jeremy Lim Inst. of Astronomy & Astrophysics, Academia Sinica, **Taiwan**
Witold Maciejewski Obserwatorium Astronomiczne Uniwersytetu Jagiellonskiego, **Poland**
Paul Martini Harvard-Smithsonian Center for Astrophysics, **USA**
Karin Menendez-Delmestre California Institute of Technology, **USA**
George Miley Sterrewacht Leiden, **The Netherlands**
Julio F. Navarro CIAR and Guggenheim Fellow, Department of Physics and Astronomy, University of Victoria, **Canada**
Masafumi Noguchi Tohoku University, **Japan**
Stephen C. Odewahn HET, University of Texas, **USA**

Jan Palous Astronomical Institute, Academy of Sciences of the Czech Republic, **Czech Republic**
Isabel Perez Kapteyn Institute, University of Groningen, **The Netherlands**
Daniel Pfenniger Geneva Observatory, University of Geneva, **Switzerland**
Michael Pohlen Instituto de Astrofisica de Canarias, **Spain**
Ivanio Puerari INAOE, **Mexico**
Jean-Loup Puget Institut d'Astrophysique Spatiale, Universite Paris Sud, **France**
Michael Regan Space Telescope Science Institute, **USA**
Brigitte Rocca-Volmerange Institut d'Astrophysique de Paris, **France**
Heikki Salo Division of Astronomy, Dept. of Phys. Sciences, Univ. of Oulu, **Finland**
Marc S. Seigar Joint Astronomy Centre, Hilo, **USA**
William Sheehan Willmar, MN, **USA**
Kartik Sheth California Institute of Technology, **USA**
Frank Shu National Tsing Hua University, Taiwan, **ROC**
Philip B. Stark Dept. of Statistics, University of California, **USA**
Thomas Steiman-Cameron Astronomy Dept., Indiana Univ., **USA**
Margarita Valdez-Gutierrez Departamento de Astronomia, Universidad de Guanajuato, **Mexico**
Catherine Vlahakis School of Physics and Astronomy, Cardiff University, **UK**
Rogier Windhorst Department of Physics and Astronomy, Arizona State University, **USA**
Guy Worthey Astronomy Program, Washington State University, **USA**
Chi Yuan Inst. of Astronomy & Astrophysics, Academia Sinica, **Taiwan**
Xiaolei Zhang US Naval Research Laboratory, **USA**
Almudena Zurita Universidad de Granada, Dept. de Fisica Teorica y del Cosmos, **Spain**

Teachers from the **Nirvana Secondary School** (Lenasia), **Cliffview Primary School** (Fairlands) and the **S G Mafaesa School** (Kagiso) participated in our outreach to schools, during the Venus Transit.

Students Participation

The students below also formed part of the Venus Transit video link-up to Cambridge University (UK): Nehal Badal, Fabian Francis, Deepika Goolab, Pauline Kgongoane, Navitha Latchman, Lester Mancasa, Solomon Matlmong, Ashvira Moodley, Muhanganei Munyai, Lindiwe Mwanda, Nisha Naka, Kashmira Naran, Ntsieni Nemanashi, Michael Nobu, Meera Ooka, Emelia Pillay,

Aifheli Rabambi, Mpho Rabambi, Kashmira Rawjee, Lebogang Seloi, Petunia Shuping,

Staff from the Nirvana Secondary School (Lenasia), Cliffview Primary School (Fairlands) and the S G Mafaesa School (Kagiso) who came to the Pilanesberg with students, included: Sue le Sueur, Yvonne Kekana (Wits Marketing), Angsha Nathoo, Pauline Kgongoane, Berendien van Aswegen.

Figure 1. Our group photograph at the Pilanesberg National Park.

Astrophysics and Space Science Library

Volume 316: ***Civic Astronomy - Albany's Dudley Observatory, 1852-2002,*** by G. Wise
Hardbound ISBN 1-4020-2677-3, October 2004

Volume 315: ***How does the Galaxy Work - A Galactic Tertulia with Don Cox and Ron Reynolds,*** edited by E. J. Alfaro, E. Pérez, J. Franco
Hardbound ISBN 1-4020-2619-6, September 2004

Volume 314: ***Solar and Space Weather Radiophysics - Current Status and Future Developments,*** edited by D.E. Gary and C.U. Keller
Hardbound ISBN 1-4020-2813-X, August 2004

Volume 313: ***Adventures in Order and Chaos – A Scientific Autobiography,*** by G. Contopoulos
Hardbound ISBN 1-4020-3039-8, December 2004

Volume 312: ***High-Velocity Clouds,*** edited by H. van Woerden, U. Schwarz, B. Wakker
Hardbound ISBN 1-4020-2813-X, September 2004

Volume 311: ***The New ROSETTA Targets- Observations, Simulations and Instrument Performances***, edited by L. Colangeli, E. Mazzotta Epifani, P. Palumbo
Hardbound ISBN 1-4020-2572-6, September 2004

Volume 310: ***Organizations and Strategies in Astronomy 5,*** edited by A. Heck
Hardbound ISBN 1-4020-2570-X, September 2004

Volume 309: ***Soft X-ray Emission from Clusters of Galaxies and Related Phenomena***, edited by R. Lieu and J. Mittaz
Hardbound ISBN 1-4020-2563-7, September 2004

Volume 308: ***Supermassive Black Holes in the Distant Universe,*** edited by A.J. Barger
Hardbound ISBN 1-4020-2470-3, August 2004

Volume 307: ***Polarization in Spectral Lines***, by E. Landi Degl'Innocenti and M. Landolfi
Hardbound ISBN 1-4020-2414-2, August 2004

Volume 306: ***Polytropes – Applications in Astrophysics and Related Fields,*** by G.P. Horedt
Hardbound ISBN 1-4020-2350-2, September 2004

Volume 305: ***Astrobiology: Future Perspectives,*** edited by P. Ehrenfreund, W.M. Irvine, T. Owen, L. Becker, J. Blank, J.R. Brucato, L. Colangeli, S. Derenne, A. Dutrey, D.

Despois, A. Lazcano, F. Robert
Hardbound ISBN 1-4020-2304-9, July 2004
Paperback ISBN 1-4020-2587-4, July 2004

Volume 304: ***Cosmic Gammy-ray Sources,*** edited by K.S. Cheng and G.E. Romero
Hardbound ISBN 1-4020-2255-7, September 2004

Volume 303: ***Cosmic rays in the Earth's Atmosphere and Underground***, by L.I, Dorman
Hardbound ISBN 1-4020-2071-6, August 2004

Volume 302:***Stellar Collapse,*** edited by Chris L. Fryer
Hardbound, ISBN 1-4020-1992-0, April 2004

Volume 301: ***Multiwavelength Cosmology***, edited by Manolis Plionis
Hardbound, ISBN 1-4020-1971-8, March 2004

Volume 300:***Scientific Detectors for Astronomy,*** edited by Paola Amico, James W. Beletic, Jenna E. Beletic
Hardbound, ISBN 1-4020-1788-X, February 2004

Volume 299: ***Open Issues in Local Star Fomation,*** edited by Jacques Lépine, Jane Gregorio-Hetem
Hardbound, ISBN 1-4020-1755-3, December 2003

Volume 298: ***Stellar Astrophysics - A Tribute to Helmut A. Abt,*** edited by K.S. Cheng, Kam Ching Leung, T.P. Li
Hardbound, ISBN 1-4020-1683-2, November 2003

Volume 297: ***Radiation Hazard in Space,*** by Leonty I. Miroshnichenko
Hardbound, ISBN 1-4020-1538-0, September 2003

Volume 296: ***Organizations and Strategies in Astronomy, volume 4,*** edited by André Heck
Hardbound, ISBN 1-4020-1526-7, October 2003

Volume 295: ***Integrable Problems of Celestial Mechanics in Spaces of Constant Curvature,*** by T.G. Vozmischeva
Hardbound, ISBN 1-4020-1521-6, October 2003

Volume 294: ***An Introduction to Plasma Astrophysics and Magnetohydrodynamics,*** by Marcel Goossens
Hardbound, ISBN 1-4020-1429-5, August 2003
Paperback, ISBN 1-4020-1433-3, August 2003

Volume 293: ***Physics of the Solar System,*** by Bruno Bertotti, Paolo Farinella, David Vokrouhlický
Hardbound, ISBN 1-4020-1428-7, August 2003
Paperback, ISBN 1-4020-1509-7, August 2003

Volume 292: ***Whatever Shines Should Be Observed,*** by Susan M.P. McKenna-Lawlor
Hardbound, ISBN 1-4020-1424-4, September 2003

Volume 291: ***Dynamical Systems and Cosmology***, by Alan Coley
Hardbound, ISBN 1-4020-1403-1, November 2003

Volume 290: ***Astronomy Communication,*** edited by André Heck, Claus Madsen
Hardbound, ISBN 1-4020-1345-0, July 2003

Volume 287/8/9: ***The Future of Small Telescopes in the New Millennium,*** edited by Terry D. Oswalt
Hardbound Set only of 3 volumes, ISBN 1-4020-0951-8, July 2003

Volume 286: ***Searching the Heavens and the Earth: The History of Jesuit Observatories,*** by Agustín Udías
Hardbound, ISBN 1-4020-1189-X, October 2003

Volume 285: ***Information Handling in Astronomy - Historical Vistas,*** edited by André Heck
Hardbound, ISBN 1-4020-1178-4, March 2003

Volume 284: ***Light Pollution: The Global View,*** edited by Hugo E. Schwarz
Hardbound, ISBN 1-4020-1174-1, April 2003

Volume 283: ***Mass-Losing Pulsating Stars and Their Circumstellar Matter,*** edited by Y. Nakada, M. Honma, M. Seki
Hardbound, ISBN 1-4020-1162-8, March 2003

Volume 282: ***Radio Recombination Lines,*** by M.A. Gordon, R.L. Sorochenko
Hardbound, ISBN 1-4020-1016-8, November 2002

Volume 281: ***The IGM/Galaxy Connection,*** edited by Jessica L. Rosenberg, Mary E. Putman
Hardbound, ISBN 1-4020-1289-6, April 2003

Volume 280: ***Organizations and Strategies in Astronomy III,*** edited by André Heck
Hardbound, ISBN 1-4020-0812-0, September 2002

Volume 279: ***Plasma Astrophysics , Second Edition,*** by Arnold O. Benz
Hardbound, ISBN 1-4020-0695-0, July 2002

Volume 278: ***Exploring the Secrets of the Aurora,*** by Syun-Ichi Akasofu
Hardbound, ISBN 1-4020-0685-3, August 2002

Volume 277: ***The Sun and Space Weather,*** by Arnold Hanslmeier
Hardbound, ISBN 1-4020-0684-5, July 2002

Volume 276: ***Modern Theoretical and Observational Cosmology,*** edited by Manolis Plionis, Spiros Cotsakis
Hardbound, ISBN 1-4020-0808-2, September 2002

Volume 275: ***History of Oriental Astronomy,*** edited by S.M. Razaullah Ansari
Hardbound, ISBN 1-4020-0657-8, December 2002

Volume 274: ***New Quests in Stellar Astrophysics: The Link Between Stars and Cosmology,*** edited by Miguel Chávez, Alessandro Bressan, Alberto Buzzoni,Divakara Mayya
Hardbound, ISBN 1-4020-0644-6, June 2002

Volume 273: ***Lunar Gravimetry,*** by Rune Floberghagen
Hardbound, ISBN 1-4020-0544-X, May 2002

Volume 272:***Merging Processes in Galaxy Clusters,*** edited by L. Feretti, I.M. Gioia, G. Giovannini
Hardbound, ISBN 1-4020-0531-8, May 2002

Volume 271: ***Astronomy-inspired Atomic and Molecular Physics,*** by A.R.P. Rau
Hardbound, ISBN 1-4020-0467-2, March 2002

Volume 270: ***Dayside and Polar Cap Aurora,*** by Per Even Sandholt, Herbert C. Carlson, Alv Egeland
Hardbound, ISBN 1-4020-0447-8, July 2002

Volume 269: ***Mechanics of Turbulence of Multicomponent Gases,*** by Mikhail Ya. Marov, Aleksander V. Kolesnichenko
Hardbound, ISBN 1-4020-0103-7, December 2001

Volume 268: ***Multielement System Design in Astronomy and Radio Science,*** by Lazarus E. Kopilovich, Leonid G. Sodin
Hardbound, ISBN 1-4020-0069-3, November 2001

Volume 267: ***The Nature of Unidentified Galactic High-Energy Gamma-Ray Sources,*** edited by Alberto Carramiñana, Olaf Reimer, David J. Thompson
Hardbound, ISBN 1-4020-0010-3, October 2001

Volume 266: ***Organizations and Strategies in Astronomy II,*** edited by André Heck
Hardbound, ISBN 0-7923-7172-0, October 2001

Volume 265: ***Post-AGB Objects as a Phase of Stellar Evolution,*** edited by R. Szczerba, S.K. Górny
Hardbound, ISBN 0-7923-7145-3, July 2001

Volume 264: ***The Influence of Binaries on Stellar Population Studies,*** edited by Dany Vanbeveren
Hardbound, ISBN 0-7923-7104-6, July 2001

Volume 262: ***Whistler Phenomena - Short Impulse Propagation,*** by Csaba Ferencz, Orsolya E. Ferencz, Dániel Hamar, János Lichtenberger
Hardbound, ISBN 0-7923-6995-5, June 2001

Volume 261: ***Collisional Processes in the Solar System,*** edited by Mikhail Ya. Marov, Hans Rickman
Hardbound, ISBN 0-7923-6946-7, May 2001

Volume 260: ***Solar Cosmic Rays,*** by Leonty I. Miroshnichenko
Hardbound, ISBN 0-7923-6928-9, May 2001

Volume 259: ***The Dynamic Sun,*** edited by Arnold Hanslmeier, Mauro Messerotti, Astrid Veronig
Hardbound, ISBN 0-7923-6915-7, May 2001

Volume 258: ***Electrohydrodynamics in Dusty and Dirty Plasmas- Gravito-Electrodynamics and EHD,*** by Hiroshi Kikuchi
Hardbound, ISBN 0-7923-6822-3, June 2001

Volume 257: ***Stellar Pulsation - Nonlinear Studies,*** edited by Mine Takeuti, Dimitar D. Sasselov
Hardbound, ISBN 0-7923-6818-5, March 2001

Volume 256: ***Organizations and Strategies in Astronomy,*** edited by André Heck
Hardbound, ISBN 0-7923-6671-9, November 2000

Volume 255: ***The Evolution of the Milky Way- Stars versus Clusters,*** edited by Francesca Matteucci, Franco Giovannelli
Hardbound, ISBN 0-7923-6679-4, January 2001

Volume 254: ***Stellar Astrophysics***, edited by K.S. Cheng, Hoi Fung Chau, Kwing Lam Chan, Kam Ching Leung
Hardbound, ISBN 0-7923-6659-X, November 2000

Volume 253: ***The Chemical Evolution of the Galaxy***, by Francesca Matteucci
Paperback, ISBN 1-4020-1652-2, October 2003
Hardbound, ISBN 0-7923-6552-6, June 2001

Volume 252: ***Optical Detectors for Astronomy II***, edited by Paola Amico, James W. Beletic
Hardbound, ISBN 0-7923-6536-4, December 2000

Volume 251: ***Cosmic Plasma Physics***, by Boris V. Somov
Hardbound, ISBN 0-7923-6512-7, September 2000

Volume 250: ***Information Handling in Astronomy***, edited by André Heck
Hardbound, ISBN 0-7923-6494-5, October 2000

Volume 249: ***The Neutral Upper Atmosphere***, by S.N. Ghosh
Hardbound, ISBN 0-7923-6434-1, July 2002

Volume 247: ***Large Scale Structure Formation***, edited by Reza Mansouri, Robert Brandenberger
Hardbound, ISBN 0-7923-6411-2, August 2000

Volume 246: ***The Legacy of J.C. Kapteyn***, edited by Piet C. van der Kruit, Klaas van Berkel
Paperback, ISBN 1-4020-0374-9, November 2001
Hardbound, ISBN 0-7923-6393-0, August 2000

Volume 245: ***Waves in Dusty Space Plasmas,*** by Frank Verheest
Paperback, ISBN 1-4020-0373-0, November 2001
Hardbound, ISBN 0-7923-6232-2, April 2000

Volume 244: ***The Universe***, edited by Naresh Dadhich, Ajit Kembhavi
Hardbound, ISBN 0-7923-6210-1, August 2000

Volume 243: ***Solar Polarization***, edited by K.N. Nagendra, Jan Olof Stenflo
Hardbound, ISBN 0-7923-5814-7, July 1999

Volume 242: ***Cosmic Perspectives in Space Physics***, by Sukumar Biswas
Hardbound, ISBN 0-7923-5813-9, June 2000

Volume 241: ***Millimeter-Wave Astronomy: Molecular Chemistry & Physics in Space,*** edited by W.F. Wall, Alberto Carramiñana, Luis Carrasco, P.F. Goldsmith
Hardbound, ISBN 0-7923-5581-4, May 1999

Volume 240: ***Numerical Astrophysics,*** edited by Shoken M. Miyama, Kohji Tomisaka,Tomoyuki Hanawa
Hardbound, ISBN 0-7923-5566-0, March 1999

Volume 239: ***Motions in the Solar Atmosphere,*** edited by Arnold Hanslmeier, Mauro Messerotti
Hardbound, ISBN 0-7923-5507-5, February 1999

Volume 238: ***Substorms-4,*** edited by S. Kokubun, Y. Kamide
Hardbound, ISBN 0-7923-5465-6, March 1999

For further information about this book series we refer you to the following web site:
www.springeronline.com

To contact the Publishing Editor for new book proposals:
Dr. Harry (J.J.) Blom: harry.blom@springer-sbm.com